Ethnobotanical Wisdom and Microbial Studies on MEDICINAL PLANTS

Ethnobotanical Wisdom and Microbial Studies on MEDICINAL PLANTS

Editors
D.R. Khanna
Ashutosh Gautam
R. Bhutiani
Gagan Matta

Associate Editors
Anil Dhiman
Vikas Singh

2011
DAYA PUBLISHING HOUSE®
Delhi - 110 002

ISBN 9789351240396

Published by : **Daya Publishing House®**
A Division of
Astral International Pvt. Ltd.
– ISO 9001:2008 Certified Company
4760-61/23, Ansari Road, Darya Ganj,
New Delhi - 110 002
Phone: 23245578, 23244987
Fax: (011) 23260116
e-mail : dayabooks@vsnl.com
website : www.dayabooks.com

Laser Typesetting : **Classic Computer Services**
Delhi - 110 035

Printed at : **Chawla Offset Printers**
Delhi - 110 052

PRINTED IN INDIA

Preface

Ayurvedic system of medicine, which deals mainly with the use of plants and their products, has been a preferable system for the last many centuries. Today's, it is also getting much importance from the developed countries because of almost no-side effects action of its medicines. On the same line, Ethnobotany, that deals with the folk use of plants, especially as medicine, is also becoming popular in the country. But both the systems need the standardization of the drugs ultimately. Hence, more researches are needed in these fields.

The present work is the compilation of various papers contributed by learned community from different parts of the country working for betterment of herbal-based medicine system. Ethnomedicinal uses of many medicinal plants in various parts of the country, viz., - *Tridax procumbens, Crotolaria sericea, Mutela occidentalis, and Nyctanthes arbor-tristis* etc. have been discussed in details in various papers. Many antimicrobial and antibacterial activities of various plant speices – *Mimusops elengi, Pongamia pinnata, Vitex negundo, Abutilon indicum Veronica cineria, Cascabela thevetia, Psidium guajava,* etc. have been experimented and documented by various contributors in their papers. In addition, there are other papers covering various aspects of medicinal plants and their uses ranging from their cultivation, development and bio-evaluation of them. Isolation and determination of some chemical consitutents have also been discussed in details in some papers. This volume also contains some *in vitro* studies of various toxic constituents of different plant species and their effects on certain organisms. Some new records of plant species and their hosts have also been described in this volume.

It is hoped that this work will serve as a supplement for the standardization of various plant drugs. It is also hoped that in the present context, when in every part of the world, the emphasis is giving on herbal – based products to overcome the deadly diseases, this volume will be of immense use, particularly in opening new vistas of the wonderful world of plant species.

We thank all the contributors who have devoted their precious time in contributing their papers for this volume. Lastly but not least, the thanks are also due to M/S Daya Publishing House, New Delhi in bringing out this volume in time with nice presentation.

Editors

Contents

Ethnobotanical Wisdom and Microbial Studies on Medicinal Plants (2011) *Pages* **1–7**
Editors: **D.R. Khanna, Ashutosh Gautam, R. Bhutiani, Gagan Matta, Anil Dhiman**
Published by: **DAYA PUBLISHING HOUSE, NEW DELHI**

Chapter 1

Effect of Refinery Effluent on Seed Germination Seedling Growth at Nursery Stage of *Vigna radiata*

Sharmila Upadhyaya, Seema Bhadauria and Ekta Bhadauria
Department of Botany, Raja Balwant Singh College, Agra

ABSTRACT

The present investigation deals with the effect of refinery effluent on germination, shoot/root length, leaf, pods and seed yield of *Vigna radiata*. The seeds were grown in pots and treated with various concentrations of refinery effluent. Concentrations 0.5 and 1.0 per cent and above were found to be deleterious while treated effluent with 0.01, 0.05 and 0.1 per cent concentrations were found to be suitable for irrigation in comparison to groundwater. A field experiment was also laid down and observations were interpreted that treated effluents can be effectively used for irrigation.

Keywords: *Effluent, Refinery, Germination.*

Introduction

Petroleum refinery unavoidably generates a large volume of oil sludge, organics like n-alkanes-paraffins, olefins, aromatics, asphaltics, Phenols and Polynuclear aromatic hydrocarbons (Atlas, 1981; Grob, 1983) which have both lethal and Sub lethal effects on a wide variety of marine organism (Chet and Mitchell, 1976), reduced photosynthetic rate in algae (Parker and Menzel 1974), reduced resistance to environmental stress in crabs accumulation of hydrocarbons in fatty tissues of fishes which get transferred to higher trophic levels including humans (Shailubhai 1986). Therefore, the concern over chronic hydrocarbon inputs in environment is ecologically mendatory and disposal of oil sludges and waste hydrocarbons is a major challenge prevailing in the petroleum industry.

Many prior attempts had been made to assess the characteristics of petroleum refinery wastes (Kale *et al.*, 1985) and all reports suggest the researchers for regular assessment of this waste in order to obtain judicious data before effluents are used for irrigation. The deliberately cause deleterious effect in agroecosystems (Shailubhai 1986).

Study Area and Methodology

Mathura oil refinery is located near Bad (Mathura District) in Uttar Pradesh processing about 15 lakhs tonnes of crude oil. The oil refinery discharges its pretreated effluents into Yamuna river through a 13.5 km long drain which is constructed with bricks and high slanting sides to avoid ground seepage of effluents. At about 0.5 km away from outlet of the drain, a pumping station has been installed which continuously transfer effluents to the irrigation canal distributing effluent and water to agricultural fields.

Experimental Layout

An experiment on Moong was conducted in the experimental field developed for this purpose, adjacent to the drain carrying treated effluent on Delhi-Agra Highway No. 2 one experiment on Moong was also conducted in pots at the experimental Net House of Botany Deptt. Raja Balwant Singh College, Agra. The study was conducted in a split plot design where main plot treatments effluent discharged from the refinery.

Design of the Experiment	*Split Plot*
Main plot treatment:	Irrigants
	(i) Groundwater
	(ii) Treated effluent
Concentrations:	0.01 per cent
	0.05 per cent
	0.1 per cent
Replications:	Three
Seed rate:	20 kg/ha
Fertilizer:	$N_{10}P_3O$ K35
Date of sowing:	22.4.2000
Irrigations (Four):	8.05.2000
	22.5.2000
	08.6.2000
	20.6.2000
Samplings:	30.40 and 50 DAS

The following growth, characteristics were studied:

0. Germination
1. Shoot length (cm) per plant
2. Root length (cm) per plant
3. Root nodule number per plant
4. Leaf number per plant
5. Shoot fresh weight (g) per plant
6. Shoot dry weight (g) per plant

Crops and Inputs

As mentioned earlier Moong *Vigna radiata* was selected for these studies. It was based on the fact that these crops are grown locally by the farmers and leguminous crops maintain the seed fertility. In field experiment each plot was irrigated with 2,500 liters of irrigant each time calculated in accordance with the size of the plot and local requirements. A 90° v-notch weir box was used for measuring the flow using the following formula.

$$Q = 1417\ H^{2.5}$$

where,

Q: Discharge, 1/s and H height of irrigant over the apex of the notch cm.

Analytical Techniques

Irrigant and soil samples were analysed following the Standard Methods of Singh, 1983 while analysis for heavy metal content was done with the help of Atomic Absorption Spectro photometer.

Statistical Analysis

All the data for growth was analysed statistically.

Results and Discussion

Most of the growth characteristics, studied at different intervals, exhibited a favourable response to the treated effluent as compared with groundwater (Table 1.1–1.6). The application of the treated effluent favoured the vegetative growth of the plants, observed at 35, 42, 49 and 56 days after sowing. The respective leaf number per plant was 25.2, 13.0, 16.1 and 13.6 per cent and shoot dry weight per plant was 33.3, 26.7, 38.7 and 27.1 per cent more than groundwater application.

Table 1.1: Effect of Treated Effluent (TE) and Groundwater (GW) on Shoot Length (cm) per Plant at Three Concentrations at Three Stages of Growth (Mean of three replicates)

Concentration	*Days After Sowing*								
	30			*40*			*50*		
	TE	*GW*	*Mean*	*TE*	*GW*	*Mean*	*TE*	*GW*	*Mean*
0.01	28.65	26.11	27.38	39.10	36.88	37.99	43.55	39.44	
0.05	27.77	27.66	27.72	39.00	37.89	38.44	41.50		
0.1	31.07	27.22	29.14	44.30	41.55	42.93	47.22	42.06	
								44.64	
								54.77	49.00
								51.89	
Mean	29.16	26.99		40.80	38.77		48.51	43.50	

CD at 5 per cent

	30	*40*	*50*
Irrigants	1.10	0.68	3.03
Concentration	N.S.	1.05	0.82
Irrigants × Concentration	N.S.	N.S.	N.S.
Concentration × irrigants	N.S.	N.S.	N.S.

N.S.: Not significant.

Table 1.2: Effect of Treated Effluent (TE) and Groundwater (GW) on Shoot Length (cm) per Plant at Three Concentrations at Three Stages of Growth (Mean of three replicates)

Concentration	*Days After Sowing*								
	30			*40*			*50*		
	TE	*GW*	*Mean*	*TE*	*GW*	*Mean*	*TE*	*GW*	*Mean*
0.01	13.33	13.22	13.27	17.89	15.67	16.78	22.22	19.33	20.77
0.05	12.77	13.33	13.05	20.85	23.77	22.31	25.78	21.89	23.83
0.1	14.78	14.22	14.50	23.85	20.22	20.03	31.33	27.99	29.66
Mean	13.83	13.58		20.86	19.88		26.44	23.07	

			CD at 5 per cent
	30	*40*	*50*
Irrigants	N.S.	N.S.	2.56
Concentration	N.S.	N.S.	1.26
Irrigants × Concentration	N.S.	N.S.	N.S.
Concentration × irrigants	N.S.	N.S.	N.S.

N.S.: Not significant.

Table 1.3: Effect of Treated Effluent (TE) and Groundwater (GW) on Shoot Length (cm) per Plant at Three Concentrations at Three Stages of Growth (Mean of three replicates)

Concentration	*Days After Sowing*								
	30			*40*			*50*		
	TE	*GW*	*Mean*	*TE*	*GW*	*Mean*	*TE*	*GW*	*Mean*
0.01	8.00	9.00	8.50	18.55	16.00	17.28	19.55	18.22	18.89
0.05	9.00	10.78	9.89	21.00	18.89	19.94	22.17	19.33	20.75
0.1	13.78	8.89	11.28	25.20	22.66	23.93	26.89	25.44	26.16
Mean	10.22	9.55		21.58	19.18		22.87	20.99	

			CD at 5 per cent
	30	*40*	*50*
Irrigants	N.S.	1.21	1.65
Concentration	0.93	1.02	1.26
Irrigants × Concentration	2.00	N.S.	N.S.
Concentration × irrigants	1.62	N.S.	N.S.

N.S.: Not significant.

Table 1.4: Effect of Treated Effluent (TE) and Groundwater (GW) on Shoot Length (cm) per Plant at Three Concentrations at Three Stages of Growth (Mean of three replicates)

Concentration	*Days After Sowing*								
	30			*40*			*50*		
	TE	*GW*	*Mean*	*TE*	*GW*	*Mean*	*TE*	*GW*	*Mean*
0.01	13.67	14.67	14.17	18.11	16.33	17.22	20.55	19.22	19.89
0.05	14.33	13.22	13.78	19.00	16.44	17.72	24.22	20.55	22.38
0.1	16.00	15.44	15.72	21.11	18.33	19.72	28.00	23.55	25.78
Mean	14.67	14.44		19.40	17.03		24.26	21.10	

CD at 5 per cent

	30	*40*	*50*
Irrigants	N.S.	2.03	1.42
Concentration	0.98	0.86	1.12
Irrigants × Concentration	N.S.	N.S.	N.S.
Concentration × irrigants	N.S.	N.S.	N.S.

N.S.: Not significant.

Table 1.5: Effect of Treated Effluent (TE) and Groundwater (GW) on Shoot Length (cm) per Plant at Three Concentrations at Three Stages of Growth (Mean of three replicates)

Concentration	*Days After Sowing*								
	30			*40*			*50*		
	TE	*GW*	*Mean*	*TE*	*GW*	*Mean*	*TE*	*GW*	*Mean*
0.01	10.22	10.55	10.39	18.86	17.33	18.00	23.96	20.94	22.45
0.05	10.06	9.22	9.62	19.55	16.88	18.22	27.33	23.18	25.26
0.1	10.22	10.44	10.33	22.66	22.66	22.66	33.77	29.32	31.54
Mean	10.17	10.17		20.29	18.95		28.35	24.47	

CD at 5 per cent

	30	*40*	*50*
Irrigants	N.S.	1.25	2.37
Concentration	N.S.	0.60	0.73
Irrigants × Concentration	N.S.	N.S.	N.S.
Concentration × irrigants	N.S.	N.S.	N.S.

N.S.: Not significant.

Table 1.6: Effect of Treated Effluent (TE) and Groundwater (GW) on Shoot Length (cm) per Plant at Three Concentrations at Three Stages of Growth (Mean of three replicates)

Concentration	*Days After Sowing*								
	30			*40*			*50*		
	TE	*GW*	*Mean*	*TE*	*GW*	*Mean*	*TE*	*GW*	*Mean*
0.01	3.59	3.48	3.53	6.21	5.87	6.04	8.25	6.95	7.60
0.05	3.35	3.08	3.21	6.51	5.75	6.13	9.32	7.86	8.59
0.1	10.69	3.48	3.44	7.70	6.23	6.96	11.50	9.88	10.69
Mean	3.45	3.34		6.81	5.95		9.69	8.22	

			CD at 5 per cent
	30	*40*	*50*
Irrigants	N.S.	0.77	0.76
Concentration	N.S.	0.20	0.45
Irrigants × Concentration	N.S.	0.54	N.S.
Concentration × irrigants	N.S.	0.35	N.S.

N.S.: Not significant.

The irrigation of Moong treated effluent enhanced the vegetative growth of the plants as compared with groundwater. This luxuriant growth resulted in the improvement of all the characteristics determining the reproductive growth in gram contrary to the 1 per cent above in a similar pattern.

The decrease in leaf No. per plant was observed at 0.5, 1.0 and 1.5 per cent refinery effluent and found that leaf number and shoot dry weight also decreased at this concentration.

The leaf number and shoot dry weight increased by 13.9 and 14.2 per cent at 40 and 14.9 and 17.8 per cent at day 50 respectively. Most of the interactions between irrigant and Moong were not significant (Tables 1.1–1.6). Contrary to the growth, irrigation with treated effluent decreased seed yield in Moong which was 9.4 per cent lower in comparison to the groundwater. It would not be out of place to mention that the percent decrease in plant dry weight of gram due to the ill effect of effluent irrigation was only 3.3 per cent in comparison with groundwater above 0.1 per cent. This shows that the plant dry weight is not adversely affected to the same extent as the other two concentration. All the interaction values between irrigant and concentration were not significant for plant growth attributing parameters.

References

Atlas, R.M., 1980. Microbial degradation of petroleum hydrocarbons: An environmental perspective. *Microbiological Reviews*, 45(1): 180–209.

Chet, J. and Mitchell, R., 1976. Petroleum hydrocarbons inhabit decomposition of organic matter in sea water. *Nature*, 261: 308–309.

Grob, R.C. (ed), 1983. *Chromatographic Analysis of Environment*. Marcel Dekker, Inc (En) N.V.U.S.A., pp. 724.

Kale, C., Raman, K.J., Mishra, R.P., Sharma, C.N., Raman, A., Dixit, P.C. and Swaminathan, B.V., 1985. Wastewater utilizations scheme for Mathura refinery: Planning and investigations. In: *Proc. Nat. Seminar Poll. Control Environ. Manag.*, Nagpur, 16–18th March, p. 37–46.

Singh, D., 1983. *Soil and Water Testing Methods: A Laboratory Manual.* IARI, New Delhi.

Parker, P.L. and Menzel, D., 1974. *National Science Foundation.* JDOE. CF.

Shailubhai, K., 1986. Treatment of petroleum industry oil sludge in soil. *Trends in Biotechnology*, 4(8): 202–206.

Ethnobotanical Wisdom and Microbial Studies on Medicinal Plants (2011) *Pages* 8–13
Editors: D.R. Khanna, Ashutosh Gautam, R. Bhutiani, Gagan Matta, Anil Dhiman
Published by: DAYA PUBLISHING HOUSE, NEW DELHI

Chapter 2

Phenol Degradation by a Bacterium *Pseudomonas putida* in the Presence of Chromium

D.S. Malik[1] and Amrita Malik[2]
[1]Department of Zoology and Environmental Science,
Gurukul Kangri University, Haridwar
[2]Environmental Microbiology Laboratory,
Industrial Toxicology Research Centre, Lucknow

ABSTRACT

Phenol and Phenolic compounds are well known components in aqueous waste of many different pollution generating industries. These compounds must be regarded as hazardous to the environment and are resistant to biological degradation. Degradation of Phenol in the presence of hexavalent chromium was studied using phenol degrading bacteria, *Pseudomonas putida*. The bacteria was able to degrade phenol up to 80 per cent in the absence of chromium but the degradation was increased up to 81.67 per cent in the presence of 25mg/l initial concentration of chromium (VI). The degradation was declined further up to 21.67 per cent, when concentration of Chromium (VI) was applied 50 mg/l.

Keywords: *Phenol degradation, Pseudomonas putida, Hexavalent chromium.*

Introduction

Phenol is a common constituent of effluents from polymeric resin production, oil refining, coking plants, textile, pulp and paper mills and tannery industries. The discharged effluents contain phenolic compounds ranging 6-200 mg/L, however their admissible limit is only 3 mg/L in the receiving water bodies (Jossens 1978). Phenol can be toxic to fish at concentrations of 5 mg/L and gives an objectionable taste to drinking water at far lower concentrations (Throop 1975). The persistence of phenolic compounds in the aquatic and terrestrial environment can be injurious to the health of humans and cause allergic dermatitis, skin irritation, cancer and mutation (Buikema *et al.*, 1979, Zitomer and Speece 1993 and Singh and Viraraghavan 1996).

Recent literature on the method of degradation of phenol and their compounds from industrial effluents focus on microbial degradation process. *Pseudomonas putida* has an ability to degrade a wide variety of aromatic compounds. The present study has revealed the potential of phenolic compounds degradation through microbial degrading bacterium *Pseudomonas putida* in the presence or absence of Chromium as the wide spread use of chromium and its compounds by various industries has led to the release of this element into environments. Being mutagenic, carcinogenic and teratogenic, hexavalent chromium is about 100-fold more toxic than the trivalent form (Petrilli and Flora 1977).

Materials and Methods

A pure culture of *Pseudomonas putida* was obtained from Institute of Microbial Technology, Chandigarh (UT).The bacterial strain (*Pseudomonas putida*) was maintained on nutrient agar medium, stored at 4 °C and sub cultured every week. The culture media used to perform the growth experiments were LB (Luri Bertani) media containing peptone 10gm/L, NaCl 10gm/L, yeast extract 5 gm/L, agar 2 per cent and LB (Luria broth) having above composition without agar. The complete growing medium were sterilized in an autoclave at 121 °C for 30 minutes and subsequently cooled. Phenol solution and chromium were added to the sterilized medium to make the solution of desired concentration. In each experiment the initial pH of the medium was raised to 7.0 by addition of concentrated NaOH before sterilization. The LB (Luria broth) media (25 mL) was dispensed into a sterile 150 ml flask fitted with a cotton plug. Aromatic compound (Phenol) and Chromium (VI) as chromate (K_2CrO_4) were added as sole electron donors for chromium (VI) reduction. Harvested cells of *P. putida* were placed into each experiment before incubation in the dark on a rotator shaker at 140 RPM and at 30 °C temperature.

The viable cells of *P. putida* were counted by using haemocytometer at the desired time intervals. The cell growth was measured by the optical density (OD at 699 nm) taken at regular intervals. Hexavalent chromium was determined calorimetrically by using spectrophotometer at 540 nm by reaction with diphenylcarbazide in acid solution (APHA 1989). The phenol concentration was determined calorimetrically using a spectrophotometer at 500 nm by reaction with 4-amino-antipyrine in the presence of potassium ferricyanide (APHA 1989).

Results and Discussion

In the culture, *Pseudomonas putida* has oxidized the phenol and used Cr (VI) as an electron acceptor. The oxidation of phenol by *P. putida* has initialized the energy flow of the culture and served as a primary energy source for the strain while Cr (VI) reduction occurred only as result of metabolism (Shen and Wang 1995). The effect of different concentration of chromium (VI) mixed with bacterial culture (*Pseudomonas putida*) growth has shown in Figure 2.1 and observation revealed that growth was inhibited up to incubation period of 6 hrs in the culture media containing chromium (VI). The chromium concentration (50 mg/L) in culture medium has exhibited more inhibitory phenomenon than culture of initial chromium concentration (25 mg/L). At the 24 hrs incubation period, the culture of chromium concentration (25 mg/L) was found growth promoters, where as chromium concentration (50 mg/L) has shown considerable inhibitory effect with increasing incubation periods.

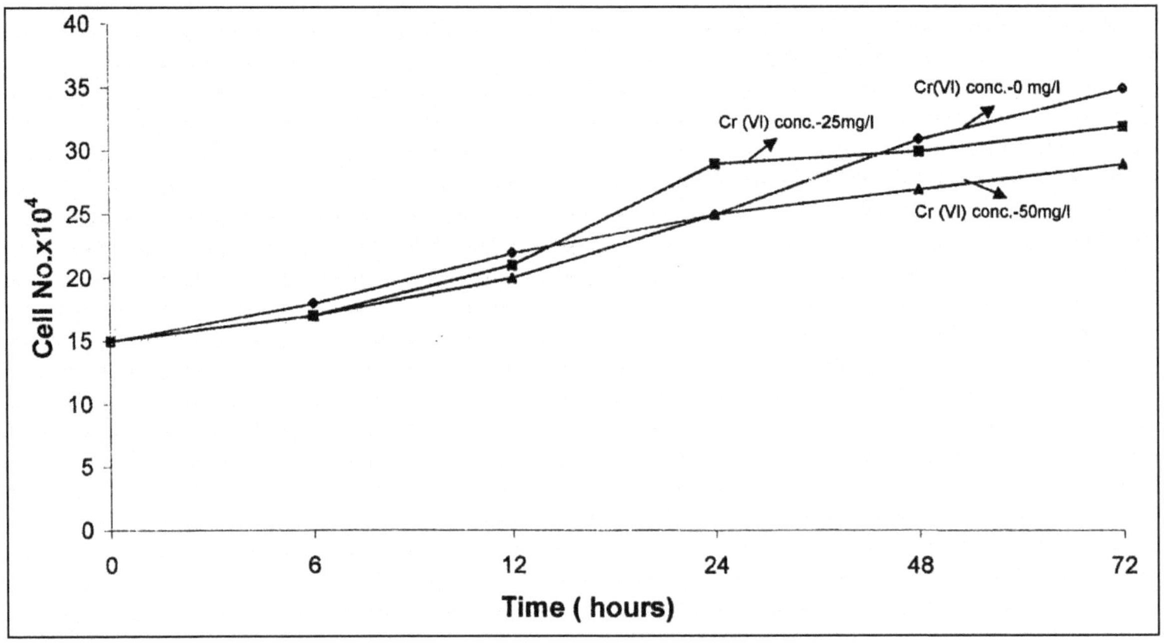

Figure 2.1: Growth Pattern of *Pseudomonas putida* at Different Concentration of Cr (VI)

The sudden change in the growth of *P. putida* at the incubation period of 24 hrs can be explained with changing pH (Figure 2.2). In the culture medium, the changing pattern of pH has observed after 12 hours exposure period. This change in pH may be due to the presence of some metabolites released by *P. putida* in the culture (Shen and Wang 1995).

In the experiments chromium (VI) reduction occurred in the culture of *P. putida* using phenol as a carbon source, the results have observed that Cr (VI) reduction was greater in the culture containing Chromium concentration (50 mg/L) (Figure 2.3). The Chromium reduction suddenly increased up to 60 per cent in the culture containing Chromium concentration (25 mg/L) after 12 hrs incubation time. The effect of Cr (VI) and phenol concentrations on Cr (VI) reduction and phenol degradation were also investigated by (Shen and Wang 1995 and Annadurai *et al.*, 1999). The phenol degradation patterns have shown in Figure 2.4 and the results occurred as Chromium (VI) with concentration of 25 mg/L have shown maximum rate of phenol degradation in between 24-48 hours incubation time, while the chromium (VI) concentration (50 mg/L) has suppressed the phenol degradation up to many fold in comparison to control.

When *Pseudomonas putida* was grown in the culture to degrade the phenol, the fluids turned yellow transiently and the yellow material accumulated in the culture was obtained maximum with the maximum phenol degradation (Molin and Nilsson 1985).

In the present study, the results have shown that *Pseudomonas putida* has degraded the phenol compound up to 80 per cent in the absence of Chromium (VI). The degradation level has enhanced up to 81.67 per cent in the presence of 25 mg/L initial concentration of Chromium (VI). The simultaneous Chromium reduction was observed 68 per cent and 70 per cent in initial Chromium concentration used as 25.0 and 50.0 mg/L respectively.

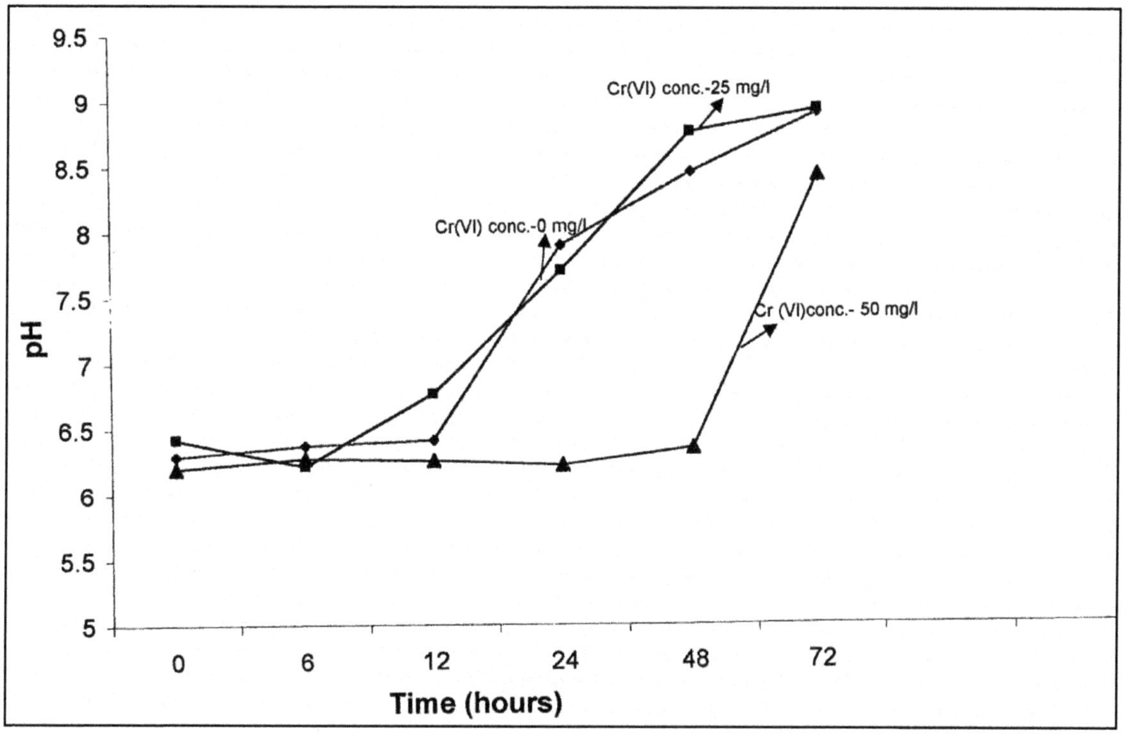

Figure 2.2: Change in pH at Different Cr (VI) Concentration with Time

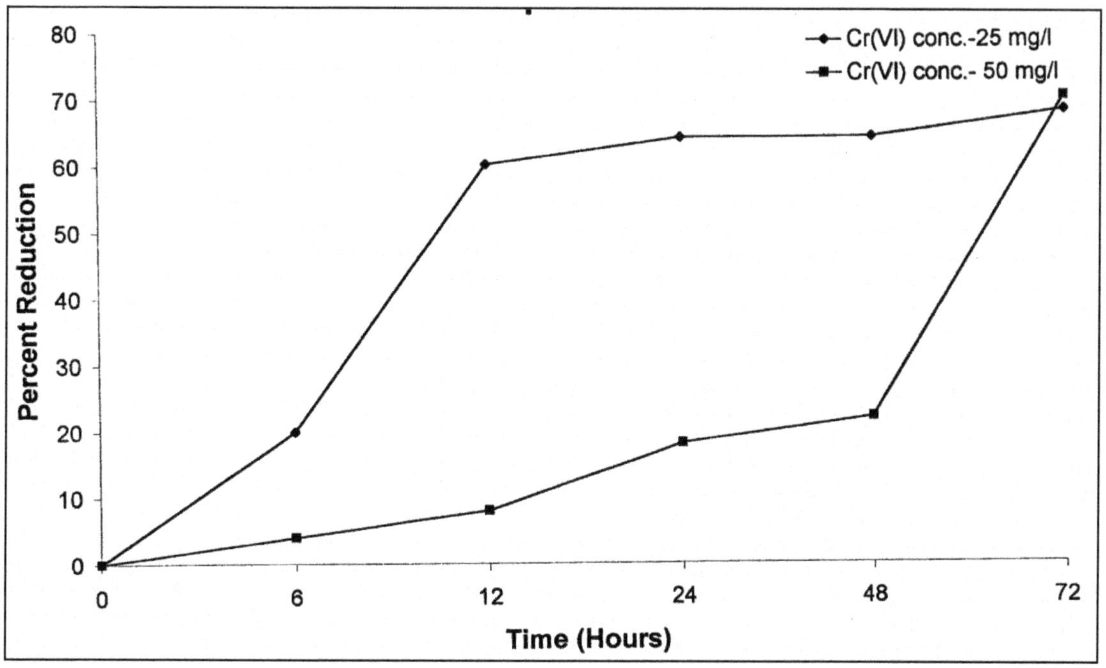

Figure 2.3: Chromium Reduction at Different Time in Presence of Phenol

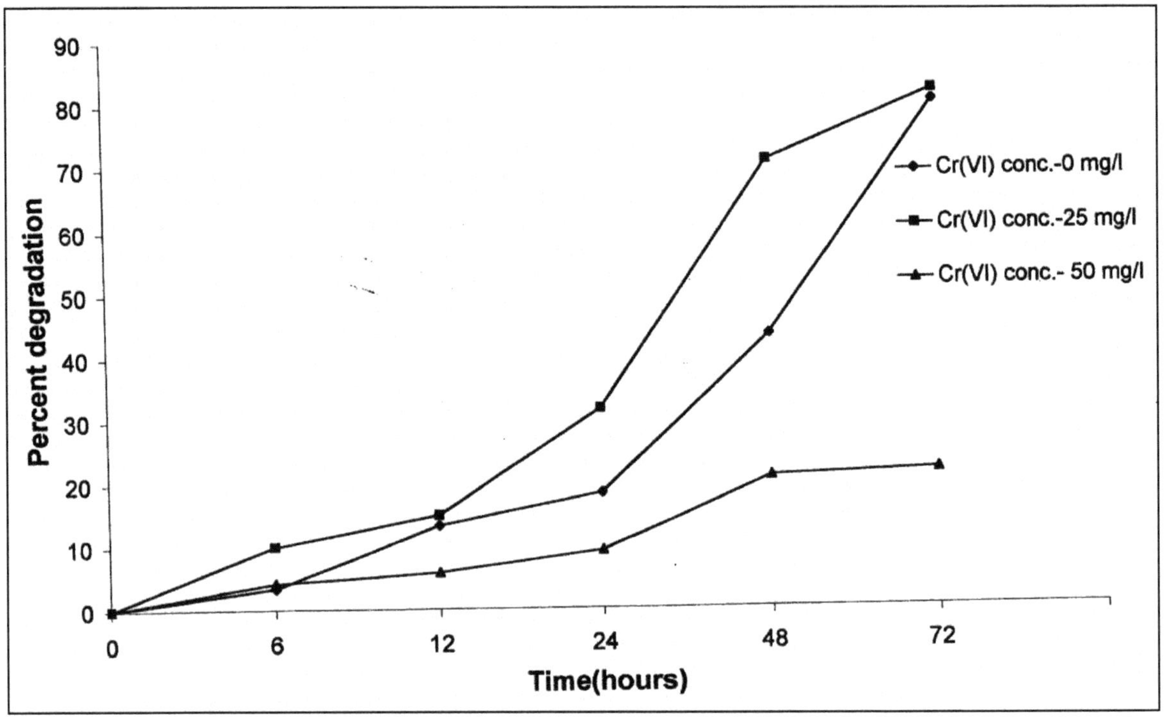

Figure 2.4: Phenol Degradation with Different Cr (VI) Concentration

References

Annadurai, G., Rajesh Babu, S., Siva Kumar, T. and Murugesan, T., 1999. Degradation of Phenol by a mixed culture of *Pseudomonas putida* (NICM 2174) and *Pseudomonas pictorum* (NICM 2074) adsorbed on chitosan. *J.E.P.*, 20(7): 493–498.

APHA, 1989. *Standard Methods for the Examination of Water and Wastewater*, 17th edn. American Public Health Association, Washington D.C.

Buikema, A.C.J., Mc Ginna, M.J. and Carius, J., 1979. Phenolics in aquatic ecosystems: A selected review of recent literature. *Marine Env. Res.*, 2: 87–245.

Jossens, L., 1978. Thermodynamics of multi-solute adsorption from dilute aqueous solution. *Chem. Engg. Sci.*, 33: 1097–1102.

Molin, G. and Nilsson, I., 1985. Degradation of phenol by *Pseudomonas putida* ATCC 11172 in continuous culture at different ratios of biofilm surface to culture volume. *App. Environ. Microbiol.*

Petrilli, F.L. and Flora, S. De., 1977. Toxicity and mutagenicity of hexavalent chromium on *Salmonella typhimunium*. *App. Environ. Microbiol.*, 33: 805–809.

Shen, H. and Wang, Y., 1995. Simultaneous chromium reduction and phenol degradation in a culture of *Escherichia coli*. ATTCC 33456 and *Pseudomonas putida* DMP–1. *App. Environ. Microbiol.*, p. 2754–2758.

Singh, K.S. and Viraraghavan, T., 1996. Biodegradation potential of mono, di and trichlorophenols: Review. *Indian J. Environmental Protection,* 17(1): 1–9.

Throop, W.M., 1975. Alternative methods of phenol wastewater control. *J. Hazardous Materials,* 1: 319–329.

Zitomer, D.H. and Speece, R.W., 1993. Sequential environmental for enhanced biotransformation of aqueous contaminants. *Env. Sci. Tech.,* 27: 227–245.

Ethnobotanical Wisdom and Microbial Studies on Medicinal Plants (2011) *Pages* **14–18**
Editors: **D.R. Khanna, Ashutosh Gautam, R. Bhutiani, Gagan Matta, Anil Dhiman**
Published by: **DAYA PUBLISHING HOUSE, NEW DELHI**

Chapter 3

Rearing and Feeding of Larvae of *Bombyx mori* Linn. in Laboratory Conditions

Smita Shrivastava, Hemant Saxena and Anita Shrivastava
Pest Control Research Lab, P.G. Department of Zoology,
S.S.L. Jain P.G. College, Vidisha, M.P.

ABSTRACT

Lepidoptera larvae are good experimental materials to observe growth phenomenon. In the present study relationship between quality and consumption of leaves and larval rearing have been depicted.

Keywords: *Lepidoptera, Larvae, Silkworm.*

Introduction

Lepidopteran larvae are known as more or less continuous feeders hence, a deficiency in the amount of food required for reaching its full potential will be manifested in various ways (Waldbaver 1968). Shortening of feeding periods for larvae of different insects have resulted in the prolongation of larval periods and decrease in final body weight (Mc Girimis and Kasting 1969), which in turn may affect reproductive success or resulting adults particularly where adult females do not feed. Srivastava *et al.*, 1982 reported extension of larval period up to 2 days and decrease in the fecundity to the extent of 32 per cent of Eri silk moth, *Philosamia ricini*.

Materials and Methods

Lepidoptera larvae are good experimental materials to observe growth phenomenon. The growth and reproduction are separate phases in the life cycle of Silkmoth. Providing nutritional mulberry leaves to the larvae is an essential component of sericulture solubility, as to get good quality and improved cocoon and Silk thread. The larval feeding especially 4th and 5th in star are very important because the larva only feed on leaves and grow. The whole phenomenon of larval feeding completes within few days only and one has not to wait for months and years to complete experiment, their smaller size makes them manageable objects and the entire experimentation can be conducted by controlled situation with petridishes, wooden trays and other material.

The growth of silk moth larvae in relation to food consumption and in relation to ecological factors such as light, temperature and humidity have been studied in detail with reference to the growth and production of cocoon solubility as to make the problem more result oriented.

Taxonomic Position of Silkmoth

Mulberry silk worm, *Bombyx mori* is a homometabolous insect and passes through four morphologically stages in its cycle *viz.* egg, larva, pupa and adult.

In first stage, the embryo grows and develops into a larva's. The second is a vegetative Stage, in which the larva takes nutrients *i.e.* mulberry leaves. Mulberry leaves are the sole food for larvae in commercial sericulture and the quality and quantity of the mulberry leaf feed during rearing decide the success of Silkworm crop. Hence, choice of mulberry leaves suitable for healthy growth of Silk worm is one of the most important factor in sericulture. The 4th is metamorphic stage in which the larva becomes pupa and then in to the adult. The fourth stage of cycle is known as reproductive, stage in which moth mates and the female moth lay eggs producing next generation. It is only the larval stage that the silk worm takes foods, grows enormously and accumulates nutrition for the moth stage, the life cycle of silk moth was under laboratory condition and following stages were studies by Chawky methods the eggs were procured from "Sericulture Centre" district Raisen, Madhya Pradesh.

Laboratory Culture of Silkmoth Eggs and Larvae

Silkworm eggs are brought from sericulture centre were sterilized by dipping in 2 per cent formaline solution for about 10 minutes and dried in shade before incubation. The incubation period as recorded for bivoltine and hibernating bivoltine eggs in room with temperature 25° C, RH 80 per cent + 5 per cent and keeping the photoperiod L:D 16:8. The data recorded for proper incubation of eggs are given in the Table 3.1. The data show that average incubation period varies between 11-12 days. It was also observed that two days before hatching, the colour of the eggs start changing into lighter shade with a blue point, the very next day colour of eggs changed into blue. This is the pigmentation stage. Eggs were kept in dark room or covered with black paper or cloth. The following technical cares need to be followed as emerged from the experimentation in the present study.

(A) Maintenance of Temperature

1. 20-28°C is the optimum range of temperature for Silk worm rearing.
2. 25°C is temperature which favored healthy growth of late instar larvae.
3. It was noticed that maximum optimum range of temperature for different instar as mentioned in Table 3.2 varies from 20 to 28 °C.

Table 3.1: Showing Range of Temperature and Humidity Required during Incubation Period Days of Incubation

		Days of Incubation											
		1	2	3	4	5	6	7	8	9	10	11	12
Hibernated	Temperature					25 °C						26 °C	
	Humidity			75 %				80 %				85 %	
Acid Treated Bivoltive Egs	Temperature		15 °C				24 °C					26 °C	
	Humidity				75 %		80 %					85 %	

4. It was observed that for young larvae upto 3rd instar a slightly higher range of temperature is required where as in last stages *i.e.* for 4th and 5th instar a little less as shown in the Table 3.2.

Table 3.2: Temperature and Humidity Required during Early Instar Rearing

Particulars	*1st Instar*	*2nd Instar*	*3rd Instar*
Temperature	28 °C	27 °C	26 °C
Humidity	85-90 per cent	85 per cent	80 per cent

(B) Humidity

It was noticed that relative humidity needed, for ideal rearing of silk worm larvae ranged between 80 to 90 per cent for younger instar, a slightly higher humidity was needed while for older, less humidity is required as shown is Table 3.2. It was maintained by using Honey well Humido-state in the insectary.

(C) Light

Silkworm larvae attains maximum growth in dimlight and was found to avoid strong light. The minimum photoperiod of light required for larval rearing was noticed to be 16:14 hours night/day. This was maintained by automatic switch 'On and Off' device available in the insectory. All these conditions were maintained in the insectery established long back at Pest Control Research Laboratory at Vidisha.

(D) Spacing of worm's in the rearing beds

This is another important aspect of sericulture. Optimum rearing bed over from brushing to the end of larval stage is important to attain full larval growth and yield of good quality cocoon. The data collected for proper spacing have been cited in Table 3.3 with spacing, regular cleaning was also felt necessary. Cleaning was done during 2nd instar, first on the next day of resumption of feeding and the next before the worms prepare for second moult in the 3rd instar also. Two cleaning is sufficient, first after second moult and second before setting for third moult. The larvae reared on good quality leaves were found grown to 10 thousands time more, than the weight of the newly born larvae within 24 to 25 days.

Table 3.3: Showing Spacing during Rearing of Larval Instar of *Bombyx mori*

Age of Larvae	*Area Required for Rearing*		*Increase in Spacing during Each Instar (Times)*
	In Beginning (Sqft)	*At the End (Sqft)*	
1st Instar larvae	4	15	3 ½
2nd Instar larvae	15	41	3
3rd Instar larvae	41	105	2 ½
4th Instar larvae	105	210	2
5th Instar larvae	210	400	About 2

Results and Discussion

A detail study was carried out as regards the feeding and rearing of *Bombyx mori* in the insectary maintained at Pest Control Research laboratory of the Institution. It was noticed that mulberry leaf feed during rearing decides the success of silk worm crop. The similar views have been expressed in several manual of Silk worm rearing specially (F.A.O. Bulletin 1988). It has been observed that young larval instar needs more care and attention because they are more susceptible to diseases. Of the total leaf consumed, 6.36 per cent was found to be utilized by early stages *i.e.* from 1st to 3rd instar larvae. For the proper growth and metamorphosis of the larvae of silk moth, proper feeding is required as mentioned in the Tables 3.1–3.5. Besides proper feeding to the larvae, proper spacing in rearing beds and clearing is also needed to be taken care of. Larvae reared, on in crowded condition showed poor growth. Rajan *et al.*, 1996 have mentioned that rearing of young larval instar of *B.mori* needs special attention. They have mentioned further that not giving proper spacing to the larvae may affects cocoon quantity and yield of fibers. Similar views has been expressed by Saxena 2000.

Table 3.4: Contents of Starch and Water in Mulberry Leaves

Contents	*Harvesting Time*	
	06=00	*18=00*
Sugar	0.490	0.921
Starch	0.707	2.043
Water	82.11	74.01

Table 3.5: Showing Frequency of Bed Cleaning during Rearing of *Bombyx mori*

Instar	*Before Moulting*	*Intermediate Period*	*After Moulting*
1st Instar	1	–	–
2nd Instar	1	–	1
3rd Instar	1	1	1
4th Instar	–	1–2	1
5th Instar	–	Once a day	

In the present study relationship between quality and consumption of leaves and larval rearing have been depicted in the Tables 3.4 and 3.5. It took about 24 to 26 days for completing the larval cycle in *Bombyx mori* under controlled laboratory conditions which are slightly different (< 2 days) as mentioned by Nath *et al.*, 1990.

Acknowledgement

We are highly thankful to Dr. Y.P. Saxena, Dy. Director, Sericulture for permission to use the facilities at Sanchi Farm and to Mr. Soni, Technical Assistant and other persons at Sanchi Sericulture for various facilities.

References

FAO, 1988. FAO Agriculture Service Bulletin 1988. *Silk Worm Rearing*, F.A.O. Organization United Nations by Pungchaun, 73/2: 1–83.

Mc Girimis, A.J. and Kasting, R., 1969. Effect of under feeding and artificial diet on growth, development and comparison of *B. mori. Can. J. Zool.*, 37: 259–266.

Nath, T.N., Shivnath and Subbarao, G., 1990. Effect of food deprivation on dry matter utilization, shell and cocoon weight and fecundity of multivoltine silk worm, *Bombyx mori* Linn. *Ind. J. Seri.*, 29(2): 174–181.

Rajan, R.K., Muroga, A. and Datta, R.K., 1996. *Manual on Young Silk Worm Rearing*. CSRTI, Mysore, pp. 1–24.

Saxena, Hement, 2000. *Ph.D. Thesis*, Barkatullah University, Bhopal, pp. 149.

Srivastava, D., Mishra, S.D. and Poomia, F.S., 1982. Effect of food deprivation on larval weight and larval fecundity of silk worm. *Indian. J.Seri.*, 21: 11–15.

Waldbauer, G.P., 1968. The consumption and utilization of food by insects. *Advances Insect Physiol.*, 5: 229–288.

Ethnobotanical Wisdom and Microbial Studies on Medicinal Plants (2011) *Pages* **19–22**
Editors: **D.R. Khanna, Ashutosh Gautam, R. Bhutiani, Gagan Matta, Anil Dhiman**
Published by: **DAYA PUBLISHING HOUSE, NEW DELHI**

Chapter 4

Effect of *Catharanthus roseus* (*L.*) (*Apocyneaceae*) Extract on Developmental Stages of *Erias favia* Stall

Sunil Kumar Dubey[1], R.C. Saxena[1], P.K. Mishra[1], Ashutosh Gautam[2] and M.L. Khare[1]
[1]Pest Control Research Laboratory, S.S.L. Jain College, Vidisha, M.P.
[2]India Glycols Ltd., Bazpur Road, Kashipur – 244 713, Uttranchal

ABSTRACT

Developmental defect in *Erias favia* was investigated due to aqueous extract of *Catharanthus roseus*. The extract caused significant reduction in adult emergence as well as in percentage of egg hatch in *Erias favia*.

Keywords: *Catharanthus roseus, Growth inhibition, Erias favia toxicity.*

Introduction

Botanical compounds particularly those with acute properties have been used as insecticides for many years (Jacobson and Crossby, 1971 and Rajendran and Gopalan, 1978). Some plants are excellent sources of substances disruptive of growth and behaviour (Kubo *et al*, 1983 and Kaur *et al.*, 1989). Kalyansundrum and Das (1985) and Despande *et al.* (1988) reported the larvicidal and enzymatic activity of *Catharanthus roseus* extract on mosquito larvae and on *Spodoptera litura*.

There are number of other plants whose extract caused juvenile and anti-juvenile hormone activity in dipteran and lepidopteran larvae (Saxena *et al.*, 1992 and Kubo *et al.*, 1983). *Catharanthus roseus* (L) (Apocynaceae) is sturdy perennial herb with known diverse biological activity. *Catharanthus roseus* showed developmental defects on larval stages of a Cotton pest of lepidoptera, *Erias favia* which is also pest on *Hibiscus esculentus*.

Plant Description and Identification

The plant, *Catharanthus roseus* (L.) Apocynaceae was found distributed through out the country. The perennial herb collected from botanical garden at the College campus, after identification in the P.G Department of Botany of the College. A voucher specimen was deposited in the herbarium of the laboratory at No. 15.

The structure of the plant was supported earlier by Rastogi and Mehrotra (1991) to contain Catharanthine Plrytochemical study with the leaves at *Catharanthus roseus* was found to possess "Catharanthine"

Materials and Methods

Plant Material

Fresh leaves of *Catharanthus roseus* collected in large quantities were thoroughly washed in tap water, shade dried and powdered (40 to 60 mesh). The powdered materials were extracted with water using the cold percolation method of Harborne (1984). The extract was concentrated using a vaccum evaporator and the residue was dissolved in acetone to make the desired concentrations.

Bioassays

Erias favia (Lepidoptera: Noctuidae) were cultured in glass jars capped with muslin cloth and rubber band in the insectary maintained at temp. + 1°C, RH 75 per cent and 14:10 hr. L:D Photoperiod Insects were fed on fresh seeds of *Hibiscus esculentus* ped soaked in drinking water. A piece of Whatman filter paper was kept in each jar for egg laying, eggs and pupae from the laboratory culture stock were used for experimental bioassays.

Treatment of Eggs

The freshly laid eggs (0-12 hr) were treated using in contact method. A film of extract was prepared by spreading 0.5 ml of different concentrations 1.0, 0.5, 0.25 and 0.05 per cent on petridish. The solvent was allowed to evaporate by rotating the peridishes. There were three replicates for each treatment. Percentage corrected mortality was calculated by Abbott's formula (1925).

Treatment of Pupae

Freshly ecdysed pupae (0-24 hr old) were treated with the different concentration of the extract topically by a Hamilton microlitre syringe in 5/ml doses. A control with acetone alone was also run separately. The pest performance parameters studied included the hatching success of eggs pupa period, adult emergence and mortality.

Results and Discussion

Hatch was considerably reduced when caped in contact with different concentration at the water extract of *Catharanthus roseus* (Tables 4.1 and 4.2). No egg hatched was recorded at the higher concentration, and at lower concentration it got reduced to 46, 22 and 6 per cent respectively. The delayed pupation and the longevity of adults also found to be 3.1 days against 7.3 days in control as reported by Garcia and Rambold (1984) who observed delayed pupation in *Rhodnius prolixus* by the treatment of Aazadirachtin. The inhibited egg development in female *Locusta migratoria* reported by Rambold and Siebr (1980) is quite similar as noticed in the egg of *Erias favia*. Tischler *et al.* (1989) described the effect of ecdysteroid on the growth of the flight muscles in *Manduca sexta*. They have mentioned that growth, development and sexual maturation largely regulated by ecdysome and JH. The results of present study also suggest that *Catharanthus roseus* could be a new success of biopesticidal compound for suppressing best population at an early stage of their development.

Table 4.1: Effect of Water Extract of *Cathranthus roseus* on Inactive Development Stages (Egg) of *Erias favia* Stall

Dose in Per cent	*No. of Egg Treated*	*Per cent Egg Hatching*	*Incubation Period (in days)*	*Per cent Mortality*	*Per cent Corrected Mortality*
1.0	100	–	–	100	100.0
0.5	100	–	–	100	100.0
0.25	100	–	–	100	100.0
0.01	100	6	4–5	90	92.0
0.05	100	22	4–5	72	75.0
0.025	100	46	4–6	54	50.0
Control	100	92	4–6	8	–

0.5 wg. Dose of each concentration was used.

First three concentration caused 100 per cent mortality of the eggs (0–12 hr).

Table 4.2: Effect of Water Extract of *Cathranthus roseus* on Inactive Development Stages (Egg) of *Erias favia* on Topical Application

Dose mg/pupa	*Conc. N (Per cent)*	*No. of Pupa Treated*	*Per cent in Death within Population (a)*	*Per cent Died Out in the Puparium (b)*	*Per cent Mortality at Pupal Stage (a+b)*	*Per cent Corrected Mortality*	*Per cent Adult Emergence*	*Adult Longevity (Days)*	*Pupal Period (Days)*
5	1.0	25	100.0	–	100.0	100	–	–	–
5	0.5	25	90.0	10.0	100.0	100	–	–	–
5	0.25	25	72.3	24.6	97.0	100	3.0	3.7	6.5
5	0.85	25	64.0	26.0	90.0	100	10.0	3.7	6.5
5	Control acetone alone	25	4.0	–	4.0	100	94.0	7.1	4.5

Adult longevity and pupal period is mentioned for living adult only.

Acknowledgement

Financial assistance received from MAPCOST (Project No. B/50/91, Bhopal) (M.P) India is gratefully acknowledged. Thanks are due to Dr. R.C. Saxena for guidance and correcting the manuscript.

References

Abbot's, W.S., 1925. A method of computing the effectiveness of an insecticide. *Journal of Economic Entomology*, 18: 265–270.

Deshpande, S.G., Joseph, M. and Sharma, P.N., 1988. Insect growth and development inhibition properties of *Cathranthus roseus*. *International Tropical Agriculture*, 6: 287–290.

Garcia, E.S. and Rembold, H., 1984. Effects of Azadirachtin on ecdysis of *Rhodnium prolixus*.

Harborne, J.B., 1984. *Phytochemical Methods*, 2nd Edn. John Willey and Sons, London, pp. 257.

Jacobson, M. and Crossby, D.G., 1971. *Naturally Occurring Insecticides*. Marcel Dekker Inc., New York, pp. 585.

Kalyansundrum, M. and Das, P.K., 1985. Larvicidal and synergistic activity of a extract for mosquito. *Indian Journal of Medical Research*, 82: 19–23.

Kaur, A., Thakur, S.S. and Sabitaraja, S., 1989. *Chrysanthemum indicum* an effective growth and development inhibitor of *Dysdercus similes*. *Journal of Environment Biology*, 10(40): 373–377.

Kubo, I.J., Klocke, A. and Assano, S., 1983. Effect of ingested phytoecdyseriod on the growth and development of two lepidopteroid larvae. *Journal of Insect Physiology*, 29(4): 307–316.

Rajendran, B. and Gopalan, M., 1978. Note on the insecticidal properties of certain plant extracts. *Indian Journal of Agriculture Science*, 9: 295–297.

Rembold, H. and Sieber, K.P., 1980. Effects of Azadirachtin on oocyte development in *Locusta migartoria* migratoriodes. Max planck Institute fur Biochemic, Martinsried federal Republic of Germany prec; *Ist Int. Neem Conf.*, Rottach Engern, pp. 75–80.

Rastogi and Mehrotra, 1991. Phytochemical study with the leaves of *Catharanthus-roseus* was found to possess "Catharantine" *Indian Medicinal Plant PID*, CSIR, New Delhi, India, pp. 90–91.

Saxena, R.C., Dixit, O.P. and Sukumaran, P., 1992. Laboratory assessment of indigenous plant extracts for anti-juvenile hormone activity in *Culex quinquefasciastus*. *Indian Journal of Medical Research*, 95: 204–206.

Tischler, E., Mark Paul, Clock, Shirly Hodsolen, Susan Mcready and Minwu, 1989. Ecdysteroids influence growth of the dorsolongitudinal flight muscles in the tobacco horn worn (*Manduca sexta*). *J. Insect. Physiol.*, 35(12): 1017–1022.

Ethnobotanical Wisdom and Microbial Studies on Medicinal Plants (2011) *Pages* **23–29**
Editors: **D.R. Khanna, Ashutosh Gautam, R. Bhutiani, Gagan Matta, Anil Dhiman**
Published by: **DAYA PUBLISHING HOUSE, NEW DELHI**

Chapter 5

Solid State Fermentation of Wheat for the Production of Cellulase

R.K. Tenguria[1], Krishna Jha[2] and Pariti Satalkar[1]
[1]Motilal Vigyan Adarsh Mahavidyalaya, Bhopal
[2]Central Institute of Agricultural Engineering, Nabibagh, Bhopal

ABSTRACT

Solid state fermentation of wheat straw for cellulase production by *Phanerochaete chrysosporium* has been studied. The most optimum moisture level for solid state fermentation was found to be 40 per cent while the best enzyme yield was obtained at an initial pH of 4.5 and incubation temperature of 30°C. Supplementation @ 0.2 per cent (W/W) was found to have enhanced the yield of the enzyme by three folds. With all conditions standardized an enzyme yield of 78 IU pergds could be obtained.

Keywords: *Cellulase, Phanerochaete chrysosporium, Fungus.*

Introduction

Utilization of lignocellulosic wastes is gaining considerable interest in most developing countries due to their availability in abundance. Almost five decades have passed since the pioneering work of Reese on fungal degradation of cellulosic substrates (Mandels and Sternberg, 1976, Esterbrauer *et al.*, 1991). In view of relative abundance of wheat straw, its biodegradation has been carried out by several fungi (Zadrazil and Brunnert, 1980: Zafar *et al.*, 1989). *Phanerochaete chrysosporium* has been reported to be a potent lignin degrading white rot fungi (Faison and Kirk 1985). In the present investigation an attempt has been made to utilize wheat straw for the production of cellulose by solid state fermentation.

Materials and Methods

Organism and Cultural Conditions

Phanerochaete chrysosporium (HHB 103737 S) was obtained from Prof. Kirk's laboratory of Forest Research Laboratory, Canada. It was grown and maintained on malt extract agar slants and repeated transfers were performed at 30 days intervals. The culture was preserved in refrigerator at 4°C. The medium for cultivation was prepared according to Kirk *et al.* (1978) with 0.22 g per litre of ammonium tartarate as the nitrogen source. As per the requirements the Erlenmeyer flask of 100 ml capacity was added with 4.5 ml of the basal medium containing 30 mm 2,2- dimethyl succinate to buffer the cultures at 4.3-4.5 and 0.5 ml of fungal mycelium. The fungal mycelium was obtained from a Roux bottle run in batch. Such an inocculums was used to obtain rapid primary growth of the culture (Ulmer *et al.*, 1983).

Preparation of Starter Culture

Starter culture of the test mould was prepared according to the standard method as described by Tao and Zuohu, (1997). Conidiospores were aseptically harvested from a 15-day old culture slant by suspending in 10 ml sterile distilled water containing 0.01 per cent Tween 80. The conidiophores suspension was suitably diluted to obtain a population of approximately 10^7 per ml One ml culture was used per 100 g substrate for all experimental purposes.

Solid State Fermentation

All fermentations were carried in autoclavable LDPE bags (30*25 cm: Himedia). The bags were perforated at 1.5 cm distance throughout and filled with the test materials. The raw materials *viz.* wheat straw was ground in Waring blender and 100 g of each were then loodely filled in the perforated autoclavable LDPE bags in triplicate. All the bags were then autoclaved at 15 psi for 30 minutes.

Optimization for Moisture Content

Moisture levels in solid state fermentation depends upon the nature of substrate, the organism and the type of end product (Ramesh and Lonsane1990). Since the nutrient concentration is inversely proportional to the quantity of water present in the substrate, an increase in the salt concentration resulting in high osmotic potential, will have an adverse effect on the growth and productivity. This preliminary experiment was thus undertaken to optimize the moisture content for each organism and substrate used in this study.

Fifty gram of substrate was transferred into the each three perforated autoclavable LDPE bags (Himedia, India) for testing of each substrate as well as organism. Each of the three bags were added with water at the rate of 25, 50, and 75 ml water. The bags were then autoclaved at 15 psi for 30 minutes. After cooling to room temperature the materials were inoculated with 2.5 ml of starter culture containing 1.0 X 10^7 cells or spores per ml. The bags were then incubated at 28° C. The growth of the organism was measured by standard plate count method.

Effect of Nitrogen Supplementation

Carbon : Nitrogen ratio is an important factor in determining the growth behaviour and productivity. Thus nitrogen in form of ammonium, nitrate and protein were supplied through addition of ammonium sulphate, urea, sodium nitrate and casein. The test substrate was tested for effect of nitrogen supplementation on enzyme yield. Three concentration *viz.* 0.2, 0.5, 1.0 per cent (as N) were added to the substrates and were inoculated with the test microorganisms. A control (without N) was also run simultaneously.

Effect

For the optimization of pH the wheat straw was soaked in water of different pH of raw material *i.e.* 3, 4.5, 6.0 and 7.0. The materials were then loosely packed in auloclavable LDPE bags and inoculated. Other fermentation conditions were same as for other experiments.

Optimization of Incubation Temperature

For experimentation the inoculated bags were incubated at different temperatures *viz.* 25°, 30°, 37° and 40° C.

Determination of Cellulose Activity

Sample Extraction and Analysis

One gram sample in triplicate were withdrawn aseptically at 5 days intervals and extracted in 10 ml citrate buffer (0.05 M; pH 4.8) using pestle mortar. The content was shaken for 10 min on a rotatory shaker and finally centrifuged. The supernatant was collected and the residue was washed twice and again centrifuged. All the supernatant were finally pooled and cellulase activity was determined.

Filter Paper (FPase) Activity

FPase activity was determined using 1.0 cm X 6.0 cm of Whatman No. 1 filter paper strip (in 1 ml of 100 mM sodium citrate buffer, pH 5.0) as substrate and incubating it with 1.0 ml of enzyme extract for 1 hr at 500° C. The reaction was terminated by boiling the contents for 2 min. The released reducing sugar content was measured by Nelson's (1944) using glucose as standard.

Carboxy Methyl Cellulose (CMCase) Acidity

CMCase activity was determined by using the IUPAC method (IUPAC, 1987). One ml of extract was added with 1 ml of Carboxymethyl cellulose (1 per cent prepared in 0.05N; pH 4.8). The contents were incubated at 500° C for 30 min. and finally the reaction was terminated by adding 3 ml of 1 mM sodium carbonate solution. The released reducing sugar was estimated by Nelson's method (1944).

One unit of cellulase activity was CMCase expressed as micromoles of glucose released per min per gram of culture.

Determination of Reducing Sugars

Two gram samples were extracted in ten ml of 70 per cent aqueous ethanol by finely grinding in pestel mortar. The contents were then centrifuged at 5000 rpm for 10 minutes. The pellets were washed thoroughly in same solution twice and again followed by centrifugation. All the supernatants were pooled and evaporated to dryness using flash evaporated. The contents were re-dissolved in 5 ml of 70 per cent aqueous ethanol and used for estimation of carbohydrates using the standard method according to Nelson (1944).

Results and Discussion

Moisture levels in solid state fermentation depends upon the nature of substrate, the organism and the type of end product (Ramesh and Lonsane, 1990; Smits, *et al.*, 1997). The results show that the most optimum moisture level was 40 per cent (Figure 5.1). Since the nutrient concentration is inversely proportional to the quantity of water present in the substrate, an increase in the salt concentaration resulting in high osmotic potential, will have an adverse effect on the growth and productivity. Our results are in accordance with those of Jha *et al.* (1995) who have also shown 40 per cent as the most

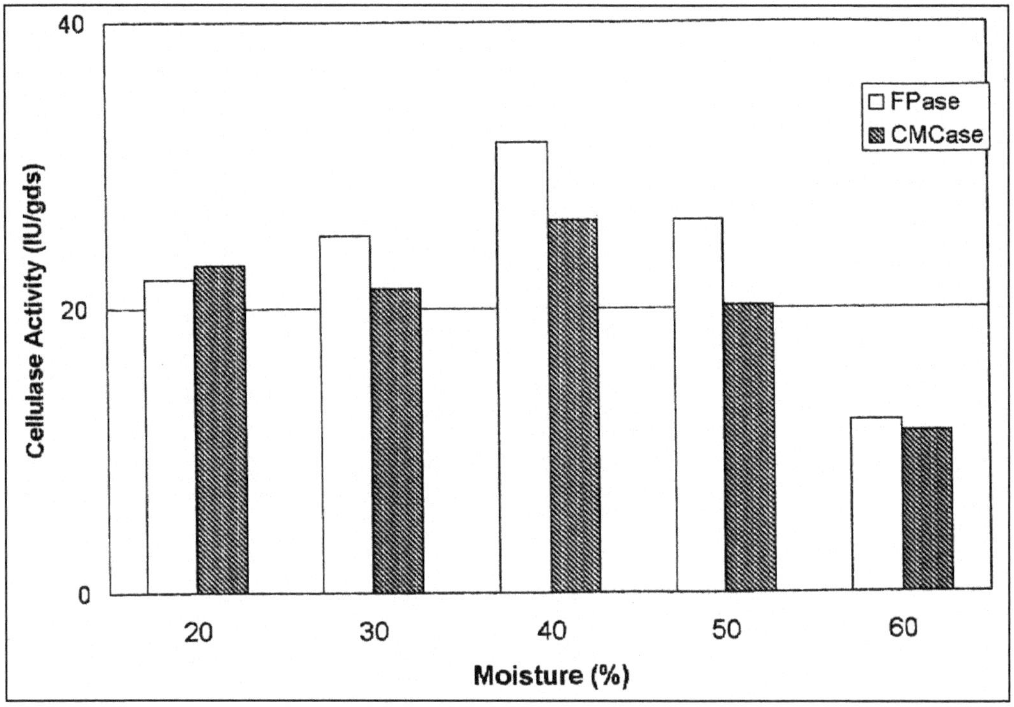

Figure 5.1: Effect of Moisture Level on the Production of Cellulase

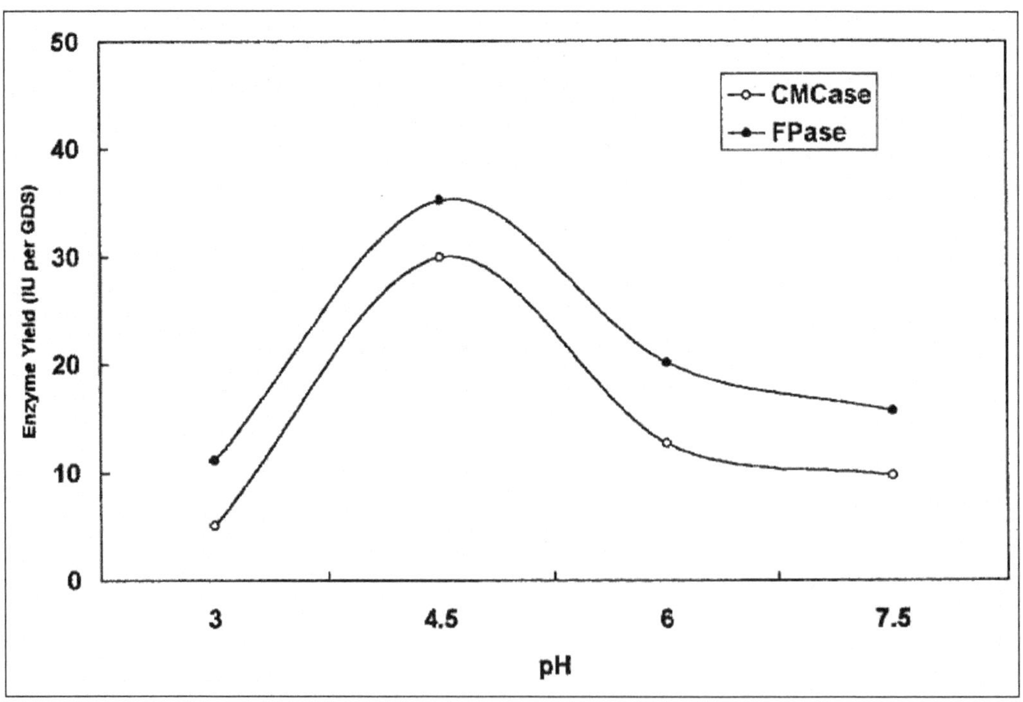

Figure 5.2: Effect of pH on the Cellulase Production

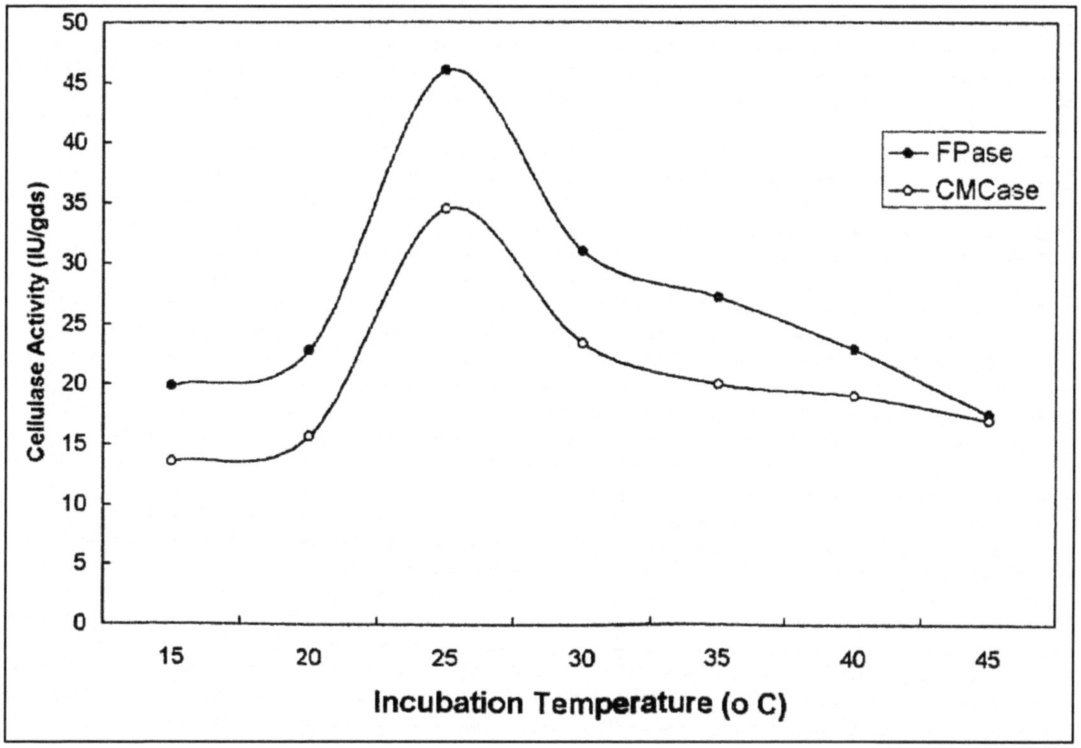

Figure 5.3: Effect of Incubation Temperature on Cellulase Activity

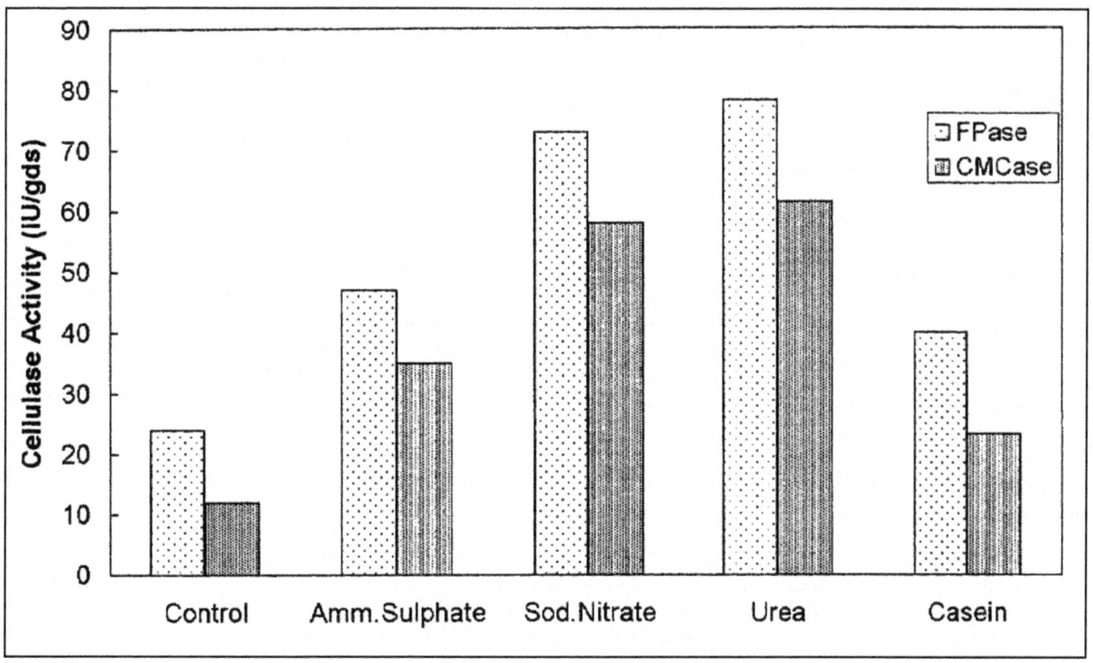

Figure 5.4: Effect of Nitrogen Sources on Cellulase Activity

optimum initial moisture content for the cellulase production from soyahulls. Zafar *et al.* (1989) and Smits *et al.* (1997) have however reported a minimum moisture of 60 per cent in solid state fermentation. It may however be mentioned that lower initial moisture is always better as it greatly facilitates handling and extraction of the enzyme at pilot scale.

While studying the physiological characters of *Phanerochaete chrysosporium* grown on wheat straw it was seen that the high cellulase yield of 31.5 IU per g dry (FPase) substrate was obtained at an initial pH of 4.5 (Figure 5.2). Further, the temperature optimization experiment indicated that 25° C was the most optimum temperature for best cellulase activity (Figure 5.3).

Effect of supplementation of various nitrogen *viz.* Ammonium sulphate, sodium nitrate, urea and casein was also studied. The results indicate that all the compounds used showed stimulatory effect on the cellulase activity (Figure 5.4). Among different nitrogen sources supplemented, urea was found to be the best nitrogen source for cellulase production. Our results are in accordance with that of Nigam *et al.* (1988) who have also shown urea to be the best nitrogen sources in the fermentation of sugarcane bagasse and soyabean huls, respectively.

The over all results indicates that under all conditions standardized, the cellulase activity of 74.9 IU per g dry substrate could be obtained in 6 days and thus concludes that wheat straw can be successfully used for the production of cellulase at pilot scale.

Acknowledgement

The authors are grateful to the Head, Department of Botany, Principal, Motilal Vigyan Mahavidyalaya, Bhopal and Project Director, SOU Centre and Director, Central Institute of Agricultural Engineering, Bhopal who have kindly permitted to undertake this work at their laboratories.

References

Esterbrauer, H., Steriner, W., Labodova, J., Hermann, A. and Hayn, M., 1991. Production of *Trichoderma* cellulase in laboratory and pilot scale. *Bioresource Technology*, 36: 51–65.

Faison, B.D. and Kirk, T.K., 1985. Factors involved in regulation of ligninase activity in *Phanerochaete chrysosporium*. *App. Environ. Microbiol.*, 49: 299–304.

IUPAC (International Union for Pure and Applied Chemistry), 1987. Measurement of cellulase activity. *Pure and Applied Chem.*, 59: 257–268.

Jha, Krishna, Khare, S.K. and Gandhi, A.P., 1995. Solid state fermentation of syahull for the production of cellulase. *Bioresource Technol.*, 54: 321–322.

Kirk, T.K., Schulz, E., Connors, W.L., Lorenze, L.F. and Zeikes, J.G., 1978. Influence of culture parameters on lignin metabolism by *Phanerochaete chrysosporium*. *Arch. Microbiol.*, 117: 277–285.

Mandels, M. and Sternbeg, D., 1976. Recent advances in cellulase technology. *Ferment. Technol.*, 54: 267–286.

Nelson, N., 1944. A photometric adaptation of Somogyi method for determination of glucose. *J. Biol. Chem.*, 15: 375.

Nigam, Poonam, Pandey, A. and Prabhu, K.A., 1988. Fermentation of Bagasse by submerged fungal cultures: Effect of nitrogen sources. *Biological Wastes*, 23: 313–317.

Ramesh, M.V. and Lonsane, B.K., 1990. Critical importance of moisture content of the edium in alpha-amylase production by *Bacillus licheniformis* M–27 in solid state fermentation system. *Appl. Microbiol. Biotechnol.*, 33: 501–505.

Smits, J.P., Rinzema, A., Tramper, J., Sxhlosser, E.E. and Knol, W., 1997. Accurate determination of process variable in a solid-state fermentation system. *Process Biochemistry*, 31: 669–678.

Tao, S., Beihui, L. and Zuohu, L., 1997. Enhanced cellulase production fed-batch solid state fermentation of *Trichoderma viride* SL-1. *Chem. Technol. Biotechnol.*, 69: 429–432.

Ulmer, D.C., Leisola, J., Puhakka, J. and Fiechter, I., 1983. *Phanerochaete chrysosporium*: Growth pattern and lignin degradation. *Eur. J. Appl. Microbiol. Biotechnol.*, 18: 153–157.

Zadrazil, F. and Brunnert, H., 1980. The influence ammonium nitrate supplementation on digestion and *in vitro* digestibility of straw colonized by higher fungi. *Eur. J. Microbiol. Biotechnol.*, 9: 37–44.

Zafar, S.I., Sheeraz, Q. and Abdullah, N., 1989. Degradation of lignocellulosic component of wheat straw–*Coriolus versicolor* Solid state fermentation under nitrogen starved conditions. *Biological Wastes*, 27: 67–70.

Ethnobotanical Wisdom and Microbial Studies on Medicinal Plants (2011) *Pages* **30–35**
Editors: **D.R. Khanna, Ashutosh Gautam, R. Bhutiani, Gagan Matta, Anil Dhiman**
Published by: **DAYA PUBLISHING HOUSE, NEW DELHI**

Chapter 6

Phytotoxic Effects of *Eucalyptus globulus* Leaf Extract on Seed Germination and Seedling Growth of Vicia

R.B. Singh[1] and U.C. Das[2]
[1]Cytogenetic Laboratory, P.G. Department of Botany, M.S. College, Motihari- 845 401
[2]Organic Laboratory, Department of Chemistry, S.N.S. College, Motihari - 845 401

ABSTRACT

Eucalyptus globulus L.(Myrtaceae) stared as antisocial tree for social forestry (Kohli, 1987) is commonly transplanted through out India on roadsides, bare lands and boundaries of crop fields without considering its allelopathic effects on crop plants, especially cereals and pulses. The aerial parts, especially leaves of this tree contain volatile tarpenes (Del Moral and Muller, 1970), water soluble toxins and growth inhibitors (Grummer and Beyer, 1960 and Al-Mousawi, Al-Naib, 1976 and Pandhy *et al.*, 2000). Toxic effects of different concentrations of leaf extract of *Eucalyptus globulus* (*viz.*, 5, 10, 15, 20 and 25 per cent) was studied on percentage of seed germination,their survival, growth of radical and plumule and establishment of secondary roots on three species of Vicia-*V. narborensis*, *Vicia benghalensis*. and *V. faba*.In the present investigation each concentration showed inhibitory effects of the above parameters which gradually increases with increasing concentrations of the leaf extract. Delayed germination was also noticed in the treated seeds. *Vicia narborensis* proved to be more sensitive to these toxic leachates than the other two species. The above findings clearly showed that the transplantation of *Eucalyptus species* on boundaries of crop plants should be totally banned.

Introduction

In plant communities, many plant species interact with each other, thus, regulate the growth of neighbouring plants by releasing certain chemicals, stimulators or inhibitors resulting into an allelopathic action (Putnam and Tang, 1986), Such Chemic Substances which interfere the physiological processes and life-cycle of another plant species in communities are referred as allelopathic substances and the phenomenon as Allelopathy. These allelochemic substances are present in various plant parts and are released into the environment by various processes such as leaching, volatilization, root exudation and decomposition of plant residues (Rice, 1984). Recently several works have been carried out on the allelopathic effects between crops and crops,between weeds and crops and between weeds and herbaceous plants (Eyini *et al.*, 1989, Chandel and Mehta, 1990, Singh, 1999, Pandhy *et al.* 2000), but interactions on the tree crops are still scanty.

Eucalyptus released several volatile tarpenes (Del Moral and Muller, 1970), water soluble toxins and growth inhibitors (Grummer and Beyer, 1960, Al-Mousawi and Al-Naib, 1976, Pandhy, 2000) and volatile materials (Baker, 1966) which in one or other way inhibit the growth of crop plants.

Considering the allelopathic effect of *Eucalyptus globulus* on several important crop plants and termed as antisocial tree for forestry (Kohli, 1987), it was planned to see the phytotoxic effects of *Eucalyptus globulus* on seed germination, seedling growth and its survival on three species of Vicia *Viz.-V. faba, V. narborensis* and *V. benghalensis.*

Materials and Methods

Fresh and matured leaves of *Eucalyptus globulus* (Myrtaceae) were plucked, washed thoroughly, shade dried and then soaked in distilled water for 48 hrs, (100 gms in 1 liter of distilled water). The leachate so obtained was considered as Stock solution. Further dilution was made by adding adequate amount of distilled water into the stock solution to obtain 5, 10, 15, 20 and 25 per cent concentrations of the Leachate.

Seeds of *Vicia faba* and *V. benghalensis* were procured from local registered suppliers while seeds of *V. narborensis* was obtained with the courtesy of Dr. A.K. Singh, the then Professor of Botany, M. S. College, Motihari.

100 healthy seeds of each species of *Vicia* were taken, thoroughly washed in distilled water, then kept for soaking in different test tubes containing various concentrations of the leachate. After soaking for 24 hr. seeds were taken out, washed with distilled water and then allowed to germinate in different pertridishes lined with moist filter paper. A control set in distilled water in identical condition was also maintained for comparison. The daily record of seed germination and their mortality was noted upto a week, while length of radical, plumule and number of secondary roots were noted after seven days.

Results and Discussion

As it reveals from Tables 6.1–6.3 that the percentage of seed germination and its survival is dose dependent. Increase of leachate concentration decreases the percentage of seed germination and survival of seedlings. It was also observed that increase of concentration delayed the process of germination and inhibited the growth of seedlings in all the test cultivars of Legumes. *V. narborensis* proved to be more sensitive than other two species of Vicia.

Table 6.1: Effect of Leaf Leachate of *Eucalyptus globulus* on Seed Germination and Seedling Growth of *Vicia faba*

Leaf Leachate Conc. (per cent)	*Germination (per cent)*	*Periods Taken into Germination (in hrs.)*	*Survival (per cent) (mm)*	*Length of Plumule and Radical (after a week)*		*No. of Sec. Roots (after a week)*
				Radical (mm)	*Plumule (mm)*	
Control	100	48–60	95.00	55	81.5	18
5	98	48–60	91.84	51	78.4	16
10	90.00	48–72	90.00	40.5	72.6	13
15	90.00	60–84	87.7	46.2	66.4	08
20	86.00	60–84	84.88	42.4	60.5	06
25	82.00	72–84	80.48	38.5	54.2	06

Table 6.2: Effect of Leaf Leachate of *Eucalyptus globulus* on Seed Germination and Seedling Growth of *V. narborensis*

Leaf Leachate Conc. (per cent)	*Germination (per cent)*	*Periods Taken into Germination (in hrs.)*	*Survival (per cent) (mm)*	*Length of Plumule and Radical (after a week)*		*No. of Sec. Roots (after a week)*
				Radical (mm)	*Plumule (mm)*	
Control	94	48–60	94.68	51.4	73.2	20
5	90	48–72	90.00	50.8	72.8	18
10	86	72–84	84.88	46.00	72.4	12.
15	80	72–84	80.00	39.02	65.4	08
20	70	84–96	75.71	37.3	58.1	04
25	65	84–108	72.30	35.2	44.4	02

Table 6.3: Effect of Leaf Leachate of *E. globulus* on Seed Germination and Seedling Growth of *V. benghalensis*

Leaf Leachate Conc. (per cent)	*Germination (per cent)*	*Periods Taken into Germination (in hrs.)*	*Survival (per cent) (mm)*	*Length of Plumule and Radical (after a week)*		*No. of Sec. Roots (after a week)*
				Radical (mm)	*Plumule (mm)*	
Control	96	48–60	100	62.24	120	32
5	90	48–60	90.00	60.05	115	27
10	90	48– 72	87.77	54.22	105	27
15	85	48–84	81.17	49.20	94	20
20	75	60–84	80.00	45.45	91	12
25	70	60–96	78.57	40.20	82	09

Delayed germination with increasing doses of the leachate might be due to presence of more phytotoxins which arrest the protease enzyme activity or action of hydrolytic enzymes (Shrivastava *et al.*, 1972, Singh and Saxena, 1991), Saxena and Singh (1987) are of opinion that seed germination is mostly controlled by osmotic potential which regulates the action of hydrolytic enzymes. Mishra and Srivastava (1978) and Khosh-Khui and Bassiri (1979) reports that the toxic substances present in the leachate damaged the enzyme system involved in the metabolic and repair mechanism or their degradation products in excess. Similar opinions have also been made by Nicollier *et al.* (1985), Singh and Singh (1997), Jha *et al.* (1998), Singh (1999) and Singh and Singh (2001)while Choe (1972) reports that indigenous hormones are denatured or leached out to the surround medium by exogenous application of plant regulating chemicals. In contrast there are several reports that the secondary metabolites such as α and β pinene, Limonine, Linolol, P-cimene, Citronella, α–terpenol, geraniol, citral etc. present in leaves of *Eucalyptus* are of toxic nature, hence might have inhibited the process of germination resulting decline in percentage as well as delayed in the germination process. Similar reports have also been made by Srinivasan *et al.*(1990) on Piegeonpea and Soyabean, Pandhy *et al.* (2000) on finger millet and Devasagayam and Ebenezar (1996) on arable crops.

Reduction in seedling growth as observed in present investigation has also been noticed by several co-workers such as by Singh and Nandlal (1993) on wheat, mustard and chick pea treated with *Eucalyptus rostrata* by Jaikumar *et al.* (1990) on ground nut; Pandhy *et al.* (2000) on Ragi treated with *E.globulus* and by Singh and Singh (2001) on *Vicia faba* treated with *Cannabis* extract and by Singh (1999) on *Ageratum conyzoides* treated with *Parthenium hysterophorus*, Tripathi *et al.* (1981) and Mukherjee and Sahai (1985) have reported that the allelopathic actions differ from plant to plant. Gleissman and Muller (1978) is of opinion that allelochemics of plant might be incorporated into the soil by leaching of the substance from shoot canopy due to dissolving action of rain water and they might have arrested the physiological, biochemical and cytological activities of seedlings leading to poor seedling growth, similar opinion has also been made by Pandhy *et al.* (2000). While Kohli (1990) is of opinion that reduction in seedling growth by the allelochemics might be due to inhibition and/or checking of protein,nucleic acid and carbohydrate synthesis. The opinion of Kohli (1990) seems to be more authentic as these allelochemics have greater amount of terpenoids, flavonoids and phenolics, which are more toxic to the various physiological and metabolic processes related to the seedling growth and has directly or indirectly inhibited the seedling growth and number of secondary roots. Decrease in the length of radicals and plumules at higher doses may safely be attributed to the toxic effects of the allelochemicals or their degradation products on the root tipcells, that retarded cell divison, thus,decline the length of primary and secondary roots and number of secondary roots. Similar findings have also been made by Shastree *et al.* (1989), Bansal *et al.* (1992), Singh and Singh (1997), and Singh (1999).

References

AL-Mouswi, A.H. and Al-Naib, F.A.G., 1976. Volatile growth inhibitors produced by *Eucalyptus microthea*. *Bull. Biol. Res. Center,* 7: 17–23.

Baker, H.G., 1966. Volatile growth inhibitors produced by *Eucalyptus globulus*. Madrono. S. Francisco. 18: 207–211.

Bansal, G.L., Nayar, M. and Bedi, Y.S., 1992. Allelopathic effect of *Eucalyptus machromlyncha* and *E. youmanni* on seedling growth of wheat and radish. *Indian. J. Agri. Sci.*, 62(2): 771–772.

Chandel, Y.S. and Mehta, P.K., 1990. Nematicidal properties of leaf extracts of wild sage (*Lantana camara*). *Indian. J. Agri. Sci.*, 60: 781–782.

Choe, H.T., 1972. Effect of presoaking seeds of *Pisum sativum* in GA_3, IAAand kinetin solution on seedling growth. *Hort. Sci.*, 7: 476–478.

Del Moral, R. and Muller, C.H., 1970. The allelopathic effect of *E. camaldulensis*. *American Midland Naturalist*, 83: 254–282.

Devasagayam, M.M. and Ebenezar, E.G., 1996. Allelopathic effect of *Eucalyptus* on Arable crops. *J. Ecotoxicol. Envirom. Moit.*, 6(1): 73–75.

Eyini, M., Jaikumar, M.and Pannirselvam, S., 1989. Allelopathic effect of bamboo leaf extract on the seedling of groundnut. *Trop. Ecol.*, 30: 138–141.

Gleissam, S.R. and Muller, C.H., 1978. The allelopathic mechanism of dominance in Brasken in Southern California. *J. Chem. Ecol.*, 4: 337–362.

Grummer, G. and Beyer, H., 1960. The influence exerted by species of Camelina on flax by means of toxic substances. In: *The Biology of Weeds*, (Ed.) J.L. Harper. Blackwell Scientific, Oxford, pp. 153–157.

Jaikumar, M., Eyini, M. and Pannirselvam, S., 1990. Allelopathic effect of *Eucalyptus globulus* in groundnut and corn. *Comp. Physiol. Ecol.*, 15: 109–113.

Jha, P.K., Choudhary, S.K. and Prasad, A., 1998. Effects of phytosaponin (*Mollugo saponin*) on the seed germination and seedling growth of *Glycine max* L. *Bio. Journal*, 10(1 and 2): 83–85.

Khosh-Khui, M. and Bassiri, 1979. Inhibition of seedling growth by wild myrtle (*Myrtus communis*). *Weed. Res.*, 19: 45–50.

Kohli, R.K., 1987. *Eucalyptus* an antisocial tree for social forestry, Social for Rural Development, Solan. *Indian Society of Tree Scientists*, pp. 235–241.

Kohli, R.K., 1990. Allelopatic potential of *Eucalyptus*, project report MAB-DOEN Project, India, p. 199.

Mishra, G.P. and Srivastava, A., 1978. A germination inhibitor from *Mellilotus indica*. *Bot. Soc. Univ. of Sagar*, 25/26: 54–56.

Mukherjee, U. and Sahai, R., 1985. Allelopathic effect of *Indigofera enneaphylla* on the seed gernination and seedling establishment of *Alysicarpus monilifer*. *J. Ind. Bot. Soc.*, 64: 169–171.

Nicollier, G. F., Pope, D.F. and Thompson, A.C., 1985. *The Chemistry and Allelopathy*. American Chemical Society, Washington D.C., pp. 207–218.

Pandhy, B., Patnaik, P.K. and Tripathy, A.K., 2000. Allelopathic potential of *Eucalyptus* leaf: Litter leachates on germination and seedling growth of finger millet. *Allelopathy Journal*, 7: 69–78.

Putnam, A.R. and Tang, C.S., 1986 *Allelopathy: State of the Sciences in the Science of Allelopathy*. John Wiley and Sons, Inc., p. 1–17.

Rice, E.L., 1984. *Allelopathy*. Academic Press, Oriando, F.L., 1: 131.

Saxena, O.P. and Singh, G., 1987 Osmotic priming studies in some vegetable seeds. *Acta. Hort.*, 215: 201–207.

Shastree, N.K., Pathak, S. and Islam, M.S., 1989. Studies on the allelopathic competence of *Argemone maxicana*. *Bio. Journal*, 1(2): 33–38.

Singh, R.B., 1999. Allelopathic effect of *Ageratum conyzoides* on seed germination and seedling growth of *Parthenium hysterophorus*. *J. Environ. and Poll.*, 6 (2 and 3): 169–172.

Singh, G. and Saxena, O.P., 1991. Physiological and Biochemical studies associated with instant germination in tomato. *Proc. Intern. Seed Symp.*, Jodhpur, pp. 313–317.

Singh, R. B. and Singh, A.K., 1997. Stimulating and toxic effects of *Cannabis extracts* on seed germination and seedling growth of Pea (*Pisum sativum*). *Flora and Fauna*, 3(2): 107–108.

Singh, R.B. and Singh, S., 2001. Toxic effects of *Cannobis sativa* extracts on *Vicia faba*. *J. of Environ. and Poll.*, 8(1): 75–78

Singh, R.B. and Nandlal, D.P.S., 1993. Allelopathic effects of *Eucalyptus* and *Leucana* Leaf Litter on the germination and seedling growth of some fodder crops. *Forage Res.*, 19: 13–16.

Srinivasan, K., Ramason, M. and Shantha, R., 1990. Tolerance of pulse crops to allelochemicals of tree species. *Indian J. Pulses. Res.*, 3: 40–44.

Srivastava, A.K., Sharma, K. and Ahuja, K.L., 1972. Effect of salt stress on seed germination and survival of soyabean seedlings. *Trop. Ecol.*, 13: 27–31.

Tripathi, R.S., Singh, R.S. and Rai, J.N.P., 1981. Allelopathic potential of *Eupatorium adenophorum*: A dominant ruderal weed of Meghalaya. *Proc. Nat. Sci. Acad.*, 47: 458–465.

Ethnobotanical Wisdom and Microbial Studies on Medicinal Plants (2011) *Pages* **36–41**
Editors: **D.R. Khanna, Ashutosh Gautam, R. Bhutiani, Gagan Matta, Anil Dhiman**
Published by: **DAYA PUBLISHING HOUSE, NEW DELHI**

Chapter 7

Pharmacological and Ethnomedicinal Studies of *Tridax procumbens* Linn Family (Compositeae) for Anti-asthmatic Activities

Sapan Patel, R.K. Tenguria and P.K. Mishra
Department of Botany, M.V.M., Bhopal, M.P.

ABSTRACT

Tridax procumbens linn. of family- *Compositeae* (*Asteraceae*) is found plentifully in the plain areas near rivers and on marshy land. The villagers use this plant in bronchial asthma, dysentery and diarrhoea. Leaf juice is generally used to kill the bed bug in cote and wooden furniture and to protect the woods from termites etc. The whole herb juice in decoction is used by the villagers to check the hemorrhage of wounds.The detailed chemistry of the insecticidal compound was carried out, which revealed the structure of the compound. In the present communication, an effort has been made to screen the *Tridax procumbens,* a common plant used in folk medicines.

Introduction

The practices of traditional medicine are based on hundreds of years of belief and observations which create the development and spread of modern medical system. In some countries traditional medicines remain an integral part of formal health system and exist at equal footing with modern medicine. The methods of practices of traditional medicine may appear to be numerous and dissimilar but they all represent variations of three basic activities, faith healing, hygiene measures and drug therapy. Traditional medicine plays an important role in health care in India.

Survey of certain pharmacologically active phyto-constituents in wild plants, which are commonly used in folk medicine, have attracted our attention during the past few decades. The results of these studies have proved to be of much significance for their commercial exploitation. Many reports are available now on the phytochemical screening of plants of a particular geographical region. However, only few are concerned with plants, used by rural folks and tribal peoples.

Materials and Methods

Tridax procumbens L.(Compositeae) is a common weed found throughout India and used for a variety of medicinal purposes. The whole herb was collected from the field of village Chiocholi(Gavasne Forest) district Betul, India.

During months of Feb-March samples were (A voucher specimen is preserved in the Herbarium) shade-dried and powered to a fine mesh extraction and purification: Air dried powdered material was soxhlated in acetone for 24 hours and solvent was evaporated under reduced pressure in vacuum evaporator. A dark green semi-solid substance was obtained which accounted for 0.01 per cent of the total dried powdered material. The crude extract was diluted with acetone to made 10 per cent stock solution, dilutions were made from this stock solution. The crude extract was initially tested on TCLP plate of Silica Gel "G" (0.22mm) where total eight spots were obtained using Benzene: Methanol (9:1). The crude extract was then poured into an open glass column (65cm x 4cm) and eluted with C_6H_6: MeOH (9:1) and about 7 fraction were collected. Some fractions overlapped when subjected to TLC.They were once again eluted using the same solvent system. The different fractions were collected in the small glass vials and were kept in refrigerator. Out of seven fractions obtained, fraction FR_5 (Brown colour) was analyzed spectroscopically for active constituents.

Results

Purification, Isolation and Structural Elucidation of Biologically Active Compound

The purification and isolation of biologically active compound from crude extract of *Tridax procumbens* was done by column chromatography and TLC. Table 7.1 indicates the results for column chromatography of petroleum, ether, acetone and methanol extracts of *Tridax*. A total of nine fractions were obtained. The solvent system used for column chromatography are Benzene : P. Ether (4 : 1), P. Ether : Acetone (9 : 1) and CCl_4 : CH_3COOH (5 : 1).The detail of the percentage yield, weight of each and colour characteristic have been mentioned in Table 7.1.

TLC of Crude Extract

The three crude extracts obtained through soxhletion were preliminarily examined on TLC plate using silica Gel "G" and different solvent systems as indicated in Table 7.2. For P.Ether extract Benzene: P.Ether (2:3) gave two spots and $CHCl_3$,: Ether(4:1) gave three spots.

In acetone extract of three different combinations using Benzene:Ether (4:1) four spots were obtained while in Benzene: Methanol (9:1) eight spots were obtained.

For methanol extract four combinations were used, which gave three and one each.In the last three combinations with different revalues and different behaviour pattern as already indicated in Table 7.2.

Further Purification

The column purification fractions were further identified for their purity on TLC Plates. Fraction FR, using C_6H_6: P. Ether (2:3) gave three spots thus it seems to be mixture of compounds whereas FR_2,

Table 7.1: Column Chromatography of *Tridax procumbens* Linn.

Plant Extract	*Solvent System Power (50 ml)*	*Fraction Obtained Dried (mg)*	*Weight of Fraction (mg)*	*Per cent Yield of Characteristics*	*Colour*
Tridax P. Ether	Benzene:P. Ether	FR_1 TPP	500	0.10	Light Green
production	2:1	FR_2 TPP	15	0.03	Light yellow
Extract	$CHCl_3$:Ether	FR_3 TPP	11	0.02	Yellowish
	4:1				
	Benzene:methane	FR_4 TPP	60	0.12	Green Yellow
	9:1				
Tridax	Benzene:Acetone	FR_5 TPP	620	0.12	
procumbens	4:1	FR_6 TPP	25	0.13	Brown
		FR_7 TPP	10	0.05	Milky Colour
	P. Ether:Acetone	FR_8 TPP	110	0.02	Light Yellow
		FR_9 TPP	320	0.022	Green
Tridax	CCl_4:CH_4COOH			0.064	Yellow
procumbens	5:1				

Table 7.2: TLC of Crude Extract of *Tridax procumbens* Linn.

Plant Extract	*Solvent System Used in TLC*	*No. of Spots*	*Behaviour*		*Rf Value of Each Spot*
			Visible Light	*UV Light*	
Tridax procumbens	Benzene:P.Ether	2 Spots			
	2:3	Spot-I	Invisible	Blue	0.25
		Spot-II	Brown	Yellow	0.35
P. Ether Extract	$CHCl_3$. Ether	3 Spots			
	4:1	Spot-I	Dark yellow	Fluorescent	0.18
		Spot-II	Yellow	Brown	0.25
		Spot-III	No spot	Fluorescent	0.12
Tridax procumbens	Benzene: Ether	7 Spots			
		Spot-II	Green	Dark blue	0.44
Crude		Spot-III	Light Yellow	Green	0.64
Acetone		Spot-IV	Invisible	Fluorescent	0.94
Extract		Spot-V	Visible	Light Yellow	0.85
		Spot-VI	Invisible	Fluorescent	0.94
		Spot-VII	Dark Green	Black	0.92
	Chloroform : Ether	4 Spots			
	4:1	Spot-I	Blue	Green	0.07]
		Spot-II	Yellow	Green	0.271
		Spot-III	Invisible	Fluorescent	0.875
		Spot-IV	Yellow	Green	1.000

Contd...

Table 7.2–Contd...

Plant Extract	*Solvent System Used in TLC*	*No. of Spots*	*Behaviour*		*Rf Value of Each Spot*
			Visible Light	*UV Light*	
	Benzene : Methanol	8 Spots			
	9:1	Spot-I	Green	Black	0.07
		Spot-II	Yellow	Fluorescent	0.]5
		Spot-III	Yellow	Invisible	0.26
		Spot-IV	Yellow	Red	0.50
		Spot-V	Green	Blue	0.75
		Spot-VI	Green	Black	0.84
		Spot-VII	Yellow	Green	0.96
		Spot-VIII	Green	Green	1.00
	Benzene : Methanol	3 Spots			
	9:1	Spot-I	Dark Green	Black	0.92
		Spot-II	Green		1.00
		Spot-III	Yellow	Green	0.96
	Chloroform : Ether	1 Spot			
	4:1	Spot-I	Green	Reddish Black	1.00
	Benzene: Acetone	1 Spot			
	4:1	Spot-I	Light Green	Black	0.75
	Benzene : P. Ether	1 Spot			
		Spot-I	Yellow	Green	0.24

gave no spot presuming it to be a purified compound. With Benzene: Methanol (9:1) also FR and FR_2, gave no further splitting hence was thought to be purified. Erections five to nine were tested using $CHCl_3$: Ether (4:1) in which fractions 3coded FR_6, FR_8 and FR_9, gave no further spot thus assumed to be purified whereas fractions coded FR_5 and FR_7 gave two spots each indicating a further need of their purification. The characteristics of each spot in UV and visible light along with their Rf values have been already mentioned in Table 7.3.

Isolation and Structural Elucidation of Compound

Isolation and structural elucidation of the purified fraction over IR,UV, HNMR and mass spectrum were carried out using spectrophotometer at R.S.I.C, Chennai. The IR spectrum revealed the absorption at 3010 cm^{-1}, (C=C-H), 2940 cm^{-1} (CH_3), 1990 cm^{-1} (Amides), 1450 cm^{-1} (C-H-1), 1370 cm^{-1}, 1210 cm^{-1}, 750 cm^{-1} (monosubstituted aromatic ring) and 660 cm (monosubstituted aromatic ring).

The UV spectra showed absorption at maximum, 412 nm, H-NMR showed intense peak at S 0.945 (7-C Methyl Group), S.170 (6H), S. 4.27 (3H), S 7.20 (Orthopteron of the aromatic ring). The $CHCl_3$, soluble fraction of the methanolic extract of the whole hert, *Tridax procumbens* which yielded a flavones glycoside (Figure 7.1)

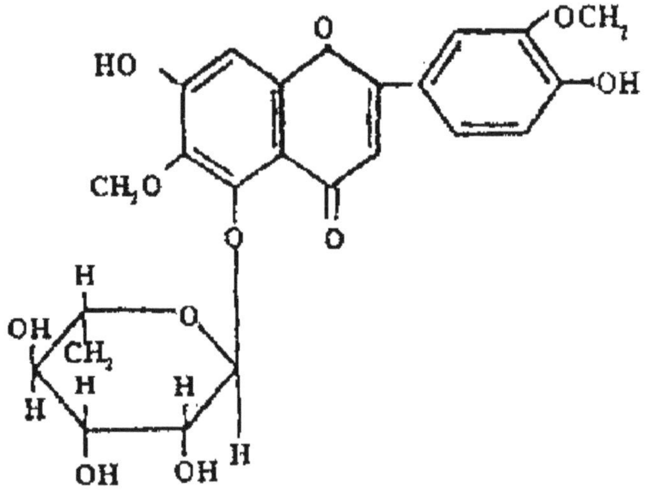

Figure 7.1: Trihydroxy-6,3 D Di methoxy Flavone 5-0-α-1-rhmnoyranoside

Table 7.3: TLC of Purified Compound of *Tridax procumbens* Linn.

Solvent	*Purified Fraction*	*No. of Spots*	*Colour Characterization in*		*Rf Value of Each Spot*
			Visible Light	*UV Light*	
Benzene : P. Ether	FR_1, TPP	3 Spots			
9:1		Spot-I	Brown	Invisible	0.10
		Spot-II	Invisible	Yellow	0.25
		Spot-III	Invisible	Blue	0.35
	FR_2, TPA	No Spot	No Spot	No Spot	Nil
	FR_3, TPA	No Spot	No Spot	No Spot	Nil
	FR_4, TPA	No Spot	No Spot	No Spot	Nil
Chloroform : Ether	FR_5, TPA	2 Spot			
		Spot-I	Dark Yellow	Fluorescent	1.12
		Spot-II	Creamy	Invisible	0.56
		No Spot	No Spot	No Spot	Nil
		2 Spots			
		Spot-I	Yellow	Brown	0.18
		Spot-II	No Spot	Fluorescent	0.25
		No Spot	No Spot	No Spot	Nil
		1 Spot			
		Spot-I	Invisible	Fluorescent	0.12

The present survey work will be useful in the commercial utilization of *Tridax procumbens* as a medicinal plant for curing dysentry, diarrhoea and to check the haemorrhage of wounds. Sinha and

Dogra (1985) also reported pharmacological study of the medicinal plants and mentioned that it would be quite helpful for commercial utilization of the medicinal plants in the country. Yadav and Kumar(1996) also reported similar compound in *Tridax procumbens* Linn.

Medicinal Use

Leaves are used for curing bronchial catarrh, dysentry, diarrhoea and haemorrage of wounds. Leaf juice is also insecticidal and piscicidal.

References

Bhattacharjee, A.K. and Das, A.K., 1969. *Econ. Bot.*, 23: 274.

Kapoor, L.O., Singh, A., Kapoor, S.L. and Shrivastava, S.N., 1969. *Lloydia*, 38: 221.

Persinos, G.J. and Quimbly, M.W., 1967. *J. Phaeon Sci.*, 56: 1512.

Rawson, N., 1990. *Can. Pharm. J.*, 51: 493.

Smolenski, S.J., Sillnis, H. and Fransworth, N.R.R., 1975. *Uoydia*, 38: 22.

Sinha, S.K.P. and Dogra, J.V.V., 1985. *Int. J. Crude Drug Res.*, 23(2): 77.

Wall, M.E., Krtder, M.H., Krewson, C.F., Redd, J.J., Williaman, C.D.S. and Gentry, U.S., 1954. *J. Amer. Pharm. Assoc. Sci.*, 13: 1.

Yadav, R.N. and Kumar, S., 1996. *Janpr.*, 199B(1): 147.

Ethnobotanical Wisdom and Microbial Studies on Medicinal Plants (2011) *Pages* **42–44**
Editors: **D.R. Khanna, Ashutosh Gautam, R. Bhutiani, Gagan Matta, Anil Dhiman**
Published by: **DAYA PUBLISHING HOUSE, NEW DELHI**

Chapter 8

Antibacterial Activity of *Mimusops elengi* (Bakul)

Prabhat[1], Navneet[1] and Sri Krishna[2]
[1]Department of Botany and Microbiology, [2]Department of Chemistry,
Gurukul Kangri University, Haridwar

ABSTRACT

Mimusops elengi (Linn.) extracts were evaluated for antibacterial activity against human pathogenic bacterial strain of Gram positive and Gram negative bacteria. The methanolic extract showed the maximum activity against *Streptococcus mutans, Staphylococcus aureus* and *Bacillus subtilis* (16 mm) whereas petroleum ether extract showed the minimum activity against *Streptococcus mutans* (9 mm) by well diffusion method. The use of *M. elengi* extracts as a potential antibacterial agent and the treatment of dental caries has been suggested.

Keywords: *Antibacterial activity, M. elengi, Dental caries, CFU (Colony forming Unit).*

Introduction

The medicinal plants have been evaluated for possible antimicrobial activity and to get remedy from a variety of ailments due to microorganisms. *Mimusops elengi* is a large glabrous ever green tree. It belongs to family Sapotaceae (Kirtikar and Basu, 1984). It is widely distributed throughout the greater parts of India. The bark and fruit enjoy a considerable reputation in Indian medicine as an astringent and tonic and are used in the treatment of diarrhoea and dysentry (Niranjan *et al.*, 1995). Several chemical substances from the plant such as saponins,steroids, terpenoids and alkaloids have been reported (Misra and Mitra, 1967)and isolated (Satyanarayana *et al.*, 1997). The leaf extract of plant showed antibacterial activity against *B. anthracis, B. mycoides, B. subtilis, Salmonella typhi* and *Staphylococcus aureus* (Kapoor *et al.*, 1969).

However, there is no report of antimicrobial activity of *Mimusops elengi* against dental caries bacteria. Therefore, the antibacterial activity of *M. elengi* against dental caries bacteria and other pathogens have been studied.

Materials and Methods

The material was collected from the plants present in the campus of Gurukul Kangri University, Haridwar, Uttaranchal. They were shade dried at room temperature and then powdered by using blender. The 100 gm. of powdered plant material was loaded in soxhlet assembly and extracted by sucessively in four different solvents *i.e.* petroleum ether, acetone, methanol and water. The polarity of the solvents would leach out compounds soluble in the particular solvent.

A total of 11 bacterial cultures were used in the screening. Muller Hinton Agar media (Himedia No. M-173) was used to carry out antimicrobial studies. Inoculam of each organism was prepared by inoculating a loopful growth from freshly prepared culture in to respective broth media. The inoculam was further diluted in sterile normal saline solution to provide 10^5 CFU/ml.

0.1 ml of approximately diluted broth culture of test bacteria was evenly mixed in Muller Hinton Agar. Wells of 8 mm diameter were punched into agar with sterilized cork borer and each well was filled with 45 µl (100 mg/ml) of plant extracts solvent for blank and antibacterial drug (ampicillin 100 µg/ml) for positive control. The plates were incubated for 24 hrs. at 37 °C.

The antibacterial activity was evaluated by measuring the inhibition zone diameter (Ahmad *et al.*, 1998).

Table 8.1: Antimicrobial Activity of *M. elengi* Extracts in Different Solvent

Pathogens	*Inhibition Zone (mm)*				
	Extracts				*Antibiotic*
	*Pet. Ether***	*Acetone***	*Methanol***	*Water***	*Ampicillin 100 mg/ml*
Staphylococcus aureus	14	15	16	15	24
S. epidermidis	10	13	15	14	23
Streptococcus mutans	9	12	16	14	25
S. sanguis	10	11	13	12	23
S. salivarius	10	13	15	14	25
Bacillus subtilis	13	14	15	13	22
B. megnetherium	10	11	13	12	24
Lactobacillus acidophillus	12	13	15	14	20
Escherichia coli	10	11	13	12	23
Klebsiella pneumoniae	14	14	16	15	22
Micrococcus luteus	10	11	13	12	21

*: Tested by well diffusion method

**: Solvents did not show any zone of inhibition

Results and Discussion

The antibacterial activity of extracts of *Mimusops elengi* against various test organisms at the concentration of 100 mg/ml were determined as presented in Table 8.1. The plant extracts were effective against both Gram +ve and Gram -ve bacteria. The extracts were found to be less effective as compared to ampicillin. The methanolic extract is effective as compared to other extracts because the antibacterial compounds (triterpenoid, saponin, glycosides) (Sen *et al.*, 1995; Sahu *et al.*, 2001) leached in more quantity. In general the extracts were highly inhibitory to, *K. pneumoniae, S. mutans* and *S. aureus* but the methanolic extracts shows maximum zone of inhibition 16 mm against *K. pneumoniae, S. mutans* and *S. aureus*.

A variety of constituents have been isolated from *Mimusops elengi* they are saponin, pentacyclic triterpenes, mimusopgenone, steroidal glycosides. These organic compounds shows the antibacterial activity against *S. aureus* (Kapoor *et al.*, 1969; Scalbert, 1991).

It is expected that the nature and presence of more than one active plant constituents may be responsible for enhanced antimicrobial activity in the crude extracts. The results encourage that the screening of medicinal plants hopefully will provide valuable substances to be exploited in the disease management of not only human and animals but also of plants as bacteriocides.

References

Ahmad, I., Mehmood, Z. and Mohammad, F., 1998. Screening of some Indian medicinal plants for their antimicriobial properties. *Journal of Ethanopharmacology*, 62: 183–188.

Kapoor, L.D., Singh, A., Kapoor, S.L. and Srivastava, S.N., 1969. Survey of Indian medicinal plants for saponins, alkaloids and flavonoids. *Lloydia*, 32: 297–302.

Kirtikar, K.R. and Basu, B.D., 1984. *Indian Medicinal Plants*, 2nd Edn. Indian press, Allahabad.

Mishra, G. and Mitra, C.R., 1967. Constituents of fruits and seeds of *M. elengi*. *Phytochemistry*, 6: 1309–1315.

Niranjan, P.S., Kaike, K., Jia, Z. and Nikaido, T., 1995. Novel triterpenoid saponins from *Mimusops elengi*. *Tetrahediiron*, 51: 13435–13446.

Sahu, N.P., Mandal, N.B., Banerjee, S. and Siddique, K.A.I., 2001. Chemistry and biology of triterpenes and saponins from seeds of *M. elengi*. *Journal of Herbs, Spices and Medicinal Plants*, 8: 23–29.

Satyanarayana, T., Rao, D.P.C. and Singh, B.S., 1997. Antibacterial activity of six medicinal plant extracts. *Indian Drugs*, 14: 209–214.

Scalbert, A., 1991. Antimicrobial properties of tannins. *Phytochemistry*, 30: 3875–3883.

Sen, S., Sahu, N.P. and Mahato, S.B.C., 1995. Pentacyclic triterpenoids from *M. elengi*. *Phytochemistry*, 38: 205–207.

Ethnobotanical Wisdom and Microbial Studies on Medicinal Plants (2011) *Pages* **45–51**
Editors: **D.R. Khanna, Ashutosh Gautam, R. Bhutiani, Gagan Matta, Anil Dhiman**
Published by: **DAYA PUBLISHING HOUSE, NEW DELHI**

Chapter 9

Bioremediation Using Artificial Constructed Wetland in Combination with Efficient Microbial Cultures

N. Sharma[1], P.S. Minhas[1], P.K. Joshi[1] and Navneet[2]
[1]**Central Soil Salinity Research Institute, Karnal**
[2]**Department of Botany and Microbiology, Gurukul Kangri University, Haridwar**

ABSTRACT

The paper highlights the use of constructed wetland in combination with effective microbial cultures (Bacterial and fungal) for the removal of BOD, COD and faecal coliform from untreated sewage water of STP2 of Karnal city (India). The constructed wetland consists of iron drums of 220-litre capacity with 60 litres of average loading rate having 85-cm length and filled with 35-cm coarse sand, 10-cm pebbles of 2-5 mm diameter, 35 cm gravel from top to bottom, was installed at CSSRI, Karnal in the state of Haryana. Drums in triplicate were planted with *Phragmite* and inoculated with bacterial cultures SWB1 (*Alcaligenes cupidus*, MTCC 6850), SWB19 (*Enterobacter intermedius*, MTCC 6849) and fungal culture SWF1 (*Aspergillus flavus*, MTCC 6589) added at an interval of one month having 1 litre of culture broth and control without inoculation. Constructed wetland was found to be influenced by season, retention time and inoculated microorganisms.Inoculated *Phragmites* showed better BOD, COD and faecal coliform reduction as compared to uninoculated *Phragmites* because of higher biomass buildup. The system being easy to operate and low cost, can provide an economical viable solution for wastewater management.

Keywords: *BOD, COD, Faecal coliform, Constructed wetland, Phragmites, Eichhornia, Typha, Alcaligenes cupidus, Enterobacter intermedius, Aspergillus flavus.*

Introduction

Bioremediation is the use of microorganisms or plants to detoxify an environment, mostly by transforming or degrading, pollutants. Four basic techniques may be used: (1) stimulation of the activity of indigenous microorganisms by the addition of nutrients, regulation of redox conditions, optimizing pH conditions; (2) inoculation of the sites with microorganisms of specific biotransforming abilities; (3) application of immobilized enzymes; and (4) use of plants (phytoremediation) to remove, contain, or transform pollutants.*In situ* bioremediation involves the use of organisms to remove pollutants at the site of contamination. Often, these organisms are indigenous to the area and may even be adapted for growth on the chemical contaminants in that particular environment. An alternative to the enhancement of bioremediation by indigenous microorganisms is the use of an inoculum of an appropriate pure or mixed culture of degrading microorganisms to effect removal of the undesired compound (s) (Gibson and Sayler, 1992). Constructed wetland aim to control systematically and optimize the ability of a wetland system to remove or transform wastewater pollutants and in many cases to also create an aesthetic environment for the development of wildlife and social objectives. In recent years interest has increased in wastewater treatment through constructed wetlands because of their low cost and energy requirement (Gersberg *et al.*,1986).Several investigators have reported that wetlands may act as efficient water purificiation system and nutrient sink. Wetlands remove aquatic pollutants through bacterial transformation and physio-chemical processes like adsorption, precipitation and sedimentation.

Constructed wetlands are of low cost, simple to operate and are more suitable for treatment of domestic wastewater. However, insufficient information is available on the design and operation of wetlands in the country. Therefore laboratory investigation was carried out to optimize conditions for bioremediation using microbes in combination with aquatic plants best found under lab conditions using constructed wetlands.

Materials and Methods

Modified form (Wood, 1995) of wetland was constructed using iron drums of 220 lit capacity having 85-cm length and filled with 35-cm coarse sand, 10-cm pebbles of 2-5 mm diameter, 35 cm gravel from top to bottom. Drums in triplicate were planted with *Phragmite* and inoculated with SWB1, SWB19, consortium of SWB1 and SWB19, SWF1 and control without inoculation. Other treatments comprised of: no plantation, *Eichhornia* and *Typha* both uninoculated. Where SWB1 and SWB19 were bacterial cultures *Alcaligenes cupidus* (MTCC 6850) and *Enterobacter intermedius* (MTCC 6849) while SWF1 was fungal culture *Aspergillus flavus* (MTTC 6589). It was down flow system with 60 litres of average loading rate. It was loaded regularly from inlet of STP2 sewage water of Karnal city having average BOD (161mg/l) and COD (340 mg/l). Efficient bacterial and fungus cultures were added at an interval of one month. Inoculum level was maintained at 1 litre culture broth per 60 litre of loaded sewage. Sampling was done in two different seasonal temperatures.

1. Sampling was done in slightly moderate seasonal temperature (January-March, 2004) employing retention time of 72 hr. (January-February), 48 hr. (February-March) and 24 hr. (March) respectively for per cent BOD and COD reduction.
2. Sampling was made at high seasonal temperature (May-June, 2004) for retention time of 24 hr., 48 hr. and 72 hr. respectively for per cent BOD and COD reduction. Faecal coliform count was also observed after retention time of 72 hr. Influent and effluent samples were collected regularly and BOD, COD and faecal coliform were analysed as per Standard Methods of APHA (1985).

Results and Discussion

Sub samples for total outlet were collected after retention time of 24 hr during March, May and June (Figures 9.1a and 9.1b). The mean per cent BOD, COD for different treatments having moderate seasonal temperature showed maximum per cent BOD reduction by *Typha* (68.4 per cent) followed by SWB19 inoculated *Phragmite sp.* (64.9 per cent). Whereas maximum mean COD reduction was observed in *Phragmite sp.* (61.4 per cent) inoculated with consortium and SWF1 inoculated. For samples at high seasonal temperature maximum per cent BOD reduction was observed in *Typha* (69.6 per cent) followed by SWB19 inoculated *Phragmite sp.* (68.0 per cent) similarly per cent COD reduction was maximum in *Typha sp.*, 67.5 per cent. Sub samples for total outlet were collected after retention time of 48 hr during February-March and May-June respectively (Figures 9.2a and 9.2b). During moderate seasonal temperature the mean per cent BOD reduction was observed maximum for SWB1 inoculated *Phragmite*

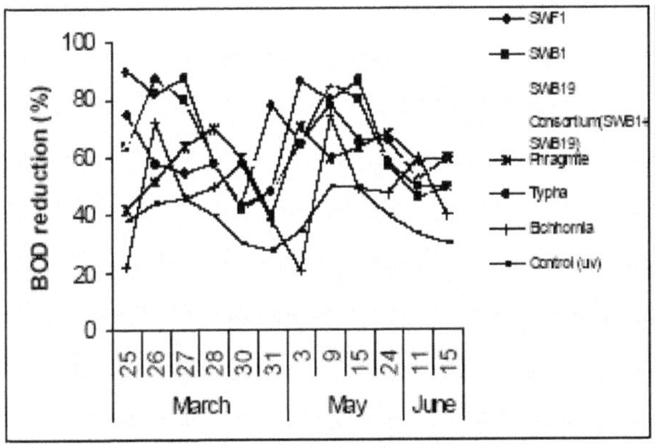

Figure 9.1a: Per cent BOD* Reduction in Different Treatments of Wetland having Retention Time (24hr)

*** Values are mean of three replicates**

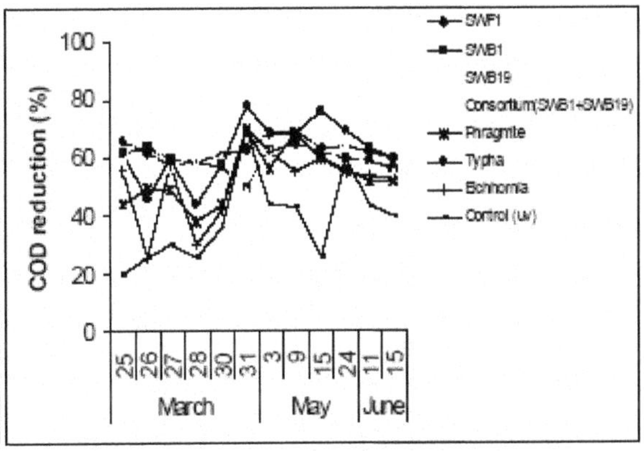

Figure 9.1b: Per cent COD* Reduction in Different Treatments of Wetland having Retention Time (24hr)

*** Values are mean of three replicates**

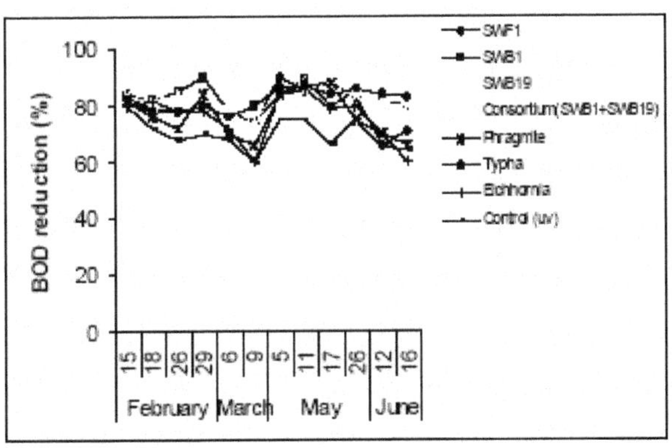

Figure 9.2a: Per cent BOD* Reduction in Different Treatments of Wetland having Retention Time (48hr)

*** Values are mean of three replicates**

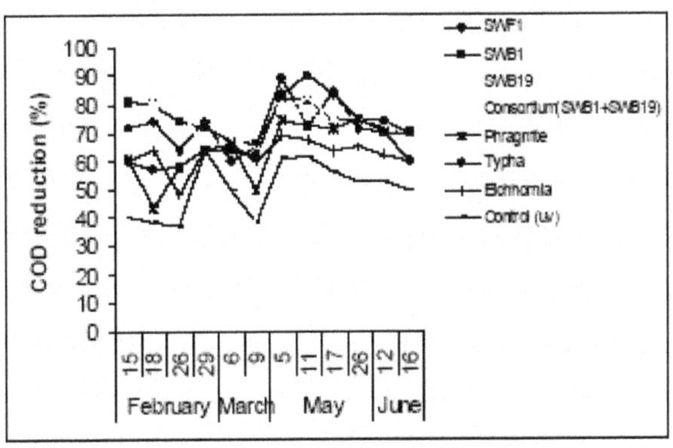

Figure 9.2b: Per cent COD* Reduction in Different Treatments of Wetland having Retention Time (48hr)

*** Values are mean of three replicates**

sp. (82.8 per cent) followed by SWB19 (82.7 per cent) and maximum per cent COD reduction was observed in consortium inoculated *Phragmite sp.* (76.9 per cent) followed by SWB1 and SWB19 inoculated *Phragmite sp.* (75.5 per cent). For samples at high seasonal temperature (extreme summer) maximum mean percent BOD reduction was observed in consortium inoculation (85.5 per cent) followed by *Typha sp.* (85.0 per cent) and COD percent reduction was maximum in SWB1 inoculation, 79.5 per cent.Sub samples for total outlet of constructed wetland were collected after retention time of 72 hr during January-February and May-June respectively (Figures 9.3a and 9.3b). The mean percent BOD reduction was maximum in consortium inoculation (46.3 per cent) followed by SWF1 inoculated *Phragmite sp.* (45.5 per cent) and COD reduction observed maximum in SWB19 inoculation (50.2 per cent) during moderate seasonal temperature. While during high seasonal temperature maximum per cent BOD reduction observed in *Typha sp.* (88.6 per cent) followed by SWB19 inoculation (86.5 per cent) and also COD reduction maximum in *Typha sp.* (83.6 per cent) followed by consortium inoculated *Phragmite* sp. 83.8 per cent.Thus it is concluded that mean per cent reduction (BOD and COD) increased with increase in seasonal temperature with same retention time. However per cent BOD and COD reduction increased with increase in retention time when subjected to same seasonal temperature. Percent BOD and COD reduction increased from 24 hr. retention time to 72 hr. retention time for outlet samples at high seasonal temperature (May-June, 2004). Long retention time and an extensive surface area in contact with the flowing water provides for effective removal of particulate and organic matter as reported by Wood (1995). Reduction in BOD and COD with three days retention during January-February was less as compared to two days retention in February-March. Similarly, for one day retention time in end of March, values of reduction were slightly less than two days but more than three days retention time in month of January-February.This indicates that bioremediation is a temperature dependent process. Maximum reduction obtained in summer can be easily ascribed to ideal temperature available for oxidation process (Trivedy and Nakate, 2002). Wetlands performance is affected by rainfall, temperature (Juwarkar *et al.*, 1995). Heritage *et al.* (1995) suggested improvement in BOD reduction over the spring and summer. It might be due to temperature increase and increased plant growth over the period. Season is a significant factor in the removal of BOD (Kuehn and Moore, 1995). Faecal coliform reduction by constructed wetland after retention time of 72 hr. during high seasonal temperature (May-June, 2004) was tried (Table 9.1). The mean inflow of faecal coliform were 95×10^5 per 100 ml which reduced maximum in SWF1 inoculated Phragmites (71×10^2/100 ml) followed

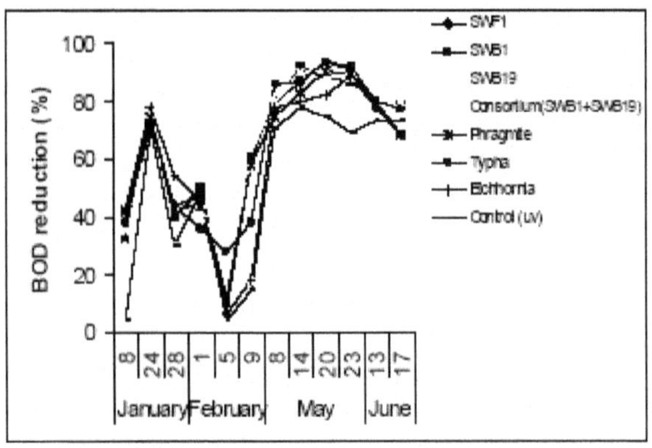

Figure 9.3a: Per cent BOD* Reduction in Different Treatments of Wetland having Retention Time (72hr)

*** Values are mean of three replicates**

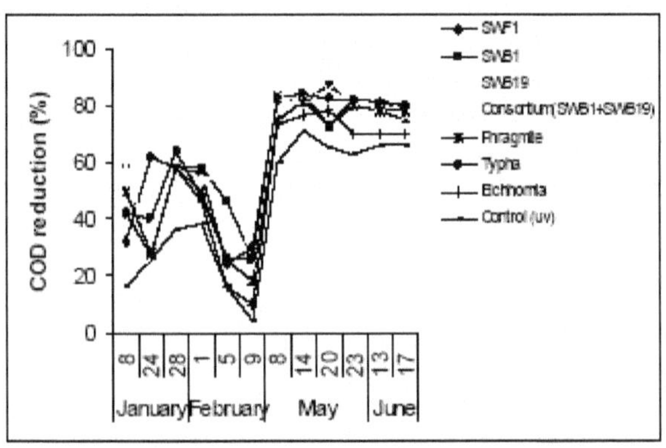

Figure 9.3b: Per cent COD* Reduction in Different Treatments of Wetland having Retention Time (72hr)

*** Values are mean of three replicates**

by SWB19 inoculated *Phragmite* ($12x10^3$/100 ml). Unvegetated control observed reduction having mean outflow value of $48x10^4$/100 ml. This reduction in bacterial load may be because of (1) the bacteria are sedimented or trapped in the root hairs of wetland plants (2) wetland plants may have the capacity to secrete a chemical substance which could have bactericidal or bacteriostatic effects. Mandi (1994) reported faecal coliform reduction of 98.6 per cent and 78 per cent during summer and winter respectively by *Eichhornia crassipes*. Bavor *et al.* (1988) cited an average value of 1 million counts per 100 ml in the effluent from 9 trickling filter in NSW. Chick and Mitchell (1995) reported marked reduction in faecal coliform count from mean inflow value of 12 million counts per 100 ml to a mean for all VFWs of less than 150,000.

Table 9.1: Faecal Coliform Count in Miniature Wetland Inlet and Outlet, having Retention Time of 72 h (May–June 2004)

Plant Type	*Inoculation @ 1.01/601*	*Faecal Coliform (MPN/100 ml) Range*	*Faecal Coliform (MPN/100 ml) (Mean)**	
			Inlet	*Outlet*
		$5x10^3$–$8x10^7$	$95x10^5$	
Phrafimite	SWB1	$6x10^3$–$7x10^4$	–do–	$22x10^3$
–do–	SWB 19	$4x10^3$–$5x10^4$	–do–	$12x10^3$
–do–	Consortium	$3x10^3$–$4x10^4$	–do–	$15x10^3$
–do–	SWFI	$2x10^3$–$5x10^4$	–do–	$71x10^2$
–do–	Nil	$2x10^3$–$5x10^4$	–do–	$29x10^3$
Typha	Nil	$6x10^2$–$11x10^5$	–do–	$28x10^3$
Eichhornia	Nil	$8x10^4$–$8x10^5$	–do–	$23x10^4$
Unvegetated (UV) Control		$14x10^4$–$13x10^5$	–do–	$4 8x10^4$

Inoculated *Phragmites* showed better BOD and COD and Faecal coliform reduction as compared to uninoculated *Phragmite* because of higher biomass of inoculated *Phragmite* compared to uninoculated *Phragmite* (Table 9.2). Ramesh *et al.* (1990) reported substitution of mango rhizosphere soil with dominant bacterial isolate resulted in maximal improvement of height and biomass of ber seedling. Fungal treatment alone or in combination with bacteria was less effective. Further studies will be required to optimize these conditions for effective treatment of wastewater under field conditions.

Table 9.2: Harvested Fresh Biomass of Miniature Wetland

Plant Type	*Inoculation @1.01/601*	*Range of Biomass (kg)*	*Average* Biomass (kg)*
Phragmite	SWB1	3.5–4.0	3.6
–do–	SWB19	2.75–4.5	3.6
–do–	Consortium	3.5–4.5	4.0
–do–	Cow dung	3.5–4.0	3.6
–do–	SWFI	3.5–4.0	3.91
–do–	Nil	3.0–3.0	3.0
Typha	Nil	4.0–6.0	5.0
Eichhornia	Nil	1.25–1.75	1.5

Conclusion

The results showed that constructed wetland with an average loading rate of 60 litre sewage in combination with effective bacterial cultures SWB1 (*Alcaligenes cupidus*, MTCC 6850), SWB19 (*Enterobacter intermedius*, MTCC 6849) and fungal culture SWF1 (*Aspergillus flavus*, MTCC 6589) added at an interval of one month having 1 litre of culture broth in *Phragmites* showed increased BOD,COD reduction with retention time of 24hr,48hr and 72hr and faecal coliform reduction with 72hr retention time (May-June 2004) as compared to uninoculated *Phragmites* because of increased biomass of culture supplemented Phragmites. Other treatments comprised of: no plantation, *Eichhornia* and *Typha* both uninoculated. Mean per cent reduction (BOD and COD) increased with increase in seasonal temperature with same retention time. However per cent BOD and COD reduction increased with increase in retention time when subjected to same seasonal temperature. BOD and COD reduction with three days retention during January-February 2004 were less as compared to two days retention in February-March,2004. Similarly, for one day retention time in end of March,2004 values of reduction were slightly less than two days but more than three days retention time in month of January-February 2004. Faecal coliform reduction by constructed wetland after retention time of 72 hr. during high seasonal temperature (May-June, 2004) was tried. The mean inflow of faecal coliform were $95x10^5$ per 100 ml which reduced maximum in SWF1 inoculated *Phragmites* ($71x10^2$/100 ml) followed by SWB19 inoculated *Phragmite* ($12x10^3$/100 ml). Unvegetated control observed reduction having mean outflow value of $48x10^4$/100 ml. The constructed wetland seems to be cost- effective alternative to conventional treatment processes which involves huge cost and its efficacy further gets improved by addition of efficient microbial cultures. Wetland performance is affected by temperature, retention time etc. Since it is not site specific, the system can be implemented near the wastewater source.

References

APHA, 1985. *Standard Methods for the Examination of Water and Wastewater*, 16th Edn. APHA, AWWA, WPCF, Washington, D.C., p. 1268

Bavor, H.J., Roser, D.J., McKenzie, S.A. and Breen, P.F., 1988. Joint study on sewage treatment using shallow lagoon-aquatic plant systems. *Final Technical Report for the Water Research Laboratory*, Hawkesbury Agricultural College, CSRIO Centre for Irrigation and Freshwater Research and Sydney Water Board, 2: 243.

Chick, A.J. and Mitchell, D.S., 1995. A pilot study of vertical flow wetlands at Coffs harbour, New South Wales, Australia. *Water. Sci. Technol.*, 32(3): 103–109.

Gersberg, R.M., Elkins, B.V., Lyen, S.R. and Goldman, C.R., 1986. Role of aquatic plants in wastewater treatment by artificial wetlands. *Wat. Res.*, 20: 363–398.

Gibson, D.T. and Sayler, G.S., 1992. *Scientific Foundations of Bioremediation: Current Status and Future Needs*. The Amer. Acad. Microb., Washington, D.C.

Heritage, A., Pistillo, P., Sharma, K.P. and Lantzke, I.R., 1995. Treatment of primary settled urban sewage in pilot scale vertical flow wet lands and filters: Comparison of four emergent macrophytes over a 12 month period. *Water Sci. Technol.*, 32(3): 295–304.

Juwarkar, A.S., Oke, B., Juwarkar, A. and Patnaik, S.M., 1995. Domestic wastewater treatment through constructed wetland in India. *Water Sci. Technol.*, 32(3): 291–294.

Kuehn, E. and Moore, A.J., 1995. Variability of treatment performance in constructed wetlands. *Water Sci. Technol.*, 32(3): 241–250.

Mandi, L., 1994. Marrakesh wastewater purification experiment using vascular aquatic plants *Eichhornia crassipes* and *Lemna gibba*. *Water Sci. Technol.*, 29(4): 283–287.

Ramesh, R., Goel, R. and Johri, B.N., 1990. Influence of Rhizosphere manipulation on the establishment of Ber (*Zizyphus mauritiana* Lam.). *Indian J. Microbiol.*, 30(3): 343–347.

Trivedy, R.K. and Nakate, S.S., 2002. Treatment of hospital waste and sewage in hyacinth ponds. In: *Low Cost Wastewater Treatment Technologies*, (Eds.) R.K. Trivedy and S. Kaul. ABD Publisher, Jaipur, pp. 132–163.

Wood, A., 1995. Constructed wetlands in water pollution control: Fundamentals to their understanding. *Water Sci. Technol.*, 32(3): 21–29.

Ethnobotanical Wisdom and Microbial Studies on Medicinal Plants (2011) *Pages* **52–64**
Editors: **D.R. Khanna, Ashutosh Gautam, R. Bhutiani, Gagan Matta, Anil Dhiman**
Published by: **DAYA PUBLISHING HOUSE, NEW DELHI**

Chapter 10

Biomonitoring and Interspecies Comparison of Metal Precipitation through Bryophytes at Petrol Pumps on Kumaon Hill

Dinesh K. Saxena* and Md. Saiful-Arfeen**
Department of Botany, Bryological Science Division,
Bareilly College, Bareilly – 243 005

ABSTRACT

The number of pollution sources raises heavy metal contamination of the environment and become increasingly higher in Kumaon hill. This problem particularly manifests itself around the petrol pump. The objective of our investigation was to measure the pollution load of Pb, Zn, Cu and Cd in near by area of Kumaon hill. Keeping this in mind present investigation was undertaken to study the adverse effect of intensity of these metal load on the ecology and human population in near by area of respective study site. Moss sample were collected from pure and undisturbed habitat in Mukteswar and were transplanted on four petrol pumps and one control site. Analysis of harvested transplants shows a significant seasonal increase in the metal in moss, which could be due to automobile exhaust and petroleum product consumption.

Keywords: *Biomonitoring, Metal, Interspecies comparison, Petrol pump, Ecology.*

E-mail: *dineshsaxenabc@rediffmail.com; **saifulafeen123@yahoo.com.

Introduction

The significance of heavy metal deposition in different cryptogam at petrol pump can be looked from different viewpoint. Because their potential to bioaccumulate heavy metal from respective source varies from species to species and also their contribution in environmental pollution rescue is increasing. Due to no other source of transport, dangerously petroleum consumption and automobile exhaust are the major source to put pollution load on the atmospheric condition in the near by area of petrol pumps (Krishnayya and Bedi 1986; Huang *et al.*, 1994; Xuolong *et al.*, 1994; Saxena *et al.* 2000a).

Monitoring of air contaminant at petrol pumps is necessary to determine upon locality of the near by population. Bryophytes have a long history to use as biological indicators of air quality. High deposition of heavy metal through dry deposition is a specific biology of cryptogams in biomonitoring studies (Saxena and Saxena 2000; Ruhling *et al.*, 1987; Ferguson *et al.*, 1984).Despite the large number of publication dealing with interaction of cryptogam with heavy metal (Nash 1989; Meenk and Tuba 1992; Saxena and Saxena 2000), detail knowledge on environmental condition of the petrol pump of Kumaon hill is in need of more attention.

Indirect or transplant bag method is best-suited technique to study the atmospheric condition (*i.e.* heavy metal) of petrol pumps (Saxena *et al.* 2000, Kirchoff and Rudolph 1989). The main objective of present work was to gain understanding of the metal accumulation potential of three different experimental species *i.e. Racomitrium crispulum, Bryum cellulare* and *Plagiochasma appendiculatum* and pollution load at different petrol pumps of Kumaon hill.

Materials and Methods

The area of our investigation is located on the Western Himalaya *i.e.* Kumaon hill. Catchment areas are the four petrol pumps of Almora, Nainital, Ranikhet and Pithoragarh of Kumaon hill. The moss species *Racomitrium crispulum, Plagiochasma appendiculatum, Bryum cellulare* was applied in our investigation. The seasonal and annual transplantation were started in year 2003 with the start of winter season (*i.e.* in the month of November). Seasonal transplantation was carried out for the fixed period of four months (*i.e.* 120 days, which represents the exposure of only one season and there is no intermittent mixing of seasonal harvesting of transplant). Sampling and transplantation were carried out by the method of Saxena and Saxena (2000). The catchment site of same height were chosen, to avoid the misleading of data. Samples were digested and analyzed using HNO_3 and $HClO_4$ (4:1 V/V) analysis by atomic absorption spectrophotometer (Model no. E 4139) by following the method of Bengston *et al.* (1982).

The per cent heavy metal loading of each location in proportion to control were calculated using the formula of Tuba and Csintalan (1993). Grodzinska *et al.* (1993) put forwarded the formula for PI (Pollution Index) value, on the basis of each metal, total number of metals, total amount of metals and number of sampling stations. The analysis of experimental heavy metal (*i.e.* Pb, Zn, Cu and Cd) for every study area were carried out in triplicate and data were presented with ± S.E. Significance (*) differs from control site (Pd" 0.05).

Results

Tables 10.1–10.3 show the result of metal analysis of three different species *i.e. Plagiochasma appendiculatum, Bryum cellulare* and *Racomitrium crispulum* harvested periodically after a fixed period of exposure in Kumaon hill. Experimental site *i.e.* petrol pumps of Almora, Nainital, Ranikhet and Pithoragarh, are highly populated area of Kumaon hill. Data obtained after every seasonal exposure

Table 10.1: Seasonal Variation in Metal Precipitation in Winter, Summer and Rain in Year 2003–04 in Moss *Plagiochasma appendiculaum*. Each value is mean of 3 replicates±S.E. Significance (*) differs from control (Pd"0.05).

Site No.	Catch-ment Area	Winter Exposure				Summer Exposure				Rain Exposure			
		Pb	*Zn*	*Cu*	*Cd*	*Pb*	*Zn*	*Cu*	*Cd*	*Pb*	*Zn*	*Cu*	*Cd*
1.	Mukteswar Forest	32.42±0.12	3.15±0.02	22.4±0.15	1.124±0.21	41.25±0.24	10.49±0.6	40.3±0.54	6.32±0.22	41.5±0.14	93.95±0.65	8.59±0.25	2.66±0.01
2.	Almora	318.5*±0.24	25.2*±0.05	46.93*±0.3	0.62*±0.05	298.4*±3.5	173.2*±5.6	50.33*±0.6	1.9*±0.05	33.9*±0.3	18.1*±0.42	41.6*±0.3	ND
3.	Ranikhet	42.04*±0.52	42.1*±0.32	62.1*±0.56	1.3*±0.2	54.9*±1.25	56.6*±0.98	83.4*±1.25	2.2*±0.05	37.7*±0.1	36.8*±0.25	47.6*±0.5	ND
4.	Nainital	211.05*±0.3	59.07*±0.2	44.2*±0.58	2.42*±0.2	335.4*±7.5	60.93*±0.8	69.72*±1.0	3.6*±0.09	33.9*±0.2	14.2*±0.84	32.0*±0.2	1.9*±0.05
5.	Pithoragarh	47.3*±0.35	62.1*±0.5	26.3*±0.25	0.12*±0.02	57.2*±0.22	72.1*±0.65	32.4*±0.36	0.3*±0.01	52.8*±0.5	ND	25.5*±0.9	ND

Table 10.2: Seasonal Variation in Metal Precipitation in Winter, Summer and Rain in Year 2003–04 in Moss *Bryum Cellulare*. Each value is mean of 3 replicates±S.E. Significance (*) differs from control (Pd"0.05).

Site No.	Catch-ment Area	Winter Exposure				Summer Exposure				Rain Exposure			
		Pb	*Zn*	*Cu*	*Cd*	*Pb*	*Zn*	*Cu*	*Cd*	*Pb*	*Zn*	*Cu*	*Cd*
1.	Mukteswar Forest	32.42±0.12	3.15±0.02	22.4±0.15	1.124±0.21	41.25±0.24	10.49±0.6	40.3±0.54	6.32±0.22	41.5±0.14	93.95±0.65	8.59±0.25	2.66±0.01
2.	Almora	318.5*±0.24	25.2*±0.05	46.93*±0.3	0.62*±0.05	298.4*±3.5	173.2*±5.6	50.33*±0.6	1.9*±0.05	33.9*±0.3	18.1*±0.42	41.6*±0.3	ND
3.	Ranikhet	42.04*±0.52	42.1*±0.32	62.1*±0.56	1.3*±0.2	54.9*±1.25	56.6*±0.98	83.4*±1.25	2.2*±0.05	37.7*±0.1	36.8*±0.25	47.6*±0.5	ND
4.	Nainital	211.05*±0.3	59.07*±0.2	44.2*±0.58	2.42*±0.2	335.4*±7.5	60.93*±0.8	69.72*±1.0	3.6*±0.09	33.9*±0.2	14.2*±0.84	32.0*±0.2	1.9*±0.05
5.	Pithoragarh	47.3*±0.35	62.1*±0.5	26.3*±0.25	0.12*±0.02	57.2*±0.22	72.1*±0.65	32.4*±0.36	0.3*±0.01	52.8*±0.5	ND	25.5*±0.9	ND

Table 10.3: Seasonal Variation in Metal Precipitation in Winter, Summer and Rain in Year 2003–04 in Moss *Racomitrium crispulum*. Each value is mean of 3 replicates per cent S.E. Significance (*) differs from control (Pd" 0.05).

Site No.	Catchment Area	Winter Exposure				Summer Exposure				Rain Exposure			
		Pb	*Zn*	*Cu*	*Cd*	*Pb*	*Zn*	*Cu*	*Cd*	*Pb*	*Zn*	*Cu*	*Cd*
1.	Mukteswar Forest	35.42±0.21	3.15±0.03	23.65±0.25	0.321±0.54	61.25±0.15	6.49±0.28	32.85±0.24	0.40±0.02	39.62±0.2	ND	9.12±0.24	ND
2.	Almora	318.5*±9.25	342*±8.645	5.43*±0.55	2.57*±0.05	224.4*±5.3	173.2*±0.5	59.46*±0.4	3.6*±0.32	35.8*±0.1	8.79*±0.24	50.81*±0.5	ND
3.	Ranikhet	46.32*±0.55	43.16*±4.5	69.3*±0.13	1.00*±0.05	53.4*±0.65	53.26*±0.9	88.9*±0.54	1.9*±0.05	35.8*±0.5	ND	42.73*±0.6	ND
4.	Nainital	22.10*±0.54	59.07±0.5	63.48*±9.5	0.3*±0.004	33.54*±2.2	68.93*±0.9	84.74*±0.4	4.7*±0.14	14.6*±0.5	32.9*±0.33	52.74*±0.5	1.25*±0.66
5.	Pithoragarh	42.64*±0.41	60.26*±0.2	20.4*±0.24	0.25*±0.06	57.7*±0.24	78.2*+0.84	25.46*+1.5	0.92*±0.2	43.3*±1.5	ND	18.5*±0.5	ND

transplant shows comparative variability in seasonal deposition as well as in interspecies variation in metal accumulation. There is a high accumulation of lead observed *i.e.* 211 ppm in winter exposure in both bryophyte *Plagiochasma appendiculatum* and *Racomitrium crispulum* in Nainital. *Racomitrium crispulum* shows a significant high value of Cu *i.e.* 142 ppm in Nainital under comparative study with *Bryum cellulare* and *Plagiochasma appendiculatum* in winter season. There is no such significant variability in the Cd content in any of the catchment area of the Kumaon hill and same is observed in all the experimental species.

Summer is the season for the higher vehicular input in the tourist place *i.e.* in the present experimental sites. A significant high value of Pb were observed in summer in *Plagiochasma appendiculatum* (335.84 ppm) in Nainital, while the highly populated area of Kumaon hill *i.e.* Almora shows maximum value of Pb input, which were observed in all the three experimental cryptogams. There is no such significant difference in Cd content all the three seasons; while it is striking that rain exposed experimental bryophytes transplant impart Cd in non-detectable limit. There is a positive correlation between metal accumulation and vehicular input in the catchment site (data not shown).

Significant variability in the metal content of annual transplant and summation of seasonal transplant, treated as annual fall out data was observed only in the transplant of *Plagiochasma appendiculatum* (Tables 10.4–10.6).

Undertaken metals *i.e.* Pb, Zn, Cu and Cd accumulation are the index of vehicular pollution and were useful to indices the vehicular input, in particular area. Per cent metal loading at each location signifies that the study area is highly loaded with experimental metal. While the inter species comparison implicit that *Racomitrium crispulum* is much better accumulator than *Bryum cellular* and *Plagiochasma appendiculatum* respectively. It is in doubt that the undertaken control site Mukteswar forest is free from metal load and need more attention in further studies. The site Almora which were observed to be highly loaded with metal is justifiable through per cent loading value (Figure 10.1–10.3).

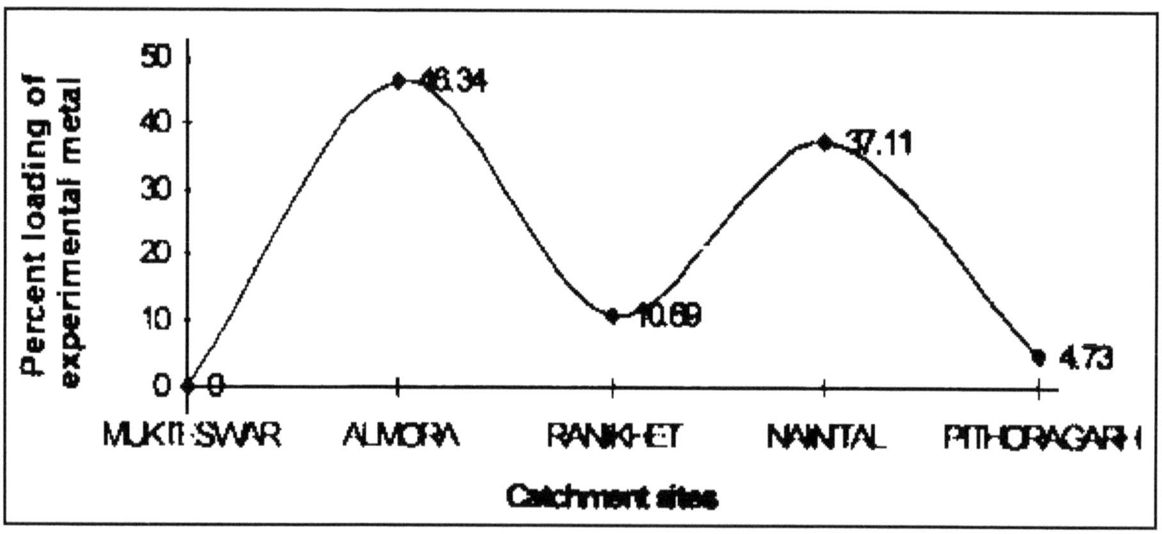

Figure 10.1: Per cent Metal Loading in *Plagiochasma appendiculatum* in the Experimental Catchment Area of Kumaon Hill in Year 2003–04

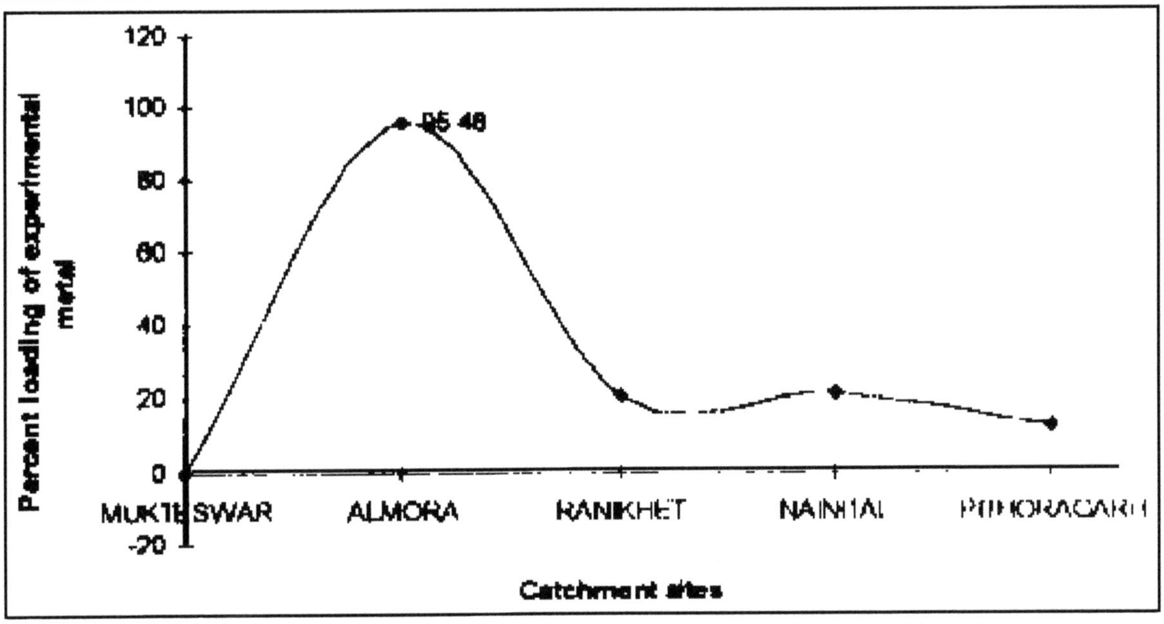

Figure 10.2: Per cent Metal Loading in *Bryum cellulare* Hook. in the Experimental Catchment area of Kumaon Hill in Year 2003–04

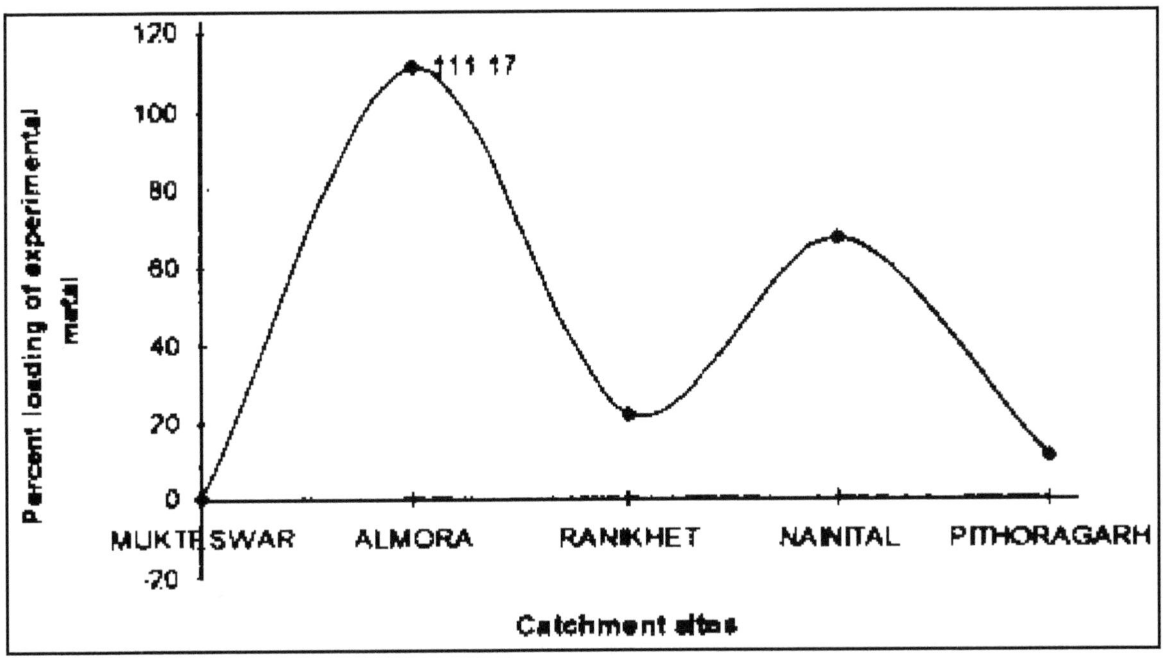

Figure 10.3: Per cent Metal Loading in *Racomitrium crispulum* in the Experimental Catchment Area of Kumaon Hill in Year 2003–04

Table 10.4: Variation in Metal Precipitation Value in Annual Transplant and Summation Value of Three Seasons in Year 2003–04 in Moss *Plagiochasma appendiculatum*. Each value is mean of 3 replicates per cent S.E. Significance (*) differs from control (Pd" 0.05).

Site No.	Catchment Area	Summation of Seasonal Transplant Data Treated as Annual Precipitation				Annual Exposure of Moss Bags Metal Data			
		Pb	Zn	Cu	Cd	Pb	Zn	Cu	Cd
1.	Mukteswar Forest	115.17±0.2	107.59±1.22	71.29±2.33	10.11±1.25	95.17±0.65	79.61±3.25	52.64±2.65	7.11±0.32
2.	Almora	650.86*±9.6	216.62*±6.42	138.95*±1.35	2.58*±0.35	425.01*±8.32	198.6*±4.2	99.64*±4.65	3.58*±0.63
3.	Ranikhet	134.67*±2.3	135.51*±2.35	193.18*±5.65	3.5*±0.33	101.94*±4.35	100.2*±2.3	159.67*±2.65	2.7*±0.36
4	Nainital	580.49*±5.6	134.29*±2.35	145.97*±3.65	7.96*±0.62	420.16*±5.65	98.64*±2.2	111.94*±6.35	6.46*±0.55
5.	Pithoragarh	157.33*±1.2	134.2*±1.65	84.23*±4.25	0.44*±0.03	133.21*±4.94	98.66*±3.2	89.64*±5.656	0.39*+0.05

Table 10.5: Variation in Metal Precipitation Value in Annual Transplant and Summation Value of Three Seasons in Year 2003–04 in Moss *Bryum cellulare*. Each value is mean of 3 replicates± S.E. Significance (*) differs from control (Pd" 0.05).

Site No.	Catchment Area	Summation of Seasonal Transplant Data Treated as Annual Precipitation				Annual Exposure of Moss Bags Metal Data			
		Pb	Zn	Cu	Cd	Pb	Zn	Cu	Cd
1.	Mukteswar Forest	136.29±0.3	9.64±0.34	65.43±0.74	0.72±0.33	131.24±0.1	8.94±0.32	62.23±0.233	0.7*±0.001
2.	Almora	578.74*±0.3	524.08*±0.5	115.71*±0.51	0.171*±0.35	572.24*±5.25	520.7*±0.2	111.11*±2.33	5.24*±0.04
3.	Ranikhet	135.57*±0.5	96.42*±0.51	200.93*±0.69	2.96*±0.098	130.52*±3.25	95.02*±0.6	193.54*±4.22	2.96*±0.3
4.	Nainital	70.25*±0.5	160.90*±0.73	200.98*±0.54	6.25*±0.25	69.25*±2.54	158.9*±0.3	192.98*±0.35	6.75*±0.32
5.	Pithoragarh	143.74*+0.7	138.46*+0.54	64.36*+0.254	1.17*+0.5	139.74*±0.65	135.4*±0.6	59.37*+0.35	1.34*+0.03

Table 10.6: Variation in Metal Precipitation Value in Annual Transplant and Summation Value of Three Seasons in Year 2003–04 in Moss *Racomitrium crispulum*. Each value is mean of 3 replicates±S.E. Significance (*) differs from control (Pd" 0.05).

Site No.	*Catchment Area*	*Summation of Seasonal Transplant Data Treated as Annual Precipitation*				*Annual Exposure of Moss Bags Metal Data*			
		Pb	*Zn*	*Cu*	*Cd*	*Pb*	*Zn*	*Cu*	*Cd*
1.	Mukteswar Forest	128.17±0.3	17.48±2.65	75.10±2.34	0.68±0.02	124.47±0.21	14.48±4.35	73.17±1.22	1.25±0.02
2.	Almora	1760.29*±8.6	537.37*±6.35	153.12*±0.25	0.32*±0.02	755.79*±9.66	529.7*±2.3	150.12*±0.5	1.49*±0.05
3.	Ranikhet	133.54*±3.2	102.27*±4.35	225.46*±0.55	2.3*±0.23	125.44*±0.55	99.27*±5.3	220.46*±2.35	2.3*±0.05
4.	Nainital	500.49*±7.6	138.48*±2.325	327.00*±6.35	7.17*±0.35	489.47*±4.2	135.4*±0.5	320.4*±0.512	7.07*±0.04
5.	Pithoragarh	109.23*±4.2	159.29*±2.35	74.80*±2.1	0.9*±0.025	100.24*±2.32	149.7*±0.5	77.80*±4.5	0.9*±0.03

The percent decrease in the metal content in winter season with respect to summer in *indirect* biomonitoring study, through experimental bryophytes imparts that Cu and Cd value decrease up to the range of 80-90 per cent. While there is no such significant difference in metal concentration observed in case of Pb and Zn in all the three species of bryophytes in winter with respect to summer (Figure 10.5). There is an up to 100 per cent decrease in Cd were observed in rainy season under comparative study with summer (Figure 10.6). We can not over rule the role of rain in pollutant leaching. Same could be the reason for decrease in Pb, Zn and Cu value up to 70-80 per cent in rainy season in all the three experimental cryptogams.

There is a statistically significant correlation found between metal accumulation, an increment in their source type and intensity of source with respect to PI (pollution index) value. The PI value were observed highly positive at Almora *i.e.* + 4.965 in *Bryum cellulare* signifies that the pollutant intensity among other petrol pump catchment area, while the site of Pithoragarh petrol pump shows negative value – 2.4406, -2.48094, -1.60022 in *Racomitrium crispulum, Plagiochasma appendiculatum, Bryum cellulare* respectively (Figure 10.4).

Discussion

We have seen that the study and comparison of the moss transplant of petrol pumps of different site in Kumaon hills give more authentic picture of atmospheric surrounding of petrol pumps and near by areas. The experimental data of seasonal and annual exposure demonstrate that all the three experimental bryophytes have positive potential to accumulate and retain heavy metal pollutant of petrol pumps. However, their efficiencies vary from species to species.

Present study also demonstrate that seasonal transplant of moss bag technique for liverwort is more reliable than annual transplant one. Chlorosis, thallus degradation, ageing, and low surface area by mass ratio could be the reason for lower metal input value in annual transplant of *Plagiochasma appendiculatum*. Ranikhet is relatively clean site as compared to Almora and Nainital, though that the catchment area is in the heart of the city. This further supported by lower per cent metal loading.

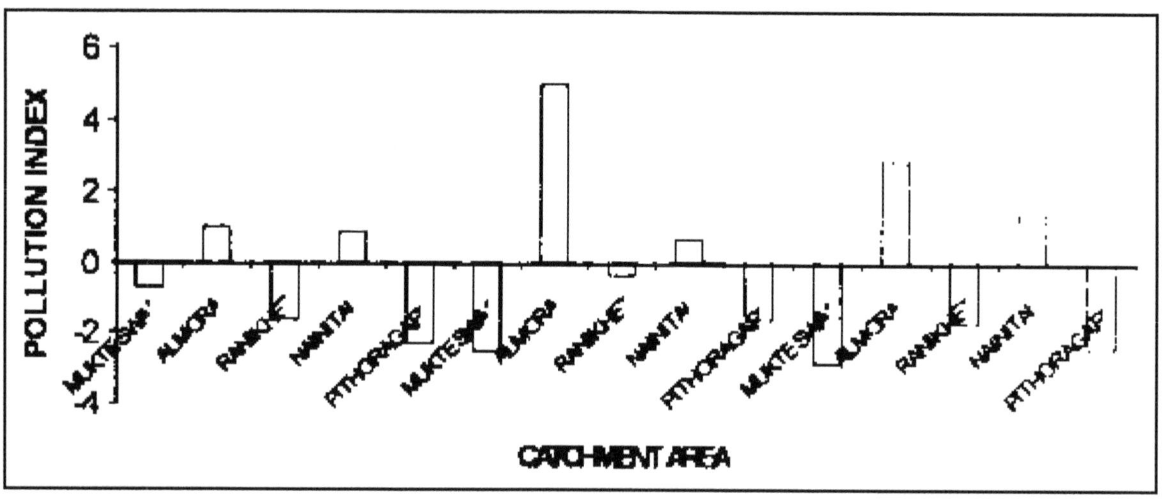

Figure 10.4: Pollution Index Value by Undertaken Bryophytes in the Experimental Catchment Area in Year 2003–04

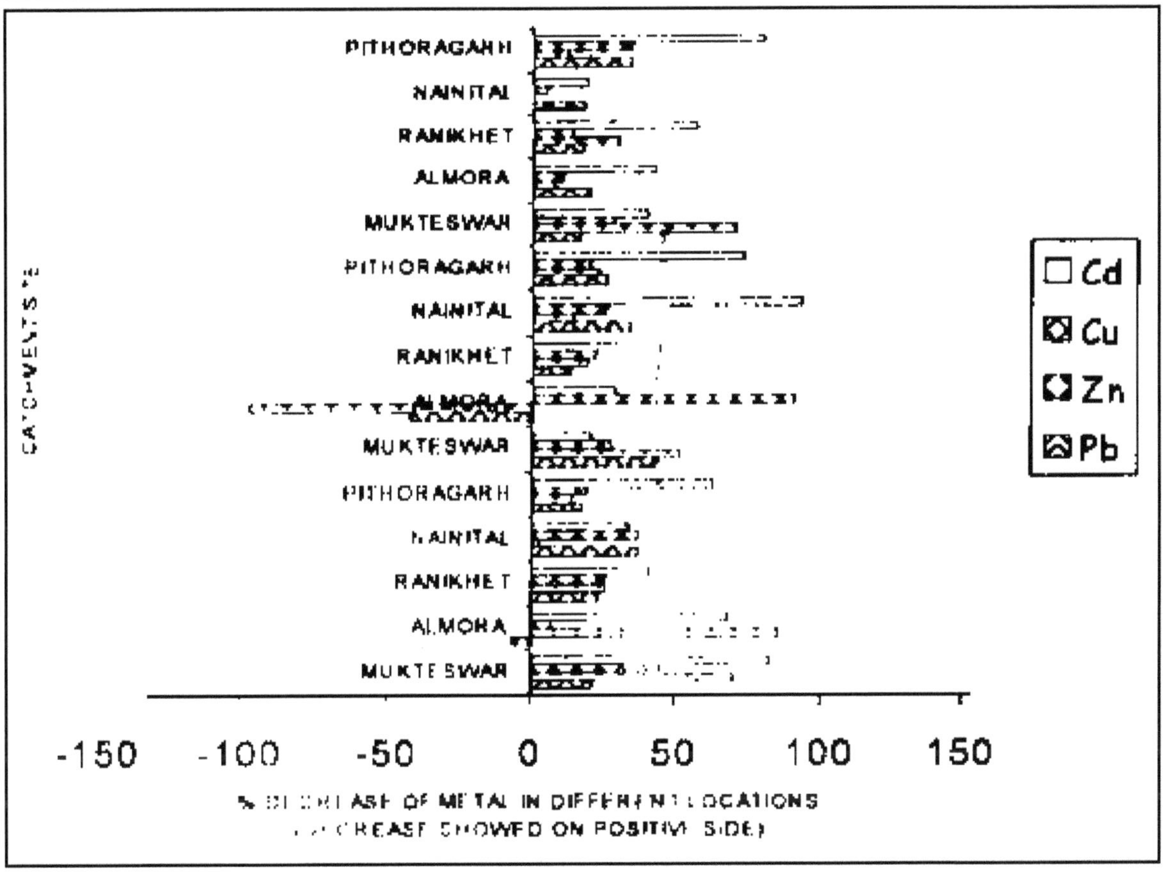

Figure 10.5: Trend of Per cent Decrease in Metal Precipitation in Winter with Respect to Summer in Experimental Catchment Areas in Experimental Bryophytes in Year 2003–04

Cu industry in Kumaon region is also a probable additional reason for high metal input value in Almora. Kerosene mixing is also one of the addition facts for increment in Cu in the petrol pump transplants, as $CuSO_4$ is mixed as an impurity in kerosene. Perhaps Cd is also present as a mining impurity in petrol. High input of these two metals (*i.e.* Cu and Cd) in summer season in all the three experimental bryophytes were reasonable.

The relative cleanness of control site *i.e.* Mukteswar is over ruled by the metal concentration observed in respective transplant in all the three bryophytes. Though their concentration is low and comparable with the other study areas. Increment in the use of Cu, Zn and Cd fertilizers could be the reason for this high value of metal in Mukteswar. Per cent loading value of Nainital and Almora signifies that there is an abrupt increment of consumption of diesel and petrol. This is a matter of serious concern, as the ecosystem flora may not be able to bear the burden of increasing metal level and other pollutant, released on the consumption of petrol and diesel.

Genetic constituent may be assumed to have considerable significance in respect to heavy metal uptake and excretion. The different uptake capacity of heavy metal and retention / excretion by interspecies survey shows variability (Ruhling *et al.*, 1987; Ross 1990; Birg *et al.*, 1995). This need

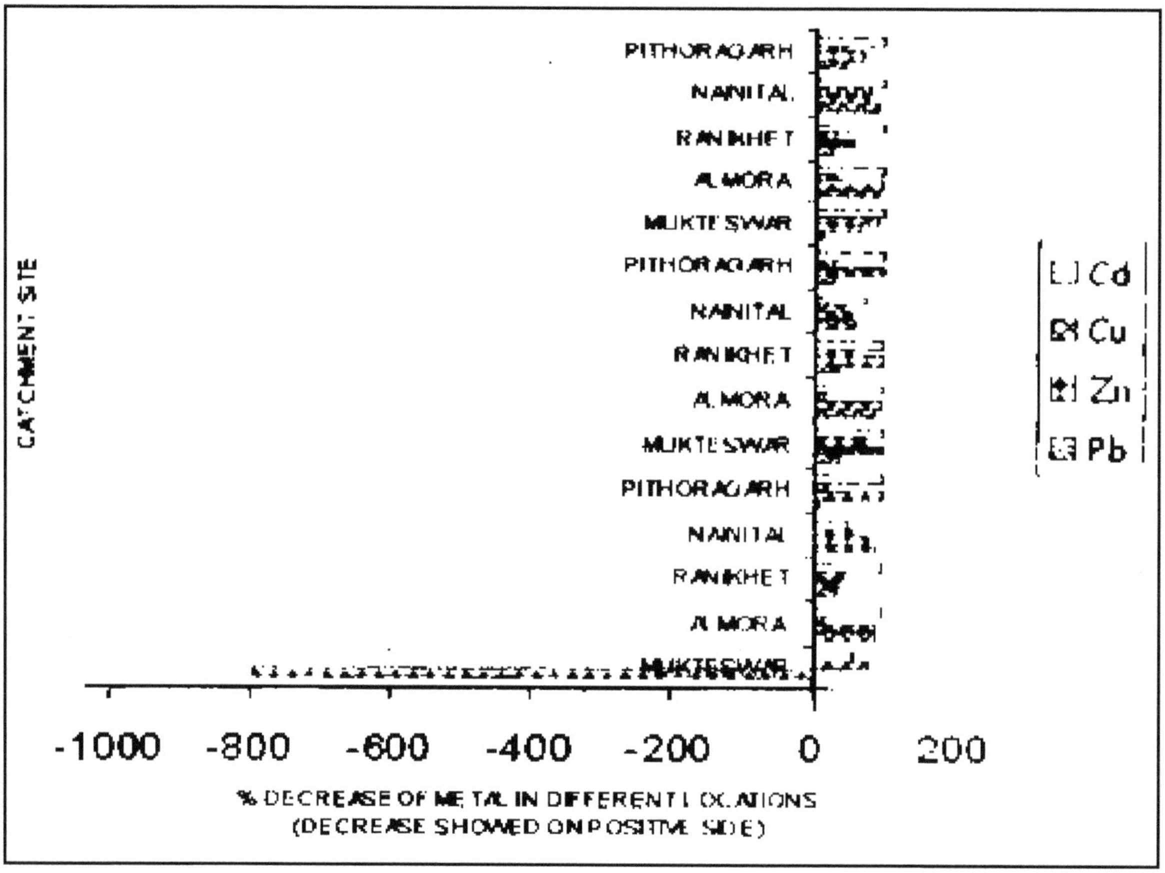

Figure 10.6: Trend of Per cent Decrease in Metal Precipitation in Rain with Respect to Summer in Experimental Catchment Areas in Experimental Bryophytes in Year 2003–04

further harmonization and research in Kumaon region. Research is in continuation by present authors in the survey area of Kumaon hill. Accumulation of heavy metal especially lead through various cryptogams is well known study and several authors described the moss capability to accumulate vehicular pollution (Ruhling and Tyler, 1968; Onianwa and Eunyomi, 1983; Ferguson *et al.*, 1984; Makinen 1987; Ross, 1990; Steinnes *et al.*, 1994; Saxena 1995).

High Pb value *i.e.* 600–800 mg / L in petrol is a direct evidence of their high concentration in bryophyte transplanted at petrol pumps. Study has also confirmed that the relationship between the environmental quality and metal accumulation in bryophytes. Present investigation confirms that *Bryum cellulare* had a lower potential to monitor the metal, while *Racomitrium crispulum* has the maximum capability to absorb the vehicular pollution on petrol pumps and might be due to this they maintained their physiology throughout the year of transplantation. This was confirmed by the non-significant difference in annual transplant metal data and summation of the seasonal transplant metal data which were treated as annual precipitation data. Seasonal and annual transplantation data of *Plagiochasma appendiculatum* further recommends that biomonitoring by liverworts is better through seasonal way.

Acknowledgements

Present work is supported with financial assistance from Department of Biotechnology, (Ministry of Science and Technology) Govt. of India. Our sincere thanks are also due to Principal, Bareilly College, Bareilly, Prof S.C Srivastava, Professor and Head, Dept. of Botany, Lucknow University, Lucknow and Prof. Zoltan Tuba, Szent Istvan University, Hungary. We express our sincere thanks to Prof. Pocs for providing specimens for comparison and identification of bryophyte sample.

References

Bengston, C., Folkeson, L. and Goransson, A., 1982. Growth reduction and branching frequency in *Hylocomium splendens* near a foundry emitting Cu and Zn. *Lindergia*, 8: 129–138.

Birg, T., Royset, O. and Steinnes, E., 1995. Moss (*Hylocomium splendens*) used as biomonitor of atmospheric trace element deposition, estimation of uptake efficiencies *Atmos. Environ.*, 29A: 353–360.

Ferguson, N.P., Robinson, R.N., Press, M.C. and Lee, J.A., 1984. Element concentration in five *Sphagnum* species in relation to atmospheric pollution. *Journal of Bryology*, 13: 107–114.

Grodzinska, K., Godzik, B. and Sjarek, G., 1993. Heavy metals and sulphur in moss from Southern Spitsbergen. *Frogmenta Floristica Et Geobotanica*, 2(2): 699–708.

Huang, X., Olmez, I., Aras, N.K. and Gordon, G.E., 1994. Emission of trace elements from motor vehicles. *Atmospheric Environment*, 28: 1385–1391.

Kirchhoff, M. and Rudolph, H., 1989. A sandwich technique for the continuous monitoring of air pollution with the bryophyte *Sphagnum*. *Journal of Hottori Botanical Laboratory*, 67: 423–431.

Krishnayya, N.S. and Bedi, S.J., 1986. Effect of automobile lead pollution on *Cassia tora* L. and *Cassia occidentalis* L. *Environmental Pollution*, 40: 221–226.

Makinen, A., 1987. *Sphagnum* moss-bag in air pollution monitoring in the city of Helsinki. *Sym. Biol. Hung.*, 35: 755–776.

Meenks, J.L. and Do Tuba, Z., 1992. In: *Biological Indicators in Environmental Protection*, (Ed.) M. Kovacs. Akademia Kiado, Budapest, p. 65–75.

Nash, T.H., 1989. In: *Metal Tolerance in Plants: Evolutionary Aspects*, (Ed.) J. Shaw. CRC Press, BocaRaton, Florida, p. 119–131.

Onianwa, P.C. and Eunyomi, A., 1983. Trace metal levels in some Nigerian mosses used as indicators of atmospheric pollution. *Environment Pollution*, 5: 71–81.

Ruhling, A. and Tyler, G., 1968. An ecological approach to the lead problem. *Bot. Notiser.*, 122: 248–259.

Ruhling, A., Rasmussen, L., Pilegaard, K., Makinen, A. and Steinnes, E., 1987. Survey of atmospheric heavy metal deposition in the Nordic countries in 1985-monitored by moss analyses. *NORD*, 21: 1–44.

Ross, H.B., 1990. On the use of mosses (*Hylocomium splendens* and *Pleurosium schreberi*) for the estimating atmospheric trace metal deposition. *Water Air and Soil Pollution*, 50: 63–76.

Saxena, D.K., 1995. Biomonitoring of lead with the help of bryophytes in Nainital. In: *Proceeding of National Conference on Bryology*, NBRI, Lucknow.

Steinnes, E., Hanssen, J.E., Rambaek, J.P. and Vogt, N.B., 1994. Atmospheric deposition of trace elements in Norway; temporal and special trends studied by moss analysis. *Water, Air Soil Pollut.*, 74: 121–140.

Saxena, D.K. and Saxena, A., 2000. Uptake of metals in *Plagiochasma* and their use in pollution monitoring. *Geophytology*, 28(1 and 2): 129–137.

Saxena, D.K., 2001. Biomonitoring of Pb, Ni, Cr, Hg with the help of bryophytes in Nainital. *Perspective in Indian Bryology*, p. 155–170.

Saxena, D.K., Saxena, Anuj and Kamakshee, 1998. Use of *Plagiochasma* in regional mapping of metal fall out. In: *4th International Symposium and Exhibition on Environmental Contamination in Central and Eastern Europe*. Warsaw, Poland.

Saxena, D.K., Saxena, Anuj and Srivastava, H.S., 2000a. Biomonitoring of metal precipitation at petrol pumps and their effect on moss *Sphagnum cuspidatum* Hoffm. *J. of Env. Studies and Policy*, 3(2):95–102.

Tuba, Z. and Csintalan, Z., 1993. *The Use of Moss Transplantation Technique for Biomonitoring of Heavy Metal Pollution*, (Ed.) Bernd Markert. VCH Publishers Inc., New York, USA, 17: 403–411.

Xuolong, H.I., Olmer, I., Aras, N.K. and Gordon, G.E., 1994. Emission of trace elements from motor vehicles: Potential marker elements and source composition profile. *Atmospheric Environment*, 28: 1385–1392.

Ethnobotanical Wisdom and Microbial Studies on Medicinal Plants (2011) *Pages* **65–70**
Editors: **D.R. Khanna, Ashutosh Gautam, R. Bhutiani, Gagan Matta, Anil Dhiman**
Published by: **DAYA PUBLISHING HOUSE, NEW DELHI**

Chapter 11

Isolation and Identification of Food Spoiling Bacteria

Bharti Chauhan
Department of Zoology, H.N.B Garhwal University, Srinagar, U.A.

ABSTRACT

Raw fruits and vegetables have been known to serve as vehicles of human disease for at least a century. In the present study a survey of different hospitals and clinics of Srinagar was done to investigate the cases of food borne diseases in the valley. It was found that *Staphylococcus aureus* was predominant in banana and grapes, which is known to cause Staphyloccocal food poisoning.

Keywords: *Bacterial contamination, Food spoilage, Staphylococcus aureus.*

Introduction

Food is one of the most important requirements for life of single cell microbes to human. Except plants and some algae most of the organisms are dependent to others for their nutritional requirements and use varied food materials. Microbes are present almost everywhere in the environment (except fire) and use a vast range of food substances. All microbes comes under different categories on different characters they have, they may be harmful, beneficial or neutral. Among these under harmful category we consider pathogenic microbes which can cause diferent disease in humans, animals as well as in plants.

In humans the most common source of disease is due to spoiled food which is result of microbial growth, mainly by bacteria and their byproducts. There are more than 250 food borne diseases, many of which cause other symptoms other than vomiting and diarrhoea. Bacterial contamination is most common cause of food poisoning, followed by viruses and parasites. Among the common bacterial species, *Camphylobacter, Salmonella, E.coli, Staphylococcus aureus, Shigella, Clostridium perfringens, Botulinum cereus* are encountered most frequent. In the present study a survey of different hospitals and clinics of Srinagar was done to investigate the cases of food borne diseases in the valley.

Materials and Methods

Sample Collection

Vegetables, fruits and meat samples were collected from the roadside market and were packed in presterile autoclavable polybags to bring in lab. Sample were washed with 100 ml sterile distill water for 2-3 times. 10 ml of this water was then taken into presterile tube and was considered as test samples.

Isolation and Identification of Bacterial Species

Isolation and identification of bacterial species responsible for food spoilage was done following the standards methods of APHA(1998).

Results

The results of the present study are given in Tables 11.1, 11.2, 11.3 and 11.5 while the data collected from Govt. Hospital for food and water related disorders is given in Table 11.4. During the course of study total number of colonies obtained from different samples were 607. In case of banana (Orange 54, yellow 13, white 47), in case of grapes (Orange 60, yellow 03, creamy 150) while in case of tomato (Orange 3, yellow 7, creamy 150) and in case of meat (yellow 20, white 100) number of colonies were found (Table 11.1).On the basis of different parameters *viz.* morphology colour and size 11 groups were prepared. One representative of each group was taken for further assay.

Table 11.1: Morphology and Number of Bacterial Colonies in Different Food Samples

Samples	*Orange*	*Yellow*	*White*	*Creamy*	*Total*
Banana	54	13	47	–	114
Grapes	60	3	–	150	213
Tomato	3	7	–	150	160
Meat	–	20	100	–	120

Gram staining reveals that all the representative isolates were 11 gram positive (Table 11.2). Cells which were found round shaped *i.e.* Coccus and clustered like grapes were *Staphylococcus* sps. while those which were coccus in shape but individual in arrangement were of *Micrococcus* Sps. Series of biochemical test were performed for each bacterial colony. Catalase test (positive 9, negative 2), mannitol test (positive 9, negative 2)

Bile esculin test and 6.5 per cent NaCl (positive 1, negative 1) Table 11.3 and Figure 11.1 (a–c).

For determining spoilage rate fresh food articles (Banana, meat and tomato) were brought from market. Bacterial colony present on the surface of these food articles cultured on nutrient agar medium.These food articles were then kept at room temperature (40° C) for determining the bacterial species responsible for spoilage. After 24 hour meat sample showed bacterial growth, produced bad odour and lesion appereance. While in case of banana all the above conditions were seen after 48 hrs.

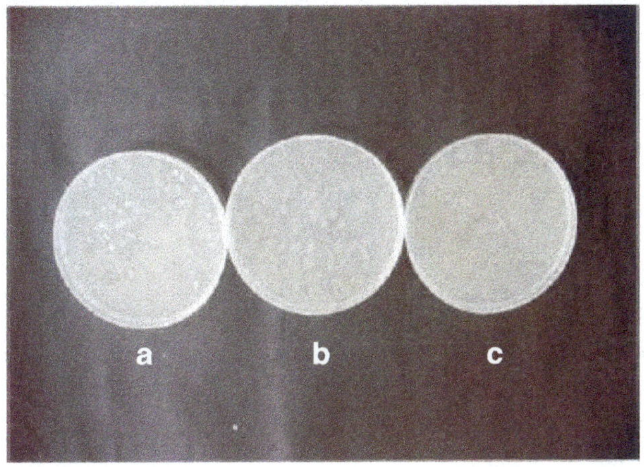

Figure 11.1: Slide Showing Gram Staining Results

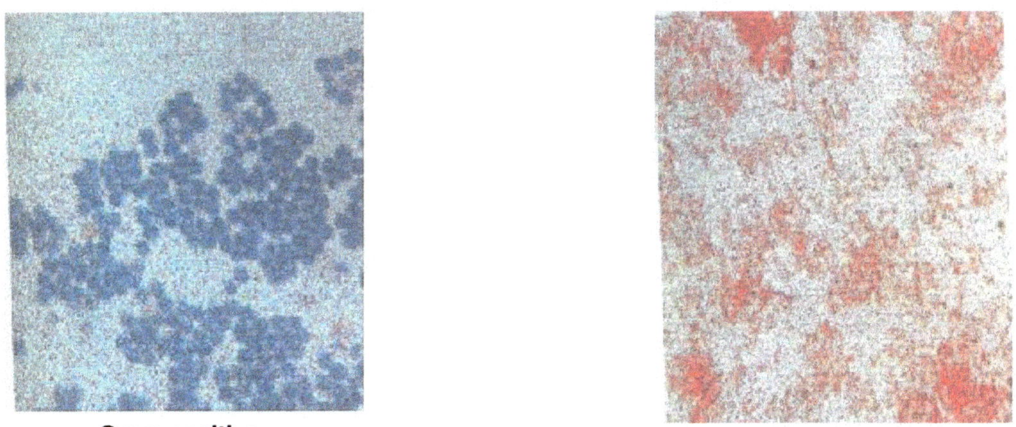

Gram positive **Gram negative**

Figure 11.2: Slide Showing Gram Staining Results

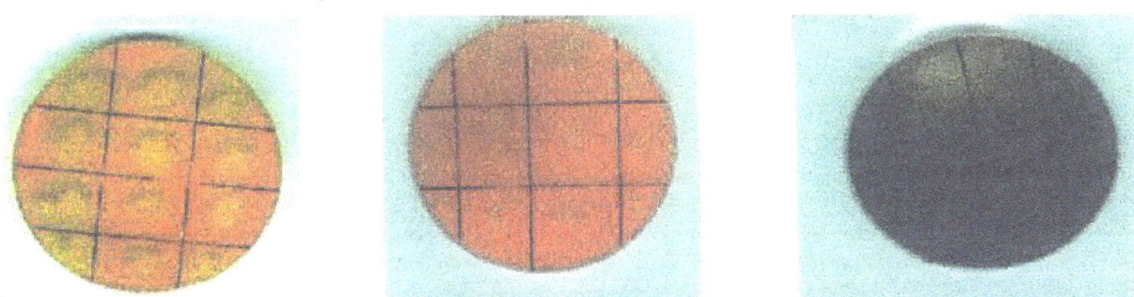

Mannitol positive test
Presence of yellow zone around bacterial colony

Mannitol negative test
Absence of yellow zone

Positive bile esculin test
Light brown colour changes to blackish brown

Figure 11.3: Evaluation of Isolate for Different Biochemical Test

Tomato relatively showed only bacterial growth and foul odour on 3rd day.These materials ware taken and were treated as sample for obtaining bacterial growth of media plates.

Table 11.2: Number of Bacterial Colony Showing Gram-Positive and Gram-Negative Test

Sample	*Colony*	*Gram Reaction*	*Shape*	*Arrangement*	*Inference*
Banana	Orange	Positive	Coccus	Chain like	*Streptococcus*
	Yellow	Positive	Coccus	Chain like	*Streptococcus*
	White	Positive	Coccus	Grape Bunch like	*Staphylococcus*
Grapes	Creamy	Positive	Coccus	Grape Bunch like	*Staphylococcus*
	Orange	Positive	Coccus	Grape Bunch like	*Staphylococcus*
	Yellow	Positive	Coccus	Grape Bunch like	*Staphylococcus*
Tomato	Yellow	Positive	Rod	Individual	*Bacillus*
	Creamy	Positive	Coccus	Grape Bunch like	*Staphylococcus*
	Orange	Positive	Coccus	Grape Bunch like	*Staphylococcus*
Meat	Yellow	Positive	Coccus	Grape Bunch like	*Staphylococcus*
	Creamy	Positive	Coccus	Grape Bunch like	*Staphylococcus*

Stain+ve: Positive test; Stain-ve: Negative test; Coccus: Round; Bacilus: Rod; Individual single; Irregular: Grapes like.

Table 11.3: List of Biochemical Test and Organism Identified

Sample	*Isolate*	*G.R.*	*Shape*	*Pigment*	*Biochemical Test*				*Inference*
					Catalse	*Mannitol*	*Bile Test*	*NACL Test*	
Banana	B.B.O1	+	Coccus	Orange	–	–	+	–	May be *S. bovis*
	B.B.C2	+	Coccus	Creamy	–	–	+	+	*E. faecalis*
	B.B.Y3	+	Coccus	Yellow	+	+	–	–	*S. aureus*
Grapes	G.B.C4	+	Coccus	Creamy	+	+	–	–	*S. aureus*
	G.B.O5	+	Coccus	Orange	+	+	–	–	*S. aureus*
	G.B.Y6	+	Coccus	Yellow	+	+	–	–	*S. aureus*
Tomato	T.B.Y7	+	Rod	Yellow	–	–	–	–	*B. cereus*
	T.B.C8	+	Coccus	Creamy	+	+	–	–	*S. aureus*
	T.B.O9	+	Coccus	Orange	+	+	–	–	*S. aureus*
Meat	M.B.Y10	+	Coccus	Yellow	+	+	–	–	*S. aureus*
	M.B.C11	+	Coccus	Creamy	+	+	–	–	*S. aureus*

B.B.O1: Banana bacterial orange isolate 1; B.B.C2: Banana bacterial orange isolate 2; B.B.Y3: Banana bacterial orange isolate 3; G.B.C4: Banana bacterial orange isolate 4; G.B.O5: Banana bacterial orange isolate 5; G.B.Y6: Banana bacterial orange isolate 6; T.B.Y7: Banana bacterial orange isolate 7; T.B.C8: Banana bacterial orange isolate 8; T.B.O9: banana bacterial orange isolate 9; M.B.Y10: Banana bacterial orange isolate 10; M.B.C11: Banana bacterial orange isolate 11; G.R.: Gram reaction.

Table 11.4: Food and Water Related Disorder of the Year 2004 Collected from Government Hospital Srinagar

Disorder	Case	Age Groups					Recorded in Months											
		0–15	15–30	30–45	45–60	60–75	J	F	M	A	MY	JN	JU	AU	S	O	N	D
Colitis	26	6	9	5	5	1	–	–	2	–	–	3	12	5	3	3	–	–
Diarrhoea	12	4	3	1	1	3	1	–	3	2	–	3	3	–	–	–	–	–
Dysentry	30	5	10	10	3	2	–	1	1	–	3	5	15	2	1	2	–	–
Gastroentritis	10	–	3	5	2	–	2	–	–	4	1	3	–	–	–	–	–	–
Acute abdomine	15	–	9	3	2	1	5	2	–	–	3	2	1	–	–	2	–	–

Table 11.5: Rate of Spoilage of Different Food Samples

Samples	*Temp.*	*Days*	*B.G.*	*Odour*	*Lesions*
Banana	40	2ND	+	+	+
Meat	40	24 HRS	+	+	+
Potato	40	3 RD	+	+	–

After incubation same bacterial colonies were observed which shows that same bacterial species are responsible for spoilage of food.

Discussion

Raw fruits and vegetables have been known to serve as vehicles of human disease for at least a century. In1899 Morse linked typhoid infection to eating celery. Warry(1903) attributed an outbreak of typhoid fever to eating watercress grown in soil fertilized with sewage. Pixley (1913) recorded two cases of typhoid from eating uncooked rhubarb which was grown in soil known to have been fertilized with typhoid excreta. Melick (1917) recovered typhoid bacilli from mature lettuce and raddish harvested from soil that has been inoculated at the time seeds were planted. Similiarly in the present *Staphylococcus aureus* was predominant in banana and grapes, which is known to cause Staphyloccocal food poisoning.

Staphylococcus aureus is known to be carried in the nasal passages of healthy food handlers and has been detected on raw produce (Abdelnoor *et al.*, 1983). According to Hauschild, 1992 cases of botulism that have been linked to fresh produce are very rare. De Rover (1998) reported outbreaks involving cooked/processed vegetable products 9 *e.g.* garlic in oil, mushrooms).

Pao and Brown (1998) studied human pathogens associated with the surface of citrus from seven commercial packing house. They found no generic *E. coli* on fruit. In the present study orange samples which were serially diluted and plated were contaminated with *S. aureus, S. bovis, B. cereus* and *E. facecalis*.

References

Abdelelnoor, A.M., Batshoun, R., Roumani, B.M., 1983. The bacterial flora of fruits and vegetables inlebanon and the effect of washing on the bacterial content. *Zbl. Bakt., I. Abt. Orig. B.*, 177: 342–349.

APHA, AWWA, WPCF, 1998. *Standard Methods for the Examination of Water and Wastewater*, 17th Edn. American Public Health Association, Washington, D.C.

De Rover, C., 1998. Microbial safety evaluations and recommendations on fresh produce. *Food Control*, 9(6): 321–347.

Hauschild, A.H.W., 1992. Epidemiology of foodborne botulism. In: *Clostridium botulinum: Ecology and Control in Foods*, (Eds.) A.H.W. Hauschild and K.L. Dodds. Marcel Dekker, Inc., New York.

Melick, C.O., 1917. The possibility of typhoid infection through vegetables. *J. Infect. Dis.*, 21: 28-38.

Pao, S. and Brown, G. E., 1998. Reduction of microorganisms on citrus fruit surfaces during packinghouse procedures. *J. Food Prot.*, 61(7): 903-906.

Pixley, C., 1913. Typhoid fever from uncooked vegetables. *New York Med. J.*, 98: 328.

Warry, J.K., 1903. Enteric fever spread by water-cress. *Lancet.*, 12: 1671.

Ethnobotanical Wisdom and Microbial Studies on Medicinal Plants (2011) *Pages* **71–77**
Editors: **D.R. Khanna, Ashutosh Gautam, R. Bhutiani, Gagan Matta, Anil Dhiman**
Published by: **DAYA PUBLISHING HOUSE, NEW DELHI**

Chapter 12

Effect of Heavy Metal Stress (Pb, Ni and Cd) on Protein and Fat Content of the Moss *Thuidium cymbifolium*

D.K. Saxena and Harinder Kaur
Department of Botany, Bareilly College, Bareilly

ABSTRACT

In the present study individual effect of Pb, Ni and Cd respectively was studied on protein and fat content of the moss *Thuidium cymbifolium*. A decrease in protein content was noticed in moss treated with Ni and Cd for 3 days, whereas for Pb an increase of 18 per cent was noticed at 0.1 M conc. However following an exposure period of 6 days, a significant increase in protein content was noticed for all metals, which again declined after 15 days of treatment. A significant increase in fat content of the moss (136 per cent and 269 per cent at 0.1 and 0.2 M Pb) was noted after 3 days. Conversely the increase in Cd ranged between 1-27 per cent during 72 hrs treatment that further increased to 77 per cent at 0.2 M Cd concentration following 6 days of exposure. After 15 days of Pb, Ni and Cd treatment a more or less decreasing trend was noticed with increasing concentration. Results revealed that *T. cymbifolium* shows heavy metal sensitivity as evidenced by change in protein and fat content.

Keywords: *T. cymbifolium, Protein, Fat, Heavy metals.*

Introduction

Bryophytes are an important part of many forest ecosystems that are very effective accumulators of heavy metals (Raeymaekers, 1970). Due to their high accumulation capacity these are used quite extensively in assessing atmospheric deposition rate of heavy metals (Tyler, 1990). However physiological and biochemical responses of bryophytes to heavy metal stress is not yet fully understood. In most studies decrease in growth rate (Margot, 1980; Klein and Bliss, 1984) and changes in photo-pigment content (Saxena *et al.*, 2002) has been used as the measure of the toxic effect. However, the fate of protein and neutral lipids (fats) under heavy metal stress has received little attention till date.

Both protein and lipids are sensitive indicators of stress and act as a key substrate for metabolism (Peters, 1973). Therefore the aim of this study was to examine the potential toxicity of Pb, Ni and Cd on fat and protein content of the moss *Thuidium cymbifolium* (Doz. and Molk.) . *T. cymbifolium* is a pleurocarpous feather moss; so far no literature is available on the use of this moss as a test plant to study phyto-toxic effect of heavy metals. Therefore this species was chosen for the present work.

Materials and Methods

Samples of moss *Thuidium cymbifolium* was collected from Mukteshwar (alt. ca. 2290 m) in polystyrene bags from a uniform area of 50 X 50 cm. After collection the samples were sorted in lab to remove extraneous material (adhering bark, other moss species, soil particles) as much as possible. It was then thoroughly cleaned and washed first with running tap water and finally with distilled water. Dilute solutions of heavy metals $CdCl_2$, $Pb(CHCOO_3)_2$ and $NiSO_4$ were used to determine plant sensitivities to metal toxicity.

Moss samples were treated with different concentrations (0.01, 0.1 and 0.2 M) of metals for 3, 6 and 15 days respectively. After treatment period plant tissues were used for physiological analyses. All the treatments had three replications.

Protein Analysis

Oven dry samples were digested in digestion mixture (0.5 g of $CuSO_4$ and 9.5 g of Na_2SO_4) with 5 ml of conc. H_2SO_4 for 5 hrs. It was estimated by Kjeldahl method (Jackson, 1962).

Fat Analysis

Oven dry samples were transferred on Whatman paper and percentage fat was estimated with the help of Soxhlet apparatus (Sankaran, 1965).

Results

The protein of the moss exposed to heavy metal stress (Pb, Ni and Cd) is shown in Figure 12.1. After 3 days of lead treatment, the crude protein in moss plants treated with 0.01 and 0.1 M Pb increased by 18 per cent and 5 per cent respectively in comparison to untreated plants while a drop follows to 19 per cent at 0.2 M concentration. The crude protein increased further to about 65 per cent and 22 per cent at 0.01 and 0.2 M concentrations respectively, whereas at 0.2 M Pb treatment the protein content was comparable to control plants. Again after 15 days of treatment the crude protein of the treated moss plants, was found to be well below the untreated plants.

On the contrary, if we focus on the impact of Ni treatment after 72 hrs on the moss plants an overall reduction in crude protein of the plant at all concentrations was noticed. However following an exposure period of 6 days, a significant increase in protein content (30-45 per cent) compared to control plants was detected, whereas a noteworthy decrease was noticed after 15 days. Likewise Ni,

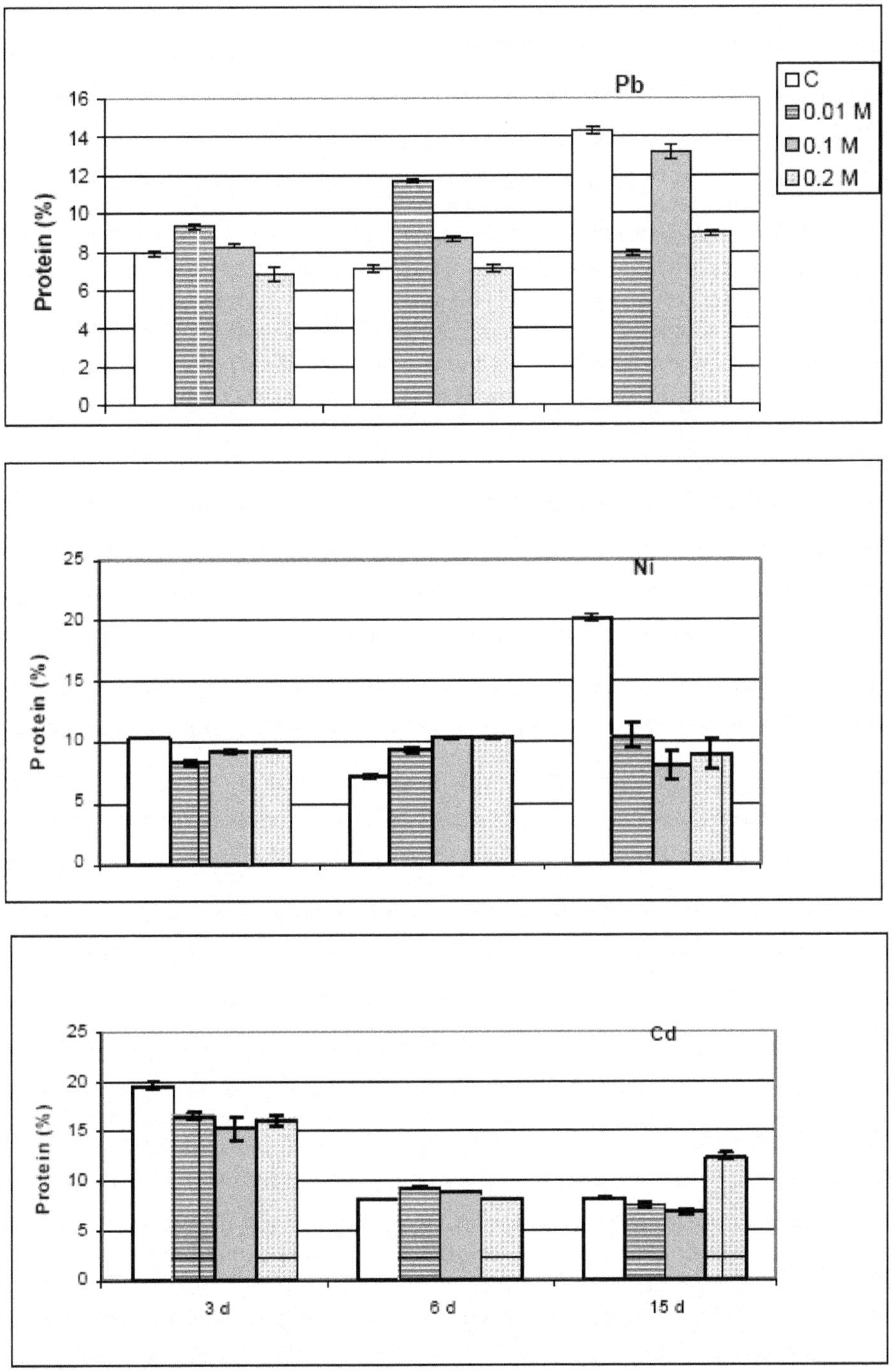

Figure 12.1: Effect of Different Metals (Pb, Ni, Cd) Concentrations on Protein (per cent) Content of Moss *Thuidium cymbifolium* during Different Days of Exposure (Data were means of three replications±S. E).

short-term treatment of Cd, caused a significant reduction at all treatments in crude protein compared to control. While after 6 days of treatment a very slight increase in crude protein at 0.01 and 0.1 M Cd concentrations was observed. After 15 days of Cd treatment in 0.2 M concentration a remarkable increase of 49 per cent relative to control plants was noticed, whereas at 0.01 and 0.1 M concentrations significant lower values were measured.

It is clearly apparent that three days of Pb treatment caused a significant increase in fat content (136 per cent and 269 per cent at 0.1 and 0.2 M Pb) of the moss with respect to control (Figure 12.2.). After 6 days of treatment, the decline in fat proportion ranged between 27-45 per cent relative to control that further decreased after 15 days of treatment. A slightly different trend was noted for Ni treatment, a more or less equivalent fat content was observed in moss plants in comparison to control after 72 hrs. However, after 6 days, a linear increase in fat content was noticed with increasing metal concentration. After 15 days an increase of almost 22 per cent was observed at 0.01 M Ni concentration relative to control that decreased gradually to 16 per cent at highest Ni concentration (0.2 M).

Figure 12.2 indicates that Cd treatment during 72 hrs of treatment induced an alteration in fat content. The percentage increase in fat ranged between 1-27 per cent with respect to control. A further increase in fat composition of moss plants in comparison to untreated plants was noticed, after 6 days of treatment. The maximum increase recorded was 77 per cent at 0.2 M concentration of Cd. However after 15 days, phytotoxicity of Cd was severe, as evident by decrease of fat content.

Discussion

The total protein content of moss *Thuidium cymbifolium* was found varying with metal supplied, concentration of metal and duration of exposure period. A striking increase in the protein content of the moss plants was detected after 6 days of metal (Pb, Ni and Cd) treatment. A comparable crude protein relative to control or slight increase was reported by several workers after metal exposure in a range of plants (Vogeli-Lange and Wagner, 1990; Srivastava *et al.*, 2002; Gupta *et al.*, 2003). The increase in protein content could be due to the direct increase of mRNA and protein synthesis (Singh *et al.*, 1997; Hirt *et al.*, 1989). Further this relative increase in the protein content could also be due to recycling of nitrogen from porphyrins and apoproteins of pigment-protein complexes in the synthesis of new protein molecules (Matile *et al.*, 1990). A common feature of many environmental stresses is the cellular dehydration, which serves as the signal for the accumulation of stress induced proteins such as dehydrins in a wide variety of plants, in response to a wide variety of environmental stresses (Wisniewski *et al.*, 1996).

Long term treatment of Cd, Ni and Pb caused a decrease in protein content may be due to the breakdown of soluble proteins or due to the increased activity of proteins or other catabolic enzymes which were activated and destroyed the proteins (Rai *et al.*, 1998). It is well known that metals interfere with number of metabolic processes associated with normal development, especially synthesis of proteins (Vierstra, 1993). Reduced protein content has also been reported by Saxena *et al.* (2002).

In present study a considerable increase in fat content of the moss plants were noted, when exposed to Pb and Cd levels. Ni exposure, on the other hand, caused small alterations in the fat content of the moss. This result is in consonance with Pham *et al.* (1985) and Sakaki *et al.* (1994), who reported that storage lipids present in low levels in leaf tissues, increase within plants that undergo some environmental stresses. It is therefore, probable that this significant increase in fat content may have an adaptive character in heavy metal pollution.

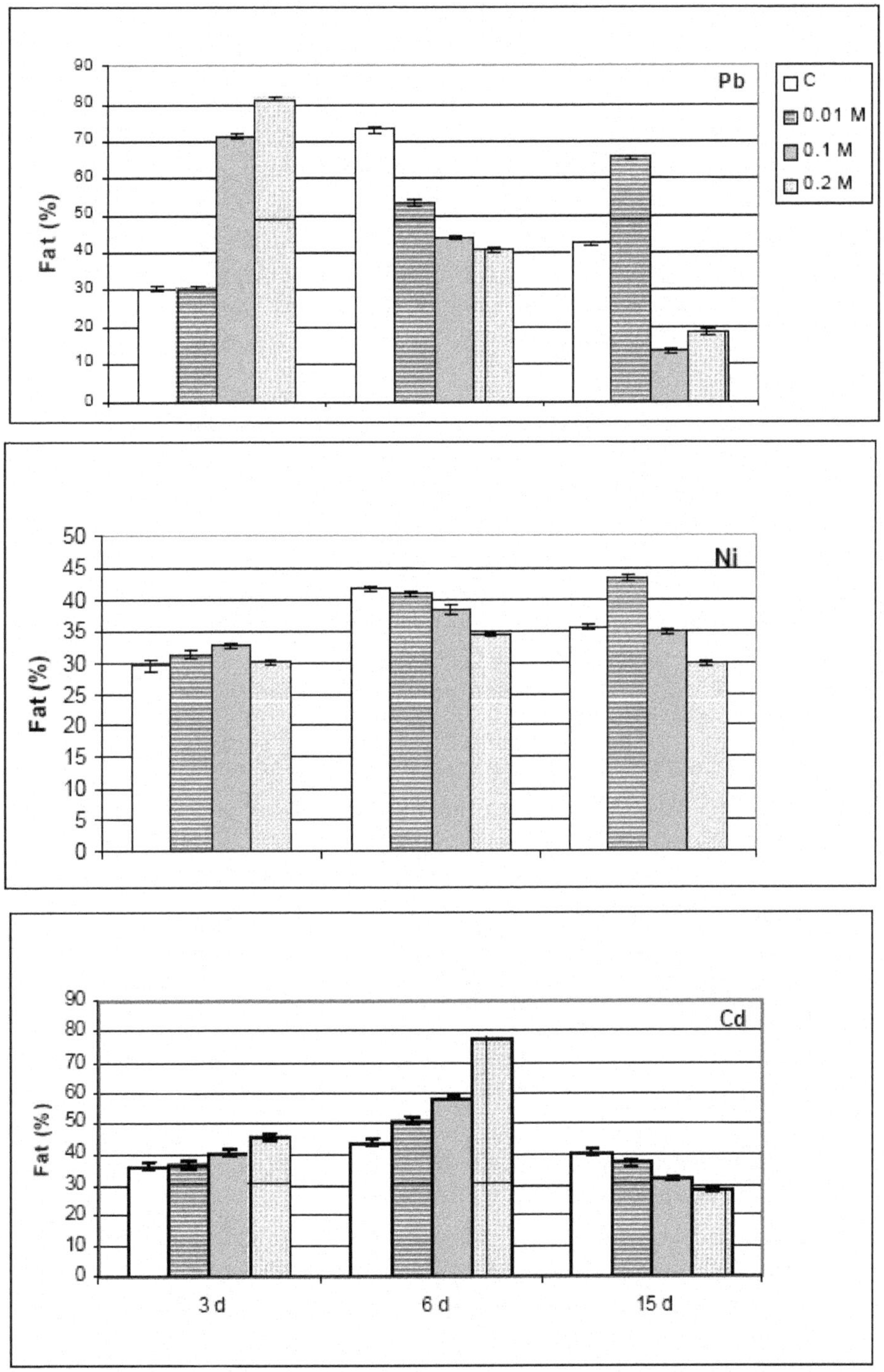

Figure 12.2: Effect of Different Metals (Pb, Ni, Cd) Concentrations on Fat (per cent) Content of Moss *Thuidium lymbifolium* during Different Days of Exposure (Data were means of three replications±S. E).

Moreover, acute exposure to Pb, Ni and Cd caused significant decrease in the fat content. Reduction in the fat level that serves as depots of metabolic fuel may be attributed to increased hydrolysis of stored fat during stress. At high external metal concentration, the energy requirement for active metal extrusion increases the ATP demand by 10-30 per cent of the normal ATP turnover (Losch and Kohl, 1999). Therefore under sub lethal toxic ion stress, this increased ATP demand is generally met by an increased supply from enhanced catabolic process. This might explain the decrease in the fat content of the moss under severe metal stress.

The experiments reported here reveal that the moss *T. cymbifolium* shows heavy metal sensitivity. Both protein and fat content of the moss varied with supply of metal, its concentration and duration of exposure. Furthermore, such in vitro experiments are indeed necessary to investigate standard responses of plants thought to be bio-indicators.

Acknowledgement

This is a part of PhD work submitted to M J P Rohilkhand University. Author is grateful to Director NBRI, Lucknow, for providing permission to consult library; Director, Mukteshwar for providing stay facility during the field trips. Our sincere thanks are also to Principal, Bareilly College, Bareilly for providing facility to carry out above work.

References

Gupta, D.K., Rai, U.N., Singh, A. and Inouhe, M., 2003. Cadmium accumulation and toxicity in *Cicer arietinum* L. *Poll. Res.*, 22(4): 457–463.

Hirt, H., Casari, G. and Barta, A., 1989. Cadmium enhanced gene expression in suspension cells of tobacco. *Planta*, 179: 414–420.

Jackson, M.L., 1962. *Soil Chemical Analysis*. Prentice Hall, Inc., England.

Klein, R.M. and Bliss, M., 1984. Decline in surface coverage by mosses on camels hump mountain, Vermont: possible relationship to acidic deposition. *Bryologist*, 87: 128–131.

Losch, R. and Kohl, K.I., 1999. Plant respiration under the influence of heavy metals. In: *Heavy Metal Stress in Plants*, (Eds.) M.N.V. Prasad and J. Hagemeyer, p. 139–156.

Margot, J., 1980. Vegetative multiplication of cryptogams, an experimental test of pollution by heavy metals. In: *Proceedings of the 3rd International Conference on Bioindicators Deteriorisationis Regionis*. Liblice, Sept. 1977, 391–401, Praha.

Matile, P., Duggelin, T., Schellenbug, M., Rentschm, D., Bortlik, K., Peisker, C. and Thomas, H., 1990. How and why chlorophyll is broken down in senescent leaves. *Plant Physiol. Biochem.*, 27: 595–604.

Peters, M., 1973. Metabolism of carbohydrates. In: *Review of Physiological Chemistry*, 14th edn, (Ed.) H.A. Herpin. Large Medical Publication, p. 232–267.

Pham, Thi A.T., Borrel-Flood, C., da Silva, J.V., Justin, A.M. and Mazliak, P., 1985. Effects of water stress on lipid metabolism in cotton leaves. *Phytochem.*, 24: 723–727.

Raeymaekers, G., 1970. Effects of simulated acidic rain and lead on biomass, nutrient status, and heavy metal content of *Pleurozium schreberi* (Brid.) *Mitt. J. Hattori Bot. Lab.*, 63: 219–230.

Rai, U.N., Gupta, M., Tripathi, R.D. and Chandra, P., 1998: Cadmium redulated nitrate reductase activity in *Hydrilla verticillata* (I. f.) Royle. *Water Air Soil Pollut.*, 106: 171–177.

Sakaki, T., Tanaka, K. and Yamada, M., 1994. General metabolic changes in leaf lipids in response to ozone. *Plant Cell Physiol.*, 35: 53–62.

Sankaran, A., 1965. *A Laboratory Manual for Agriculture Chemistry*. Publication, Bombay.

Saxena, D.K., Kamakshi and Saxena, A., 2002: Reduction in chlorophyll and protein in *Marchantia polymorphaa* L. as indicator of air pollution: A study in Nainital hills. In: *Role of Plant Tissue Culture in Biodiversity Conservation and Economic Development*, (Eds.) S.K. Nandi, L.M.S. Palni and A. Kumar. Gyanodaya Prakashan, Nanital, pp. 623–636.

Singh, R.P., Tripathi, R.D., Sinha, S.K., Maheshwari, R. and Srivastava, H.S., 1997. Response of higher plants to lead contaminated environment. *Chemosphere*, 34(11): 2467–2493.

Srivastava, S., Chakravarty, B. and Srivastava, P.S., 2002. Plant's responses to cadmium: *Linum* as a model system. In: *Role of Plant Tissue Culture in Biodiversity Conservation and Economic Development*, (Eds.) S.K. Nandi, L.M.S. Palni and A. Kumar. Gyanodaya Prakashan, Nanital, pp. 201–219.

Tyler, G., 1990. Bryophytes and heavy metals: a literature review. *Botanical J. Linnean Society*, 104: 231–253.

Vierstra, R.D., 1993. Protein degradation in plants. *Annu. Rev. Phys. Pl. Mol. Biol.*, 44: 385–410.

Vogeli-Lange, R. and Wagner, G.L., 1990. Subcellular localization of cadmium and cadmium binding peptides in Tobacco leaves. *Plant Physiol.*, 92: 1086–1093.

Wisniewski, M., Close, T.J., Artlip, T. and Arora, R., 1996. Seasonal patterns of dehydrins and 70 kDa heat-shock proteins in bark tissues of eight species of woody plants. *Physiol Plant.*, 96: 496–505.

Ethnobotanical Wisdom and Microbial Studies on Medicinal Plants (2011) *Pages* **78–80**
Editors: **D.R. Khanna, Ashutosh Gautam, R. Bhutiani, Gagan Matta, Anil Dhiman**
Published by: **DAYA PUBLISHING HOUSE, NEW DELHI**

Chapter 13

Antibacterial Potential of Karanja *Pongamia pinnata* (Vent.)

P.K. Sagar[1], Navneet[2] and R.K. Paliwal[3]
[1]Drug Testing Laboratory, Rishikul Ayurvedic Pharmacy, Haridwar
[2]Department of Microbiology, [3]Department of of Chemistry,
Gurukul Kangri University, Haridwar

ABSTRACT

The *in vitro* antibacterial activity of seed and root extracts of *Pongamia pinnata* has been investigated against *Escherichia coli, Staphylococcus aureus, Salmonella typhi, Klebsiella pneumoniae, Pseudomonas sp.* and *Hygrobacterium sp.* The plant shows strong activity against *E. coli, S. aureus, S. typhi* and no activity against *K. pneumoniae* in seed extract. The root extract was found to be effective against *E.coli, S. aureus, S. typhi, K. pneumoniae* and was not effective against *Pseudomonas sp.* and *Hygrobacterium sp* large as the one produced by one unit concentration of the antibiotic oxytetracycline. The ethylalcohol (95 per cent) extract gave a light yellow colour and orange red colour in methanol (95 per cent) extract. The ethyl alcoholic extract of seeds shows a higher inibitory effect on *S. aureus* followed by *E. coli*, and *S. typhi* (Jain and Aggarwal, 1978).

Keywords: *Antibacterial effect, Sensitivity, Effective inhibitory zone, Plant extract, Pongamia pinnata (vent.)*

Introduction

The evaluation of plant extract for their antibacterial activity have been known for more than seventy years (Machat and Kankel, 1920). The microorganism have developed resistance to many antibiotics. This has created an immense clinical problems in the treatment of disease, therefore, there is a need to develop an alternative to these drugs for the treatment of disease. The medicinal herbs represent a rich source of antibacterial activity. Keeping this in view an attempt has been made to study the antibacterial activity of alcoholic seeds and root extracts in alcohol of *Pongamia pinnata* (Dhar *et al.*, 1968).

Method-I

100 gm of the roots (coarse form) were kept in ethyl alcohol in soxhlet extractor for 72 hrs. The extract in the form of orange-red colour was collected and fresh ethyl alcohol (95 per cent) was again added and kept in the soxhlet extractor for next 72 hrs. The procedure was repeated till colour of decoction become light. All decoctions were mixed and ethyl alcohol (95 per cent) was separated by vacuum distillation (Jain and Aggarwal, 1978).

Method-II

For the preparation of seed extract, the seeds were first washed 2-3 times with tap water and then with sterilized distilled water. 100 gm. seeds of *Pongamia pinnata* were crushed in blender resulting in the formation of a paste, which was mixed in 250 ml of absolute alcohol (Ghanaksha and Kaushik, 1999). Alcoholic extract so prepared was allowed to evaporate at room temperature until 80 ml of it was left. This extract was squeezed through double layer musline cloth and filtered through Whatman filter paper No-42 and was centrifuged at 5000 r.p.m. for 20 minutes and then sterilized by passing through 0.2 micron disposable filter.

Maller Hinton Agar (Hi media No. M173) media was used to test antimicrobial activity against *E. coli, S. aureus, S. typhi* and *K. pneumoniae* by disc diffusion method (Ananthanarayana and Panikar, 1996). 5 mm diameter discs are charged with appropriate concentration of the extract and standard antibiotic oxytetracycline (one unit concentration), distilled water and absolute alcohol served as control. The plates were incubated at 37°C for 24 hours. The antimicrobial activity was evaluated by measuring the diameter of inhibition zone in mm.

Table 13.1: The Antibacterial Effect of *P. pinnata* Undiluted Alcoholic Seed Extract

Organism	*Inhibition Zone in mm*				*Effective Zone of Inhibition (in mm)*
	Antibiotic Zone	*Undilute Extract Zone (A)*	*Control Alcohol Zone (B)*	*Distilled Water Zone (C)*	*(A–B)*
Escherichia coli	15.0	14.0	5 mm	Nil	9.0
S. aureus	17.2	15.0	5 mm	Nil	10.0
Salmonella typhi	30.0	13.0	5 mm	Nil	8.0
K. pneumoniae	35.0	Nil	5 mm	Nil	Nil

Results and Discussion

The data presented in Table 13.1 shows that alcoholic seed extract of *P. Pinnata* substantially inhibited the growth of *E.coli, S. aureus* and *S. typhi* but not inhibited the growth of *K. pneumoniae* as indicated by size of zone of inhibition. Thus the seed extract of the plant shows broad spectrum antibacterial activity. Similar results of biological activity of plant against fungi and bacteria were reported by Mehta *et al.* (1993) and Ahmad *et al.* (2001). The seed extract was more effective against Gram +ve.

- 100 per cent: Concentration of root extract
- 50 per cent: Concentration of root extract
- 25 per cent: Concentration of root extract
- Antibiotic: (Oxytetracyline 1 unit conc.)

Table 13.2: The Antibacterial Effect of *P. pinnata* Root Extract

Organism	Inhibition Zone in mm						Effective Zone of Inhibition (in mm)		
	Antibiotic Zone (Oxytetra Cycline I Unit Conc.)	Undilute Extract Zone (A)			Control Alcohol Zone (B)	Distilled Water Zone	(A–B)		
		100%	50%	25%	mm	mm	100%	50%	25%
Escherichia coli	35	17	15	15	5	Nil	12	10	10
Pseudomonas sp.	33	Nil	Nil	Nil	5	Nil	Nil	Nil	Nil
Hygrobacterium sp.	37	Nil	Nil	Nil	5	Nil	Nil	Nil	Nil
Salmonella typhi	21	12	10	Nil	5	Nil	7	5	Nil
K. pneumoniae	30	12	11	Nil	5	Nil	7	6	Nil

Acknowledgement

Extends my sincere thanks to Dr. Vishal Upadhyaya (Microbiologist), G.K. University for providing advance research studies and encouragement from time to time.

References

Ahmad, I., Zaiba, Beg, A.Z. and Mehmood, Z., 2001. Antimicrobial potency of selected medicinal plants with special interest in activity against phytopathogenic fungi. *Indian Vet. Med. J.*, 23 299–306.

Ananthanarayan, R. and Jayaram Paniker, C.K., 1996. *Text Book of Microbiology*, 5th Edn. *Orient Longman*, pp. 578–580.

Dhar, M.L., Dhar, M.M., Dhawan, B.N., Mehrotra, B.N. and Ray, C., 1968. Screening of Indian plants for biological activity Part I. *Indian J. Exp. Biol.*, 6: 232.

Ghanaksha, A. and Kaushik, P., 1999. *Antibacterial effect Rhyncostylis retusa Bt* (*Orchidaceae*): A study. *Plant Science*, 12(11): 593–598.

Jain, P.C. and Agrawal, S.C., 1978. Notes on the activity of some odlori ferous: Organic compounds against some Keratinophilic fungi. *Trans. Mycol. Soc., Japan*, 19: 197.

Machat, D.I. and Kundel, W.M., 1920. *Proc. Soc. Exp. Bio. Med.*, 19: 68.

Mehta, B.K., Shitut, S. and Wankhade, H., 1993. *In vitro antimicrobial efficiency of Triphala Fitoterpia*, 64: 371–372.

Ethnobotanical Wisdom and Microbial Studies on Medicinal Plants (2011) *Pages* **81–83**
Editors: **D.R. Khanna, Ashutosh Gautam, R. Bhutiani, Gagan Matta, Anil Dhiman**
Published by: **DAYA PUBLISHING HOUSE, NEW DELHI**

Chapter 14

Crotalaria sericea: A Miraculous Ethnomedicinal Plant for Snake Bite in North Western Tarai Forest of U.P.– A New Report

Babli Singh and T.P. Mall
Post Graduate Department of Botany, Kisan P.G. College, Bahraich, U.P.

ABSTRACT

During the taxonomic and ethnomedicinal survey for the Flora of the Dudwa Tiger Reserve on Nov 24, 2006 along with M.Sc. Botany students the authors were introduced by their guide Sri Baddal Ram Rana, a local Tharu tribe, to a miraculous ethnomedicinal plant, locally named as Van Sanai. The author were told by Mr. Rana that the leaf juice of the plant has a miraculous effect in case of snake bite. If a tea spoonful of leaf juice is anyhow administered to the patient through mouth, the effect of snake bite is neutralized within a very short time. Mr. Rana is resident of village Muen Nuchani (a tharu village) P.O. Parsia, P.S. Chandan Chauki, Distt. Kheri Lakhimpur. His statement was later confirmed by Mr. Kewal Singh Rana resident of the same village and also working as guide over there, Mr. Mihi Lal Dangaura, a witchery (an elderly Jadu Tona specialist) resident of village Balera P.O. Dhuskiya, Distt. Kheri Lakhimpur and Mr. Rahul Kumar Singh resident of village and Post Semra Hardoo Distt. Kushinagar (U.P.).

Van Sanai was found in very few number in sporadic manner at Kharighat bank of Suheli river in Sonaripur range of Dudwa National Park. According to Sri Y.P. Shukla, The Chief wild life warden, Dudwa Tiger Reserve, the habitat of Van Sanai is a low land grass land vegetation having water logging for about 4-6 months starting from pre monsoon to upto retaining monsoon period. Van Sanai is identified as *Crotalaria sericea* Retz (Fabaceae) with the help of FUGP, 1994 and latter confirmed by Prof. S.K. Singh Retd. Prof. and Head of Botany, DDU Gorakhpur University, Gorakhpur.

The same plant but Single in number was observed by the author near Nishangara Forest Rest House in Kakarha Forest range of Katarniaghat wild life reserve of Distt. Bahraich on Jan 05, 2007 during survey. Since the plants are very few in number can be considered as endangered species, they are at the brink of extinction but miraculous and ethnomedicinally very important hence the plant needs conservation.

Keywords: *Ethnomedicine, Snake bite.*

Introduction

India is one of the twelve mega biodiversity countries of the world having rich vegetation with a wide variety of plants with medicinal value. Plants with medicinal value enjoyed the highest reputation in the indigenous system of medicine all over the world (Mehrotra, 1989). The two great epics- The Ramayana and Mahabharata which have originally written in Sanskrit language also have mention of numerous plant names and their role in the life of people of those days (Rao, 1989). Dudwa Tiger Reserve, which is one of the hot spot of biodiversity, situated in Kheri district of U.P. (India). The region is full of flora, fauna and Tharu tribal population. The indigenous traditional knowledge of medicinal plants which is mostly undocumented and transmitted orally from generation to generation, is fast disappearing from the face of world due to advent of modern civilization. The present study was under taken with a aim to collect and document the traditional indigenous knowledge of the plants and their ethnomedicinal use before its lost for ever.

Dudwa Tiger Reserve lies on the India Nepal border in the foot hills of the Himalaya and the plain of 'Tarai' region. The study area lies between 28° 30′ 60″ N and 80° 41′0″ E and comprises 884 km^2 of Kheri district, Uttar Pradesh. The Tharus of Kheri district, live in villages situated in the vicinity of the Dudwa National Park and adjoining the territory of Nepal. The district is bounded on the East by the district Bahraich, on the South by Sitapur and Hardoi, on the West by Shahjahanpur and Pillibhit district and on the North by the territory of Nepal separated by the river Mohan. There are 41 Tharu villages in Kheri district, occupying an area of 8,194 hectare in the vicinity of Dudwa National Park. The district is inhabited mainly by the Rana Tharus, Danguria Tharus and Kathurias Tharu population.

Methodology

The survey was made on Nov 24, 2006 in Dudwa Tiger Reserve with M.Sc. Botany students to collect the ethnomedicinal plant and their ethnomedicinal use, under the guidance of Sri Baddal Ram Rana and Sri Kewal Singh Rana (both a local Tharu tribe) resident of village Muen Nuchani (a Tharu village) P.O. Parsia, P.S. Chandan Chauki, Distt. Kheri Lakhimpur. Questionnaire method was adopted for documentation of folk indigenous knowledge about medicinal plants and herbs (Shah and Khan, 2007). During the field survey, the plants were collected and their use by inhabitants were recorded in the field books. All the collected plants were pressed, dried, preserved, mounted and identified through the available taxonomic literature and latter confirmed by Prof. S.K. Singh Retd. Prof. and Head of Botany, D.D.U. Gorakhpur University, Gorakhpur. All the collected and preserved plant specimens were deposited in the Herbarium maintained in the department.

Results and Discussion

After the survey of Dudwa Tiger Reserve about 55 plant species belonging to different family were collected. Out of these species the present study deals with an important medicinal plant of *Crotalaria sericea* Retz. (Fabaceae) locally called as Van Sanai, collected at Kharighat bank of Suheli river in Sonaripur range of Dudwa National Park. Our guide, Mr. Rana told us about the plant that the leaf juice of plant is very effective in case of snake bite. His statement was latter confirmed by Mr. Mihi Lal Dangaura a witchery resident of village Balera P.O. Dhuskiya, Distt. Kheri Lakhimpur and Mr. Rahul Kumar Singh resident of Village and Post Semra Hardoo Distt. Kushinagar (U.P.). It is an robust under shrub, 3-4 ft. high, branches stout, striated, nearly glabrous. Leaves shortly stalked, 2.6 in. long, oblong, lanceolate, acute, glabrous, above finely silky beneath, stipules large, leafy, persistent. Racemes terminal, often one foot long, bracts ovate, leafy, persistent, pedicels longer than the calyx. calyx ½ in. long nearly glabrous. Corolla 4/5 in. much exerted, yellow colour. Flower during the cold season.

The man has deep involvement with plants and forest not only through history, culture, religion and philosophy but also through their livelihood. The local used of plants as a cure is as old as human civilization (Behera, 2006). Forest are the reservoir of medicinal plants and play a vital role in the economy of the Tharus. The tribal people of Kheri district who are residing in villages are very poor, illiterate and mostly dependent on the forest wealth and ambient vegetation for their food, clothing, oil, fibre, housing and medicine. They utilize root, rhizome, stem, leaf, flower, fruit and seed of medicinal plant in different ailments such as fever, cough, bronchitis, asthma, rheumatism, earache, toothache, leucorrhoea and as antidote to snake bite and scorpion sting. Medicinal plants have several curative properties due to the presence of various chemical substances of different composition. *Crotalaria sericea* Retz. have several important medicinal value but since the plant is found very few in number in the study area hence the conservation and regeneration of the plant is necessary. It is hoped that chemical analysis of the plants and their pharmacotherapeutics will provide much needed lead for further research and new drug development.

Acknowledgments

The authors are thankful to Sri Mohd. Ahsan, Chief Wildlife warden, Uttar Pradesh, Lucknow for giving his permission for scientific research for the exploration of floral vegetation of the territory under his command and to Dr. S.P. Singh, Principal, Kisan P.G. College, Bahraich for providing necessary laboratory and library facilities.

References

Behera, K.K., 2006. *Ethnomedicinal Plants Used by the Tribals of Similipal Bioreserve, Orissa, India: A Pilot Study.*

Mehrotra, B.N., 1989. Collection and processing of plants for biological screening. In: *Methods and Approaches in Ethnobotany*, (Ed.) S.K. Jain. Society of Ethnobotanists, Lucknow, p. 25–37.

Rao, R.R., 1989. Methods and techniques in ethnobotanical study and research: Some basic considerations. In: *Methods and Approaches in Ethnobotany*, (Ed.) S.K. Jain. Society of Ethnobotanists, Lucknow, p. 13–23.

Shah, M.G. and Khan, M.A., 2007. *Check list of Medicinal Plants of Siran Valley Mansehra, Pakistan.*

Ethnobotanical Wisdom and Microbial Studies on Medicinal Plants (2011) *Pages* **84–89**
Editors: **D.R. Khanna, Ashutosh Gautam, R. Bhutiani, Gagan Matta, Anil Dhiman**
Published by: **DAYA PUBLISHING HOUSE, NEW DELHI**

Chapter 15

Effect of Leaf Leachates of Potato Plants on Germination of *Phytophthora infestans* Sporangia and Zoospores

Navneet and Deepika
Department of Botany and Microbiology,
Gurukula Kangri University, Haridwar

ABSTRACT

Leaf leachates of two cultivars of potato *i.e.* Kufri Chandramukhi (KCM) and Kufri Jyoti (KJ) contain nine amino acids. Total sugar content of KCM and KJ leachates varied between 0.10 and 0.37 μg per sq. cm and 0.02 to 0.23 μg per sq. cm. respectively. The sugars exuded more by KCM. Total phenol content of KCM and KJ leachates varied between 1.00 to 2.0 μg per sq. cm and 1.05 to 3.16 μg per sq. cm respectively. The percentage germination of *P. infestans* sporangia in KCM and KJ cultivars was more in mature leaf leachates as compared to young leaves, moreover it was less in KJ cultivar as compared to KCM. The germ tube length of *P. infestans* zoospores in leaf leachates was inhibited in both the cultivars except in mature leaf leachate of 94 days old Kufri Chandramukhi.

Introduction

Leaf leachates play a vital role in stimulating or inhibiting the growth of propagules settled down on leaf surface. The nature of leaf leachates varies from plant to plant and is influenced by the age of the organ.This study deals with the composition of leaf leachates of two cultivars of potato *i.e.* Kufri

Chandramukhi (KCM) and Kufri Jyoti (KJ) and their role in germination of *Phytophthora infestans* sporangia and zoospores.

Material and Method

Leaf leachates of young leaves (second leaf from top) and mature leaves (fifth/sixth leaf from top) of each variety were collected in sterilized plastic boxes by the method of Godfrey (1975) in 22nd, 222rd and 5th sampling the age of plants were 42, 67 and 94 days respectively. Amino acids were qualitatively analysed by unidirectional ascending paper chromatography (Block *et al.*, 1958). Total sugar contents of leachates were estimated by anthrone reagent test (Morris, 1948). Total phenols were estimated by the method of Farkas and Kiraly (1962).Germination of *P. infestans* sporangia were estimated by placing 0.02 ml of leachate plus 0.02 ml of sporangial suspension in one concavity of sterilized double concavity slide. The other concavity served as control was inoculated by 0.02 ml of sterilized distilled water plus 0.02 ml of sporangial suspension. The slides were incubated at 8 °C for 1–1/2 hours, 2 hours and 2-1/2 hours. Five non overlapping fields were observed for percentage sporangial germination.Germination of *P. infestans* zoospores were estimated by placing 0.02 ml of leaf leachate plus 0.01 ml of zoospore suspension in one concavity and the other concavity which served as control was inoculated with 0.02 ml of sterilized distilled water and 0.01 ml of zoospores suspension. The slides were incubated at 25 °C for hours. The germ tube length was measured. The present inhibition in germ tube length was calculated by using the formula suggested by Vincent (1947).

Results

Nine amino acids *viz.*. L-glutamic acid, DL-serine, DL- threonine, L-ornithine monohydrochloride, L-arginine monohydrochloride, DL-α-alanine, L-proline, L-cystine and DL-valine were detected in the leaf leachates of two cultivars of potato *i.e.* KCM and KJ. Two amino acids, DL-serine and L-arginine monohydrochloride were exclusively detected in the leachates of KJ cultivar. While the rest seven were common in both the cultivars. Four amino acids *viz.* DL-threonine, Dl- α-alanine, L-proline and DL-valine and five amino acids *viz.* DL-serine, DL-threonine, L-arginine monohydrochloride, L-cystine and DL-valine were detected from young leaves of KCM and KJ cultivars respectively.

Total sugars and phenol content of leachates of young and mature leaves of KCM and KJ cultivars of potato were given in Table 15.1. Total sugar content of KCM leachates varied between 0.10 and 0.37 µg per sq. cm. The lowest amount of sugars was exuded by leaves of 25 sampling. In 22nd and 25th sampling the sugar content was more in young leaf leachates as compared to mature ones. Similarly the total sugar content of KJ leaf leachates varied between 0.02 to 0.23 µg per sq. cm. The maximum amount of sugar was present in the leachates of mature leaves of 22nd sampling. The sugar content was more in the leachates of young leaves of the 22nd and 222rd sampling. On comparing the sugar content of leaf leachates of cultivars it was found that sugars was exuded more by KCM.

Total phenol content of KCM leachates varied between 1.00 to 2.0 µg per sq. cm the phenols were present more in young leaf leachates of 22nd and 222rd sampling as compared to mature leaf leachates, while the reverse was true for 25th sampling. The phenol content of KJ leachates varied between 1.05 to 3.16 µg per sq. cm. In all the samplings total phenol content was more in young leaf leachates and the maximum amount was found in the young leaf leachates of 25th sampling. On comparing the phenolic content of KCM and KJ leachates we found that in 22nd and 222rd sampling more phenols were exuded by young leaves of KCM and mature leaves of KJ. In the 25th sampling phenolic content was much higher in young leaves of KJ cultivar as compared to other samples of leaf leachates.

Table 15.1: Total Sugars and Total Phenols in the Leaf Leachates of Two Cultivars of *Solanum tuberosum*

Sampling Number		Sugars µg/cm²		Phenol µg/cm²	
		KCM	KJ	KCM	KJ
II	Young	0.37	0.20	2.80	1.84
	Mature	0.12	0.23	1.09	1.50
III	Young	0.20	0.11	1.91	1.22
	Mature	0.31	0.02	1.00	1.05
IV	Young	0.14	0.22	1.23	3.16
	Mature	0.10	0.15	1.44	1.21

The percentage germination of *P. infestans* sporangia in the leachates of KCM and KJ cultivars of *Solanum tuberosum* at different time intervals is given in Figure 15.1. After 1-1/2 hours percent sporangial germination was found enhanced except young leaf leachates of 22nd sampling when compared to control in both the cultivars, germination. While, sporangial germination was inhibited when compared

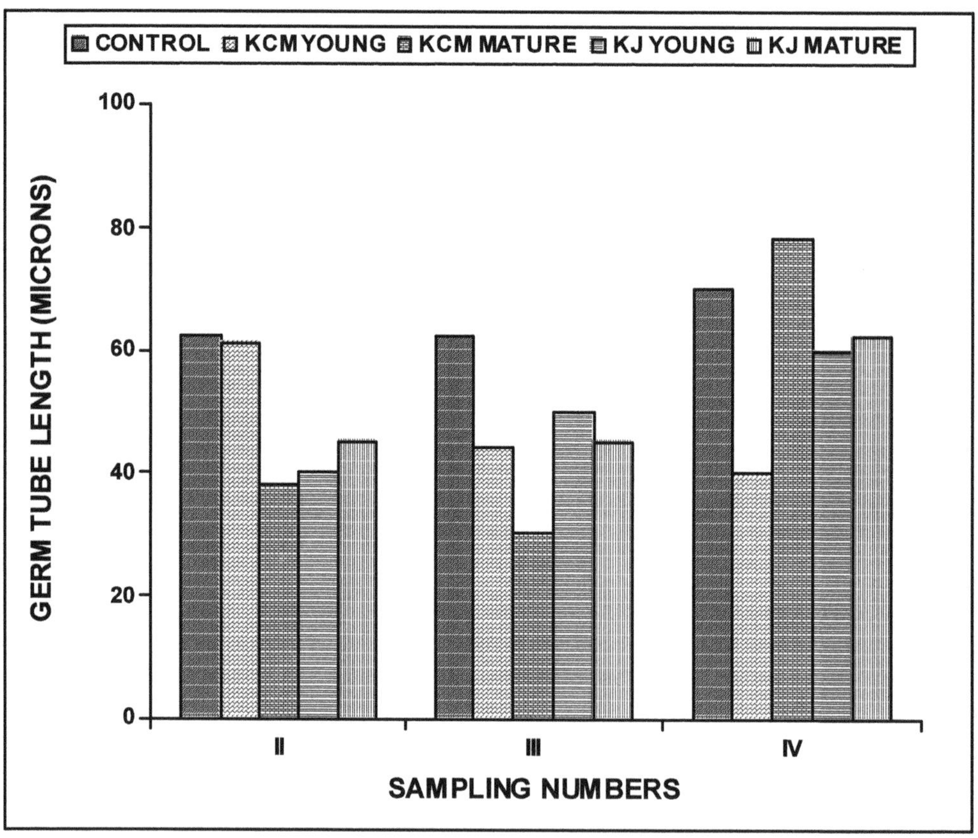

Figure 15.1: Growth of Germ Tube Length in Control and Leachates of Kufri Chandramukhi and Kufri Jyoti Cultivars

Godfrey, B.E.S., 1974. Phylloplane mycoflora of braken, *Pteridium aquilinum*. *Trans. Br. Mycol. Soc.*, 62: 305–311.

Link, K.P. and Walker, J.C., 1933. The isolation of catechol from pigmented onion scales and its significance in relation to disease resistance in onions. *J. Boil. Chem.*, 100: 379–383.

Morris, D.L., 1948. Quantitative determination of carbohydrates with Dreywood's anthrone reagent. *Science*, p. 254–255.

Mishra, R.R. and Tewari, R.P., 1978. Studies on pyllosphere fungi: Germination behaviour of certain fungi in leaf extract and exudates of wheat and barley. *Acta Bot. Indica*, 6: 21–30.

Purnell, T.J., 1971. Effects of pre-inoculation washings of leaves with water on subsequent infections by *Erisyphe cichoracearum*. In: *Ecology of Leaf Surface Microorganisms*, (Eds.) T.F. Preece and C.H. Dickinson. Academic Press, London, pp. 269–275.

Sadasivam, K.V., Rangaswami, G. and Prasad, N.N., 1976. Studies on pyllosphere microflora of tapioca (*Manihot utilissima* Pohl.). *Zbl. Bakt. Abt. Bd.*, 131: 632–643.

Singh, P.N., Sindhu, I.R. and Gupta, K., 1986. Effect of leaf exudates and extract of spinach on some phylloplane fungi. *Acta Bot. Indica*, 14: 104–110.

Shepherd, C.J. and Mandryk, M., 1963. Germination of conidia of *Peronospora tabacina*. II. Germination *in vivo*. *Aust. J. Soil. Sci.*, 16: 77– 87.

Sharma, J.K. and Sinha, S., 1971. Effect of exudates of *Sorgum* varieties varying in susceptibility and maturity on the germination of conidia of *Colletotrichum graminicola* (Ces.) Wilson. In: *Ecology of Leaf Surface Microorganisms*, (Eds.) T.F. Preece and C.H. Dickinson. Academic Press, London, pp. 597–601.

Singh, S.A. and Gupta, P.K.S., 1983. Effect of leaf exudates and extracts on conidial germination of rice leaf scald fungus, *Rhynchosporium oryzae*. *Indian J. Plant Pathol.*, 1: 109–113.

Schneider, R.W. and Sinclair, J.B., 1975. Inhibition of conidial germination and germ tube growth of *Cercospora canescens* by cow pea leaf diffusates. *Phytopathology*, 65: 63–65.

Vincent, J.M., 1947. Distortion of fungal hyphae in the presence of certain inhibitors. *Nature*, (London) 159: 850.

Wang, S.C. and Pinckard, J.A., 1973. Spore germination of *Diplodia gossypina* in the presence of carbohydrates and phenolic compounds in relation to boll rot of cotton. *Phytopathology*, 63: 1181–1185.

Ethnobotanical Wisdom and Microbial Studies on Medicinal Plants (2011) *Pages* **90–93**
Editors: **D.R. Khanna, Ashutosh Gautam, R. Bhutiani, Gagan Matta, Anil Dhiman**
Published by: **DAYA PUBLISHING HOUSE, NEW DELHI**

Chapter 16

Dyeing of Wool and Cotton Fibres with Fruit Rind of *Juglans regia* as Natural Dyes, and Standardization of Ancient Dyeing Procedure

J.S. Jangwan[1], Maneesha Dobhal[1] and S.C. Sati[2]
[1]Department of Chemistry, H.N.B. Garhwal University Campus, Badhshanithaul Tehri, Garhwal
[2]Department of Chemistry, H.N.B. Garhwal University Campus, Srinagar, Garhwal

ABSTRACT

Vegetable dyes and their uses are known in ancient time more than 2000 years ago. It was found that the discovery of synthetic dyes reduce the use of natural dyes and consequently the export of same was affected in large extent. However due to non toxic nature, eco-friendly behaviour and properties to impart variety of colours once again natural dyes are getting more attention of chemists and dyers. The present study deals with optimization of dyeing and mordanting of wool and cotton samples with fruit rind of *Juglans regia* and standardization of ancient dyeing procedure and also the fastness studies of dyed samples toward light and washing.

Introduction

The *Juglans regia* (Ankhrot) belong to family Juglandaceac is a large tree. Flowers are green, male and female appearing with the leaves. The bark of *Juglans regia* (Ankhrot) were known to contain substances that could be used for colouring the fabrics, cloths, utensils, and household implements.

The colouring matter presents in the bark of *Juglans regia* is a mixture of Lignins, Tannins, Anthocyanins and Xanthones. The bark is credited with astringent properties. The fresh bark powder and seed coat of the plant are used as biopesticide in agriculture. (Gaur, 1999 and Lemmens and Wulijarni-Soetijipto, 1991).

Materials and Methods

The fruit rinds of the plant were collected from the forests of Ranichauri Distt.- Tehri Garhwal, Uttrakhand. During rainy season, and dried in shade. All the chemicals used as mordant were of L.R/ A.R and BDH fine chemicals. The white wool and cotton samples were freed from traces of protein, gums and oil impurities by washing it with detergent or soap for a required periods, followed by kneading, squeezing and rinsing with lot of water, till it was free from traces of detergent. It was then dried in shade and ironed before dyeing and mordanting, the cotton and wool samples were soaked for an hour. The optical density was recorded by using digital spectrophotometer.

The optimum concentration of dye material and the time for extraction of the dye and dyeing of cotton and wool were found out by taking different concentration of dye material (2-12gm) in different beakers.

In order to find out optimum time for dyeing one gram of cotton and wool fibers was added to seven beakers and dyed for 30, 45, 60, 75, 90, 105, and 120 minutes. The samples were stirred occasionally to obtain an even dyeing. The dyed samples were taken out from beakers and dried in shade. The evenness of dye, depth of shade and overall appearance are evaluated by a panel of judges. The highest percentage rating was calculated by Judgment Table 16.2.

Number of shades were obtained by mordanting the cotton and wool with different concentration of mordants such as chrome, alum, copper sulphate, ferrous sulphate, stannous chloride etc. and also using the pre-mordanting (mordanting before dyeing) Post-mordanting (mordanting after dyeing) and simultaneous mordanting (dyeing and mordanting together in same pot.)

Analysis of Samples Towards Light and Washing Fastness

The dyed samples of cotton and wool to be tested for the light fastness were cut into specimens of 3x 6cm, numbered and mounted on a cardboard frame. The blue wool and cotton sample (five of each) standard having a light fastness rating from low to high (1 low-5 high cotton, 1 low-5 high wool) were stapled on cardboard strip. A strip of thick black chart paper was pinned in such a manner that half of it lies on the specimen and the standards thereby leaving the samples half exposed and half covered (Bird, 1972).

A strip of thick black paper was put inside the fluorescent lamp. The specimens and standards were checked at regular intervals. The samples were compared with blue wool and cotton standards and rated. The different methods of mordanting produce good range of colours having various shades ranging from dark brown to brownish black, black to greenish, brown to black. The premordanting method gave the best results/colours with Stannous Chloride and Ferrous Sulphate. A variety of shades was also obtained with post and simultaneous mordanting process.

The washing fastness of sample was done by marking the specimens of 5 x 7 cm size and placed between two layers of the fabric of the same size. These layers were seen from a complex specimen. The soap or detergent solution was prepared by dissolving 10 gram of soap or detergent in 1000ml of water. Each composite sample was treated with soap solution in Launderometer for 30-40 minutes at a temperature of 60 °C. The samples were then removed and rinsed thoroughly under running tap water, dried and ironed.

The rating of washing fastness was done on the basis of the staining of the fabrics. Washing brought a considerable change in the colour of dyed sample Chrome, Alum and Stannous chloride mordanted samples showed very poor fastness. Cotton was stained with $SnCl_2$ whereas wool was noticeably stained with Alum (Sati *et al.*, 2003).

Ferrous Sulphate mordanted samples has fair washing fastness while Copper Sulphate and Chrome mordanted samples have poor to fair washing fastness. Chrome and $SnCl_2$ showed slight stain on wool.

Results and Discussion

From the data of Table 16.1 and 16.2 it is evident that 8gm of rind power was optimum to give the maximum optical density of dye liquor while 75 minutes of extraction time and dyeing time was found to be optimum. Best shades of colours were obtained by using 12 gm of Alum, 5 gm of Chrome, 8 gram of $CuSO_4$, 10 gram of $SnCl_2$ and $FeSO_4$ respectively per 100gm of wool and cotton.

Table 16.1: Optimum Concentration of Dye Materials

Conc. of Dye material g/100 ml of Water	*Optical Density*	*Percentage Rating*
2 gm	0.081	52
4 gm	0.097	55
6 gm	0.121	60
*8 gm	*0.137	*67
10 gm	0.128	59
1 2gm	0.123	62

Wavelength 320 mn.

* Selected sample.

The optimum time for extraction of dye was found out by extracting the optimum concentration of dye material at 80°C for 30, 45, 60, 75, 90,105 and 120 minutes and the optical density of the liquor was measured (Table 16.2).

Table 16.2: Optimum Time for Extraction of Dye and Dyeing

Time for Extraction of Dye (min.)	*Conc. of Dye Material (gm)*	*Optical Density*	*Per cent Absorption*	*Per cent Rating*
30	8	0.142	0.72	
45	8	0148	0.75	
60	8	0.153	0.82	
*75	8	*0.162	*0.84	*0.84
90	8	0.160	0.83	
105	8	0.161	0.82	
120	8	0.161	0.82	

Wavelength 320 mn.

* Selected sample.

In present study it was observed that cotton samples was noticeable stained with Stannous Chloride using post mordanting technique, while wool was noticeably stained with Alum using same procedure. The washing fastness of both sample of cotton and wool was found fairly good.

Efforts was also made to mordant the samples with naturally occurring mordants of plants origin like bark of *S. racemosa*, seed rind of *P. granatum*, bark of *M. esculenta* etc. which produces a number of beautiful shades on wool and cotton samples.

The whole study points out that the dye extracted from fruit rind of *Juglans regia* can be use to obtain from array of colours either alone or in combination with mordants (natural/synthetic) with different fabrics. The optimization of mordanting has not only rendered the dyed wool with good wet fastness but has also helped in achieving eco-friendly processing parameters in terms of effluent within tolerance limit for the residual heavy metal content.

References

Bird, C.L., 1972. *Theory* and *Practice of Wool Dyeing*, 4th edn. Society of Dyers and Colourist, 4 Edn.

Gaur, R.D., 1999. *Flora of District Garhwal*. Transmedia, Srinagar Garhwal.

Lemmens, R.H.M.J. and Wulijarni-Soetijipto, N., 1991. *Dye and Tennin Producing Plants Vol 3: Plant Resources of South East Asia.*

Sati, S.C., Rawat, U., Sati, O.P. and Srivastav, B., 2003. *Colourage*, 44.

Ethnobotanical Wisdom and Microbial Studies on Medicinal Plants (2011) *Pages* **94–96**
Editors: **D.R. Khanna, Ashutosh Gautam, R. Bhutiani, Gagan Matta, Anil Dhiman**
Published by: **DAYA PUBLISHING HOUSE, NEW DELHI**

Chapter 17

Antibacterial Activity of Nirgundi (*Vitex negundo* Linn.)

Sanjay[1], Navneet[1] and M.M. Tiwari[2]
[1]Department of Botany and Microbiology,
[2]Faculty of Engineering and Technology,
Gurukul Kangri University, Haridwar – 249 404, Uttarakhand

ABSTRACT

The antibacterial activity of Nirgundi (*Vitex negundo*) extracts were tested against bacterial dental infections (*Staphylococcus aureus, Streptococcus mutans, S. sanguis, S. salivarius* and *Lactobacillus acidophillus*) and some other pathogens (*Bacillus subtilus, E. coli*).The aqueous, methanolic and petroleum ether extracts of plants were tested for their antibacterial activity using well diffusion method at the sample concentration of 200 mg/ ml. The methanolic extract showed maximum activity as compared to other extracts. The methanolic extract was most active against *S. salvarius* followed by *Staphylococcus aureus, Streptococcus mutans, S. sanguis* and *Lactobacillus acidophillus*.

Keywords: *Vitex negundo, Dental plaque, Antibacterial activity.*

Introduction

The different systems of medicine practiced in India, Ayurveda, Siddha, Unani, Amchi and local health traditions, utilize large number of plants for the treatment of human diseases. Most of these medicinal plants have been identified and their uses are well documented and described by different authors (Nadkarni, 1976, Dastur, 1995, Saradamma, 1990, Jain, 1991, Kirtikar and Basu, 1991), but the efficacy of many of these plants are yet to be verified. In order to evaluate the efficacy of these plants by scientific investigations, an interdisciplinary programme was started with the aim of screening plants for their antimicrobial activity. Selection of the plants from the literature were made on the basis of their common use in the treatment of infectious diseases like fever, bronchitis, ulcer, diarrhoea, dysentery and skin disease. The medicinal herbs represent a rich source of antibacterial and antifungal activity (Ahmad *et al.*, 1998). *Vitex negundo* has shown activity against *Escherichia coli, Klebsiella aerogenes, Proteus vulgaris, Pseudomonas aerogenes* (Perumal *et al.*, 1998). But there is no record of its antibacterial property against dental infections caused by *Streptococcus mutans, S. sanguis, S. salivarius, S. aureus* and *Lactobacillus acidophillus*. In this study the antibacterial activity of *V. negundo* is presented along with references to their traditional uses.

Materials and Methods

Plant Material and Extracts

The plant was collected from the foot hills of Shivalik range of Himalayas in Haridwar and identified at Botanical Survey of India, Dehradun, Uttarakhand. The plant was shade dried at room temperature and then powdered plant material was loaded in soxhlet assembly and extracted in four different solvents such as petroleum ether, acetone, methanol and water. The polarity of the solvents would leach out compounds soluble in particular solvents.The extracts were concentrated by using vacuum evaporator.

Antibacterial Assay

The well diffusion or cup plate method is used for antibacterial activities. The 8mm diameter wells were punched in the agar and filled with extracts and respective solvents for control and standard antibiotic ampicillin (100 mg / ml) was used as positive control, the plate were incubated at 37 °C for 24 hours. The antibacterial activity was evaluated by measuring the diameter of inhibition zone in mm.

Results and Discussion

The plant shows broad spectrum antimicrobial activity (Table 17.1) *i.e.* the methanol, water, acetone and petroleum ether extracts were active against both gram positive and gram negative bacteria. The *V. negundo* extracts were found to be less effective as compared to ampicillin. The methanolic extracts exhibited the highest degree of antibacterial activity as compared to aqueous, acetone and petroleum ether extracts. The extracts were highly inhibitory to *S. mutans* and *S. salivarius*, but the maximum inhibition zone 23 mm was attained with methanol extract of *V. negundo*.

The medicinal plants are used by large proportion of the Indian population because there is a true improvement of disease condition after herbal treatment with no harmful side effects and high cost of other forms of treatment. *V. negundo* extracts show broad spectrum of action against microbes causing dental caries. Ampicillin is more effective against microorganisms because it is used in a pure form as compared to crude extract of *V. negundo* in different solvents.

The methanolic extracts exhibited the highest degree of antimicrobial activity as more compounds *i.e.* alkaloids, steroids, terpenoides and glycosides are extracted by methanol in comparison to petroleum ether and acetone.

Further work is going on to locate the active principle from the various extracts and these preliminary results of this investigation indicates that *V. negundo* has a wide spectrum of antibacterial activity.

Table 17.1: Antibacterial Activity of *Vitex negundo* Extracts in Different Solvents

Microorganisms	*Inhibition Zone (mm)**				
	Extracts				*Antibiotic*
	*Pt. Ether*** 100 mg/ml*	*Acetone*** 100 mg/ml*	*Water*** 100 mg/ml*	*Methanol*** 100 mg/ml*	*Ampicillin 100 mg/ml*
Staphylococcus aureus	20	21	22	20	24
Streptococcus mutans	19	20	22	21	25
S. salivarius	21	23	22	23	26
S. sanguis	17	17	20	19	21
L. acidophillus	18	17	19	21	23
B. subtilus	9	10	12	II	21
E. coli	13	14	17	16	22

References

Ahmad, I., Mehmood, Z. and Mohammad, F., 1998. Screening of some Indian medicinal plants for their antimicrobial properties. *Journal of Ethnopharmacology*, 62: 183–193.

Dastur, J.F., 1995. *Medicinal Plants of India and Pakistan*. D. B. Taraporevala, Bombay.

Jain, S.K., 1991. *Dictionary of Indian Folk Medicine and Ethnobotany*. Deep Publications, New Delhi.

Kirtikar, K.R. and Basu, B.D., 1991. *Indian Medicinal Plants*, 2nd edn. Bishen Singh Mahendra Pal Singh, Dehradun, India.

Nadkarni, K.M., 1976. *India Materia Medica*. Popular Prakashan, Bombay.

Perumal Samy, R., Ignacimuthu, S. and Sen, A., 1998. Screening of 34 Indian medicinal plants for antibacterial properties. *Journal of Ethnopharmacology*, 62: 173–182.

Saradamma, I., 1990. *All India Co-ordinated Research Project on Ethnobiology*. Final technical report. RRI Drug research, CCRAS, Government of India, Poojapura, Trivandrum, Kerala, India.

Valsaraj, R., Pushpangadan, P., Smith, V. W., Adsersen, A. and Nyman, V., 1997. Antimicrobial screening of selected medicinal plants from India. *Journal of Ethnopharmacology*, 58: 75–83.

Ethnobotanical Wisdom and Microbial Studies on Medicinal Plants (2011) *Pages* **97–99**
Editors: **D.R. Khanna, Ashutosh Gautam, R. Bhutiani, Gagan Matta, Anil Dhiman**
Published by: **DAYA PUBLISHING HOUSE, NEW DELHI**

Chapter 18

Isolation and Structure Determination of New Anthraquinone from the Flowers of *Tagetes erecta*

J.S. Jangwan[1], M. Dobhal[1] and S.C. Sati[2]
[1]Department of Chemistry, HNB Garhwal University Campus Badhshaithaul, Tehri Garhwal
[2]Department of Chemistry, HNB Garhwal University Campus, Srinagar, Garhwal

ABSTRACT

Tagetes erecta belongs to family asteraceae is found plentifully in most of the places of India, either cultivated in gardens or wild. The villagers use this plant in bronchial asthma, anusthans, and in different ceremony. The pastes of flowers were often applied on wound and cuts and leaves juice dropped in otalgia. *Tagetes erecta* are richest source of yellow dyes in ancient arts of Rajasthan and Orissa. The present communication deals with the isolation and structure elucidation of a new anthraquinone together with quercitin and loganic acid. The structure of compound was described with the help of spectral data and chemical studies.

Introduction

The family asteraceae is one of the largest family of plant kingdom, which includes about 1100 genera and 30,000 species. About 157 genera and 900 species are reported in India. *Tagetes erecta*(Gainda), is an annual, sparingly-branched, aromatic herb. Branches angular, ribbed, leaves pinnate, oblong, acute, base decurrent, flowers yellow or orange. Flowering throughout the year. It is most commonly cultivated in gardens in major parts of India. The pastes of flowers were often applied on wound and cuts and leaves juice dropped in otalgia (Gaur, 1999)

Most of the species of Tagetes were analysed for their essential oil compositions (Chawdhury, 2001). *Tagetes patula* is widely distributed in montane and sub montane Himalayan zone,Its leaves powder is used as an insect repellent and paste used in skin ailments. Flowers of *Tagetes erecta* are richest source of yellow dyes in ancient arts of Rajasthan and Orissa (Lemmens and Wulijarni 1991).

Principle constituents isolated from flowers of *T. minuta* are anthocyanins and it derivatives. Some long chain fatty acids, aromatic hydrocarbons and phenyl acetaldehydes are extracted from floral extract of *T. erecta*. The present study describe the isolation and structure elucidation of a new anthraquinone glycoside together with quercitin and loganic acid.

Materials and Methods

Collection of Plant Material

The flowers of *Tagetes erecta* was collected from the Pasulok Barrage Rishikesh of District, Dehradun, (Garhwal) in the month of August. The identity of the plant was confirmed by Dr. P.K. Uniyal, Department of Botany, H.N.B. Garhwal University Campus, Badshahithaul, Tehri Garhwal (U.A.) and the voucher specimen is available in the herbarium of Plant Identification Laboratory of Botany Department.

Extraction and Isolation

The air-dried and coarsely powdered flowers of the plant were defatted with light petroleum in a soxhlet. The defatted mass was exhaustively extracted repeatedly with 90 per cent aqueous EtOH, until the extractive became colourless. All the extracts were mixed and concentrated under reduced pressure using rotatory vacuum evaporator.

The concentrated extract was adsorbed on Silica gel and fractionated through column chromatography using the solvent system of chloroform-methanol (95:5). The polarity of solvent was gradually increased by addition of methanol. Repeated column chromatography afforded compounds MT-1 (1.5 gm), MT-2 (0.5 gm),and MT-3 (0.7gm), with some other inseparable compounds.Compound MT-2 and MT-3 were identified as quercetin and loganic acid by their reported with an authentic sample.

M.P.were uncorrected.UV spectra were taken in MeOH.^{1}H-NMR spectra were taken using TMS as internal standard and $CDCl_3$ and CD_3OD as solvent, all the signals are expressed as values downfield from TMS.CC was carried out on silica gel (60-120 mesh, Merck, eluting solvent $CHCl_3$: MeOH).

Results and Discussion

The ethanolic extract of flowers of *Tagetes erecta* on repeated column chromatography over silica gel afforded compounds MT-1,MT-2 and MT-3.Compounds MT-2 and MT-3 were identified as quercetin and loganic acid by comparison with authentic sample and reported data (Potter and Thomas, 1995).

Compound MT-1

It was crystallized from methanol as yellow powder.

Melting point	:	177-178 °C
Molecular formula	:	$C_{15}H_{10}O_4$
Molecular weight	:	254 amu

U.V (max MeOH) nm	: 242(sh), 279, 310, 414.
I.R (maxKBr) cm^{-1}	: 3240, 3040, 1895, 1660, 1625, 1450, 1250, 1190, 915
EI-MS (m/z)	: 255[M+H]+254[M]+,236[M-H_2O]+,226[M-CO]+,225[M-CHO]+152, 141,115.
^{1}H-NMR ($CDCl_3$, δ ppm)	: 13.19-8.27(1H,s),8.22-8.30(2H,m),7.74-7.80(2H,m),7.27(1H,s),2.23 (3H,s).
^{13}C-NMR ($CDCl_3$, δ ppm)	: 142.2(C-1), 142.8(C-2), 124.2(C-3), 124.1(C-4), 112.7(C-5), 151.3(C-6), 123.2(C-7), 158.2(C-8), 180.2(C-9), 179.2(C-10), 138.8(C-11), 113.3(C-12), 114.1(C-13), 139.8 (C-14), 29.8(CH_3).

Compound MT-1 was crystallized from methanol as yellow powder. Its UV spectrum showed absorption bands at 242,279,310 and 414 nm, and IR spectrum displayed characteristic peaks for carbonyl group at 1625 cm^{-1}. Its IR spectrum also showed a bands at 3450, 1662 and 1630cm^{-1} for free hydroxyl, unchelated and chelated carbonyl groups, respectively. The EI mass spectrum of compound displayed the peak at m/z 254 [M]+,calculated for molecular formula $C_{15}H_{10}O_4$. The ^{1}H-NMR spectrum of compound showed a peri hydroxyl group at 13.19 and two meta-coupled protons at 8.22 (2H,m) and 7.27(2H,m),and a singlets for methyl group at 2.23 (3H,s).

The ^{13}C-NMR data of the compound indicate the presence of fifteen carbon atoms. The downfield signals in ^{13}C-NMR appeared at 142.2, 142.8 and 153.16 were assigned for substituted C-1, C-3 and C-2 carbons respectively. Thus on the basis of above spectral findings the compound MT-1 was identified as 1, 3-dihydroxy-2-methyl anthraquinone (Figure 18.1). It was further confirmed by comparison of its data with that of reported compound (Thomson, 1971 and Wijnsma and Verpoorte, 1986).

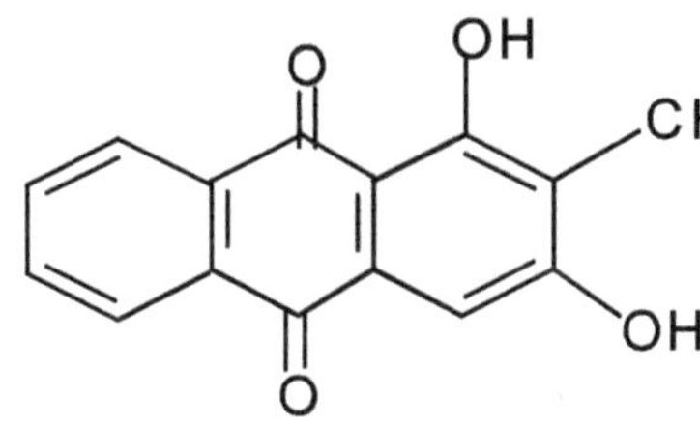

Compound MT- 1

Figure 18.1

References

Chawdhury, A.R., 2001. *J. Essential Oil Bearing Plants*, 4(1): 9–12.

Lemmens, R.H.M.J. and Wulijarni, N., 1991. *Dye and Tannin Producing Plants*, 3.

Gaur, R.D., 1999. *Flora of District Garhwal*. Trans Media, Srinagar Garhwal, pp. 592.

Potter, L., and Thomas, J., 1995. *J. of Essential Oil Res.*, 7: 347.

Thomson, R.H., 1971. *Naturally Occurring Quinones*. Chapman and Hall, London.

Wijnsma, R. and Verpoorte, R., 1986. *Prog. Chem. Org. Nat. Prod.*, 49, 79.

Ethnobotanical Wisdom and Microbial Studies on Medicinal Plants (2011) *Pages* **100–102**
Editors: **D.R. Khanna, Ashutosh Gautam, R. Bhutiani, Gagan Matta, Anil Dhiman**
Published by: **DAYA PUBLISHING HOUSE, NEW DELHI**

Chapter 19

Investigation of Ethnomedicinal Properties of *Mutella occidentales* Used by Tribal Peoples of Bhandara District Vidarbha Region, Maharashtra, India

D.D. Ramteke[1], S.B. Zade[2] and C.J. Khune[1]
[1]Department of Botany and Microbiology,
[2]P.G. Department of Zoology and Sericulture,
M.B. Patel College Sakoli, Dist. Bhandara, Maharashtra

ABSTRACT

Use of faunal wealth for medicinal purposes have been employed since dawn of civilization for protecting life of man by combating various ailments. Ancient tribal societies around the world have learned to utilize their neighbourhood wealth for curative as well as offensive purposes. The knowledge developed in their human laboratories through century's perfectly documented from generation to generation as domestic practices. The present investigation deals with study of ethnomedicinal use of *Mutella occidentales* an insect from order hemiptera.

Keywords: *Ethnomedicinal, Mutella occidentales, Vidarbha.*

Introduction

Today the system of allopathic is still unknown to the people inhabiting in interior areas. Most of the peoples depend on Ayurvedic, Unani and traditional folk medicines to cure various human problems. In India we have rich tradition of medicine and health case for safety and efficacy. Tribal and rural peoples in the area of Vidarbha region of Maharashtra State uses the natural wealth for maintaining their health. The local practitioners use wild plants as well as some animals for ethnomedicinal purposes. The present study deals with ethnomedicinal uses of *Mutella occidentales* an invertebrate insect from order Hemiptera. It is beautiful colored insect, which is smooth, small, shiny, and hairy in appearance measuring about 12mm in length. The insect is known by various names by tribal and innate peoples of district like Birbahuti, Indrawadhu, Vershakit etc. It is burrowing insect and can be observed during commencement of monsoon and comes out from soil after first showers during rainy season in month of June. It is present in mountains buried in soil and sandy loamy soils of gardens. After touching it becomes inactive and gets contracted. The present communication was done about ethnomedicinal use of *Mutella occidentales*. It is generally sold by unani and ayurvedic practitioners as "Kirme mukhmukh" or indrawadhu. The insect has several applications in rural and tribal health care.

Materials and Method

In the present investigation ethnomedicinal properties of *Mutella occidentales* were studied. The tribal people of district widely use this insect for curing various ailments. It is collected during monsoon season from various localities like mountains, gardens etc. where soil is rich in humus content. It is used for various ethnomedicinal purposes to cure diseases. It is preserved by drying and store for long time use wherever necessary. Generally it is eaten directly with fresh water. The dried insect is eaten directly or in powdered form. Some times other formulations are prepared by mixing together with honey, ghee, yogurt, or jaggery. It is also taken by mixing it with "Paan or vida". The present investigation has been done in Bhandara district of Vidarbha region of Maharashtra state. The study includes important medicinal animal, which is under study. It also include survey, conservation along with their utilization with repect to economic ethnomedicinal properties.

Results and Discussion

The innate people of Bhandara district of Vidarbha region are poor and mostly depend on agriculture. During their daily work they observe nature carefully and their folk knowledge is used for treatment of various ailments. They are nature loving and used natural wealth carefully.

Tribal peoples of Bhandara District of Vidarbha region of Maharashtra State commonly use *Mutella occidentales*. The insect is used ethno medicinally as toner for various parts of body to tone up. The dried insect is mixed with milk and applies on affected part of body. The peoples of District widely used this insect on inflammation. It is applied with Indian wax; it is generally boiled with Indian wax and applied on affected part. It is very effective on paralysis, joint pains and heart diseases and used internally by the tribal peoples. It is generally used of stimulator and used in improvement of mortality. Special oil is prepared by boiling it with coconut or sesame oil for massage. It is also used to cure acne and pimples by girls and women's of this region.

Acknowledgements

The authors are very thankful to Dr.R.D.Singh Principal, M.B.Patel College Sakoli for his precious help during research work.

References

Anantkrishnan, T.N., 1969. *Indian Thisanoptera.* CSIR Zool. Monograph, 171 pp.

Anantkrishnan, T.N., 1973. *Biology and Control.* Macmillan India Pvt. Ltd., 120 pp.

Anderson, S.O. and Weiss-fog, T., 1964. *Advances in Insect Physiology*, p. 1–65.

Ayyar, T.V.R., 1963. *Handbook of Econonomic Entomology*. Govt. Press, Chennai, 22–40 pp.

Bursel, E., 1970. *An Introduction to Insect Physiology*. Academic Press, N.Y., pp. 1–94.

Busvine, J.R., 1966. *Insects and Hygiene*. Methuen and Co. Ltd., London, pp. 1–95.

Clark, L.R., P.W., Geier, R.D., Hughes, R.F. and Morish, 1966. *The Ecology of Insect Population*. Methuen and Co. Ltd., London, p. 12–56.

Davey, K.G., 1965. *Reproduction in the Insects*. Freeman and Co., p. 1-50.

Imes, Rick, 1996. *Insects: The Practical Guide to Entomology*. Quantum Books Ltd., London.

Ethnobotanical Wisdom and Microbial Studies on Medicinal Plants (2011) *Pages* **103–108**
Editors: **D.R. Khanna, Ashutosh Gautam, R. Bhutiani, Gagan Matta, Anil Dhiman**
Published by: **DAYA PUBLISHING HOUSE, NEW DELHI**

Chapter 20

Antibacterial Activity of *Abutilon indicum* Linn.

Amit Kumar[1], Navneet[2], Sanjay[2] and Prabhat[2]
[1]Regional Drug Testing Laboratory,
Ministry of Health and Family Welfare, Government of India, Chandigarh
[2]Department of Botany and Microbiology,
Gurukul Kangri University, Haridwar

ABSTRACT

The *in vitro* antibacterial activity of leaves, stems, roots and seeds of *Abutilon indicum* Linn. have been investigated against human pathogenic bacteria *i.e. Klebseilla pneumoniae, Staphylococcus aureus, Staphylococcus epidermidis, Streptococcus mutans, Bacillus megnetherium* and *Escherichia coli*. The ethanolic extract showed the maximum antibacterial activity followed by petroleum ether and aqueous against all pathogens.

Introduction

Plants have a great potential for producing new drugs of great benefit to mankind. There are many approaches to the search for new biologically active principles. Many efforts have been done to discover new antibacterial compounds from various kinds of sources such as soil, microorganisms, animals and plants, one of such resources is folk medicine and systematic screening of them may results in the discovery of novel effective compounds (Parekh and Chanda,, 2006).The microbes slowly become resistant to drugs and some genetic changes take place in these microbes in course of time. The resistant strains are continuously appearing, imposing the need for a permanent search and development of new drugs, there is a need to develop an alternative to these drugs for the treatment of microbial diseases. Some medicinal herbs represent a rich source of antibacterial activity (Prabhat *et al*, 2005). In present study the *in vitro* antibacterial effect of different parts of *Abutilon indicum* Linn against *Klebsiella pneumoniae, S. aureus, B. megnetherium, Escherichia coli, S. mutans* and *S. epidermidis* have been studied.

Materials and Methods

The different parts (leaves, stems, roots, seeds) of medicinal plant *Abutilon indicum* Linn. were collected from the local areas of Meerut and were identified by Botanical Survey of India (BSI), Dehradun. All parts of the plant were powdered by using grinder and extracted with petroleum ether, ethyl alcohol and water and the obtained extracts were dried by using vacuum rotatory evaporator under reduced pressure.

Test Organisms

Staphylococcus aureus MTCC 1144, *Staphylococcus epidermidis* MTCC 435, *Streptococcus mutans* MTCC 890, *Bacillus megnetherium* MTCC 453, *Escherichia coli* MTCC 433 and *Klebsiella pneumoniae* MTCC 109 were used.

Antimicrobial Assay

Antimicrobial study was carried out by cup plate method (Prabhat *et al.*, 2005). The medium used for assay was Muller Hinton Agar media (Hi media No. 173). The fresh culture of each organism was prepared by inoculating a loopful growth from respective media. The culture was mixed in Muller Hinton agar well of 6mm diameter were punched into agar with sterilized cork borer and each well was filled with 45 micro liter (100mg/ml) of plant extracts, solvents for blank and antibiotic (Ampicillin 100mg/ml) for positive control. The plates were incubated at 37 °C for 24 hrs. After incubation, a clear zone of inhibition was formed around the disc. It was measured with the help of zone reader. The average zone of inhibitions were noted. The zone includes the size of cup (6mm) and per cent of potency was calculated from the mean values as (Reyes *et al.*, 1997).

$$\text{Percentage of potency} = \frac{C - T}{C} \times 100$$

where,

C= positive control, T= Test

Results and Discussion

Results of antibacterial activities of the isolated materials obtained through petroleum ether, ethanol and water are summarized in Table 20.1–20.4. *Abutilon indicum* has an important place in the Indian system of medicine and the most parts of the plant are used therapeutically (Ahmad *et al.*, 1999). The different parts *i.e.* (leaves, stems, roots and seeds) were extracted in petroleum ether, ethyl alcohol and water. Most of the extracts showed prominent antibacterial activity against microorganisms *i.e. S. aureus, S. epidermidis, K. pneumoniae, B. megnetherium, S. mutans* and *E. coli*, at the concentration of 100mg/ml. as given in Table 20.1–20.4. The ethanolic extract showed the maximum activity followed by petroleum ether and water against all pathogens. The petroleum ether extract of leaves is highly active against *S. aureus* followed by, *S. epidermidis, B. megnetherium* and *E. coli*, as compared to other solvents. The percentage of potency of leaves extract against all bacteria in comparison to antibiotic was maximum in ethanol followed by petroleum ether and water.

Table 20.1: The Percentage of Potency of Plant *Abution indicum* Leaves Extract and Antibiotic (Ampicillin) Against Pathogens

Sl.No.	Pathogens	Zone of Inhibition (Antibiotic 100 mg/ml)			Zone of Inhibition (Extra 100 mg/ml)			Per cent of Potency		
		P. Ether	E. Alcohol	Water	P. Ether	K Alcohol	Water	P. Ether	E. Alcohol	Water
1.	*S. aureus*	16	18	19	No Zone	17	16	–	5.55	15.78
2.	*S. epidermidis*	17	19	18	9	14	16	47.05	26.31	11.11
3.	*K. pneumoniae*	15	17	17	No Zone	13	–	–	23.52	–
4.	*B. megnetherium*	16	19	20	10	12	15	37.50	36.84	25.00
5.	*S. mutans*	15	18	20	No Zone	12	14	–	33.33	30.00
6.	*E. coli*	15	17	20	10	11	13	33.33	35.29	35.00

Table 20.2: The Percentage of Potency of Plant *Abution indicum* Stem Extract and Antibiotic (Ampicillin) Against Pathogens

Sl.No.	Pathogens	Zone of Inhibition (Antibiotic 100 mg/ml)			Zone of Inhibition (Extra 100 mg/ml)			Per cent of Potency		
		P. Ether	E. Alcohol	Water	P. Ether	K Alcohol	Water	P. Ether	E. Alcohol	Water
1.	*S. aureus*	16	18	19	9	13	14	43.75	27.77	26.31
2.	*S. epidermidis*	17	19	18	10	14	13	41.17	26.31	27.77
3.	*K. pneumoniae*	15	17	17	10	13	12	33.33	23.52	29.41
4.	*B. megnetherium*	16	19	20	11	12	13	31.25	36.84	35.00
5.	*S. mutans*	15	18	20	10	11	11	33.33	38.88	45.00
6.	*E. coli*	15	17	20	10	12	12	33.33	29.41	40.00

Table 20.3: The Percentage of Potency of Plant *Abution indicum* Roots Extract and Antibiotic (Ampicillin) Against Pathogens

Sl.No.	Pathogens	Zone of Inhibition (Antibiotic 100 mg/ml)			Zone of Inhibition (Extra 100 mg/ml)			Per cent of Potency		
		P. Ether	E. Alcohol	Water	P. Ether	K Alcohol	Water	P. Ether	E. Alcohol	Water
1.	*S. aureus*	16	18	19	14	15	13	12.50	16.66	31.57
2.	*S. epidermidis*	17	19	18	13	17	16	23.52	10.52	11.11
3.	*K. pneumoniae*	15	17	17	11	14	13	26.66	17.64	23.52
4.	*B. megnetherium*	16	19	20	10	13	15	37.50	31.57	25.00
5.	*S. mutans*	15	18	20	11	12	14	26.66	33.33	30.00
6.	*E. coli*	15	17	20	10	12	13	33.33	29.41	35.00

Table 20.4: The Percentage of Potency of Plant *Abution indicum* Seeds Extract and Antibiotic (Ampicillin) Against Pathogens

Sl.No.	Pathogens	Zone of Inhibition (Antibiotic 100 mg/ml)			Zone of Inhibition (Extra 100 mg/ml)			Per cent of Potency		
		P. Ether	E. Alcohol	Water	P. Ether	K Alcohol	Water	P. Ether	E. Alcohol	Water
1.	*S. aureus*	16	18	19	11	14	13	31.25	22.22	31.57
2.	*S. epidermidis*	17	19	18	10	15	13	41.17	21.05	27.77
3.	*K. pneumoniae.*	15	17	17	9	13	15	40.00	23.52	11.76
4.	*B. megnetherium*	16	19	20	10	12	13	37.50	36.84	30.00
5.	*S. mutans*	15	18	20	10	11	12	33.33	38.88	40.00
6.	*E. coli*	15	17	20	11	12	12	26.66	29.41	40.00

* Zone of inhibition mm.

The petroleum ether extract of stem is highly active against *S. aureus*, followed by *S. epidermidis, K. pneumoniae, E. coli, S. mutans* and *B. megnetherium* as compared to other solvents. The percentage of potency of stem extract in comparison to antibiotic was much in petroleum ether and followed by water and ethyl alcohol. The root extracts of plant *Abutilon indicum* is highly active against *S. epidermidis* followed by *K. pneumoniae, S. aureus, B. megnetherium* and *S.mutans*.The percentage of potency of petroleum ether extract of roots is maximum and followed by water and ethyl alcohol. The water extract of seeds of *A. indicum* is highly active against *K. pneumoniae* followed by *S. aureus* and *B. megnetherium.*

The crude extracts (100mg/ml) of each part were used for determination of their potency against pathogens as compared with antibiotic (100mg/ml).The extracts based on inhibition zone diameter has described as low (12-18 mm) moderate (19-22 mm) and strong activity (23-38 mm) (Ahmad *et al*, 1999). In our study all parts of plant showed low to moderate activity against pathogens. The ethanolic extract is highly effective against pathogens because more organic compounds were leached in this solvent. Although water is reported by the traditional healers and herbalists to be the most commonly used solvent for extracting the active compound due to its easy availability. Screening of medicinal plants to detect antimicrobial activity has clearly demonstrated that alcohol is better solvent as compared to water and petroleum ether.

The various parts of *Abutilon indicum* were tested for their antimicrobial activity against large number of bacteria (*S. aureus, S. epidermidis, K. pneumoniae, B. megnetherium, S.mutans* and *E. coli*. The water extracts of leaves were devoid of any antibacterial activity against *K. pneumoniae, E. coli* and *B. cereus* etc., but in our study it was active against both gram positive and gram negative bacteria. The above results showed antibacterial activity of this plant and are correlated with the traditional uses of this plant in ayurvedic system of medicine. Further work is required before they can be recommended as therapeutic agent. Our result shows that, the petroleum ether extract is highly active against *S.aureus, B. megnetherium* and *E. coli*. The chemical analysis of petroleum ether extract from roots of the plant showed the presence of caprylic, palmitic, myristic and oleic acid. Geeta and Gupta, 1983 also reported that essential oils obtained from *A. indicum* exhibited antibacterial activity against *B. subtilis, P. vulgaris* while it is inactive against *S. pullorum, S. typhimurium* and *Klebsiella spp.*

Aqueous extract of the plant parts *i.e.* leaves stems, roots and seeds showed antibacterial activity against *S. epidermidis, B. megnetherium* followed by *S. aureus, K. pneumoniae, S.mutans* and *E. coli*. The water extract contain carbohydrates, amino acids, saponins and flavanoids. They have been found *in vitro* to be effective antimicrobial substances. This activity is probably due to their ability to complex with extracellular and soluble proteins and to complex with bacterial cell wall for quinones. More liphophilic flavanoids may also disrupt microbial membranes (Tsuchiya *et al.*, 1996). Naqvi *et al.*, 1991 found that aqueous extract of the plant at concentration of 5 to 10 mg/ml showed activity against *S. aureus, S. viridans, C. diphtheria E. coli, S. typhi, S. paratyphi A* and *B., S. flexneri.*Valsraj *et al.* (1997) reported that 80 per cent of alcoholic extracts of the roots of *A. indicum* showed no activity against *E. coli, P. aeruginosa, B. subtilis* and *S. aureus*. In our study ethyl alcohol extract showed good activity against *E. coli* and *S. aureus*. The root of the plant was extracted in 95 per cent ethyl alcohol and antimicrobial studies were carried out by cup plate method. The gallic acid was present during chemical analysis of the roots. Sato *et al.*, 1995 reported that gallic acid is effective against *S. aureus*.

Mehta *et al.*, 1997 while studying ethanolic and acetone extract of the roots exhibited significant antibacterial activity against *E. coli, Proteus* and *P. aeruginosa*. The hexane extract was active against *P.aeruginosa* and *S. aureus*. The chloroform and ethyl acetate extract showed good activity against most of the bacteria.

Most of the Ayurvedic practitioners use aqueous extracts. This substantiate the view that aqueous extract contain most of the polar and nonpolar compounds *i.e.* flavonoids, saponins, sugars etc and thus the aqueous extract is antibacterial in action. So, efficacy and toxicity studies remains to be tested before further exploitation of the drug for human treatment.

References

Ahmad, I.Z., Beg, A.Z. and Mehmood, Z., 1999. Antimicrobial potency of selected medicinal plants with special interest in activity against phytopathogenic fungi. *Indian Vet. Med. Journal*, 23: 299–306.

Geeta, A. and Gupta, A.K., 1983. Chemical investigation of essential oil of *Abutilon indicum. Perfum. Flav.*, 18: 39.

Mehta, B.K., Negi, R., Kotra, S., and Mall, O.P., 1997. Antimicrobial activity of *Abutilon indicum. Fitotropia*, 68: 273–274.

Naqvi, S.A.H., Khan, M.S.V. and Vohra, S.B., 1991. Antimicrobial, antifungal and anthelmintic investigations on Indian Medicinal Plants. *Fitotropia*, 62: 221–228.

Parekh, J. and Chanda, S., 2006. *In vitro* antimicrobial activities of extracts of *Launaea procumbens* Roxb. (Labiateae), *Vitis venifera* L. (Vitaceae) and *Cyperus rotundus* L. (Cyperaceae). *African Journal of Biomedical Research*, 9: 89–93.

Prabhat, Navneet, and Shrikrishna, 2005. Antimicrobial activity of apamarga (*A. aspera*). *Indian Academy of Science Letters*, 28(11–12): 379–381.

Prabhat, Navneet, and Shrikrishna, 2005. Antimicrobial activity of *Mimusops elengi* (Bakul). *Environment Conservation Journal*, 6(2): 59–61.

Reyes, C.R., Quiroz, V.R.I., Jiménez, E.M., Navarro, O.A., Cassani, H.J., 1997. Antifungal activity of selected plant secondary metabolites against *Corilus versicolar*. *Journal of Tropical Forest Products*, 3: 110–113.

Sato, Y., Odetani, H., Syngyouchi, K., Ohtsubo, T., Kihara, M., Shibata, H. and Higuti, T., 1995. Flavones with antibacterial activity and carcinogenic bacteria. *Journal of Ethnopharmacology*, 48: 161–164.

Tsuchiya, H., Sato, M., Miyazaki, T., Fujiwara, S., Tanigaki, S., Ohyama, M., Tanaka, T., Takase, I. and Namikawa, I., 1996. Comparative study on the antibacterial activity of phytochemical flavonones against methicillin resistant *S. aureus*. *Journal of Ethnopharmacology*, 5: 27–54.

Valsraj, R., Pushpangadan, P., Smith, U.W., Adserson, A. and Nynan, U., 1997. Antimicrobial screening of selected medicinal plants from India. *Journal of Ethnopharmacology*, 58: 75–83.

Ethnobotanical Wisdom and Microbial Studies on Medicinal Plants (2011) *Pages* **109–115**
Editors: **D.R. Khanna, Ashutosh Gautam, R. Bhutiani, Gagan Matta, Anil Dhiman**
Published by: **DAYA PUBLISHING HOUSE, NEW DELHI**

Chapter 21

Impact of Treated and Untreated Distillery Effluent on *Momordica charantia* L. var. Jaunpuri Super Special

Raj Kumar Rampal and Preeti Dogra
Department of Environmental Sciences,
University of Jammu, Jammu, J&K

ABSTRACT

The present study has been made to evaluate the impact of treated and untreated distillery effluent from M/s Devans Modern Breweries Ltd. Talab Tillo, Bohri, Jammu on seed germination, growth parameters and yield of Bitter Gourd (*Momordica charantia* L. var. Jaunpuri Super Special). From the analysis of data on the impact of distillery effluent on various quantitative morphological features of *Momordica charantia*, it can be concluded that the Untreated Distillery Effluent is detrimental for the growth of plants whereas Treated Distillery Effluent without dilution is also harmful for agriculture.

Introduction

Industrial activity has expanded so much all over the world today that it has become a matter of great concern of the deteriorating environment. With the rapid growth of industries in the country, the pollution of Environment by industrial waste has increased tremendously. However, the wastewater from industries can act as a useful resource in agriculture thus preventing environmental hazard. Use of wastewater in agriculture is gaining importance now-a-days, because of its value as a potential irrigant and a nutrient donor. Use of wastewater for irrigation makes it possible to conserve the limited water resources for crop production and also prevent pollution of water bodies, as soil is a very good sink. Also, application of some effluents to agricultural land may promote the growth of crops and conserve water and nutrients. A large number of wastes *viz.* dairy waste, food processing waste, tannery waste, pulp and paper mill waste etc. have been successfully used for irrigation of crops. In addition to providing large quantities of water, some effluents contain considerable amounts of nutrients, which may prove beneficial for plants.

Distillery waste is amongst the worst pollutants produced by industries both in magnitude and in strength. Although distillery effluent contains many constituents, which are phytotoxic at higher concentrations, many of these constituents are, however essential for the growth and development of plants.

The effect of distillery effluent on different plant communities has been investigated by various workers (Sahai and Srinivatava, 1986; Banerjee *et al.*, 2004; Chandra *et al.*, 2004).

In the present study, attempt has been made to assess the impact of treated and untreated distillery effluent from M/s Devans Modern Breweries Ltd. Talab Tillo, Bohri, Jammu on seed germination, growth parameters and yield of Bitter Gourd (*Momordica charantia* L. var. Jaunpuri Super Special).

Materials and Methods

To carry out these investigations, 13 Experimental Sets each comprising of 10 polythene bags were prepared – One Control Set (without any treatment to seed as well as soil); Four Seed Treatment Sets (*i.e.* one set each having seeds soaked in 100 per cent Treated Distillery Effluent for 24 hrs. and 48 hrs. and one set each having seeds soaked in 100 per cent Untreated Distillery Effluent for 24 hrs. and 48 hrs.); Eight Soil Treatment Sets- *i.e.* Four soil treatment sets having treatment with Treated Distillery Effluent at 25 per cent, 50 per cent, 75 per cent and 100 per cent concentration and four sets having treatment with Untreated Distillery Effluent at 25 per cent, 50 per cent, 75 per cent and 100 per cent. Each of the eight soil treatment set was given respective concentration of respective effluent at interval of 5 days during first month, at interval of 10 days during second month, at interval of 15 days during third month. In impact study, the seed germination, root and shoot length, number of leaves per plant, number of tendrils per plant, wet weight of root and shoot, dry weight of root and shoot, number of flowers, number of fruits and quantitative morphology of fruits (dry weight of fruit, dry weight of seed and number of seeds per fruit) were studied at regular intervals. Physico-chemical characteristics like pH, electrical conductivity, BOD, COD, TDS, TSS and Volatile solids of effluents were also studied.

Results and Discussion

The physico-chemical parameters of Untreated Distillery Effluent and Treated Distillery Effluent are represented in Table 21.1. The comparison of physico-chemical analysis of Treated and Untreated Distillery Effluent is supported by the work of Chandra *et al.* (2004) *i.e.* they also observed acidic nature of raw effluent and higher values of BOD, COD, TDS and TSS of raw effluent as compared with that of treated effluent. BOD value (25mg/l) of the Treated Distillery Effluent was found to be less than the BOD value of 30mg/l which is prescribed as Minimum National Standard for surface water disposal of liquid effluents form fermentation industry (Table 21.2).

The seed treatment with Treated Distillery Effluent for 24 hours had a stimulatory effect on percentage seed germination (96 per cent) as compared with that of the Control Set (88-92 per cent) whereas seed treatment with Treated Distillery Effluent for 48 hours and with Untreated Distillery Effluent for 24 hours as well as 48 hours inhibited percentage seed germination (*i.e.* 81 per cent, 79 per cent and 70 per cent respectively) as compared with that of the Control Set (Table 21.3). This observation partly finds support from the work of Sahai and Srinivastava (1986), Dayama (1987), Rajendrababu (1987), Sahai and Srivastava (1988) and Gupta (1991) on different crops using different types of industrial effluents. They observed that industrial effluent has affected the seed germination more severely when effluent treatment was given to seeds and less severely when effluent treatment was given to soil.

Table 21.1:- Physico-Chemical Parameters of Distillery Effluent

Parameter	*Untreated Effluent*	*Treated Effluent*
Colour	Brown	Colourless
Odour	Unpleasant	Odourless
Temperaturo (°C)	75	40
pH	6.97	8.26
Electrical Conductivity (μMho)	2.78	2.09
BOD (mg/l)	2300	25
COD (mg/l)	4200	80
Total Dissolved Solids (mg/l)	1450	769
Total Suspended Solids (mg/l)	575	230
Volatile Solids (mg/l)	415	280

Table 21.2: Minimum National Standards for Liquid Effluents from Fermentation Industry

Parameters	*Concentration not to Exceed mg/l*
Maltry	
BOD, 5 days, 20°C:	30
Suspended Solids:	100
Brewery	
BOD, 5 days, 20°C:	30
Suspended Solids:	100
Distillery:	100 for land disposal, 30 for
BOD, 5 days, 20°C:	surface water disposal

Babu and Chakrabarti (2002).

The analysis of the data further revealed that seed treatment with Treated Distillery Effluent for 24 hrs (Set II) and 48 hrs (Set III) inhibited the average root length (31.3 cm and 31.8 cm. respectively), whereas the seed treatment with Untreated Distillery Effluent for 24 hrs (Set VIII) and 48 hrs (Set IX) increased the average root length (32.6 cm and 33.0 cm respectively) as compared with that (32.4 cm) of the Control Set at 60 days after thinning. The soil treatment with Treated Distillery Effluent was observed to exhibit stimulatory effect on average root length with increasing concentration from 25 per cent to 75 per cent, but it had inhibitory effect at 100 per cent concentration. The soil treatment with Untreated Distillery Effluent exhibited stimulatory effect on average root length at 25 per cent concentration (Set X) and then it exhibited inhibitory effect on average root length with increasing concentration from 50 per cent to 100 per cent (Table 21.3). Patel *et al.* (1990) also observed that root length of mustard plant increased with pharmaceutical factory effluent up to 40 per cent and higher concentrations had inhibitory effect. Kumar *et al.* (1990) and Gupta and Nathawat (1992) also observed decrease in root length of various plants with increasing concentration of different industrial effluents.

The analysis of the data further revealed that seed treatment with 100 per cent Treated Distillery Effluent and soil treatment with 25 per cent, 50 per cent and 75 per cent Treated Distillery Effluent had a stimulatory effect on the average shoot length, whereas soil treatment with 100 per cent Treated Distillery Effluent had a inhibitory effect on the shoot length as compared with shoot length of the Control Set. The seed treatment with 100 per cent Untreated Distillery Effluent as well as soil treatment with 25 per cent Untreated Distillery Effluent was also observed to exhibit stimulatory effect on the shoot length whereas soil treatment with 50 per cent, 75 per cent and 100 per cent Untreated Distillery Effluent exhibited inhibitory effect on the shoot length of *Momordica charantia* as compared with that of the Control Set (Table 21.3).

The study further revealed that Treated Distillery Effluent treatment to seeds in Sets –II (soaked for 24 hrs) and III (soaked for 48 hrs); Untreated Distillery Effluent treatment to seeds in Sets –VIII (soaked for 24 hrs) and IX (soaked for 48 hrs) as well as the soil treatment with Treated Distillery Effluent up to 75 per cent and the 25 per cent Untreated Distillery Effluent treatment to soil exhibited stimulatory effect on average number of leaves as well as tendrils per plant as compared with those of the Control Set, whereas soil treatment with 100 per cent Treated Distillery Effluent and Untreated Distillery Effluent above 25 per cent exhibited inhibitory effect on average number of leaves as well as tendrils per plant as compared with those of the Control Set (Table 21.3). Banerjee *et al.* (2004) also observed increased growth rate of *Casurina equisetifolia* with distillery effluent up to 60 per cent.

The analysis of data regarding dry weight of root revealed that seed treatment with treated as well as Untreated Distillery Effluent and soil treatment with Treated Distillery Effluent at 100 per cent as well as Untreated Distillery Effluent at all concentrations inhibited the average dry weight of root. But soil treatment with Treated Distillery Effluent stimulated dry weight with corresponding increase in effluent concentration from 25 per cent to 75 per cent as compared with that of the Control Set (Table 21.3). Ramana *et al.* (2002) have reported a similar pattern of increase in dry weight with application of distillery effluent.

The critical analysis of data further revealed that seed treatment with treated as well as Untreated Distillery Effluent and the soil treatment with Treated Distillery Effluent from 25 per cent to 75 per cent and Untreated Distillery Effluent at 25 per cent exhibited stimulatory effect on dry weight of shoot as compared with that of the Control Set. But the soil treatment with 100 per cent Treated Distillery Effluent and Untreated Distillery Effluent at 50 per cent, 75 per cent and 100 per cent concentration had inhibitory effect on dry weight of shoot as compared with that of the Control Set (Table 21.3). Dayama (1987), Patel *et al.* (1990) and Sundaramoorthy and Kunjithapatham (2000) also observed decrease in average dry weight with increase in effluent concentration while working on different plants using different industrial effluents.

The cumulative number of flowers per set in seed treatment sets of Treated Distillery Effluent –II (soaked for 24 hrs) and III (soaked for 48 hrs) and the soil treatment sets of Treated Distillery Effluent exhibited an increase in cumulative number of flowers at 25 per cent, 50 per cent and 75 per cent, but in all the Untreated Distillery Effluent Sets and Treated Distillery Effluent at 100 per cent, these exhibited low value as compared with that of the Control Set (Table 21.3).

Table 21.3: Impact of Treated and Untreated Distillery Effluent on Seed Germination and Quantitative Morphological Features of *Momordica charantia* L. var. Jaunpuri Super Special

Parameters	*Name of Experimental Set*												
	Control Set	*Treated Distillery Effluent Sets*						*Untreated Distillery Effluent Sets*					
	I	*II*	*III*	*IV*	*V*	*VI*	*VII*	*VIII*	*IX*	*X*	*XI*	*XII*	*XIII*
%age of seed germination	90	96	81	92	91	88	89	79	70	90	92	88	89
Avg. root length (cm)	32.4± 6.08	31.3 ± 5.61	31.8 ± 5.56	33.3±5.01	39.1± 4.10	42.6±9.83	20.4±3.47	32.6± 4.52	33.0±5.15	33.1±5.80	19.9±2.25	19.3±1.36	16.7±0.53
Avg. shoot length (cm)	99.1±32.77	105.3±28.98	107.0±26.85	115.2±24.42	128.5±10.56	151.4±33.65	87.5±13.44	106.7±30.12	100.5±32.41	100.8±32.66	80.2±21.80	70.6±16.86	65.9±15.12
Avg. no of leaves /plant	39.6±4.72	43.3.±5.13	44.0±4.58	42.6±8.14	46.3±14.01	63.0±19.92	39.3±2.52	44.6±6.02	40.6±6.43	41.3±5.86	34.6±5.51	33.3±6.03	31.0±4.58
Avg. no of tendrils/plant	24.0±7.21	28.0±8.72	29.0±3.05	26.3±3.05	31.3±6.66	40.3±12.58	23.3±6.11	30.0±6.93	25.6±8.14	26.0±7.94	21.6±3.05	21.0±3.00	20.6±2.52
Avg. dry wt. of root (gm)	0.88±0.04	0.77±0.05	0.86±0.04	0.92±0.01	1.20±0.47	1.47±0.42	0.83±0.04	0.87±0.06	0.87±0.05	0.86±0.03	0.78±0.11	0.72±0.10	0.66±0.01
Avg. dry wt. of shoot (gm)	3.72±1.27	4.13±1.59	4.24±1.56	4.01±1.33	4.46±1.42	4.88±0.80	3.17±0.87	4.35±1.66	3.76±1.11	3.99±1.27	2.51±0.99	1.79±0.55	1.11±0.01
Cumulative no. of flowers per set	83(75)	87(78)	84(78)	86(75)	93(82)	105(98)	75(68)	82(79)	76(71)	65(57)	59(53)	56(49)	47(45)
Cumulative no. of fruits per set	3	3	—	5	2	7	2	—	—	1	1	1	—
Avg. dry wt. of fruit (mg)	4766.7	4831.8	—	5178.8	5082.6	5985.9	4295.5	—	—	4081.3	3511.3	3013.6	—
Avg. dry wt. of seed (mg)	87.5	98.0	—	131.6	110.1	160.3	79.2	—	—	71.3	46.5	31.6	—
Avg. no. of seeds/fruit	16.33	16.33	—	17.14	16.8	18.0	16.5	—	—	16.0	15.5	15.0	—

Set I: Control Set (Without any treatment to soil and seed); Set II: Seeds soaked in 100 per cent Treated Effluent for 24hrs before sowing; Set III: Seeds soaked in 100 per cent Treated Effluent for 48hrs before sowing; Set IV: 25 per cent Treated Effluent Treatment to soil; Set V: 50 per cent Treated Effluent Treatment to soil; Set VI: 75 per cent Treated Effluent Treatment to soil; Set VII:100 per cent Treated Effluent Treatment to soil; Set VIII: Seeds soaked in 100 per cent Untreated Effluent for 24hrs before sowing; Set IX: Seeds soaked in 100 per cent Untreated Effluent for 48hrs before sowing; Set X: 25 per cent Untreated Effluent Treatment to soil; Set XI: 50 per cent Untreated Effluent Treatment to soil; Set XII: 75 per cent Untreated Effluent Treatment to soil; Set XIII: 100 per cent Untreated Effluent Treatment to soil.

Seed treatment Sets- III (having seeds soaked in 100 per cent treated effluent for 48 hrs), VIII (having seeds soaked in 100 per cent untreated effluent for 24 hrs) and IX (having seeds soaked in 100 per cent untreated effluent for 48 hrs) exhibited no fruit formation. The cumulative number of fruits in soil treatment sets of Treated Distillery Effluent exhibited a random pattern thereby showing no relation with the effluent concentration *i.e.* the cumulative number of fruits was observed to be 5, 2, 7, and 2 in Sets – IV (25 per cent), V (50 per cent), VI (75 per cent) and VII (100 per cent) respectively (Table 21.3).

The Untreated Distillery Effluent Sets- X (25 per cent), XI (50 per cent) and XII (75 per cent) exhibited one fruit each whereas no fruit formation was observed in Set XIII (100 per cent). The average dry weight of fruit was observed to be more (4831.8 mg) in Seed Treatment Set II (having seeds soaked in 100 per cent treated effluent for 24 hrs) than that (4766.6 mg) of the Control Set. The average dry weight of fruit in soil treatment sets of Treated Distillery Effluent exhibited no relation with increasing effluent concentration but exhibited higher values of 5178.8 mg at 25 per cent, 5082.6 mg at 50 per cent, 5985.9 mg at 75 per cent and lower value of 4295.5 mg at 100 per cent as compared with that (4766.7 mg) of the Control Set (Table 21.3).

The average dry weight of fruit in Untreated Distillery Effluent Sets recorded lower value with increase in the effluent concentration *i.e.* of 4081.3 mg at 25 per cent, 3511.3 mg at 50 per cent and 3013.6 mg at 75 per cent. The average dry weight of seed exhibited a higher value of 98.0 mg in Seed Treatment Set II (having seeds soaked in 100 per cent treated effluent for 24 hrs) as compared with that (87.5 mg) of the Control Set. The average dry weight of seed in soil treatment sets of Treated Distillery Effluent exhibited higher values of 131.6 mg at 25 per cent, 110.1 mg at 50 per cent, 160.3 mg at 75 per cent and lower value of 79.2 mg at 100 per cent whereas in Untreated Distillery Effluent Sets, the average dry weight of seed exhibited lower value of 71.3 mg at 25 per cent, 46.5 mg at 50 per cent and 31.6 mg at 75 per cent (Table 21.3). The average number of seeds per fruit in the Seed Treatment Set II (having seeds soaked in 100 per cent treated effluent for 24 hrs) remained same (16.33) as that of the Control Set.

The average number of seeds in soil treatment sets of Treated Distillery Effluent exhibited higher value of 17.14 at 25 per cent, 16.8 at 50 per cent and 18.0 at 75 per cent and lower value of 16.5 at 100 per cent and Untreated Distillery Effluent Sets exhibited lower value of 16.0 at 25 per cent, 15.5 at 50 per cent and 15.0 at 75 per cent (Table 21.3).

Conclusion

From the above analysis, it can be concluded that Treated Distillery Effluent is better in physico-chemical characteristics as compared with Untreated Distillery Effluent *i.e.* BOD value of Treated Distillery Effluent was observed to be 25 mg/l which is less than 30 mg/l as prescribed in Minimum National Standards for surface water disposal of liquid effluents from fermentation industry. From the analysis of data on the impact of distillery effluent on various quantitative morphological features of *Momordica charantia*, it can be concluded that the Untreated Distillery Effluent is detrimental for the growth of plants whereas Treated Distillery Effluent without dilution is also harmful for agriculture.

But the Treated Distillery Effluent at 25 per cent -75 per cent concentration was observed to have stimulatory effect on the growth of plants. Therefore, it can be recommended that Treated Distillery Effluent after diluting to 25 per cent *i.e.* 75 per cent concentration can be used for irrigation of agricultural land. Rampal and Dorjey (2001) also concluded that 50 per cent concentration of untreated foam industry effluent favoured growth and yield of *Lens esculenta*. Chandra *et al.* (2004) also made similar conclusion while working on the impact of treated and untreated distillery effluent irrigation on soil micro flora, growth, total chlorophyll and protein contents of *Phaseolus aureus* L.

References

Babu, N.R. and Chakrabarti, S.P., 2002. Control and management of environmental pollution in India: A review. In: *Environmental Pollution*, (Ed.) B.D. Sharma. Anmol Publications Pvt. Ltd., New Delhi, p. 34–62.

Banerjee, A.C., Bajwa, I. and Bahal, K.K., 2004. Effect of distillery effluents on growth of *Casurina equisetifolia*. *Poll. Res.*, 23(1): 179–182.

Chandra, R., Kumar, K. and Singh, J., 2004. Impact of anaerobically treated and untreated (raw) distillery effluent irrigation on soil microflora, growth, total chlorophyll and protein contents of *Phaseolus aureus*. *J. Environ. Biol.*, 25(4): 381–386.

Dayama, O.P., 1987. Influence of dyeing and textile water pollution on nodulation and germination of gram (*Cicer arietinum* L.). *Acta Ecologica*, 9(1 and 2): 34–37.

Gupta, A. and Nathawat, G.S., 1992. Effect of textile effluent on germination and seedling growth of *Pisum sativum* var. RPG–3. *Acta Ecologica*, 13(2): 109–112.

Gupta, R., 1991. Toxic effects of mercury on seed germination of bean and mustard. *Comp. Physio. Eco.*, 16(1): 43–45.

Kumar, K.T.R., Kumar, P., Patel, B. and Pushjalata, 1990. Effect of chemical factory effluent on germination and growth of Guar, *Cyamopsis tetragonoloba* L. Taub. var. PNB. *Adv. Plant Sci.*, 3(1): 34–42.

Patel, B., Kumar, P. and Kumar, K.T.R., 1990. Effect of pharmaceutical factory effluent on germination, dry matter accumulation and crop productivity of mustard plant *Brassica juncea* L. *Polln. Res.*, 9: 113–119.

Rajendrababu, S., 1987. Impact of potassium chromate contaminated well and reservoir water on seed germination and plant growth. *Poll. Res.*, 6(3 and 4): 97–99.

Ramana, S., A.K. Biswas and Singh, A.B., 2002. Effect of distillery effluents on some physiological aspects in Maize. *Bioresour. Technol.*, 84(3): 295–297.

Rampal, R.K. and Dorjey, P., 2001. Studies on the effect of foam industry effluent on *Lens esculenta* Moench var. Malika. *Indian J. Environ. Prot.*, 21(1): 14–17. *Polln. Contl.*, 17(1): 165–169.

Sahai, R. and Srivastava, C., 1988. Effect of fertilizer factory waste on seed germination, seedling growth and pigment content of plants. *J. Indian Bot. Soc.*, 67(3 and 4): 309–311.

Sahai, R. and Srinivastava, N., 1986. Effect of distillery waste on the seed germination, seedling growth and pigment content of *Cajanus cajan* L. *J. Indian Bot. Soc.*, 65(2): 208–211.

Sundaramoorthy, P. and Kunjithapatham, J., 2000. Effect of paper mill effluent on seed germination and seedling growth of six varieties of groundnut (*Arachis hypogea*). *J. Ecotoxico. Environ. Monit.*, 10(1): 53–57.

Ethnobotanical Wisdom and Microbial Studies on Medicinal Plants (2011) *Pages* **116–119**
Editors: **D.R. Khanna, Ashutosh Gautam, R. Bhutiani, Gagan Matta, Anil Dhiman**
Published by: **DAYA PUBLISHING HOUSE, NEW DELHI**

Chapter 22

In vitro Antibacterial Activity of Sahdevi (*Vernonia cineria* Less)

Prabhat[1], Navneet[1], Pramod[2] and Chhaya[2]
[1]Department of Botany and Microbiology, [2]Department of Chemistry, Gurukul Kangri University, Haridwar

ABSTRACT

Sahdevi (*Vernonia cineria* less) were evaluated for antibacterial activity against human pathogenic bacterial strains *i.e. Staphylococcus aureus, Klebsiella pneumoniae, Salmonella typhi* and *Escherichia coli*. The methanolic extracts of the plant parts showed the significant activity followed by water and petroleum ether extracts. The methanolic extract of roots showed maximum activity against *K. pneumoniae* (18mm) minimum activity against *S. typhi* (14mm) by well diffusion method. The use of plant for its antibacterial activity and treatment of fever has been suggested.

Keywords: *Vernonia cineria, Antibacterial, Fever.*

Introduction

Microorganisms are closely associated with the health and welfare of human beings. Human cannot escape a dynamic on going association with microbes. In Indian system of medicine, the plants are used as a curative in cough, asthma, dropsy, bronchitis, colic, vomiting, leprosy, cardiac disorders, anemia and vitiated conditions of kappa and vata and general debility (Prabhat *et al.,* 2005). The microorganisms have developed resistance to many antibiotics. This has created immense problem for the treatment of microbial diseases, therefore, there is a need of an alternative to these antibiotics for the treatment of diseases. One of the methods to reduce this resistance is by using medicinal plants. The plants are known to produce a variety of phytoconstituents that have antimicrobial properties (Prabhat and Navneet, 2007). The unripe fruits of *Mimusops elengi* showed antibacterial activity against *Bacillus anthrasis, B. subtilis, Salmonella typhi, Staphylococcus aureus, S. sanguis* and *S. mutans* (Kapoor *et al.*, 1989; Prabhat *et al.*, 2005). In the present study, we have selected different parts *i.e.*roots, leaves, stem and whole plant of *Vernonia cineria* against *Klebsiella pneumoniae, Salmonella typhi, Staphylococcus aureus* and *E. coli.*

Materials and Methods

The plants were collected from the foot hills of Shivalik range in Haridwar, Uttarakhand. The root, stem, leaves were separated by peeling. These were washed by running tap water to remove the adhering unwanted material and cut into small pieces, dried at room temperature and then powdered by using blender. The powdered plant materials were loaded in soxlet assembly and extracted in three different solvents *i.e.* petroleum ether, methanol and water for 72 hrs by successive method. At the end of each extract, it was passed through Whattman filter paper No. 40 and the filtrates were evaporated under reduced pressure. Muller Hinton agar media (Hi media No. 173) was used to test to antimicrobial activity against *Escherichia coli*, gram negative (MTCC 739), *Staphylococcus aureus*, gram positive (MTCC 1144), *Salmonella typhi*, gram negative, (MTCC 109) and *Klebsiella pneumoniae* (MTCC) by agar well cup plate method (Ahmad *et al.*, 1998). 8 mm diameter wells were punched in the agar and filled with abstracts of respective solvents for control and standard antibiotic ampicillin (100 mg/ml) was used as positive control. The plates were incubated at 37° C for 24 hours. The antimicrobial activity was evaluated by measuring the diameter of inhibition zone in mm.

Results and Discussion

The plant parts showed broad spectrum antimicrobial activity (Table 22.1) *i.e.* the petroleum ether, methanol and water extracts were active against both gram positive and gram negative bacteria. The *Vernonia cineria* extracts were found to be less effective as compared to ampicillin. The antibacterial activity of plant (*V. cineria*) parts *i.e.* leaves, roots, stems, whole plant and antibiotic ampicillin in all the three solvents against *Escherichia coli, Salmonella typhi, Staphylococcus aureus* and *Klebsiella pneumoniae* at concentrations of 100 mg/ml is given in Tables 22.1–22.4. The zone of inhibitions obtained in crude extracts of different parts of plants varied against pathogens while it was same in case of ampicillin. The results revealed that antibiotic is more effective as compared to crude extracts. The methanolic extracts showed the highest activity against *Klebsiella pneumoniae* followed by *Escherichia coli, Salmonella typhi* and *Staphylococcus aureus* as compared to other extracts.

The extracts of roots showed maximum activity followed by whole plant, leaves and stem. The antibacterial therapy is going through crises due to the rapidly increasing development of resistance to existing agents. Antibiotic resistance has increased substantially in the recent years and is posing an increasing therapeutic problem. The use of plants as primary health remedies is quite common due to their pharmacological properties. The plant produced a variety of phytoconstituents that have antibacterial activity. These compounds includes flavonoids, phenols, phenolic glycosides and glycerinates etc. the natural antimicrobial compounds in plants can inhibit the growth of bacteria by means of unknown mechanisms other than that of known antibiotics and for this reason the search for new antibiotics must continue. The present study has shown that the plant and its parts are potentially a rich source of antibacterial agents. The plant extracts inhibited the growth of all pathogens. The antibacterial activity of active extracts of parts were observed in decreasing order roots>whole plant> leaves> stem. The crude extracts (100mg/ml) of each part was used for determination of their potency against pathogens and compared with antibiotic (100mg/ml). The extent of antimicrobial activity of the extracts based on inhibition zone diameter has been described as low (10-14), moderate (15-20) and strong (21-26) by Ahmad *et al.* (1999). In our study, root extracts showed moderate activity against all pathogens and followed by whole plant, leaves and stem. The methanolic extracts of each part is highly effective against all pathogens because more phytoconstituents were leached in this solvent. Our findings have validated the use of these medicinal plants for the treatment of microbial infections. It seems important to recommend that further studies using isolated constituents instead of whole extracts must be done in this field. Health foundations have to increase their funding of these studies and research to help saving the lives of many peoples. This will also offer a great help in facing the emergence and spread of antimicrobial resistance.

Table 22.1: The Percentage of Potency of Plant *Vernonia cineria* Roots Extract and Antibiotic (Ampicillin) Against Pathogens

Sl.No.	*Human Pathogens*	*Zone of Inhibition (Antibiotic 100 mg/ml)*			*Zone of Inhibition (Extract 100 mg/ml)*			*Per cent of Potency*		
		Pt. Ether	*Methanol*	*Water*	*Pt. Ether*	*Methanol*	*Water*	*Pt. Ether*	*Methanol*	*Water*
1.	*K. pneumoniae*	15	19	18	–	16	14	–	26.32	22.22
2.	*S. typhi*	14	18	17	–	14	14	–	22.22	17.64
3.	*S. aureus*	14	18	20	11	15	15	21.43	16.67	25.00
4.	*E. coli*	14	19	20	12	14	15	14.29	26.32	25.00

Table 22.2: The Percentage of Potency of Plant *Vernonia cineria* Stems Extract and Antibiotic (Ampicillin) Against Pathogens

Sl.No.	*Human Pathogens*	*Zone of Inhibition (Antibiotic 100 mg/ml)*			*Zone of Inhibition (Extract 100 mg/ml)*			*Per cent of Potency*		
		Pt. Ether	*Methanol*	*Water*	*Pt. Ether*	*Methanol*	*Water*	*Pt. Ether*	*Methanol*	*Water*
1.	*K. pneumoniae*	15	19	18	10	12	1	33.33	36.84	44.44
2.	*S. typhi*	14	18	17	9	13	12	35.71	27.78	29.41
3.	*S. aureus*	14	18	20	11	14	14	21.42	22.22	30.00
4.	*E. coli*	14	19	20	10	13	10	28.57	26.32	50.00

Table 22.3: The Percentage of Potency of Plant *Vernonia cineria* Leaves Extract and Antibiotic (Ampicillin) Against Pathogens

Sl.No.	*Human Pathogens*	*Zone of Inhibition (Antibiotic 100 mg/ml)*			*Zone of Inhibition (Extract 100 mg/ml)*			*Per cent of Potency*		
		Pt. Ether	*Methanol*	*Water*	*Pt. Ether*	*Methanol*	*Water*	*Pt. Ether*	*Methanol*	*Water*
1.	*K. pneumoniae*	15	19	18	9	14	13	40.00	26.32	27.78
2.	*S. typhi*	14	18	17	10	13	10	28.57	27.78	41.18
3.	*S. aureus*	14	18	20	10	13	12	28.57	27.78	40.00
4.	*E. coli*	14	19	20	11	11	12	21.43	42.11	40.00

Table 22.4: The Percentage of Potency of Plant *Vernonia cineria* Whole Plant Extract and Antibiotic (Ampicillin) Against Pathogens

Sl.No.	*Human Pathogens*	*Zone of Inhibition (Antibiotic 100 mg/ml)*			*Zone of Inhibition (Extract 100 mg/ml)*			*Per cent of Potency*		
		Pt. Ether	*Methanol*	*Water*	*Pt. Ether*	*Methanol*	*Water*	*Pt. Ether*	*Methanol*	*Water*
1.	*K. pneumoniae*	15	19	18	9	15	13	40.00	21.05	27.78
2.	*S. typhi*	14	18	17	8	14	10	35.71	22.22	41.18
3.	*S. aureus*	14	18	20	10	12	14	28.57	27.78	30.00
4.	*E. coli*	14	19	20	11	13	13	21.42	31.58	35.00

References

Ahmad, I., Mehmood, Z. and Mohammad, F., 1998. Screening of some Indian medicinal plants for their antimicrobial properties. *Journal of Ethnopharmacology*, 62: 183–193.

Ahmad, I., Beg, A.Z. and Mehmood, Z., 1999. Antimicrobial potency of selected medicinal plants with special interest in activity against phyto pathogenic fungi. *Indian Veterinary Medical Journal*, 23: 299–306.

Kapoor, L.D., Singh, A., Kapoor, S.L. and Srivastava, S.N., 1989. Survey of Indian medicinal plants for saponins, alkaloids and flavonoids. *Llodia*, 32: 297–302.

Prabhat, Navneet and Shrikrishna, 2005. Antibacterial activity of *Mimusops elengi* (Bakul). *Environment Conservation Journal*, 6(2): 59–61

Prabhat and Navneet, 2007. *Antibacterial activity of medicinal plants against dental infections.* In: *Medicinal Plants: Conservation, Cultivation and Utilization*, (Eds.) D.R. Khanna *et al.* Daya Publishing House, New Delhi, pp. 139–146.

Ethnobotanical Wisdom and Microbial Studies on Medicinal Plants (2011) *Pages* **120–122**
Editors: **D.R. Khanna, Ashutosh Gautam, R. Bhutiani, Gagan Matta, Anil Dhiman**
Published by: **DAYA PUBLISHING HOUSE, NEW DELHI**

Chapter 23

Rajnigandha: A New Host of *Fusarium* sp. from Bahraich–A New Report

Nishat Asif*, D.P. Singh and T.P. Mall
P.G. Department of Botany, Kisan P.G. College, Bahraich – 271 801, U.P.

ABSTRACT

Polianthes tuberosa L.(Rajnigandha) is an important medicinal plant possessing diuretic emetic and antigonorrhoeic properties.Plants of *P.tuberosa* growing in pots were found suffering from a leaf spot disease.Oval to elongated spots were produced throughout the leaf blade.The spots remained border in the center and pointed at the ends. Light brown spots are surrounded by dark brown border. Spots may also coalesce to form large irregular spots. Isolation from the disease tissue yielded a species of *Fusarium*. Pathogenicity test when carried out by spray inoculation on healthy potted plants of *P. tuberosa,* proved positive. On the bases of cultural characters and morphology of vegetative and reproductive bodies, the isolated species from *P. tuberosa* was identified as *Fusarium solani*. Search of the available literature revealed that *P. tuberosa* is a new host for *F. solani* which is being first reported from Bahraich district.

Introduction

Rajnigandha (*Polianthes tuberosa* L., Eng.–Tuberosa; Hindi–Rajnigandha, Gulshaba, Family–Agavaceae) is a herb native to Mexico grown as an ornamental plant for its white, fragment flowers. It is an important ethnomedicinal plant possessing diuretic emetic and antigonorrhoeic properties. The plants is grown in pots in family gardens, parks, nurseries and government gardens. In November

* Corresponding Author: E-mail: cute_nishat@rediffmail.com.

2006 during our routine gardening activities in morning we observed that some plants were suffering from some foliar spots. The number of spots and size per leaf was gradually increasing significantly. The observation resulted a curiosity to know about the microbe responsible for the leaf spot. The disease on Rajnigandha was a foliar spot disease so it generate a curiosity as the leaves are known the suitable habitat which provide ample surface area and nutritional supply to the fungal pathogen for its overall growth.

Keeping this view in our mind, authors surveyed the gardens of housing colonies, parks, nurseries and government gardens upto November 2006 and infected specimens were collected and gone through for the detailed study for the disease symptom.

Materials and Methods

The collected specimens were pressed and dried by routine herbarium technique as described by Jain and Rao (1978). The fresh infected leaves were collected, hand cut section and scrap mount were prepared of infected parts in lactophenol and cotton blue as described by Kamal *et al.* (2003).

For pathogenicity test the isolated fungus was cultured on PDA and its pathogenicity test was done on healthy potted plants by spray inoculation of sporulating mass in aqueous solution as described by McCallum and Tekauz (2002).

Results and Discussion

The infected plants showed oval to elongated spots produced throughout the leaf blade. The spots were remained broader in the centre and pointed at the ends. Light brown spots are surrounded by dark border. Spots merge to form a large irregular patch. The microscopic examination of slides prepared showed that the causal organism of the leaf spot is a fungus having microconidia 2.5-5 micron; macroconidia 5-14 micron in size. The consultation of monographs showed that the tested fungus is *Fusarium solani* (Mart.) Sacc.

The pathogenicity test by spray method was positive. Latter the fungus was conformed by Prof. Kamal, Emeretus Professor in Botany, D.D.U. Univ. of Gorakhpur, Gorakhpur (U.P.).

During pathogenicity test the first spot appeared after 10 days of spray which latter covered about 60 per cent area of leaf blade within 15 days in form of irregular patches. This system is latter followed by infection on leaf sheath and spike.

The severity of spot causes defoliation. Leaf spots are numerous,large and irregular which cause a considerable reduction in the photosynthetic area of leaf. The inhabiting fungi interfere with the physiology of the host and host as well as pathogen produces toxins which may degrade the ethnomedicinal quality of the plant. The fungal pathogen reduces the productivity of host. Ahir *et al.* (2006) has also reported *F.solani* on Rajnigandha leaf blades only causing less damage.

So some important strategies must be adopted for the conservation of *Polianthes tuberosa* from plant pathogen considering its importance in garden as well as its use in green herbal medicine system. The plants are still under observation and further study will be done to save the plant from loss.

Acknowledgement

The authors are thankful to Sri Mohd. Ahsan, Chief Wildlife Warden, Uttar Pradesh, Lucknow for giving his permission for scientific research for the exploration of floral vegetation of territory under his command and to Dr. S.P. Singh, Principal, Kisan P.G. College, Bahraich for providing necessary library and laboratory facilities.

References

Ahir, R.R., Tak, S.K., Singh, N. and Verma, O.P., 2006. Two new diseases of medicinal plants caused by *Fusarium solani*. *J. Mycol. Pl. Pathol.*, 36(1): 18.

Jain, S.K. and Rao, R.R., 1978. *A Handbook of Field and Herbarium Methods*. Today and Tomorrow Printers and Publishers, New Delhi, p. 33–58.

Kamal, Sharma, Nidhi and Singh, P.N., 2003. Three new taxa of *Corynespora* causing foliar blight in forest plants of North Eastern Uttar Pradesh. *J. Mycol. Pl. Pathol.*, 33(1): 26.

McCallum, B.D. and Tekauz, A., 2002. Influence of inoculation method and growth stage on *Fusarium* head blight in barley. *Can. J. Plant Pathol.*, 24: 77–80.

Ethnobotanical Wisdom and Microbial Studies on Medicinal Plants (2011) *Pages* 123–129
Editors: D.R. Khanna, Ashutosh Gautam, R. Bhutiani, Gagan Matta, Anil Dhiman
Published by: DAYA PUBLISHING HOUSE, NEW DELHI

Chapter 24

Synergistic Effect of *Pongamia pinnata* Bark and *Tamarindus indica* Fruit Extract Against Aflatoxin Producing Fungi *i.e. Aspergillus flavus*

Harish Chandra[1], O.P. Sidhu[2], Jatin Srivastava[3], Nishant Rai[1], Sachin Chauhan[1], Ajay Singh[4] and A.R. Nautiyal[5]

[1]Department of Microbiology, [4]Department of Chemistry, Gayatri College of Biomedical Science, Ballupur Chowk Dehradun

[2]National Botanical Research Institute (CSIR), Rana Pratap Marg, Lucknow, U.P.

[3]Department of Environmental Sciences C.S.J.M. University, Kanpur

[5]HAPTRC, HNB Garhwal University, Srinagar, Garhwal

ABSTRACT

Aflatoxin is the most toxic of many naturally occuring toxins produced by fungi. *Aspergillus flavus* and *Aspergillus parasiticus* are the major casual organisms. These co-exist with and grow on almost all crops. *Pongamia pinnata* (Karanj) and *Tamarindus indica* (Imli) and other tree species producing non edible oil were screened for their possible antifungal activity. Methanolic fraction were assayed to control the fungi and significant reduction in fungus growth was observed when applied synergistically than the individual plant extract.It was found that Combination of both extract were more effective than the individual extract when tested alone i.e 62 per cent inhibition of fungal growth as comapared to *Pongamia* bark alone (33.20 per cent) and *Tamarindus indica* alone (34.12 per cent).

Keywords: Aflatoxin, Tamarindus indica, Pongamia pinnata.

Introduction

Aflatoxin are the secondary metabolite *i.e.* the metabolite not required during the growth of microorganism and it was mainly produced by the fungi *Aspergillus flavus* and *Aspergillus parasiticus* and in some cases by *Aspergillus nomius*. The most important group of toxigenic Aspergilli are the Aflatoxigenic molds, *A. flavus*, *A. parasiticus* and the recently described but much less common species *A. nomius* all of which are classified in *Aspergillus* section Flavi (Gams *et al.*, 1985). Although these three species are closely related and shares many similarities a number of characteristics may be used in their differentiation. *A. flavus* is widely distributed in nature but *A. parasiticus* is less wide spread, the actual extent of its occurrence being complicated by the tendency for both species to be reported indiscriminately as *A. flavus*. In a wide ranging survey of the mycoflora of commodities in Thailand. *A. flavus* was the most common species in the peanut and second most common (after *Fusarium moniliforme*) in corn. *A. nomius* was reported from both commodities. Soyabeans, mung bean, sorghum and other commodities also contained considerable population of *A. flavus* but *A. parasiticus* was found.

A. flavus and *A. parasiticus* have strong affinity with nuts and oil seeds, corn, peanuts and cotton seed are the most important crops invaded by these mold and in many instances, invasion takes place before harvest not during storage. Peanuts are invaded while still in the ground if the crops suffer drought stress or related factor (Cole *et al.*, 1982; Pitt *et al.*, 1991; Sanders *et al.*, 1981). In corn insect damage to developing kernels allow entry of Aflatoxigenic molds but invasion can also occur through the silks of developing ears (Lilehoj *et al.*, 1980) cotton seeds invaded through nectaries. (Klich *et al.*, 1984). Cereals and spices are common substrate for *A. flavus* (Pitt *et al.*, 1991), but aflatoxin production in these commodities is almost always a result of poor drying, handling or storage and aflatoxin levels are rarely significant.

Significant amount of aflatoxin can occur in peanuts, corn and other nuts and oil seeds particularly in some tropical countries where crops may be grown under marginal condition and where drying and storage facilities are limited.

Aspergillus flavus can produce Aflatoxin B_1, B_2 and cyclopiazonic acid, but only a proportion of isolates are toxigenic. *A. parasiticus* produces Aflatoxin B_1, B_2, G_1 and G_2 but not cyclopiazonic acid, and almost all the isolate are toxigenic. *A. nomius* is morphologically similar to *A. flavus*, but like *A. parasiticus* produces B and G aflatoxin without cyclopiazonic acid. Because these species appear to uncommon, it has been little studied, so the potential toxigenicity of isolates is not known and practical importance of this species is hard to access.

There is an immense potential of active fractions from many biodiversity resources available in the country. The proposed study focuses on a systematic study with respect to antifungal potential of bioactive compounds of these resources in relation to fungal infestation and aflatoxin production in high risk groundnut, its oil and cake. *Pongamia pinnata* Pierre (Leguminosae) is commonly known as Karanja. It is distributed throughout Western Ghats and chiefly found in tidal forests of India (Krishnamurthi, 1969). Different parts of the plant have been used in traditional medicines for bronchitis, whooping cough, rheumatic joints and to quench dipsia in diabetes (Kirtikar *et al.*, 1995). Previous phytochemical examination of this plant indicated the presence of furanoavones, furanoavonols, chromenoavones, avones, and furanodiketones (Talapatra *et al.*, 1980; Murty *et al.*, 1944; Rangaswami *et al.*, 1942; Sharma *et al.*, 1973; Pathak *et al.*, 1983; Toshiyuki *et al.*, 1992). In the present communication, we describe the isolation and characterization of three new furano-avonoid glucosides, pongamosides A-C (1–3), and a new avonol glucoside pongamoside D (4). Tamarind (*Tamarindus indica* L.) belongs to the family Leguminosae and grows naturally in many tropical and sub-tropical regions. In Thailand, two types of Tamarind are found in abundance, the so-called sweet and sour varieties. Tamarind is an important food resource for the Thai population. The flower and leaf are eaten as vegetables, while the germ obtained from the seed is used for manufacturing Tamarind gum, which is well known as a component of jelly (Phakruschaphan, 1982). Tamarind seeds are also reported to contain phenolic antioxidants, such as 2-hydroxy-30, 40-dihydroxyacetophenone,methyl 3,4-dihydroxybenzoate, 3,4-dihydroxyphenyl acetate and epicatechin (Tsuda *et al.*, 1994). Extracts exhibit antioxidant potential by reducing lipid peroxidation *in vitro* (Tsuda *et al.*, 1993, 1994) and anti-microbial activity (De *et al.*, 1999). Pumthong (1999) described the antioxidant activity of extracts of Tamarind pericarp, and reported the presence of mainly polymeric tannins and oligomeric procyanidins but these were not identified or quantitated. From this stand point it was of interest to compare the polyphenolic content in methanolic extracts of Tamarind pericarp and seeds, utilizing the methods of Owen *et al.* (2000a, 2003b). Furthermore, the antioxidant capacity of the methanolic extracts was determined using the hypoxanthine/xanthine oxidase (HX/XO) and 2-deoxyguanosine assays (Owen *et al.*, 2000a, 2003b).

Materials and Methods

Collection and Extraction of Plant Materials

Pongamia bark were collected from the campus of Banthara Research Station of National Botanical Research Institute (CSIR), Lucknow and Fruits of *Tamarindus* were purchased from the local market of Lucknow. Pongamia bark were then air dried to a uniform moisture level. Fruits of *T. indica* obtained were extracted in methanol using Polytron homogenizer (PT 6100 KINEMETICA). Methanolic extracts collected by filtration (Whatman No. 1) were concentrated under vacuum at low temperature using Rotary Evaporator of Heidolph, Switzerland. The residue then dissolved in 15 per cent Ethyl alcohol to get the desired concentration for the activities.

Maintenance of Fungal Strains

The strain of *Aspergillus flavus* MTCC2799 were obtained from Microbial Type Culture Collection from IMTECH, Chandigarh. The culture was maintained at 4 ± 1°C on Slants of Potato Dextrose Agar (PDA).

Antifungal Assay

Antifungal activity of *P.pinnata* and *T. indica* was tested against aflatoxin producing fungal strains of *A. flavus* obtained from IMTECH, Chandigarh, India.

Preparation of Inoculum

The spore suspension was prepared as described by Fan and Chen (1999). *A. flavus* was grown on PDA (HiMedia) slant for 5–7 days at 25 ± 3°C and the spore were harvested by adding 10 ml of sterile water and aseptically dislodging the spore with a sterile inoculating loop. This was diluted to obtain desired concentration of spore suspension.

Agar Well Diffusion Method

The Potato Dextrose Agar media (HiMedia) was cooled down up to 40-45 °C after autoclaving and added desired amount or concentration of plant extract. Shaked it very well and poured the media in Petriplates. After solidifying the media, three wells of 8mm diameter were made in each Petriplates. Without addition of plant extract were used as controls. 40 µL of spore suspension contained 18×10^4 spores mL^{-1} of *A. flavus* were added to each wells and incubated at 28°C for 5 days. Fungal growth of both the treated and untreated control plates was measured at every 24 hrs for 5 days (Perea *et al.*, 1990).

The percentage of inhibition was calculated using the following formula (Rasooli and Abyaneh, 2004)

$$I = \frac{C - T}{C} \times 100$$

where,

I: Percentage inhibition

C: Radial Growth in control

T: Radial growth in treatment

Results and Discussion

Production of toxin are linked to fungal growth and the environment in which the grains / cereals are stores (especially Relative Humidity and Temperature). Fungal growth and subsequent mycotoxin product in stored grain can be inhibited by physical method (aeration, modified atmosphere, etc.) or by fungistatic of which is propionic acid; acetic acid and sorbic acid are the most common used (Paster *et al.*, 1988). Report by several authors (Morozumi, 1978; Azzouz and Bullerman, 1982; Bahk and Marth, 1983; Bullerman *et al.*, 1977; Yin and Cheng, 1998; Hitokoto *et al.*, 1980) supports the fact that the extract of certain spices and herbs of medicinal importance exhibit antifungal property. These natural antifungal agents can be potential exploited in controlling the growth of fungi consequently inhibiting aflatoxin formation (Yin and Cheng, 1998; Grayer and Harborne, 1994).

Methanolic extract of *Pongamia* bark were tested for their antifungal activity against aflatoxin producing fungi. Fungal growth was significantly reduced in all the treatments as compared to that of control. Antifungal activity varied significantly among different treatments (Table 24.1). Methanolic extract of bark inhibited fungal growth from 26.44 per cent to 33.20 per cent.

Table 24.1: Efficacy of Bark of *P. pinnata* Against Aflatoxin Producing Fungal Strain *A. flavus*

Plant Parts Used	*Concentration (in ppm)*	*Radial Dia in mm After 72 hrs*	*Percentage Inhibition in per cent*
Bark	Control	30.21 ± 1.436	—
	500	22.22 ± 0.780	26.44
	1000	21.78 ± 1.080	27.9
	1500	20.35 ± 0.800	32.6
	2000	20.18 ± 0.980	33.20

Different concentrations i.e 500ppm, 1000ppm, 1500ppm and 2000ppm of methanolic extract of tamarind fruits were tested for their efficacy against aflatoxin producing fungi. All the concentrations tested were found to decrease fungal growth as compared to that of control. Percentage inhibition of fungal growth ranged from 11.60 to 34.12 per cent at 500ppm and 2000ppm concentration, respectively (Table 24.2). Another experiment was also set to study the synergistic effect of both extract. Percentage inhibition of fungal growth ranged from 39.75 to 62.72 per cent at 500ppm and 2000ppm concentration, respectively it was found that at 2000 ppm *i.e.* percentage inhibition increased approximately 50 times higher than individual extract.

Table 24.2: Efficacy of Polar Fraction of *Tamarindus indica* Against *A. flavus*

Plant Parts Used	*Concentration (in ppm)*	*Radial Dia in mm After 72 hrs*	*Percentage Inhibition in per cent*
Fruit	Control	30.21± 1.436	0
	500	26.70 ± 0.025	11.6
	1000	25.03 ± 0.095	17
	1500	23.13 ± 0.145	23.4
	2000	19.9 ± 0.295	34.12

Table 24.3: Combined Effect of Methanolic Extract *Tamarindus* Fruit and *Pongamia* Bark Against Aflatoxin Producing *A. flavus*

Plant Parts Used	*Concentration (in ppm)*	*Radial Dia in mm After 72 hrs*	*Percentage Inhibition in per cent*
Fruit + Bark	Control	30.21± 1.436	0
	500	18.20 ± 0.010	39.75
	1000	16.13 ± 0.085	46.6
	1500	12.45 ± 1.055	58.78
	2000	9.75 ± 0.470	62.72

In conclusion, the present study reports the antifungal properties of *Pongamia pinnata* and *T.indica* extracts, which can be commercially exploited and applied to foodsystems. Further studies are needed on the isolation and characterization of individual compounds to elucidate their different antifungal.

References

Azzour, M.A. and Bullerman, L.B., 1982. Comparative antimycotic effects selected herbs, spices, plant components and commercial antifungal agents. *Journal of Food Protection*, 45: 1298–1301.

Bahk, J. and Marth, E.H., 1983. Aflatoxin production is inhibited by selected herbal drugs. *Mycopathologia*, 8: 129–134.

Bullerman, L.B., Lieu, F.Y. and Seier, S.A., 1977. Inhibition of growth and aflatoxin production by cinnamon and clove oils, cinnamic aldehyde and eugenol. *Journal of Food Science*, 42: 1107–1109.

Cole, R.J., Hill, R.A., Blankenship, P.D., Sanders, T.H. and Garren, H., 1982. Influence of irrigation and drought on invasion of *Aspergillus flavus* in corn kernel and peanut pods. *Dev. Ind. Microbiol.*, 23: 299–326.

Fan, J.J. and Chen, J.H., 1999. Inhibition of aflatoxin producing fungi by Welsh onion extracts. *Journal of Food Protection*, 62: 414 – 417.

Gams, W., Christensen, M., Onions, A.H.S., Pitt, J.I. and Samson, R.A., 1985. Intrageneric taxa of *Aspergillus*. In: *Advances in Penicillium and Aspergillus Systematics*, (Eds.) R.A. Samson and J.I. Pitt. Plenum Press, New York, NY, pp. 55–62.

Grayer, R.J. and Harborne, J.B., 1994. A survey of antifungal compounds from higher plants. *Phytochemistry*, 37: 19–42.

Hitokoto, H., Morozumi, S., Wanke, T., Sakai, S. and Kurata, H., 1980. Inhibitory effects of spice on growth and toxin production of toxigenic fungi. *Applied Environmetal Microbiology*, 39: 818 –822.

Kirtikar, K.R. and Basu, B.D., 1995. *Indian Medicinal Plants*, 2nd edn. International Book Distributors, Dehradun, India, 1: 830.

Klich, M.A., Thomas, S.H. and Mellon, J.E., 1984. Field study on the mode of entry of *Aspergillus flavus* into cotton seeds. *Mycologia*, 76: 665–669.

Krishanamurthi, A., 1969. *The Wealth of India, Vol. VIII*. Publication and Information Directorate, CSIR, New Delhi, India, pp. 206.

Lillehoj, E.B., Kwolek, W.F., Horner, E.S., Wild strom, N.W., Josephson, L.M., Franz, A.O. and Catalano, E.A., 1980. Aflatoxin contamination of preharvest corn: Role of *Aspergillus flavus* inoculum and insect damage. *Cereal Chemistry*, 57: 255–257.

Morozumi, S., 1978. Isolation, purification and antibiotic activity of O-Methoxycinnm aldehyde from cinnamon. *Applied Environmental Microbiology*, 36: 577–583.

Murthy, P.B.R. and Seshadri, T.R., 1944. Chemical examination of the flowers of *Pongamia glabra* and a note on the glycoside component of *Butea frondosa* flowers. *Proceeding of Indian Academy of Sciences*, 20A: 279–291.

Owen, R.W., Giacosa, A., Hull, W.E., Haubner, R., Spiegelhalder, B. and Bartsch, H., 2000a. The antioxidant/anticancer potential of phenolic compounds isolated from olive oil. *European Journal of Cancer*, 36: 1235–1247.

Owen, R.W., Giacosa, A., Hull, W.E., Haubner, R., Wurtele, G., Spiegelhalder, B. and Bartsch, H., 2000b. Olive oil consumption and health: The possible role of antioxidants. *Lancet Oncology*, 1: 107–112.

Paster, N., Juven, B.J. and Harshemesh, H., 1988. Antimicrobial activity and inhibition of aflatoxin B1 formation by olive plant tissue constituents. *Journal of Applied Bacteriology*, 64: 293–297.

Pathak, V.P, Saini, T.R, Khanna, R.N., 1983. *Glabra chalcone* a chromenochalcone from *Pongamia glabra*. In: *Proceeding of Indian Academy of Sciences*, 16A: 319–322.

Perea, C., Paul, M. and Bazerque, P., 1990. Antibiotic assay by Agar well diffusion method. *Acta Biol. Med. Exp.*, 15: 113–115.

Phakruschaphan, T., 1982. Comparison of peeling and extraction methods in the production of Tamarind seed gum. *The Kasetsart Journal of Natural Sciences*, 16(2): 74–81.

Pitt, J.I., Dyer, S.K. and Mc Cammon, S., 1991. Systemic invasion of developing peanut plants by *Aspergillus flavus*. *Letter Appl. Microbial.*, 13: 16–20.

Pumthong, G., 1999. Antioxidative activity of polyphenolic compounds extracted from seed coat of *Tamarindus indica* Linn. *Ph.D.Thesis*, Chiangmai University, Thailand.

Rangaswami, S., Rao, J.V. and Seshadri, T.R., 1942. Kanugin: A crystalline compound of the root of *Pongamia glabra*. *Proceeding of Indian Academy of Sciences*, 16A: 319–322.

Rasooli, I. and Abyaneh, M.R., 2004. Inhibitory effect of Thymes oil on the growth and Aflatoxin production by *Aspergillus parasiticus*. *Food Control*, 15: 479–483.

Sanders, T.H., Hill, R.A., Cole, R.J. and Blankenship, 1981. Effects of drought on the occurrence of *Aspergillus flavus* in maturing peanuts. *J. Am. Oil. Chem. Soc.*, 58: 966A–970A.

Sharma, P., Seshadri, T.R. and Mukerjee, S.K., 1973. Some synthetic and Natural analogues of glabra Chromene. *Indian journal of Chemistry*.11: 965–986.

Talapatra, S.K., Mallik, A.K. and Talapatra, B., 1980. Pongaglabol: A new hydroxyfuranoflavone and aurantimide acetate, a dipeptide rom the flower of *Pongamia glabra*. *Phytochemistry*, 19: 1199–1202.

Tosiyuki, T., Munekaza, I., Kaone, Y., Yuku, F. and Mizuo, M., 1992. Flavonoids in root bark of *Pongamia pinnata*. *Phytochemistry*, 31(3): 993–998.

Tsuda, T., Osawa, T., Makino, Y., Kato, H. and Kawakishi, S., 1993. Screening for antioxidant activity of edible pulses. *Bioscientific Biotechnological Biochemistry*, 57: 1606–1608.

Tsuda, T., Watanabe, M., Ohshima, K., Yamamoto, A., Kawakishi, S. and Osawa, T., 1994. Antioxidative components isolated from the seed of Tamarind (*Tamarindus indica L.*). *Journal Agricultural and Food Chemistry*, 42: 2671–2674.

Yin, M. C. and Cheng, W.S., 1998. Inhibition of *Aspergillus niger* and *Aspergillus flavus* by some herbs and spices. *Journal of Food Protection*, 61: 123–125.

Ethnobotanical Wisdom and Microbial Studies on Medicinal Plants (2011) *Pages* **130–150**
Editors: **D.R. Khanna, Ashutosh Gautam, R. Bhutiani, Gagan Matta, Anil Dhiman**
Published by: **DAYA PUBLISHING HOUSE, NEW DELHI**

Chapter 25

Moss Bag Technique for Monitoring of Metal Precipitation

Dinesh K. Saxena[1]*, Kajal Srivastava[2] and Shivom Singh[1]
[1]Bryology Division, Department of Botany, Bareilly College, Bareilly, U.P.
[2]SRF, Ministry of Environment and Forest, Government of India)

ABSTRACT

Moss *Barbula vinealis* and *Rhodobryum roseum* used as biomonitor of metal precipitation at Nainital during 2004–2005 to examine deposition of Zn, Cu, Cd, Pb at Nainital area, around point sources in all the four directions. Moss bags were transplanted at 8 sites for fixed exposure time in sampling seasons (summer, monsoon and winter) and were harvested periodically after 4 months of exposure. An increase in amount of metals in 2004–2005 reflects an increase in metals in air. High metallic load was observed in locations in proximity of higher traffic density. From the result, it is concluded that local sources in Nainital, especially due to enhanced tourism during summer, contributes to elevated metal deposition in comparison to winter and monsoon season. Active monitoring by 4 months of exposure of mosses gives reliable results on metal contamination. Study also aim at assessing the suitability of both mosses as a biomonitor for metal deposition. Study confirms that in Nainital, where due to higher vehicular traffic, wear and tear of vehicular parts and beside it increasing tourist activity, high level of Zn, Pb was measured in moss transplant bags.

Bioaccumulation ability in these two mosses was evaluated statistically using Dunkun's Multiple Range Test and was presented on contour maps obtained from SURFER program.

Keywords: *Element concentration, Barbula vinealis, Rhodobryum roseum, biomonitoring.*

* Corresponding Author: E-mail: dinesh.botany@gmail.com.

Introduction

Activities of man and the uncontrolled development of large and small cities and urbanization during the recent past and especially on hills resulted in the contamination of ecosystem. Combustion of fuel, increasing tourist activity and heavy traffic load contribute a considerable amount of pollutants to air. In recent decades the number and intensity of anthropogenic sources, such as waste burning, fertilizers, vehicle emissions, agricultural and sewage sludge, have increased the overall environmental element concentration (Bargagli, 1998). This fact seems to be true for the city of Nainital where there are large numbers of vehicles and other human activities increased with time. Therefore, monitoring of air contaminants of study area is necessary to determine impact upon ecosystem and control measure requires for abatement of their sources in investigated region.

Bryophytes have been used as terrestrial biomonitors and bioindicators of air pollution worldwide as well as by this laboratory and are recognized as more sensitive to pollution than other plants (Fernandez *et al.*, 2000). Among them mosses have tremendous ability to absorb metal as dry fallout from the atmosphere (Saxena, 2001). They have capability to receive and accumulate chemical substances predominantly from surrounding atmosphere (Fernandez *et al.*, 2004). Due to this accumulation potential they have been preferred in present study.

In present work, active monitoring (moss bags) has been used to determine total level of atmospheric deposition of heavy metals in Nainital city. This technique is very useful especially in such polluted areas where wild growing mosses are lacking (Makholm and Miladenoff, 2005).

Study is an attempt to use moss *Barbula vinealis* and *Rhodobryum roseum* to investigate the spatial distribution of the metals (Cu, Zn, Pb and Cd) contamination in Nainital city during Monsoon, summer and winter seasons with the help of moss bag (transplant) technique during different seasons of 2004 to 2005.

Materials and Methods

Study Area

Nainital is located at an altitude of 5900 ft. on Kumaon hills and connected by road only. It is situated in area of 11.7 Sq. Km. and surrounded by hills from all sides except east, which has only entry in the city (Figure 25.1). The climate was quite cold between October–April and mild warm through May–June followed by monsoon rain till September. The average rainfall measured 80" and relative humidity range from 85 to 90 per cent in the months of July and August. The maximum and minimum temperatures were recorded 27°C to 10°C in summer and 15°C to 3°C in winter respectively.

Sampling

Moss *Barbula vinealis* and *Rhodobryum roseum* were collected from the forest cover of Mukteshwar (unpolluted area), situated at an altitude of 2300 meters. A complete green patch of moss was transplanted in nylon bags at 8 study sites of investigated areas and sufficient amount of the same moss was taken for digestion to determine the baseline concentration of each season. Each moss bag was suspended 6 feets above the ground in triplicate. These moss bags were transplanted cross section-wise in all the four directions at the distance of 100m and 300m and were harvested after an exposure period of four months *i.e.* first week of November (for monsoon monitoring), March (for winter monitoring) and July (for summer monitoring).

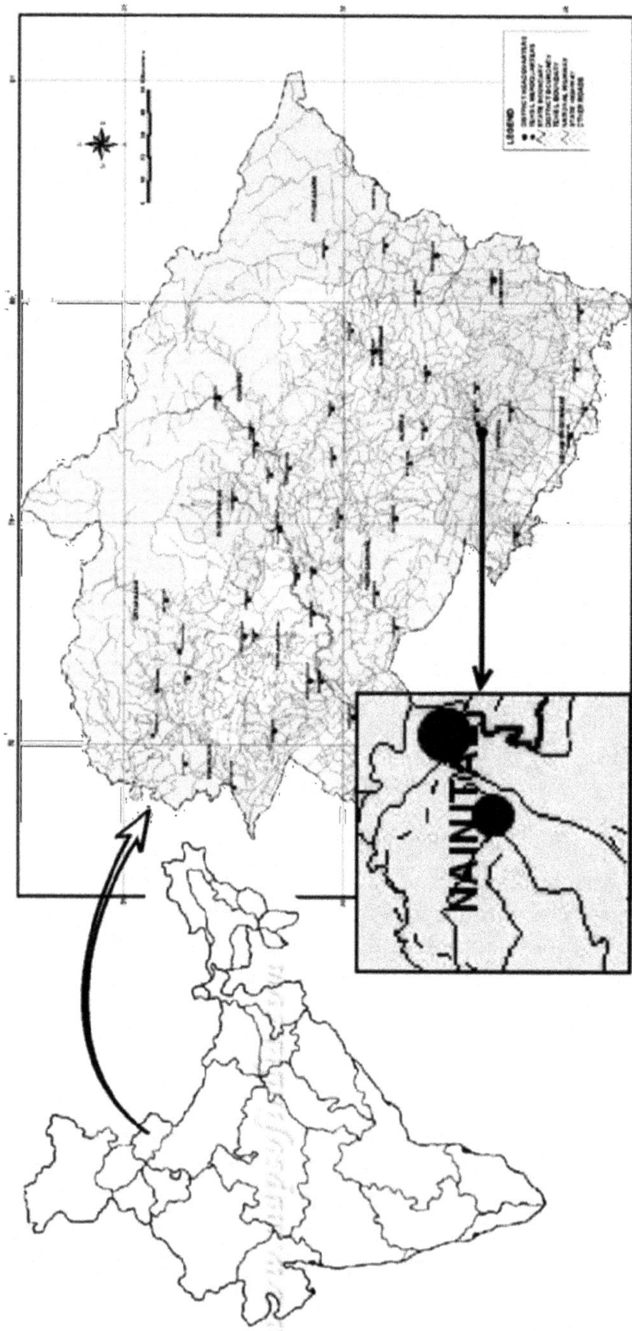

Figure 25.1: Map of Monitoring Sites (Nainital) of Kumaon Hills, India

Metal Analysis

Upon return to the laboratory, harvested moss samples were oven dried at 40°C for 24 hours. Prior to analysis, adhering substrate and litter was removed by hand, great care being taken to avoid metal contamination. Triplicate samples were digested with concentrated HNO_3 and $HClO_4$ in ratio of 5:2 v/v on a hot plate. The digestion was completed after all organic material had disappeared. The extract obtained was filtered and the filtrate was made up to a final volume of 50 ml and fraction was quantitatively analyzed by atomic absorption spectrophotometer. Suitable blank were used to check for possible contamination during extraction.

Data Analysis

Samples were collected in triplicate to conduct the statistical analysis. Value represented as mean ± standard error (Snedecor and Cochran, 1967). ANOVA revealed significant differences in the metal concentration at different distances and seasons (for $p<0.01$, $p<0.05$) utilizing Dunkun's Multiple Range test (Karmer, 1956). Catographic representation of the results was performed with the program package Surfer (Golden Software Inc., U. S. A.).

Results

Mean concentration (mg g^{-1} dry wt.) of metal ions were detected in *Rhodobryum roseum* and *Barbula vinealis* exposed during 2004-2005 transplant period at 8 monitoring stations of Nainital city (Tables 25.1 and 25.2). From analysis of moss transplant, it appears that different concentration of metal accumulation was observed which was related to different pollution sources, its level in the study sites and seasonal variations. A positive correlation was observed in all four metals in each season.

Rhodobryum roseum

In view of seasonal affect maximum concentration of zinc, lead, copper and cadmium were measured 2.056, 1.585, 0.785 and 0.064 mg g^{-1} dry wt. respectively at 100m in south direction. However, minimum concentration was measured in west direction of Nainital at 300m for Zn (0.709 mg g^{-1} dry wt.), Pb (0.477 mg g^{-1} dry wt.), Cd (0.008 mg g^{-1} dry wt.), whereas, in Cu (0.325 mg g^{-1} dry wt.) was lowest in north (300m).

Results were highly significant different for values ($p \leq 0.05$) of Zn, Pb and Cu in each site and for all metals undertaken, while, Cd value was non-significant at north direction between winter and summer season. An identical trend was observed in both transplant belongs to west direction (monsoon and winter season).

Barbula vinealis

The maximum concentration of Zinc was measured as 2.055 mg g^{-1}dry wt. at 100m (south) and minimum as 1.408 mg g^{-1} dry wt.at 300m (west), whereas, in case of lead, higher value was analyzed 1.846 mg g^{-1} dry wt. at 300m (east) and lowest at 0.753 mg g^{-1} dry wt. at 300m (west). In copper concentration was as high as 1.229 mg g^{-1} dry wt. in south direction (100m) and decline up to 0.416 mg g^{-1} dry wt. in west (300m). Cadmium was detected quite low in all samples, the maximum and minimum values were measured as 0.070 mg g^{-1} dry wt. at 300m (south) and 0.013 mg g^{-1} dry wt. at 100 m in north direction respectively.

After calculating ANOVA Zn, Pb and Cu show almost significant different values in different seasons. Cd shows almost non-significant results. However, results of eight study sites show significant different values for other metals Zn, Pb and Cu in all three seasons in comparison to control.

Table 25.1: Metals Concentration (mg g^{-1} dry wt.) in Transplants of *Rhodobryum roseum* at Different Sites of Nainital City During Monsoon, Winter and Summer Season (2004–2005)

(a)

Sites	*Rhodobryum roseum*							
	Monsoon 2004				*Winter 2005*			
	Zn	*Pb*	*Cu*	*Cd*	*Zn*	*Pb*	*Cu*	*Cd*
Control	0.418±0.004	0.268±0.005	0.207±0.004	0.005±0.002^{ABC*}	0.582±0.001	0.399±0.006	0.312±0.001^{DEF*}	0.024±0.052^{GHIJa*}
South, 100m	1.030±0.008	0.668±0.004	0.412±0.006	0.026±0.005^{A*}	1.205±0.003	0.776±0.003	0.525±0.013	0.034±0.026
South, 300m	1.354±0.002	0.925±0.014	0.456±0.003	0.023±0.002	1.575±0.006	1.125±0.014	0.577±0.004	0.044±0.029
North, 100m	1.183±0.007	0.844±0.012	0.378±0.004	0.009±0.0061	1277±0.009	0.958±0.004	0.484±0.006^{D}	0.024±0.027^{Gb}
North, 300m	0.880±0.002	0.723±0.013	0.325±0.009	0.022±0.002	1.010±0.008	0.804±0.027	0.454±0.002^{E}	0.039±0.043^{c}
East, 100m	0.980±0.003	0.771±0.002	0.453±0.008	0.020±0.003*	1.134±0.003	0.950±0.014	0.604±0.002	0.034±0.021*
East, 300m	1.207±0.004	0.903±0.026	0.449±0.011	0.009±0.001	1.476±0.009	1.003±0.002	0.506±0.003*	0.029±0.038^{Hf}
West, 100m	0.376±0.011	0.706±0.003	0.348±0.004	0.017±0.0211	1.156±0.004	0.837±0.064	0.518±0.011	0.026±0.0329^{gI}
West, 300m	0.709±0.005	0.477±0.004	0.414±0.003^{de}	0.003±0.001^{h}	0.828±0.046	0.704±0.004	0.221±0.166^{Fd*}	0.033±0.038^{Jh*}

(b)

Sites	*Rhodobryum roseum*			
	Summer 2005			
	Zn	*Pb*	*Cu*	*Cd*
Control	0.758±0.052	0.488±0.046	0.403±0.002	0.026±0.035^{KLa}
South, 100m	2.056±0.018	1.585±0.003	0.785±0.003	0.064±0.003
South, 300m	1.659±0.005	0.954±0.023	0.762±0.015	0.054±0.001
North, 100m	1.728±0.045	1.286±0.035	0.623±0.024	0.027±0.013^{Kb}
North, 300m	1.277±0.038	1.023±0.002	0.529±0.004	0.055±0.021^{c}
East, 100m	1.428±0.026	1.123±0.021	0.756±0.035	0.046±0.033*
East, 300m	1.627±0.002	1.295±0.026	0.668±0.041	0.030±0.036^{Lf}
West, 100m	1.390±0.003	1.125±0.003	0.556±0.015	0.046±0.031
West, 300m	1.228+0.004	1.025±0.024	0.678±0.047^{e*}	0.050±0.055*

* $p \leq 0.05$

Values are represented as mean ± standard error.

With in vertical columns of different seasons, values superscripted with same small alphabets are not significantly different at $P < 0.01$.

With in horizontal rows, values with same Capital alphabets are not significantly different at $P < 0.01$ in comparison to control.

Table 25.2: Metals Concentration (mg g^{-1} dry wt.) in Transplants of *Barbula vinealis* at Different Sites of Nainital City During Monsoon, Winter and Summer Season (2004–2005)

(a)

Sites	*Barbula vinealis*							
	Monsoon 2004				*Winter 2005*			
	Zn	*Pb*	*Cu*	*Cd*	*Zn*	*Pb*	*Cu*	*Cd*
Control	0.479±0.023	0.370±0.020	0275±0.002	ND	0.675±0.014	0.470±0.012	0.395±0.003	0.005±0.014^{a*}
South, 100m	1.355±0.102^{b}	0.954±0.023^{cd}	0.505±0.003	0.018±0.052^{fg}	1.531±0.009^{b}	1.417±0.241^{ce}	0.666±0.015	0.038±0.005^{fh}
South, 300m	1.333±0.017	1.030±0.087	0.704±0.023	0.025±0.046^{i*}	1.653±0.007	1.346±0.021	0.884±0.006	0.047±0.0111
North, 100m	1.226±0.006	0.896±0.017	0.457±0.006	0.013±0.072^{kl}	1.378±0.121	1.052±0.016	0.585±0.072	0.025±0.002^{km}
North, 300m	1.151±0.027	0.792±0.001	0.429±0.012	0.025±0.002^{o*}	1278±0.052	0.977±0.012	0.766±0.013	0.040±0.011^{no}
East, 100m	1.300±0.028	0.846±0.035	0.568±0.179	0.017±0.069^{p}	1.578±0.012	1.260±0.017	0.636±0.023	0.030±0.005^{pq*}
East, 300m	1.380±0.011	1.076±0.029	0.517±0.007	0.030±0.087^{rs}	1.704±0.023	1.430±0.008	0.629±0.011	0.045±0.011^{rt}
West, 100m	1.178±0.005	0.876±0.005	0.526±0.010	0.018±0.005^{t}	1.379±0.011	1.093±0.001	0.723±0.013	0.033±0.006^{t*}
West, 300m	1.048±0.003	0153±0.075	0.466±0.015	0.014±0.055^{uv}	1.254±0.013	0.994±0.003	0.676±0.009	0.029±0.005^{uw}

(b)

Sites	*Barbula vinealis*			
	Summer 2005			
	Zn	*Pb*	*Cu*	*Cd*
Control	0.775±0.005	0.576±0.010	0.435±0.012	0.015±0.012^{aAB*}
South, 100m	2.055±0.012	1.481±0.052^{de}	1229±0.011	0.064±0.018^{gh}
South, 300m	1.805±0.029	1.630±0.008	0.931±0.016	0.070±0.011^{j*}
North, 100m	1.714±0.008	1.421±0.011	0.829±0.058	0.032±0.007^{lmA}
North, 300m	1.501±0.007	1.204±0.002	0.955±0.104	0.062±0.0690^{o*}
East, 100m	1.774±0.013	1.519±0.058	0.746±0.010	0.060±0.077^{q}
East, 300m	2.004±0.023	1.846±0.098	0.795±0.002	0.058±0.052st
West, 100m	1.618±0.052	1.307±0.024	0.954±0.053	0.053±0.053*
West, 300m	1.453±0.075	1.126±0.041	0.873±0.088	0.036±0.075^{vwB}

* $p \leq 0.05$

Values are represented as mean ± standard error.

With in vertical columns of different seasons, values superscripted with same small alphabets are not significantly different at $P < 0.01$.

With in horizontal rows, values with same Capital alphabets are not significantly different at $P < 0.01$ in comparison to control.

Metal Distribution Pattern

The distribution mapping pattern of undertaken metals (Cu, Zn, Cd, Pb) show maximum concentration in south and south east direction, whereas, minimum concentration was observed in west direction of transplant sites (in most of the cases)at Nainital during different seasons in both the mosses (Figures 25.2–25.25). The reason could be due to higher vehicular load as bus station and petrol pump is located in south side of Nainital city. In addition, at some places samples get direct exposure while in other areas not. We cannot ignore even meteorological factor and direction of prevailing wind, which may be the other cause for variation in distribution pattern.

Discussion

The metal concentrations of the analyzed samples varied greatly, depending on the environmental precipitation and correlated well with the distribution of emission sources. In different seasons, metal precipitation was not constant and recorded maximum during summer followed by winter and lowest in monsoon season. The more significant seasonal trend is reflected by the fact that consumption of gasoline peaks during summer due to many fold increased in tourist activity and also due to decrease in growth during dry summer period (Gerdol *et al.*, 2002). However, in monsoon tourist activity decreases and pollutants leach out. Further more, increase in growth and biomass occurs more rapidly during rains and thus reduces the metal percentage in leaves in proportion to biomass. Present observation are ample documented by Tables 25.1 and 25.2 confirm that statistically only Pb, Zn and Cu showed any degree of positive correlation for the whole data set.

The study confirms that there are different levels of contaminants accumulation in the transplants. It is evident from analytical parameters that high amount of metals (Pb, Cu, Cd, Zn) was detected in south direction in moss *Rhodobryum roseum* and *Barbula vinealis* due to the reason that site is near bus station and petrol pump, dry deposition of metals spewed out from automobile (Imperato *et al.*, 2003), the only source of transportation connecting the foot hill (Haldwani) to Nainital. On the contrary, Pb was measured maximum at 100m (north) in *Barbula vinealis* may be due to traffic density, proximity to other roads, precipitation, meteorological factors, and finally by direction of prevailing wind. The lowest ratios of metal were found in the transplants, west direction along with north site during monsoon season. This reveals that mosses exposed to traffic are relatively high (south) compared with those having low traffic volume and human interference (west).

This is further confirmed for Pb and Zn both elements were found high in concentration. This may be due to busy roads, motorways, traffic density, use of fertilizers for growth of crops and orchards (Saxena and Saxena, 2000). Besides, it is an integral constituent of lead as steel and automobile industries. The likeliest source of the Zn from vehicles is engine and particularly tires wear, along with very little from exhaust emission (Pearson *et al.*, 2000). The main source of Pb is vehicular emission. It is also found associated with anthropogenic variables of the environment.

Since, Cu is an integral part of discarded metallic waste and used in fertilizers etc.On the contrary it is also used as fungicides and pesticides, practices in agriculture (Gerdol *et al.*, 2000; Otvos *et al.*, 2003). Higher values in domestic waste were reported due to their improper dispersal in atmosphere. In addition, higher concentration of Cd was due to automobile exhausts (Stefano and Bononi, 2000). The contamination may also come from use of metallic or plastic pipes, sewage sludge, abrasion of automobile tires and from domestic wastes in urban areas (Grodzinska and Szarek-Lukaszewska, 2001).

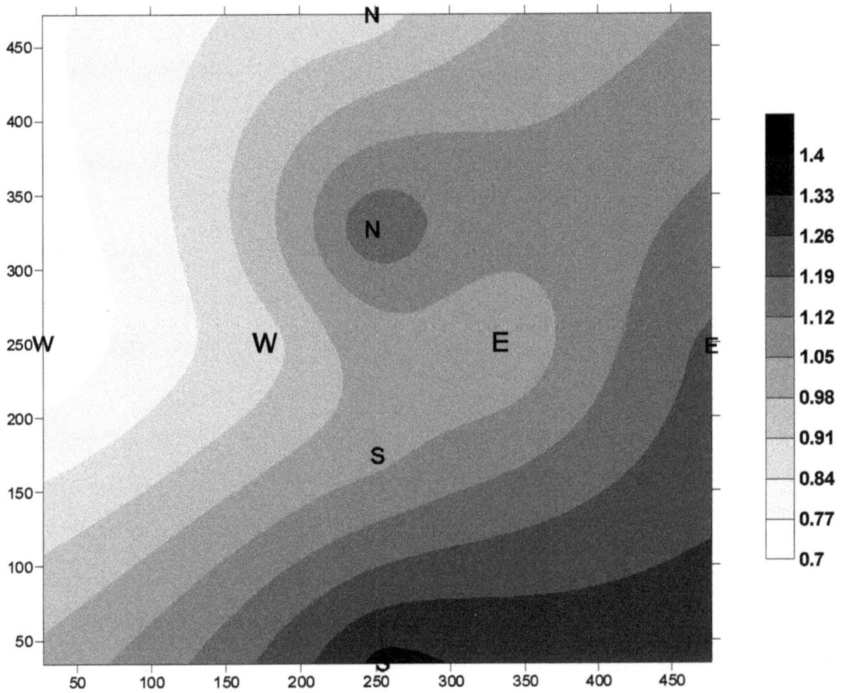

Figure 25.2: Distribution Map of Zn Content *R. roseum* (mg g^{-1}) in Monsoon at Nainital

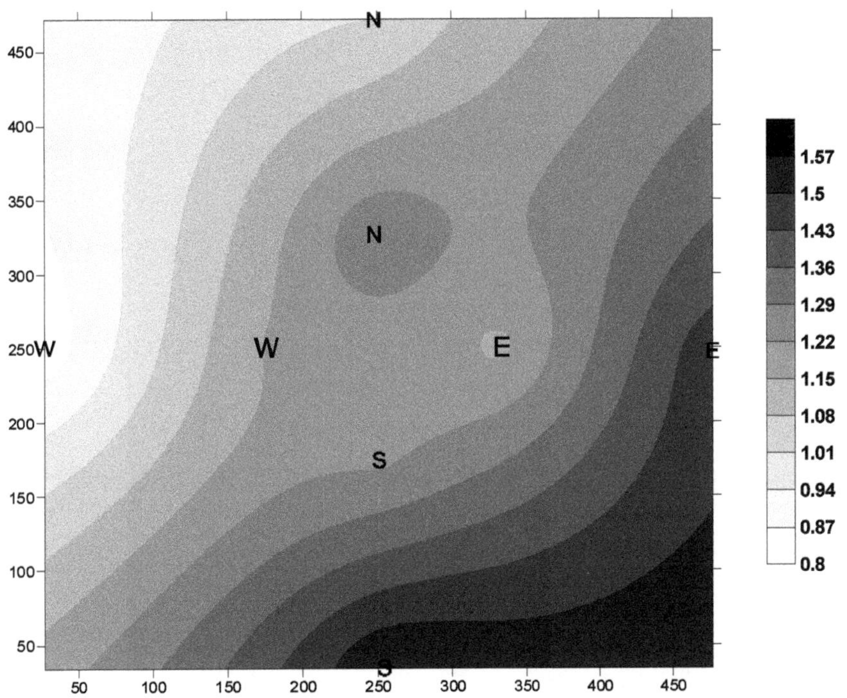

Figure 25.3: Distribution Map of Zn Content *R. roseum* (mg g^{-1}) in Winter at Nainital

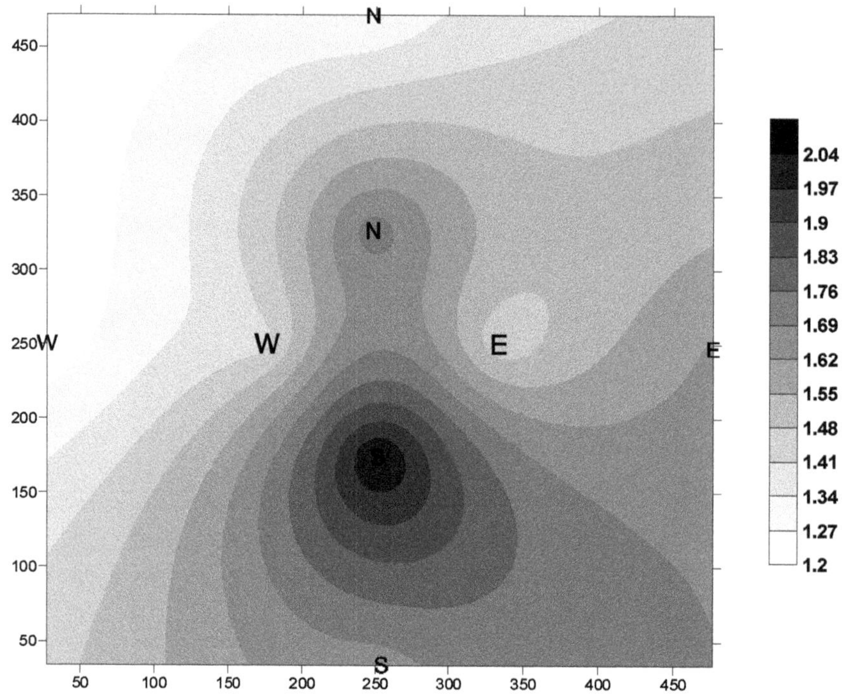

Figure 25.4: Distribution Map of Zn Content *R. roseum* (mg g^{-1}) in Summer at Nainital

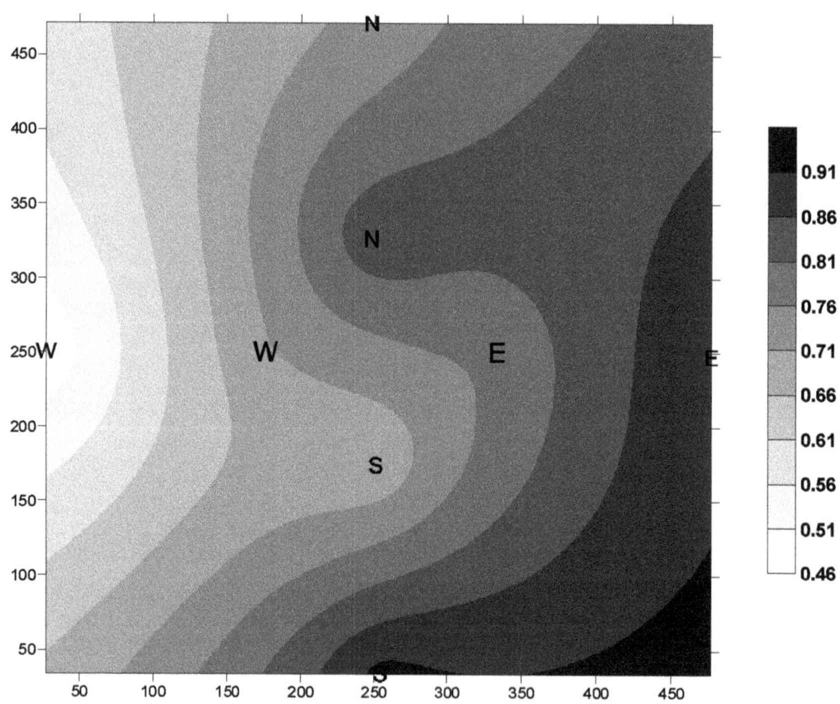

Figure 25.5: Distribution Map of Pb Content *R. roseum* (mg g^{-1}) in Monsoon at Nainital

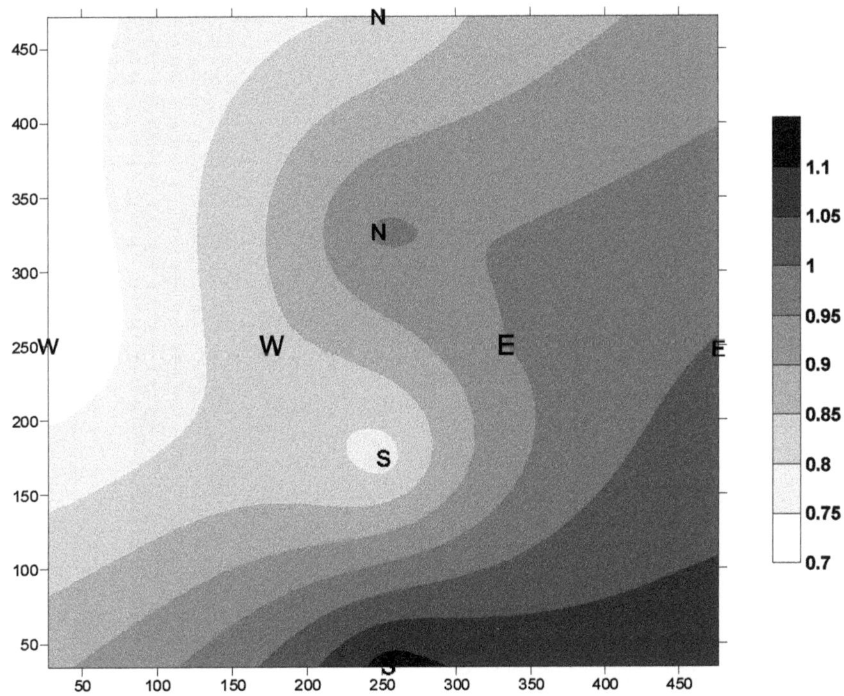

Figure 25.6: Distribution Map of Pb Content *R. roseum* (mg g⁻¹) in Winter at Nainital

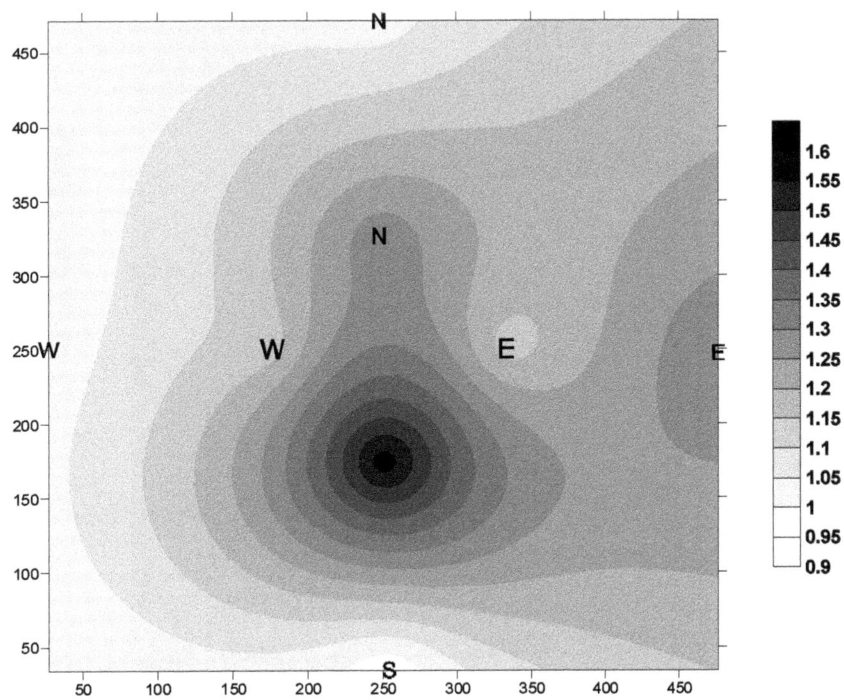

Figure 25.7: Distribution Map of Pb Content *R. roseum* (mg g^{-1}) in Summer at Nainital

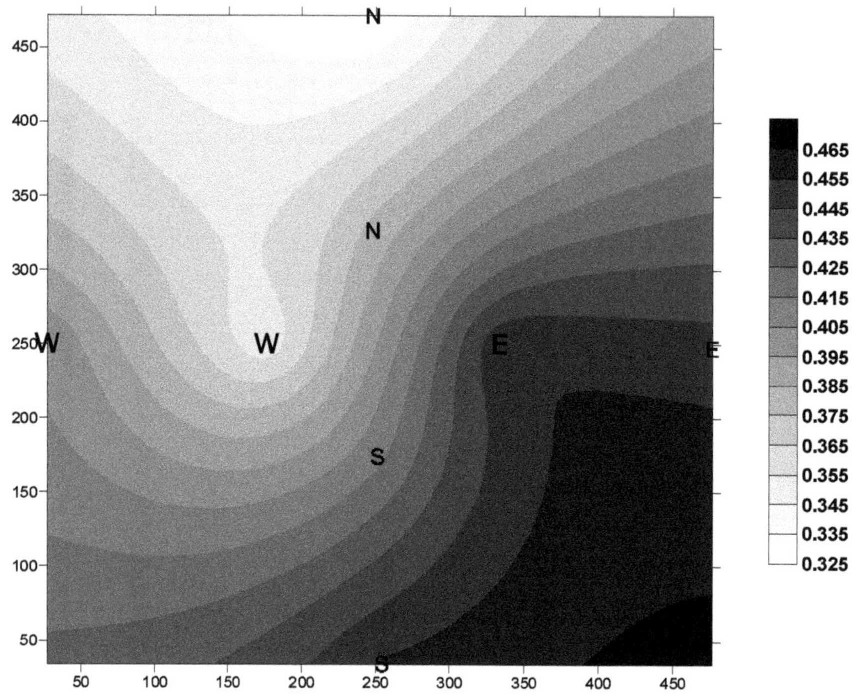

Figure 25.8: Distribution Map of Cu Content *R. roseum* (mg g^{-1}) in Monsoon at Nainital

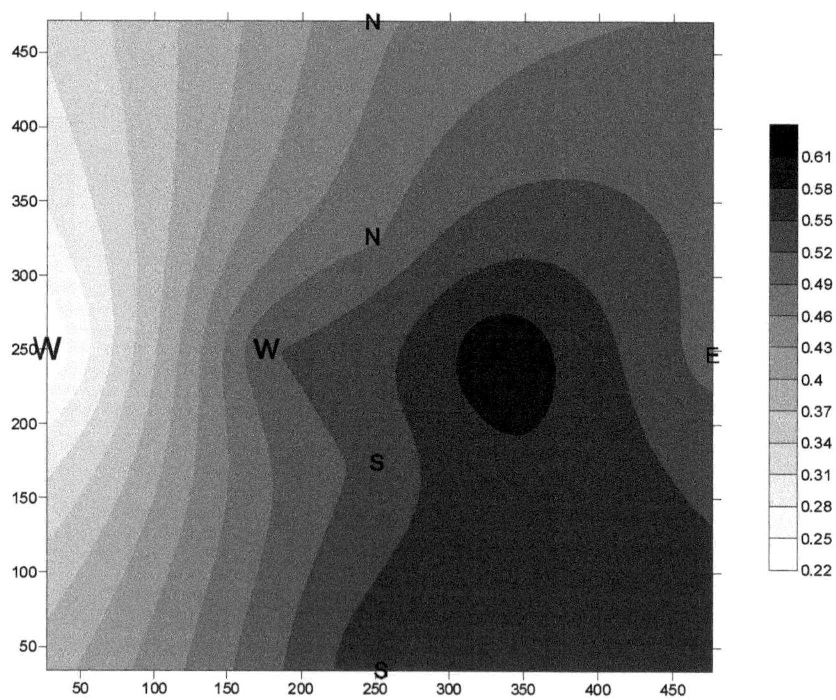

Figure 25.9: Distribution Map of Cu Content *R. roseum* (mg g^{-1}) in Winter at Nainital

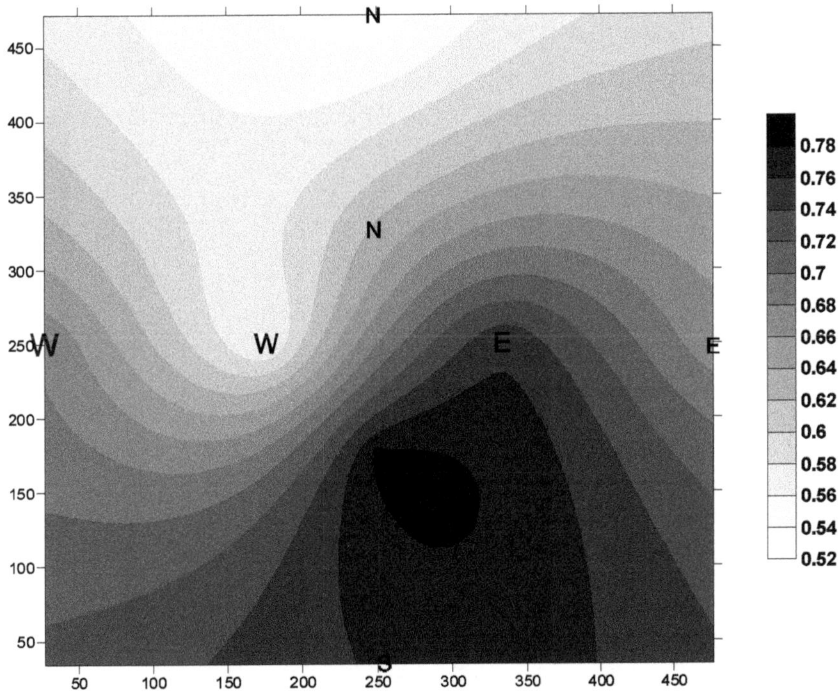

Figure 25.10: Distribution Map of Cu Content *R. roseum* (mg g^{-1}) in Summer at Nainital

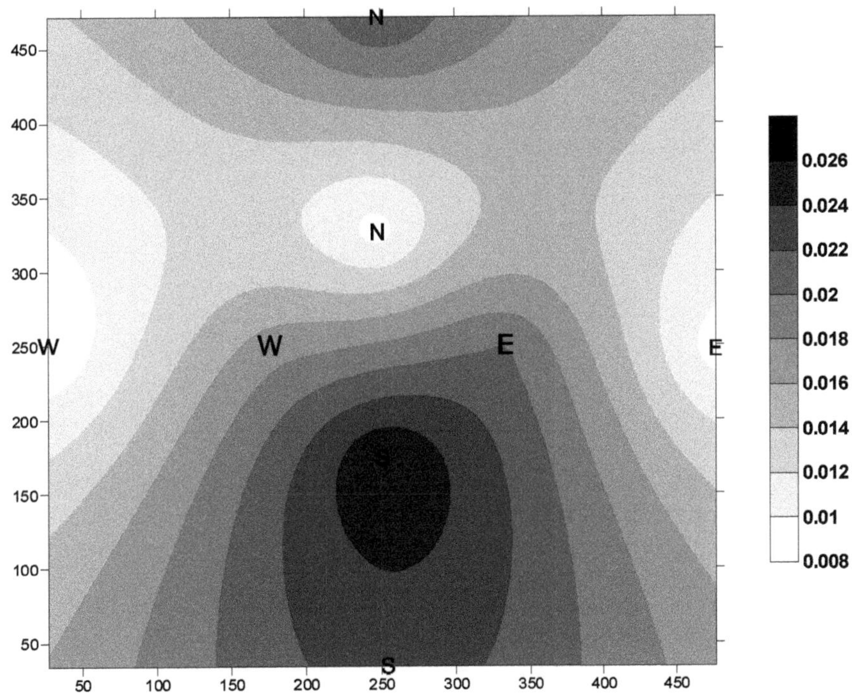

Figure 25.11: Distribution Map of Cd Content *R. roseum* (mg g^{-1}) in Monsoon at Nainital

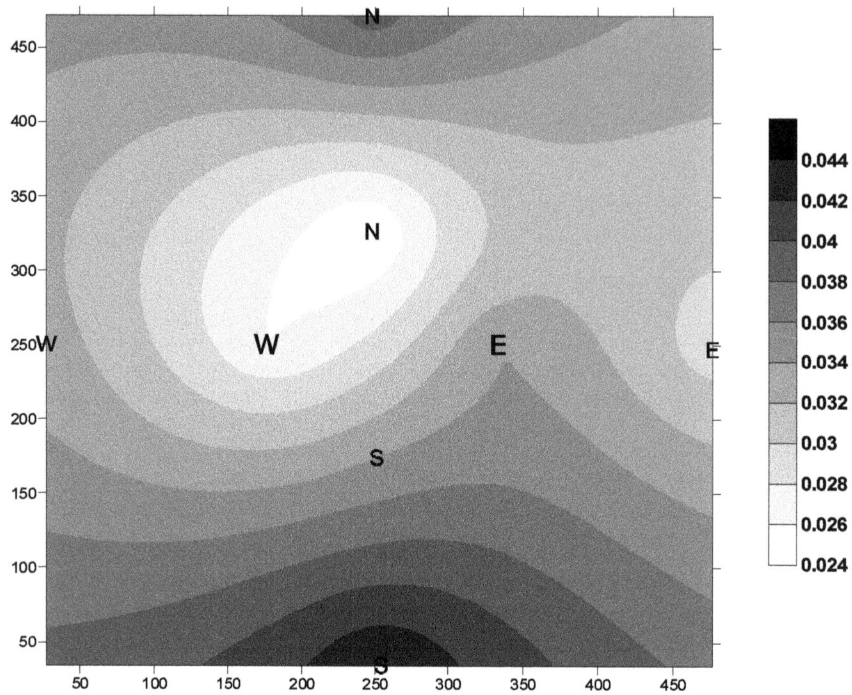

Figure 25.12: Distribution Map of Cd Content *R. roseum* (mg g⁻¹) in Winter at Nainital

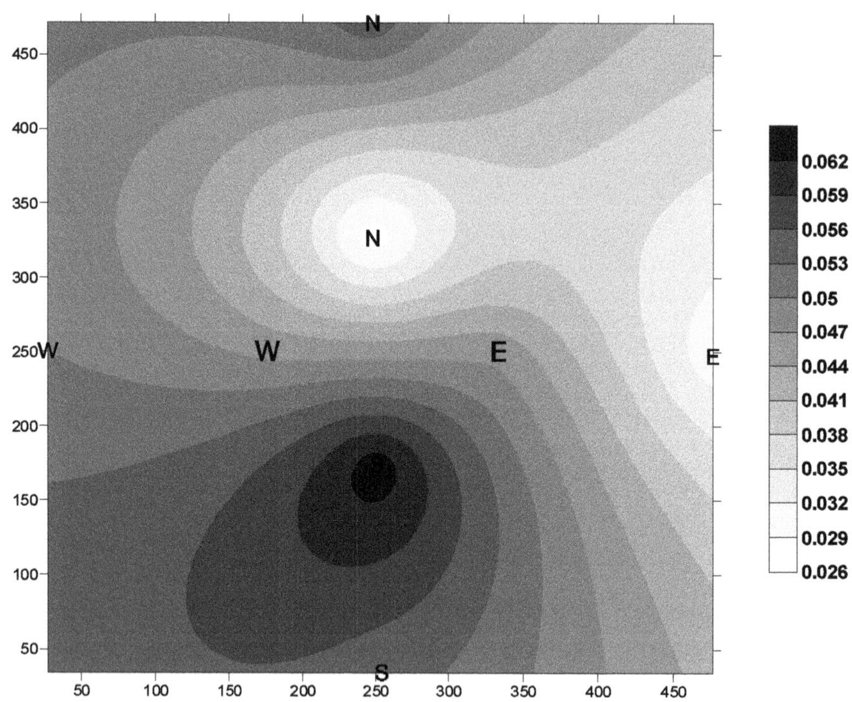

Figure 25.13: Distribution Map of Cd Content *R. roseum* (mg g^{-1}) in Summer at Nainital

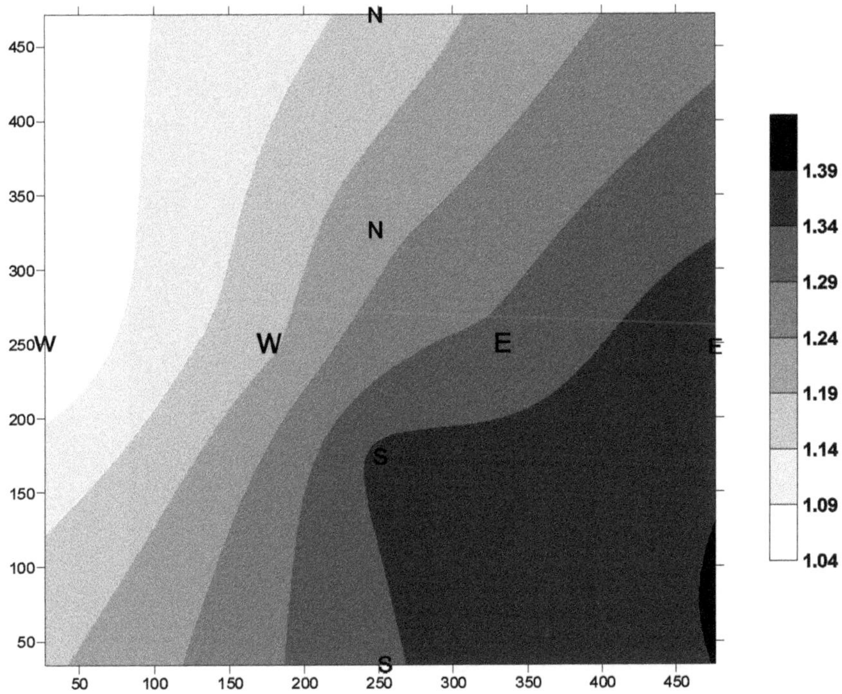

Figure 25.14: Distribution Map of Zn Content *B. Vinealis* (mg g⁻¹) in Monsoon at Nainital

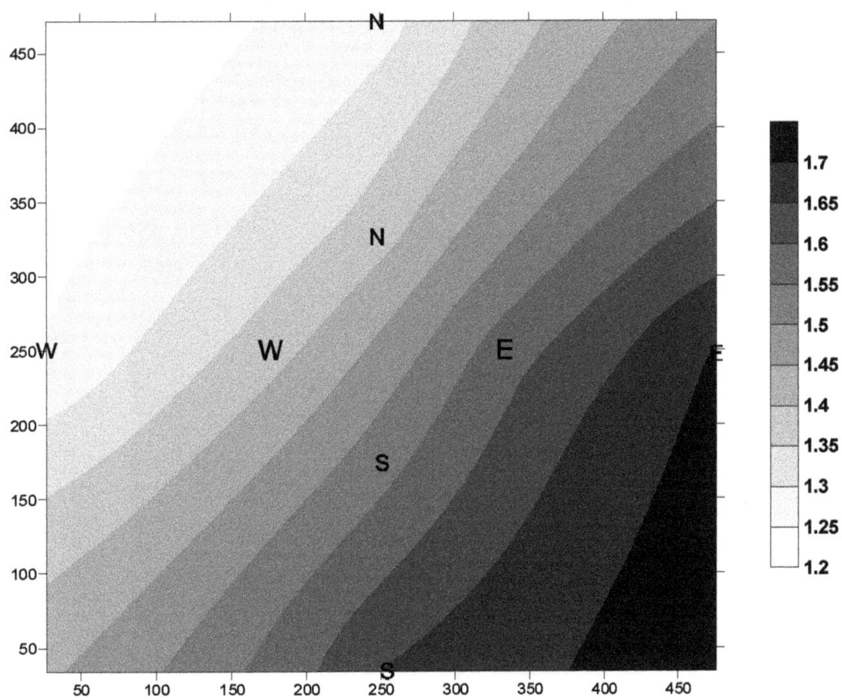

Figure 25.15: Distribution Map of Zn Content *B. Vinealis* (mg g⁻¹) in Winter at Nainital

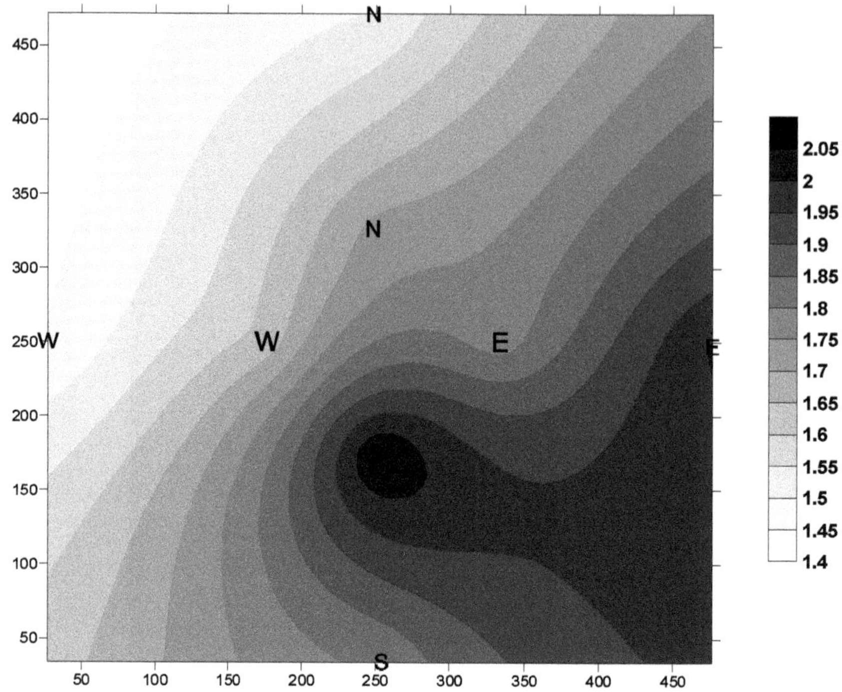

Figure 25.16: Distribution Map of Zn Content *B. Vinealis* (mg g^{-1}) in Summer at Nainital

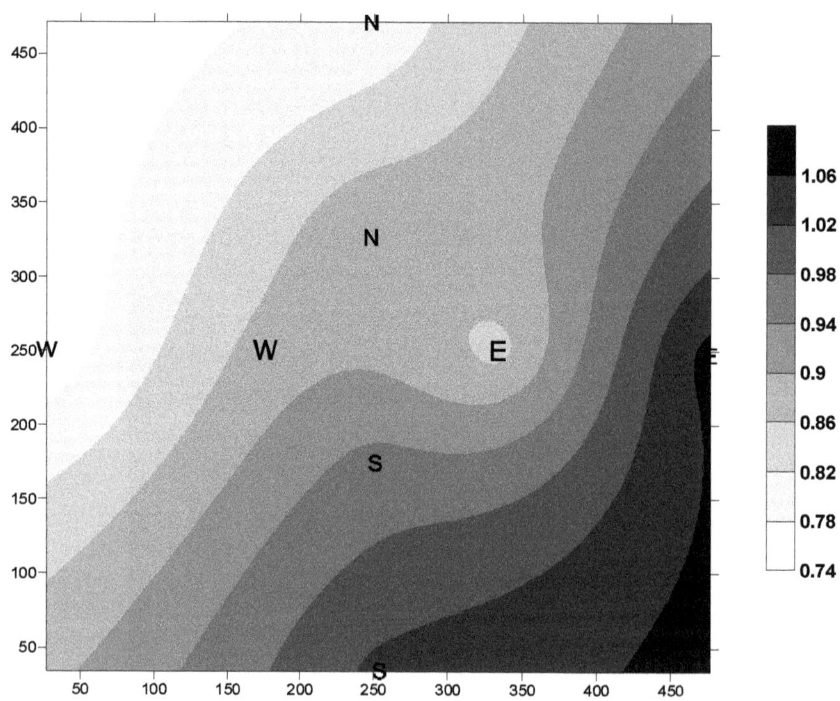

Figure 25.17: Distribution Map of Pb Content *B. Vinealis* (mg g^{-1}) in Monsoon at Nainital

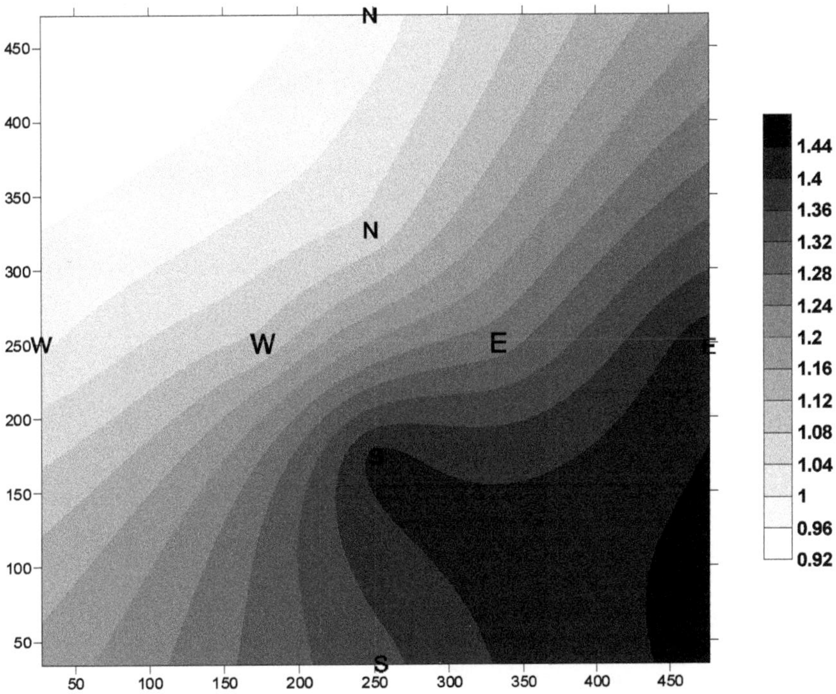

Figure 25.18: Distribution Map of Pb Content *B. Vinealis* (mg g^{-1}) in Winter at Nainital

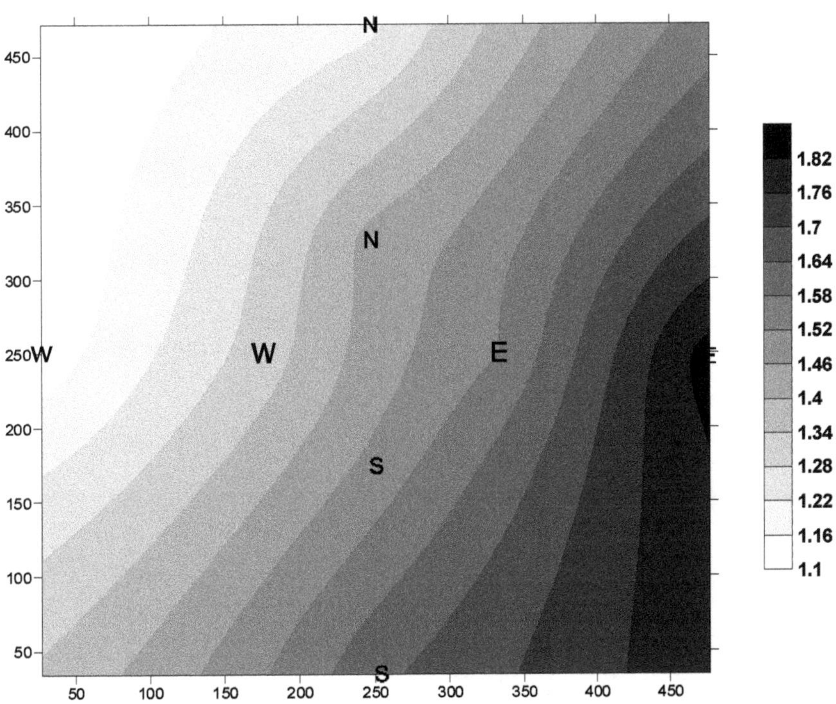

Figure 25.19: Distribution Map of Pb Content *B. Vinealis* (mg g^{-1}) in Summer at Nainital

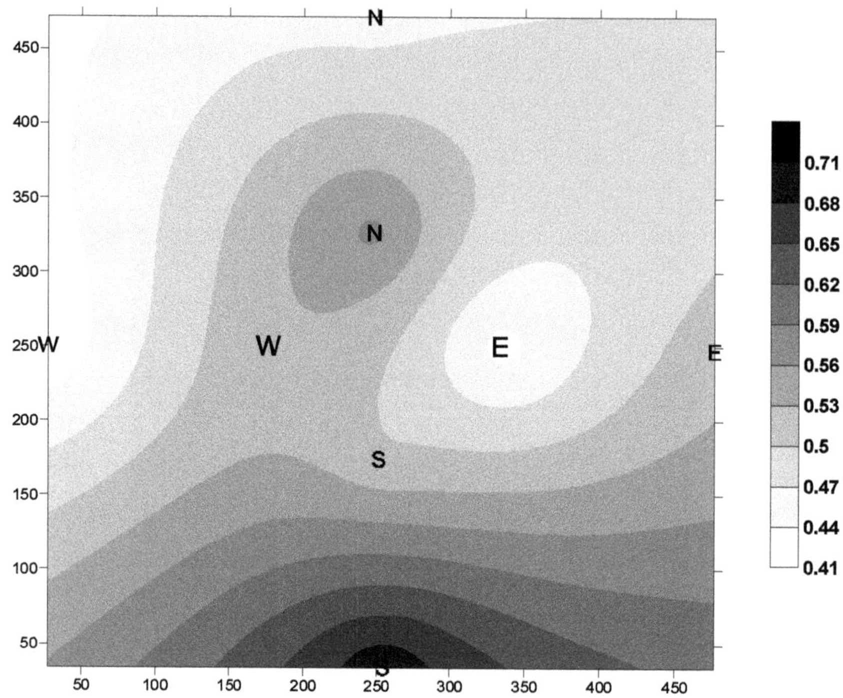

Figure 25.20: Distribution Map of Cu Content *B. Vinealis* (mg g⁻¹) in Monsoon at Nainital

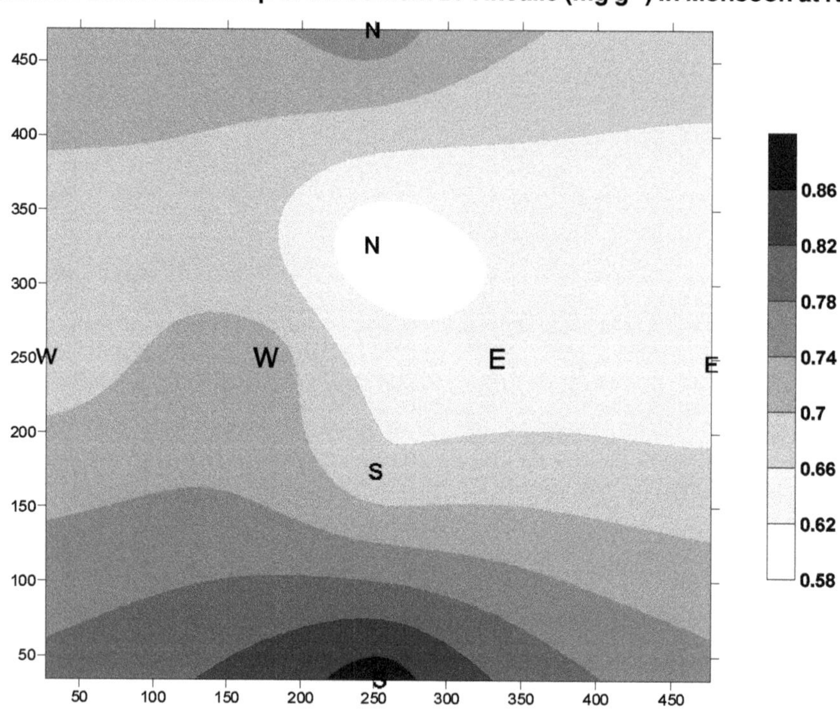

Figure 25.21: Distribution Map of Cu Content *B. Vinealis* (mg g^{-1}) in Winter at Nainital

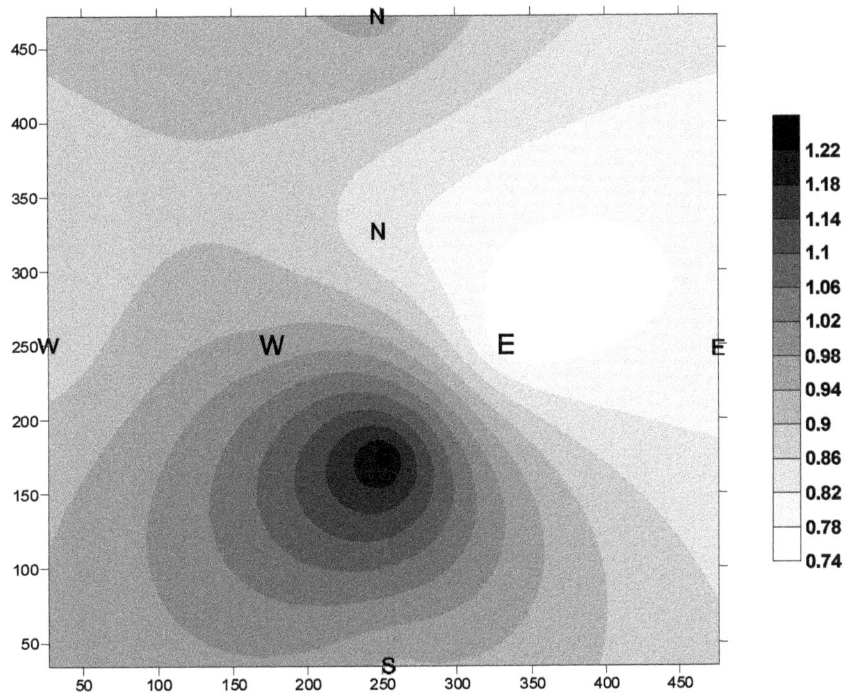

Figure 25.22: Distribution Map of Cu Content *B. Vinealis* (mg g^{-1}) in Summer at Nainital

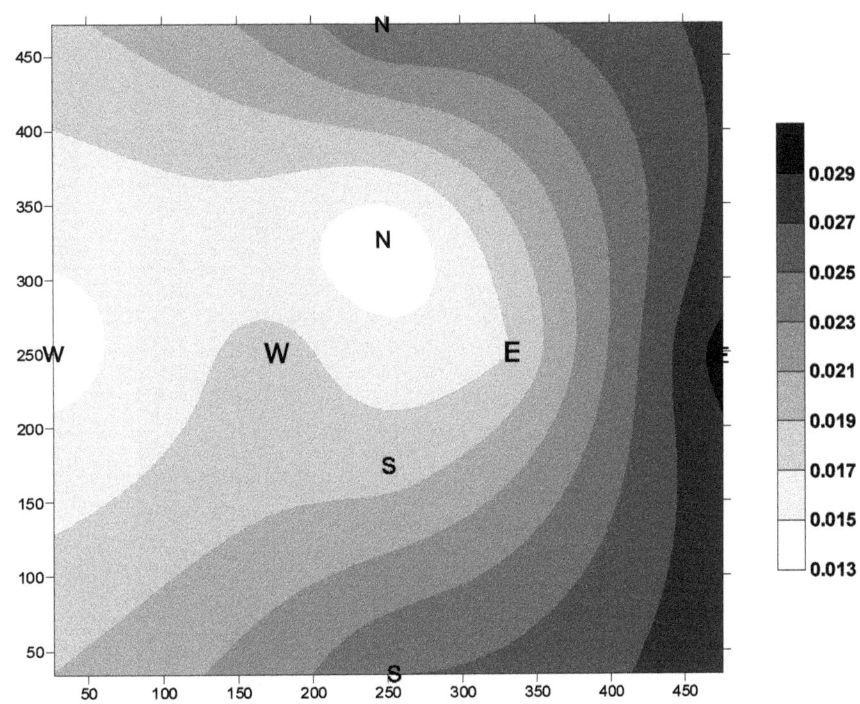

Figure 25.23: Distribution Map of Cd Content *B. Vinealis* (mg g^{-1}) in Monsoon at Nainital

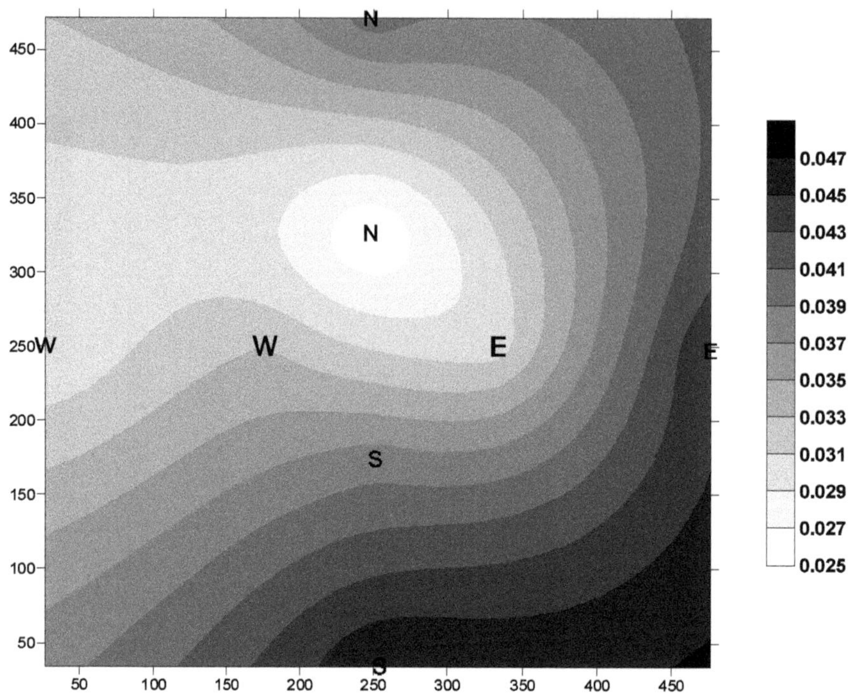

Figure 25.24: Distribution Map of Cd Content *B. Vinealis* (mg g⁻¹) in Winter at Nainital

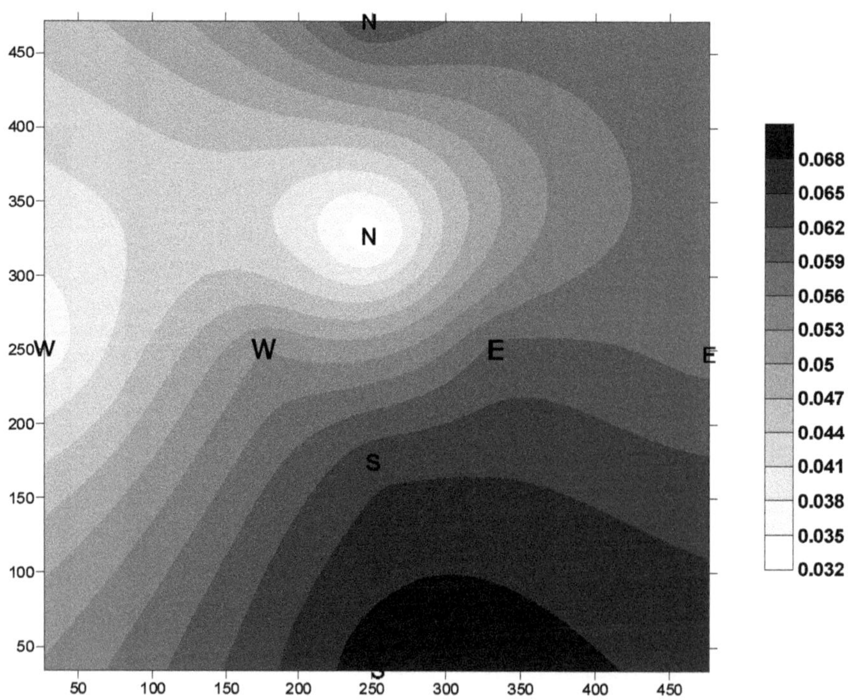

Figure 25.25: Distribution Map of Cd Content *B. Vinealis* (mg g^{-1}) in Summer at Nainital

Study successfully demonstrated the regional gradients of metals. Mosses *Rhodobryum roseum* and *Barbula vinealis* were very common in urban areas and were often the only species near busy main road, an indication of their general tolerance to atmospheric pollution in Nainital.

Acknowledgements

The authors are thankful to Ministry of Environment and Forest (Govt. of India) for their financial support. We wish to express our gratitude to Dr. R. D. Sharma of C. A. R. I., Izatnagar, Bareilly for his support. Authors are also grateful to Prof. Zoltan Tuba, Hungary for their informative suggestions. Help rendered by Bareilly College, Bareilly is gratefully acknowledged.

References

Bargagli, R., 1998. *Trace Elements in Terrestrial Plants: An Ecophysiological Approach to Biomonitoring and Biorecovery*. Springer Verlag, Berlin, New York, pp. 324.

Fernandez, J.A., Aboal, J.R. and Carballeria, A., 2000. Use of native and transplanted mosses as complimentary techniques for biomonitoring mercury around an industrial facility. *Science of the Total Environment*, 256(2–3): 151–156.

Fernandez, J.A., Aboal, J.R. and Caballeira, A., 2004. Identification of pollution sources by means of moss bags. *Ecotoxicology and Environmental Safety*, 59(1): 76–83.

Grodzinska, K. and Szarek-Lukaszewska, G., 2001. Response of mosses to the heavy metal deposition in Poland: An overview. *Environmental Pollution*, 114: 443–451.

Gerdol, R., Bragazza, L. and Marchesini, R., 2002. Element concentrations in forest moss *Hylocomium splendens*: variation associated with altitude, net primary production and soil chemistry. *Environmental Pollution*, 116: 129–135.

Gerdol, R., Bragazza, L., Marchesini, R., Alber, R., Bonetti, L. and Lorenzovi, G., 2000. Monitoring of heavy metal deposition in Northern Italy by moss analysis. *Environmental Pollution*, 108: 201–208.

Imperato, M., Admo, P., Naimo, D., Arienzo,M., Stanzione, D., Violant, P., 2003. Spatial distribution of heavy metals in urban soils of Naples city Italy. *Environmental Pollution*, 124: 247–256.

Karmer, C.Y., 1956. Extension of multiple range tests to group means with unequal numbers of replication. *Biometrics*, 12: 307–310.

Makholm, M. M. and Miladenoff David, J., 2005. Efficacy of a biomonitoring (moss bag) technique for determining element deposition trends on a mid-range (37 Km) scale. *Environmental Monitoring and Assessment*, 104: 1–18.

Otvos, E., Pazmandi, T. and Tuba, Z., 2003. First national survey of atmospheric heavy metal deposition Hungary by the analysis of mosses. *Science of the Total Environment*, 309: 151–160.

Pearson, J., Wells, D.M., Seller, K.J., Bennett, A., Soares, A., Woodall, J. and Ingroyille, M.J., 2000. Traffic exposure increase natural 15 N and heavy metal concentrations in mosses. *New Phytology*, 147: 317–326.

Saxena, D.K. and Saxena, A., 2000. Uptake of metals in Plagiochasma and their uses in pollution monitoring. *Geophytology*, 28(1 and 2): 129–137.

Saxena, D.K., 2001. Biomonitoring of Pb, Ni, Cr, Hg with the help of bryophytes in Nainital. *Perspective in Indian Bryology*, pp. 155–170.

Snedecor, G.W. and Cochran, W.G., 1967. *Statistical Methods.* Iowa State University, U.S.A. Oxford and I.B.H. Publishing Co., New Delhi.

Stefano, L. and Bononi, I.L., 2000. Lichens and mosses as biomonitors of trace elements in areas with thermal springs and fumarole activity (Mt. Amiata, Central Italy). *Chemosphere*, 41: 1333–1336.

Ethnobotanical Wisdom and Microbial Studies on Medicinal Plants (2011) *Pages* **151–154**
Editors: **D.R. Khanna, Ashutosh Gautam, R. Bhutiani, Gagan Matta, Anil Dhiman**
Published by: **DAYA PUBLISHING HOUSE, NEW DELHI**

Chapter 26

A Study *in vitro* Antibacterial Potential of *Riccia sp.* (Bryophyte) Plant Extract

P. Kaushik and Vishal Upadhyay
Department of Botany and Microbiology,
Gurukula Kangri Vishwavidyalaya, Haridwar – 249 404

ABSTRACT

Effect of alcoholic extract of *Riccia sp.* (Bryophyte) has been tested for its antibacterial activity against bacterial strains of *Escherichia coli, Sarcina lutea* and *Vacillus megaterium*. All the plant parts showed antibacterial property. Effective inhibitory zone produced by undiluted extract was nearly as large as that produced by 1 : 10000 concentration of the antibiotic oxytetracyclin.

Keywords: *Bryophyte, Riccia, Antibacterial.*

Introduction

The bryophytes comes in lower group of plants due to its least medicinal value for humans but except some of their physiological ecology. Evaluation of plant extract for their antimicrobial activity has been done by several workers. Ansari 1995 studied effect of alcoholic extract of shoot extract of *Pholiciota articulate* Lindl., against *Bacillus subtilis, Staphylocoecus aureus* and *Klebsiella pneumoniae*. Information regarding drug plants and micro-organisms dates back to Vedic period (Kaushik 1983, 1985 and 1988), Antimicrobial activity of various plants parts of *Aerya persica* has been tested against human pathogenic bacterial strains and pathogenic fungal species by Gehlot and Bohra 1998, earlier literature on antibacterial potential of plants has been reviewed.

Investigations into the folkloric antimicrobial activity of *Landolphia owrrience* against *Escherichia coli, Bacillus subillis* and *Candida albicans* (Ebi and Ofoe fule 1997), chemical composition and antimicrobial activity of *Croton urucurana Baillon* (Euorbiaceae) against *Staphylococcus aureus* and *Salmonella typhinuriun* and antimicrobial properties of *Piper aduncum* against the AIDS related pathogens. *Candida albicans, Cryptococcus neoformans* and *Cobacterium intracellulare* have already been studied. Evaluation of antioxidative and antimicrobial activity of essential oils of *Salvia officinalis* and *Thymus vulgaris* was seen by Battistuta *et al.* (1996). Recently three species of orchids and garlic were studied for the presence of alkaloids and antibacterial potential by Kaushik and Kishore (1997).

A large nmnber of bryophytes have been used in China (Ding 1980), Europe (Gamier *et al.*, 1969) and North America (Ando and Matsuo, 1984). Banerjee and Sen (1979) reported antibiotic activity against some bryophytes and Dikshit *et al.* (1982) reported antifungal activity of some bryophytes against some human pathogens.

Materials and Methods

For the preparation of plant extract, plant parts were first washed 2-3 times with tap water then again with sterilized distilled water. Finally the surface sterilization was done with 90 per cent alcohol, 100 gm of thalli of *Riccia* sp. were crushed in ware blender resulting the formation of a paste, which was mixed in 250 ml of absolute alcohol. This method was also, adopted by Ghanaksha and Kaushik (1999). Alcoholic extract so prepared was allowed to evaporate at room temperature until 80 ml was left. This extract was squeezed through double layer musline cloth and filtered through Whattman filter paper no. 42 and was centrifuged at 5000 r.p.m. for 20 minutes and then sterilized by passing through 0.2 microne disposable filters for the evaluation of antibacterial effect of extracts on growth on *Escherichia coli, Sardna lutea* and *Bacillus migaterium* organisms.

Ananthanarayana and Paniker (1996) gave the disc diffusion method using filter paper discs, 5 mm in diameter, charged with appropriate concentration of the extract. The discs are stored dry in cold. A suitable dilution of a broth culture or a broth suspension of the test bacterium is flooded on the surface of a solid medium (nutrient agar). The plate is titled to ensure uniform spreading and the excess broth pipetted off. Inoculation may also be performed by spreading with swabs. After drying the plates (37°C for 30 minutes), extract discs are applied with sterile forceps. After overnight incubation. the degree of sensitivity is determined by measuring the zone of inhibition which is influenced by variety of factors, such as diffusibility of the drug or extract, the nature and composition of medium, its thickness, and presence of inhibitory or stimulatory substances, pH and time of incubation. It is also necessary to check the potency of the discs periodically using as control a standard bacterium of known sensitivity, such as *Staphylococcus aureus* oxford strain (NCTC,6571). Disc with antibiotic Oxytetracyclin (1 : 1000) concentration, distilled water and absolute alcohol served as control. Several replicates of the experiment were set.

Supporting Data on the Nature of Extract and its Chemical Characterization

This study is concerned with an examination of the antibiotic activity of 52 species (in 40 genera) of the bryophytes. The plants were extracted in water, methanol, ethanol, ether and acetone and tested against 12 microorganisms, including 3 Gram positive (*Staphylococcus aureus*, apenicillin-sensitive and *penicillin-resistant strain, Sarcina lutea* and *Bacillus subtilis*) 5 Gram negative (*Escherichia coli, Salmonella typhi, Vibrio cholera, Klebsiella pneumoniae* and *Pseudomonas aeruginosa*) one acid fast bacterium (*Mycobacterium phlei*) and 3 fungi (*Curvularia lunata, Aspergillus niger* and *Helminthosporium oryzae*). Solubility data and antibiotic spectra of the active plants indicate the occurrence of a variety of antibiotic substances among bryophytes. The following work had been done on the bryophytes (Banerjee and Sen, 1979).

Table 26.1

Plants	*Locality Altitude*	*Test Organisms*	*Solubility of Active Principles*	*Insoluble in*
Anthoceroserectus	Mussorie 2000 m	Mp 14-15, St 10-12 Pa 18-21	Me, E, Ac	W, Et
Notothylas indica	Kalyani 9 m	St 12-15,	Me, E, Ac	W, Et
Campylopus lactus	Shillong 1400 m	St 10-13, Ve 12-20	Me, E, Et	W, Ac
Philonotis falcata	Mussorie 2100 m	Bs-12	Me	W, E, Et, Ac
Rhodobryum giganteum	Darjeeling 1650 m	Mp 11-12	Me, E, Ac	W, Et
Mnium sp.	Mussorie 2100 m	Mp 14-16, Bs-10 St 10-11, Ve-10	Me, Bw, Ac	E, Et
Marchantia sp.	Pachmarhi 1100 m	Mp 12, Sa-11, St 18-19 Ve-12	Me, E	W, Et, Ac
Marchantia sp.	Jalpaiguri 1100 m	St 16-17, Pa 12-14	Me, E	W, Et, Ac
M. palenea bertol	Darjeeling 1650m	Sa 15-30, SaR 16-25 Bs 15-18, Mp 15-18 St 14-19, Ve 15-20 Kp 20-22	Me, E, Et, Ac	W
M. Palmata Vees	Darjeeling 22503 m	Sl 14-16, Mp 14-16 St 12-13, Kp 13-14	Me, E, Ac	W, Et
Riccia discolor	Kalyani 9 m	S1 18	Me	W, E, Et, Ac
Reboulia hemisphaerica	Darjeeling 2200 m	Sa 10-12, Bs 15-17 St 10, Kp 18-20	Me, E, Ac	W, Et
Targionia hypophyta	Darjeeling 1650m	Mp 14, Kp 20	Me	W, E, Et, Ac
Asterella sp.	Pachmarhi 1100 m	Sl 20	Me	W, E, Et. Ac

Antibiotic activity of bryophyte extracts against bacteria indicate the diameter of inhibition zone in mm when the active principle was assayed by agar cup method. Inhibition zone represents difference between control value and that of the extract. S. *autrus* (Sa), Penicillin-resistant *S. aureus* (SaR), *S. lutea* (Sl), *B. sutilis* (Bs), *M. phlei* (Mp), *E. coli* (Ec), *S. typhi* (St), *V. cholerae* (Vc), *K. pneumoniae* (Kp), *P. aeruginosa* (Pa), *Acetone* (Ac), Ethanol (E), Ether (Et), Methanol (Me), Water room temperature (W).

Results and Discussion

The present investigation showed that the alcoholic extract of Riccia (Bryophyte) substantially inhibited the growth of *Escherichia coli, Sarcina lutea* and *Bacillus megaterium* were nearly same when observed at room temperature.

An interesting feature of bryophytes is that they relatively free from attack by parasitic microorganisms, there are, however a few reports of infection of bryophytes by fungi but the infected parts constitute only a small fraction of the large number of bryophyte solubility far (Bapna, 1962), by the methanol extract of the Indian liver wort *Marchantia polymorpha*, two new cyclic bis (bibenzyls) isomarchantin C and isoriccardin C, and a new phenanthrene, derivative, 2-hydroxyl 3,7-

dimethoxyphenanthrene, were isolated together with the previously known cyclic bis (bibenzyls) marchantin A, C, D and E riccardin C and perrottetin E and their structure were established by extensive 'II NMR' spectroscopic examination isomarchantin C, isoriecardin C, marchantimn C and G and riccardin C were also isolated from the *Indian M. palmate*. The two *Marchantia* sp. are chemically quite similar (Asakawa 1984). *Marchantia polymorpha* is a rich source of cyclic bis (bibenzyls) and possess a number of interesting biological activities *e.g.* diuretic (Garnier *et al.*, 1969), antifungal (Asakawa 1984) and dermatitic.

References

Ananthanarayana, R. and Jayaram Paniker, C.K., 1996. *Textbook of Microbiology*, 5th edn. Orient Longman, pp. 578–580.

Ansari, M.M., 1995. Control of sheath blight of rice by plant extract. *Indian Phytopath.*, 48(3): 268–270.

Asakawa, Y.J., 1984. *Hattori Bot. Lab.*, 56: 125.

Banerjee, R.D. and Sen, S.P., 1979. Antibiotic activity of bryophytes. *The Bryologist*, 82(2): 141–153.

Bapna, K.R., 1962. Some fungi on *Riccia* sp. *The Bryologist*, 65: 47–51.

Battistuta, F., Candido, E., Ciola, L., Giomo, A., Comi, G.I., Conte, L.S. and Zironi, R., 1996. Evaluation of antioxidative and antimicrobial activity of essential oils of *Salvia officinalis* and *Thymus vulgaris*, *Atti convegno internazionale, Coltivazione miglor amento dipiante officinali*, Trento, Italy, 2–3, Gingno. 1994. P.B. Trento, Italy, instituto sperimentale per 1. Assestamento Forestale e per 1. Alplcoltura, p. 481–486.

Ebi, G.C. and Ofoefule, S.I., 1997. Investigation into Folkloric antimicrobial activities of *LandoJphia awrrience*. *Phytotherapy Research*, 11(2): 149–151.

Garnier, G., Bezaniger, B.I. and Debraux, G., 1969. *Resources Medicinales de Laboratory Flora Francaile*. Vol 1: 78, Vigot Freres, Paris.

Gehlot, D. and Bohra, A., 1998. Antimicrobial activity of various plant parts extract of *Aerva persico*. *Ad. Plant Sci.*, 11(1): 109–111.

Ghanaksha, A. and Kaushik, P., 1999. Antibacterial effect of *Rhyncostylis retusa* Bl. (Orchidaceae): A study *in vitro*. *Adv. Plant Sci.*, 12(11): 593–598.

Kaushik, P., 1983. *Ecological and Anatomical Marvels of the Himalayan Orchids*. Today and Tomorrow's Printer's and Publisher, New Delhi.

Kaushik, P., 1985. Glimpses of medical botany in Atharvareda. *The Vedic Path.*, 48: 64–67.

Kaushik, P., 1988. *Indigenous Medicinal Plants including Microbes and Fungi*. Today and Tomorrow's Printer's and Publisher, New Delhi, pp. 243.

Kaushik, P. and Nand Kishore, 1997. *In vitro* antibacterial activity of *Cleisostoma micranthum* K and P. *Advances in Plant Sciences*, 9(11): 81–84.

Kaushik, P. and Upadhyay, Vishal, 1999. Antibacterial activity of Garlic (*Allium sativum*) plant extract. In: *Sustainable Ecosystem and Environment*, (Eds.) D.R. Khanna, Ashutosh Gautam and Aditya Goutam, pp. 129–132.

Ethnobotanical Wisdom and Microbial Studies on Medicinal Plants (2011) *Pages* **155–159**
Editors: **D.R. Khanna, Ashutosh Gautam, R. Bhutiani, Gagan Matta, Anil Dhiman**
Published by: **DAYA PUBLISHING HOUSE, NEW DELHI**

Chapter 27

A Comparison of *Phyllosphere mycoflora* of Healthy and Diseased Leaves of Potato

Navneet
Department of Botany and Microbiology,
Gurukul Kangri University, Haridwar – 249 404, Uttranchal

ABSTRACT

Phyllosphere mycoflora of healthy and late blight infected leaves of same age of two cultivars of potato *i.e.* Kufri jyoti and Kufri chandramukhi were carried out on 12th of January, 1989. In 1988-89 crop season late blight of potato appeared on 6th of January, 1989. Qualitatively the *Phyllosphere mycoflora* of healthy and diseased leaves of both the cultivars consists of 27 fungal species. 14 and 17 fungi belonging to 9 and 10 genera respectively were isolated from healthy and diseased leaves of Kufri chandramukhi. While 18 fungi belonging to 12 and 10 genera respectively were isolated from healthy and diseased leaves of Kufri jyoti by dilution plate, leaf washing and leaf impression method. Quantitatively the *Phyllosphere mycoflora* of diseased leaves was rich as compared to healthy ones. The density of *P. infestans* sporangia/cm^2 was 4604 and 7056 on KJ and KCM leaves respectively. *Alternaria alternata, A. tenuissima, Aspergillus niger, Cladosporium cladosporioides, Epicoccum nigrum, Mucor hiemalis* and *Penicillium* were associated with both healthy and diseased leaves of both the cultivars. Diseased leaves support more *Aspergilli* and fast growing fungi as compared to healthy ones, *Bipolaris indica* and *Rhizopus oryzae* were restricted to diseased leaves of both the cultivars while *Aspergillus terreus* to diseased Kufri chandramukhi leaves and *Trichoderma longibrachiatum* to diseased Kufri jyoti leaves only. *Curvularia lunata, Nigrospora sphaerica, Phoma glomerata* and *Periconia saraswatipurensis* were isolated from healthy leaves of Kufri jyoti and *Periconia echinocholae* from healthy leaves of Kufri chandramukhi only.

Keywords: *Mycoflora, Phyllosphere, Kufri jyoti, Kufri chandramukhi, Potato.*

Introduction

Leaf surface provides a suitable habitat for the growth, reproduction and multiplication of microorganisms. The epiphytic microorganisms on the leaf surface play an important role in controlling the incidence of plant disease. They utilize the energy source present on the leaf surface which ultimately restrict the growth of the pathogen. Cook and Baker 1983 have written it clearly that any pre-emption of a critical nutrient by non-pathogenic members of the phylloplane flora will probably be reflected in less growth by the necrotrophic parasites and hence less damage of the leaf. The present study deals with the *Phyllosphere mycoflora* of healthy and diseased leaves of potato.

Materials and Methods

The phyllosphere studies were conducted on two potato cultivars Kulfri chandramukhi (KCM) and Kufri jyoti (KJ) which were sown according to randomized block design on 20th of October, 1988. *Phyllosphere mycoflora* was determined by three methods namely, dilution plate method (Waksman 1927), leaf washing and leaf impression method on 12th of January, 1989.

Results and Discussion

In 1988-89 crop season late blight of potato appeared on 6th of January, 1989. Comparative studies of healthy and late blight infected leaves of same age were carried out on 12th of January, 1989. Qualitatively the *Phyllosphere mycoflora* of healthy and diseased leaves of both the cultivars consist of 27 fungal species. 14 and 17 fungi belonging to 9 to 10 genera respectively were isolated from healthy and diseased leaves of Kufri chandramukhi. While 18 fungi belonging to 12 and 10 genera respectively were isolated from healthy and diseased leaves of Kufri jyoti by dilution plate, leaf washing and leaf impression method. Quantitatively the *Phyllosphere mycoflora* of diseased leaves was rich as compared to healthy ones. The density of *P. infestans* sporangia/cm^2 was 4604 and 7056 on KJ and KCM leaves respectively. *Alternaria alternata, A. tenuissima, Aspergillus niger, Cladosporium cladosporiodes, Epicoccum nigrum, Mucor hiemalis* and *Penicillium* were associated with both healthy and diseased leaves of both the cultivars. Diseased leaves support more *Aspergilli* and fast growing fungi as compared to healthy ones, *Bipolaris indica* and *Rhizopus oryzae* were restricted to diseased leaves of both the cultivars while *Aspergillus terreus* to diseased Kufri chandramukhi leaves and *Trichoderma longibrachiatum* to diseased Kufri jyoti leaves only. *Curvaularia lunata, Nigrospora sphaerica, Phoma glomerata* and *Periconia saraswatipurensis* were isolated from healthy leaves of Kufri jyoti and *Periconia echinocholate* from healthy leaves of Kufri jyoti and *Periconia echinocholate* from healthy leaves of Kufri chandramukhi only (Table 27.1).

Maximum number of fungal species was isolated by leaf impression method followed by leaf washing and least by dilution plate method. Leaf impression method was used for qualitative estimation of mycoflora as this method provided data about surface distribution of saprophytic leaf population (Dickinson 1971, Sharma *et al.*, 1974). The combination of all these techniques gave sufficient data regarding the number, nature and occurrence of fungi on leaves. The deuteromycetous forms occur in larger proportions than other groups and constituted a major portion of the total fungal species. Consistent presence of *Cladosporium cladosporioides, Alternaria alternata, Aspergillus niger* and *Penicillium eyelopium* on leaf surface is more or less in accordance with the observation of Sharma and Mukherji 1972, Narula and Mehrotra 1981, Bansal *et al.*, 1988. Prevalence of these forms was in all probability due to their fast growing and high sporulating abilities and also their capacity to utilize a wide range of substances present on leaf surface.

Table 27.1: Distribution of *Phyllosphere mycoflora* of Healthy and Diseased Leaves of KCM and KJ Cultivars of *Solanum tuberosum* Employing Various Methods of Isolation

Sl.No.	Microorganisms	Kufri Chandramukhi						Kufri Jyoti					
		Healthy			Diseased			Healthy			Diseased		
		DP	LW	LI	DP	LW	LI	DP	LW	LI	DP	LW	LI
1.	*Alternaria alternata*	3.84	1.32	+	20.0	1.89	+	16.66	1.50	+	21.42	4.84	+
2.	*A. tenuissima*	–	–	+	–	–	+	–	–	+	–	–	+
3.	*Aspergillus candida*	–	0.52	–	–	0.53	+	–	–	–	3.57	0.60	–
4.	*A. flavus*	–	–	–	3.83	0.26	–	5.55	1.20	+	–	1.20	–
5.	*A. niger*	19.23	–	–	6.66	0.53	+	11.11	1.50	+	7.14	2.11	–
6.	*A. versicolor*	–	–	–	–	–	–	–	–	–	10.74	–	–
7.	*A. terreus*	–	–	–	–	–	+	–	–	–	–	–	–
8.	*Bipolaris indica*	–	–	–	–	–	–	–	–	–	–	0.60	–
9.	*Cladosporium cladosporioides*	26.92	1.86	+	33.33	3.78	+	33.33	4.54	+	32.14	6.66	+
10.	*C. herbarum*	–	–	–	–	–	–	5.55	–	–	3.57	1.20	–
11.	*Curvularia lunata*	–	–	–	–	–	–	–	0.30	–	–	–	–
12.	*Epicoccum nigrum*	11.53	1.06	+	–	0.53	+	–	1.20	+	–	0.30	+
13.	*Fusarium equiseti*	3.84	?	+	–	–	+	–	0.60	+	–	–	–
14.	*Gilmaniella humicola*	7.69	0.26	–	3.33	–	–	–	–	–	–	0.30	–
15.	*Mucor hiemalis*	–	–	+	–	–	+	–	0.60	+	–	0.60	+
16.	*Nigrospora sphaerica*	–	–	–	–	–	–	–	–	+	–	–	–
17.	*Penicillium citrinum*	3.84	–	–	6.66	–	–	–	–	–	7.14	–	+
18.	*P. cyloplum*	19.23	1.32	+	26.66	1.62	+	22.22	1.50	–	10.71	2.11	+
19.	*P. oxalicum*	3.84	1.60	–	–	1.08	–	–	0.60	–	–	–	–
20.	*P. purpurogenum*	–	–	–	–	–	+	5.55	–	–	–	–	+

Contd...

Table 27.1–Contd...

Sl.No.	Microorganisms	Kutri Chandramukhi						Kutri Jyoti					
		Healthy			Diseased			Healthy			Diseased		
		DP	LW	LI	DP	LW	LI	DP	LW	LI	DP	LW	LI
21.	*Periconia echinocholae*	–	0.52	–	–	–	–	–	–	–	–	–	–
22.	*P. saraswatipurensis*	–	–	–	–	–	–	–	0.60	–	–	–	–
23.	*Phoma glomerata*	–	–	–	–	–	–	–	–	+	–	–	–
24.	*Rhizopous oryzae*	–	–	–	3.33	–	–	–	–	–	3.57	–	–
25.	*Trichoderma koningii*	–	–	–	–	–	–	–	0.30	–	–	–	
26.	*T. longibrachiatum*	–	–	–	–	–	–	–	–	–	–	0.60	+
27.	Sterile form	–	0.26	–	–	–	–	–	0.30	–	–	–	–
	Total	9	9	7	8	8	12	7	13	10	9	12	9

DP: Dilution Plate Method (Figures indicate percentage of abundance); LW: Leaf Washing Method (Figures indicate density *i.e.* hundred/gm dry weight); LI: Leaf Impression Method; +: Present: –: Absent.

References

Bansal, B.K., Navneet and Mehrotra, R.S., 1988. *Phyllosphere mycoflora* of wheat in relation to leachates and its resistance to *Helminthosporium blight*. *Indian Phytopathology*, 41: 398–405.

Cook, R.J. and Baker, K.F., 1983. *The Nature and Practice of Biological Control of Plant Pathogens*. The American Phytopathological Society, St. Paul. Minnesota, 539 pp.

Dickinson, C.H., 1971. Cultural studies of leaf saprophytes. In: *Ecology of Leaf Surface Microorganisms*, (Eds.) T.F. Preece and C.H. Dickinson. Academic Press, London, pp. 129–137.

Narula, K.L. and Mehrotra, R.S., 1981. *Phylloplane microflora* of *Colocasia esculenta* in relation to *Phytohthara colocasiae*. *Geobios*, 8: 152–156.

Sharma, K.R., Bahera, M. and Mukerji, K.G., 1974. A comparison of three techniques for the assessment of phylloplane microbes. *Trans. Mycol. Soc., Japan*, 15: 223–233.

Sharma, K.R. and Mukherji, K.G., 1972. Succession of fungi on cotton leaves. *Ann. Inst. Pasture*, 122: 425–454.

Sharma, K.R. and Mukherji, K.G., 1976. Microbial ecology of *Scsamum orientale* L. and *Gossypium hirsutum* L. In: *Microbiology of Aerial Plant Surfaces*, (Eds.) C.H. Dickinson and T.F. Preece. Academic Press, London, pp. 375–390.

Waksman, S.A., 1927. *Principles of Soil Microbiology*. Williams and Wilkins, Baltimore, pp. 897.

Ethnobotanical Wisdom and Microbial Studies on Medicinal Plants (2011) *Pages* **160–164**
Editors: **D.R. Khanna, Ashutosh Gautam, R. Bhutiani, Gagan Matta, Anil Dhiman**
Published by: **DAYA PUBLISHING HOUSE, NEW DELHI**

Chapter 28

Citric Acid Production from Alkali Treated Wheat Straw by Solid State Fermentation

R.K. Tenguria[1], Krishna Jha[2] and Priti Satalkar[1]
[1]Motilal Vigyan Adarsh Mahavidyalaya, Bhopal
[2]Central Institute of Agricultural Engineering, Nabibagh, Bhopal

ABSTRACT

Citric acid production from alkali treated wheat straw by solid state fermentation has been attempted. Alkali treatment was found to have increased the lignocellulose hydrolysis to the extent of 25 per cent. Almost 7.11 g of citric acid production could be attained from 100g alkali treated straw on dry weight basis. Various workers have reported prefermentation to enhance the citric acid yield by four folds from soy residue.

Keywords: *Citric acid, Alkali treatment, Wheat straw.*

Introduction

Lignocellulosic wastes are considered as potential source for production of many industrial products such as citric acid (Panda *et al.*, 1984; El-Abayad *et al.*, 1992; Khare *et al.*, 1995), Enzymes (Elshafei *et al.*, 1990; Hayashi *et al.*, 1992; H. Kittikun and Tani, 1986), Vitamins (Kinoshita *et al.*, 1985), Production of ethanol from lignocellulosic wastes, particularly for mixing with petrol holds a great promise. Out of the various important constituents of straw, lignin is known to adversely affect the hydrolysis of cellulose (Fan *et al.*, 1987; Christakopoulos *et al.*, 1991). Several investigations have thus been carried out on physical and chemical treatments of lignocellusosic wastes to influence their hydrolysis (Thompson *et al.*, 1992; Christakopoulos *et al.*, 1991). Alkali treatment is one of the most frequently used chemohydrolysis methods for increasing free sugar and digestibility of lignocellulosic substrates (Aiello *et al.*, 1996). In the present investigation an attempt has been made to study the citric acid production from wheat straw treated with alkali.

Materials and Methods

Preparation of Starter Culture

Citric acid producing *Aspergillus niger* (NRRL 330) culture was obtained from Northern Regional Research Laboratory, Peoria, USA. It was grown and maintained on malt extract agar (Himedia, India). The culture was preserved in refrigerator at 4°C. The medium for cultivation was prepared with 0.22 g per litre of ammonium tartrate as the nitrogen source. As per the requirements the Erlenmeyer flask of 100 ml capacity was added with 4.5 ml of the basal medium containing 30 mM 2,2 dimethyl succinate to buffer the cultures at 4.3–4.5 and 0.5 ml of fungal mycelium. Conidiospores were aseptically harvested from a 14 days old culture slant by suspending in 10 ml sterile distilled water containing 0.01 per cent Tween 80. The conidiospore suspension was diluted to obtain a population of approximately 10^7 per ml. One ml culture was used per 100 g substrate for all experimental purposes.

Alkali Treatment of Wheat Straw

The dried straw was treated with sodium hydroxide @ 5g/100g dry substrate for 30 min. at ambient temperature and at autoclaving temperature of 121°C. After the alkali treatment the substrate was repeatedly washed with dilute hydrochloric acid until the final pH was 4.5. Excess water was drained until the final moisture of the substrate was 40 per cent.

Fermentation for Citric Acid Production

50 g of treated straw was loosely packed in perforated autoclavable LDPE bags and sterilized at 15 psi for 30 min. After cooling they were inoculated with 1 ml spore suspension of *Aspergillus niger* containing 1.0×10^5 spores per ml. A control with untreated straw runs simultaneously.

Determination of Reducing Sugars

2 gram samples were extracted in 10 ml of 70 per cent aqueous ethanol by finely grinding in pestel mortar. The contents were then centrifuged at 5000 rpm for 20 minutes. The pellets were washed thoroughly in same solution twice again followed by centrifugation. All the supernatants were pooled and evaporated to dryness using flash evaporated. The contents were re-dissolved in 5 ml of 70 per cent aqueous ethanol and used for estimation of carbohydrates using the standard method according to Nelson (1944).

Determination of Cellulose, Hemicellulose and Lignin

Fresh as well as treated, 1 g samples were suspended in 10 ml of distilled water to measure the pH. The remaining sample was dried and grind to fine powder using pestle mortar to pass through 20 mesh screen. Cellulose, hemicellulose and lignin were detennined by fibre analysis using the method of Van Soest (1963) and Van Soest and Wine (1967). The method for cellulose digestibility was adapted from McQueen and Van Soest (1975). Triplicate samples (50 mg) were incubated in 5 ml of a solution of cellulase extract at 40°C for 72 hours without agitation. After incubation the reducing sugar content were determined by Nelson's method (1944).

Determination of Citric Acid

Two grams substrate was extracted in 10 ml distilled water. Supernatant was collected by centrifugation at 5000 rpm for 20 min. citric acid was estimated following the pyridine-acetic anhydride method (Marier and Boulets, 1958).

Results and Discussion

The results in Figures 28.1 and 28.2 show the effect of alkaline hydrolysis at ambient and 121°C temperatures. The changes in reducing sugar, cellulose and hemicellulose indicate that chemical hydrolysis of lignocellulosic wheat straw is far more efficient method for release of fermentable sugars than microbial or enzymatic method. Pretreatment of wheat straw with 1 M NaOH for different periods showed that conversion of cellulose was complete in 6 hours. Alkali pretreatment also resulted in delignification of the substrate thus making the cellulose more accessible.

The alkali pretreated substrate was further inoculated with *Aspergillus niger* for citric acid production. The results in Figure 28.3 indicates almost 8 fold increase in the citric acid yield. While in untreated wheat straw only 0.94 g citric acid was produced per 100 g dry straw, the alkali treated straw at 121°C was found to have produced as high as 7.11 g citric acid per 100 g dry substrate. Khare *et al.* (1994) have reported a maximum yield of 5.1 g per 100 g dry substrate using *Aspergillus terreus*

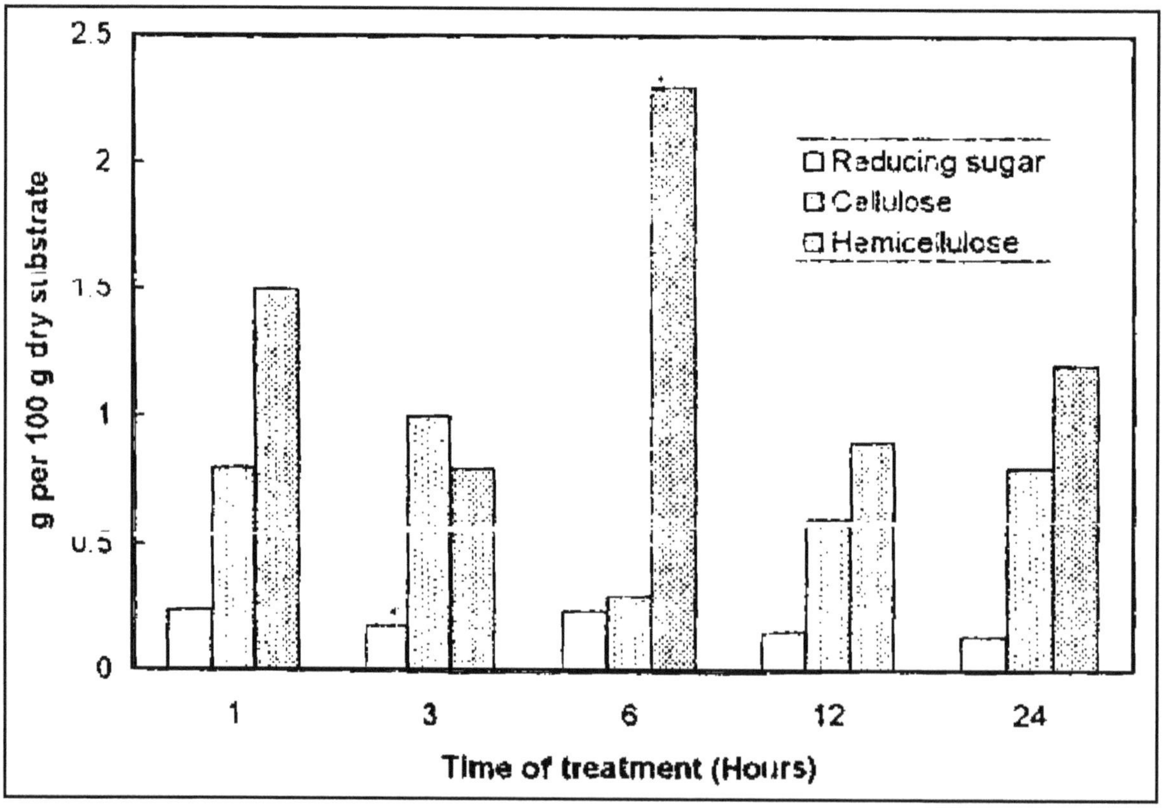

Figure 28.1: Effect of Alkali Treatment on Straw Hydrolysis at Ambient Temperature

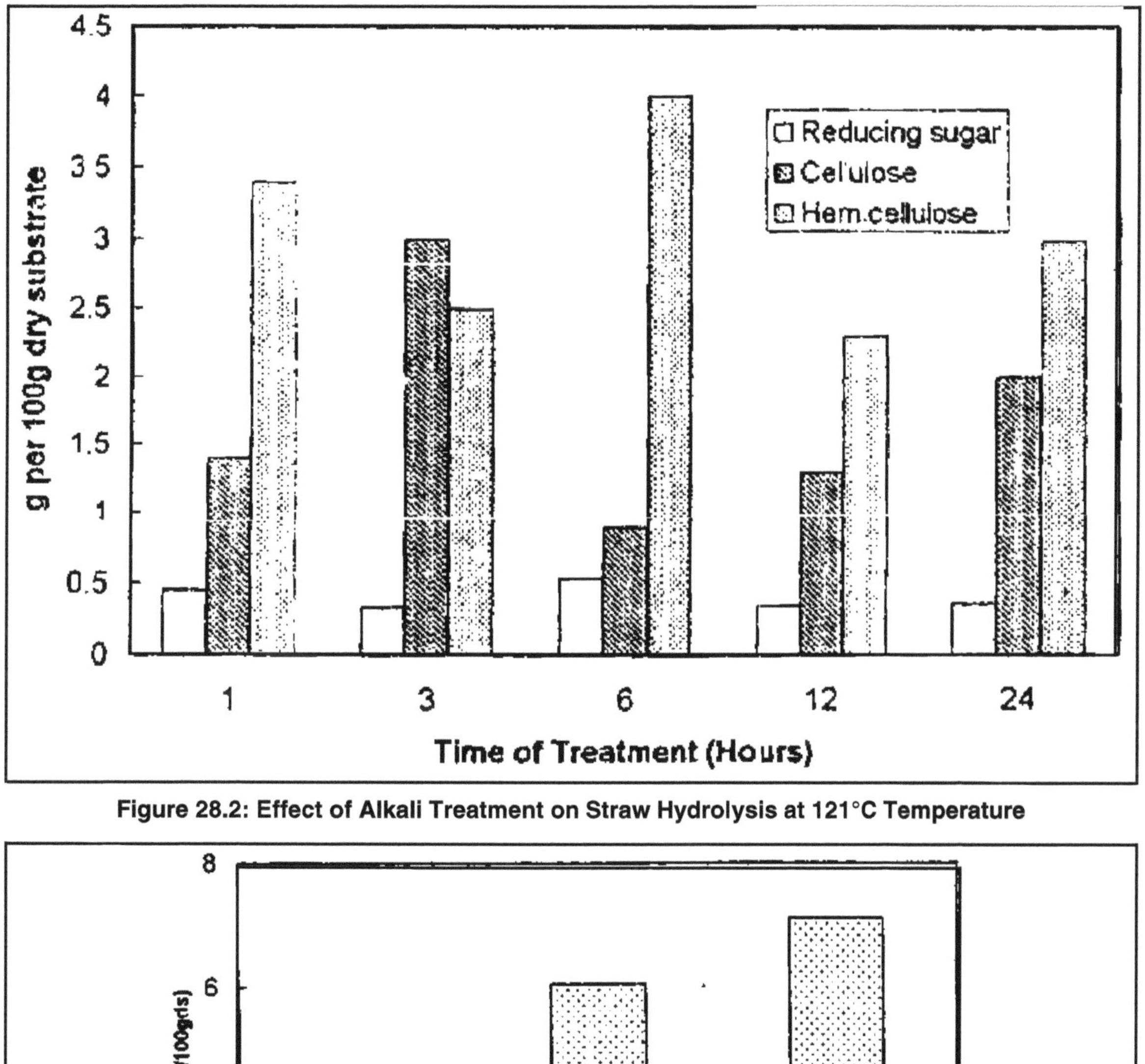

Figure 28.2: Effect of Alkali Treatment on Straw Hydrolysis at 121°C Temperature

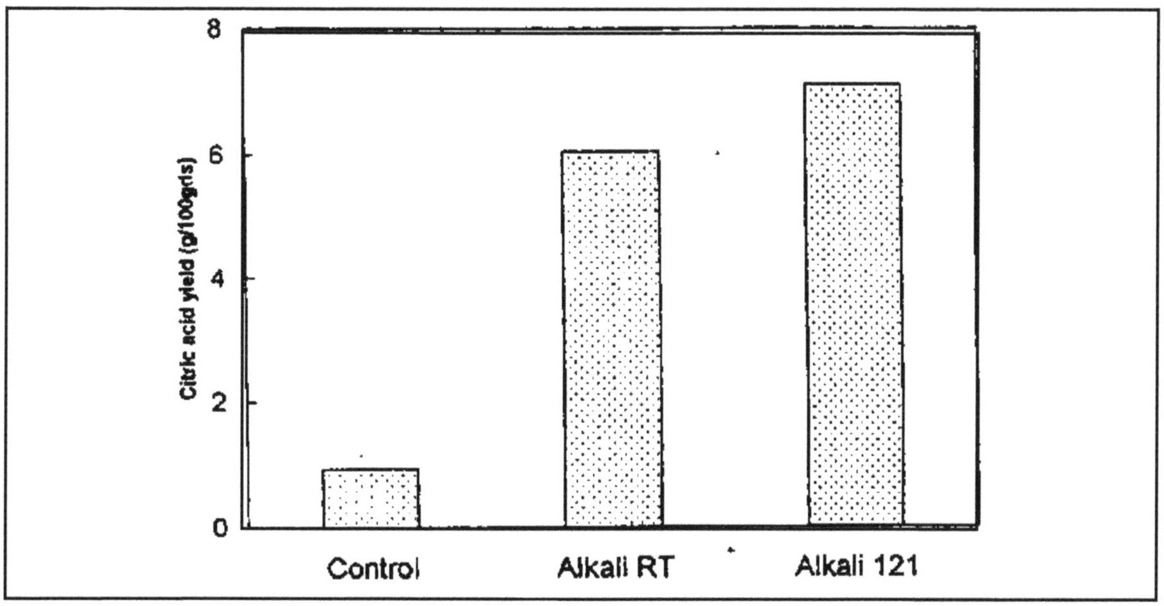

Figure 28.3: Citric Acid Production with Alkali Treated Wheat Straw

fermented soy residue as substrate. The study thus concludes that chemo thermal treatment is a better option for release of fermentable free than enzymatic treatment or microbial prefermentation.

Acknowledgements

The authors are grateful to the Head, Department of Botany, Principal, Motilal Vigyan Mahavidyalaya, Bhopal and Project Director, SPU Centre and Director, Central Institute of Agricultural Engineering, Bhopal who have kindly permitted to undertake this work at their laboratories.

References

Aiello, C., Ferrer, A. and Ledesma, A., 1996. Effect of alkaline treatments at various temperatures on cellulose and biomass production using submerged sugarcane bagasse fermentation with *Trichoderma reesi* QM 9414. *Biores. Technol.*, 57: 13–18.

Christakopoulos, P., Koullas, D.P., Kekos, D., Koukois, E.G. and Macris, B.J., 1991. Direct ethanol conversion of pretreated straw by *Fusarium oxysporum*. *Biores. Technol.*, 35: 207–300.

El-Abayad, M.S., Hamissa, F.A. and Gad, A.S., 1992. Treatment of beet molasses for citric acid production by a potent strain of *Aspergillus niger* van Tieghem. *Biores. Technol.*, 39: 19–197.

Fan, L.T., Gharpurey, M.M. and Lee, Y.H., 1987. *Cellulose Hydrolysis*. Springer-Verlag, Berlin.

H–Kittikun, A. and Tani, Y., 1986. Preliminary study on microbial production of lipase using soyabean residue from tofu making process. *Ann. Rep. IC Biotech.*, 9: 275–276.

Khare, S.K., Jha, Krishna and Gandhi, A.P., 1994. Single cell protein production by solid state fermentation okara (soy residue). *Poll. Res.*, 13: 417–420.

Kinoshita, S., H-Kittikun, A. and Pinthong, R., 1985. Production of riboflavin from waste of tofu (soy curd). *Ann. Rep. IC Biotech.*, 8: 322–324.

Marier, J.R. and Boulets, M.A., 1958. Direct determination of citric acid in milk with amlysis improved pyridine acetic anhydride method. *J. Dairy Sci.*, 41: 1683–1692.

McQueen, R.E. and Van Soest, P.J., 1975. Fungal cellulase and hemicellulase prediction of forage digestility. *J. Dairy Sci.*, 58: 1482–1491.

Nelson, N., 1944. A photometric adaptation of Somogyi method for determination of glucose. *J. Biol. Chem.*, 15: 375.

Panda, T., Kundu, S. and Majumdar, S.K., 1984. Studies on citric acid production by *Aspergillus niger* using treated sugar cane molasses. *Proc. Biochem.*, 19: 183–187.

Thompson, D.N., Chen, H.C. and Grethiein, H.E., 1992. Comparison of petreatment methods on the basis of available surface area. *Biores. Technol.*, 39: 155–163.

Van Soest, P.J. and Wine, R.H., 1967. Use of detergents in analysis of fibrous feeds. IV. Determination of plant cell wall constituents. *J. Assoc. Off. Agric. Chem.*, 50: 50.

Ethnobotanical Wisdom and Microbial Studies on Medicinal Plants (2011) *Pages* 165–175
Editors: D.R. Khanna, Ashutosh Gautam, R. Bhutiani, Gagan Matta, Anil Dhiman
Published by: DAYA PUBLISHING HOUSE, NEW DELHI

Chapter 29

Decolourisation of Textile-Dye-Containing Effluents Using Biofilm: A Case Study

Maihafizah Mohd. Zahari[1], Zaharah Ibrahim[1], Fauziah Othman[2], Madihah Md. Salleh[1], Nooraini Abdul Rashid[1], Adibah Yahya[1], Aishah Husin[1] and Chan Giek Far[1]

[1]Department of Biology, Faculty of Science, Universiti Teknologi Malaysia, 81300 Skudai, Johor Darul Takzim, Malaysia

[2]Microscopy and Microanalysis Unit, Laboratory of Enzyme Technology and Microbial Institute of Bioscience, Universiti Putra Malaysia, Serdang, Selangor Darul Ehsan, Malaysia

ABSTRACT

Bioremediation based on microbial technologies has been extensively used for treating coloured textile wastewater. In this research, the potential application of the indigenous and exogenous bacterial cultures, found as biofilm were investigated for colour removal. Initial study in the treatment of the textile wastewater showed that the mechanism involved in decolourisation was degradation, which was carried out *via* molecular technique involving amplification of the DNA sequence responsible for decolourisation using pure dye; Orange II. In laboratory scale experiment, the bacteria were grown as mixed culture in suspension and biofilm using shake flasks technique. Their abilities to decolourise textile wastewater were studied under semianaerobic conditions. Generally, bacteria in the form of biofilms were found to remove colour at faster rates compared to that of suspended cells. Evidence of biofilm formation during decolourisation of textile wastewater was also examined using SEM.

Introduction

The textile industry uses more than 100 billion gallons of water each year in its preparation and dyeing processes. In Malaysia, textile wastewater accounts for 22 per cent of the total volume of industrial wastewater (Rakmi, 1993). The textile wastewater has strong colour in the form of persistent organics and also variety of the other pollutants including chloride, ammonia, organic nitrogen, nitrate, phosphate and heavy metals such as Fe, Zn, Cu, Cr and Pb (McMullan *et al.*, 2001). Synthetic dyes have been used increasingly in textile and dyeing industries because of its cost effectiveness in synthesis,firmness, and variations in colour compared with natural dyes (Griffiths, 1984). Synthetic dyes also can be classified by their chromophores such as azo, antraquinone and indigo chromophores. Azo dyes used in the textile industry are constituted the largest group of over 10,000 commercial dyestuffs which account for the majority of the synthetic dyes (Villegas-Navarro *et al.*, 2001). Many studies indicated that most of the azo dyes were affected human health as they are highly toxic (McMullan *et al.*, 2001). Since azo dyes are relatively resistant to biological and chemical degradation, therefore, it makes colour removal in particular, a major interest of scientific research (Banat *et al.*, 1996). The presence of very small amount of dyes in water (less than 1 ppm for some dyes) is highly visible and affects the aesthetic merit, water transparency and gas solubility in lakes, rivers and water bodies. The removal of colour from wastewaters is often more important than the removal of the soluble colourless organic substances, which usually contribute the major fraction of the Biochemical Oxygen Demand (BOD). Methods for the removal of BOD from most effluents are fairly well established. However, dyes are more difficult to treat because of their synthetic origin and mainly complex aromatic molecular structures. Such structures are often synthesized to resist fading on exposure to sweat, soap, water, light or oxidizing agents and this renders them more stable and less amenable to biodegradation (Seshadri *et al.*, 1994). Despite the existence of a variety of chemical and physical treatment processes, bioremediation of textile effluent is still seen as an attractive solution due to its reputation as a low-cost, environment friendly, and publicly acceptable treatment technology (Banat *et al.*, 1996). A number of biological processes such as biosorption have been proposed as having potential application in removal of dyes from textile wastewater (Bustard *et al.*, 1998). Mechanisms involved in the colour removal of wastewater include adsorption, transformation or metabolisation of coloured textile effluent by bacterial cells. Microbial consortium in the form of biofilm has the ability to decolourise and metabolize dyes since the presence of intracellular mechanisms that will bring about the degradation or biosorption of dyestuffs (Watnick and Kolter, 2000). Biofilm usually composed of a mixed microbial population of cells growing on surfaces surrounded by exopolysaccharides (EPS). Consequently, in most natural environments, biofilm is the prevailing microbial lifestyle. The EPS secreted by various types of bacterial strains serve mainly to protect the bacteria against desiccation and predation, as well as to assist in adhesion to surfaces. The formation of EPS generally occurs in two forms: capsule, which is tightly bound to the surface of bacterium and slime, which is only loosely attached to the bacterium (Abrahamson *et al.*, 1996). EPS are also considered as the most immediate interfacial boundary between the bulk aqueous phase and the bacterial cells. EPS generally consist of a wide variety of macromolecular compounds including acidic polysaccharides and proteins, as well as lipids (Fang *et al.*, 2002). This study aimed at investigating the potential application of biofilm consisting of selected indigenous and exogenous bacteria in the bioremediation of colour from textile wastewater.

Most Recent Advances

Studies carried out at the research laboratories have resulted in the isolation of various mixed bacterial cultures capable of growth on several kinds of azo (Amaranth), diazo (Remazol Red) and reactive dyes (Synazol Blue, Synazol Yellow), both under aerobic and anaerobic conditions (Mohd Zahari *et al.*, 2004, Husin *et al.*, 2003). Furthermore, they also were capable of growing in filter sterilized textile wastewater (fstw) supplemented with glycerol 0.5 per cent (Mohd Zahari *et al.*, 2004). 26 different types of pure cultures were isolated from wastewater and biofilm formed in textile treatment ponds. Most of them were gram-negative bacteria, non-motile and produced exopolymeric substances (EPS). Referring to Figure 29.1, it was clearly shown that different concentration (mg/L) SF Red 3BS showed maximum absorption peak of 517.5 nm and 542.5 nm in fstw. The selection of SF Red 3BS was based on it being one of the commonest dye used in all steps of dye processing. However, the absorbance value of each concentration for this dye at both wavelengths is not shown. These isolates were further screened for colour removal using synthetic dye; SF Red 3BS (CDM + 100ppm dye) and filter sterilized textile wastewater (fstw) in separate experiments (Table 29.1).

Table 29.1: Screening on Colour Removal by Potential Isolates in SF Red 3BS in a Synthetic Medium (CDM) and Filter Sterilized Textile Wastewater (FSTW)

Types of Sample	*CDM + 100 ppm Dye After 3 Days Incubation*		*FSTW After 12 Hours Incubation*	
	Residual Colour (ADMI Unit)	*Colour Removal (per cent)*	*Residual Colour (ADMI Unit)*	*Colour Removal (per cent)*
Control	9320	0	406	0
Bacillus cereus	5840	37	146	64
Aeromonas hydrophila	740	92	28	93
Aeromonas caviae	800	91	51	87
Enterobacter aerogenes	1100	88	27	93
Citrobacter freundii	800	92	84	79
Shigella flexneri	1000	89	112	72

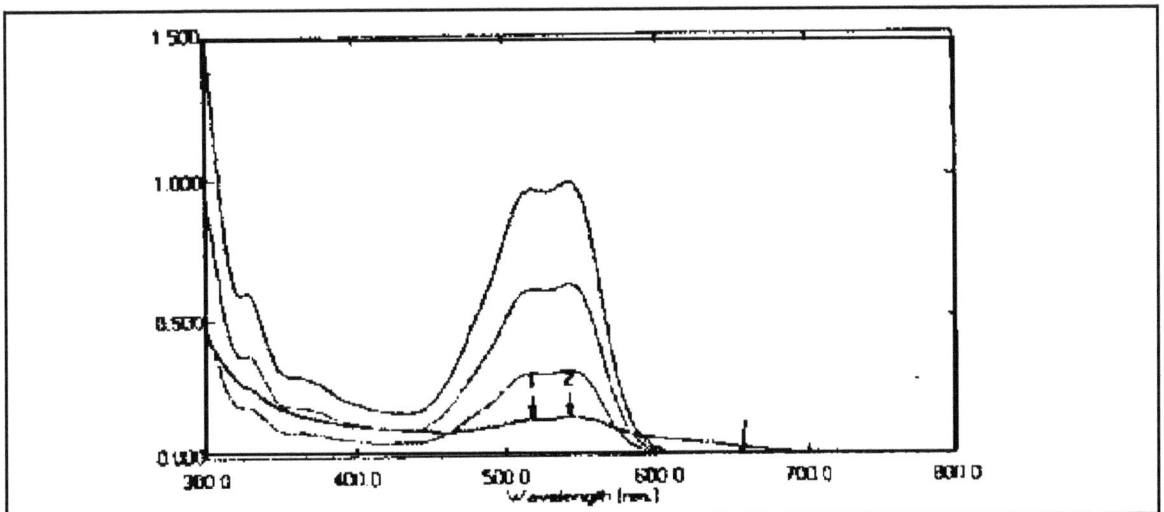

Figure 29.1: Spectrum of Filter Sterilized Textile Wastewater (FSTW) and Pure Dye Used for Screening (____) Filter Sterilized Textile Wastewater (IX); (.......) SF Red 3BS (100 mg/l); (____) SF Red 2BS (50 mg/l); (____) SF Red 3BS (25 mg/l); at two maximum peaks: (1) 517.5 nm; (2) 542.5 nm.

According to Figure 29.1, filter sterilized textile wastewater (FSTW) brought the maximum absorbance at 517.5 nm and 542.5 nm. It should be noted that, FSTW as originated from real wastewater, collected from the textile treatment pond with initial colour concentration of 519.0 ADMI units. In particular, colour intensity of FSTW was lesser than the real wastewater by 21.8 per cent; that is 405.9 ADMI units. The wastewater was further filtered using 0.2 µm pores filter membrane to remove indigenous microorganisms (culturable and unculturable microorganisms) by using of 0.2µm pores filter membrane. However, by doing so, some of the other wastewater parameters such as COD, TSS and colour intensity were generally decreased.

Numerous bacteria capable of dye decolourisation have been reported. Referring to Table 29.1 above, it was shown that from the 26 bacteria isolated, six bacteria were capable of decolourisation over 50 per cent; these were later identified as *Bacillus cereus, Aeromonas hydrophila, Aeromonas caviae, Enterobacter aerogenes, Citrobacter freundii* and *Shigella flexneri* using the 16S rRNA analysis. Effort to isolate bacterial cultures capable of degrading azo dyes started in 1970's with reports of a *Bacillus subtilis*, then *Aeromonas hydrophila* followed by the *Bacillus cereus* (Banat *et al.*, 1996). Similar observation achieved by Husin *et al.* (2003) where *Aeromonas hyrophila* and *Bacillus cereus* in the form of biofilm had shown the same ability to treat coloured textile wastewater. Banat *et al.* (1996) had also reported that isolation of such microorganisms proved to be a difficult task. Extended periods of adaptation in chemostat cultures are needed to maintain the performance of the isolates. As an example, a gram-positive bacteria; *Bacillus cereus*, showed a decrease of colour removal from 96 per cent (Husin *et al.*, 2003) to only 64 per cent after 12 h of incubation. Stock cultures were preserved at -80°C with addition of 12.5 per cent glycerol into FSTW. Even though they were isolated from enrichment cultures previously, they were maintained in FSTW as sole carbon sources in semi-anaerobic conditions for over a year (Husin *et al.*, 2003). This might cause in disruption of intracellular function due to cell dwarfing in very limited nutrient condition (Yu *et al.*, 2001).

An investigation into the efficiency of growth and decolourisation for these cultures, exogenous and selected indigenous, concluded they were facultative, with an ability to grow under both aerobic and anaerobic conditions in FSTW, but with highest growth rate and decolourisation under semi-anaerobic conditions in FSTW; 37°C, with shaking (100 rpm), with initial pH of 7.4. Growth and decolourisation of two mixed culture were enhanced in FSTW supplemented with 0.5 per cent (v/v) sterile glycerol. Further test using the shake flasks technique in separate experiment, including two types of mixed culture; selected indigenous isolates (*Bacillus cereus, Aeromonas hyrophila and Aeromonas caviae*) and selected exogenous isolates (*Enterobacter aerogenes, Citrobacter freundii* and *Shigella flexneri*) were compared to see any differences in their efficiencies to decolourise filter sterilized textile wastewater (FSTW).

Mechanism of Decolourisation

Following screening study for dye decolourisers, 6 of the isolates (ANB1, ANB2, ANB3, Cb01,Cb02 and Cb15) were selected to further determine their mechanism of decolourisation via molecular probing. This involved amplification of two genes (I*sfA* and *ssu*D) for enzyme synthesis of aromatic sulphonates degradation (Quadroni *et al.*, 1999). For this purpose, genomic DNA of the most potential bacteria, *Aeromonas caviae* was extracted using Promega Wizard Genomic DNA Purification Kit. The genomic DNA and suitable primers were used for the amplification of the desired gene. The sample of genomic DNA (1 µg) was mixed with 10 pmol of forward and reverse primers and 2x PCR Master Mix (25 µL). Four sets of different primers were used to amplify the genes of interest (Table 29.2). Polymerase Chain Reaction (PCR) was performed for 25 cycles 94°C for 1 min, 50°C for 1 min and 72°C for 2 min. PCR products were observed by agarose gel electrophoresis.

Table 29.2: Different Types of Primer for Amplifying Genes Involved in Dye Degradation

Primer	*Tm (°C)*	*Expected Size (bp)*	*References*
lsfA gene			
LsfA for			Quadroni *et al.*
5'-CAA GCT CAA GGA CCA GTT CG	59.4	639	(1999)
LsfArev			
5'-C AG CGA AGG CAC GAT TAC C	58.8		
ssuoperon			
Ssuprofor2			Kahnert *et al.*
5'-AAG AGC TCC CCA AAG GTT ATC GCG	64.4	5300	(2000)
Ssuprorevl			
5'-TCG TGC AAG CGC TCT TCC	58.2		
ssuF gene			
SsuF for			Quadroni *et al.*
5'-AGC CAT CAA CGT TCG TAA CC	57.3	216	(1999)
SsuFrev			
5'-AGC TTG GCG ATG GAG ACC	58.2		
ssuDgene			
EE24for			Kahnert *et al.*
5'-CAT CTG GAA GCT TAC TCA ACT G	58.4	1220	(2000)
EE27rev			
5'-ATA ACC AAG CTT TCA CTG GCG	57.9		

Bacterial decolourisation of azo dyes could either be due to azo reduction and/or desulphonation. The disappearance of colour is due to the reductive cleavage of azo bond(s), which is catalysed by enzymes such as flavin reductase and quinone reductase (Russ *et al.*, 2000). For decolourisation of sulphonated azo dyes, the release of the sulphonic group may be required to decolourise the azo dye since desulphonation results in the destabilization of the benzene ring structure (Kertersz and Wietek, 2001). Since decolourisation of azo dyes by the all six selected isolates above may involve desulphonation and/or azo reduction, primers of genes for desulphonation (Table 29.2) were used to amplify the I*sfA* gene and *ssu*D gene. The amplification of the I*sfA* gene and *ssu*D gene yielded fragments of ~700 bp and ~1400 bp respectively from bacterium *Aeromonas caviae*. This might signify the possibility of desulphonation by this bacterium, which is related to decolourisation of the sulphonated azo dyes (result not shown).

Proposed Mechanism on Decolourisation

Several researchers from the University of Stuttgart, Germany had experimentally proven that azo dye reduction *in vivo* may involve the role of bacterial cytoplasmic and extracellular 'azoreductases'. A gram-negative isolate *Sphingomonas* sp. BN6, which has recently, through molecular analysis, been shown to represent a novel species named *Sphingomonas xenophaga* after its ability to "eat foreign compound". It was demonstrated that an extracellular mechanism of dye reduction existed in addition to the nonspecific cytoplasmic enzymes which functioned as azoreductases by transferring electrons via reduced flavin groups to the dye molecule and thus bringing about a purely chemical reduction. Refering to Figure 29.2, Keck *et al.* (1997) demonstrated that certain quinone-based compound generated during metabolism of specific substrate acted as mediators shuttling redox equivalents to azo dye molecules from the bacterial membrane. Russ *et al.* (2000) had investigated if cytoplasmic enzymes played any role in dye decolourisation *in vivo*, a flavin reductase [NAD(P)H: flavin oxidoreductase] was cloned and overexpressed in both *E. coli* and *S. xenophaga*. In cell extracts, the strains with overexpressed flavin reductase demonstrated elevated azoreductase activity; however in whole cell studies these strains showed little improvement in their dye decolourising capabilities. Russ *et al.* (2000) also provided evidence that reports of aerobic azoreductases could be explained by the isolates in which such a phenomenon was described having elevated flavin reductase activities.

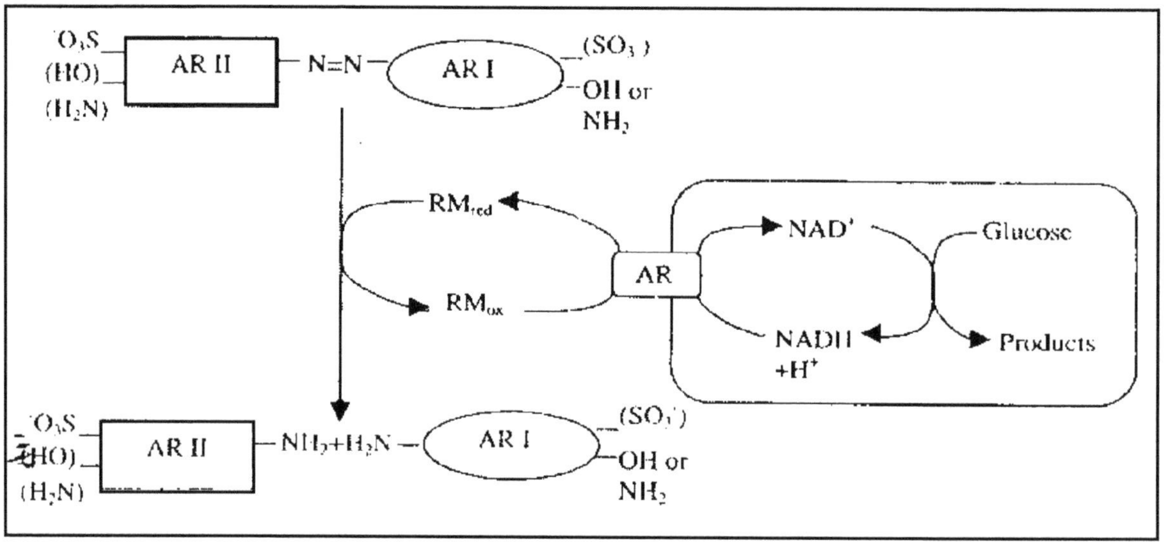

Figure 29.2: Proposed Mechanism for the Redox-mediafor-dependent Reduction of Azo Dyes by *Sphingomonas xenophaga* BN6. (AR: Azoreductase, RM: Redox mediator). Reproduced from Keck *et al.* (1997).

Decolourisation with Bacterial Biofilms: Anaerobic-Aerobic Biodegradation of Dyes

Experiments on the use of selected bacteria for colour removal were conducted using shake flasks technique. Real wastewater (Non-FSTW) and filter sterilized textile wastewater (FSTW) with initial concentration 405.9 ADMI units and 519.0 ADMI Units respectively, were inoculated with either mixed culture of the exogenous or selected indigenous bacteria. For Set (A) experiment, the FSTW consisted of *Bacillus cereus: Aeromonas caviae: Aeromonas hydrophila*, at the ratio of 1:1:1 (v/v). However, for Set (B) experiment, the bacteria involved were *Enterobacter aerogenes: Citrobacter freundii: Shigella flexneri* also at the ratio of 1:1:1 (v/v). Both sets of experiments were carried out under varying conditions to compare the effectiveness of suspended cells and biofilm in the treatment of textile wastewater. The percentage of decolourisation by 11 different systems after 12 h of incubation by both exogenous and indigenous biofilm is presented in Table 29.3.

After 12 hours of incubation under optimized condition (37°C, 100 rpm, pH 7.4, supplemented with 0.5 per cent glycerol), both the exogenous (Set B; Systems 4 and 10) and indigenous biofilms (Set

A; System 10) achieved almost complete decolourisation (~97 per cent) compared to suspended cells (Systems 5 and 11) which below 80 per cent. From the results obtained, it is important to note that decolourisation of the wastewater containing indigenous microbes (System 6), that containing culturable and unculturable microorganisms; showed only about 1.8 per cent of decolourization compared to the wastewater containing selected bacteria (System 8 and 9) in suspension (29 per cent for Set A; 19 per cent for Set B) and in the form of biofilm (46 per cent for Set A; 48 per cent for Set B). This strongly indicated that the use of selected bacteria capable of colour removal played a significant role in enhancing decolourisation of the wastewater in the absence of added glycerol.

Table 29.3: Rate of Decolourization (Per cent h^{-1}) in Shake Flasks Experiment

Number of System	*System Components*	*Rate of Decolourisation per Hour (0–12h)*		*Rate of Decolourisation per Hour (0–12h)*	
		A	*B*	*A*	*B*
1.	FSTW	0.00		0.00	
2.	FSTW + biofilm	3.74	2.28	2.52	4.25
3.	FSTW + suspended cell	1.58	b.90	1.92	0.92
4.	FSTW +biofilm + glycerol	4.15	8.19	2.22	0.08
5.	FSTW + suspended cell + glycerol	4.52	5.09	1.01	2.75
6.	non-FSTW	0.15		0.01	
7.	non-FSTW + glycerol	1.46		0.58	
8.	non-FSTW + biofilm	3.87	4.04	2.59	4.42
9.	non-FSTW+ suspended cell	2.40	1.62	3.75	1.84
10.	non-FSTW + biofilm + glycerol	8.12	8.08	0.23	0.22
11.	non-FSTW + suspended cell+ glycerol	6.56	5.98	0.33	0.35

Non-FSTW: Wastewater contained indigenous bacteria (Initial concentration: 519.0 ADMI Unit) FSTW: Filter Sterilized Textile Wastewater (Initial concentration: 405.9 ADMI Unit).

A: Mixed indigenous culture (*Bacillus cereus : Aeromonas caviae : Aeromonas hydrophila*)

B: Mixed exogenous culture (*Enterobacter aerogenes* : *Citrobacter freundii* : *Shigella flexneri*)

During the decolourisation study, the rate of colour removal was monitored with time over a period of 24 h. In general, it was found that there were two distinct linear regions that corresponded to two different rates of decolourisation (data not shown, calculated rates shown in Table 29.3). During the first 12 h of incubation, the rate of decolourisation was higher of biofilm bacteria; System 10 (Set A and B) and System 4 (Set B) was higher (1.2-1.6 times) than those in suspension (System 11). This may be attributed to the presence of extracellular polymers (EPS), which provide diffusion barrier against high inhibitory concentrations of toxic substances, allowing better survival of the bacteria in the form of biofilm. In addition, structural feature of biofilm provide both aerobic and anaerobic zones which were both required for complete mineralisation of the dye. Besides that, biofilms also offer higher solids retention times necessary to prevent washout of adapted microorganism (Jiang and Bishop, 1994). Accordingly, control experiment showed the addition of a meager amount of glycerol (0.5 per cent) contributed a medial effect on decolourisation; rate of decolourisation from System 7 (1.46 per cent h-1) which incorporates the glycerol, was higher than System 6 (0.15 per cent h-1).

As for the suspended cells in the presence of indigenous (System 11), the addition of glycerol showed increased rate of decolourisation to about 1.5-1.7 times more than that without glycerol (System 9) during the first 12h of incubation. Glycerol may be used as an alternative carbon source for growth and may also act as an electron donor for the colour reduction (Nigam *et al.*, 1996 b) during the first 12h of incubation.

Figure 29.3(a): Biofilm Form by the Potential Indigenous; *Bacillus cereus, Aeromonas caviae* and *Aeromonas hydrophila* in FSTW

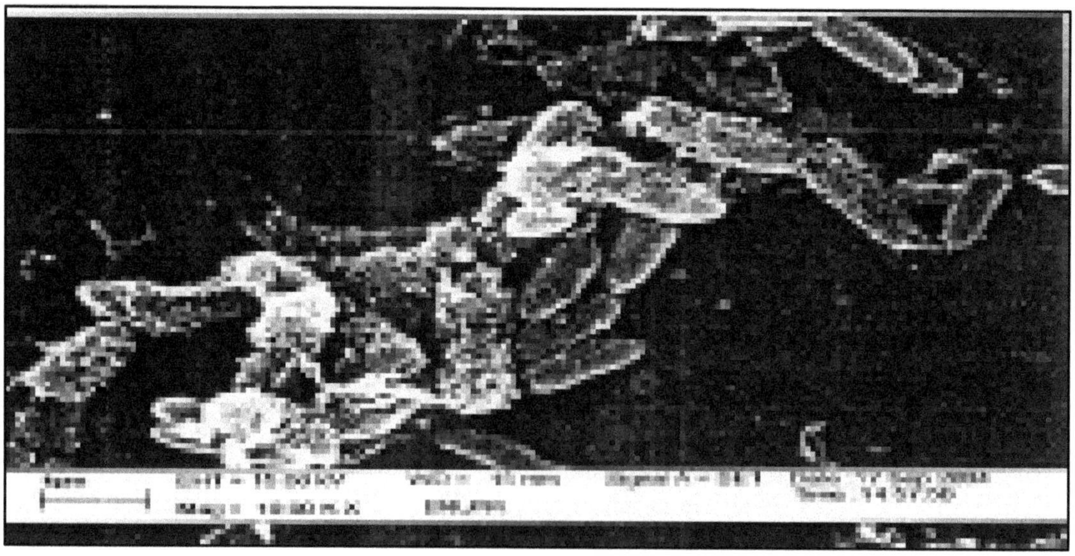

Figure 29.3(b): Biofilm Form by the Potential Indigenous; *Enterobacter aerogenes, Citrobacter freundii* and *Shigella flexneri* in FSTW

Experiments using selected indigenous and exogenous microbes in FSTW, showed a similar profile on decolourisation. Rates of colour removal by potential microbes in the formed of biofilm (System 2) were much higher (about 2.5- 4.6 times) compared to suspended cells (System 3) during the first 12 h of incubation. The rate of decolourisation remained unchanged after 12 h. The low efficiency of potential microbes in decolourisation may be related to the low population of cells present and low tolerance towards the toxic elements in the wastewater. Furthermore, dye chromophore act as the sole source of nutrient (carbon and energy source), bacterial growth is dependent on dye degradation, which in turn may contribute to a longer adaptation period and slower rate of decolourisation. The presence of bacteria in the form of biofilm is advantageous since the biofilm provides aerobic and anaerobic zones, which may facilitate complete mineralisation of dyes (Jiang and Bishop, 1994).

From the work that has been carried out, the best system for dye removal of the textile wastewater is System 10, which indigenous microorganisms contained selected bacteria in real wastewater (or non-fstw) in the form of biofilm. Closer examination of the biofilm clusters using SEM revealed that most of the bacteria were attached to the support matrices by means of extracellular strands. The morphology of biofilm formed by the indigenous and exogenous during the treatment of textile wastewater is shown in Figure 29.3(a) and 29.3(b) respectively. At some stage of growth, weblike structure termed as "bioweb" comprising of EPS was found as part of biofilm (Paulsen *et al.*, 1997). This morphology was reported to facilitate quicker plugging of pore throats by trapping floating biofilm fragments and other detritus. Meanwhile, both figures also illustrated the population of cells growing on surface of support matrix enveloped in a matrix of exopolymers. Electron microscopy studies demonstrated that the biofilm is not found as simple layers of bacteria, but are enveloped in the exopolymers. In this manner, biofilm enhances the removal of toxic organic compounds such as those found in textile wastewater (Fang *et al.*, 2002).

Bioremediation of textile-dye-containing effluent using biofilm had worked effectively either in the presence of great potential indigenous microbes such as *Bacillus cereus*, *Aeromonas caviae*, *Aeromonas hydrophila* or selected exogenous bacteria *Enterobacter aerogenes*, *Citrobacter freundii* and *Shigella flexneri* in the wastewater. From the work had been carried out, it was also suggested that a great potential of decolouriser bacteria in the form of biofilm system provide a complete decolourisation with only hours of exposure.Biofilm structure and spatial distribution influence all biofilm properties, including fixed cell activity. Biofilm structure and formation can be observed by electron microscopy techniques, enhanced in some cases by image analysis. Mechanism study on decolourisation may take advantage of biofilm processes in textile wastewater treatment system. Therefore, further studies are needed to develop simpler and more precise analytical methods for bacterial biomass correlated with enzyme activity estimation and a better knowledge of the composition and function of the biofilm.

References

Abrahamson, M., Lewandoski, Z., Geesey, G., Skjak-Braek, G., Strand, W. and Christensen, B.E, 1996. Development of an artificial biofilm to study the effects of a single microcolony on mass transport. *Journal of Microbiological Methods*, 26: 161–163.

Banat, I.M., Nigam, P., Singh, D. and Marchant, R, 1996. Microbial decolorisation of textile-dye containing effluents: A review. *Bioresource Technology*, 58: 217–227.

Bustard, M, McMullan, G. and McHale, A.P, 1998. Biosorption of textile dyes by biomass derived from *Kluyveromyces marxianus* IMB3. *Bioprocess Engineering*, 19: 427–430.

Fang, H.H.P., Xu, L.C. and Chan, K.Y, 2002. Effect of toxic metals and chemicals on biofilm and biocorrosion. *Water Research*, 36: 4709–4716.

Griffiths, J, 1984. Developments in the light absorption properties of dyes: Color and photochemical reactions. In: *Developments in the Chemistry and Technology of Organic Dyes*, (Ed.) J. Grifliths. Society of Chemistry Industry, Oxford, pp. 1–30.

Husin, A., Ibrahim, Z., Md Salleh, M., Abdul Rashid, N.A. and Yahya, A., 2003. Biological treatment of textile wastewater using biofilm. In: *IWA Conference on Environmental Biotechnology: Advancement on Water and Wastewater Application in the Tropics*, pp. 14.

Jiang, H. and Bishop, P.L., 1994. Aerobic biodegradation of azo dyes in biofilms. *Water Science Technology*, 29: 525–530.

Kahnert, A., Vermeij, P., Wietek, C., James, P., Leisinger, T. and Michael, A.K., 2000. The *ssu* Locus Plays a key role in organosulphur metabolism in *Pseudomonas putida* S-313. *Journal of Bacteriology*, pp. 2869–2878.

Keck, A., Klein, J., Kudlich, M., Stolz, A., Knackmuss, H.J. and Mattes, R., 1997. Reduction of azo dyes by redox mediators originating in the naphtalenesulphonic acid degradation pathway of *Sphingomonas* sp. Strain BN6. *Applied in Environmental Microbiology*, 6: 3684–3690.

Kertesz, M.A. and Wietek, C., 2001. Desulphurisation and desulfonation: Applications of sulphur-controlled gene expression in bacteria. *Applied Microbiology and Biotechnology*, 57: 460–466.

McMullan, G., Meehan, C., Conneely, A., Kirby, N., Robinson, T., Nigam, P., Banat, I.M., Marchant, R., Smyth, W.F., 2001. Microbial decolourisation and degradation of textile dyes. *Applied Microbiology and Biotechnology*, 56: 81 –87.

Mohd Zahari, M., Ibrahim, Z., Md Salleh, M.,Yahya, A., Abdul Rashid, N.A., Syed Muhammad Rafeii, S.N. and Zakaria, Z., 2004. *In situ* bioremediation of textile effluent sing biofilm. *Water and Environmental Management Series*, p. 93–100.

Nigam, P., McMullan, G., Banat, I.M. and Marchant, R., 1996b. Decolourization of effluent from the textile industry by a microbial consortium. *Biotechnology Letters*, 18(1): 117–120.

Paulsen, J.E., Oppen, E. and Bakke, R., 1997. Biofilm morphology in porous media: A study with microscopic and image technique. *Water Science and Technology*, 36: 1–9.

Quadroni, M., James, P., Dainese-Hatt, P. and Michael, A.K., 1999. Proteome mapping, mass spectrometric sequencing and reverse transcription–PCR for characterization of the sulfate starvation–induced response in *Pseudomonas aeruginosa* PA01. *European Journal in Biochemistry*, 266: 986–996.

Rakmi, A.R., 1993. Characterisation and prospects for biological treatment of textile finishing wastewater. In: *Waste Management in Malaysia: Current Status and Prospects for Bioremediation*, (Eds.) B.G. Yeoh, K.S. Chee, S.M. Phang, I. Zaid, I. Azni, and M. Maketab. Ministry of Science, Technology and the Environment, Malaysia, p. 99–108.

Russ, R., Rau, J. and Stolz, A., 2000. The function of cytoplasmic flavin reductases in the reduction of azo dyes by bacteria. *Applied in Environmental Microbiology*, 66(4): 429–1434.

Seshadri, S., Bishop, P.L. and Agha, A.M., 1994. Anaerobic/aerobic treatment of selected azo dyes in wastewater. *Waste Management*, 15: 127– 137.

Villegas-Navarro, A., Ramirez, M.Y., Salvader, S.M.S. and Gallador, J.M., 2001. Determination of wastewater LC_{50} of the different processes stages of the textile industry. *Ecotoxicological and Environmental Safety*, 48: 56–61.

Watnick, P. and Kolter, R., 2000. Mini review: Biofilm, city of microbes. *Journal of Bacteriology*, 182(10): 2675–2679.

Yu, J., Wang, X. and Yue, P.L., 2001. Optimal decolourisation and kinetic modelling of synthetic dyes by *Pseudomonas* strains. *Water Research*, 35(15): 3579–3586.

Ethnobotanical Wisdom and Microbial Studies on Medicinal Plants (2011) *Pages* **176–184**
Editors: **D.R. Khanna, Ashutosh Gautam, R. Bhutiani, Gagan Matta, Anil Dhiman**
Published by: **DAYA PUBLISHING HOUSE, NEW DELHI**

Chapter 30

Effect of Herbicide (Basalin EC) on Pollen Germination and Tube Growth of Twelve Hours Stored Pollen of Five Cultivars of Apocynaceae: A Critical Review*

S.A. Salgare
Salgare Research Foundation Pvt. Ltd.,
Prathamesh Society, Shivaji Chowk, Karjat – 410 201, Maharashtra

ABSTRACT

All the concentrations (10^{-17}-10^{-2} -10^{-3}, 1, 5, 10, 20-20-100 mg/ml) of basalin EC tried suppressed the germination of pollen of F-72 series of pink-flowered cultivar of *Catharanthus roseus*. Basalin EC stimulated the germination of pollen and tube growth in either sets in all the 5 cultivars of the Apocynaceae. Stored pollen shows the decrease in the germinability of pollen as well as tube growth in control and treated sets in all the 5 cultivars of the Apocynaceae.

Keywords: *Palynology, Toxicology, Environmental Sciences.*

* Further evidence of a criticism of Banerji and Gangulee (1937), Sudhakaran (1967-Ph.D.Thesis), Dharurkar (1971–Ph.D. Thesis), Berg (1973), Brandt (1974), Vick and Bevan (1976), Rasmussan (1977), Navara, Horvath and Kaleta (1978), Mhatre (1980-Ph.D. Thesis), Mhatre, Chaphekar, Ramani Rao, Patil, Haldar (1980), Shetye (1982-Ph.D. Thesis) and Giridhar (1984 -Ph.D. Thesis).

Introduction

The use of vegetation as biological indicator of environmental quality has a long history dating back to the miners canary, to the recognition about 100 years ago. Recent studies have shown the feasibility of using natural vegetation for monitoring pollution (Berg, 1973; Brandt, 1974; Rasmussan, 1977; Navara *et al.*, 1978).

Materials and Methods

Pollen of successive flowers (*viz.* F, F-24, F-48, F-72 series *i.e.* open flowers and the flower buds which require 24, 48, 72 hours to open respectively) of 5 cultivars of the Apocynaceae *e.g.* red-, pink- and white-flowered cultivars of *Nerium odorum* Soland and pink- and white-flowered cultivars of *Catharanthus roseus* (L.) G. Don. were collected at the stage of the dehiscence of anthers in the open flowers and stored at room temperature (21-31°C) having RH 59 per cent and in diffuse laboratory light at the department of botany, Govt. Institute of Science, Mumbai. Pollen viability was tested by using 2,3,5-triphenyl tetrazolium chloride (Hauser and Morrison, 1964). Germination of stored pollen grains of successive flowers was made with 2 hours intervals for the first 12 hours in the optimum concentrations of sucrose as well as in the optimum concentrations of sucrose supplemented with the optimum concentrations of basalin EC (Table 30.1). However, the present investigation is restricted only with the pollen stored 12 hours at the room temperature (Table 30.1). Observations were recorded 24 hours after incubation. For each experiment a random count of 200 grains was made to determine the percentage of pollen viability and germination. For measurement of length of pollen tubes 50 tubes were selected randomly and measured at a magnification of 100x.

Results and Discussion

Pollen viability is a subject that has a great deal of practical as well as theoretical interest. In the present investigation even the different cultivars of the same species show the variations in the percentage of their pollen viability (Table 30.1). Reduced pollen viability has been interpreted as an indication of suspected hybridity in wild populations. Nevertheless, variations in pollen viability may affect the breeding systems of the species concerned, and if the pollen viability can be altered by the environment, then the breeding system itself may be under some degree of environmental control.

Potentiality of the germinability of pollen is noted only in F series of pink- and white-flowered cultivars of *Nerium odorum*. Both of them are single-flowered cultivars (Salgare, 1983). Potentiality of the germinability of pollen was recorded in F and F-24 series of *Physalis minima* and *Solanum xanthocarpum* (Ram Indar, 1981), in red-flowered (double-flowered) cultivar of *Nerium odorum* and in white-flowered cultivar of *Catharanthus roseus* (Salgare, 1983), in all the 5 cultivars of *Petunia grandiflora* (Sharma, 1984), in all the 5 cultivars of *Solanum melongena* (Singh, 1985) and in all the 5 cultivars (light-violet-, pink-, violet- and white-violet-flowered cultivars) of *Petunia axillaris* except for white-flowered cultivar (Salgare, 1986a). Pollen germination *in vitro* culture of sucrose was noted in F, F-24 and F-48 series of *Brunfelsia americana* and in violet-flowered form of *Datura fastuosa* (Ram Indar, 1981), in all the 3 cascades (Sharma, 1984) and in white-flowered cultivar of *P. axillaris* (Salgare, 1986a). However, it was the pollen of white-flowered form of *D. fastuosa* (Ram Indar, 1981) and pink-flowered cultivar of *C. roseus* (Salgare, 1983) showed their germination *in vitro* culture of sucrose in all the 4 series (F, F-24, F-48, F-72 series) investigated. Potentiality of the germinability of pollen in all the 4 series investigated was also noted by Salgare (1986g) in 3 Leguminous crops *viz. Cyamopsis tetragonoloba* Var. Pusa Navbahar – gawar, *Phaseolus aureus* Var. J-781- mung and *Phaseolus mungo* Var. T-9- urid. Sebastian (1987) observed the germination of pollen of one of the Leguminous crops *i.e. Vigna mungo* Type 9, of

Table 30.1: Effect of Basalin EC on Pollen Germination and Tube Growth of Twelve Hours Stored Pollen of Five Cultivars Apocynaceae

Species		pgtgsaps								pgtg 12 haps				
		C				T				HC	C	T	C	T
	Series	PV	SC	G	μm	HC	G	HC	μm	HC	G	G	μm	μm
N. odorum pink-flowered	F	80	50	35	1485	10^{-15}	38	10^{-17}	1490	10^{-15}	Ng_1	25	Ng_1	386
N. odorum red-flowered	F	74	20	20	1250	10^{-15}	30	10^{-17}	1450	10^{-15}	Ng_1	28	Ng_1	640
N. odorum white-flowered	F	62	50	20	675	10^{-15}	30	10^{-17}	1000	10^{-15}	Ng_1	29	Ng_1	516
C. roseus pink-flowered	F	90	20	60	1575	10^{-15}	78	10^{-17}	1867	10^{-15}	45	78	967	1256
C. roseus white-flowered	F	88	20	40	1256	10^{-15}	76	10^{-17}	1280	10^{-15}	28	70	110	483
N. odorum red-flowered	F-24	74	20	6	485	10^{-15}	23	10^{-17}	490	10^{-15}	Ng_1	20	Ng_1	320
C. roseus pink-flowered	F-24	90	50	28	240	10^{-15}	70	10^{-17}	1025	10^{-15}	Ng_1	16	Ng_1	600
C. roseus white-flowered	F-24	88	50	16	248	10^{-15}	76	10^{-17}	407	10^{-15}	6	40	80	272
C. roseus pink-flowered	F-48	90	50	14	95	10^{-15}	60	10^{-17}	390	10^{-15}	Ng_1	Ng_1	Ng_1	Ng_1
C. roseus pink-flowered	F-72	90	80	10	65	Ng_2	Ng_2	Ng_2	Ng_2	Ng_2	1	Ng_2	15	Ng

C: in control sets pollen germination and tube growth; G: Germination of pollen in per cent; HC: Optimum concentrations of herbicide in mg/ml; Ng_1 and Ng_2: No germination of pollen after 12 and 24 hours of sowing respectively; pgtgsaps: Pollen germination and tube growth in the sets, sets soon after pollen storage; pgt12 haps: Pollen germination and tube growth in the sets, sets 12 Hours after pollen storage at room temperature; SC: Optimum concentrations of sucrose in per cent; PV: pollen viability in per cent; T: In treated sets pollen germination and tube growth; mm: Pollen tube length in mm.

Uttar Pradesh in all the 4 series investigated *in vitro* culture of sucrose. Gawde (1988) noted the germinability of pollen of 2 Leguminous crops *viz. Vigna unguiculata* Var. Pusa Barsati – cowpea and *Vigna radiata.* Var. Pusa Baisakhi of Delhi in all the 4 series investigated. Johri and Chowdhury (1957) stated that in *Citrullus colocynthis,* where pollen grains 'mostly remained attached in tetrads', satisfactory germination is observed.

Salgare (1983) observed the germination of pollen of F-72 series of pink-flowered cultivar of *Catharanthus roseus in vitro* culture of sucrose. However, Palathingal (1990) failed to germinate the pollen of F-72 series of pink-flowered cultivar of *C. roseus* in Brewbaker and Kwack's (1963) culture medium. This proves that the culture medium is also having the bearing on the germination of pollen. This also points out that Brewbaker and Kwack's (1963) culture medium is not ideal for pollen culture.

As a rule the percentage of pollen germination is always less than the pollen viability. However, Banerji and Gangulee (1937) and Dharurkar (1971) reported higher percentage of pollen germination than the pollen viability in *Eichhornia crassipes.* The claim of Banerji and Gangulee (1937) and Dharurkar (1971) is challenged by Salgare (1986c, 95, 2000b, 2006d) who stated that the observations of Banerji and Gangulee (1937) and Dharurkar (1971) are exaggerating.

All the concentrations (10^{-17}-10^{-2}-10^{-3}, 1, 5, 10, 20-20-100 mg/ml) of basalin EC tried suppressed the germination of pollen of F-72 series of pink-flowered cultivar of *C. roseus* (Table 30.1). Pollen of F series of duet, F-24 series of white cascade, duet and sonata and F-48 series of all the 3 cascades did not germinate when treated with 10^{-17} mg/ml of basalin EC. All of them are the cultivars of *Petunia grandiflora* (Sharma, 1984). Even the lowest concentration (10^{-17} mg/ml) of basalin EC tried suppressed the germinability of pollen of F-24 series of light-violet- and violet-flowered cultivars of *Petunia axillaris* (Salgare, 1986a). This proves that the pollen of the said series are highly sensitive and acts as an ideal indicators of pollution. Thus it is confirmed that the pollen development and activity are more sensitive indicators of adverse factors in the botanical environment and the use of an entire vascular plant (Berg, 1973; Brandt, 1974; Rasmussan, 1977; Navara *et al.,* 1978; Mhatre, 1980; Mhatre *et al.,* 1980; Shetye, 1982 and Giridhar, 1984) as an indicator of pollution is a very crude method and rather a wrong choice. There is no evidence of any entire vascular plant exhibiting this much degree of sensitivity. This is confirmed in the present critical review. This was already proved earlier by Salgare (1983, 84, 85a-c, 86a, d-g, 2000a, 01a-b, 05a-c, 06b, f), Salgare and Sebastian (1986), Salgare and Phunguskar (2002), Salgare and Singh (2002, 06a-b) and Salgare and Pathak (2005) and his Research Group (Ram Indar, 1981; Sharma, 1984; Singh, 1985; Sebastian, 1987; Gawde, 1988 and Palathingal, 1990) in their extensive work.

Basalin EC stimulated the germination of pollen and tube growth in either sets in all the 5 cultivars of the Apocynaceae. Stored pollen caused decrease in the germinability of pollen and tube growth in control as well as treated sets in all the 5 cultivars of the Apocynaceae (Table 30.1). Stored pollen of F series of all the 3 cultivars and F-24 series of red-flowered cultivar of *Nerium odorum* and F-24 and F-48 series of pink-flowered cultivar of *Catharanthus roseus* failed to germinated *in vitro* culture of sucrose after 10 hours of their sowing. However, they were found germinated *in vitro* culture of sucrose supplemented by basalin EC (Table 30.1). This proves that the herbicide extended the longevity of the stored pollen.

In many instances due to hyper- or hypo-nutrition the percentage of germination and length of the tube are considerably reduced. Bursting of pollen also increases and occasionally the pollen tubes were observed to eject their content. In addition to this various pollen tube deformities *viz.* 'bloating' or 'bulla' formation resulting in the swelling of the tip of the pollen tube were also observed. In the pollen tubes that grew in the coiled or *zig-zag* manner the wall was not straight. *Catharanthus roseus* though characterized by the presence of monosiphonous condition at a low frequency bisiphonous and trisiphonous condition was also recorded in the present investigation along with the branched pollen tubes. In this connection it should be pointed out that Sudhakaran (1967) stated that in *Vinca rosea* L. [*Catharanthus roseus* (L.) G. Don.] besides pollen grains which produced single pollen tube, it has also been noticed that tetraploid grains frequently produce more than one pollen tube. Pollen tubes are branched quite frequently. Aberrations of this type in the pollen tube development are not observed in diploid pollen tubes, but quite frequently met with the pollen grains of irradiated plants. Salgare (1983) made it very clear that Sudhakaran (1967) had failed to trace out the branched pollen tubes and polysiphonous condition which is fairly common even in diploid pollen grains. Apart from this Sudhakaran (1967) was not able to report the various types of pollen tube deformities either with diploid or tetraploid grains. Present investigation as well as the extensive work of Salgare (1983, 86b, 2006a, c, e, g) proved that the observations of Sudhakaran (1967) are superficial and misleading.

References

Banerji, I. and Gangulee, H.C., 1937. Spermatogenesis in *Eichhornia crassipes* Solms. *J. Indian Bot. Soc.*, 16: 289–296.

Berg, H., 1973. Plants as indicators of air pollution. *Toxicol.*, 1: 79–89.

Brandt, C.C., 1974. Plants as indicators of air quality. In: *Indicators of Environmental Quality*, (Ed.) W.A. Thomas. Plenum Press, New York, pp. 101–107.

Brewbaker, J.L. and Kwack, B.H., 1963. The essential role of Ca ion in pollen germination and tube growth. *Amer. J. Bot.*, 50: 859–865.

Dharurkar, R.D., 1971. Effects of herbicides on the cytomorphology of *Eichhornia crassipes* Solms (Mart.). *Ph.D. Thesis*, Univ. Bombay.

Gawde, Suwarna, 1988. Effect of industrial pollution on crop physiology–I. *Ph.D. Thesis*, Univ. Bombay.

Giridhar, B.A., 1984. Study of interactions between industrial air pollutants and plants. *Ph.D. Thesis*, Univ. Bombay.

Hauser, E.J.P. and Morrison, J.H., 1964. The cytochemical reduction of notroblue tetrazolium as an index of pollen viability. *Amer. J. Bot.*, 51: 748–752.

Johri, B.M. and Chowdhury, Chhaya Roy, 1957. A contribution to the embryology of *Citraullus colocynthis* Schare. and *Melothria maderaspatana* Cong. *New Phytol.*, 56: 51–60.

Mhatre, G.N., 1980. Studies in responses to heavy metals in industrial environment. *Ph.D. Thesis*, Univ. Bombay.

Mhatre, G.N., Chaphekar, S.B., Ramani Rao, I.V., Patil, M.R. and Haldar, B.C., 1980. Effect of industrial pollution on the Kalu river ecosystem. *Environ. Pollut. Series A*, 23: 67–78.

Navara, J., Horvath, I. and Kaleta, M., 1978. Contribution to the determination of limiting of values of SO_2 for vegetation in the region of Bratislava. *Environ. Pollut. Series A*, 16: 249–262.

Palathingal, Trisa, 1990. Evaluation of industrial pollution of Bombay by pollen–I. *M.Phil. Thesis*, Univ. Mumbai.

Ram Indar, 1981. Effect of herbicides on plants of the Solanaceae–I. *M.Sc. Thesis*, Univ. Bombay.

Rasmussan, L., 1977. Epiphytic bryophytes as indicators of changes in the background levels of air borne metals from 1951 to 1975. *Environ. Pollut. Series A*, 14: 34–45.

Salgare, S.A., 1983. Pollen physiology of Angiosperms–I. *Ph. D.Thesis*, Univ. Bombay.

Salgare, S.A., 1984. Further evidence of a criticism on the hypothesis of Mhatre (1980), Mhatre, Chaphekar, Ramani Rao, Patil, Haldar (1980) and Shetye (1982). In: *Proc. 4th Indian Palyno. Conf.*, Held on March 19–21, at Andhra Univ., Visakhapatnam, Abstract No. S III–06.

Salgare, S.A., 1985a. A criticism on the findings of Mhatre (1980), Mhatre, Chaphekar, Ramani Rao, Patil, Haldar (1980) and Shetye (1982). In: *Proc. 2nd Ann. Conf. of Nat. Environ. Sci. Acad.*, Held on May 25–27, at Awadh Univ., Faizabad, Abstract No. 2.

Salgare, S.A., 1985b. Further evidence of a criticism of the findings of Mhatre (1980), Mhatre, Chaphekar, Ramani Rao, Patil, Haldar (1980), Shetye (1982) and Giridhar (1984). In: *Proc. Nat. Symp. on Environ. Sci. and Warm Water Aquculture-Finfish*, Held on November 6–9, at M.J.K.(P.G.) College, Bettiah, pp. 1–2, Abstract No. 2.

Salgare, S.A., 1985c. A criticism on the findings of Mhatre (1980), Mhatre, Chaphekar, Ramani Rao, Patil, Haldar (1980) and Shetye (1982). In: *Proc. Nat. Symp. on Environ. Sci. and Warm Water Aquculture-Finfish*, Held on November 6–9, at M.J.K.(P.G.) College, Bettiah, pp. 52, Abstract No. 73.

Salgare, S.A., 1986a. Effect of herbicides on pollen physiology of *Petunia axillaris* BSP. *Ph.D. Thesis*, World Univ.

Salgare, S.A., 1986b. A criticism on Ph.D. Thesis of Sudhakaran (1967) entitled, 'Cytogenetic studies in *Vinca rosea* Linn.' In: *Proc. of 1st Nat. Symp. on Environ. Biol.*, Held on December 30–31, at Deptt. Zool. and Microbiol., S. K. Univ., Anantapur, Abstract No. 5.

Salgare, S.A., 1986c. A criticism on Ph.D.Thesis of Dharurkar (1971) entitled, 'Effect of herbicides on the cytomorphology of *Eichhornia crassipes* Solms (Mart.). *Ph.D. Thesis*, Univ. of Bombay. In: *Proc. of 1st Nat. Symp. on Environ. Biol.*, Held on December 30–31, at Deptt. Zool. and Microbiol., S. K. Univ., Anantapur, Abstract No. 6.

Salgare, S.A., 1986d. A criticism on Ph.D. Thesis of Giridhar (1984) entitled, 'Study of interactions between industrial air pollutants and plants.' In: *Proc. 1st Nat. Symp. on Environ. Biol.*, Held on December 30–31, at Deptt. Zool. and Microbiol., S.K. Univ., Anantapur, Abstract No.14.

Salgare, S.A., 1986e. A criticism on the hypothesis of Berg (1973), Brandt (1974), Vick and Bevan (1976), Rasmussan (1977), Navara, Horvath and Kaleta (1978), Mhatre (1980), Mhatre, Chaphekar, Ramani Rao, Patil, Haldar (1980), Shetye (1982) and Giridhar (1984). In: *Proc. 1st Nat. Symp. on Environ. Biol.*, Held on December 30–31, at Deptt. Zool. and Microbiol., S.K. Univ., Anantapur, Abstract No.15.

Salgare, S.A., 1986f. A criticism on Ph.D. Thesis of Shetye (1982) entitled, 'Effect of heavy metals on plants.' In: *Proc. 1st Nat. Symp. on Environ. Biol.*, Held on December 30–31, at Deptt. Zool. and Microbiol., S.K. Univ., Anantapur, Abstract No.16.

Salgare, S.A., 1986g. Pollen physiology of successive flowers. *D.Sc. Thesis*, Marathwada Univ.

Salgare, S.A., 1995. Interesting observations on the physiology of pollen of *Eichhornia crassipes*. Solam. (Mart). In: *Proc. 13th Nat. Symp. on Life Sci., Proceeded with Ann. Session of Indian Soc. of Life Sci.*, Held on December 30–31, 1995 and January 1, 1996 at CH. Charan Singh Univ., Meerut, Abstract No. 83.

Salgare, S.A., 2000a. Additional evidence of a criticism on the hypothesis of Berg (1973), Brandt (1974), Vick and Bevan (1976), Rasmussan (1977), Navara, Horvath and Kaleta (1978), Mhatre (1980–Ph.D. Thesis), Mhatre, Chaphekar, Ramani Rao, Patil, Haldar (1980), Shetye (1982–Ph.D. Thesis), and Giridhar (1984–Ph.D. Thesis): A Critical Review. In: *Proc. 1st Nat. Conf. on Recent Trends in Life Management*, Held on October 22–23, at Bipin Bihari (P.G.) College, Jhansi, pp. 64–65, Abstract No. 92.

Salgare, S.A., 2000b. A criticism on the findings of Banerji and Gangulee (1937) and Dharurkar (1971–Ph.D. Thesis). *Him. J. Environ. Zool.*, 14: 159–160.

Salgare, S.A., 2001a. Monitoring of herbicide (nitrofen) toxicity by using pollen as indicators: A critical review–I. *Website www.microbiologyou.com*. U.S.A's Publications (35 pages).

Salgare, S.A., 2001b. Monitors of pollution – A criticism on the hypothesis of Berg (1973), Brandt (1974), Vick and Bevan (1976), Rasmussan (1977), Navara, Horvath and Kaleta (1978), Mhatre (1980 – Ph.D. Thesis), Mhatre, Chaphekar, Ramani Rao, Patil, Haldara (1980), Shetye (1982 – Ph.D. Thesis) and Giridhar (1984 – Ph.D. Thesis): A critical review–V. *Biojournal*, 13: 39–44.

Salgare, S.A., 2005a. Monitoring of herbicide (simazine) toxicity – A criticism on the hypothesis of Berg (1973), Brandt (1974), Vick and Bevan (1976), Rasmussan (1977), Navara, Horvath and Kaleta (1978), Mhatre (1980–Ph.D. Thesis), Mhatre, Chaphekar, Ramani Rao, Patil, Haldar (1980), Shetye (1982–Ph.D. Thesis) and Giridhar (1984–Ph.D. Thesis) – A critical review. *Him. J. Environ. Zool.* 19: 69–71.

Salgare, S. A. 2005b. Monitoring of herbicide (dalapon) toxicity by using pollen as indicators – Pollen of some cultivars of Apocynaceae and further evidence of a criticism on the hypothesis of Berg (1973), Brandt (1974), Vick and Bevan (1976), Rasmussan (1977), Navara, Horvath and Kaleta (1978), Mhatre (1980–Ph.D. Thesis), Mhatre, Chaphekar, Ramani Rao, Patil, Haldar (1980), Shetye (1982–Ph.D. Thesis) and Giridhar (1984–Ph.D. Thesis): A critical review. *Flora and Fauna*, 11: 49–50

Salgare, S.A., 2005c. Monitoring of herbicide (basalin EC) toxicity – A criticism on the hypothesis of Berg (1973), Brandt (1974), Vick and Bevan (1976), Rasmussan (1977), Navara, Horvath and Kaleta (1978), Mhatre (1980–Ph.D. Thesis), Mhatre, Chaphekar, Ramani Rao, Patil, Haldar (1980), Shetye (1982–Ph.D. Thesis) and Giridhar (1984–Ph.D. Thesis): A critical review. *Internat. J. Biosci. Reporter*, 3: 298–300.

Salgare, S.A., 2006a. Further evidence of a criticism of the findings of Sudhakaran (1967–Ph.D.Thesis) and Katre and Ghatnekar (1978)*. *Internat. J. Biosci. Reporter*, 4: 19–20.

Salgare, S.A., 2006b. Monitoring of herbicide (2,4–D) toxicity by using pollen as indicators – Pollen of Apocynaceae and further evidence of a criticism of the hypothesis of Berg (1973), Brandt (1974), Vick and Bevan (1976), Rasmussan (1977), Navara, Horvath and Kaleta (1978), Mhatre (1980–Ph.D. Thesis), Mhatre, Chaphekar, Ramani Rao, Patil, Haldar (1980), Shetye (1982–Ph.D. Thesis) and Giridhar (1984–Ph.D. Thesis): A critical review. *Internat. J. Plant Sci.*, 1: 134–136.

Salgare, S.A., 2006c. Effect of gramoxone on pollen germination and tube growth of twelve hours stored pollen of Apocynaceae and further evidence of a criticism of the hypothesis of Sudhakaran (1967)–A critical review – II. *Plant Archives*, 6: 389–390.

Salgare, S.A., 2006d. Further evidence of a criticism of the findings of Banerji and Gangulee (1937) and Dharurkar (1971)*. *Internat. J. Biosci. Reporter*, 4: 169–170.

Salgare, S.A., 2006e. Effect of acrolein on pollen germination and tube growth of stored pollen of Apocynaceae and further evidence of a criticism of the hypothesis of Sudhakaran (1967): A critical review. *Him. J. Environ. Zool.*, 20: 137–139.

Salgare, S.A., 2006f. Monitoring of herbicide (basalin EC) toxicity by using pollen as indicators – Pollen of *Petunlia axillaris* and further evidence of a criticism of the hypothesis of Berg (1973), Brandt (1974), Vick and Bevan (1976), Rasmussan (1977), Navara, Horvath and Kaleta (1978), Mhatre (1980–Ph.D. Thesis), Mhatre, Chaphekar, Ramani Rao, Patil, Haldar (1980), Shetye (1982–Ph.D. Thesis) and Giridhar (1984–Ph.D. Thesis): A critical review. *Him. J. Environ. Zool.*, 20: 269–271.

Salgare, S.A., 2006g. Effect of fernoxone on pollen germination and tube growth of twelve hours stored pollen of Apocynaceae and further evidence of a criticism of the hypothesis of Sudhakaran (1967). *Asian J. Bio Sci.*, 1: 160.

Salgare, S.A. and Phunguskar, K.P., 2002. Monitors of pesticide (alphamethrin) toxicity by using pollen as indicators – Pollen of pink-flowered *Catharanthus roseus* – A criticism on the hypothesis of Berg (1973), Brandt (1974), Vick and Bevan (1976), Rasmussan (1977), Navara, Horvath and Kaleta (1978), Mhatre (1980–Ph.D. Thesis), Mhatre, Chaphekar, Ramani Rao, Patil, Haldara (1980), Shetye (1982–Ph.D. Thesis) and Giridhar (1984–Ph.D. Thesis): A critical review – III. *Biojournal*, 14: 1–4.

Salgare, S.A. and Pathak, Sanchita, 2005. Monitoring of heavy metal (bismuth nitrate) toxicity by using pollen as indicators–Pollen of white-flowered *Catharanthus roseus* and further evidence of a criticism on the hypothesis of Berg (1973), Brandt (1974), Vick and Bevan (1976), Rasmussan (1977), Navara, Horvath and Kaleta (1978), Mhatre (1980–Ph.D. Thesis), Mhatre, Chaphekar, Ramani Rao, Patil, Haldar (1980), Shetye (1982–Ph.D. Thesis), and Giridhar (1984–Ph.D. Thesis): A critical review. *Flora and Fauna*, 11: 69–70.

Salgare, S.A. and Singh, Sanju, 2002. Monitors of heavy metal (ferrous sulphate) toxicity by using pollen as indicators–Pollen of white-flowered *Catharanthus roseus*–A criticism on the hypothesis of Berg (1973), Brandt (1974), Vick and Bevan (1976), Rasmussan (1977), Navara, Horvath and Kaleta (1978), Mhatre (1980–Ph.D. Thesis), Mhatre, Chaphekar, Ramani Rao, Patil, Haldara (1980), Shetye (1982–Ph.D. Thesis) and Giridhar (1984–Ph.D. Thesis): A critical review–II. *Biojournal*, 14: 5–7.

Salgare, S.A. and Singh, Sanju, 2006a. Monitoring of heavy metal (copper sulphate) toxicity by using pollen as indicators–Pollen of white-flowered *Catharanthus roseus* and further evidence of a criticism on the hypothesis of Berg (1973), Brandt (1974), Vick and Bevan (1976), Rasmussan (1977), Navara, Horvath and Kaleta (1978), Mhatre (1980–Ph.D. Thesis), Mhatre, Chaphekar, Ramani Rao, Patil, Haldar (1980), Shetye (1982–Ph.D. Thesis), and Giridhar (1984–Ph.D. Thesis): A critical review – I. *Him. J. Environ. Zool.*, 20: 276–278.

Salgare, S.A. and Singh, Sanju 2006b. Monitoring of heavy metal (cobalt nitrate) toxicity by using pollen as indicators – Pollen of pink-flowered cultivar of *Catharanthus roseus* and further evidence of a criticism of the hypothesis of Berg (1973), Brandt (1974), Vick and Bevan (1976), Rasmussan (1977), Navara, Horvath and Kaleta (1978), Mhatre (1980–Ph.D. Thesis), Mhatre, Chaphekar, Ramani Rao, Patil, Haldar (1980), Shetye (1982–Ph.D. Thesis) and Giridhar (1984–Ph.D. Thesis): A critical review – II. *Him. J. Environ. Zool.*, 20: 283–284.

Salgare, S. A. and Sebastian, Theresa 1986. Further evidence of a criticism of the hypothesis of Berg (1973), Brandt (1974), Vick and Bevan (1976), Rasmussan (1977), Navara, Horvath and Kaleta (1978), Mhatre (1980–Ph.D. Thesis), Mhatre, Chaphekar, Ramani Rao, Patil, Haldar (1980), Shetye (1982–Ph.D. Thesis) and Giridhar (1984–Ph.D. Thesis). In: *Proc. 1st Nat. Symp. on Environ. Biol.*, Held on December 30–31, at Deptt. Zool. and Microbiol., S. K. Univ., Anantpur. Abstract No.15.

Sebastian, Theresa, 1987. Effect of herbicides on crop physiology–I. *Ph.D. Thesis*, Univ. Bombay.

Sharma, R.I., 1984. Effect of herbicides on plants of the Solanaceae–II. *Ph.D. Thesis*, Univ. Bombay.

Shetye, R.P., 1982. Effect of heavy metals on plants. *Ph.D. Thesis*, Univ. Bombay.

Singh, S.R., 1985. Effect of herbicides on pollen physiology of Brinjal. *M.Sc. Thesis*, Univ. Bombay.

Sudhakaran, I.V., 1967. Cytogenetic studies in *Vinca rosea* Linn. *Ph.D. Thesis*, Univ. Bombay.

Ethnobotanical Wisdom and Microbial Studies on Medicinal Plants (2011) *Pages* **185–192**
Editors: **D.R. Khanna, Ashutosh Gautam, R. Bhutiani, Gagan Matta, Anil Dhiman**
Published by: **DAYA PUBLISHING HOUSE, NEW DELHI**

Chapter 31

Monitoring of Herbicide (MH) Toxicity by Using Pollen as Indicators–Pollen of Five Cultivars of *Petunlia axillaris* BSP.: A Critical Review*

S.A. Salgare
Salgare Research Foundation Pvt. Ltd.,
Prathamesh Society, Shivaji Chowk, Karjat – 410 201, Maharashtra

ABSTRACT

Germination of pollen of F series of white-flowered and white-violet-flowered and F-24 series of white-violet-flowered cultivars of *Petunia axillaris* was noted even in 100 mg/ml MH.

Keywords: *Physiology of pollen, Palylnology, Toxicology, Environmental sciences.*

Introduction

Extensive use of herbicides leaves behind residues which contaminate our environment. It is primarily needed to work out some simple system for the evaluation of the toxicity of herbicides.

* Further evidence of a criticism of Banerji and Gangulee (1937), Dharurkar (1971–Ph.D. Thesis), Berg (1973), Brandt (1974), Vick and Bevan (1976), Rasmussan (1977), Navara, Horvath and Kaleta (1978), Mhatre (1980–Ph.D. Thesis), Mhatre, Chaphekar, Ramani Rao, Patil, Haldar (1980), Shetye (1982–Ph.D. Thesis) and Giridhar (1984–Ph.D. Thesis).

Materials and Methods

Pollen of successive flowers (*viz.* F, F-24, F-48, F-72 series *i.e.* open flowers and the flower buds which require 24, 48, 72 hours to open respectively) of 5 cultivars of *Petunia axillaris* BSP. *e.g.* light-violet-, pink-, violet-, white- and white-violet-flowered cultivars were collected at the stage of the dehiscence of anthers in the open flowers. Germination of pollen grains of successive flowers was studied by standing-drop technique in the optimum concentrations of sucrose (acts as control) as well as in the optimum concentrations of sucrose supplemented by the different concentrations (10^{-17}-10^{-2}-10^{-3}, 1, 5, 10, 20-20-100 mg/ml) of Maleic Hydrazide (MH) (1,2-dihydropyridazine, 3-6-dione) (Table 31.1). The cultures were then transferred to a moist filter chamber, stored at room temperature (21.9-32.2°C) having RH 58 per cent and in diffuse laboratory light. Observations were recorded 24 hours after incubation. For each experiment a random count of 200 grains was made to determine the percentage of pollen germination. For measurement of length of pollen tubes, 50 tubes were selected randomly and measured at a magnification of 100x.

Results and Discussion

Potentiality of pollen germinability was noted in F and F-24 series of all the 5 cultivars of *Petunia axillaris* and in F-48 series of white-flowered cultivar of *P. axillaris*. Thus the potentiality of pollen germinability in *P. axillaris* was recorded in 11 out of 20 series investigated (Table 31.1). Potentiality of the germinability of pollen is noted only in F series of pink- and white-flowered cultivars of *Nerium odorum*. Both of them are single-flowered cultivars (Salgare, 1983). Potentiality of the germinability of pollen was recorded in F and F-24 series of *Physalis minima* and *Solanum xanthocarpum* (Ram Indar, 1981), in red-flowered (double-flowered) cultivar of *N. odorum* and in white-flowered cultivar of *Catharanthus roseus* (Salgare, 1983) and in all the cultivars (light-violet-, pink-, violet- and white-violet-flowered cultivars) of *Petunia axillaris* except for white-flowered cultivar (Salgare, 1986a). Pollen germination *in vitro* culture of sucrose was noted in F, F-24 and F-48 series of *Brunfelsia americana* and in violet-flowered form of *Datura fastuosa* (Ram Indar, 1981) and in white-flowered cultivar of *P. axillaris* (Salgare, 1986a). However, it was the pollen of white-flowered form of *D. fastuosa* (Ram Indar, 1981) and pink-flowered cultivar of *C. roseus* (Salgare, 1983) showed their germination *in vitro* culture of sucrose in all the 4 series (F, F-24, F-48, F-72 series) investigated. Potentiality of the germinability of pollen in all the 4 series investigated was also noted by Salgare (1986d) in 3 Leguminous crops *viz. Cyamopsis tetragonoloba* Taub. var. Pusa Navbahar – gawar, *Phaseolus aureus* Roxb. var. J-781- mung and *Phaseolus mungo* Roxb. var. T-9- urid. Sebastian (1987) observed the germination of pollen of one of the Leguminous crops *i.e. Vigna mungo* (L.) Hepper Type 9, of Uttar Pradesh in all the 4 series investigated *in vitro* culture of sucrose. Gawde (1988) noted the germinability of pollen of 2 Leguminous crops *viz. Vigna unguiculata* (L.) Walp. var. Pusa Barsati – cowpea and Baisakhi of Delhi in all the 4 series investigated. Johri and Chowdhury (1957) stated that in *Citrullus colocynthis,* where pollen grains 'mostly remained attached in tetrads', satisfactory germination is observed.

Salgare (1983) observed the germination of pollen of F-72 series of pink-flowered cultivar of *Catharanthus roseus in vitro* culture of sucrose. Palathingal (1990) stated that the pollen of F-72 series of pink-flowered cultivar of *Catharanthus roseus* did not germinate in Brewbaker and Kwack's (1963) culture medium. This proves that the culture medium is also having the bearing on the germination of pollen. This points out that Brewbaker and Kwack's (1963) culture medium is not ideal for the pollen culture.

Table 31.1: Inhibitory Effect of MH on Pollen Germination and Tube Growth of Successive Flowers of Five Cultivars of *Petunia axillaris* BSP

Cultivars			*iocs*				*rchi*		*pgtgstch*				
	Series	*PV*	*SC*	*PG*	*TG*	*V/O*	*RCPG*	*RCTG*	*HC*	*PG*	*TG*	*V/O*	*TC*
Light-violet	F	76	50	32	30	0.09	10^{-17}–40	10^{-17}–40	40	88.57	66.67	0.03	40
Pink	F	93	50	28	35	0.11	10^{-17}–40	10^{-17}–40	40	96.67	75.00	0.03	60
Violet	F	80	50	25	38	0.11	5–60	10^{-17}–60	60	92.86	66.67	0.03	80
White	F	95	30	34	80	0.24	10^{-17}–100	60–100	NWO	NWO	NWO	NWO	NWO
White-violet	F	90	30	30	325	0.88	5–100	10^{-17}–100	NWO	NWO	NWO	NWO	NWO
Light-violet	F-24	76	30	25	45	0.14	10^{-5}–60	10^{-17}–60	60	89.29	80.00	0.03	80
Pink	F-24	93	10	16	30	0.09	10^{-13}–40	10^{-17}–40	40	92.86	60.00	0.03	60
Violet	F-24	80	60	25	30	0.09	10^{-17}–10^{-13}	10^{-17}–60	60	64.29	60.00	0.29	80
White	F-24	95	10	26	30	0.09	10^{-3}–20	10^{-17}–20	20	91.67	71.43	0.29	40
White-violet	F-24	90	30	30	210	0.57	10^{-17}–100	10^{-17}–100	NWO	NWO	NWO	NWO	NWO
White	F-48	95	10	13	40	0.12	10^{-7}–60	10^{-17}–60	60	80.00	77.78	0.29	80

HC: Concentrations of herbicide in mg/ml; iocs: In optimum concentrations of sucrose germination of pollen and tube growth; Ng: No germination; NOW: Not worked out; PG: Pollen germination in per cent; pgtgstch: Pollen germination and tube growth in sub-toxic concentrations of herbicide; PV: Pollen viability in per cent; rchi: Range of concentrations of herbicide for inhibition of pollen germination and tube growth; rcpg: Range of concentrations of herbicide for inhibition of pollen germination; rctg: Range of concentrations of herbicide for inhibition of pollen tube growth; SC: Optimum concentrations of sucrose in per cent; TG: Pollen tube growth in μm; V/O: *In vitro* tube length in compare to *in vivo* in per cent.

Even the lowest concentration (10^{-17} mg/ml) of MH tried suppressed the germinability of pollen of F-24 series of red-flowered cultivar of *Nerium odorum* and F-48 and F-72 series of pink-flowered cultivar of *Catharanthus roseus* (Salgare, 1983). Singh (1985) stated that the germinability of pollen of F series of brinjal round and F-24 series of brinjal long, muktakeshi and round was prevented even by the lowest concentration (10^{-17} mg/ml) of MH. All of them are the cultivars of *Solanum melongena*. This proves that the pollen of the said series are highly sensitive and acts as an ideal indicators of pollution. Thus it is confirmed that the pollen development and activity are more sensitive indicators of adverse factors in the botanical environment and the use of an entire vascular plant (Berg, 1973; Brandt, 1974; Rasmussan, 1977; Navara *et al.*, 1978; Mhatre, 1980; Mhatre *et al.*, 1980; Shetye, 1982 and Giridhar, 1984) as an indicator of pollution is a very crude method and rather a wrong choice. There is no evidence of any entire vascular plant exhibiting this much degree of sensitivity. This is confirmed in the present critical review (Table 31.1) and was also proved earlier by Salgare (1983, 84, 85a-c, 86a-d, 2000, 1a-b, 05a-c, 2006), Salgare and Sebastian (1986), Salgare and Phunguskar (2002), Salgare and Sanju Singh (2002) and Salgare and Pathak (2005) and Salgare's Research Group (Ram Indar, 1981; Sharma, 1984; Singh, 1985; Sebastian, 1987; Gawde, 1988 and Palathingal, 1990) also supports the present findings.

Inhibition in the germination of pollen was caused by MH in 13, 7, 4, 6, 11 series of Solanaceae (Ram Indar, 1981), Apocynaceae (Salgare, 1983), *Petunia grandiflora* (Sharma, 1984), *Solanum melongena* (Singh, 1985) and *Petunia axillaris* (Table 31.1) respectively.

The widest range of concentrations of MH found to be 10^{-17}- 10, 10^{-5}-100, 10^{-17}-40, 10^{-17}-100 mg/ml which inhibited the germination of pollen of *Petunia axillaris* (in F-24 series of violet-flowered cultivar) (Salgare, 1986a-Table 31.1), Apocynaceae (in F series of pink-flowered cultivar of *Catharanthus roseus* (Salgare, 1983), *Petunia grandiflora* (in F series of pink and white cascades) (Sharma, 1984) and brinjal (in F series of all the five cultivars of brinjal except for brinjal round) (Singh, 1985) respectively.

Sub-toxic concentration of MH caused as high as 96.67 per cent, 98.59, 75.00, 96.77, 96.00 per cent inhibition in the pollen germination of *P. axillaris* (in F series of pink-flowered cultivar) (Table 31.1), Solanaceae (in F series of violet-flowered form of *Datura fastuosa* (Ram Indar, 1981), Apocynaceae (in F series of white-flowered cultivar of *Nerium odorum* and F-24 series of white-flowered cultivar of *C. roseus* (Salgare, 1983), *P. grandiflora* (in F series of pink cascade) (Sharma, 1984) and brinjal (in F-24 series of brinjal small) (Singh, 1985) respectively.

Ratio between the series and inhibition caused by MH (in sub-toxic concentration) in the germination of pollen is as (Inhibition is represented in the form of percentage):

F:F-24:F-48:F-72 = 92.70:84.53:80.00:00.00 in *Petunia axillaris* (Table 31.1-Salgare, 1986a)

F:F-24:F-48:F-72 = 89.70:93.84:81.74:70.58 in Solanaceae (Ram Indar, 1981)

F:F-24:F-48:F-72 = 75.00:75.00:00.00:00.00 in Apocynaceae (Salgare, 1983)

F:F-24:F-48:F-72 = 94.28:96.77:00.00:00.00 in *Petunia grandiflora* (Sharma, 1984)

F:F-24:F-48:F-72 = 00.77:78±4.56:00.00:00.00 in brinjal (Singh, 1985)

This shows that MH caused maximum inhibition in the germination of pollen of F series of *Petunia axillaris* (Table 31.1), F-24 series of Solanaceae (Ram Indar, 1981), F and F-24 series of Apocynaceae (Salgare, 1983), F-24 series of *Petunia grandiflora* (Sharma, 1984) and of F-24 series of brinjal (Singh, 1985).

MH inhibited the pollen tube growth in 13, 7, 4, 6, 11 series of Solanaceae (Ram Indar, 1981), Apocynaceae (Salgare, 1983), *Petunia grandiflora* (Sharma, 1984), *Solanum melongena* (Singh, 1985) and *Petunia axillaris* (Table 31.1) respectively.

The widest range of the concentrations of MH tried, found to be 10^{-17}- 100, 10^{-17}-100, 10^{-17}-100, 10^{-17}- 40, 10^{-17}-100 mg/ml which inhibited the pollen tulbe growth of *Petunia axillaris* (in F and F-24 series of white-violet-flowered cultivar) (Salgare, 1986a-Table 31.1), Solanaceae (in white-flowered form of *Datura fastuosa*) (Ram Indar, 1981), Apocynaceae (in F series of pink- and red-flowered cultivars of *Nerium odorum* and pink-flowered cultivar of *Catharanthus roseus* and F-24 series of red-flowered cultivar of *N. odorum*)(Salgare, 1983), *Petunia grandiflora* (in F series of pink and white cascades) (Sharma, 1984) and brinjal (in F series of all the cultivars of *Solanum melongena* except for brinjal round) (Singh, 1985) respectively.

Sub-toxic concentration of MH caused as high as 80.00 per cent inhibition in the pollen tube growth of *Petunia axillaris* (in F-24 series of light-violet-flowered cultivar) (Salgare, 1986a-Table 31.1). However, the maximum inhibition in the pollen tube growth *viz.* 96.14, 77.78, 94.59, 78.72 per cent) was reported by Ram Indar (1981) in Solanaceae (in F series of *Brunfelsia americana*), Salgare (1983) in Apocynaceae (in white-flowered cultivar of *Nerium odorum*), Sharma (1984) in *Petunia grandiflora* (in F series of red cascade) and Singh (1985) in *Solanum melongena* (in F-24 series of brinjal small) respectively

Ratio between the series and inhibition caused by MH (in sub-toxic concentration) in the pollen tube growth is as (Inhibition is represented in the form of percentage):

F:F-24:F-48:F-72 = 69.45:67.86:77.78:00.00 in *Petunia axillaris* (Table 31.1-Salgare, 1986a)

F:F-24:F-48:F-72 = 89.70:93.84:81.74:70.58 in Solanaceae (Ram Indar, 1981)

F:F-24:F-48:F-72 = 75.00:75.00:00.00:00.00 in Apocynaceae (Salgare, 1983)

F:F-24:F-48:F-72 = 94.28:96.77:00.00:00.00 in *Petunia grandiflora* (Sharma, 1984)

F:F-24:F-48:F-72 = 00.77:78±4.56:00.00:00.00 in brinjal (Singh, 1985)

This shows that MH caused maximum inhibition in the pollen tube growth of F-48 series of *Petunia axillaris* (Table 31.1), F-24 series of Solanaceae (Ram Indar, 1981), F and F-24 series of Apocynaceae (Salgare, 1983), F-24 series of *Petunia grandiflora* (Sharma, 1984) and F-24 series of brinjal (Singh, 1985).

Tube length *in vitro* culture (sucrose + MH) of MH (in sub-toxic concentration) is 0.03 per cent (in F series of light-violet-, pink-, and violet-flowered and F-24 series of light-violet- and pink-flowered cultivars of *Petunia axillaris*) in *P. axillaris* of the tube length found *in vivo* is the longest of all the cultivars investigated (Table 31.1).

References

Berg, H., 1973. Plants as indicators of air pollution. *Toxicol.*, 1: 79–89.

Brandt, C.C., 1974. Plants as indicators of air quality. In: *Indicators of Environmental Quality*, (Ed.) W.A. Thomas. Plenum Press, New York, pp. 101–107.

Brewbaker, J.L. and Kwack, B.H., 1963. The essential role of Ca ion in pollen germination and tube growth. *Amer. J. Bot.*, 50: 859–865.

Gawde, Suwarna, 1988. Effect of industrial pollution on crop physiology–I. *Ph.D. Thesis*, Univ. Bombay.

Giridhar, B.A., 1984. Study of interactions between industrial air pollutants and plants. *Ph.D. Thesis*, Univ. Bombay.

Johri, B.M. and Chowdhury, Chhaya Roy, 1957. A contribution to the embryology of *Citraullus colocynthis* Schare. and *Melothria maderaspatana* Cong. *New Phytol.*, 56: 51–60.

Mhatre, G.N., 1980. Studies in responses to heavy metals in industrial environment. *Ph.D. Thesis*, Univ. Bombay.

Mhatre, G.N., Chaphekar, S.B., Ramani Rao, I.V., Patil, M.R. and Haldar, B.C., 1980. Effect of industrial pollution on the Kalu river ecosystem. *Environ. Pollut. Series A*, 23: 67–78.

Navara, J., Horvath, I. and Kaleta, M., 1978. Contribution to the determination of limiting of values of SO_2 for vegetation in the region of Bratislava. *Environ. Pollut. Series A*, 16: 249–262.

Palathingal, Trisa, 1990. Evaluation of industrial pollution of Bombay by pollen–I. *M.Phil. Thesis*, Univ. Mumbai.

Ram Indar, 1981. Effect of herbicides on plants of the Solanaceae–I. *M.Sc. Thesis*, Univ. Bombay.

Rasmussan, L., 1977. Epiphytic bryophytes as indicators of changes in the background levels of air borne metals from 1951 to 1975. *Environ. Pollut. Series A*, 14: 34–45.

Salgare, S.A., 1983. Pollen physiology of Angiosperms–I. *Ph. D.Thesis*, Univ. Bombay.

Salgare, S.A., 1984. Further evidence of a criticism on the hypothesis of Mhatre (1980), Mhatre, Chaphekar, Ramani Rao, Patil, Haldar (1980) and Shetye (1982). In: *Proc. 4th Indian Palyno. Conf.*, Held on March 19–21, at Andhra Univ., Visakhapatnam, Abstract No. S III–06.

Salgare, S.A., 1985a. A criticism on the findings of Mhatre (1980), Mhatre, Chaphekar, Ramani Rao, Patil, Haldar (1980) and Shetye (1982). In: *Proc. 2nd Ann. Conf. of Nat. Environ. Sci. Acad.*, Held on May 25–27, at Awadh Univ., Faizabad, Abstract No. 2.

Salgare, S.A., 1985b. Further evidence of a criticism of the findings of Mhatre (1980), Mhatre, Chaphekar, Ramani Rao, Patil, Haldar (1980), Shetye (1982) and Giridhar (1984). In: *Proc. Nat. Symp. on Environ. Sci. and Warm Water Aquculture-Finfish*, Held on November 6–9, at M.J.K.(P.G.) College, Bettiah, pp. 1–2, Abstract No. 2.

Salgare, S.A., 1985c. A criticism on the findings of Mhatre (1980), Mhatre, Chaphekar, Ramani Rao, Patil, Haldar (1980) and Shetye (1982). In: *Proc. Nat. Symp. on Environ. Sci. and Warm Water Aquculture-Finfish*, Held on November 6–9, at M.J.K.(P.G.) College, Bettiah, pp. 52, Abstract No. 73.

Salgare, S.A., 1986a. Effect of herbicides on pollen physiology of *Petunia axillaris* BSP. *Ph.D. Thesis*, World Univ.

Salgare, S.A., 1986b. A criticism on Ph.D. Thesis of Sudhakaran (1967) entitled, 'Cytogenetic studies in *Vinca rosea* Linn.' In: *Proc. of 1st Nat. Symp. on Environ. Biol.*, Held on December 30–31, at Deptt. Zool. and Microbiol., S. K. Univ., Anantapur, Abstract No. 5.

Salgare, S.A., 1986c. A criticism on Ph.D.Thesis of Dharurkar (1971) entitled, 'Effect of herbicides on the cytomorphology of *Eichhornia crassipes* Solms (Mart.). *Ph.D. Thesis*, Univ. of Bombay. In: *Proc. of 1st Nat. Symp. on Environ. Biol.*, Held on December 30–31, at Deptt. Zool. and Microbiol., S. K. Univ., Anantapur, Abstract No. 6.

Salgare, S.A., 1986d. A criticism on Ph.D. Thesis of Giridhar (1984) entitled, 'Study of interactions between industrial air pollutants and plants.' In: *Proc. 1st Nat. Symp. on Environ. Biol.*, Held on December 30–31, at Deptt. Zool. and Microbiol., S.K. Univ., Anantapur, Abstract No.14.

Salgare, S.A., 2000. Additional evidence of a criticism on the hypothesis of Berg (1973), Brandt (1974), Vick and Bevan (1976), Rasmussan (1977), Navara, Horvath and Kaleta (1978), Mhatre (1980–Ph.D. Thesis), Mhatre, Chaphekar, Ramani Rao, Patil, Haldar (1980), Shetye (1982–Ph.D. Thesis), and Giridhar (1984–Ph.D. Thesis): A Critical Review. In: *Proc. 1st Nat. Conf. on Recent Trends in Life Management*, Held on October 22–23, at Bipin Bihari (P.G.) College, Jhansi, pp. 64–65, Abstract No. 92.

Salgare, S.A., 2001a. Monitoring of herbicide (nitrofen) toxicity by using pollen as indicators: A critical review–I. *Website www.microbiologyou.com. U.S.A's Publications (35 pages).*

Salgare, S.A., 2001b. Monitors of pollution – A criticism on the hypothesis of Berg (1973), Brandt (1974), Vick and Bevan (1976), Rasmussan (1977), Navara, Horvath and Kaleta (1978), Mhatre (1980 – Ph.D. Thesis), Mhatre, Chaphekar, Ramani Rao, Patil, Haldara (1980), Shetye (1982 – Ph.D. Thesis) and Giridhar (1984 – Ph.D. Thesis): A critical review–V. *Biojournal*, 13: 39–44.

Salgare, S.A., 2005a. Monitoring of herbicide (simazine) toxicity – A criticism on the hypothesis of Berg (1973), Brandt (1974), Vick and Bevan (1976), Rasmussan (1977), Navara, Horvath and Kaleta (1978), Mhatre (1980–Ph.D. Thesis), Mhatre, Chaphekar, Ramani Rao, Patil, Haldar (1980), Shetye (1982–Ph.D. Thesis) and Giridhar (1984–Ph.D. Thesis) – A critical review. *Him. J. Environ. Zool.* 19: 69–71.

Salgare, S. A. 2005b Monitoring of herbicide (dalapon) toxicity by using pollen as indicators – Pollen of some cultivars of Apocynaceae and further evidence of a criticism on the hypothesis of Berg (1973), Brandt (1974), Vick and Bevan (1976), Rasmussan (1977), Navara, Horvath and Kaleta (1978), Mhatre (1980–Ph.D. Thesis), Mhatre, Chaphekar, Ramani Rao, Patil, Haldar (1980), Shetye (1982–Ph.D. Thesis) and Giridhar (1984–Ph.D. Thesis): A critical review. *Flora and Fauna*, 11: 49–50

Salgare, S.A., 2005c. Monitoring of herbicide (basalin EC) toxicity – A criticism on the hypothesis of Berg (1973), Brandt (1974), Vick and Bevan (1976), Rasmussan (1977), Navara, Horvath and Kaleta (1978), Mhatre (1980–Ph.D. Thesis), Mhatre, Chaphekar, Ramani Rao, Patil, Haldar (1980), Shetye (1982–Ph.D. Thesis) and Giridhar (1984–Ph.D. Thesis): A critical review. *Internat. J. Biosci. Reporter*, 3: 298–300.

Salgare, S.A., 2006. Monitoring of herbicide (2,4–D) toxicity by using pollen as indicators – Pollen of Apocynaceae and further evidence of a criticism of the hypothesis of Berg (1973), Brandt (1974), Vick and Bevan (1976), Rasmussan (1977), Navara, Horvath and Kaleta (1978), Mhatre (1980–Ph.D. Thesis), Mhatre, Chaphekar, Ramani Rao, Patil, Haldar (1980), Shetye (1982–Ph.D. Thesis) and Giridhar (1984–Ph.D. Thesis): A critical review. *Internat. J. Plant Sci.*, 1: 134–136.

Salgare, S.A. and Phunguskar, K.P., 2002. Monitors of pesticide (alphamethrin) toxicity by using pollen as indicators – Pollen of pink-flowered *Catharanthus roseus* – A criticism on the hypothesis of Berg (1973), Brandt (1974), Vick and Bevan (1976), Rasmussan (1977), Navara, Horvath and Kaleta (1978), Mhatre (1980–Ph.D. Thesis), Mhatre, Chaphekar, Ramani Rao, Patil, Haldara (1980), Shetye (1982–Ph.D. Thesis) and Giridhar (1984–Ph.D. Thesis): A critical review – III. *Biojournal*, 14: 1–4.

Salgare, S.A. and Pathak, Sanchita, 2005. Monitoring of heavy metal (bismuth nitrate) toxicity by using pollen as indicators–Pollen of white-flowered *Catharanthus roseus* and further evidence of a criticism on the hypothesis of Berg (1973), Brandt (1974), Vick and Bevan (1976), Rasmussan (1977), Navara, Horvath and Kaleta (1978), Mhatre (1980–Ph.D. Thesis), Mhatre, Chaphekar,

Ramani Rao, Patil, Haldar (1980), Shetye (1982–Ph.D. Thesis), and Giridhar (1984–Ph.D. Thesis): A critical review. *Flora and Fauna*, 11: 69–70.

Salgare, S.A. and Singh, Sanju, 2002. Monitors of heavy metal (ferrous sulphate) toxicity by using pollen as indicators–Pollen of white-flowered *Catharanthus roseus*–A criticism on the hypothesis of Berg (1973), Brandt (1974), Vick and Bevan (1976), Rasmussan (1977), Navara, Horvath and Kaleta (1978), Mhatre (1980–Ph.D. Thesis), Mhatre, Chaphekar, Ramani Rao, Patil, Haldara (1980), Shetye (1982–Ph.D. Thesis) and Giridhar (1984–Ph.D. Thesis): A critical review–II. *Biojournal*, 14: 5–7.

Salgare, S. A. and Sebastian, Theresa 1986. Further evidence of a criticism of the hypothesis of Berg (1973), Brandt (1974), Vick and Bevan (1976), Rasmussan (1977), Navara, Horvath and Kaleta (1978), Mhatre (1980–Ph.D. Thesis), Mhatre, Chaphekar, Ramani Rao, Patil, Haldar (1980), Shetye (1982–Ph.D. Thesis) and Giridhar (1984–Ph.D. Thesis). In: *Proc. 1st Nat. Symp. on Environ. Biol.*, Held on December 30–31, at Deptt. Zool. and Microbiol., S. K. Univ., Anantpur. Abstract No.15.

Sebastian, Theresa, 1987. Effect of herbicides on crop physiology–I. *Ph.D. Thesis*, Univ. Bombay.

Shetye, R.P., 1982. Effect of heavy metals on plants. *Ph.D. Thesis*, Univ. Bombay.

Singh, S.R., 1985. Effect of herbicides on pollen physiology of Brinjal. *M.Sc. Thesis*, Univ. Bombay.

Ethnobotanical Wisdom and Microbial Studies on Medicinal Plants (2011) *Pages* **193–202**
Editors: **D.R. Khanna, Ashutosh Gautam, R. Bhutiani, Gagan Matta, Anil Dhiman**
Published by: **DAYA PUBLISHING HOUSE, NEW DELHI**

Chapter 32

Whether Optimum Pollen Germination and Tube Length Attained in the Same Growth Medium (Sucrose + Basalin EC) by Five Cultivars of *Petunia axillaris* BSP.: –A Critical Review*

S.A. Salgare
Salgare Research Foundation Pvt. Ltd.,
Prathamesh Society, Shivaji Chowk, Karjat – 410 201, Maharashtra

ABSTRACT

The widest range of concentration of basalin EC was proved to be 10^{-17}-10^{-11} mg/ml which stimulated the germination of pollen of *Petunia axillaris* BSP., while 10^{-17}-10^{-15} mg/ml was confirmed as the widest range of concentration which stimulated the tube growth. Pollen of F-24 series of pink-flowered cultivar showed the highest stimulation (142.86 per cent) in the germination of pollen. Basalin EC produced maximum stimulation (66.67 per cent) in the pollen tube growth of *P. axillaris* (in F series of violet-flowered cultivar).

Keywords: *Physiology of pollen, Palylnology, Toxicology, Environmental sciences.*

* Further evidence of a criticism of Banerji and Gangulee (1937), Sudhakaran (1967-Ph.D.Thesis), Dharurkar (1971–Ph.D. Thesis), Nair, Nambudiri and Thomas (1973), Berg (1973), Brandt (1974), Vick and Bevan (1976), Rasmussan (1977), Navara, Horvath and Kaleta (1978), Mhatre (1980–Ph.D. Thesis), Mhatre, Chaphekar, Ramani Rao, Patil, Haldar (1980), Shetye (1982–Ph.D. Thesis) and Giridhar (1984–Ph.D. Thesis).

Introduction

The residual matter of the herbicides left over in the soil, may, further, proves to be an important factor in the growth of the subsequent crops.

Materials and Methods

Pollen of successive flowers (*viz.* F, F-24, F-48, F-72 series *i.e.* open flowers and the flower buds which require 24, 48, 72 hours to open respectively) of 5 cultivars (light-violet-, pink-, violet-, white- and white-violet-flowered) of *Petunia axillaris* BSP were collected soon after the dehiscence of anthers in the open flowers. Germination of pollen grains was studied by standing-drop technique in the optimum concentrations of sucrose which acts as control as well as in the optimum concentrations of sucrose supplemented with the wide range of concentrations (10^{-17}-10^{-2}-10^{-3}, 1, 5, 10, 20-20-100 mg/ml) of basalin EC. Pollen grains were incubated soon after the dehiscence of anthers. The cultures then transferred to a moist filter chamber, stored at room temperature (29.3-32.5°C) having RH 64 per cent and in diffuse laboratory light. The experiments were run in triplicate and average results were recorded. Observations on the germination of pollen and tube growth were recorded 24 hours after incubation. For each experiment a random count of 200 grains was made to determine the percentage of pollen germination. For measurement of length of pollen tubes, 50 tubes were selected randomly and measured at a magnification of 100x.

Results and Discussion

Potentiality of the germinability of pollen was noted only in F series of pink- and white-flowered cultivars of *Nerium odorum*. Both of them are single-flowered cultivars (Salgare, 1983). Potentiality of the germinability of pollen was recorded in F and F-24 series of *Physalis minima* and *Solanum xanthocarpum* (Ram Indar, 1981), in red-flowered (double-flowered) cultivar of *Nerium odorum* and in white-flowered cultivar of *Catharanthus roseus* (Salgare, 1983) and in all the five cultivars (pink, red and white cascades, duet and sonata) of *Petunia grandiflora* (Sharma, 1984) in all the cultivars (light-violet-, pink-, violet- and white-violet-flowered cultivars) of *Petunia axillaris* except for white-flowered cultivar (Salgare, 1986a). Pollen germination *in vitro* culture of sucrose was noted in F, F-24 and F-48 series of *Brunfelsia americana* and in violet-flowered form of *Datura fastuosa* (Ram Indar, 1981), in all 3 cascades (Sharma, 1984) and in white-flowered cultivar of *P. axillaris* (Table 32.1). However, it was the pollen of white-flowered form of *D. fastuosa* (Ram Indar, 1981) and pink-flowered cultivar of *C. roseus* (Salgare, 1983) showed their germination *in vitro* culture of sucrose in all the 4 series (F, F-24, F-48, F-72 series) investigated. Potentiality of the germinability of pollen in all the 4 series investigated was also noted by Salgare (1986f) in 3 Leguminous crops *viz. Cyamopsis tetragonoloba* Taub. Var. Pusa Navbahar – gawar, *Phaseolus aureus* Roxb. Var. J-781- mung and *Phaseolus mungo* Roxb. Var. T-9- urid. Sebastian (1987) observed the germination of pollen of one of the Leguminous crops *i.e. Vigna mungo* (L.) Hepper Type 9, of Uttar Pradesh in all the 4 series investigated *in vitro* culture of sucrose. Gawde (1988-Ph.D.Thesis) noted the germinability of pollen of 2 Leguminous crops *viz. Vigna unguiculata* (L.) Walp. Var. Pusa Barsati – cowpea and *Vigna radiata* (L.) Wilczek. Var. Pusa Baisakhi of Delhi in all the 4 series investigated. Johri and Chowdhury (1957) stated that in *Citrullus colocynthis*, where pollen grains 'mostly remained attached in tetrads', satisfactory germination is observed.

Salgare (1983) observed the germination of pollen of F-72 series of pink-flowered cultivar of *Catharanthus roseus in vitro* culture of sucrose. Palathingal (1990) stated that the pollen of F-72 series of pink-flowered cultivar of *C. roseus* did not germinate in Brewbaker and Kwack's (1963) culture medium. This confirms that Brewbaker and Kwack's (1963) culture medium is not perfect. This also proves that

Table 32.1: Stimulatory Effect of Basalin EC on Pollen Germination and Tube Growth of Successive Flowers of Five Cultivars of *Petunia axillaris* BSP.

Cultivars			*iocs*				*rchi*		*pgtgstch*				
	Series	*PV*	*SC*	*PG*	*TG*	*V/O*	*RCPG*	*RCTG*	*HC*	*PG*	*TG*	*V/O*	*TC*
Light-violet	F	76	50	32	30	0.09	10^{-17}–40	10^{-17}–40	40	88.57	66.67	0.03	40
Pink	F	93	50	28	35	0.11	10^{-17}–40	10^{-17}–40	40	96.67	75.00	0.03	60
Violet	F	80	50	25	38	0.11	5–60	10^{-17}–60	60	92.86	66.67	0.03	80
White	F	95	30	34	80	0.24	10^{-17}–100	60–100	NWO	NWO	NWO	NWO	NWO
White-violet	F	90	30	30	325	0.88	5–100	10^{-17}–100	NWO	NWO	NWO	NWO	NWO
Light-violet	F-24	76	30	25	45	0.14	10^{-5}–60	10^{-17}–60	60	89.29	80.00	0.03	80
Pink	F-24	93	10	16	30	0.09	10^{-13}–40	10^{-17}–40	40	92.86	60.00	0.03	60
Violet	F-24	80	60	25	30	0.09	10^{-17}–10^{-13}	10^{-17}–60	60	64.29	60.00	0.29	80
White	F-24	95	10	26	30	0.09	10^{-3}–20	10^{-17}–20	20	91.67	71.43	0.29	40
White-violet	F-24	90	30	30	210	0.57	10^{-17}–100	10^{-17}–100	NWO	NWO	NWO	NWO	NWO
White	F-48	95	10	13	40	0.12	10^{-7}–60	10^{-17}–60	60	80.00	77.78	0.29	80

HC: Concentrations of herbicide in mg/ml; iocs: In optimum concentrations of sucrose germination of pollen and tube growth; Ng: No germination; NOW: Not worked out; PG: Pollen germination in per cent; pgtgstch: Pollen germination and tube growth in sub-toxic concentrations of herbicide; PV: Pollen viability in per cent; rchi: Range of concentrations of herbicide for inhibition of pollen germination and tube growth; rcpg: Range of concentrations of herbicide for inhibition of pollen germination; rctg: Range of concentrations of herbicide for inhibition of pollen tube growth; SC: Optimum concentrations of sucrose in per cent; TG: Pollen tube growth in µm; V/O: *In vitro* tube length in compare to *in vivo* in per cent.

the culture medium is also having the bearing on the germination of pollen. This pointed out that Brewbaker and Kwack's (1963) culture medium is not ideal for pollen culture of successive flowers.

Even the lowest concentration (10^{-17} mg/ml) of basalin EC tried found to be toxic for the germination of F-24 series of light-violet- and violet-flowered cultivars of *Petunia axillaris* (Table 32.1), F-72 series of pink-flowered cultivar of *Catharanthus roseus* (Salgare, 1983) and F series of duet, all the 3 cultivars of cascades of F-48 series and except for pink and red cascades all the cultivars investigated of F-24 series of *Petunia grandiflora* (Sharma, 1984). This proves that the pollen of the said series are highly sensitive and acts as an ideal indicator of pollution. Thus it is confirmed that the pollen development and activity are more sensitive indicators of adverse factors in the botanical environment and the use of an entire vascular plant (Berg, 1973; Brandt, 1974; Rasmussan, 1977; Navara *et al.*, 1978; Mhatre, 1980; Mhatre *et al.*, 1980; Shetye, 1982 and Giridhar, 1984) as an indicator of pollution is a very crud method and rather a wrong choice. There is no evidence of any entire vascular plant exhibiting this much degree of sensitivity. This is also confirmed in the present critical review (Table 32.1). This was already proved earlier by the extensive work of Salgare (1983, 84, 85a-c, 86a, c-e, 2000, 1a-b, 05b, d-e, 06a), Salgare and Sebastian (1986), Salgare and Phunguskar (2002), Salgare and Singh (2002) and Salgare and Pathak (2005) and Salgare's Research Group (Ram Indar, 1981; Sharma, 1984; Singh, 1985; Sebastian, 1987; Gawde, 1988 and Palathingal, 1990) also supports the present findings.

It was the only pollen of F series of pink-flowered cultivar of *Petunia axillaris* showed their germination even in 100 mg/ml basalin EC (Table 32.1). This proves that the pollen of the said series are very resistant.

The widest range of concentration of basalin EC was proved to be 10^{-17}-10^{-15} mg/ml which stimulated the germination of pollen of *Petunia axillaris* (Table 32.1), while Salgare (1983) and Sharma (1984) confirmed 10^{-17}-100 and 10^{-17}-10^{-11} mg/ml basalin EC as the widest ranges of concentrations for Apocynaceae (in F-24 and F-48 series of white- and pink- flowered cultivars of *Catharanthus roseus* respectively) (Salgare, 1983) and in *P. grandiflora* (in F series of red cascade) (Sharma, 1984) respectively.

Basalin EC stimulated the germination of pollen of 1, 8, 2 series of *Petunia axillaris* (Table 32.1) (Salgare, 1986a), Apocynaceae (Salgare, 1983) and *P.grandiflora* (Sharma, 1984) respectively.

Ratio between the series and stimulation produced by basalin EC (in an optimum concentration) in the germination of pollen in the number of series is as:

F:F-24:F-48:F-72 = 0:1:0:0 in *Petunia axillaris* (Table 32.1-Salgare, 1986a)

F:F-24:F-48:F-72 = 4:3:1:0 in Apocynaceae (Salgare, 1983)

F:F-24:F-48:F-72 = 1:1:0:0 in *Petunia grandiflora* (Sharma, 1984)

This shows that basalin EC stimulated the germination of pollen in maximum number of series of F and F-24 series of Apocynaceae (Salgare, 1983) and *P. axillaris* (Table 32.1) (Salgare, 1986a) respectively, while in an equal number of series of F and F-24 series of *P. grandiflora* (Sharma, 1984).

An optimum concentration of basalin EC produced 142.86 per cent stimulation in the germination of pollen of F-24 series of pink-flowered cultivar of *Petunia axillaris* (Table 32.1) (Salgare, 1986a), while as high as 375.00 per cent and 75.00 per cent in Apocynaceae (in F-24 series of white-flowered cultivar of *Catharanthus roseus*) (Salgare, 1983) and in *Petunia grandiflora* (in F series of red cascade) (Sharma, 1984) respectively.

Ratio between the series and stimulation produced by basalin EC (in an optimum concentration) in the germination of pollen is as (Stimulation is represented in the form of percentage):

F:F-24:F-48:F-72 = 00.00142.86:00.00:00.00 in *Petunia axillaris* (Table 32.1-Salgare, 1986a)

F:F-24:F-48:F-72 = 55.00:269.44:328.57:00.00 in Apocynaceae (Salgare, 1983)

F:F-24:F-48:F-72 = 75.00:25.00:00.00:00.00 in *Petunia grandiflora* (Sharma, 1984)

This shows that basalin EC produced maximum stimulation in the germination of pollen of F, F-25 and F-48 series of *Petunia grandiflora* (Sharma, 1984), *P. axillaris* (Table 32.1) (Salgare, 1986a) and Apocynaceae (Salgare, 1983) respectively.

Basalin EC stimulated the pollen tube growth of *Petunia axillaris* (Table 32.1) (Salgare, 1986a), Apocynaceae (Salgare, 1983) and *P. grandiflora* (Sharma, 1984) in 2, 6, 1 series respectively.

Ratio between the series and stimulation produced by basalin EC (in an optimum concentration) in the pollen tube growth in the number of series is as:

F:F-24:F-48:F-72 = 2:0:0:0 in *Petunia axillaris* (Table 32.1-Salgare, 1986a)

F:F-24:F-48:F-72 = 3:2:1:0 in Apocynaceae (Salgare, 1983)

F:F-24:F-48:F-72 = 1:0:0:0 in *Petunia grandiflora* (Sharma, 1984)

This shows that basalin EC stimulated the pollen tube growth in maximum number of series of F series of all cases investigated.

The widest range of concentration of basalin EC was proved to be 10^{-17}-10^{-13} and 10^{-17}-80 mg/ml which stimulated the pollen tube growth of *P. axillaris* (in F series of light-violet-flowered cultivar) (Table 32.1) (Salgare, 1986a) and Apocynaceae (F-48 series of pink-flowered cultivar of *Catharanthus roseus*) (Salgare, 1983) respectively. The pollen tube length of F series of red cascade was stimulated by basalin EC in the range of 10^{-17}-10^{-7} mg/ml basalin EC (Sharma, 1984).

An optimum concentration of basalin EC produced as high as 66.67 per cent and 327.08 per cent stimulation in the pollen tube growth of *Petunia axillaris* (in F series of violet-flowered cultivar) (Salgare, 1986a) and Apocynaceae (in F-24 series of pink-flowered cultivar of *Catharanthus roseus*) (Salgare, 1983) respectively. An optimum concentration of basalin EC produced maximum stimulation (158.78 per cent) in the pollen tube growth of *Petunia grandiflora* (in F series of red cascade) (Sharma, 1984).

Ratio between the series and stimulation produced by basalin EC (in an optimum concentration) in the pollen tube growth is as (Stimulation is represented in the form of percentage):

F:F-24:F-48:F-72 = 58.34:00.00:00.00:00.00 in *Petunia axillaris* (Table 32.1-Salgare, 1986a)

F:F-24:F-48:F-72 = 27.58:242.04:310.53:00.00 in Apocynaceae (Salgare, 1983)

F:F-24:F-48:F-72 = 158.78:00.00:00.00:00.00 in *Petunia grandiflora* (Sharma, 1984)

This shows that basalin EC produced maximum stimulation in the pollen tube growth of F series of both the species of *Petunia* (Salgare, 1986a and Sharma, 1984) and F-48 series Apocynaceae (Salgare, 1983).

Tube length *in vitro* culture (sucrose + basalin EC) of basalin EC (in an optimum concentration) is 0.14, 13.18, 1.99 per cent in *Petunia axillaris* (in F series of light-violet- and violet-flowered cultivars) (Table 32.1) (Salgare, 1986a), in Apocynaceae (in F series of red-flowered cultivar of *Nerium odorum*) (Salgare, 1983) and *P. grandiflora* (in F series of red cascade) (Sharma, 1984) respectively of the tube length found *in vivo* is the longest of all the cultivars investigated.

However, 10^{-17}-10^{-13} mg/ml basalin EC proved to be the widest range of concentrations which stimulated the tube growth of *P. axillaris*. It was the pollen of F-24 series of pink-flowered cultivar of *P. axillaris* which showed the highest stimulation in the germination of pollen. Basalin EC produced as high as142.86 per cent stimulation in the germination of pollen of successive flowers of *P. axillaris*. However, 66.67 per cent stimulation proved to be the highest produced by the herbicide in the pollen tube growth of successive flowers of *P. axillaris* (Table 32.1) (Salgare, 1986a). It should be pointed out that horticulturists and plant breeders often failed to get the fertile seeds in spite of all the care taken during artificial pollination. Unless sterility is the main cause, failure of seed setting may be due to slow growth of the pollen tube or its early degeneration in the style. As a rule, the length of the pollen tube obtained *in vitro* is significantly shorter than that *in vivo*. Consequently, one of the main problems is to obtain *in vitro* germination and tube length comparable to that *in vivo*. Tube length *in vitro* (in an optimum concentration of sucrose supplemented by an optimum concentration of basalin EC) is 0.14 per cent in F series of light-white- and violet-flowered cultivars of *P. axillaris* of the tube length found *in vivo* is the longest of all the cultivars investigated of *P. axillaris* (Table 32.1) (Salgare, 1986a). Tube length *in vitro* is 13.18 per cent in F series of red-flowered cultivar of *Nerium odorum* of the tube length found *in vivo* is the longest of all the cultivars investigated of Apocynaceae (Salgare, 1983). This proves that though basalin EC stimulated the pollen tube growth, however, it can not be produced as long as that found *in vivo*. However, the use of the herbicide as the growth substance is very economical.

It should be pointed out that in a few cases the length of the tubes in cultures does equal that in nature (Knight, 1917; Schoch-Bodmer, 1921; Brink, 1924; Branscheidt, 1929, 30; Ehlers, 1951; Vasil, 1960).

Pollen germination and tube elongation are two distinct processes differing in their sensitivity to different concentrations of the herbicide was also confirmed with the present work (Table 32.1) (Salgare, 1986a). However, Nair, Nambudiri and Thomas (1973) stated that it has been significant that the optimum percentage of germination and tube length were attained in the same growth medium. However, with the present work (Table 32.1) as well as previous extensive work of Salgare (1979, 83, 86a-b, e, 2004, 05a, c, 06b-c), Salgare and Bindu (2002, 05) and Salgare and Tessy Mol Antony (2005a, b) it could be concluded that the observations of Nair, Nambudiri and Thomas (1973) are superficial and misleading.

References

Berg, H., 1973. Plants as indicators of air pollution. *Toxicol.*, 1: 79–89.

Brandt, C.C., 1974. Plants as indicators of air quality. In: *Indicators of Environmental Quality*, (Ed.) W.A. Thomas. Plenum Press, New York, pp. 101–107.

Branscheidt, P., 1929. Die Befruchtungs verhaltnisse beim Obst und der Rebe. *Gartenbauwiss*, 2: 158–270

Branscheidt, P., 1930. Zur Physiologie der Pollenkeimung und ihere experimentellen Beeinflussung. *Planta*, 11: 368-453.

Brewbaker, J.L. and Kwack, B.H., 1963. The essential role of Ca ion in pollen germination and tube growth. *Amer. J. Bot.*, 50: 859–865.

Brink, R.A., 1924 The physiology of pollen. III. Growth *in vitro* and *in vivo*. *Amer. J.Bot.*, 11: 351–364.

Ehlers, H., 1951. Untersuchungen Zur Ernahrungsphysiologie der Pollenschlauche. *Biol. Zentralbl.*, 70: 432–451.

Gawde, Suwarna, 1988. Effect of industrial pollution on crop physiology–I. *Ph.D. Thesis*, Univ. Bombay.

Giridhar, B.A., 1984. Study of interactions between industrial air pollutants and plants. *Ph.D. Thesis*, Univ. Bombay.

Johri, B.M. and Chowdhury, Chhaya Roy, 1957. A contribution to the embryology of *Citraullus colocynthis* Schare. and *Melothria maderaspatana* Cong. *New Phytol.*, 56: 51–60.

Knight, L.I., 1917. Physiological aspects of the self-sterility of the apple. In: *Proc. Amer. Soc. Hort. Sci.*, 14: 101–105.

Mhatre, G.N., 1980. Studies in responses to heavy metals in industrial environment. *Ph.D. Thesis*, Univ. Bombay.

Mhatre, G.N., Chaphekar, S.B., Ramani Rao, I.V., Patil, M.R. and Haldar, B.C., 1980. Effect of industrial pollution on the Kalu river ecosystem. *Environ. Pollut. Series A*, 23: 67–78.

Navara, J., Horvath, I. and Kaleta, M., 1978 Contribution to the determination of limiting of values of SO_2 for vegetation in the region of Bratislava. *Environ. Pollut. Series A*, 16: 249–262.

Nair, P.K.K., Nambudiri, E.M.V. and Thomas, M.K., 1973. A note on pollen germination at various stages of development of flower buds of Balsam (*Impatians balsam*). *J. Palyno.*, 4: 29–33.

Palathingal, Trisa, 1990. Evaluation of industrial pollution of Bombay by pollen–I. *M.Phil. Thesis*, Univ. Mumbai.

Ram Indar, 1981. Effect of herbicides on plants of the Solanaceae–I. *M.Sc. Thesis*, Univ. Bombay.

Rasmussan, L., 1977. Epiphytic bryophytes as indicators of changes in the background levels of air borne metals from 1951 to 1975. *Environ. Pollut. Series A*, 14: 34–45.

Salgare, S.A., 1979. A criticism of Nair, Nambudiri and Thomas' paper entitled, 'A note on pollen germination at various stages of development of flower buds of balsam (*Impatiens balsam*)'. In: *Proc. of 66th Session of the Indian Sci. Cong.*, Held on January 3–7, at Hyderabad, Section of Bot., 3:61–62, Abstract No. 131.

Salgare, S.A., 1983. Pollen physiology of Angiosperms–I. *Ph. D.Thesis*, Univ. Bombay.

Salgare, S.A., 1984. Further evidence of a criticism on the hypothesis of Mhatre (1980), Mhatre, Chaphekar, Ramani Rao, Patil, Haldar (1980) and Shetye (1982). In: *Proc. 4th Indian Palyno. Conf.*, Held on March 19–21, at Andhra Univ., Visakhapatnam, Abstract No. S III–06.

Salgare, S.A., 1985a. A criticism on the findings of Mhatre (1980), Mhatre, Chaphekar, Ramani Rao, Patil, Haldar (1980) and Shetye (1982). In: *Proc. 2nd Ann. Conf. of Nat. Environ. Sci. Acad.*, Held on May 25–27, at Awadh Univ., Faizabad, Abstract No. 2.

Salgare, S.A., 1985b. Further evidence of a criticism of the findings of Mhatre (1980), Mhatre, Chaphekar, Ramani Rao, Patil, Haldar (1980), Shetye (1982) and Giridhar (1984). In: *Proc. Nat. Symp. on Environ. Sci. and Warm Water Aquculture-Finfish*, Held on November 6–9, at M.J.K.(P.G.) College, Bettiah, pp. 1–2, Abstract No. 2.

Salgare, S.A., 1985c. A criticism on the findings of Mhatre (1980), Mhatre, Chaphekar, Ramani Rao, Patil, Haldar (1980) and Shetye (1982). In: *Proc. Nat. Symp. on Environ. Sci. and Warm Water Aquculture-Finfish*, Held on November 6–9, at M.J.K.(P.G.) College, Bettiah, pp. 52, Abstract No. 73.

Salgare, S.A., 1986a. Effect of herbicides on pollen physiology of *Petunia axillaris* BSP. *Ph.D. Thesis*, World Univ.

Salgare, S.A., 1986b. A criticism on Ph.D. Thesis of Sudhakaran (1967) entitled, 'Cytogenetic studies in *Vinca rosea* Linn.' In: *Proc. of 1st Nat. Symp. on Environ. Biol.*, Held on December 30–31, at Deptt. Zool. and Microbiol., S. K. Univ., Anantapur, Abstract No. 5.

Salgare, S.A., 1986c. A criticism on Ph.D.Thesis of Dharurkar (1971) entitled, 'Effect of herbicides on the cytomorphology of *Eichhornia crassipes* Solms (Mart.). *Ph.D. Thesis*, Univ. of Bombay. In: *Proc. of 1st Nat. Symp. on Environ. Biol.*, Held on December 30–31, at Deptt. Zool. and Microbiol., S. K. Univ., Anantapur, Abstract No. 6.

Salgare, S.A., 1986d. A criticism on Ph.D. Thesis of Giridhar (1984) entitled, 'Study of interactions between industrial air pollutants and plants.' In: *Proc. 1st Nat. Symp. on Environ. Biol.*, Held on December 30–31, at Deptt. Zool. and Microbiol., S.K. Univ., Anantapur, Abstract No.14.

Salgare, S.A., 1986e. Pollen physiology of successive flowers. *D.Sc. Thesis*, Marathwada Univ.

Salgare, S.A., 2000. Additional evidence of a criticism on the hypothesis of Berg (1973), Brandt (1974), Vick and Bevan (1976), Rasmussan (1977), Navara, Horvath and Kaleta (1978), Mhatre (1980–Ph.D. Thesis), Mhatre, Chaphekar, Ramani Rao, Patil, Haldar (1980), Shetye (1982–Ph.D. Thesis), and Giridhar (1984–Ph.D. Thesis): A Critical Review. In: *Proc. 1st Nat. Conf. on Recent Trends in Life Management*, Held on October 22–23, at Bipin Bihari (P.G.) College, Jhansi, pp. 64–65, Abstract No. 92.

Salgare, S.A., 2001a. Monitoring of herbicide (nitrofen) toxicity by using pollen as indicators: A critical review–I. *Website www.microbiologyou.com. U.S.A's Publications (35 pages).*

Salgare, S.A., 2001b. Monitors of pollution – A criticism on the hypothesis of Berg (1973), Brandt (1974), Vick and Bevan (1976), Rasmussan (1977), Navara, Horvath and Kaleta (1978), Mhatre (1980 – Ph.D. Thesis), Mhatre, Chaphekar, Ramani Rao, Patil, Haldara (1980), Shetye (1982 – Ph.D. Thesis) and Giridhar (1984 – Ph.D. Thesis): A critical review–V. *Biojournal*, 13: 39–44.

Salgare, S.A., 2004. Whether optimum pollen germination and tube length attained in the same growth medium (sucrose + 2,4–D) by some species of Apocynaceae – A criticism to the hypothesis of Nair, Nambudiri and Thomas (1973): A critical review – IV. *Him. J. Env. Zool.*, 18: 53–56.

Salgare, S.A., 2005a. Whether optimum pollen germination and tube length attained in the same growth medium (sucrose + 2,4,5–T) by some cultivars of Apocynaceae and further evidence of a criticism to the hypothesis of Nair, Nambudiri and Thomas (1973): A critical review. *Flora and Fauna*, 11: 42–44.

Salgare, S.A., 2005b. Monitoring of herbicide (simazine) toxicity – A criticism on the hypothesis of Berg (1973), Brandt (1974), Vick and Bevan (1976), Rasmussan (1977), Navara, Horvath and Kaleta (1978), Mhatre (1980–Ph.D. Thesis), Mhatre, Chaphekar, Ramani Rao, Patil, Haldar (1980), Shetye (1982–Ph.D. Thesis) and Giridhar (1984–Ph.D. Thesis) – A critical review. *Him. J. Environ. Zool.* 19: 69–71.

Salgare, S.A., 2005c. Whether optimum pollen germination and tube length attained in the same growth medium (sucrose + nitrofen)–A criticism to the hypothesis of Nair, Nambudiri and Thomas (1973): A critical review. *Him. J. Environ. Zool.*, 19: 93–95.

Salgare, S.A. 2005d. Monitoring of herbicide (dalapon) toxicity by using pollen as indicators – Pollen of some cultivars of Apocynaceae and further evidence of a criticism on the hypothesis of Berg (1973), Brandt (1974), Vick and Bevan (1976), Rasmussan (1977), Navara, Horvath and Kaleta (1978), Mhatre (1980–Ph.D. Thesis), Mhatre, Chaphekar, Ramani Rao, Patil, Haldar (1980), Shetye

(1982–Ph.D. Thesis) and Giridhar (1984–Ph.D. Thesis): A critical review. *Flora and Fauna*, 11: 49–50

Salgare, S.A., 2005e. Monitoring of herbicide (basalin EC) toxicity – A criticism on the hypothesis of Berg (1973), Brandt (1974), Vick and Bevan (1976), Rasmussan (1977), Navara, Horvath and Kaleta (1978), Mhatre (1980–Ph.D. Thesis), Mhatre, Chaphekar, Ramani Rao, Patil, Haldar (1980), Shetye (1982–Ph.D. Thesis) and Giridhar (1984–Ph.D. Thesis): A critical review. *Internat. J. Biosci. Reporter*, 3: 298–300.

Salgare, S.A., 2006a. Monitoring of herbicide (2,4–D) toxicity by using pollen as indicators – Pollen of Apocynaceae and further evidence of a criticism of the hypothesis of Berg (1973), Brandt (1974), Vick and Bevan (1976), Rasmussan (1977), Navara, Horvath and Kaleta (1978), Mhatre (1980–Ph.D. Thesis), Mhatre, Chaphekar, Ramani Rao, Patil, Haldar (1980), Shetye (1982–Ph.D. Thesis) and Giridhar (1984–Ph.D. Thesis): A critical review. *Internat. J. Plant Sci.*, 1: 134–136.

Salgare, S.A., 2006b. Whether optimum pollen germination and tube length attained in the same growth medium (sucrose + sodium arsenite) and further evidence of a criticism of the hypothesis of Nair, Nambudiri and Thomas (1973): A Critical Review. *Internat. J. Plant Sci.*, 1: 132–133.

Salgare, S.A., 2006c. Whether optimum pollen germination and tube length attained in the same growth medium (sucrose + atrataf 50w)–A criticism to the hypothesis of Nair, Nambudiri and Thomas (1973): A critical review. *Internat. J. Biosci. Reporter*, 4: 145–146.

Salgare, S.A. and Bindu, 2002. Whether optimum pollen germination and tube length attained in the same growth medium (sucrose + calcium chloride) by pink-flowered *Catharanthus roseus*–A criticism to the hypothesis of Nair, Nambudiri and Thomas (1973): A critical review–I. *Biojournal*, 14: 21–23.

Salgare, S.A. and Bindu, 2005. Whether optimum pollen germination and tube length attained in the same growth medium (sucrose + calcium chloride) by white-flowered *Catharanthus roseus* and further evidence of a criticism on the hypothesis of Nair, Nambudiri and Thomas (1973): A critical review. *Flora and Fauna*, 11: 183–184.

Salgare, S.A. and Phunguskar, K.P., 2002. Monitors of pesticide (alphamethrin) toxicity by using pollen as indicators – Pollen of pink-flowered *Catharanthus roseus* – A criticism on the hypothesis of Berg (1973), Brandt (1974), Vick and Bevan (1976), Rasmussan (1977), Navara, Horvath and Kaleta (1978), Mhatre (1980–Ph.D. Thesis), Mhatre, Chaphekar, Ramani Rao, Patil, Haldara (1980), Shetye (1982–Ph.D. Thesis) and Giridhar (1984–Ph.D. Thesis): A critical review – III. *Biojournal*, 14: 1–4.

Salgare, S.A. and Pathak, Sanchita, 2005. Monitoring of heavy metal (bismuth nitrate) toxicity by using pollen as indicators–Pollen of white-flowered *Catharanthus roseus* and further evidence of a criticism on the hypothesis of Berg (1973), Brandt (1974), Vick and Bevan (1976), Rasmussan (1977), Navara, Horvath and Kaleta (1978), Mhatre (1980–Ph.D. Thesis), Mhatre, Chaphekar, Ramani Rao, Patil, Haldar (1980), Shetye (1982–Ph.D. Thesis), and Giridhar (1984–Ph.D. Thesis): A critical review. *Flora and Fauna*, 11: 69–70.

Salgare, S.A. and Singh, Sanju, 2002. Monitors of heavy metal (ferrous sulphate) toxicity by using pollen as indicators–Pollen of white-flowered *Catharanthus roseus*–A criticism on the hypothesis of Berg (1973), Brandt (1974), Vick and Bevan (1976), Rasmussan (1977), Navara, Horvath and Kaleta (1978), Mhatre (1980–Ph.D. Thesis), Mhatre, Chaphekar, Ramani Rao, Patil, Haldara (1980),

Shetye (1982–Ph.D. Thesis) and Giridhar (1984–Ph.D. Thesis): A critical review–II. *Biojournal*, 14: 5–7.

Salgare, S.A. and Antony, Tessy Mol, 2005a. Whether optimum pollen germination and tube length attained in the same growth medium (sucrose + potassium chloride) by pink-flowered *Catharanthus roseus* and further evidence of a criticism to the hypothesis of Nair, Nambudiri and Thomas (1973): A critical review. *Flora and Fauna*, 11: 61–62.

Salgare, S.A. and Antony, Tessy Mol, 2005b. Whether optimum pollen germination and tube length attained in the same growth medium (sucrose + sodium oxalate) by white–flowered *Catharanthus roseus* and further evidence of a criticism to the hypothesis of Nair, Nambudiri and Thomas (1973): A critical review. *Flora and Fauna*, 11: 75–76.

Salgare, S. A. and Sebastian, Theresa 1986. Further evidence of a criticism of the hypothesis of Berg (1973), Brandt (1974), Vick and Bevan (1976), Rasmussan (1977), Navara, Horvath and Kaleta (1978), Mhatre (1980–Ph.D. Thesis), Mhatre, Chaphekar, Ramani Rao, Patil, Haldar (1980), Shetye (1982–Ph.D. Thesis) and Giridhar (1984–Ph.D. Thesis). In: *Proc. 1st Nat. Symp. on Environ. Biol.*, Held on December 30–31, at Deptt. Zool. and Microbiol., S. K. Univ., Anantpur. Abstract No.15.

Schoch-Bodmer, H., 1921. Reservestoffe bei einigen anemophilen Pollenarten. *Vjschr. Naturf. Gas. Zurich*, 66: 339–346.

Sebastian, Theresa, 1987. Effect of herbicides on crop physiology–I. *Ph.D. Thesis*, Univ. Bombay.

Shetye, R.P., 1982. Effect of heavy metals on plants. *Ph.D. Thesis*, Univ. Bombay.

Singh, S.R., 1985. Effect of herbicides on pollen physiology of Brinjal. *M.Sc. Thesis*, Univ. Bombay.

Ethnobotanical Wisdom and Microbial Studies on Medicinal Plants (2011) *Pages* **203–206**
Editors: **D.R. Khanna, Ashutosh Gautam, R. Bhutiani, Gagan Matta, Anil Dhiman**
Published by: **DAYA PUBLISHING HOUSE, NEW DELHI**

Chapter 33

Folicolous Fungi of Medicinal Plants in North Western Tarai Region of Uttar Pradesh

D.P. Singh and T.P. Mall
Post Graduate Department of Botany,
Kisan P.G. College, Bahraich, U.P.

ABSTRACT

The North Western Tarai belt of U.P. is next only to Eastern and Western ghats as one of the hot spots of Biodiversity in general and the diversity of fungal organisms inhabiting point leaves in particular. Keeping the above view in mind a survey trip was organized for Bhinga forest village, Shrawasti on October 29, 2006 and Dudwa Tiger Reserve, Kheri Lakhimpur on November 23 and 24, 2006 for collection and documentation about ethnomedicinal plants as well as Folicolous fungi infecting the medicinal plants. During one survey more than fifty five ethnomedicinal plants were collected. Out of these plants about 25 plants species represented by 20 Genera of sixteen families have been found infected with sixteen different fungal species. *Cercospora* sp. and *Stenella* sp. are recorded to infected four plants each. *Cercospora* was recorded on *Cassia fistula, Momordica* sp., *Clerodendron* and *Lagerstraemia* where as *Stellella* sp. was recorded on *Celastrus panicuatus, Litisea grutinosa, Teliocora acuminate, Ficus sacra. Phoma* sp. was found on *Glycosmis pentaphylla, Teliocorpa* sp. and *Mallotus philipinsis. Zygosporium* was recorded on *Mallotus philipinsis* and *Sterculia balen.* Remaining fungus were recorded only on single host *viz. Leptozyphium* on *Diosphros, Stellella cassiae* on *Cassia fistula, Alternaria* sp. on *Ichnocarpus frutiscens, Sirosporium* on *Carica carandus, Passalora* sp. on *Calotropis procera,* uredial stage of rust on *Tectona grandis, Pseudocercospora* on *Ziziphus, Meliola* sp. on *Mallotus philipinsis, Acremonium zonatum* on *Teliocora* sp. *Fuligomyces* on *Litisea chinensis; Cladosporium* on *Ficus carica* and *Cephaloros* sp. on *Justicia* sp.,

Teliocora sp. had been recorded to be infected by three different fungus *viz., Phoma* sp., *Stenella* sp. and *Acremonium zonatum* where as *Mallotus philippinsis* was found infected with *Zygosporium, Phoma* sp. and *Meliola* sp.

Keywords: *Folicolous fungi, Medicinal plants, Ethnomedicine.*

Introduction

India is inhabited a wide variety of about 563 tribal communities who dwell in forest areas and depend on surrounding resources for their livelihood. Plants with medicinal properties enjoyed the highest reputation in the indigenous system of medicines all over the world. The state of Uttar Pradesh is a rich repository of ethnomedicinal plants diversity having large number of tribal communities, inhabiting in remote village and mostly in village of forest areas. Though there is nothing in this universe, which is non-medicinal, which can not be made use for many purposes and by many modes (Shankar *et al.,* 2000). In tribes since the information about plants is passed from one generation to another orally as well as they have their own names for all the plants they recognize, it is necessary to document the names of the plants and their way of uses. The leaves provide a very suitable habitat for the growth and development of endophytic fungal pathogens by providing ample surface area and nutrient supply. Such leaf inhabiting fungi are known as "Folicolous" or "Foliar" and the invaded areas of the leaf as leaf spot or leaf lesions. The fungi interfere with the physiology of the host and develop a characteristic spot and produce toxins.

The North-Western Tarai belt of U.P. is next only to Eastern and Western Ghats as one of the hot spot for Biodiversity in general and the diversity of fungal organisms inhabiting plant leaves in particular.

Keeping the above views in mind a survey trip was organized for Bhinga Forest Range, Shrawasti on October 29, 2006 and Dudwa Tiger reserve, Kheri Lakhimpur on November 23 and 24, 2006 for collection and documentation about ethnomedicinal plants as well as Folicolous fungi infecting the medicinal plants.

Materials and Methods

Method of collection of ethnomedicinal plant and recording its data was followed as described by Jain, 1989. The selected specimens were pressed and dried by routine herbarium technique as described by Jain and Rao (1978). The specimens were processed and given a field note number and field note data showing name of the plant with family, local name, locality of collection, the plant parts used in different ailment, route of administration, form of medicines, mode of administration and any other special feature regarding ethnomedicinal uses and date of collection. Plants were identified with the help of different flora available in the departmental library. The laboratory processing for fungus were done by scrap mount, Collodion preparation, Squash preparation and land cut section preparation. For identification of foliar fungi collected, a thorough survey of literature was done by going through publication made all over the world and consulting the different Mycological papers, Mycological Memoirs and other relevant Mycological monographs.

Results and Discussion

During the survey more than fifty ethnomedicinal plants were collected. Out of these plants about 25 plants species represented by 20 genera of 16 families have been found infected with sixteen different fungal species (Table 33.1).

Cercospora sp. and *Stenella* sp. are recorded to infected four plants each. *Cercospora* was recorded on *Cassia fistula, Momordica* sp., *Clerodendron* and *Lagerstraemia* where as *Stenella* sp. was recorded on *Celastrus paniculatus, Litisea grutinosa, Teliocora acuminate, Ficus sacra, Phoma* sp. was found on

Table 33.1: Ethnomedicinal Plants Infected with Folicolous Fungi

Sl.No.	*Name of Ethnomedicinal Plant*	*Family*	*Name of Folicolous Fungus*
1.	*Carisa carandus* L.	Apocynaceae	*Sirosporium* sp.
2.	*Cassia fistula* L.	Caesalpinaceae	*Stenella cassia* sp.
3.	*Calotropis procera* R.Br.	Asclepiadaceae	*Passalora* sp.
4.	*Clerodendron* sp. Linn.	Verbenaceae	*Corenospora* sp.
5.	*Celastrus paniculatus* Willd.	Apocynaceae	*Stenella* sp.
6.	*Diosphyros gurke*	Ebenaceae	*Leptozyphium* sp.
7.	*Dolichos lablab* Linn.	Fabaceae	*Cercospora* sp.
8.	*Ficus sacra*	Moraceae	*Stenella* sp.
9.	*Ficus carica* Linn.	Moraceae	*Cladosporium* sp.
10.	*Glycosmis pentaphylla* Correa	Rutaceae	*Phoma/Phomopsis* sp.
11.	*Ichnocarpus frutiscens* R. Br.	Apocynaceae	*Alternaria* sp.
12.	*Justicia* sp. Linn.	Acanthaceae	*Cephaloros* sp.
13.	*Lagerstroemia paruiflora* Linn.	Lythraceae	*Cercospora* sp.
14.	*Litisea grutinosa* Lamk.	Lauraceae	*Stenella* sp.
15.	*Litisea chinensis* Lamk.	Lauraceae	*Fuligomyces* sp.
16.	*Mallotus philippinsis* Mull Arg.	Euphorbiaceae	*Phoma* sp.
17.	*Mallotus philippinsis* Mull Arg.	Euphorbiaceae	*Zygosprium* sp.
18.	*Mallotus philipinsis* Mull Arg.	Eupharbiaceae	*Meliola* sp.
19.	*Momordica* sp. Roxb.	Sapotaceae	*Cercospora* sp.
20.	*Sterculia balen* L.	Sterculiceae	*Zygosporium* sp.
21.	*Tectona grandis* Linn.	Verbenaceae	*Uredo* sp.
22.	*Telicora* sp. Coleber	Menispermaceae	*Phoma* sp.
23.	*Telicora acuminata* Coleber	Menispermaceae	*Stenella* sp.
24.	*Teliocora* sp. Coleber	Menispermaceae	*Acremonium zonatum*
25.	*Ziziphus* sp. L.	Rhamnaceae	*Pseudocercospora* sp.

Glycosmis pentaphylla, Teliocora sp. and *Mallotus philippinsis. Zygosporium* was recorded on *Mallotus philipinsis* and *Sterculia balen.* Remaining fungus were recorded only on single host *viz. Leptozyphium* on *Diosphros, Stenella cassiae* on *Cassia fistula, Alternaria* sp. on *Ichnocarpus frustiscens, Sirosporium* on *Carica carandas; Passalora* sp. on *Calotropis procera;* uredial stage of rust on *Tectona grandis, Pseudocercospora* on *Ziziphus; Meliola* sp. on *Mallotus philipinsis; Acremonium zonatum* on *Teliocora* sp; *Fuligomyces* on *Litisea chinensis; Cladosporium* on *Ficus carica* and *Cephaloros* sp. on *Justicia* sp. *Teliocora* sp. had been recorded to be infected by three different fungus *viz., Phoma* sp., *Stenella* sp. and *Acremonium zonatum* where as *Mallotus philippinsis* was found infected with *Zygosporium, Phoma* sp. and *Meliola* sp.

The ethnic and rural people of India have preserved a large bulk of traditional knowledge of medicinal use of plants growing around them. India is one of the twelve mega biodiversity countries of the world having vegetation with a wide variety of plants with medicinal value. Herbal medicines have good value in treating many diseases including infectious diseases. Ethnomedicinal knowledge can not only save lives of many, it is also important from humanitarian point of view in that in long run this knowledge may help to identify important medicinal uses that can help in curing health care around the world. So it is important to protect the traditional knowledge from disappearing and documenting the indigenous knowledge.

The folicolous fungal pathogens interfere with the manufacturing rate of food and other valuable substances by damaging the photosynthetic elements of living leaves; bringing about qualitative and quantitative dearrangements in the living tissue of the host in various ways. Several leaf spot pathogens are known to produce toxins of various kinds, some of which are host specific. When the leaf spots are numerous or are large, there is a considerable reduction in the photosynthetic area of the leaf. In addition some times rapid defoliation occurs due to such infections. As such the productivity of the host plants is substantially reduced. The weeds and forest plants serve as reservoirs of leaf spot pathogens which on getting opportunity may spread to agricultural and horticultural plants. The destruction caused by these enemies of leaves is a serious problem before us.

References

Jain, S.K., 1989. *Methods and Approaches in Ethnobotany.* Society of Ethnobotanists, I.ucknow.

Jain, S.K. and Rao, R.R., 1978. *A Handbook of Field and Herbarium Methods.* Today and Tomorrow Printers and Publishers, New Delhi, p. 33–58.

Shankar, D., Ved, D.K. and Geetha, V.G., 2000. A green pharmacy: Indian health traditions. *The Hindu*, Special Issue with the Sunday Magazine October 8, p. 1–2.

Ethnobotanical Wisdom and Microbial Studies on Medicinal Plants (2011) *Pages* **207–209**
Editors: **D.R. Khanna, Ashutosh Gautam, R. Bhutiani, Gagan Matta, Anil Dhiman**
Published by: **DAYA PUBLISHING HOUSE, NEW DELHI**

Chapter 34

Studies on the Natural Infection of Mosaic Disease of Brinjal (*Solanum melongena* L.) in Tarai Region of Eastern U.P.

Alka Gautam, R.K. Shukla, Nafees Ahmed Ansari, G.P. Srivastava and J.P. Tewari
Botany Department, M.L.K. (P.G.) College, Balrampur, U.P.

ABSTRACT

Brinjal crops were surveyed for the incidence of mosaic disease in Devipatan mandal of Tarai region, U.P. On the basis of symptoms, host range, physical properties, sap transmission, insects transmission different hosts and seed transmissions, three different virus isolates were characterized as PVY, TRSV and TMV.

Keywords: *Disease incidence, Brinjal, Mosaic disease.*

Introduction

Brinjal or eggplant (*Solanum melongena* L.) is one of the most commonly cultivated vegetable crops in India, especially in eastern Uttar Pradesh and is presently affected by a number of viral diseases. During routine survey of Brinjal growing fields various types of Mosaic and ring spot symptoms were noticed viral disease characterized by chlorotic, vein banding, oak leaf pattern, yellow blotches, slight puckering and stunting were frequently observed in crop (var. Pusa Purple long and Pusa Purple round) around Balrampur district of Devipatan mandal. One isolates among these was transmitted by aphids readily. The identification of various mosaic diseases of brinjal is an objective of the present study. This communication deals with the results of the studies pertaining to the identification of various diseases, conducted at Post Graduate Department of Botany, Balrampur, U.P.

Materials and Methods

The cultures of these diseases were maintained on the brinjal var. Pusa Purple long and Pusa Purple round in glass house conditions. Based on the symptoms they were grouped into three categories and labeled as virus A, B and C. The infected leaves with chlorotic rings and oak patterns with yellow patches, virus C infected leaves exhibited mild mosaic symptoms with necrotic spots on stem and leaves. The mechanical sap transmission studies of these three viruses were conducted by macerating the infected leaf material in phosphate buffer pH-7 (0.05M) and inoculated to the test plants *Solanum melongena* L.

Aphid transmission studies were conducted by using *Aphis gossypii* Olov and *Myzus peersicae* Sulz. During the transmission studies the aphids were given two hour fasting followed by 30 min. each acquisitation and inoculation periods and brinjal was used as a test plant. Seed transmission tests were also conducted by collecting the seeds from the fruits of infected plants and were raised in the glass house.

Results and Discussion

Based on the host range and physical properties, aphid and seed transmission viruses A, B and C were identified as potato Y(PVY), tobacco ring spot (TRSV) and tobacco mosaic virus (TMV). For quick identifications of these three viruses, a key having a set of four differential hosts (*Physalis floridana* Rydb., *Cucumis sativus* L., *Phaseolus vulgaris* var. Pinto and *Vigna unguiculata* (Linn.) Walp sub. Sp. *Cylindrica* van-Eseltine) was developed and their reaction are given (Table 34.1). Only virus A was transmitted by both the aphids and the transmission was 70 per cent. The results indicated that only viruses B and C were transmitted and the percentage of transmission was 12 and 41 respectively.

From India, Sastry *et al.* (1974) reported a strain of PVY which had thermal in activation point between 55–60°C, dilution and end point between 1/1000 to 1/10000, and ageing *in vitro* for 24 hours. The present PVY isolate is different with this isolate in physical properties (Table 34.2). Tobacco ring spot virus under study differs with the report of Sastry and Nayudu (1976) and also Sastry (1982) both in host range and physical properties. Hence, it is new strain TRSV occurring on brinjal, which is not reported earlier in Eastern U.P. The third virus C, TMV differs with the report of Naqvi and Mahmood (1976, 1978) in having higher thermal inactivation point and also in host range and this being different reported from Tarai region of eastern U.P.

In the present study, *Physalis floridana* Rydb., *Cucumis sativus* L., *Phaseolus vulgaris* var. and *Vigna unguiculata* (Linn) Walp sub. Sp. *Cylindrica* van-Eseltine were found to be good differential hosts for PVY, TRSV and TMV strains infecting brinjal. *Physalis floridana* produced necrotic local lesions only with PVY. *Cucumis sativus* L., and *Vigna unguiculata* (Linn) reacted to TRSV and produced mosaic mottling followed by chlorotic rings and reddish brown necrotic local lesions respectively. *Phaseolus vulgaris* var. Pinto produced mosaic mottling and severe mosaic mottling with TRSV and TMV on inoculation. With the help of these four differential hosts, the three viruses under study can easily be identified during the germplasm screening in the breeding programme.

Table 34.1: Host Range of Viruses Infecting Brinjal

Host Range	*Reaction*		
	Virus-A	*Viruse-B*	*Virus-C*
Solanum melongena L.	SMM	Ch.R.	MM
Capsicum annum L.	MM	Ch.R., NLL	MM
Physalis floridana Rydb.	NLL		–
Datura stramonium L.	–	NLL, Ch.R.,	MM
Datura metel L.	MM	–	MM
Phaseolus vulgaris var. plato	SMM	MM	MM
Lablab niger L.	–	NLL	NLL
Nicotiana tabacum var. WB(L)	MM, VN	NLL	MM
Nicotiana glutinosa L.	MM	NLL, Ch.R.	NLL
Cucumis sativus L.		NLL, Ch.R.	MM
Vigna ungulculata (Linn.) Walp sub.sp. *Cylindrica* van-Eseltine		NLL	Ch.R.
Chenopodium amaranticolor Coste and Reyn.	NLL	NLL	NLL

SMM: Severe mosaic mottling, Ch.R.: Chlorotic ring; MM: Mosaic mottling; NLL: Necrotic local lesions; VN: Velnal necrosis; –: No reaction.

Table 34.2: Physical Properties of Infecting Brinjal

Physical Properties	*Virus-A*	*Virus-B*	*Virus-C*
Thermal inactivation point	60–65°C	65–70°C	90–95°C
Dilution end point	1/1000	1/1000	1/10,0000
	10000	1/10000	1/10,00,000
Ageing *in vitro*	5 days	10 days	95 days

Acknowledgements

The authors are grateful to Head, Department of Botany, M.L.K. (P.G.) College, Balrampur, U.P. for providing necessary facilities.

References

Naqvi, Q.A. and Mahmood, K., 1976. Studies on brinjal mild mosaic virus. *Indian Phytopath.*, 29: 138–142.

Naqvi, Q.A. and Mahmood, K., 1978. Purification and properties of brinjal mild mosaic virus. *Phytopath. Z.*, 93: 86–89.

Sastry, K.S., 1982. Studies on the identification of mosaic disease of brinjal (*Solanum melongena*) in Karnataka. *Curr. Sci.*, 51: 568–569.

Sastry, K.S. and Naydu, M.V., 1976. Ring spot symptoms of eggplant incited by tobacco ring spot virus. *Phytopath. Medit.*, 15: 60–62.

Sastry, K.S., Sastry, K.S.M. and Singh, S.J., 1974. A mosaic disease of eggplant (*Solanum melongena* L.). *Phytopath. Medit.*, 13: 176–178.

Ethnobotanical Wisdom and Microbial Studies on Medicinal Plants (2011) *Pages* **210–213**
Editors: **D.R. Khanna, Ashutosh Gautam, R. Bhutiani, Gagan Matta, Anil Dhiman**
Published by: **DAYA PUBLISHING HOUSE, NEW DELHI**

Chapter 35

Deviation in Nitrogen Metabolism of Ridge Gourd (*Luffa acutangula* Roxb.) Infected with Three Strains of Watermelon Mosaic Virus

Deepika Shukla, R.K. Shukla, Nafees Ahmad Ansari, G.P. Srivastava and J.P. Tewari
Botany Department, M.L.K. (G.P.) College, Balrampur – 271 2001, U.P.

ABSTRACT

The effect of three strains of Watermelon mosaic virus (WMV) infection on the nitrogen metabolism of *Luffa acutallgula* leaves were studied and found that the infected levels of total nitrogen, total protein, nitrate and nitrate nitrogen and some of free amino acids (aspartic, proline, serine and cysteine) were higher in comparison to healthy plants, but ammonical nitrogen, threonine, alanine and lysine were low.

Keywords: *Ridge gourd, Nitrogen metabolism, WMV.*

Introduction

A virus disease found to be most prevalent in eastern D.P. on *Luffa actangula* Roxb. was characterized by severe mosaic and reduction in leaf size, distortion and chlorotic spotting and considerable loss in the yield of crop. The causal agent was earlier reported to be a strain of watermelon mosaic virus (Shukla *et al.*, 2004) Physiological disturbances which are induced by the viral infection have been reviewed by many workers (Bawden, 1959; Diener, 1963; Yarwood, 1967). In India, work on physiology of virus infected plants is studied for a very few diseases (Sadashivan, 1963). The present investigation was therefore, undertaken to study the deviation in nitrogen metabolism of *Luffa acutangula* leaves infected by watermelon mosaic virus.

Materials and Methods

Luffa acutangula (Roxb.) plants were used as host and three strains of Watermelon mosaic virus (WMV) *viz.*, WMVVB, WMVMM and WMVC were used as pathogen. Five day old plants were taken into four groups, each containing fifteen seedlings. Seedlings of the first, second and third groups were inoculated mechanically by usual method with all three strains of Watermelon mosaic virus, while seedling of the fourth group served as healthy control. A random method of composite samplings was used for estimation and extraction were obtained from fifty days old healthy and diseased plants of uniform size. Three replicates were taken for each estimation and the average of same is reported in the results. Total nitrogen was determined in day samples by ash analysis method of Snell and Snell (1949.) Protein nitrogen and non-protein nitrogen by method of Pregal (1945). Nitrate nitrogen and nitrite nitrogen by Humphries (1956), ammonical nitrogen by Stronganov (1964) and tree amino acid was determined according to Draper (1963).

Results and Discussion

It is evident from the result given in Tables 35.1 and 35.2 that total nitrogen and protein were higher in diseased leaves, maximum in WMVVB followed by WMVMM and WMVC strain. The level of nitrate nitrogen was lower in the diseased leaves. Nitrite nitrogen was higher in diseased leaves but ammonical nitrogen was lower, in diseased samples, whereas protein nitrogen was higher in diseased leaves. Non-protein nitrogen was reduced in diseased leaves. Among free amino acids, aspartic serine, glutamic acid, cystine and proline were higher, but threonine, alanine, glycine and lysine were lower in diseased leaves.

Table 35.1: Influence of Three Strains of Watermelon Mosaic Virus Infection on Nitrogenous Fractions (µg/100 µg dry wt.) of *Luffa acutangula* Roxb. Leaves

Nitrogen Fraction	*Healthy*	*Diseased*			*S.D.*
		WMVVB	*WMVMM*	*WMVC*	
Total N_2	3.50	4.08	3.95	3.85	0.219
Protein	11.20	13.90	13.50	13.24	1.112
Nitrate N	0.128	0.142	0.138	0.133	0.005
Nitrite N	15.44	16.82	16.52	16.26	0.613
Ammonical N	0.142	0.126	0.122	0.119	0.011
Protein N	5.60	6.41	5.81	5.63	0.429
Non-protein N	2.85	3.95	3.77	3.64	0.537

It is now almost generally agreed the host viral infection of plants should be regarded as change in the nitrogen metabolism of the host (Bawden 1959). The increased levels of different nitrogenous fractions in virus infected plants is in accordance with previous observations (Diener, 1963; Harman *et al.*, 1970). Ammonical nitrogen decreased in virus infected leaves. Similar observations were found by Commoner and Dietz. (1952). Nambiar (1966) reported that the increased activity of nitrate reductase may be one of the reasons for increased nitrite nitrogen, as developed here due to viral infection.

Table 35.2: Influence of Three Strains of Watermelon Mosaic Virus Infection on Nitrogenous Fractions (µg/100 µg dry wt.) of *Luffa acutangula* Roxb. Leaves

Amino Acids	*Healthy*	*Diseased*			*S.D.*
		WMVVB	*WMVMM*	*WMVC*	
Aspartic	20.8	292.5	286.7	281.5	102.615
Glutamic	196	228.4	216.3	211.8	97.81
Threonine	46.1	16.2	15.2	14.8	14.422
Serine	11.8	32.4	31.7	31.5	9.304
Alanine	72.4	23.9	22.8	21.8	26.75
Glycine	26.3	18.5	17.5	16.3	4.418
Proline	11.1	14.5	14.0	13.8	1.705
Cysteine	50.8	60.6	61.4	60.8	5.132
Lysine	30.9	24.8	23.5	22.6	3.628

Luffa acutangula Roxb. leaves infected with three strains of Watermelon mosaic virus reveals pronounced changes in the free amino acid pool. Infected leaves in comparison to healthy leaves contain higher levels of some amino acids (aspartic, glutamic, serine, praline and cystine) and lower levels of others (threonine, alarnine, glycine and lysine). Similar results have been reported by Mohanty and Sridhar (1982) in RTV infected rice plants.

Selman *et al.* (1961) observed that increased nitrogenous fractions may be due to check of growth and normal protein synthesis. Goodman *et al.* (1967) and Commoner and Dietz. (1953) reported the active utilization of non-protein complex for the synthesis of virus during nitrogen deficient conditions results in the decrease of some amino acids.

Acknowledgements

The authors are thankful to Dr. S.V. Singh, Head, Dr. O.P. Mishra, Principal of the college for providing facilities.

References

Bawden, F.C., 1959. Physiology of virus diseases. *Ann. Rev. Pl. Physiol.*, 10: 253–299.

Commoner, B. and Dietz, P.M., 1952. Changes in non-protein nitrogen metabolism during tobacco mosaic biosynthesis. *J. Gen. Physiol.*, 35: 447–456.

Commoner, B. and Dietz, P.M., 1953. Relationship between tobacco mosaic virus biosynthesis and the nitrogen metabolism of the host. *J. Gen. Physiol.*, 36: 807–830.

Diener, T.O., 1963. Physiology of virus infected plants. *Ann. Rev. Phytopathol.*, 1: 197–218.

Draper, S.R., 1963. Physiology of virus infected plants. *Phytochem.*, 11: 639–641.

Goodman, R.N., Kiraley, Z. and Zaitlin, 1967. *The Biochemistry and Physiology of Infectious Plants Diseases.* D. Van Nostrand, Inc. Princton, New Jersey.

Harman, G.E., Gooding, G.V. and Herbet, T.T., 1970. Effect of Tobacco mosaic virus infection in some chemical constituents of the cured tobacco. *Tobacco*, 171: 29–31.

Humphries, E.G., 1956. Mineral components and ash analysis. In: *Modern Methods of Plant Analysis*, (Eds.) K. Peach, and M.V. Tracey. Springer Verlag, Berlin, I. Gottingen, Heidelberg, pp. 468–502.

Mohanty, S.K. and Sridhar, R., 1982. Physiology of rice tungro virus disease: Proline accumulation due to infection. *Physiol. Pl.*, 56: 89–93.

Nambiar, K.K.N., 1966. Studies on pigeon pea sterility mosaic disease. *Ph.D. Thesis*, University of Madras.

Pregal, F., 1945. *Quantitative Organic Micro Analysis*. J. and Churchil, A. Ltd., London.

Sadashivan, T.T., 1963. Physiology of virus infected plants. *J. Indian Bot. Soc.*, 42: 339–357.

Selman, I.W., Brierly, M.R., Pegg. G.E. and Hill, T.A., 1961. Changes in free amino acids in tomato plants inoculated with tomato spotted wilt virus. *Ann. Appl. Biol.*, 49: 601–615.

Shukla, D., Ansari, N.A., Gautam, A., Srivastava, G.P. and Tewari, J.P., 2004. Occurrence of a mosaic disease of Ridge gourd (*Luffa acutangula* Roxb.) from eastern Uttar Pradesh. *J. Liv. World*, 11: 15–20.

Snell, E.D. and Snell, C.T., 1949. *Colorimetric Methods of Analysis*, 3rd edn. D. Van. Nostrand Co. Inc., New York, pp. 813–819.

Stroganov, N.S., 1964. Method for ammonia determination used in studies on fish metabolism. In: *Techniques for Investigation of Fish Physiology*, (Ed.) E.N. Pavlovskii. Israel Progr. Sci. Transl., Jerusalem.

Tripathi, S.C., 1989. Studies on a mosaic disease of bottle gourd (*Lagenaria vulgaris* L.). *Ph.D. Thesis*, Avadh University, Faizabad.

Yarwood, C.E., 1967. Response to parasite. *Ann. Rev. Pl. Physiol.*, 18: 413–438.

Ethnobotanical Wisdom and Microbial Studies on Medicinal Plants (2011) *Pages* **214–226**
Editors: **D.R. Khanna, Ashutosh Gautam, R. Bhutiani, Gagan Matta, Anil Dhiman**
Published by: **DAYA PUBLISHING HOUSE, NEW DELHI**

Chapter 36

Microbial Degradation of Urea from Urea Bearing Wastewaters

P.P. Pathe[1], T. Nandy[1], S.N. Kaul[1], C.V. Deshpande[1] and L. Szpyrkowicz[2]
[1]National Environmental Engineering Research Institute, Nagpur – 440 020, India
[2]Department of Environmental Sciences, Venice, Italy

ABSTRACT

Urea and ammonia are mostly present in wastewaters from nitrogenous fertilizer industries. Urea can not be oxidized by usual oxidation methods, microbiological hydrolysis of urea is necessary for removal of urea from wastewater. Biological degradation of urea is a two staged process (*i*) urea hydrolysis and (*ii*) ammonia stripping/nitrification-denitrification. Microbiological hydrolysis of urea through biohydrolyzer removes urea from fertilizer effluents, in which ureolytic bacteria *Bacillus pasteurii* converts urea into ammonia and carbon dioxide through ammonium carbonate as an intermediate product. Removal of ammonia is either by stripping or converting into nitrate on nitrification by chemoautotrophic bacteria *Nitrosomonas* sp. and further by *Nitrobacter* sp. into nitrate. On denitrification nitrate is finally converted into nitrogen gas by means of heterotrophic bacteria. Based on earlier laboratory investigations, urea bio-hydrolyser was designed, installed and commissioned in fertilizer industry as additional full scale unit to treat urea and ammonia bearing wastewaters. Ammonia stripped effluent was mixed with septic tank effluents from industry and township sewage and routed through various lagoons cultured with algae–chlorella to minimize nitrogen. Evaluation of full scale urea bio-hydrolyser and effluent treatment plant before and after modifications at ETP is discussed in this study.

Keywords: *Urea, Ammonia, Nitrogenous fertilizer industry, Microbiological treatment, Nitrification, Urea bio-hydrolysis.*

Introduction

In India, per hectare consumption of fertilizer nutrients has been steadily increasing from early 0.55 kg in 1950–51 to 68.00 kg in 1995. The production capacity of nitrogenous fertilizer in the country has boost up from a modest capacity of 85,000 tonnes in 1950–51 to 9.480 million tones in March, 1997. The projected production by the end of 2004–05 is 9.940 million tonnes (India 1995 and Monthly Review of Indian Economy 1997). Coal based nitrogenous fertilizer industries in India produced mostly ammonia, urea, nitric acid, ammonium nitrate and ammonium bicarbonate using coal and fuel oil as raw materials. Coal is converted into carbon-dioxide. Ammonia is manufactured from atmospheric nitrogen (air) and steam. Urea and ammonium bicarbonate are manufactured from carbon dioxide and ammonia. Ammonium nitrate is synthesized from nitric acid and ammonia. Therefore, the wastewater from fertilizer industry contains nitrogenous compounds *viz.*, urea, ammonia, nitrate and ammonium bicarbonate. Urea and ammonia are the major constituents of wastewater from nitrogenous fertilizer industry manufacturing ammonia and urea as the major products. The volume of wastewater from ammonia production generally varies from 4.9 to 6.7 m^3/tonne of ammonia produced with concentrations of ammonia and urea varying from 4.65 to 10.0 and 15.0 to 17.5 kg/tonne of ammonia produced, respectively. Wastewater generation from urea plant varies from 8.0 to 12.0 m^3 tonne of urea produced depending on the quantity of raw materials consumption. Ammonia and urea concentrations varying between 200–1500 mg/L and 340–20,000 mg/L, respectively (IS-9841 1981).

Urea as such is non-toxic for human, animal and aquatic life even at high concentrations. The test fish–Creek Chub in the Detroit river were observed to survive even at higher dose of more than 1000 mg/L of urea (Gillette *et al.*, 1952). However, urea under favorable conditions hydrolyses to ammonia and carbon dioxide. An ammonia concentration of 1.2 to 3.0 mg/L is toxic to fish.

Evans *et al.* (1973) reported that 4 to 6 days are required for urea degradation and 20°C with urea concentrations ranged between 1 to 15 mg/L and 14 days or more at a temperature lower than 8°C. This indicates that urea hydrolysis is a very slow process and can take place only in presence of urase, otherwise the hydrolysis may assume to be negligible.

Process of Nitrogen Removal

Nitrogen removal through microbiological treatment of wastewater containing urea involves the following unit processes:

Urea Hydrolysis

Urea hydrolysis is brought out by microorganisms mainly ureolytic bacteria *Bacillus pasteurii* as follows:

$$\underset{\text{Urea}}{O{=}C\begin{matrix} NH_2 \\ NH_2 \end{matrix}} + \underset{\text{Water}}{H_2O} \longrightarrow \underset{\text{Ammonium Carbonate}}{(NH_4)_2CO_3} \longrightarrow \underset{\text{Ammonia}}{2NH_3} + \underset{\text{Carbon-dioxide}}{CO_2} + \underset{\text{Water}}{H_2O}$$

Ammonia and ammonium ions in aqueous solution remain in equilibrium as follows:

$$NH_4 \longleftrightarrow NH_3 + H^+$$

At pH 7.0 and temperature 0 to 250°C, over 99 per cent of ammonical nitrogen is in ammonium (NH_4^+) form. Autotrophs include the nitrifying bacteria which obtain energy from oxidizing inorganic nitrogen (chemoautotroph) and algae which obtain energy from sunlight (photoautotroph).

Nitrification

Ammonia is biologically oxidized to nitrate in a two stage process by two groups of chemoautotrophic bacteria which operates in sequence. The first step is oxidizing ammonia into nitrite by members of genus *Nitrosomonas* sp. Nitrite is further oxidized to nitrate by members of genus *Nitrobacter* sp. as follows:

$$2NH_4^+ + 3O_2 \xrightarrow{\textit{Nitrosomonas}\ \text{sp.}} 2\,NO_2 + 4\,H^+ + 2H_2O + \text{Energy}$$

$$2NO_2 + O_2 \xrightarrow{\textit{Nitrobacter}\ \text{sp.}} 2NO_3 + \text{Energy}$$

Energy released in this oxidation is used in synthesizing cell materials from carbon-dioxide. It is evident from these equations that oxygen is required for nitrification process which gives rise to oxygen demand in wastewater. Liberation of hydrogen ion may result in pH drop in poorly buffered wastewater, resulting in loss of process stability.

Denitrification

Denitrification of nitrate is brought out by heterotrophic bacteria first into nitrite and then to nitrogen gas. As the energy yield is lower than that from oxygen respiration, denitrification is effective only at low dissolved oxygen concentrations. An external carbon source is necessary to obtain maximum reaction rate. Methanol can be used as one of the carbon source and the reaction can be represented as follows:

$$NO_3 + CH_3OH \longrightarrow N_2 + CO_2 + H_2O + OH$$

Urea biohydrolyser breaks urea molecule into ammonia and carbon-dioxide with the help of ureolytic bacteria *Bacillus pasteurii.* After hydrolysis the ammonia produced is stripped off. The remaining ammonia nitrogen undergoes nitrification producing nitrate. Nitrate and ammonia are utilized by algae for growth. Unutilized nitrate is converted into nitrogen gas after denitrification. Thus nitrogen free effluent is produced.

Brief Details of Nitrogenous Fertilizer Industry Under Study

The major products being produced at the coal based nitrogenous fertilizer Industry are ammonia, urea, nitric acid, ammonium nitrate, ammonium bicarbonate, coke and tar. The rated capacities of these products are 900, 1000, 200, 35, 12, 325 and 10 tonnes/d, respectively.

Total water requirement of the fertilizer Industry was around 79124 m^3/d. This includes water for process, cooling and domestic uses. Wastewater generations from the various unit processes of the industry were monitored. The total wastewater generation observed to be in the range of 34608–37944 m^3/d. Table 36.1 presents the quantity of wastewater generated from the various process units of the fertilizer industry.

This chapter gives an account of studies on a full scale urea bio-hydrolizer cum stripper as additional unit and effluent treatment plant to treat nitrogenous wastewater from fertilizer industry.

Materials and Methods

Effluent Treatment Plant

Nitrogenous fertilizer industry under study had full scale effluent treatment plant comprising of ten lagoons/ponds of various capacity of 100000 m^3. Total detention time of all ten lagoons at ETP was 25 days. As lagoons 1, 2, 3 were under repairs, wastewater from urea manufacturing plant was directly received in lagoon No. 4. Septic tank effluents from the industry and township sewage were also being discharged in this lagoon and the mixed effluents were routed through various lagoons for further treatment. Treated urea effluent was discharged into lagoon No. 9. Other combined wastewater from coke oven, ammonium bicarbonate, nitric acid and resorcinol plants was received in lagoon No. 9. Effluent from steam generation was directly discharged into treated effluent from lagoon No. 9, into natural drain, which further joins a river after traveling a distance of about 500 meters.

Table 36.1: Quantity of Wastewater Generated from Various Process Units

Sl.No.	*Process Unit*	*Quantity, m^3/day*	
		Range	*Average*
1.	Oil Gasification	1656–752	1680
2.	Rectisol	792–936	840
3.	DM		
	- Acidic	1344–1560	1440
	- Alkaline	864–1056	984
4.	Steam Generation	1584–1656	1608
5.	Compressor House	384–528	456
6.	Coal and handling house	384–576	480
7.	Urea	1800–1968	1896
8.	Coke Oven	1152–1296	1224
9.	Nitric Acid	360–528	432
10.	Ammonium Nitrate	288–432	360
11.	Ammonium Bi-carbonate	168–264	216
12.	Power Plant	9600–10800	10128
13.,	Septic Tank Effluent	14322–14592	14424
14.	Combined Effluent	34608–37944	36168

Performance evaluation of ETP was carried out for 7 days. Wastewater samples at various stages of treatment (lagoons) at ETP were collected over a period of 8 hours. These samples were analyzed for pH, TDS, SS, total and free ammonical nitrogen, total kjeldahl nitrogen, nitrite and nitrate nitrogen as per Standard Methods of APHA (1989). Urea was analysed as per procedure described Watt and Chrisp 1954. Characteristics of influent and effluent from various lagoons at ETP are presented in Table 36.2.

Treated effluent from ETP does not confirm to the standards prescribed by the MEF/State pollution Control Board for discharge of effluent into Inland Surface Water with respect to parameters- total ammonia nitrogen, free ammonia nitrogen, total kjeldahl nitrogen and nitrate nitrogen.

Based on performance evaluation of ETP and bench scale laboratory studies on urea bio-hydrolyser conducted earlier, modifications at ETP were suggested to remove urea and ammonia from urea bearing wastewater. Urea bio-hydrolyser cum stripper was designed and its installation as an additional unit before lagoon no. 1 at ETP was recommended.

Table 36.2: Characteristics of Influent and Effluent from Various Lagoons at ETP

Effluent Type	Parameters													
	pH	*TDS*	*SS*	*DO*	*COD*	*BOD*	*Total Ammonical Nitrogen*	*Free Ammonical Nitrogen*	*Total Kjedahl Nitrogen*	*Urea*	*Nitrite*	*Nitrate*	*Oil and Grease*	*Total Chromium (Cr)*	*Cyanide (Cn)*
Effluent from urea plant to lagoon 4	9.0± 0.6	668± 82	22±7	1.8± 0.2	1.42± 26	60± 15	1810± 94	722± 88	2440± 146	1350± 70	3±2	36±8	32±6	ND	ND
Lagoon 4 effluent	9.0± 0.5	792± 98	7±3	2.2± 0.3	88± 16	35±8	665± 50	348± 24	1142± 74	936 ±56	6±3	34±6	26±5	ND	ND
Lagoon 5 effluent	8.6± 0.4	786± 86	5±2	2.5± 0.4	52± 10	20±5	612± 62	264± 21	1030± 44	856± 28	8±3	30±8	19±6	ND	ND
Lagoon 6 effluent	8.6± 0.4	110± 80	4±2	3±1	44±8	15±3	511± 78	126± 13	958± 38	750± 32	9±2	32±6	16±4	ND	ND
Lagoon 7 effluent	8.5± 0.4	764± 82	3±1	3.2± 0.4	36±5	12±3	530± 46	84±8	764± 52	532± 46	10±2	31±4	10±2	ND	ND
Lagoon 8 effluent	8.4± 0.4	752± 83	3±1	3.5± 0.4	30±6	8±2	494± 46	48± 10	615± 44	366± 28	11±2	30±4	10±2	ND	ND
Lagoon 8A effluent	8.±3 11.5	728± 88	3±1	3.9± 0.4	24±4	6±2	348± 28	30±6	424± 68	274± 46	12±8	26±3	9±2	ND	ND
Factory combined effluent at lagoon 9	8.4± 0.6	918± 108	163± 23	2.0± 0.4	252± 28	120± 30	105± 26	54± 22	160± 40	12±4	20±8	42± 10	14±4	ND	ND
Lagoon 9 effluent	8.3± 0.4	848± 84	22±9	4.0± 0.3	35±8	10±4	202± 22	26±6	282± 42	84±8	14±4	22±4	6±2	ND	ND

All values except pH are in mg/l. ND: Not detectable.

Modified Effluent Treatment Plant-Full Scale Urea Bio-hydrolyser

Modified effluent treatment plant comprises a full scale urea bi-hydrolyser unit followed by ten lagoons in series. Urea bio-hydrolyser plant was installed and commissioned. The plant comprises of 16 compartments of different sizes arranged in 8 horizontal rows. Total volume of all these compartments was 6012 m^3 with detention period of 90 hours. The layout plan of urea bio-hydrolyser is shown in Figure 36.1. Pressurized steam was introduced at the bottom of compartment Nos. 11, 12, 13, 14, 15 and 16 to facilitate ammonia stripping, produced during urea hydrolysis. In addition, ammonia from hydrolyser plant was further stripped off through cascade aeration system. Further, ammonia stripped effluent was then routed through lagoons 1 to 8A to reduce ammonia from the effluent. Septic tank effluent from industry and township was being discharged into lagoon 4 to remove ammonia from wastewater. Intermittent addition of algae-chlorella in lagoons was practiced.

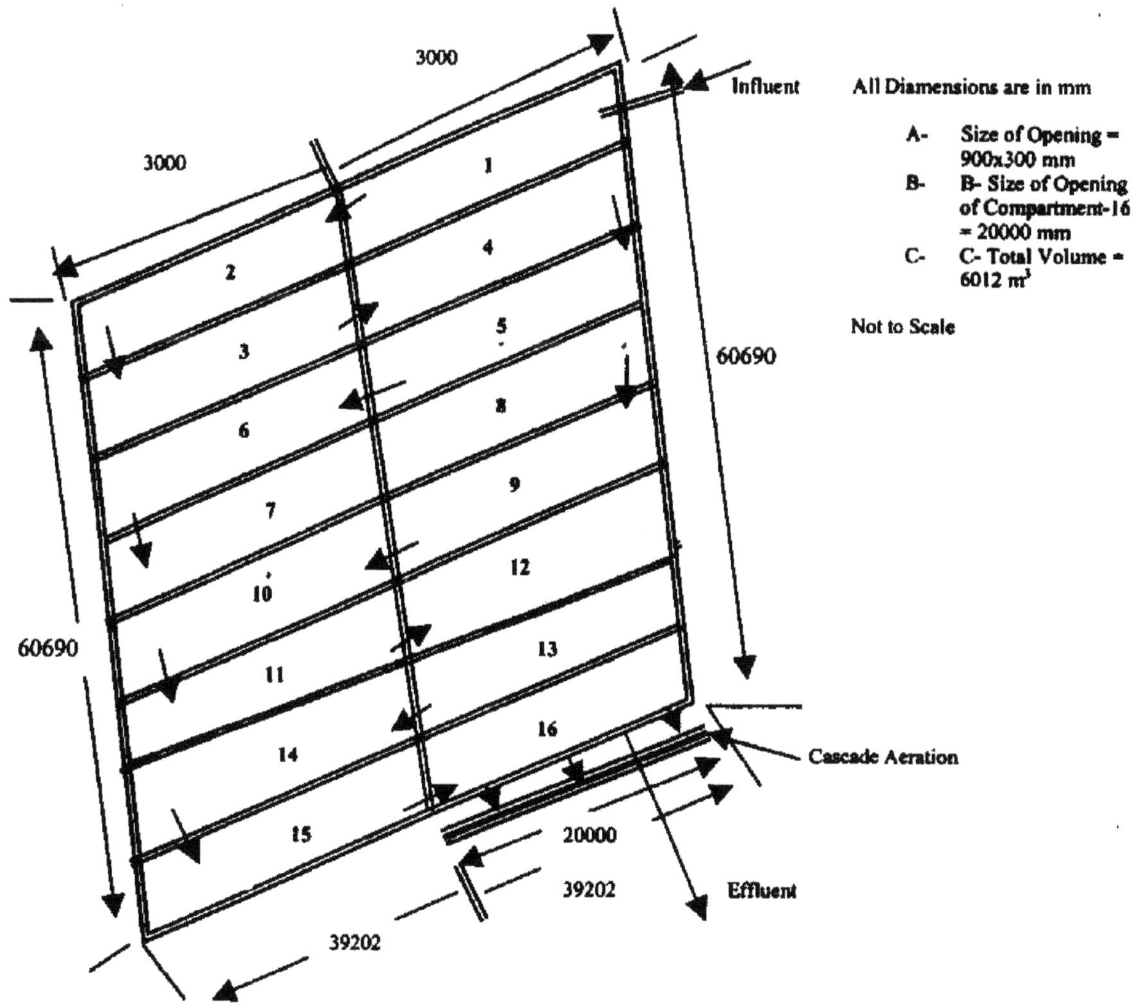

Figure 36.1: Layout Plan of Biological Urea Hydrolyser Plant

Microbial culture produced during earlier bench scale study was used for start up of urea bio-hydrolyser. Chalked paddy straw of 10 to 50 mm size was used as the carbon source for microorganisms. As urea bearing wastewater was Phosphate deficient, Super phosphate was added in requisite doses to supplement phosphate nutrient to microorganisms.

The performance of the Urea bio-hydrolyser unit and modified effluent treatment plant were monitored. Hourly wastewater samples were collected from each compartment of the urea bio-hydrolyser over a period of 8 hours for 7 days. The samples were analyzed for pH, TDS, SS, Total ammonical Nitrogen, Free ammonical Nitrogen, Total Kjeldahl Nitrogen, Nitrite Nitrogen, Nitrate Nitrogen and urea. Characteristics of influent and effluent from each compartment of urea bio-hydroliser plant are presented in Table 36.3. Hourly samples from each lagoon were also collected over a period of 8 hours for 7 days and analyzed for afore-referred parameters. Characteristics of effluent and influent from urea bio-hydroliser and effluent characteristics from various lagoons at modified effluent treatment plant are in Table 36.4. The characteristics of combined final treated effluent from the fertilizer industry being discharged into river are presented in Table 36.5.

Results and Discussion

Initial urea concentrations, 1350±70 mg/L from urea bearing wastewater was reduced to 274±46 mg/L during wastewater treatment in lagoons 4 to 8 A. Biological nitrification resulted in increase of nitrite nitrogen concentrations in effluent from lagoons. Similarly, reduction in concentration of nitrate nitrogen in effluent from lagoons indicating occurrence of denitrification process in various lagoons. High concentration of total ammonical nitrogen even after its utilization by algae in lagoons may be due to biological hydrolysis of urea in presence of sewage.

After modifications at ETP, urea by hydrolyser reduces initial concentration of urea, 1230±180 mg/L from urea bearing wastewater to substantially low concentrations, 160±52 mg/L, thereby achieving 85 to 90 per cent urea hydrolysis. The treated effluent from urea bio-hydrolyser contained free ammonia nitrogen ranging between 272 to 336 mg/L, removing around 61.3 to 62.9 per cent of free ammonia by ammonia stripping. Lagoon I received low concentrations of urea (160±52 mg/L) total Kjeldahl nitrogen (400±52 mg/L)

And total ammonia nitrogen (346±32 mg/L) for further treatment as compared to concentrations before modifications at ETP.

The urea bio-hydroliser unit was designed to accept initial urea concentrations up to 6000 mg/L to hydrolyse 90 to 95 per cent urea from urea bearing wastewater. The flow sheet of urea bio-hydroliser and removal of nitrogen from the wastewater is as follows:

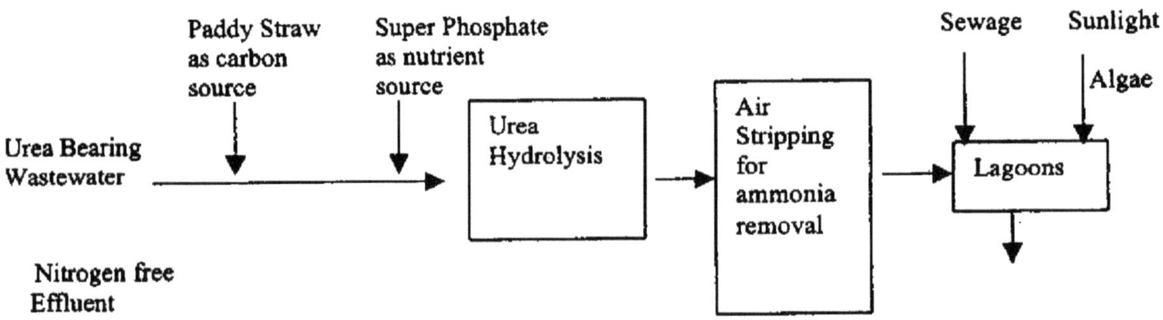

Table 36.3: Characteristics of Influent and Effluent from Each Compartment of Urea Biohydroliser Plant

Effluent Type	Parameters									
	Temperature	*pH*	*TDS*	*SS*	*Total Ammonical Nitrogen*	*Free Ammonical Nitrogen*	*Total Kjeidhal Nitrogen*	*Urea*	*Nitrate Nitrogen*	*Nitrite Nitrogen*
Compartment no. 1 influent	24±3	10.1 ±0.3	1126± 120	126±18	975 ±55	794±106	1840±168	1230±182	0.9±0.1	2.0±1.0
Compartment no. 1 effluent	25±3	10.1 ±0.3	1694 ±104	794 ±76	984±64	804±101	1794 ±152	1200±176	1.28±0.10	2.1±1.1
Compartment no. 2 effluent	25.±3	10.1±0.3	1538±102	748±84	996±58	864±94	1678±152	1200±176	1.28±0.11	2.2±1.1
Compartment no. 3 effluent	25±3	10.1±0.3	1420±96	680±92	1034±68	936±88	1476±136	854±122	1.16±0.11	2.3±1.1
Compartment no. 4 effluent	25±2	10.2±0.2	1392±90	656±74	1082±78	996±92	1282±140	628±108	1.40±0.14	2.4±1.2
Compartment no. 5 effluent	25±2	10.2±0.2	1356±92	632±68	1132±80	1074±86	1204±132	440±82	1.46±0.12	2.5±1.2
Compartment no. 6 effluent	25±2	10.3±0.2	1318±91	600±52	1158±84	1092±90	1196±124	356±64	1.52±0.10	2.6±1.2
Compartment no. 7 effluent	25±2	10.3±0.2	1250±66	580±48	1004±92	978±84	1042±116	334±58	1.64±0.11	2.7±1.1
Compartment no. 8 effluent	25±2	10.3±0.2	1222±78	528±50	820±68	798±70	886±110	310±50	1.16±0.13	2.7±1.2
Compartment no. 9 effluent	25±3	10.2±0.2	1204±62	462±38	806±52	728±46	865±112	294±56	1.88±0.12	2.8±1.2
Compartment no. 10 effluent	25±3	10.2±0.2	1178±72	370±42	712±46	641±48	830±98	278±52	2.06±0.14	2.9±1.1
Compartment no. 11 effluent	25±3	10.1±0.2	1144±64	296±38	604±50	558±42	782±84	264±64	2.18±0.18	3.0±1.1

Contd...

Table 36.3–Contd...

Effluent Type	Parameters									
	Temperature	pH	TDS	SS	Total Ammonical Nitrogen	Free Ammonical Nitrogen	Total Kjeldahl Nitrogen	Urea	Nitrate Nitrogen	Nitrite Nitrogen
Compartment no. 12 effluent	25±3	10.1±0.2	1109±70	252±32	534±38	496±44	622±76	238±56	2.26±0.20	3.2±1.1
Compartment no. 13 effluent	24±3	10.0±0.2	1090±62	224±36	478±32	442±40	512±64	222±60	2.56±0.1	3.5±1.2
Compartment no. 14 effluent	24±3	10.0±0.2	1058±74	208±32	414±30	394±42	500±5	196±70	2.64±0.26	3.7±1.2
Compartment no. 15 effluent	24±3	10.0±0.2	1028±68	196±24	382±34	346±36	442±56	118±62	2.66±0.28	3.8±1.2
Compartment no. 16 effluent	24+3	10.0±0.2	954±60	138±16	346±32	304±32	400±52	160±52	2.70±0.30	4.0±1.2

All values except pH are in mg/L. ND: Not detectable.

Table 36.4: Characteristics of Influent and Effluent from Urea-Bio-hydrolyser Plant and Various Lagoons of ETP

Effluent Type	*Parameters*														
	pH	*TDS*	*SS*	*DO*	*COD*	*BOD*	*Total Ammonical Nitrogen*	*Free Ammonical Nitrogen*	*Total Kjedahl Nitrogen*	*Urea*	*Nitrite*	*Nitrate*	*Oil and Grease*	*Total Chromium (Cr)*	*Cyanide (Cn)*
Urea bio-hydrolyser influent	10.1±0.3	1726±120	126±18	ND	332±6	200±25	975±55	794±106	1840±168	1230±182	0.90±0.10	2±1	45±10	ND	ND
Urea bio-hydrolyser effluent	10.0±0.2	994±60	138±16	0.5±0.1	258±48	145±20	346±16	304±32	400±50	160±52	2.70±0.30	4±1	40±8	ND	ND
Lagoon 1 effluent	9.9±0.2	750±22	100±8	0.6±0.2	216±32	105±16	316±20	282±30	382±48	142±46	4.00±0.80	4.0±1.0	34±9	ND	ND
Lagoon 2 effluent	9.7±0.3	700±24	74±6	1.1±0.2	186±36	85±20	288±22	264±28	367±44	104±32	6.90±0.60	5±1	32±8	ND	ND
Lagoon 3 effluent	9.6±0.3	644±16	53±7	1.1±0.3	158±24	58±13	256±28	240±24	35±38	56±16	6.80±0.70	5±1	30±7	ND	ND
Lagoon 4 effluent	9.6±0.4	604±20	35±4	2.0±4	102±12	32±6	234±24	212±22	371±34	32±8	8.15±0.85	5±1	25±5	ND	ND
Lagoon 5 effluent	9.6±0.3	584±16	25±3	2.5±0.4	62±10	28±4	208±22	194±20	320±30	20±6	9.35±0.85	6±1	19±3	ND	ND
Lagoon 6 effluent	9.5±0.4	542±14	17±3	3.1±0.3	50±7	20±2	184±20	160±16	250±26	12±5	11.50±0.5	7±2	15±2	ND	ND
Lagoon 7 effluent	9.5±0.3	504±10	10±2	3.6±0.3	38±5	15±2	162±22	124±25	198±25	8±4	12.20±0.9	9±2	11±2	ND	ND
Lagoon 8 effluent	9.4±0.3	480±12	7±2	4.0±0.3	34±6	11±3	138±20	82±18	132±24	5±2	14.00±1.0	10±3	7±2	ND	ND
Lagoon 8A effluent	9.0±0.3	462±18	5±2	4.9±0.2	30±5	7±2	112±20	60±16	108±20	2±1	16.75±0.75	12±3	5±1	ND	ND
Factory combined effluent to lagoon 9	7.6±0.7	952±96	140±8	1.9±0.2	284±54	130±118	85±16	42±12	88±16	1±1	8.95±0.95	45±8	13±2	0.07±0.02	0.10±0.03
Lagoon 9 effluent	8.1±0.2	946±84	36±6	4.7±0.3	42±8	14±4	76±6	7±2	52±12	ND	13.10±2.00	15±1	7±0.01	ND	0.03±0.03

All values except pH are in mg/l. ND: Not detectable.

Table 36.5: Characteristics of Combined Treated Effluent Discharged in to River

Sl.No.	*Parameter*	*ETP Combined Treated Effluent*	*Standard for Treated Effluent (Inland Surface Water)*	
			MoEF[1]	*SPCB*[2]
1.	pH	7.5–8.0	5.5–9.0	6.5–8.0
2.	Total dissolved solids (TDS)	750±52	–	–
3.	Suspended Solids	46±14	100	100
4.	Dissolved Oxygen Demand (DO)	4.8–5.2	–	–
5.	Chemical Oxygen Demand (COD)	54±18	250	250
6.	Biochemical Oxygen Demand (BOD_5 d, 20°C)	16.5±5	30	30
7.	Total Ammonical Nitrogen	40±5	50	75
8.	Free Ammonical Nitrogen	2.3±0.3	5	4
9.	Total Kjeldahl Nitrogen	52±12	100	100
10.	Nitrite (NO_2) Nitrogen	7.2±0.4	–	–
11.	Nitrate (NO) Nitrogen	9.0±0.8	10	10
12.	Urea	ND	–	–
13.	Total Phosphate	0.08±0.3	5	–
14.	Total Phenolics	0.3±0.1	1	–
15.	Cyanides	0.02±0.01	0.2	–
16.	Fluorides	ND	2.0	–
17.	Oil and grease	4.6±2.1	10	10
18.	Heavy Metals			
	- Iron (Fe)	0.52±0.08	3	3
	- Lead (Pb)	0.002±0.001	1	1
	- Zinc (Zn)	0.11±0.04	5	5
	- Nickel (Ni)	ND	3	3
	- Chromium	ND	2	2

All values except pH are in mg/L. ND: Not detectable.

1: Ministry of Environment and Forests, Govt. of India; 2: State Pollution Control Board.

Paddy straw served as carbon source to Uerolytic bacteria *Bacillus pasteurii* and algae utilized ammonia nitrogen for their growth. Urea is being hydrolised into ammonia which then stripped off at high pH and also by injecting steam. Up to compartment no. 6 a constant increase in ammonia concentrations and constant decrease in urea concentrations indicating brisk reaction of urea hydrolysis. Resulting in 70 to 73 per cent was removal in first 30 hours. Remaining 15 to 17 per cent urea was removed in the remaining 10 compartments of urea bio-hydrolyser plant. However, the concentrations of ammonia in these compartments showed lower values as compared to tricometric values, due to ammonia stripping at high pH of wastewater. At pH above 8.6, ammonia is released from ammonia solution in to the atmosphere. Thus, by maintaining high pH, ammonia stripping was achieved. Urea bio-hydroliser removed about 61 to 63 per cent ammonia nitrogen due to ammonia stripping. On injecting steam in latter compartments facilitate further removal of ammonia in the range of 44 to 48 per cent. The reduction of dissolved solids in the effluents from different compartments, indicating release of ammonia nitrogen into atmosphere. Increase in concentrations of suspended solids in the effluent was mostly due to addition of chopped rice straw which floats on water surface and escape through effluent.

Insignificant increase in nitrite and nitrate concentrations in effluent from various compartments of urea biohydroliser indicate that nitrification of ammonia was not taking place. The slow build up of nitrite and nitrate concentrations in the lagoons indicating nitrification of ammonia during wastewater stabilization. The urea concentration in lagoon effluent was reduced due to the presence of ureolytic bacteria present in septic tank effluent in lagoons. As algae-chlorella present in lagoons by virtue of addition of algal culture, utilize ammonia nitrogen and other nitrogen species from wastewater, thereby reduction of ammonia nitrogen was observed. The concentrations of COD and BOD was also reduction significantly. The characteristics of final treated effluent being discharged into river confirmed to the discharge standards prescribed by regulatory agencies into inland surface water.

Conclusions

Based on the studies on full scale urea bio-hydrolyser plant to treat urea bearing wastewater, it was concluded that ureolytic bacteria *Bacillus pasteurii* could hydrolyse 85 to 90 percent urea from wastewater. Urea hydrolysis process was brisk at initial period of 30 hours, hydrolyzing around 70 to 73 per cent urea in the first six compartment of the bio-hydrolyser plant. Ammonia nitrogen produced during urea bio-hydrolysis could be stripped off at pH 10.2± 0.2 without applying mechanical means. Injecting pressurized steam at the bottom of latter compartments facilitate 44 to 48 per cent ammonia removal by stripping. Paddy straw, an agricultural residue can profitable be used as an carbon source for microorganisms as against costly chemicals like methanol and other organics. Ammonia nitrogen trapped in the form of in-organics further nitrified in lagoons with the help of microorganisms into nitrate via nitrite. Algae–*chlorella* present in lagoons reduce ammonia nitrogen and nitrate nitrogen from the wastewater. Nitrate nitrogen was denitrified by microorganisms into nitrogen gas to obtain nitrogen free wastewater. BOD and COD was also reduced in the lagoons.

Urea bio-hydrolyser cum stripper reduces significant concentration of urea and ammonia in the wastewater from nitrogenous fertilizer industry to confirm standards prescribed by regulatory agencies for discharge of treated effluent into inland surface water.

References

APHA-AWWA-WPCF, 1989. *Standard Methods for the Examination of Water and Wastewater, 17th* edn. American Public Health Association, Washington, DC.

Evans, W.H., David, E.J. and Patterson, S.J., 1973. Biodegradation of urea in river water under controlled laboratory conditions. *Water Res.*, 7: 975.

Gillette, L.A., Miller, D.L. and Redman, H.E., 1952. Appraisal of chemical waste problem by fish toxicity tests. *Sewage and Industrial Waste*, 24: 1397.

India 1995. A reference Annual Published by Ministry of Information and Broadcasting. Government of India, pp. 545–546.

ISI : 9841. 1981. *Guide for Treatment and Disposal of Fertilizer Industry.* Indian Standards Institution, New Delhi.

Monthly Review of Indian Economy, 1997. Economy Intelligence Service Series, 1997, Published by Center for Monitoring Indian Economy (CMIE) Pvt. Ltd., February, pp. 58.

Watt, G.W. and Chrisp, J.D., 1954. Spectrophotometric method for determination of urea. *Anal. Chem.*, 26: 452.

Ethnobotanical Wisdom and Microbial Studies on Medicinal Plants (2011) *Pages* **227–255**
Editors: **D.R. Khanna, Ashutosh Gautam, R. Bhutiani, Gagan Matta, Anil Dhiman**
Published by: **DAYA PUBLISHING HOUSE, NEW DELHI**

Chapter 37

Environmental Significance of Helminth Parasites and their Removal in Wastewater Treatment Processes

Shanta Satyanarayana, P.R. Chaudhari and S.N. Kaul
National Environmental Engineering Research Institute,
Nehru Marg, Nagpur – 440 020, India

ABSTRACT

Epidemiological surveys have reported important risks regarding intestinal parasites because of their widespread distribution (Crompton, 1989), their high survival rate in the environment (Shuval *et al.*, 1988) and their low infective dose (Kowal, 1988; Gunnerson *et al.*, 1985). The presence of large number of helminth eggs in various effluents used for agricultural purposes has been reported by several studies (Barbier *et al.*, 1990; Schwartzbrod *et al.*, 1986). Occurrence of helminthes, a human intestinal parasite, in sewage is of considerable environmental significance. In the past, the different helminth parasites were distributed only in some pockets on earth. But the migration of people and enormous production of sewage with population explosion has resulted in widespread occurrence of these parasites. These parasites are transmitted through their resistant life stages such as eggs and larvae and are responsible for infection of intestine, illness, pneumonia, cough, eosinophilia, blood stained sputum, bowel obstruction, infection of liver, gall bladder and appendix, vitamin deficiency and also occasional death in both children and adults. Contaminated articles like vegetables, hands, utensils and dust help in distribution and transmission of parasites. Sewage treatment plant workers are also exposed to the infection by helminth parasites.

Strict standards for the number of eggs are laid down by W.H.O. for the reuse of treated wastewater. Various wastewater treatments processed differ in their efficiency in the removal of highly resistant helminth eggs. In sewage treatment, most of the helminth eggs are concentrated in the sludge, which require further treatment such as composting, aerobic digestion and anaerobic digestion. The information on all these aspects are reviewed in this study, including the approaches proposed by different workers to make the treatment more effective to make sewage and sludge suitable for safe reuse and the suitable monitoring methods developed for the enumeration of helminth parasites in wastewater and treated effluents and sludge.

Keywords: *Helminthes parasites, Life cycle, Clinical aspects, Sewage treatments, Parasite removal.*

Introduction

Numerous different parasites are of common occurrence in sewage in different life stages of their life cycle. These are viruses, bacteria, protozoa and worms. Worms are of three types-nematode, cestode and trematode. However, the common nematode (helminthes) parasites *i.e.* roundworm, hookworm, tapeworm and pinworm along with reference to some other worms are dealt with in this document.

Occurrence of helminth parasites of human origin in wastewater and their transmission during recycling of the treated wastewater are the greatest public health problems in sanitary engineering. Man is the positive host for the adult helminth parasites *e.g. Ascaris,* hookworm and tapeworm etc. The eggs of these parasitic worms and their dormant stages are discharged in large numbers in the faeces of infected people and then pass to the sewage treatment works. Contamination of crops irrigated by wastewater effluent leads to transmission of infection. Of course, intermediate hosts are also there who carry the infection further. Fishes and aquatic edible species are the carriers of nematodes when sewage is used in aquaculture ponds. The most important practical problem lies in the foolproof treatment technology, which can guaranty a safer user of effluent and sludge for the agriculture and pisciculture by employing cheaper and acceptable solution to the problem. So it is obvious that eradication of these eggs/ova from sewage effluent and sludge should be the prime motto, so that hazards from the use of sludge and sewage for agricultural and pisciculture use of effluent will be removed.

The stages of intestinal parasites found in sewage effluent and sludge are either the helminth eggs or cysts in case of protozoan. Eggs and cysts are the dormant resting stages and are highly resistant to treatments and drying, than the full-grown adults.

Helminth parasites are of concern due to their ability to resist the conditions existing during wastewater treatment. Because of their small dimensions and the density nearly equal to that of water, they are only partially removed in sedimentation processes. However, the sewage treatment processes *e.g.* oxidation ponds are largely quiescent. Nevertheless, they appear in greatly concentrated numbers in primary sludge compared with sewage and treated effluent. But still the risk prevails in the agricultural use of effluent and more so with the sludge. Hence sewage treatment and the ensuing sludge will have to be tackled equally on war footing so as to rid them of all possible pathogenic parasites.

Most of the parasites, but by no means all, are removed by settlements in the primary or secondary stages of treatment technology, the survival and distribution of eggs can be ascertained by studying the viability of these eggs, which have undergone treatment. Viability of helminth eggs assumed an important parameter in public health aspects of reuse of sewage effluents. All the above aspects are discussed in detail in this review.

Common Helminthes of Human Nature

Biological characteristics and environmental significance of three major helminthes, which are commonly prevalent in human and a few others, are discussed here. They are hookworms like *Ancylostoma duodenale, Ancylostoma* spp., *Necator americanus* and round worms like *Ascaris lumbricoides* and others like *Taenia saginata, Cysticerxcus bovis* and *Schistosoma mansoni* etc.

Occurrence and Distribution

Historically, human hookworm infections were probably confined to the Eastern Hemisphere of the earth *Necator americanus* occurring south of 20° North latitude and *Ancylostoma duodenalee* north of 20° North latitude. In the past 500 years, population migrations, mostly notably those involving Spanish and Portuguese colonization in the New world and Southern Africa led to the introduction of *Ancylostoma duodenale* in these areas as well as the introduction of *Necator americanus* to Portugal. At the same time the slave trade from Africa to North and South America and the Caribbean islands led to the widespread distribution of *Necator americanus* in the Western Hemisphere. *Ascaris* worms are very common and occur worldwide and particularly in warm climates in rural community. It is one of the most prevalent human helminthes and it infects 700–1000 million people. Children are the most vulnerable for these infections.

Biological Characteristics

A. duodenale and *N. americans* are dioecious, sexually dimorphic, round worms belonging to the phylum Nematoda, order Strongylida, superfamily Ancylostomatoidea. The adult worms are small and off white or rusty in colour. *A. duodenale* is somewhat larger than *N. americanus,* the males being 5–10 mm long and the females 10–18 mm long depending on the species. The eggs of *A. duodenale* measures 56–60 μm by 36–40 μm and those of *N. americanus* are 64–76 μm by 36–40 μm in size.

Ancylostoma ceylanicum, a hookworm of dogs, cats and other animals, can infect man and develop into the adult stage. It has been reported to be of some importance in India, Surinam, West Indies (Indonesia) and elsewhere (Banwell and Sctiad, 1978). *Ancylostoma braziliense* and *Ancylostoma caninum,* the cat and dog hookworm, rarely develop to the adult stage in man but their larvae can cause a creeping dermatitis called larvamigrans.

Ascaris lumbricoides, a nematode, is the common roundworm of man. Females are 200–400 mm in length, whereas males are 150–300 mm. The fertile eggs are ovoid and measures 45–70 μm by 33–50 μm. Man is the host for *Ascaris lumbricoides.* Pigs and dogs may disperse the undeveloped eggs of *Ascaris lumbricoides* by eating them in human faeces and excreting them later at another place.

Health Risk of Handling Sewage and Night Soil

Human excreta and the domestic sewage are the principal vehicles for the transmission and the spreading of a wide range of communicable diseases (Table 37.1). The severity of some of the diseases are such that whole of the community gets wiped out. In rural and semi-urban communities where poverty, malnutrition and poor hygienic condition prevails, the helminthic diseases are of very common occurrence in children, mostly and also elders are no exception. Besides helminthic parasites, dreaded diseases like diarrhoea, endemic malaria and cholera are the other main causes of death among the children and infants in developing countries.

These and many others diseases start their origin from an infected person to a new victim through excreta or sewage. Therefore, the collection, transport, treatment, handling and ultimate disposal of human excreta and the domestic sewage are of utmost importance in the protection of the health of whole human community throughout the universe. It is particularly very important in places like south-east Asia (Taiwan, Indonesia and Malaysia), China and African countries where the human excreta is used with/without proper treatment for fertilization of the fish ponds, agriculture and also for biogas production which are cheaper than alternative methods of disposal. Hence, in developing countries and under developed countries such reuse of excreta and domestic sewage are a welcome proposal as it supports their economic activity and food production.

Table 37.1: Some Diseases and Causative Organisms of Concern in Handling Night Soil and Sewage Sludges

Category	*Disease*	*Organisms (Where identified)*
Viral	Infective hepatitis	Adenovirus
	Gastroenteritis	Reovirus
	Respiratory illness	Enterovirus (poliovirus)
	Poliomyelitis	*Salmonella typhosa*
Bacterial	Typhoid fever	*Salmonella* sp.
	Salmonellosis	*Shigella* sp.
	Bacillary dysentery/(shigellosis)	*Vibrio cholera*
	Cholera	*Mycobacterium tuberculosis*
	Tuberculosis	*Entamoeba histolytica*
Protozoan Helminthic	Amoebiasis (amoebic dysentery)	*Ascaris lumbricoides*
	Roundworm	*Oxyaris vermicularis*
	Pin worm	*Trichuris trichura*
	Whip worm	*Taenia saginata*
	Tape worm	*Ancylostoma duodenale*
	Hook worm	*Necator americanus*

Khuder *et al.* (1998) observed that wastewater treatment workers (WWTW) are potentially exposed to a variety of infectious agents and toxic materials. The WWTW exhibited a significantly higher percentage of gastroenteritis, gastrointestinal symptoms (specifically abdominal pains) and headache.

Clinical Symptoms

Ancylostomiasis and Ascariasis are the infections of small intestine caused by hookworms and *Ascaris* respectively. Ascariasis is the infection of particular importance to those engaged in sanitation programs. Ancylostomiasis is frequently symptomless but it does produce illness and constitutes a public health problem. The most important features are anaemia and resulting weakness, debility and other consequences. Gastrointestinal pain, transient, cutaneous and pulmonary symptoms and odema may also be experienced. Hookworm is seldom recorded as a direct cause of death. Some grossly anemic individuals die of heart failure. The disease is undoubtedly a common contributory cause of death when other normally nonfatal infections attack a severely anemic and debilitated person. Ascariasis is extremely common in most part of the world because eggs of the *Ascaris* worms are very persistent in environment and difficult to be eliminated by conventional sewage or night soil treatment processes. About 85 per cent of Ascariasis infections are without any outward visible symptoms, but the presence of even few worms is potentially dangerous. The earliest symptoms are the pneumonia with severe cough, fever, eosinophilia and sometimes blood stained sputum. Infection generally occurs in small children and also in adults. In case of heavy Ascariasis infection, adult worms are passed in the faeces or by mouth. Serious complications like bowel obstruction or death due to the migration of the adult worm to the liver, gall bladder or appendix and rarely due to perforation of the intestine are also observed. Ascariasis leads to deficiency in vitamin A and C in children.

Transmission and Larval Development

Thin shelled ovoid unsegmented eggs are discharged by the adult female worms into the lumen of the small intestine. A full grown larger and stouter female of *Ancylostoma duodenale* produces approximately 3 times more number of eggs per day (10^4 to 2×10^4) than *Necator americanus* (5×10^3 to 10^4). The eggs develop rapidly in the gut and are usually at the four or eight cells stage, when excreted along with faeces. If faeces are deposited in a suitable environment, the eggs hatch in 24–48 hours to give rise to rhabdiform first stage larvae. Optimum conditions for hatching and subsequent larval development are–protection from strong sunlight; soil environment of the right particle size, denseness and structure *i.e.* ideally a light sandy loam; adequate but not excessive moisture, and both desiccation and water logging are rapidly lethal to hatched hookworm larvae. A temperature between 28°C–32°C for *Necator americanus* and between 20°C–27°C for *Ancylostoma duodenale* is ideal. Below these temperature ranges, larval development slows down and is completely arrested below 10°C and above 40°C.

Adequate quantity of decomposing organic material and microorganisms in the soil provide food supply for the developing larvae. If conditions are satisfactory, the larvae undergo two molts outside the human body on the third and fifth day of their free living existence, giving rise to third stage filariform larvae which become infective to man about 6 days after hatching from the eggs. These larvae survive normally for 3–6 weeks and have a maximum life span of 15 weeks, and larvae have vertical range of migration of about one metre in suitable soils but their lateral movement is restricted to about 0.3 meters and usually much less.

Mode of Infection

Infection of man occurs most commonly when the third stage larvae penetrate the skin usually between the toes or on feet and ankles. In the case of *Ancylostoma duodonale*, the third stage larvae can also infect man when they are ingested with unwashed raw vegetables, contaminated by the soil where they have grown.

After penetration of the skin, larvae of both hookworm species enter small veins and lymphatic vessels and are carried to the heart and then to the lungs. *N. americanus* undergoes a period of development in the lungs then moves to bronchi and trachaea, and is swallowed and reaches small intestine in 24–48 hours after the skin penetration. Two further molts occur, one at about first day and second at 13th day after reaching the small intestine. *Ancylostoma duodenale* under goes the whole of its development in the small intestine regardless of whether it enters the human body by the oral or cutaneous route. Sexually mature fertilized females of both species begin egg laying between 4 and 8 weeks after infection.

In case of *Ascaris*, the females lay upto 2,00,000 eggs per day. The unsegmented fertilized eggs are passed in the faeces. About 15 per cent of the excreted eggs are infertile and these are longer and narrower than the fertile eggs. The first stage larva in the egg must molt to produce second stage larva before the egg is infective under ideal conditions of moisture and temperature (22°C–33°C). A minimum of 10–15 days are required for about 75 per cent of freshly passed eggs to become infective. Infective eggs can survive for long periods. Survival of infective *Ascaris* eggs in soil has been reported for a period of 7 years.

Ascaris infection spreads through contaminated vegetables, infected hands, utensils, dust etc. When the infective *Ascaris* eggs are ingested through these agents, the larvae hatch in the duodenum and are carried by the blood vessels to liver, heart and lungs. They develop into adults in about 60–75 days and then live upto 1.5 years.

The Incubation Period

Incubation period in case of hookworms (the interval between infection and the development of symptoms and signs of illness) varies from a few weeks to several years depending on number of worms that parasitize an individual, his daily iron intake, body iron stores, blood losses etc.

In case of *Ascaris,* the incubation period varies from a few days to several months. The development of clinical illness depends upon the number of worms in the body and the state of health of the host. In case of many light infections in healthy people, no evidence of illness ever appears.

Prevalence of Infection

Hookworm and Ascaris infections are extremely common and occur in many countries. Perhaps over 700 million people are infected worldwide. Table 37.2 shows the prevalence of hookworm infections in fifteen countries. There are many hundred of reports on *Ascaris* prevalence in various communities. Persons having acute ascariasis infection excrete huge number of eggs up to 2,00,000 per female worm per day and hence causes considerable potential transmission. Individual egg outputs of up to about 3,00,000 per gram of faeces are reported. Assuming 100 gram of faeces per day per capita, this suggests at least 150 mature female worms in the intestine.

Life Span and Survival in the Environment

Eggs will continue to be excreted, as long as there are adult worms in the small intestine of host human body. Adult hookworms may live for upto 7 years in the case of *Ancylostama duodenale* and 15 years in *Necator americanus* (Miller 1979). These are exceptional life span however 80 per cent of worms survive for less than three years. Full-grown adult worms of *Ascaris* generally live for less than 10 months with maximum life span of up to 1.5 years.

Hookworm eggs and larvae in nature have been found mainly in soil at place where people defecate or where excreta are applied on fields. Farm workers working on fields and sanitary workers, get the percutaneous infection. Vegetables grown on sewage farms get contaminated with these parasitic eggs and become a potential health hazard. Hookworm eggs survival in environment is comparatively shorter than *Ascaris* eggs, which are much sturdier in nature and therefore provide a better indicator organism for environmental helminthology.

Hookworm eggs will tend to settle in water and eventually die in the bottom sediments. Their survival in seawater has been reported as less than 5 hours, compared to over 30 hours for *Ascaris* eggs under the same experimental conditions (Livingston, 1978). Helminth parasites are found in surface waters, night soil, sewage, vegetables grown in sewage or night soil applied farms, soils etc.

Recovery and Viability of Helminthes Eggs

Effects of wastewater treatment technology upon the survival and distribution of eggs can be ascertained by studying the viability of these eggs, which have undergone treatment. The fertilized eggs of *Ascaris* are released from the female worm, undergo cell divisions and develop into infective larval stage in which embryonic motile form can be seen inside the shell. Incubation of treated eggs in acidified water for three weeks is often used as the basis of assaying viability of *Ascaris* eggs (Pike *et al.,*

1983). Care must be taken in interpreting the results of the experiments to determine the viability of parasitic eggs.

Table 37.2: Prevalence of Hookworm Infection in Fifteen Countries

Sl.No.	Country	Location	Age (Years)	Prevalence (Per cent)	Age of Max. Prevalence (Years)
1.	Bangladesh	Rural	All ages	29	ND
2.	Columbia	Urban	All ages	35	10–14
3.	Egypt	Rural	All ages	28	20–30
4.	Gambia	Rural	All ages	23	40
		Rural	1–5	12	ND
5.	Gautemala	Urban Poor	1–5	4	ND
		Urban wealthy	1–5	0	ND
6.	Haiti	Rural	All ages	24	15–9
7.	India	Urban	All ages	8	ND
		Rural	1–11	68	ND
		Rural	All ages	44–71	ND
8.	Iran	Rural	All ages	25	11–15
		Urban	All ages	8	6–10
9.	Ivory Coast	Rural	7–14	73	ND
10.	Malaysia	Rural	6–12	43	ND
		Urban poor	4–6	5	13–15
		Urban wealthy	4–6	0	ND
11.	Papua New Guinea	Rural	All ages	68	20–29
12.	South Korea	Urban	All ages	14	30–39
		Rural	All ages	19	30–39
13.	Taiwan	Rural	All ages	52	Over 60
14.	Thailand	Rural	All ages	61	ND
15.	Zambia	Rural	All ages	49	6–10

ND: Data not available.

As the number of helminth eggs in effluent of wastewater treatment plants is normally very low (Lakshminarayana and Abdulappa, 1972; Mara and Silva, 1986; Schwartzbrod *et al.*, 1987), any method used must have a high recovery rate and/or use a suitable sample volume to allow detection to the guideline level; additionally it must also be suitable for routine use in only basically equipped laboratories.

Ayers *et al.* (1991) compared four methods for egg recovery on effluents of various qualities and their speed, ease and cost were assessed. The recovery of eggs was found to be consistently higher using the method recommended by World Health Organization (Bailenger Method) but only when 10 L samples, rather than 1 L sample, were processed.

Stien and Schwartzbrod (1988) proposed an experimental method which associates both the detection of very low numbers of *Ascaris* eggs and their viability determination from samples of wastewater after nbutanol treatment, which precipitates fertile eggs. In this study, 86 per cent of the fertile eggs separated by nbutanol was viable and 90 per cent of the non-fertile ones were non-viable. The detection of the low levels of eggs required a minimal sample volume; further study shows that 25 L was the best sample size for this method. The recoveries obtained with this method compared quantitatively with those reported by Arther *et al.* (1981) (39 per cent recovery), Reimers *et al.* (1981) (62 per cent recovery) and Teichman (1986) (33–50 per cent recovery). This method has the advantage of supplying information from water samples. Classical methods use toxic reagents that exclude any possibility of estimating viability.

Gaspard *et al.* (1996) developed a method to estimate viability of helminth eggs to evaluate the parasitic risks. The extraction of these eggs was performed with adiphasic treatment (DS 0.01 per cent + ethylacetate) coupled with a filtration on 500 µm and 100 µm sieves followed by concentration using two flotation with NaCl (d-1.19). For the culture, 3 parameters tested showed a faster egg development at 30°C in deionized water with continuous aeration, whereas organic compounds reduce this development. This culture was performed during 13, 10 and 8 days respectively to obtain *Ascaris suum, Toxacara cannis* and *Capillaria* sp. larva and 16 days for *Trichuris vulpis*, which presented a slower development. The quantification was realized by microscopic examination (100 X) after sonication and sodium hypochloride treatment (0.5° chl), the eggs reaching the larva stage were considered as viable and the per cent of viability was expressed by the ratio: nematode eggs at the larva stage/total of nematode eggs counted.

Standards for Reuse of Wastewater Effluent and Sludge

The Report published under the auspices of several international organizations (UNDP, WHO, WB), reevaluates sanitary risks posed by the agricultural use of treated wastewater and proposed new recommendations. The World Health Guidelines for the microbiological quality of treated wastewater used for crop irrigation require an arithmetic mean of 1 intestinal nematode egg per litre (IRCWD, 1985; Prost, 1988; WHO, 1989). Using epidemiological evidence and a risk assessment model, the authors specifically show the importance of strict parasitological criteria for such reuse, recommending a limit of less than one viable intestinal nematode egg per litre (Cross and Strauss, 1985; Blum and Feachem, 1985; Mara and Cairmeross, 1987; Hespanhol and Prost, 1994). However, there is no standard limitation regarding the use of sludge for agricultural purposes. According to WHO Guidelines and to the EPA Part 503 biosolid rule (Cross and Strauss, 1985), only the viability of parasitic eggs is important from public health viewpoint.

Elimination/Survival of Parasites in Wastewater Treatment

Sewage Treatment Processes

Sewage treatment works are usually designed to remove or oxidize as much decomposable matter as possible so that the resulting effluent will not deplete the dissolved oxygen in the receiving water. So different sewage treatment processes stress more on the removal of organic pollutants than the microorganisms. In the process, the final effluent may satisfy the recommended standard for general pollutants but may contain large numbers of disease producing pathogenic organisms. Helminth parasites are of concern due to their ability to resist the conditions existing during wastewater treatment.

Study of the effect of various sewage treatment processes on ova and cysts of intestinal parasites were carried out way back in 1932. Since 1932, when Hirst (1932) undertook a study of hookworms in sewage, numerous workers have evaluated treatment plants in relation to elimination of helminth parasitic eggs. First report on the parasitic helminth eggs was reported from Russia by Vasilkova (1936) and in Brazil by Amaral and Leal (1940). Later Wright *et al.* (1942) reported effects of different sewage treatment processes on the ova and cysts of intestinal parasites. They reported that ova of parasitic helminths, particularly those of *Ascaris lumbricoides* may remain viable after passing through various treatment stages of sedimentation, digestion and drying. Total of 75 samples of sludge was examined from the seventeen Army Camps with parasitic helminth ova in 27 or 36 per cent of the samples. Results are indicated in the Tables 37.3 and 37.4.

Table 37.3: Presence of Helminthes Ova in 75 Samples of Sewage Sludge

Sl.No.	*Source of Samples*	*No. of Samples Examined*	*No. of Samples Positive*	*Helminthes Ova Encountered*		
				Ascaris	*Ascaris and Trichuris*	*Ascaris, Trichuris and Hymenolepis*
1.	Raw sewage	2	1	1	–	–
2.	Primary settling tank	14	3	2	1	–
3.	Secondary settling tank	9	1	–	1	–
4.	Activated sludge return	4	1	1	–	–
5.	Primary digester	7	3	2	1	–
6.	Secondary digester	9	5	4	1	–
7.	Digester	7	1	1	–	–
8.	Dry bed	17	12	10	1	1
9.	Imhoff tank	5	0	–	–	–
10.	Wet well sump	1	0	–	–	–
	Total	75	27	21	5	1

–: Absent.

Mostly these studies reported have used eggs freshly dissected from proglottids. However, although proglottids may shed some of their eggs whilst still within the human alimentary canal, eggs are still found within proglottids recovered from faeces. Additionally the treatment of taeniasis usually involves the chemical purging of the worm trom the gut. In these circumstances whole tapeworms containing viable eggs will find their way into sewage treatment plants. The proglottids may afford some protection to the eggs when on fields (Jepsen and Roth, 1952) equally, eggs may be similarly protected by the proglottids during the sewage treatment process.

Cram 1943 studied the survival rate of helminth ova under different conditions of sewage treatment processes with inoculum of known qualities of helminth eggs and he reported their survival rates in detail. He further reported that primary settling did not remove hookworm and *Ascaris* (Table 37.5). Studies also revealed that helminth ova may pass through trickling filters and that they would survive activated sludge treatment irrespective of the efficiency of the filter.

Table 37.4: Positive Findings of Helminthes Ova in Various Types of Sewage Sludge from Different Sewage Treatment Processes

Sl.No.	*Origin of Sludge Sample*	*Ascaris*	*Trichuris*	*Hymeno-lepis*	*Hook-worm*	*Length of Time in Drying Bed*
1.	Raw sewage	+				2 samples and
	Activated Sludge Return	+				13, 62 days
	Digester	+				respectively
	Drying Bed	+				
2.	Secondary Digester	+				
3.	Primary Settling Tank	+				
	Drying bed	+	+			4 days
	Drying Bed	+				6 weeks
4.	Primary Settling Tank	+	+		?	
	Primary Settling Tank	+				
	Secondary Settling Tank	+	+		?	
	Primary Digester	+				
	Primary Digester	+	+			
	Secondary Digester	+	+			
	Drying Bed	+				4 to 6 weeks
5.	Secondary Digester (Two samples)	+				
	Drying Bed	+				3 samples
						2, 2, 22 day
	Drying Bed	+				4 hrs.
6.	Secondary Digester	+				
7.	Primary Digester	+	+	+		3 samples
						1 sample 55
	Drying Bed	+				days, others unknown

The complete treatment of sewage (primary and secondary treatment without chlorination) if properly carried out, would appear to provide a high degree of removal of helminth eggs. In contrast, primary treatment alone allows a large percentage of eggs to pass through. Tertiary treatment of secondary effluent by filtration, land treatment or lagooning will also remove the remaining hookworm eggs. Final effluent chlorination does not have much effect on hookworm and *Ascaris* eggs (Cram 1943).

Large number of hookworm, *Ascaris* eggs enter the sewage treatment plants in endemic areas. The fate of hookworm and *Ascaris* eggs during sewage treatment is similar. The major difference being that hookworm eggs are less dense and that they may hatch. Specific gravity of hookworm is 1.055 and that of *Ascaris* is 1.11. Hence hookworm eggs have a lower settling velocity and thus are less prone to removal by sedimentation process (Cram 1943). Sedimentation is the main mechanism of removal of helminth eggs during sewage treatment and therefore the reported removals for hookworm eggs are typically a little lower than those of *Ascaris* eggs.

Table 37.5: Comparative Counts of Helminthes Ova during Primary Settling

Inoculum–No. of Ova of Ascaris lumbricoides–1,500,000				*Inoculum–No. of Ova Hookworms–50,000 Toxascaris–35,000*				
Count (No. of Ova)				*Count (No. of Ova)*				
Level Time	*Top*	*¼ Down*	*Bottom*	*Level Time*	*Top Hook-worms*	*1/3 Down Hook-worms*	*Bottom*	
							Hookworms	*Toxascaris*
At start	300	208	220	10 min	30		252	918
15 min	10	0	1261	20 min	27		280	540
30 min	0	0	0	30 min	51		396	210
1 hr	0	0	5040	45 min	8			
1½ hr	0	0	406	1 hr	0		25	
1¾ hr	0	0	48	1¼ hr	2			
2 hr	0	0	24	1½ hr	3			
2¼ hr	0	0	104	2 hr	0		5	
2½ hr	0	0	0	2½ hr	3		3	
2¾ hr	0	0	22	3 hr	0		0	

Unlike Ascaris eggs, hookworm eggs may hatch during sewage treatment. Cram (1943) observed the hatching of hookworm eggs on trickling filter stone, in activated sludge tanks and in drying sludge. Larvae will tend to stay in the liquid fraction and are carried along in the effluent. But generally the eggs are concentrated in the sludge. The larvae may survive in sewage treatment process for upto 5 days, which is quite sufficient for them to be discharged in the final effluent. Removal of hookworm eggs by sewage treatment is generally quoted as percentage reduction in concentration between influent and effluent. These obscure the fact that most eggs not found in effluent have not been destroyed but are merely concentrated in the sludge. Conventional sewage treatment plants have short detention times and operate at an ambient temperature and therefore do not destroy hookworm eggs. Their role is to transfer eggs from liquid fraction (the effluent) to the solid fraction (the sludge).

Newton *et al.* (1949) studied the effects of various pilot plant treatment processes on the eggs of *T. saginata*. The results indicated that sedimentation at moderate overflow rates of 1 to 2 hours would remove the majority of the eggs but significant percentage of eggs still would appear in the effluent. In two experiments, 30 per cent and 38 per cent of the eggs passed through a trickling filter operated at slow rates. Activated sludge treatment of sewage showed little effect on the eggs of helminthic parasites except a marginal reduction in numbers in the effluent, indicating that there was no noticeable effect upon the eggs after five months of operation.

In sand filtration process at a rate of one mgd, complete (100 per cent) removal of helminth eggs were achieved in three out of four experiments and a removal efficiency of 99.6 per cent were observed in the fourth experiment. This is a positive indication that sand filtration technique would be an effective method for the removal of *Taenia saginata* eggs from primary effluents. Thus sewage treatment processes ordinarily employed with the exception of sand filtration probably are not efficient in the removal of helminth ova (Newton *et al.*, 1949). Silverman and Griffiths (1955) have critically reviewed the different sewage disposal methods with special reference to epizoology and *Cysticercus bovis.*

Bhaskaran *et al.* (1956) carried out studies on the effects of sewage treatment processes on the survival of intestinal parasites. They have reported that while septic tanks removed about 70 per cent of *Ascaris* and hookworm eggs, the trickling filter and activated sludge plant removed 81 to 100 per cent and over 90 per cent respectively (Table 37.6). Dozanska and Ivanczuk 1958 have reported influence of different sewage treatment methods on the destruction of worm eggs.

Table 37.6: Removal of Helminthes by Conventional Treatment Process

Sl.No.	*Treatment Process*	*Hours of Settling*	*Percentage Removal*	
			Hookworm	*Ascaris*
1.	Sedimentation	1.5	46	67
	2.0	75	75	
2.	Trickling filter	–	100	100
3.	Activated sludge			
	Unit I	–	81	–
	Unit II	–	96	–

Kabler (1959) has detailed the removal of pathogenic organisms by sewage treatment processes. He has discussed different treatment processes like trickling filter, activated sludge, anaerobic digestion, chlorination and sedimentation ponds (Table 37.7). Trickling filter was effective against most of the parasites; activated sludge was effective only against *Ascaris*, hookworm and *Trichuris* while anaerobic digestion destroyed the eggs of tapeworm, ova, *E. histolytica*, cysts, *Schistosoma japonivum*, *Ascaris* and *A. lumbricoides*. Chlorination was not effective.

Studies were carried out in Puerto Rico to compare the efficiency of three types of sewage treatment processes in removing eggs of *Schistosoma mansoni* and *Ascaris lumbricoides* (Rowan, 1964). The results revealed that primary sedimentation and decantation methods removed 35–75 per cent of the *Ascaris* eggs and 83 per cent of *Schistosoma* eggs, whereas secondary treatment by biological filtration or activated sludge process removed 94.7–99.8 per cent and 97.0–100 per cent of the *Ascaris* eggs respectively and 99.7 per cent of *Schistosoma* eggs. Waste stabilization ponds were also used for helminth removal and were able to eliminate hookworm eggs completely and reliably. Eggs of hookworm hatch in aerobic waste stabilization ponds and larvae were observed in the effluent (Lakshminarayana *et al.*, 1969).

Panicker and Krishnamoorthy (1978) reported hookworm and *Ascaris* removal by a variety of sewage treatment plants in India. Removal rates of hookworms were consistently less than those of *Ascaris* and this was almost certainly due to the lower specific gravity of hookworm eggs and the resultant poorer removal by sedimentation (Table 37.8). Arundel and Adolph 1980 are of the opinion that modern sewage treatment methods like activated sludge and trickling filter do not remove the eggs of *Taenia saginata* from the final effluent. While helminth eggs because of their size and density are removed by sedimentation during primary treatment of wastewater and are concentrated in sludge. Lagooning is slightly less efficient than either of digestion systems, but offers an inexpensive solution to the problem of sewage disinfection of parasitic eggs. Zhao Xi Hui (1985) has discussed the survival of parasitic pathogens in biogas fermentation process. At a retention time of 30 days, when the digested slurry is in a static state, about 95 per cent of parasitic ova are distributed in the sediments and in the scum. The Table 37.9 gives details of the survival time of pathogens in biogas digester.

Table 37.7: Percentage of Organisms Removal by Different Methods

Organisms Studied	*Trickling Filter*	*Activated Sludge*	*Anaerobic Digestion*		*Chlorination*	*Stabilization Pond*	*Remark*
Tapeworm ova	18–26	–	97	–	No effect	–	Air drying for 2 yrs removed
E. histolytica cyst	88–99	No reduction	Removed	–	–	–	–
Ascaris lumbricoides	70–76	Does not affect viability	Not effective	–	–	–	Viability not reduced after 3 months. Viability after 64 days at 20°C
Ancylostoma canninum (dog hookworm)	70–76	Do	Not effective	–	–	–	Viability not reduced after 3 months. Viability after 64 days at 20°C
Toxoscaris leonina ova (dog ascaris)	70–76	Do	Not effective	–	–	–	Viability not reduced after 3 months. Viability after 64 days at 20°C
Tapeworm ova		–	Not removed	–	–		
Schislosomo japonicum	Reduced	Excellent hatching medium	90 in 25 to 30 days	–	–	–	–
S. japonicum ova Miracidia	–	–	–	–	Killed	–	30 min residual of 3.9–11 mg/l
S. mansoni Miracidia	–	–	–	–	–	–	30 min residual of 0.2 to 4 mg/l
A. lumbricoides	–	–	Reduced	–	–	–	Removed by 1 hr settling
T. saginata	62–70	Little effect	Very slow	–	–	–	Normal eggs recovered after 6 months digestion at 75°C–85°C
Ascaris	99.8	93–98	45	–	–	–	In 105 days at 80°F to 92°F
Hookworm	100	81.5–96	–	–	–	–	–
Trichuris	–	91.8–100	–	–	–	–	–

Table 37.8: Reduction of Helminthes Ova by Sewage Treatment Process in India

Sl.No.	Process	Helminthes				
		Hookworm	*Ascaris*	*Hymenolepis*	*Trichuris*	*Taenia*
1.	Two hour sedimentation	80	96	90	90	75
2.	Complete activated sludge plant	85	98	95	100	ND
3.	Two complete trickling filter plants	82	95	80	93	ND
4.	Pilot scale biodisc plant with one hour of secondary sedimentation	50	79	60	60	ND
5.	Pilot scale oxidation ditch with secondary sedimentation	81	94	89	100	100
6.	Pilot plant aerated lagoon without secondary sedimentation	70	92	78	100	100
7.	Four waste stabilization systems					
	1	93	100	100	100	100
	2	88	100	100	100	100
	3	100	100	100	100	100
	4	100	100	100	100	100

Table 37.9: Survival Time of Pathogens in Biogas Digester

Pathogens	Thermophilic Fermentation 50–55°C Fatality		Mesophilic Fermentation 35–37°C Fatality		Ambient Fermentation 8–25°C Fatality	
	Days	Rate (Per cent)	Days	Rate (Per cent)	Days	Rate (Per cent)
Sallmonella	1–2	100	7	100	44	100
Shigella	1	100	5	100	30	100
Poliovirus			9	100		
Colities	2	10^{-1}–10^{-2}	21	10^{-4}	40–60	10^{-5}–10^{-4}
Schistoma ova	Several hours	100	7	100	7–22	100
Hookworm ova	1	100	10	100	30	90
Ascaris ova	2	100	36	98.8	100	53

Impact of wastewater treatment processes has been reported by Schwartzbrod *et al.* (1989). Different processes like activated sludge, lagoon treatment and sand filtration have been investigated for the removal of helminthes (Tables 37.10 and 37.11). Activated sludge treatment was carried out by primary decantation, followed by the biological treatment with activated sludge. After anaerobic digestion (15 to 20 days at 31°C) and conditioning with lime and ferric chloride and then the sludge is partially dehydrated. Lagoon treatment was carried out in two stabilization basins in series, retention time of 5 days in each basin (total retention of 10 days) with a flow of 6 litres/day. The number of eggs recovered from the samples varied from 160 to 340 eggs/100 g wet matter depending upon the sample type. Trematode eggs completely disappeared while nematode and cestode eggs were always found.

Activated sludge treatment seemed to have an effect particularly on the cestode eggs, which were found only in low numbers. It is concluded that among the three treatments studied, lagoon treatment was most efficient in eliminating helminthes eggs from treated wastewater. Not a single helminthes egg was found in the effluent from the stabilization basins. Two treatments, activated sludge and sand filtration, reduced the number of helminthes eggs by 77.7 per cent and 98.8 per cent respectively in their final effluent. Most of the eggs were found in the sludge and sediments from 160 to 340 eggs in 100 g wet matter.

Table 37.10: Number of Helminthes Eggs in Wastewater (Raw and Treated)

Treatment Process	*Activated Sludge*		*Lagoon Treatment*		*Sand Filtration*	
	Entry	*Exit*	*Entry*	*Exit*	*Entry*	*Exit*
Number of analyses	15	15	23	23	6	6
Number of helminthes eggs	9	2	18	0	840	40
Types of eggs recovered						
Nematode	7	1	9	0	670	7
Cestode	2	1	9	0	140	3
Trematode	0	0	0	0	30	0
Percent reduction in parasitic load	77.7		100		98.8	

Nematode: *Ascaris, Toxascaris, Trichuris*

Cestode: *Hymenolepis, Taenia*

Trematode: *Dirocoelium*

Helminthes eggs were found in all sludge and sediment samples.

Table 37.11: Number of Helminthes Eggs in Sludge and Sediment

Sample Types	*Sludge Sample at the Treatment Plant Exit (Effluent)*	*Sediment Sample at the Lagoon Basin*	*Sludge Sample After Sand Filtration*
Number of analysis	15	23	6
Average number of helminthes eggs/100g of wet matter	160	215	340
Types of eggs recovered			
Nematode	156	114	0
Cestode	4	101	170
Trematode	0	0	0

Finally it has been concluded that although all the three types of wastewater treatments investigated are effective in elimination of helminthes eggs, the lagoon treatment is by far the most efficient, even with very short retention time. This simple treatment procedure ensures optimal removal of parasites allowing the eventual reuse of treated effluent for agricultural purposes. Nevertheless, the contaminated sediment poses a problem, which this treatment method does not address.

Literature shows that the number of eggs recovered from wastewater varies considerably from 4 eggs/L (Panicker and Krishnamoorthy, 1978) to 16,000 eggs/L (Mara and Silva 1978) for ex. 62 eggs/L (Forester and Engelbrecht, 1974),360 eggs/L (Bradely and Hadiday, 1981), 10 eggs/L (Kabrick and Jewell, 1982) and 1440 eggs/100g (Arther *et al.*, 1981).

Gaspard *et al.* (1997) observed the parasites in 99 samples including urban sludges (89), lagoon sediments (3), and composts (7). The results expressed per 100g of dry matter showed a high proportion (47 per cent) of samples presenting concentrations lower than 60 eggs. Thirty eight percent of the samples were with higher concentrations ranging from 60 to 240 eggs; only 15 per cent of the analysis results indicated concentrations higher than 240 eggs, with a maximum of 898 eggs. In lagooning sediments, the concentration observed ranged from 56 to 569 eggs and the analysis of compost samples yielded average concentration of 40.8 eggs. When whole study is taken into account nematode eggs (*Toxacara, Ascaris, Capillaria, Trichuris, Ascaridia, Enterobins*) are mostly represented with 93.2 per cent whereas cestode eggs (*Taenia, Hymenolepis*) are detected only in a proportion of 6.8 per cent. The study of lime treatment impact on 10 treatment plants showed a decrease in helminthes eggs concentration in seven samples and no difference for the three others. For the egg viability, sludge from all types of treatments (nesophilic, anaerobic and aerobic digestion, composting, liming) contained viable eggs in concentrations higher than or equal to 10 eggs per 100 g of dry matter.

Aulicino *et al.* 1998 reported the presence of helminthes eggs only form 3 out of 27 total samples. Jimenez and Chavez 1998 described the application of advanced primary treatment system (APT) to remove helminthes eggs and faecal coliforms from domestic wastewater in order to obtain an effluent suitable for agricultural reuse. The system comprises a coagulation-flocculation unit with a high rate sedimentation tank and a sludge blanket (Densadeg) along with a system of sand filtration and chlorine disinfection. It was found that Densadeg reduced the helminthes eggs (HE) from 22.6 HE L.I to 1.2 HE L^{-1} with hydraulic load of 1440 M^3. M^{-2}. d^{-1} (60 mh^{-1}) in the sedimentation tank. It is necessary to filter the effluent and apply 17 mg L^{-1} of chlorine with a contact time of 1 hour.

Treatment of Parasite Laden Sludge

Little work has been done on the sludge associated organisms, but available information indicates that there are numerous different parasites in many stages of their life cycle found in sewage sludges. Sewage treatment processes which include period of sedimentation prior to biological treatment are likely to produce sludges in which the eggs of parasities have been concentrated. When this sludge is applied on pastures, these parasitic eggs remain dormant for considerable length of time. As favourable conditions return they become health hazards. To reduce the health risks to animals and humans, treatment of sludge prior to their disposal on land is desirable. Many different treatment systems are effective in reducing parasitic viability to some extent.

Golueke and Gotaas (1954) have evaluated health aspects of aerobic composting in detail. To reduce health risk of helminth parasites to animals, treatment of sludges from sewage treatment processes is subjected to further treatment by:

1. Composting
2. Aerobic digestion, and
3. Anaerobic digestion.

Aerobic/anaerobic digestion may be followed by lagooning. Sludge digestion will reduce number of parasitic helminth eggs. Of course, the destruction depends on the temperature. Little information exists on the efficacy of composting or liquid sludge storage (lagooning) in reducing the viability of tapeworm eggs.

Composting

The most practical method of parasites elimination from sewage sludges is suggested as composting where higher temperatures needed to destroy pathogens are achieved. In aerobic composting, microorganisms utilize oxygen and feed upon the organic matter, in this process a great deal of energy is evolved in the form of heat. When composting is done by pilling the wastes the inner temperature may go as high as 75°C during fermentation. Destruction of parasitic helminthes during composting solely depends on temperature Gotas (1956), Forestner (1971), Their *et al.* (1978). Biological antagonism in composting probably may also effect the destruction of pathogenic organisms and parasites (Wiley, 1962).

Composting of night soil or sludge with city garbage and woodchips or other suitable carbonaceous bulking material is highly effective in removal of hookworm eggs and larvae. Studies on hookworm elimination by composting have been reported from China (Oldt, 1926; Hou *et al.*, 1959), Srilanka (Nicholls and Gunawardana, 1939) and the USSR (Gudzhabidze and Lyubchenko, 1959). *Ascaris* being very sturdy, it is preferable to monitor *Ascaris* eggs in the final compost.

Bhoyar (1990) has reported a detail study on the fate of intestinal parasites during aerobic and anaerobic composting. Both laboratory and field surveys were carried out and the results obtained are shown in Table 37.12. He studied windrow method of aerobic composting. Samples from both surface and central core revealed good destruction of parasites (Table 37.13).

Table 37.12: Aerobic Composting at Different Temperatures and Effect on Ova Viability

Composting Time	*Composting Temperature, °C*					
	30		*40*		*50*	
	Ascaris	*Trichuris*	*Ascaris*	*Trichuris*	*Ascaris*	*Trichuris*
0	60.17	54.29	31.81	28.64	42.2	38.0
10	57.56	53.70	5.36	*	*	*
20	56.31	51.64	*	*	*	*
30	56.12	49.99	*	*	*	*
60	53.64	49.99	*	*	*	*

All values are expressed as per cent viability.

*: Nil viability.

Anaerobic composting is generally carried out in pits. It is reported that the viability of ova is more in anaerobic composting than in aerobic composting. The work on anaerobic composting by Bhoyar (1990) indicates that even after 120 days of composting, the helminthes ova present in the sludge were viable. Only after 180 days of composting, they become non-viable (Table 37.14).

Table 37.13: Parasitological Analysis of Field Composting

Composting Time in Days	*Temperature*		*Ascaris Ova (Per cent viability)*	
	Surface	*Center*	*Surface*	*Center*
0	69.0	72.66	70	–
7	58.0	62.00	55	0
14	485	60.00	19	0
21	44.5	49.50	5	0
28	30.0	35.50	0	0
35	30.0	30.00	0	0
50	28.0	28.00	0	0
60	28.0	28.00	0	0

Ambient temperature: 28°C–30°C.

Viability of *Ascaris* at the start of experiment was 78 per cent.

Table 37.14: Aerobic Composting (Pit Composting)

Composting Time (Days)	*Temperature (°C)*	*Viable Ova (Per cent)*
0	28	40
1	30	–
2	32	–
3	35	–
4	35	–
5	40	–
30	30	Ova viable
90	30	Ova viable
120	30	Ova viable
180	30	Ova non viable

Aerobic Sludge Digestion

Cram (1943) reported that *Ascaris* and hookworm ova survive the digestion and subsequent drying of sludge. Silverman and Griffiths (1955) recommended that sludge might be allowed to dry for one year to kill *T. saginata* eggs. Reports are there that heating of sludge to 103°C for 3 minutes destroyed 100 per cent of *Ascaris* and hookworm eggs. Kabrick and Jewell (1982) reported that the removal of various pathogens in thermophilic aerobic digestion was superior to anaerobic digestion. Carrington 1980, however, suggested that cold storage of wet sludge has little effect on the viability of eggs.

The effects of mesophilic anaerobic or aerobic sludge digestion on survival of eggs from roundworm *Ascaris suum, Toxacara canis, Trichuris vulpis* and *Trichuris snis* and from rat tapeworm *Hymenolepis dimenuta* were reported (Black *et al.*, 1982). In this study destruction of eggs throughout a

15 day treatment period as well as their viability after re-isolation was analysed. Studies were carried out on laboratory model digesters at 15 days HRT. *Ascaris and Trichuris* eggs were destroyed in aerobic digester as 38 per cent and 11 per cent respectively.

There have been reports on the effects of sewage sludge stabilization processes on parasite egg survival (Black *et al.*, 1982; O'Donnel *et al.*, 1984). These studies were largely concentrated upon parasitic nematodes and have been provided clear relationship between eggs, survival and treatment, especially the importance of the temperature of storage and digestion.

Storey (1987) reported the survival of tapeworm eggs, free and enclosed in proglottids during the treatment of simulated sewage sludge. In this study, the effect of various digestion processes and wet sludge storage has been tried. It has also been reported that on storing the liquid sludge at different temperatures of 4°C, ambient and 35°C, the eggs were still viable in the proglottids upto 43 per cent, 18 per cent and 4 per cent respectively at the end of the experiment. Destruction of eggs in digesters was correlated with the state of eggs before subjection to the treatment processes *i.e.* some *Ascaris* and *Trichuris* eggs were already embroyonated in the host intestinal contents or faeces and hence passed their most resistant stage. The viability of *Ascaris* and *Toxacara* eggs that survived the digestion processes were greater in anaerobically treated than in aerobically treated material. Eggs from *Hymenolepis dimunuta* were non-viable before use in the experiments. However, their removal was more in aerobic digestion than in anaerobic digestion. Storey 1987 reported a 100 per cent removal of parasitic helminth eggs at 55°C in aerobic digestion system.

Anaerobic Sludge Digestion

In the treatment of wastewater for hygienic disposal, a portion of the solids, which is removed and require further processing, is highly organic and biodegradable in nature. The segregation of such solids along with other solids produced in the treatment is subjected to separate fermentation step in an anaerobic digester for further stabilization. This process is termed as "Anaerobic sludge digestion".

As early as the eighteenth century, the formation of methane from anaerobic decomposition of organic sludges was known. However, it was just one century ago (1881) when anaerobic treatment was reported to be a useful method for reducing the solids bulk and putriscible nature of suspended organic material removed from municipal wastewater. Since then the applications of anaerobic treatment have grown steadily, as has knowledge about the chemistry and microbiology involved.

Since the seventies, more attention has been paid to pollution control and human environment. In every country, night soil and sewage disposal is being tackled as high-level priority problem to improve sanitary situation and its multifunctional benefits. Sludge digestion is a very common treatment technology used for the sewage sludge, excreta and cattle dung. In recent times, sludge digestion by biodigesters is being advocated in general for organic sludge disposal. In India, Khadi Village Industries Commission has taken up this problem and is helping rural as well as urban population in erecting biogas digesters.

Of course, sludge digestion provides valuable methane as fuel but at the same time partially or improper digestion of sludge leads to spreading of many diseases including helminthic parasite infection. These health hazards may be assessed on the basis of available information on the occurrence and survival of pathogenic organisms in raw sewage digesters and in sludge digesters and in sludge used on the agricultural fields. The hazard in the use of sewage sludge comes from contact of the sludge with the edible portion of the crops and vegetables, which are eaten raw. Inspite of the survival of some pathogens and parasites, the literature documents no disease outbreaks associated with the use of digested sewage sludges or night soil and animal wastes in crop production.

The inhabitants of most rural communities or villages in developing countries are probably already exposed to the enteric diseases endemic to their area. The introduction of anaerobic sludge digestion for sewage sludge, night soil, organic wastes and agricultural residues, therefore, should not create any new or additional health hazard, on the contrary it should reduce the present health hazards significantly. The literature available on survival of pathogenic microorganisms in the anaerobic digestion process has a wide range of values. Survival data for some of the more important enteric pathogens are shown in Table 37.15 (Methane Fermentation). These examples, together with other data available demonstrate the importance of the anaerobic sludge digestion process in the treatment of sewage sludge and night soil. With few exceptions, pathogenic enteric microorganisms are effectively killed off if the digestion time is not significantly shorter than 14 days at a temperature not significantly lower than 35°C.

Table 37.15: Dieoff of Enteric Microorganisms in Public Health Significance during Anaerobic Digestion

Organisms	*Temperature (°C)*	*Residence Time (Days)*	*Dieoff (Per cent)*
Poliovirus	15	2	98.5
Salmonella sp.	22–37	6–20	82–96
Salmonella typhosa	22–37	6	99
Mycobacterium tuberculosis	30	Not reported	100
Ascaris	29	15	90
Parasitic cysts	30	10	100

a: Time indicated is time of digestion; b: Does include ascaris.

Ascaris eggs are the major exceptions to the pathogenic microorganisms effectively killed during anaerobic digestion while other encysted helminthes are completely destroyed. *Ascaris* cysts are able to survive even after 14 days of anaerobic fermentation at 35°C. Another study reported that thermophilic digestion destroyed the *Ascaris* eggs completely. However, sludge from mesophilic digestion required additional storage and drying for about 6 months to destroy the *Ascaris* eggs completely (Wright *et al.*, 1942). Nevertheless, since the die off rate is expected to be sufficiently high say 90 per cent, anaerobic digestion of organic material for biogas production provides a public health benefit beyond that of any other treatment likely to be in use in rural areas of developing countries.

Hookworm eggs did not develop in digesting sludge but that they could survive for upto 64 days at 20°C and 41 days at 30°C (Cram, 1943). He reported that during anaerobic digestion even though the development of eggs were stopped, few eggs got embryonated probably because they were brought to the top layer by gas bubbles. Digestion for three months hardly affected the viability of eggs whereas digestion for six months destroyed about 10 per cent of the viable eggs. He also reported that eggs of *Ascaris* and hookworm were found to be extremely resistant to digestion and viable even after digestion of the sludge for a period of one year.

Newton *et al.* (1949) reported a 50 per cent mortality of *T. saginata* eggs after 60 days digestion at 24–29.5°C and 85–90 per cent after 200 days. It was assumed without much experimental study that digestion of raw sludge and subsequent drying of the digested sewage sludge could destroy all viable helminthes ova present in the raw sludge. But later studies revealed that large percentage of viable ova escape digestion process as well as sludge drying. It was observed by Hoggs (1950) that dehydration even at moisture content as low as 5.5 per cent did not destroy the viable eggs completely.

Petrik (1954) reports that sludge digestion at temperatures below 40°C does not eliminate hookworm eggs. Silverman and Guiver 1960 achieved a 100 per cent kill in 5 days at 35°C but only 40–50 per cent kill after 20 days at 4°C whereas Menshel (1963) found that *T. saginata* eggs survived aerobic digestion at 26–28°C for only 56 days.

The hookworm infections are very common to many tropical and semitropical developing countries. Studies on the survival of *Ancylostoma duodenalee* in septic tanks suggested that a few viable hookworm eggs were recovered in the septic tank effluents, although high removal rates were achieved (90 per cent) (Majumder *et al.*, 1960). Since biogas plants will similarly carry out process, the risk of indiscriminate dissemination of eggs of parasitic organism is minimized. Storage and drying after the digestion period will further minimize and perhaps eliminate the risk of parasitic eggs spreading in rural environment of developing countries.

Reyes *et al.* (1963) conducted aerobic and anaerobic digestion of nights oil and observed that at low temperatures of 25°C to 35°C of digestion, eggs were found viable and even developed when transferred to favourable conditions in both cases. When the temperature was maintained at or over 38°C for anaerobic digestion and 45°C for aerobic digestion, it was observed that nights oil could be well digested and eggs were destroyed within 20 days in the aerobic digestors and within 30 days in the anaerobic digesters. Forstner (1970) reported that *T. saginata* eggs survived upto 26 days digestion at 30°C. Elevated temperature of >50°C in the digesters has been reported to result in rapid death of helminth eggs (Silverman, 1955; Silverman and Guvier, 1960; Smith *et al.*, 1980; Owen and Crewe, 1982).

McGarry and Stainforth (1978) reported studies in China on a biogas plant which may be likened to septic tank with a long retention time. The influent contained hookworm eggs in 87 per cent of samples with an average concentration of 840 eggs per litre, the effluent contains hookworm eggs in 23 per cent samples with an average concentration of 4 eggs per litre.

Anaerobic digestion will destroy the eggs of *Taenia saginata* with continuous digestion being more effective than batch processing. This may be because of the continuous feed, which ensures a renewable carbon source for microbes. Continuous digesters also have a greater proportion of active digesting sludge and consequently a higher enzymatic activity than batch digesters. Eggs survived for longer periods in aerobic than in anaerobic digestion. This conflicts with the findings of Leftwich *et al.* (1981) on the survival of *Ascaris* and *Toxacara* at 35°C in aerobic and anaerobic digester. All the sludge digestion processes generally will reduce the number of viable helminth eggs, their efficiency being clearly dependent upon the length of treatment and exposure temperature with apparently higher death rate above 35°C.

Anaerobic sludge digestion has been reported to have destroyed only 23 per cent of *Ascaris* eggs, whereas *Trichuris* eggs in anaerobic digesters and *Toxocara* eggs in either anaerobic or aerobic digestion were not destroyed (Black *et al.*, 1982). Pike *et al.* (1983) found similar death rates at 4°C and 39°C, however these experiments were performed in saline medium where microbial activity would be expected to be low. Pike *et al.* (1983, 1988) have carried out detailed experiments to study the inactivation of ova of the parasites *T. saginata, Ascaris suum* during heated anaerobic digestion. They reported that mesophilic anaerobic digestion at 35°C has shown to reduce viability of *A. suum* ova recovered by only 35–50 per cent, but after thermophilic digestion at 49°C viability was reduced by at least 99 per cent.

However, atleast 50 to 75 per cent of the ova added could not be recovered after digestion and may have autolysed. To determine the death rate for these ova, at the temperatures of 50°C to 60 °C and subsequent mesophilic digestion, needs further studies.

A 0.4 m model steel biodigester was used for studying the survival of *Ascaris* ova and other entero pathogens. A system of three electric heat tubes, mercury thermostats and transistor relays were used to control the temperature of the digesters by Hubei Sanitation and Anti epidemic Station (Shanta, 1989). The results reported are given in Table 37.16.

Table 37.16: The Survival Rate of *Ascaris* Ova in Biodigesters

Temp. (°C)		*Days*										
		0	*1*	*2*	*3*	*4*	*5*	*6*	*7*	*8*	*10*	*12*
51±1	N	800	800	800	800	600	333					
	R	76.1	0.5	0	0	0	0					
48±1	N	800	800	578	800	600	218	800				
	R	90.8	79.3	1.3	0	0	0	0				
43±1	N	800	–	835	–	400	–	800	800	800	800	
	R	82.6	–	96.8	–	88.0	–	86.6	90.7	58.5	58.2	

N: Number of ova examined; R: Survival rate (per cent).

Shanta (1989) has reported a detailed study on the fate of *Ancylostoma duodenale* and *Ascaris lumbricoides* during mesophilic and thermophilic anaerobic digestion. Work on both laboratory and pilot plant scale studies at mesophilic range at different HRTs and organic loadings have been reported (Tables 37.17 and 37.18). A single laboratory digester was operated for a period of one month at a constant temperature of 50°C and was inoculated with known numbers of viable eggs of *Ascaris* and hookworm eggs. After every 5 minutes, the samples were withdrawn and studied for the survival of eggs till no viable eggs were found. On the basis of available information, it seems clear that using the sludge from heated digesters and thermophilic digesters as fertilizer will not pose any health risk, as complete elimination of helminth parasites can be achieved in the anaerobic thermophilic process. The results of thermophilic laboratory digesters (Table 37.19) revealed a 100 per cent removal of *Ancylostoma duodenale* after 15 minutes and 100 per cent removal of *Ascaris lumbricoides* after 45 minutes (Shanta, 1989).

Table 37.17: Per cent Reduction of Helminth Ova in Lab. Night Soil Digester

Detention Time (Days)	*Influent No. of Ova/Lit*		*Effluent No. of Ova/Lit*		*Per cent Reduction*	
	Ascaris	*Hookworm*	*Ascaris*	*Hookworm*	*Ascaris*	*Hookworm*
20	4000	58200	2080	21360	48	63.2
25	11400	18200	38324	1693	66.5	90.6
30	94110	21820	28248	1528	70	93

Anaerobic digestion of sewage sludge alongwith refuse and the fate of *Ascaris* and *Trichuris* has been reported by Bhoyar (1990). He reported that within 60 days period of digestion, reduction in viability of *Ascaris* ova was 48.15 per cent. Whereas in batch-fed digesters, where only sludge was used, 43.79 per cent reduction in *Ascaris* viability was noted. In case of *Trichuris*, 43.79 per cent reduction in *Ascaris* viability during batch digestion of sludge and reuse and 48.35 per cent reduction with only sludge was reported (Table 37.20). A slightly modified table of the results reported by Bhoyar on the continuous fed digester is shown in Table 37.21.

Table 37.18: Removal of Helminthes Ova in 18 m³ Capacity Pilot Digester

Loading (Kg VS/m³.d)	*Per cent Removal*	
	Hookworm	*Ascaris*
1.99	66±23.18	37.0±22.61
1.60	70±12.11	52.1±16.51
1.79	65.3	35.3±22.7

Table 37.19: Survival of Helminthes Ova at Thermophilic Lab. Digesters

Time (Minutes)	*Per cent Removal*	
	Ascaris	*Hookworm*
5	25	80
10	45	86
15	58	100
20	79	
25	83	
30	96	
35	100	
40	100	
45	100	

Table 37.20: *Ascaris* and *Trichuris* Ova Viability during Anaerobic Digestion (Batch Experiment)

Temperature (°C)	*Parasite*	*Period of Digestion (Days)*				
		0	*15*	*30*	*45*	*60*
30°C Reuse+ sludge	*Ascaris*	66.66	57.91	55.06	27.30	18.51
	Trichuris	58.33	50.00	46.4	19.76	14.54
30°C sludge	*Ascaris*	60.51	52.77	43.71	31.66	15.49
	Trichuris	58.94	51.79	40.44	32.75	10.59
40°C sludge	*Ascaris*	31.88	0.0	0.0	0.0	0.0
	Trichuris	28.91	0.0	0.0	0.0	0.0
50°C sludge	*Ascaris*	62.05	0.0	0.0	0.0	0.0
	Trichuris	46.42	0.0	0.0	0.0	0.0

Table 37.21: Ova Viability in Continuous Fed Anaerobic Digester

Detention Time		*Influent*		*Ova Concentration/litre Effluent*		*Ova concentration in Bottom Sediment*
		Ova Concentration	*Per cent Viability*	*Ova Concentration*	*Per cent Viability*	
Mesophilic 16 days	17	400546	46–50	540	42.85	
	18			560	40.32	
	19			480	38.24	
	20			320	45.45	100–1509 (28 per cent viable)
	21			480	41.30	
	22			580	40.00	
	23			240	42.50	
25 days	26			685	30.12	
	27			274	35.36	
	28			548	30.00	
	29			37	45.00	90-150 (30 per cent viable)
	30			137	42.65	
	31			274	41.32	
	32			548	40.54	
Thermophilic 50°C	17			60	5.00	
16 days						
	18			66	8.25	
	19			90	10.00	
	20			105	15.00	150–300 (5–10 per cent viable)
	21			110	15.5	
	22			95	20.0	
	23			105	20.25	

Conclusions

Occurrence of common nematode parasites of human intestine in sewage and sludge from wastewater treatment plant is extremely significant from public health point of view. As the dormant stages of these parasites and eggs are resistant, the conventional methods needs to be modified to increase their efficiency, not only in accumulating them in sludges of wastewater treatment plant but to kill them effectively by further treatment of sludge. Advanced primary treatment system has been advocated by some to kill the helminth parasites from wastewater. Similar such studies are required to be carried out to eradicate the public health problems. Aerobic/anaerobic digesters may also be used for the same purpose. Presently, the effluent standard with respect to presence of helminthes for agricultural reuse of wastewater is available. However, the standard for reuse of digested sludge as fertilizer is not available. Efforts are needed in the direction in evolving standard for the treated sludge so as to ensure safe reuse of digested sludge as fertilizer for enhancing the production of food crops.

Acknowledgements

Authors wish to thank Director, National Environmental Engineering Research Institute, Nagpur for giving facilities to carry out the work on helminthes and the publication of this review paper.

References

Allucino, F.A., Colombi, A., Calcaterra, E., Carene, M., Mastrantonio, A. and Ossini, P., 1998. Microbiological and chemical quality of sludges from domestic wastewater plants. *Journal of Environmental Health Research,* 8(2): 137–144.

Amaral and Leal, R.A., 1940. Influence of different sewage treatment methods on the destruction of worm eggs. *Rev. Bio. Hyg.,* S. Paulo, 11: 35–39.

Arther, R.G., Fox, J.C. and Fitzgerald, P.R., 1981. Parasite ova in anaerobically digested sludge. *J. Wat. Pollut. Control Fed.,* 53: 1334–1338.

Arundel, J.M. and Adolph, A.J., 1980. Preliminary observations on the removal of *T. saginata* eggs from sewage using various treatment processes. *Australian Vet. J.,* 56(10): 492–495.

Ayres, R.M., Slott, R., Lee, P.L., Mara, D.D. and Silva, S.A., 1991. Comparison of techniques for the enumeration of human parasitic helminth eggs in treated wastewater. *Environmental Technology,* 12: 617–623.

Banwell, J.G. and Schad, G.A., 1978. Hookworms Clinics. *Gastroenterology,* 7: 129–156.

Barbier, D., Perrine, D., Duhamel, C., Doublet, R. and Georges, P., 1990. Parasitic hazard with sewage sludge applied to land. *Appl. Environ. Microbiol.,* 56(5): 1420–1422.

Bhaskaran, T.R., Sampathkumaram, M.A., Sur, T.C. and Radhakrishan, I., 1956. Studies on the effects of sewage treatment processes on the survival of intestinal parasites. *Indian J. Med. Res.,* 44: 163.

Bhoyar, R.V., 1990. Studies on fate of intestinal parasitics during aerobic and anaerobic composting. *Ph.D. Thesis,* Nagpur University, India

Black, M.I., Scarpino, P.V., O'Donnel, C.J., Meyer, K.B., Jones, J.V. and Kaneshiro, E.S., 1982. Survival rates of parasite eggs in sludge during aerobic and anaerobic digestion. *Appl. Environ. Microbiol.,* 44: 1138–1143.

Blum, D. and Feachem, R.G., 1985. IRCWD Report No. 05/85.

Bradely, R.M. and Hadiday, S., 1981. Parasitic infection and the use of untreated sewage for irrigation of vegetables with particular reference to *Aleppo syria. Public Health Engineer,* 9: 154–157.

Carrington, E.G., 1980. *The Fate of Pathogenic Microorganisms during Wastewater Treatment and Disposal.* Technical Report T.R. 128, Water Research Centre.

Cram, E.B., 1943. The effect of various treatments processes on the survival of helminth ova and protozoan cysts in sewage. *Sewage Works Journal,* 15(6): 1119.

Crompton, D.W.T., 1989. Biology of *Ascaris lumbricoides.* In: *Ascariasis and its Prevention,* (Eds.) D.W.T. Crompton, M.C., Nesheim, Z.S. Pawlowski. Taylor and Francis, London, New York and Philadelphia, p. 9–44.

Cross, P. and Strauss, M., 1985. IRCWD Report no. 03/85.

Dozanska, W. and Ivanczuk, 1958. Influence of different sewage treatment methods on the destruction of worm eggs. *Chem. Abs.,* 129: 6934.

EPA: Environmental Protection Agency, 1994. *Alain English Guide to the EPA Part 503 Biosolid Rule.* EPA/832/R 93/003, pp. 175.

Forestner, M.J.K., 1971. The effect of sewage overlying liquor and composting on the viability of parasitic reproductive stages. In: *Water Pollution Abstract.*

Forestner, M.J., 1970. Der einflltues von abwasser, Schwemmisi und kompostierung anf die lebensfatigkeit parasitarer vermehrunsta dien. *Z. Wass. Abwass Forsch,* 3: 176–1984.

Fosster, D.H. and Engelbrecht, R.S., 1974. *Microbial Hazards of Disposing of Wastewater on Soil.* EPA/2/74/003, 217–241.

Gaspard, P., Wiart, J. and Schwarzbrod, J., 1996. A method for the viability of nematode eggs in sludge. *Environmental Technology,* 17: 415–420.

Gaspard, P., Wiart, J. and Schwartzbrod, J., 1997. Parasitological contamination of urban sludge used for agricultural purposes. *Waste Management and Research,* 15(4): 429–436.

Golucke, E.G. and Gotaas, H.G., 1954. Public health aspects of waste disposal by composting. *Amer. J. Pub. Hlth.,* 44: 339.

Gotass, H.B., 1956. *Composting, Sanitary Disposal and Reclamation of Organic Wastes.* Monograph Series 31, Geneva, World Health Organization.

Graham, H.J., 1981. *Parasites and the Land Application of Sewage Sludge.* Report Ontario Ministry of the Environment No. 110.

Gudzhabidze, S.I. and Lyubechenko, S.D., 1959. Control of Ascariasis and Ancylostomiasis by composting of organic waste material. *Meditsinskaia Parazitologiial, Parazitarnye Bolenzni,* 28: 576–578.

Gunnerson, E.G., Shuval, H.I. and Arlosoroff, S., 1985. Health effects of wastewater irrigation and their control in developing countries. In: AWWA Research Foundation (Ed.), Denver, *Future of Water Reuse,* pp. 1576–1602.

Hespanhol, I. and Prost, A.M.E., 1994. W.H.O. Guidelines and national standards for reuse and water quality. *Water Res.,* 28(1): 119–124.

Hirst, L.F., 1932. Hookworm diseases and Ceylon sewage works. *Ceylon J. Sci.,* 2: 245.

Hoggs, E.S., 1950. Parasitization, ova and cysts in digested sludge study. *Sewage and Industrial Wastes J.,* 22: 81098.

Hou, T.E., Chung, Holy, H.L. and Weng, H.E., 1959. Achievements in the fight against parasitic diseases in New China. *Chinese Medical Journal,* 79: 493–520.

IRCWD, 1985. Health Aspects of Wastewater and Excreta Use in Agriculture and Aquaculture. In: The Engelberg Report. IRCWD News No. 23, International Reference Centre for Waste Disposal, Duebendorf, Switzerland, pp. 11–18.

Jepsen, A. and Roth, H., 1952. Epizootiology of *Cysticercus* bovis resistance of the eggs of *Taenia saginata.* In: *Proc. 14th Int. Vet. Congr.* 11, London, pp. 43–50.

Jimenez, B. and Chavez, A., 1998. Removal of helminth eggs in an advanced primary treatment with sludge blanket. *Environmental Technology,* 19(11): 1061–1071.

Kabler, P., 1959. Removal of pathogenic microorganisms. *Sewage and Industrial Wastes,* 31(12): 1373.

Kabrick, R.M. and Jewell, W.J., 1982. Fate of pathogens in thermophilic aerobic sludge digestion. *Wat. Research,* 16: 1051–1060.

Khuder, S.A., Arthur, T.R., Bisesi, M. and Schaub, E.H., 1998. Prevalence of infectious diseases and associated symptoms in wastewater treatment workers. *American Journal of Industrial Medicine,* 33(6): 571–577.

Kowal, N.E., 1988. *Health Effects of Land Application of Municipal Sludge.* U.S. EPA 600/1085–015, 33–38, Cincinnati.

Lakshminarayana, J.S.S. and Abdulappa, M.K., 1969. The effect of sewage stabilization ponds on helminths. In: *Low Cost Waste Treatment,* (Ed.) C.A. Sastry. CPHERI, Nagpur, India, pp. 290–299.

Lakshminarayana, J.S.S. and Abdulappa, M.K., 1972. The effect of sewage stabilization ponds on helminths. In: *Low Cost Waste Treatment,* (Ed.) C.A. Sastry. CPHERI, Nagpur, India.

Leftwich, D.B., Reimers, R.S. and Englande, A.J., 1981. Inactivation of parasitic contaminated domestic wastewater sludges. In: *Chemistry in Wafer Reuse,* Vol. 2, (Ed.) W.J. Copper. Ann. Arbor, Science, Ann Arbor, Michigan.

Livingston, D.J., 1978. Decay of microorganisms in the marine environment. In: *Proceedings of Symposium on Disinfection of Water.* Pretoria, S. Africa National Institute for Water Research.

Majumder, N., Prakasam, T.B.S. and Suryapakasam, M.V., 1960. Critical study septic tank performance in rural area. *J. Inst. Engineers (India),* 40: 742–761.

Mara, D.D. and Cairneross, S., 1987. UNEP, WHO Report.

Mara, D.D. and Silva, S.A., 1978. Sewage treatment in waste stabilization ponds: Recent research in North East Brazil. *Progress in Water Technology,* 11: 341–344.

Mara, D.D. and Silva, S.A., 1986. *J. Tropical Med. Hyg.,* 89: 71–74.

McCarty, P.L., 1981. One hundred years of anaerobic digestion. In: *Anaerobic Digestion* 1981, (Eds.) D.E. Hughes *et al.* Elsevier Bio Medical Press Amsterdam, pp. 3–22.

McGarry, M.G. and Stainforth, J., 1978. *Compost: Fertilizer and Biogas Production from Human and Farm Wastes in the Peoples Republic of China, Ottawa.* International Development Research Centre.

Menschel, E., 1963. Beobachtunge uber Einflusse Faulturmen auf die Eier des Bandwumes *Taenia saginata. Munch. Beitr. Abwass Fisch Fhissbio,* 10: 175–179.

Miller, T.A., 1979. Hookworm infection in man. *Advances in Parasitology,* 17: 315–384.

Newton, W.L., Bennett, H.J. and Figgat, W.B., 1949. Observations on the effects of *Taenia saginata. American J. Hyg.,* 49: 166–175.

Nicholls, L. and Gunawardana, S.A., 1939. The destruction of helminth ova in night soil by composting. *Ceylon J. of Science,* Section D5: 1–9.

O'Donnel, C.H., Meyer, K.B., Jones, J.V., Benton, T., Kaneshiro, E.S., Nichols, J.S. and Schaefer, F.W., 1984. Survival of parasite eggs upon storage in sludge. *Appl. Environ. Microbiol.,* 48: 618–625.

Oldt, F., 1926. XIII Studies on the viability of hookworm eggs in stored night soil in South China. *American J. Hygiene* Monographic Series No. 7: 265–291.

Owen, R.R. and Crewe, W., 1982. Environmental parasitology: Review with particular reference to beef tapeworm. *Environ. Health,* Dec., p. 321–323.

Panicker, P.V.R.C. and Krishnamoorthy, K.P., 1978. Elimination of enteric parasites during sewage treatment processes. *Industrial Asso. for Wat. Poll. Contr. Technical Annual,* 5: 130–138.

Panicker, P.V.R.C. and Krishnamoorthy, K.P., 1981. Parasite eggs and cyst reduction in oxidation ditches and aerated lagoons. *J. Wat. Pollut. Control Fed.,* 53: 1413–1419.

Petrik, M., 1954. Utilization of nights oil, sewage and sewage sludge in agriculture. *Bulletin of the World Health Organization,* 10: 207–228.

Pike, E.B., Morris, D.L. and Carrington, E.G., 1983. The inactivation of ova of the parasites *Taenia saginata, Ascaris suum* during heated anaerobic digestion. *Wat. Poll. Cont.,* 82: 501–507.

Pike, E.B., Carrington, E.G. and Harman, S.A., 1988. Destruction of *Salmonella,* enteroviruses and ova of parasites in wastewater sludges by pasteurization and anaerobic digestion. *Wat. Sci. Tech.,* 20(11/12): 337.

Prost, A., 1988. *Revision of the 1973 WHO guidelines: A WHO scientific Group proposes revised health guidelines for the use of wastewater.* IRCWD News No. 24/25, International Reference Centre for Waste Disposal, Duebendorf, Switzerland, pp. 11.

Reimers, R.S., Little, M.D., Englande, A.J., Leftwich, D.B., Bowman, D.D. and Wilkinson, R.F., 1981. *Parasites in the Southern Sludges and Disinfection by Standard Sludge Treatment.* EPA–600/52–81–166.

Reyes, W.L., Kruse, C.W. and Batson, M.S., 1963. *Ascaris* eggs, effect of digestion of nights oil on. *Amer. J. Trop. Med. Hygiene,* 12: 46–55.

Rowan, W.B., 1964. Sewage treatment and schistosome eggs. *Am. J. Trop. Med. Hyg.,* 13: 572–576.

Schwartzbrod, J., Thevenot, M.T., Collomb, J. and Baradel, J.M., 1986. Parasitological study of waste sludge. *Environ. Letters,* 3(7): 155–162.

Schwartzbrod, J., Mathieu, C. and Thevenot, M.T., 1987. *Water Sci. Tech.,* 19(8): 33–40.

Schwartzbrel, L., Stien, J.L., Bouhoum, K. and Baleux, P., 1989. Impact of wastewater treatment on helminth eggs. *Wat. Sci. and Technol.,* 21(3): 295.

Shanta Satyanarayan, 1989. Survival of helminth ova in anaerobic digestion of night soil: A pilot plant and laboratory study. *Asian Environment,* p. 3–9.

Shuval, H.I., Adin, A., Fattal, B., Rawitz E. and Yekutiel, P., 1988. *Wastewater Irrigation in Developing Countries.* World Bank Publ. Technical Paper Series no. 5.

Silverman, P.H., 1955. The biology of sewers sand sewage treatment: The survival of the eggs of the beef tapeworm (*T. saginata*). *Advmt Sci.,* London, 12: 108–111

Silverman, P.H. and Griffiths, R.B., 1955. A review of methods of sewage disposal in Great Britain with special reference to epizoology and *Cysticercous bouis. Ann. Trop. Med. Parasitology,* 49: 436–450.

Silverman, P.H. and Guiver, K., 1960. Survival of eggs of *Taenia saginata* (the human beet) tape worm after mesophilic anaerobic digestion. *J. Proc. Inst. Sewage Purif.,* 3: 345–347.

Smith, M.R., Zinder, S.H., and Mah, R.A., 1980. Microbial methanogenesis from aselate process. *Biochem.,* 15(3): 34–39.

Stien, J.L. and Schwartzbvrod, J., 1988. Viability determination of *Ascaris* eggs recovered from wastewater. *Environmental Technology Letters,* 9: 401–406.

Storey, G.W., 1987. Survival of tapeworm eggs free and inproglottids during simulated sewage treatment processes. *Water Research,* 21(2): 199.

Teichmann, A., 1986. *Ang. Parasit.,* 27: 145–150.

Their, J.H., Bolton, V. and Storm, D.R., 1978. Helminth ova in soil and sludge from twelve urban areas. *Wat. Poll. Contr. Fed.,* 50(11): 2485.

Vasilkova, Z.G., 1936. Removal of tapeworm eggs by complete sewage treatment. *Med. Paraziloi, Parazilar BolezniMoskva,* 5: 657–673.

W.H.O (World Health Organization) 1989. *Health Guidelines for the Use of Wastewater in Agriculture and Aquaculture.* Technical Report No. 778, Geneva.

Wiley, J.S., 1962. Pathogen survival in composting municipal waste. *Wat. Poll. Cont. Fed.,* 34: 80–90.

Wright, W.H., Cram, E.B. and Nolan, M.O., 1942. Preliminary observation on the effect of sewage treatment processes on the ova and cyst of intestinal parasites. *Sewage Works J.,* 14; 1276.

Zhao, Xi Hui, 1985. The Hygienic and environmental protection effect of biogas digester in China. In: *Proc. 4th Int. Symp. on Anaerobic digestion* held in Guangzhou, China.

Ethnobotanical Wisdom and Microbial Studies on Medicinal Plants (2011) *Pages* **256–261**
Editors: **D.R. Khanna, Ashutosh Gautam, R. Bhutiani, Gagan Matta, Anil Dhiman**
Published by: **DAYA PUBLISHING HOUSE, NEW DELHI**

Chapter 38

Antibacterial Potential of *Cascabela thevatica* Leaf Extract Against *Escherichia coli*: A Study *in vitro*

Vishal Upadhyay and P. Kaushik
Department of Botany and Microbiology, Gurukul Kangri Vishwavidyalaya, Haridwar – 249 401, Uttaranchal

ABSTRACT

Cascabela thevatica leaf part was tested for antibacterial efficacy against *Escherichia coli* organism using the disc diffusion method. The crude alcoholic extract was found effective against *Escherichia coli* organism. The antibacterial activity is attributed to the presence of alkaloids, which was confirmed by gas chromatography and positive alkaloid test. The minimum inhibitory concentration (MIC) was determined for the crude extract by paper disc diffusion method. The results were compared with reference antibiotic tetracycline (one unit strength).

Keywords: *Cascabela thevatica, Escherichia coli, Apocynaceae, Leaf extract, Antibacterial, Antimicrobial activity.*

Introduction

Cascabela thevatica is evergreen and glabrous shrub or small tree with 3–6 m height. The leaves are simple, glabrous and narrowed at both ends. It is native of West Indies and Central America, now cultivated in gardens and found along roadsides throughout India. Earlier Literature shows that the seeds are abortifacient and alexeteric and are used as purgative in dropsy and rheumatism. The leaves are emetic and purgative, bark is used in different kinds of fevers (Chopra *et al.*, 1956) and the rot in the form of a plaster is applied to tumor (Chaterjee and Pakrashi, 1995). Dinda and Saba (1990), isolated lupeol, β-amyrin, α-amyrin, γ-teraxasterol and taraxasterol from the root of the plant. The viridoside has also been isolated from stem bark and L-bomesitol from leaves and stem (Chatterjee and Pakrashi, 1995).

Evaluation of plant extract for their, anti microbial activity has been done by several workers. Chakraborty and Brantner (1999) studied antibacterial steroid alkaloids from the stem bark of *Holarrhena pubescens,* antibacterial effect of *Rhynchostylis retusa* BL. and *Aerides multiflora* Roxb. (Orchidaceae) were studied by Ghanaksha and Kaushik (1999a, 1999b) and that of *Adiantum capilius veneris* Linn. Ansari (1995) studied effect of plant extract against the pathogen of leaf sheath blight or rice. Kaushik and Kishore (1991) studied the effect or alcoholic extract of shoot extract of *Pholidota articulata* Lindl., against *Bacillus subtilis, Staphylococcus aureus* and *Klebsiella pneumoniae.* Information regarding drug plants and microorganism dated back to Vedic period (Kaushik, 1983, 1985, 2000). Gehlot and Bohra (1998) have tested antimicrobial activity of various plant parts of *Aerva persica* against human pathogenic bacterial strains and pathogenic fungal species. Earlier literature on antibacterial potential of plants has been reviewed by Kaushik and Dhiman (2000). Recently three species of orchids and garlic (*Allum sativum*) studied for the presence of alkaloids and antibacterial potential by Kaushik and Kishore (1991, 1995) and Kaushik and Upadhyay (1999) and Kaushik *et al.* (2000) which showed interesting results.

Investigations into the folkloric antimicrobial activity of *Landolphia owrrience* against *Escherichia coli, Bacillus subtilis* and *Candida albicans* was done by Ebi and Ofoefule (1997); chemical composition and antimicrobial activity of *Croton urucurana* Baillon (Euphorbiaceae) against *Staphylococcus aureus* and *Salmonella typhimurium* was done by Peres *et al.* (1997); and anti-microbial properties of *Piper aduncum* against the AIDS related pathogens *Candida albicans, Cryptococcus neoformans* and *Mycobacterium intracellular* was done by Okunade (1997). The main aim of the present work is to justify the use of the plant for treatment of infectious diseases.

Materials and Methods

Plant Material

Leaves of the plant *Cascabela thevatica* were collected from the Raja Ji National Park, adjacent to Haridwar.

Extract Preparation

For the preparation of the plant leaf extract, plant leaves were first washed 2–3 times with tap water and then again with sterilized double distilled water. Finally the surface sterilization was done with 90 per cent ethyl alcohol, 100 gm leaves of plants were crushed in water blender resulting the formation of a paste, which was mixed in 250 ml of absolute ethyl alcohol. Alcoholic extract so prepared was allowed to evaporate at room temperature until 50 ml of it was left. This extract was squeezed through double layer muslin cloth and filtered through Whattman filter paper No. 42 and centrifuged at 5000 rpm for 20 minutes and then sterilized by passing through 0.2 micron disposable filters.

Selection of Test Organisms

In vitro antibacterial studies were carried out against one Gram negative *Escherichia coli,* MTCC–739. The strain were self purchased from Microbial Type Culture Collection (MTCC), Chandigarh and further developed in our lab. The bacterial culture was incubated for 24 h at 37±2°C.

Antibacterial Testing

The crude ethanolic extract was tested for its antibacterial activity as per the Bauer *et al.* (1966).

Disc Diffusion Method

Whattman filter paper No. 42 discs (δ = 5 mm) were impregnated with 100 per cent, 50 per cent and 20 per cent concentration on the surface of a solid medium (nutrient agar). After solidification of the medium, abstract discs of different concentrations are applied with sterile forceps. After 24 hours incubation, the degree of sensitivity is determined by measuring the zone of inhibition. The experiments were repeated 3 times.

Minimum Inhibitory Concentration (MIC)

Minimum inhibitory Concentration (MIC) was determined by agar diffusion method. The plant extract was serially diluted with double distilled sterilized water to obtain the desired concentrations and inoculated on agar plate containing pH-6.8 by placing Whattman filter paper No. 42 discs (δ = 5mm). After 24 hours the growth was seen. The MIC is defined as the lowest anti microbial concentration of the test compounds, which inhibit bacterial growth. The experiments were repeated 3 times.

Gas Chromatography (GC)

In phytochemical analysis gas chromatography (GC) was actively carried out. In gas chromatography (GC) HP–5 column was used. By this column GC detect the alkaloids, fatty acids and drugs etc. and solvent was ethyl alcohol it shows high peak in comparison to all other peaks which was detected by GC in the alcoholic extract (Figure 38.1).

Alkaloid Test

Considering the remarkable results with *C. thevatica* leaf extract, alkaloid test was carried out by leaf extract. For alkaloid presence test, 0.85 gm bismuth nitrate was dissolved in 50 ml. Acetic acid (20 per cent) and was labeled as solution X. 20 gm potassium iodide was dissolved in 50 ml sterilized distilled water and labeled as solution Y and refrigerated in dark up to the time of use because in the presence of air and light the solution gets oxidized. A drop of plant extract was placed in a watch glass and mixed with the equal amount of solution X and Y the presence of alkaloid was confirmed by the appearance of bright orange colour (Stahl, 1969). The presence of alkaloids in leaves of *Cascabela thevatica* confirmed antibacterial potential, as alkaloids are said to be antibacterial in nature.

Results and Discussion

The crude alcoholic leaf extract from *Cascabela thevatica* plant was submitted to an antibacterial screening, using the disc diffusion method. The results are presented as inhibition zone in diameters in Table 38.1 and as MIC respectively in Table 38.2. The test sample showed more potent inhibitory effect on gram bacteria, so the results of present investigations clearly indicate that alcoholic leaf extract of *Cascabela thevatica* possesses antibacterial property. The size of effective zone of inhibition of undiluted extract of *Cascabela thevatica* against *Escherichia coli* measured 14 mm. The MIC of *Cascabela thevatica* extract against *Escherichia coli* was on 35 per cent concentration. The inhibitory zone measured for antibiotic control is 25 mm for *E. coli.* The results indicated that the antibiotic tetracycline was more effective on *E. coli* in comparison to undiluted alcoholic extract of *Cascabela thevatica.*

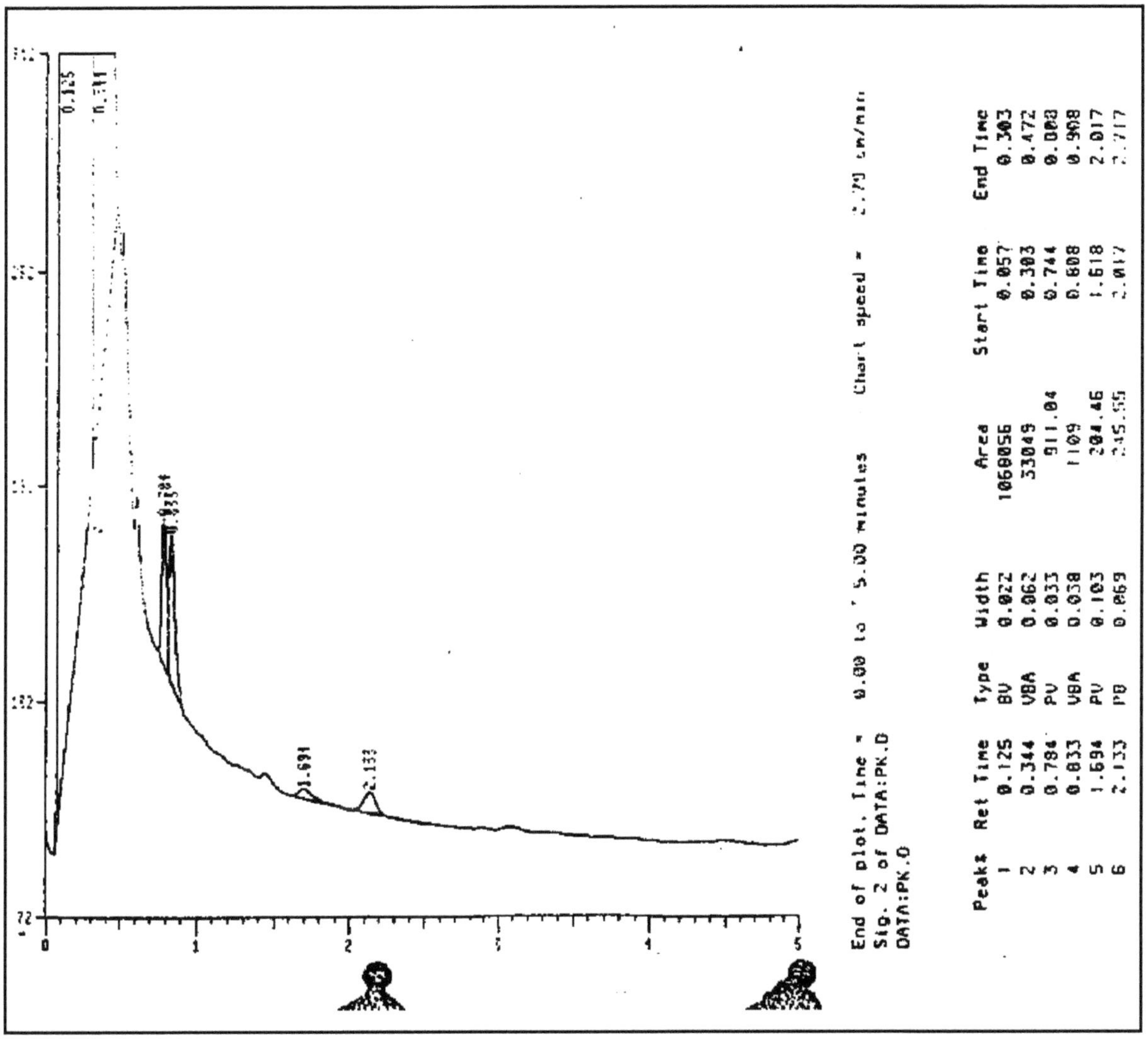

Figure 38.1: Gas Chromatography of *C. thevatica* Leaf Extract

Table 38.1: Determination of the Antibacterial Activity of *C. thevatica* Leaf Extract, and Reference Antibiotic by Means of the Disc Diffusion Method

Test Organism	*Inhibition Zone in mm*				
	Antibiotic Zone	*Undiluted Extract Zone (A)*	*Control Alcohol Zone (B)*	*Distilled Water Zone (C)*	*Effective Zone of Inhibition (A-B)*
Escherichia coli	25	14	5	Nil	9

Table 38.2: Minimum Inhibitory Concentration (MIC) of *C. thevatica* Leaf Extract

Pathogen	*Escherichia Coli*
Alcoholic extract concentration (per cent)	MIC 35

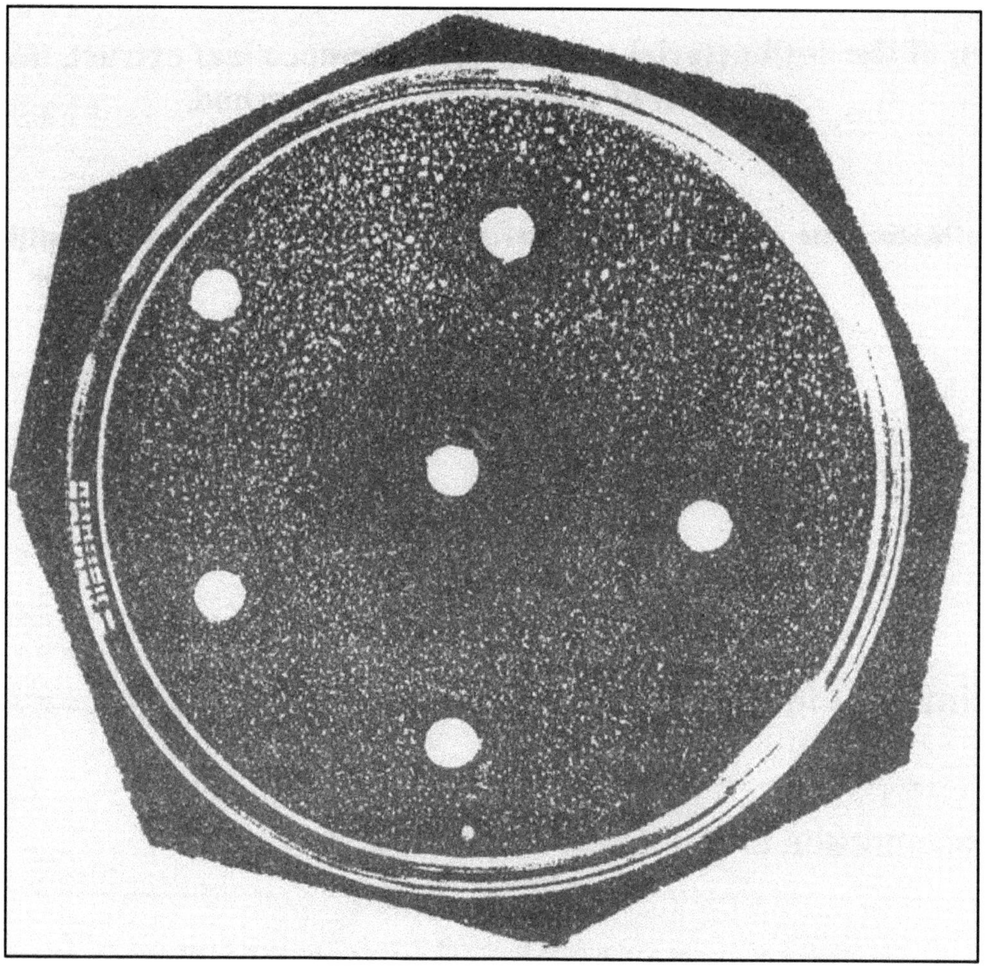

Figure 38.2: Zone of Inhibition Against *E. coli*

References

Ansari, M.M., 1995. Control of sheath blight of rice by plant extract. *Indian Phytopath.*, 48(3): 268–270.

Bauer, A.W., Kirbey, W.M.M., Sherries, J.C. and Turck, 1966. Antibiotic susceptibility testing by a standardized single disc method. *American Journal of Clinical Pathology*, 45: 493.

Chakraborty, A. and. Brantner, A.H., 1999. Antibacterial steroid alkaloids from the stem bark of *Holarrhena pubescens*. *Journal of Ethnopharmacology*, 68: 339–344.

Chatterjee, A. and Pakrashi, S.C., 1995. *The Treatise on Indian Medicinal Plants*, Volume 4, Publication and Information Directorate, C.S.I.R., New Delhi.

Chopra, R.N., Nayar, S.L. and Chopra, J.C., 1956. *Glossary of Indian Medicinal Plants.* Council of Scientific and Industrial Research, New Delhi.

Dinda, B. and Saba, S., 1990. Chemical constituents of *Plumbago zeylanica* aerial parts and *Thevatica nerrifolia* roots. *J. Indian Chem.*, 67: 88–89.

Ebi, G.C. and Ofoefule, S.L., 1997. Investigation into folkloric anti microbial activities of *Landolphia owrrience. Phytotherapy Research,* 11(2): 149–151.

Gehlot, D. and Bohra, J.A., 1998. Antimicrobial activity of various plant parts extract of *Aerva persico. Advances in Plant Sciences,* 11(1): 109–111.

Ghanaksha, A. and Kaushik, P., 1999a. Antibacterial effect of *Rhynchostylis retusa* BL. (Orchidaceae). *Advances in Plant Sciences,* 12(11): 593–598.

Ghanaksha, A. and Kaushik, P., 1999b. Antibacterial effect of *Aerides multiflora* Roxb: A study in vitro. *J. Orchid Soc., India,* 13(1–2): 765–768.

Kaushik, P., 1983. *Ecological and Anatomical Marvels of the Himalayan Orchids.* Today and Tomorrow's Printer's and Publisher, New Delhi, pp. viii+123+71 plates.

Kaushik, P., 1985. Glimpses of medical botany in Atharvaveda (Kand IV). *The Vedic Path.*, 48: 64–67.

Kaushik P., 1988. *Indigenous Medicinal Plants Including Microbes and Fungi.* Today and Tomorrow's Printer's and Publisher, New Delhi, pp. vii+243.

Kaushik, P., 2000. *Introductory Microbiology.* Emkay Publications, Delhi, pp. viii+466.

Kaushik, P. and Kishore, N., 1995. Antibacterial activity of *Dendrobium amoenum* wall Ex. Lindl: A study *in vitro. J. Orchid Soc., India,* 9(102): 33–35.

Kaushik, P. and Dhiman, A.K., 2000. *Medicinal Plants and Raw Drugs of India.* Bishen Singh Mahinder Pal Singh, Dehradun, pp. xii+623.

Kaushik, P. and Kishore, N., 1991. Antibacterial potential of *Pholidota articulata* Lindl: A study *in vitro. J. Orchid Society of India,* 5(1,2): 93–96.

Kaushik, P. and Upadhyay, Vishal, 1999. Antibacterial activity of garlic (*Allium sativum*) plant extract. In: *Sustainable Ecosystem and Environment,* (Eds.) D.R. Khanna, Ashutosh Gautam and A. Goutam. Action for Sustainable, Efficacious Development and Awareness (ASEA), pp. 129–132.

Kaushik, P., Upadyay, Vishal and Singh, Padma, 2000. Antibacterial potential of bryophytes. *The Botanica,* pp. 131–135.

Okunade, A.L., Hufford, C.D., Clark, A.M. and Lantz, D., 1997. Antimicrobial properties of constituents of *Piper aduncum. Phytotherapy Research,* 11(2): 142–144.

Peres, M.T., Moncha, A.B., Cruz Pizzelati, M.G. and Yunes, R.A., 1997. Chemical composition and antimicrobial activity of *Croton urucurana* Baillon (Euphorbiaceae). *Journal of Ethanopharmacology,* 56(3): 223–226.

Stahl, E., 1969. *Thin Layer Chromatography,* 2nd edn. MRF Ashworth Springer Verlag. Heidelber.

Ethnobotanical Wisdom and Microbial Studies on Medicinal Plants (2011) *Pages* **262–272**
Editors: **D.R. Khanna, Ashutosh Gautam, R. Bhutiani, Gagan Matta, Anil Dhiman**
Published by: **DAYA PUBLISHING HOUSE, NEW DELHI**

Chapter 39

Response of an Algal Community to Chromium

P.K. Goel and R.G. Shete
Department of Pollution Studies, Y.C. College of Science, Vidyanagar, Karad – 415 124, Maharashtra

ABSTRACT

This chapter deals with the study conducted for evaluation of the effects of chromium on the algal population by laboratory bioassay test. The results show that the algal species are affected differentially by the chromium concentrations ranging from 0.5 to 10 ppm. In general, the sensitive species to chromium include *Closterium acutum, Closterium cambricum, Closterium venus, Crucigenia fenestrata, Kirchneriella contorta, Monoraphidium capricomutum, Monoraphidium contortum, Scenedesmus acuminatus, Tetraedron trigonum, Anabaena incrassata, Chroococcus turgidus, Merismopedia tenuissima, Spirulina laxissima, Cylindrotheca gracilis* and *Melosina agranulata*. However, a few species like *Navicula viridula, Cosmarium depressum, Kirchneriella microscopica* and *Microcystis aeruginosa* were found to be tolerant to chromium, which could be exploited for their indicator value along with other species found tolerant by other workers.

Keywords: *Algae, Response, Chromium, Bioassay test.*

Introduction

Algae being autotrophic in nature have an important niche in aquatic systems. They are important link in the food chains and any change in their composition due to contamination of water can modify the biotic relationships thus changing the whole ecosystem. Many environmental contaminants like pesticides, heavy metals, variety of organic toxicants, organic matter and even nutrients can greatly influence the algal populations. The presence of pollutants have become a pervasive threat to many natural aquatic ecosystems. The pollutants can have toxic effects on almost all kinds of organisms and the biological process at the cellular, population, community and ecosystem level of organisations. The pollutants act on a number of cellular and biochemical processes critical to growth and production of algae. These processes include photosynthesis, nucleic acid production, protein and lipid biosynthesis, nitrogen fixation and uptake of nutrients.

Toxic effects on algae are difficult to generalize owing to interactions of toxicities with other physico-chemical factors, and the difference in the tolerance level of different algal species. Secondary effects of pollutants on algae may also exist due to stress on zooplankton community declining the grazing pressure. Due to this it becomes difficult to extrapolate the laboratory studies to natural field conditions (Trainor, 1984).

It has been reported that the limits of safe concentrations of various pollutants to fish and aquatic biota may well be in the toxic range of several algal and zooplankton species. It becomes increasingly important, therefore, to understand the effects of pollutants on all kind of aquatic life to manage and keep the aquatic systems healthy, as well as to maintain the algal based wastewater treatment and recycling systems (Patrick *et al.*, 1968).

Though, a vast information has been generated on the effects of various pollutants on different aquatic life forms (Cairns *et al.*, 1978; Gohash and Konar, 1980; Welch, 1980 and Goel *et al.*, 1987) relatively little is known about the effects of heavy metals on algae (Statton and Corke, 1979; Rai *et al.*, 1981; Goel and Shete, 1997). The present chapter deals with the effect of chromium on a natural algal community in laboratory conditions.

Experimental Set Up

The experiment was conducted in glass jars of 500 ml capacity having a wide mouth with surface area of 40.7 cm^2. Chromium solutions of various strengths, *i.e.* 0.5, 1, 2, 5, and 10 ppm were made from potassium dichromate using algae water after killing and removing algae by boiling and filtration. The algal water was collected from two different ponds of which one was comparatively more polluted than the other. Both the ponds were having a well developed algal community represented by a large number of genera. For running the experiment the algal waters from these two ponds were mixed in 1 : 1 ratio. The mixed algal water was acclimatized to the laboratory condition for few days in a glass aquarium tank.

For experiment, the mixtures of variously diluted chromium solutions and the algal water were taken in the ratio of 3 : 1(300 ml) of chromium solution plus 100 ml of algal water. In each set of various dilution, two replicates were taken thus employing total 12 jars with 400 ml of the above mixtures in each. The harvesting schedule of the experiment was kept as to collect 50 nil sample from each jar on first, third and sixth day from the start of the experiment. Some quantity of the original mixture was also kept to take the reading for the zero day. The collected samples were concentrated 10-fold by centrifugation, and were preserved with 4 per cent formalin for further study.

Results

The results of the experiment with regard to the change in the density of algal taxa in various concentrations of chromium are presented in Figure 39.1, Tables 39.1 and 39.2. The data indicated that a large number of species were declined in density, while a few showed a complete elimination during the six days of the experiment. However, a few concentrations of chromium were favorable to certain species that has reflected an increase of their density during the experimental period.

Table 39.1: Algae Showing Decrease (Per cent) Over the Control on 6th Day in Different Sets of Various Concentrations of Chromium

Name of the Species	*10 ppm*	*5 ppm*	*2 ppm*	*1 ppm*	*0.5 ppm*	*Control*
Class: Chlorophyceae						
Clostertum acutum	71.87	75.00	73.44	76.56	50.00	35.94
Clotertum venus	66.66	64.58	66.66	64.58	47.91	37.50
Coelastrum cambrtcum	86.44	89.83	85.59	86.44	70.34	–
Cosmarium depressum	–	–	37.50	00.00	31.25	–
Crucigenia fenestrata	38.09	61.90	69.04	69.04	40.47	28.57
Crucigenia rectangularis	19.67	75.40	89.34	78.68	79.50	65.57
Kirchneriella contorta	100.00	76.47	87.50	80.88	72.05	–
Kirchneriella microscopica	25.12	6.37	–	–	–	
Monoraphidium capricornutum	8.64	40.90	52.42	55.15	56.21	26.36
Monoraphidium contortum	74.36	67.94	62.82	69.23	46.15	
Scenedesmus acuminatus	55.55	36.11	100.00	66.66	33.33	27.77
Scenedesmus arcuatus	12.16		37.84	64.86	3.97	–
Scenedesmus quadricauda	80.85	55.32	84.04	–	13.83	69.15
Tetraedron trigonum	68.00	100.00	100.00	66.00	62.00	11.50
Tetrastrum triangulare	78.12	48.96	88.54	75.00	–	61.46
Class: Cyanophyceae						
Anabaena incrassata	69.90	88.67	81.07	88.02	81.07	70.55
Chroococcus limneticus	–	40.38	15.38	46.15	0.64	–
Chroococcus turgidus	80.85	55.31	74.47	56.38	65.95	
Merismopedia tenuissima	90.25	66.79	79.60	86.46	90.79	35.19
Microcystis aeruginosa	–	–	–	57.77	–	–
Oscillatoria limnetica	20.83	–	29.16	43.75	–	25.00
Sirulina laxissima	100.00	100.00	100.00	100.00	100.00	48.27
Class: Bacillariophyceae						
Cyclotella catenata	–	4.27	51.24	36.65	–	–
Cylindrotheca gracilis	58.06	64.51	61.29	69.35	77.41	–
Melosira granulata	48.15	62.96	100.00	66.66	74.07	–
Nitzschlia closterium	100.00	100.00	100.00	100.00	100.00	100.00

Effect of Chromium on Individual Algal Species

Chlorophyceae

The species *Closterium acutum* showed an adverse effect of chromium where it was decreased in the range of 50.0–76.56 per cent in various concentration of the metal. However, the decrease in density did not show a correlation with concentration of chromium. *Closterium venus* also showed an adverse effect with the decrease in density to the magnitude of 47.91–62.66 per cent in various concentrations of chromium. On the other hand, the control showed a decrease of only 37.5 per cent. *Closterium cambricum* was also the species that registered a quite significant decline up to 89.83 per cent in 5 ppm of chromium. The control for this species, however, showed an increase in the density by 23.0 per cent. *Cosmarium depressum* showed a mixed trend having a decline in density with 0.5 and 2 ppm chromium, and an increase in density with 5 and 10 ppm. Its density remained unchanged in 1 ppm. The increase in density was much higher than the magnitude of decline. Control also showed an increase in the density. The algal species *Crucigenia fenestrata* was found to get decreased in numbers in all the concentrations of chromium with maximum decline of 69.04 per cent in 1 and 2 ppm concentrations. The control for this showed a much lower decline of 28.57 per cent. *Crucigeniella rectangular* is also showed a decline in all the dilution with the maximum decline of 89.43 per cent in 2 ppm chromium concentration. *Kirchneriella contorta* registered a significant decline with total elimination in 10 ppm of chromium concentration. However, the control for this alga showed a marginal increase. Another species of this genus, *Kirchneriella microscopica,* on the other hand, showed a mixed trend with a slight decrease in density with 5 and 10 ppm chromium, but showing an increase in density in the lower concentrations. As the increase in the density was much higher in control with the value of 66.66 per cent, it can be infer, therefore, that this species in adversely affected by chromium to a great extent.

Table 39.2: Algae Showing Increase (Per cent) Over the Control on 6th Day in Different Sets of Various Concentrations of Chromium

Name of the Species	*10 ppm*	*5 ppm*	*2 ppm*	*1 ppm*	*0.5 ppm*	*Control*
Class: Chlorophyceae						
Coelastrum cambricum	–	–	–	–	–	23.73
Cosmarium depressum	85.41	133.33	–	0.0	–	135.42
Kirchneriella contorta	–	–	–	–	–	16.54
Kirchneriella microscopica	–	–	43.14	23.04	9.91	66.66
Monoraphidium contortum	–	–	–	–	–	98.72
Scenedesmus arcuatus	–	15.54	–	–	–	40.54
Scenedesmus quadricauda	–	–	–	18.08	–	–
Tetrastrum triangulare	–	–	–	–	1.0	–
Class: Cyanophyceae						
Chroococcus limneticus	3.84	–	–	–	–	195.51
Chroococcus turgidus	–	–	–	–	–	10.64
Microcystis aeruginosa	170.00	138.88	48.88	–	13.33	164.44
Oscillatoria limnetica	–	6.25	–	–	25.00	–
Class: Bacillariophyceae						
Cyclotella catenata	6.04	–	–	–	7.83	15.30
Cylindrotheca gracilis	–	–	–	–	–	145.16
Melostra granulata	–	–	–	–	–	166.66
Navicula virdula	216.66	229.16	16.66	91.66	806.25	304.16

Figure 39.1: Variation in Density of Phytoplankton Species in the Various Concentrations of Chromium

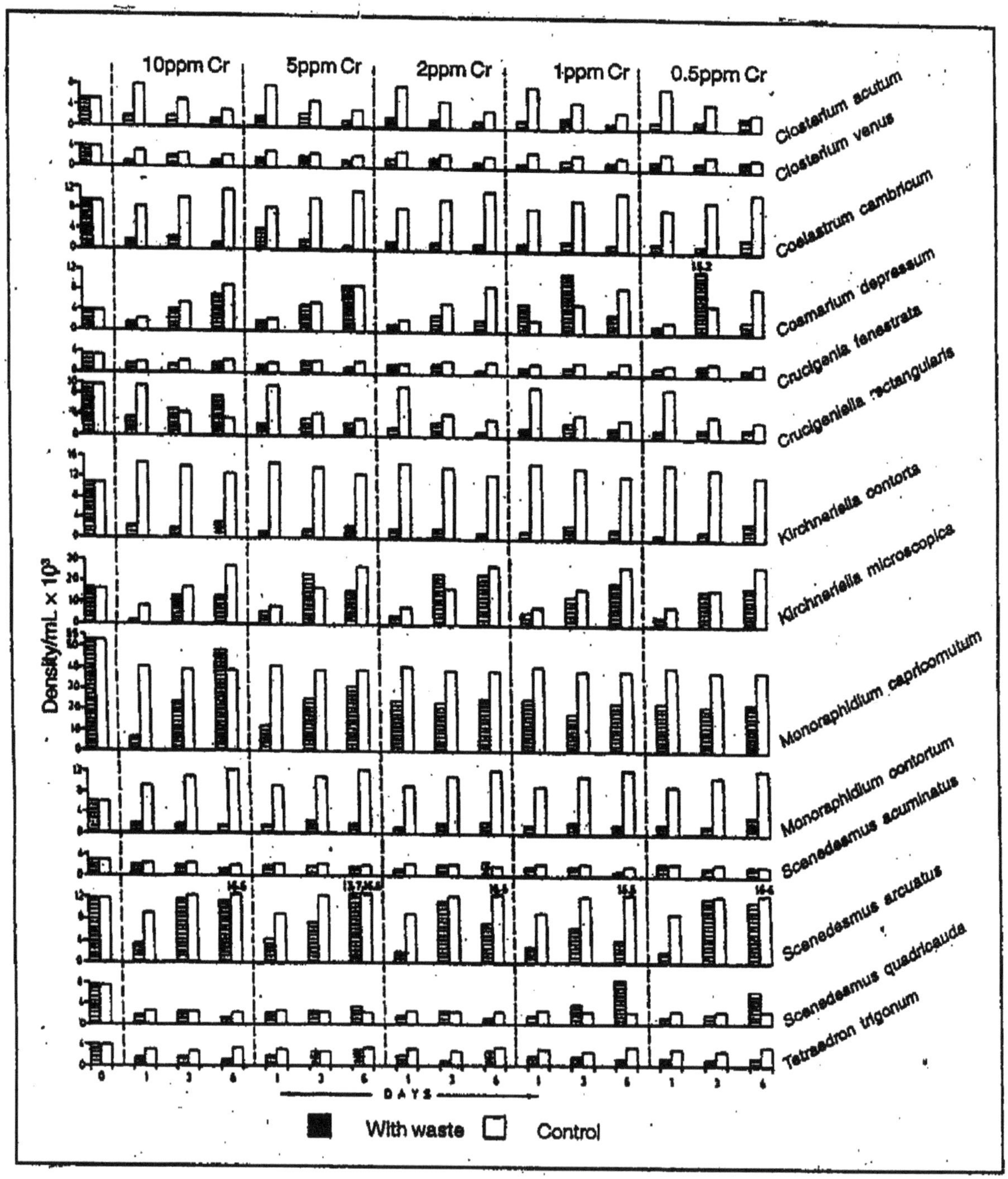

(A)

Contd...

Figure 39.1–Contd...

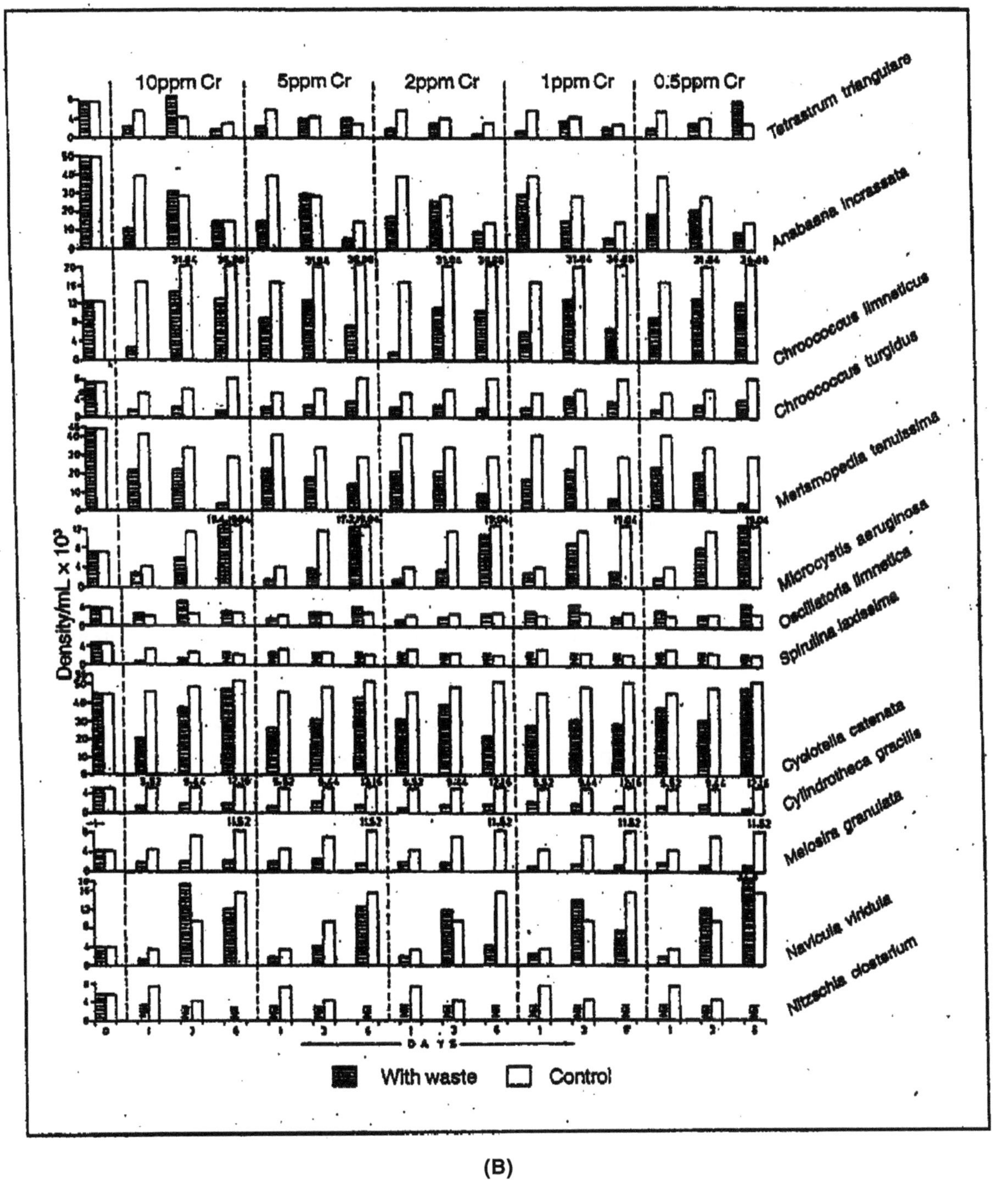

(B)

The species *Monoraphidium capricornutum* also showed adverse effect of chromium having a decline in density in all the concentrations with highest fall of 56.21 per cent in 0.5 ppm and lowest of 8.64 per cent in 10 ppm. The data indicate that the species is affected more in the lower concentrations of chromium. The second species of this genus, *M. contortum*, however, was much more affected than the former with the highest value of decline by 74.36 per cent in 10 ppm chromium. On the contrary, the control for this species showed a tremendous increase of more than 98.0 per cent. There were three species of *Scenedesmus*, all of them getting adversely affected by chromium. *S. acuminatus* was completely eliminated in 2 ppm and registered a decrease of 66.66 per cent in 1 ppm. The control for this showed a decline of only 27.77 per cent. *S. arcuatus* was not much affected as compared to the previous one with the maximum fall in density to be 64.86 per cent in 1 ppm followed by 37.84 per cent in 2 ppm. This species showed, however a slight increase in 5 ppm as well in control. *S. quadricauda* has fallen in density to the tune of 84.04 per cent in 2 ppm followed by 85 per cent in 10 ppm. This species also showed a marginal increase in 1 ppm of chromium. The control for this species showed a greater decline of 69.15 per cent.

Tetraedron trigonum showed a tremendous decline in density by chromium with even total elimination in 2 and 5 ppm concentrations. The algal species *Tetrastrum triangulare* also suffered a decline in density with chromium concentrations from 1 to 10 ppm. However, this species showed a very slight increase (104 per cent) in 0.5 ppm concentration of chromium.

Cyanophyceae

Anabaena incrassata showed a decline from 69.90 per cent to 88.67 per cent in various dilutions of chromium. However, the control also showed a substantial decline of 70.55 per cent in density. Still, the data indicate that the species is adversely affected by chromium. *Chroococcus limneticus* while showing a slight increase of 3.84 per cent in 10 ppm showed a substantial decline in the remaining dilution. The control for this, however, registered an almost 200 per cent rise in density, thus indicating that the species is greatly suppressed by chromium. Another species of this genus *C. turgidus* also showed a substantial decrease in density with all the dilutions, whereas in control it was marginally higher.

Merismopedia tenuissima was also the species that got drastically affected by chromium with a highest decline of 99.25 per cent in 10 ppm. The control for this species showed only 35.19 per cent fall.

Microcystis aeruginosa, except for 1 ppm of chromium, showed an increase in all the dilutions with maximum value of 170.0 per cent in 10 ppm chromium concentration. As the control for this also showed a higher increase of 164.44 per cent, it can be included that the species remains almost unaffected by chromium.

Oscillatoria limnetica showed a mixed trend with an increase in the density in 0.5 and 5 ppm chromium and a decrease in density in the remaining. The maximum decline was obtained (43.75 per cent) in 1 ppm while the maximum increase was 25 per cent in 0.5 ppm chromium. The control for this showed a decline in density by 25 per cent. *Spirulina laxissima* was drastically affected by chromium and showed complete elimination in all the dilutions.

Bacillariophyceae

Cyelotella catenata showed a decrease in density in 1, 2 and 5 ppm of chromium with the maximum value of decline by 51.24 per cent in 2 ppm. In 0.5 and 10 ppm of chromium, however there was a marginal increase. The control for this also showed an increase but it was not high (15.30 per cent). The overall results indicate that the species is only slightly affected. *Cylindrotheca gracilis* suffered in all concentrations of chromium with a fall in density, which was maximum in 0.5 ppm concentrations of chromium with a magnitude of decrease by 77.41 per cent. In control, on the other hand, this species showed an increase by 145.16 per cent, *Melosira granulata* also declined in density in all the concentrations of chromium with total elimination in 2 ppm followed by 74.07 per cent fall in 0.5 ppm. In control, the species showed a substantial increase of 166.66 per cent.

Nitzschia closterium was found to be destroyed completely by all the concentrations of chromium as well as in control. *Navicula viridula* showed an increase in all the chromium dilutions together with control. The maximum increase was observed to be 806.25 per cent in 0.5 ppm chromium. The increase in control was also of high magnitude (304.6 per cent). The data indicate that the low concentration of chromium might be promoting the growth of this species.

Species Adversely Affected by Chromium

The results of the experiments reveal that all most all the species except a few showed a distinct trend of getting adversely affected by different concentrations of the chromium. Table 39.3 gives the name of the species which were totally eliminated in either of dilutions or control. It was found that there were three species eliminated in each 5 and 10 ppm of chromium, 4 species in 2 ppm chromium and 2 species each in 0.5 and 1 ppm of chromium.

Table 39.3: Algal Species Showing Different Responses to Chromium

Species Totally Eliminated	*Species Adversely Affected*	*Species Promoted*	*Species with No Clear Trend*
Kirchneriella contoria	*Closterium acutum*	*Navicula viridula*	*Oscillatoria limentica*
Scenedesmus acuminatus	*Closterium venus*	*Cosmartum depressum*	
Tetraedron trigonum	*Coelastrum cambricum*	*Kirchneriella microscopica*	
Spirulina laxissima	*Crucigenia fenestrata*	*Microcystis aeruginosa*	
Nitzschia closterium	*Crucigeniella rectangularis*		
	Kirchneriella contorta		
	Monoraphidium capricormutum		
	Monoraphidium contorium		
	Scenedesmus acumtnatus		
	Scenedesmus arcuatus		
	Scenedesmus quadricauda		
	Tetraedron trigonum		
	Tetrastrum triangulare		
	Anabaena incrassata		
	Chroococcus limnetlcus		
	Chroococcus turgidus		
	Merismopedia tenuissima		
	Microcystis aeruginosa		
	Oscillatoria limnetica		
	Spirulina laxissima		
	Cyclotella catenata		
	Cylindrotheca gracilis		
	Melosira granulata		

Some of the important species which showed an adverse effect of chromium were *Closterium acutum, Closterium venus, Coelastrum cambricum, Scenedesmus acuminatus, Tetraedron trigonum, Anabaena incrassata, Chroococcus turgidus, Merismopedia tenuissima, Spirulina laxissima, Cylindrotheca gracilis* and *Melosira granulata.*

The species *Nitzschia closterium* showed a total decline of the population in all the concentrations of chromium as well as the control, hence, it is difficult to make any inference whether the metal has any adverse effect on this species or there are some other environmental conditions common to all the sets including control which are responsible for decline in density.

Species Promoted by Chromium

The species *Navicula viridula* showed a favourable response to the lower concentrations of Chromium while *Cosmarium depressum* was promoted at 5 and 10 ppm chromium and *Kirchneriella microscopica* was promoted at 0.5 and 2 ppm chromium. *Microcystis aeruginosa* also showed an increase in growth at 5 and 10 ppm of chromium.

Discussion

Most heavy metals are toxic to algae with a varying degree depending upon the tolerance of the algal species (Kanth, 1989).While a few metals in small concentrations may be essential micronutrients to algae like copper, molybdenum and zinc etc., other such as cadmium, chromium, lead and mercury are toxic at almost all concentrations. The growth of algae in low concentration of essential metals remain normal but their increased concentration, as a rule, becomes toxic or lethal.

Whatever information on effects of heavy metals is available, comes from the laboratory studies, which according to Kanth (1989) may not be fitted or extrapolated to natural habitats with large number of interacting factors. To overcome these limitations upto certain extent in the present study, the natural water enriched with chromium was used to grow the mixed population of algae rather than using single species cultures.

The studies so far conducted clearly indicate that the heavy metals pose a severe threat to the organisms in water including algae (Rai *et al.*, 1981; Patric, 1978; Whitton, 1970, 1980, 1984; Welch, 1980; Strokes, 1983; Goel and Shete, 1997). It is revealed from the studies that the algae exhibit different responses to different concentrations of heavy metals. The present experiment shows that while a large number of species (22) are sensitive to chromium, a few (four) show tolerance. Some of the extremely sensitive species to chromium include *Closterium acutum, C. venus, Kirchneriella contoria, Monoraphidium contortum, Scenedesmus acuminatus, Tetraedron trigonum, Spirulina laxissima, Cyclotella catenata* and *Melosira granulata.*

The species showed tolerance to chromium include *Navicula viridula, Cosmarium depressum, Kirchneriella microscopica* and *Microcystis aeruginosa.* Palmer (1980) gave a list of 12 indicator algal species for chromium, *i.e. Closterium acerosum, Euglena acus, E. oxyceris, E. sociabilis, E. stellata, E. viridis, Navicula atomus, N. cuspidata, Nitzschia linearis, N. palea, Stigeoclonium lenue,* and *Tetraspora* species. It seems that the species of *Euglena*, in general, are quite tolerant to chromium. Our study adds some more species to the list of indicators species of chromium.

The reduction in the growth of *Scenedesmus*, by chromium in the present study is in confirmity to the observations made by Patrick (1978) who found that 5 ppm concentration of the metal is extremely toxic to this species. Many species of *Scenedesmus* have also been reported to be inhibited by other metals like cadmium (Goel and Shete, 1997). According to Patrick (1978) many diatoms can flourish in lower concentrations of chromium, but can be replaced by blue greens at increased concentrations. It was also found in the present study that *Microcystis aeruginosa* increases substantially at 10 ppm of chromium, while *Navicula viridula* increases to several folds only at 0.5 ppm concentration.

The oxidation state of chromium is an important factor which may influence the toxicity as indicated by Patrick *et al.* (1968) who found that TLm (killing of 50 per cent organisms) for diatoms was quite low (0.59 ppm) in case of Cr^{+6} as compared to the TLm value for Cr^{+3} revealing that the former is much more toxic than the latter.

Many studies showed that several species of algae bioaccumulate different heavy metals and, thus, may be used as indicators of metal pollution (Palmer, 1980; Whitton, 1984). While the impact of metals in isolation on a single species is worth studying, there is a need to study the toxicity of metals in relation to the presence of other influencing factors common to most natural situations. For example, a decrease in oxygen level in water can lead to the rise in toxicity of chromium (Cr^{+3}) to a great extent (Goel, 1997). Considering the important status of algae, in aquatic systems, it is increasingly felt for an emphasis to understand the responses of algae to diverse pollutants specially in the natural habitats having complex interacting factors.

References

Cairns, Jr., Dickson, K.L. and Maki, A.W., 1978. *Estimating the Hazards of Chemical Substances to Aquatic Life.* (Eds.) ASTM., Philadelphia.

Goel, P.K., Trivedi, R.K. and Nakate, S.S., 1987. Effect of refined mineral oils on freshwater phytoplankton. *Environment and Ecology*, 5(3): 508–510.

Goel, P.K., 1997. *Water Pollution: Causes, Effects and Control.* New Age International, (P) Ltd., Publishers, New Delhi.

Goel, P.K. and Shete, R.G., 1997. Studies on the response of diversified algal community to different concentrations of cadmium. In: *Recent Advances in Ecobiological Research (Vol. 1)*, (Ed.) M.P. Sinha. A.P.H. Publishing Corporation, New Delhi.

Gohash, T.K. and Konar, S.K., 1980. Toxicity of chemicals in wastewaters of paper mills to worms, plankton and molluscs. *Indian J. Environ. Hlth.*, 22(4): 278–285.

Kanth, S., 1989. Algae and heavy metal pollution. In: *Environmental Risk Assessment*, (Eds.) Y.N. Sahai, R.B. Deshmukh, T.A. Mathai, and K.S. Pillai. The Academy of Environmental Biology, Muzaffarnagar, India.

Palmer, C.M., 1980. *Algae and Water Pollution.* Castle House Publication, pp. 1–119.

Patrick, R., Cairns, J.Jr. and Scheier, A., 1968. The relative sensitivity of diatoms, snails, and fish to 20 common constituents of industrial wastes. *Progressive Fish Culturist*, 30: 137–140.

Patrick, R., 1978. Effects of trace metals in the aquatic ecosystem. *Am. Sci.*, 66: 185–191.

Rai, L.C., Gaur, J.P. and Kumar, H.D., 1981. Phycology and heavy metal pollution. *Biol. Rev.*, 56: 99–151.

Statton, G.W. and Corke, C.T., 1979. The effect of cadmium ion on the growth, photosynthesis, and nitrogenase activity of *Anabaena inequalis, Chemopsphere*, 5: 277–282.

Strokes, P.M., 1983. Responses of freshwater algae to metals. In: *Progress in Phycological Research*, (Eds.) F.E. Round and D.J. Chapman. Elsevier Science Publ., 2: 87–109.

Trainer, F.R., 1984. Indicator algal assays: Laboratory and field approaches. In: *Algae as Ecological Indicators*, (Ed.) L. Elliot Shubert. Academic Press, 1: 3–14.

Welch, E.B., 1980. *Ecological Effects of Wastewater.* Cambridge University Press, Cambridge, p. 337.

Whitton, B.A., 1970. Toxicity of heavy metals to freshwater algae: A review. *Phykos*, 9: 116–125.

Whitton, B.A., 1980. Zinc in plants in rivers and streams. In: *Zinc in the Environment, Part–II: Health Effects*, (Ed.) J.P. Nriagu. John Wiley, New York, pp. 363–400.

Whitton, B.A., 1984. Algae as monitors of heavy metals in fresh waters. In: *Algae as Ecological Indicators*, (Ed.) L. Elliot Sbubert. Academic Press, 4: 257–280.

Ethnobotanical Wisdom and Microbial Studies on Medicinal Plants (2011) *Pages* 273–290
Editors: D.R. Khanna, Ashutosh Gautam, R. Bhutiani, Gagan Matta, Anil Dhiman
Published by: DAYA PUBLISHING HOUSE, NEW DELHI

Chapter 40

Recent Progress in Aflatoxin Analysis and their Inactivation

Harish Chandra[1], Parul Bishnoi[2], Jatin Srivastava[3], A.R. Nautiyal[4] and S.V.S. Chauhan[2]
[1]Department of Microbiology, Gayatri College of Biomedical Sciences, G.M.S. Road, Near Ballupur Chowk, Dehradun, Uttarakhand
[2]Institute of Life Science Dr. Bhim Rao Ambedkar University, Agra, U.P.
[3]Department of Environmental Sciences, Institute of Biologigal Science and Biotechnology, Chhatrapati Sahu Ji Maharaj University, Kanpur, U.P.
[4]High Altitude Plant Physiology Research Centre, H.N.B. Garhwal Univesity, Srinagar Garhwal, Uttarakhand

ABSTRACT

Aflatoxin is the secondary metabolite of low molecular weight produced by naturally occurring fungi mainly *Aspergillus flavus* and *Aspergillus parasiticus*. Rapid progress in the area of aflatoxin analysis and detoxification method has been made during the last few years. Simplified method, cleanup protocol and chromatographic methods have been continuously developing to make these methods more sensitive and reliable. In this review we are discussing different aflatoxin analysis methods as well as different methods of aflatoxin detoxification.

Keywords: *Aflatoxin, Aspergillus flavus, Aspergillus parasiticus.*

Introduction

Cereals, especially maize and groundnut, are the major sources of carbohydrates and proteins in Asia and are important as export products in some Asian countries. Grain quality is critically monitored to meet international standards in exports. Aflatoxin development in many stored cereal grains has constantly hampered the availability of good quality grains in Asian countries. The most important group of toxigenic Aspergilli are the Aflatoxigenic molds, *Aspergillus flavus*, *Aspergillus parasiticus* and the recently described but much less common species *Aspergillus nomius* all of which are classified in Aspergillus section Flavi (Gams *et al.*, 1985). Although these three species are closely related and share many similarities a number of characteristics may be used in their differentiation. *A. flavus* is widely distributed in nature but *A. parasiticus* is less wide spread, the actual extent of its occurrence being complicated by the tendency for both the species reported indiscriminately as *A. flavus*.

A. flavus Link ex Fries and *A. parasiticus* Speare have been identified as the major pests that produce aflatoxin and deter the quality of grain when stored. They especially affect oilseed. Edible nuts and cereals in subtropical regions throughout the world are spoiled due to inadequate storage conditions. The main cause of disease in the human is their secondary metabolite *i.e.* Aflatoxins which are carcinogenic and result in liver cancer. Liver cancer takes time to develop but the aflatoxin acts as an immunosuppressant so that affected individuals become susceptible to wide range of diseases. In the last decade, aflatoxin levels on product were found to exceed an acceptable level limit of 20 ppb stipulated in most export specifications. Aflatoxin contamination has affected maize, groundnuts, peanuts and cottonseeds in several countries including Thailand and India; and coconut in the Philippines, Sri Lanka and other Pacific countries. Aflatoxicosis, both in humans and animals, has been more prevalent in areas where maize and groundnut constitute a major part of the diet. Many private and government organizations have embarked on aflatoxin research. However, aflatoxin control is still an intricate problem. Aflatoxins have been widely detected in cereals, oilseeds, fermented beverages and milk which form basic human diet (Bilgrami *et al.*, 1983; Bullerman, 1986). Maize has been reported to be the most susceptible followed by groundnut to aflatoxin contamination. Bhat *et al.* (1997) reported that about 26 per cent maize and 21 per cent of groundnut samples collected from 11 states exceed the permissible limit of 30ppb aflatoxin per kg of material. *Prevention of Food Adulteration Act* (1954), amended in 1986, U.S. Food and Drug Administration (FDA) and the Codex Alimentarius Commission (1989) have recommended a permissible limit of aflatoxin of 30ppb/kg, 20ppb/kg and 5ppb/kg respectively.

A survey from thirty countries concluded that aflatoxins are the leading toxins present in agricultural commodities when compared to the presence of other mycotoxins (Hesseltine, 1986). Aflatoxin in animal feed presents two problems; first deleterious effect on the health of the animal ingesting the contaminated feed and second, on the health of humans consuming aflatoxin residues in foods derived from such animals.

There is great variation among strains of *A. flavus* in the quantity of aflatoxins produced (Cotty, 1989), this quantity is independent of a strain's ability to infect and colonize developing cottonseed. Simultaneous inoculation of wounded 28-to-32-day-old cotton balls with toxigenic and aflatoxigenic strains of *A. flavus* led to lower levels of aflatoxin B_1 in the cottonseed at maturity than in balls inoculated with the toxigenic strain alone. Less B_1 was detected when the aflatoxigenic strain was introduced into the wound 1 day before inoculation with a toxigenic strain than when atoxigenic and toxigenic strains were co-inoculated (Cotty, 1990). In this review we are discussing some important methods for analyzing aflatoxin and methods of inactivation by physical, chemical as well as biological means.

Sampling and Sample Preparation for Aflatoxin Analysis

Sampling is important and usually the largest source of error. This is because a small percentage of the kernels and contaminated kernels have high level of Aflatoxin and other fungal toxin (Whitakar and Wiser, 1969). With small samples, it is difficult to get contaminated kernel in the sample. Studies have shown that less than 1 kernel per 1000 (0.1 per cent) is contaminated in a lot or raw shelled peanuts. The sample has to be selected in such a way that every kernel in the lot has an equal chance of being chosen. The sample should be the accumulation of many small incremental portions taken at many different locations throughout the lot (Whitaker, 2000). The sample is ground in mill. The kernels are broken into many small pieces. Then a small sub sample of ground product is removed from the sample. The aflatoxin and other fungal toxin in replicated subsample will also vary about the sample concentration. Sample error is reduced by increasing sub-sample size and grinding the sample into finer particles.

Detection of Aflatoxin in Food Commodities

There are various methods for determination of aflatoxin in food commodities. Some of the important methods are:

Thin Layer Chromatography (TLC)

Aflatoxin has fluorescent properties *i.e.* Aflatoxin B_1 and B_2 give blue and Aflatoxin G_1 and G_2 give green fluorescence under U.V. light at 360nm. This property of fluorescence has been utilized for the detection of different types of aflatoxin by TLC. The first thin layer chromatographic (TLC) separations of aflatoxins were proposed simultaneously by Coomes and Sanders (1963). They used paper chromatography and reported that the system did not resolve aflatoxins B_1 and B_2, and the least detectable amount of B ($B_1 + B_2$) was about 0.2 µg. Broadbent *et al.* (1963) used glass plates coated with neutral alumina and reported that B_1 and B_2 were not resolved but as little as 6×10^{-3} µg of B could be detected, about a 30 fold increase in sensitivity over that attainable with paper chromatography for development. The two dimensional thin layer chromatography was proposed by Peterson and Ciegler (1967) to yield improved separation of aflatoxins from impurities in plant extracts, and for better separation of aflatoxin B_2 from G_2. Neshiem (1968) has investigated the effect of variations in developing solvents, commercial silica gels, calcium sulfate binders, gel thickness, humidity and vapor phase composition, all of which influence the separation of aflatoxins in thin layer chromatography. TLC method has detection limit of 50-100 ppb. In TLC method compare the fluorescence intensities of the spots at the Rf of B_1 in the sample with those of the B_1 standard spots and determine which of the sample spots matches one of the standards and record the corresponding aliquot volumes. If the sample spot intensity lies between two adjacent standard spots the average liquor volume of the standard spots is recorded. If the spots of the smallest volume of sample are too intense to match the standards the sample extract should be diluted and re-chromatographed. Stroka and Anklam (2000) developed miniaturized and low power consuming detector cell for the densitometeric measurement of aflatoxin on TLC plates. A UV-light emitting diode (UV-LED) with a peak emission wavelength of 370nm was used for the fluorescence excitation, while photo diode with peak sensitivity of 440 nm in combination with a 418 nm cut off filter was applied for detecting the fluorescence intensity. The resulting signal was further amplified by means of commonly used operational amplifier integrated circuit (OA) and directly converted into digital signal with simple analogue-digital converter (ADC). This signal was recorded at the serial (RS232) port of a portable PC and processed with spreadsheet program.

Pestka *et al.*, (1980) introduced HPTLC-ELISA that involves the separation of Mycotoxins in HPTLC, followed by blotting the chromatogram to nitrocellulose membrane coated with antibody, incubation with mycotoxin enzyme conjugate and finally incubation with substrate to develop the color. The only disadvantage of this technique is use of large amount of antibody.

High Performance Liquid Chromatography (HPLC)

The development of highly automated HPLC systems has afforded very precise, selective and sensitive quantification techniques for aflatoxin analysis. HPLC methods have been developed using both normal and reverse phase systems in conjunction with UV adsorption and fluorescence detection techniques. Reverse phase HPLC separations of aflatoxins are more widely used than normal-phase separations. Aflatoxin analysis using HPLC for separation and detection is quite similar to TLC because similar sampling and extraction procedures are used. The major advantages of HPLC over TLC are speed, automation, improved accuracy, and precision. Both normal-phase and reverse-phase HPLC separations have been developed for aflatoxin analysis. Early experimental work by Seitz (1975) on HPLC separations revealed that aflatoxins could be separated on normal-phase columns and detected with either a UV detector or a fluorescence detector. Seitz (1975) noted that the fluorescence detector had limited usefulness for aflatoxin B_1 and B_2 with normal phase separations. Panalaks and Scott (1977) developed a silica-gel packed flow cell for fluorometric detection of B_1, B_2, G_1, and G_2 with normal phase aflatoxin separations. A silica-gel packed cell was used by Pons (1979) and Thean *et al.* (1980) in two different HPLC methods for determination of aflatoxins. The major disadvantage of the packed cell is lack of stability. The cell needs to be repacked often and the detector signal weakens with time. The advantages of a packed cell method are that no derivative is necessary for detection and the mobile phase can be recycled. Reverse-phase HPLC separations of aflatoxins are more widely used than normal-phase HPLC separations. However, the fluorescence intensities of B_1 and G_1 are diminished in reverse-phase solvent mixtures so the derivatives B and G are diminished in reverse-phase solvent mixtures so the derivatives B and G are generally prepared before injection. Analysts should be aware that derivatives B and G are not stable in methanol, which should be used with caution, especially in the injection solvent. Acetonitrile-water mixtures do not degrade B and G rapidly and are preferred to methanol-water mobile phases. Several reverse-phase methods have been published (Stubblefield and Shotwell 1977; Hutchins and Hagler, 1983, Tarter *et al.*, 1984). Stubblefield and Shotwell (1977) found that M_1, and M_2 as well as B_1, B_2, G_1, and G_2 could be resolved and detected with a UV detector at 350 nm using reverse-phase chromatography. The methods developed by Hutchins and Hagler (1983) and Tarter *et al.* (1984) all use TFA derivatization and apparently compare favorably with other methods.

According to the procedure described by Sharma and Marquez (2001) a 50gm sample was extracted with 100 ml of methanol-water (60:40 v/v) for I minute and then filtered on Whatmann filter paper no. 4. The sample extract was concentrated to a volume of 30ml and was then diluted with 40ml PBS (pH 7.4). It was cleaned up on an immunoaffinity column by gentle syringe pressure at flow rate of 5ml/min and then the column was washed with distilled water (20 ml). Elute from each column containing the analytes were evaporated to dryness under nitrogen in 2 ml vials, the residue were derivatised with hexane/trifluoroacetic acid (300μl each) mixed and incubated for 10 min at 40 °C. The solvent were evaporated and finally the residues were dissolved in Acetonitrile and distilled water (60 and 180 μl respectively). Aflatoxins were separated isocratically on Perkin-Elmer HPLC chromatograph, connected to reverse phase C18 column particle size 5 μm, LC-10 flourescence detector (Perkin Elmer) and LCI-100 computing integrator. Measurements were made by peak area. The mobile phase was 60 per cent water, 22 per cent Acetonitrile and 18 per cent methanol (filtered through 0.2 μm

Millipore filter) at flow rate of 1 ml/min and detection was observed by fluorescence with excitation at 370 nm (λ_{em}) and emission cut off at 418 nm (λ_{em}). Quantification of each toxin was performed by measuring peak areas at their retention time, and comparing with their relevant standard calibration curve. The identity of each toxin was confirmed in all the analyzed samples by injecting sequentially sample extract and comparing the peak area ratio with their corresponding standard. HPLC has partly superceded TLC in the analysis of food for mycotoxin. Separation is usually much better than those obtained with one dimensional TLC. HPLC methods generally provide good quantitative information and the equipment employed in HPLC systems is generally automated rather easily.

Otta *et al.* (2000) developed rapid, reproducible and cost effective Over Pressure Layer Chromatography (OPLC) method for quantitative determination of Aflatoxin B_1, B_2, G_1 and G_2 in different foods. Using OPLC one can analyse ten samples simultaneously. Chiavaro *et al.* (2001) studied the effect of succinyl-β-cyclodextrin (β-CD-Su), Dimethyl-β-cyclodextrin (DINEB) and β-cylodextrin (β-Cd) on the fluorescence of B_1, B_2, G_1, G_2 and M_1. β-CD-Su promoted the largest fluorescence enhancement for AFB_1 and AFM_1 while DIMEB showed better result for AFG_1. On the basis of the Fluorescence enhancement, a RP-HPLC method for detecting aflatoxin B_1, B_2, G_1, G_2 and M_1 was developed using cyclodextrins directly dissolved in the LC elutent. Aflatoxin B_1, B_2, G_1 and G_2 were resolved using methanol: water as mobile phase to which β-CD-Su or β-Cd was added. Chromatographic response of AFB_1 and AFG_1 achieved using β-CD dissolved in the mobile phase were enhanced respectively, 8 and 12 times and 10 to 15 times with β-CD-Su. Blesa *et al.* (2003), developed method based on solid-solid phase dispersion (MSPD) extraction to determine aflatoxin B_1, B_2, G_1 and G_2 from peanut. The method used 2 gm of peanut sample, 2 gm of C18 bonded silica as MSPD sorbent and Acetonitrile as eluting solvent. The limit of quantification ranges from 0.125 to 2.5 ng/g for the four studied aflatoxin using Liquid Chromatography. Calleri *et al.* (2007) developed fully automated HPLC method with fluorescence detection for the determination of AFB_1 in aqueous solution by using anti-aflatoxin B_1 immunoaffinity monolithic disk. Polyclonal anti AFB_1 is covalently immobilized in batch on an epoxy activated monolithic Convective Interaction Media (CIM) disk (12mm x 3mm i.d) by a one step reaction. 0.96 mg of antibody were immobilized. The CIM disk was coupled through a switching valve to reverse-phase column Chromolith Performance RP-18e. The total analysis time with integrated system is 46 min and the retention time of AFB_1 is approx. 29 min.

Radio Immuno Assay (RIA)

The RIA involves the incubation of specific antibody simultaneously with unknown sample or known standard and constant amount of labeled toxin. After separation of the free and bound toxin, the radioactivity in those fraction is then determined. The toxin concentration of the unknown sample is determined by comparing the results to a standard curve, which is established by plotting the ratio of radioactivities in the bound fraction and free fraction with log concentration of unlabeled standard toxin. RIA can detect 0.25 to 0.5 ng of purified mycotoxin. The sensitivity of RIA can be improved by a simple cleanup procedure after extraction or by using radioactive markers of high specific activity (^{125}I-labeled mycotoxin) (Chu, 1991).

Enzyme Linked Immunosorbent Assay

There are two methods of ELISA used in the analysis of Aflatoxin. One type, direct ELISA, involves the use of an aflatoxin-enzyme conjugate and other system, indirect ELISA involves the use of a protein-aflatoxin conjugate and a secondary antibody to which an enzyme has been conjugated (Chu, 1986). In the direct competitive assay specific antibodies are first coated to a solid phase, including polystyrene microtiter plate (Chu, 1991). The sample solution or standard toxin is generally incubated simultaneously with enzyme conjugate or incubated separately in two steps. After washing, the amount of enzyme bound to the plate is then determined by incubation with a substrate solution containing hydrogen peroxide and appropriate oxidizable chromogen. The resulting color is measured spectrophotometrically or by visual comparison with the standard. In this assay, toxin in the sample and toxin-enzyme conjugate compete for the same binding site with the antibody coated to the solid-phase. Because the toxin-enzyme and antibody concentration are constant, the color intensity as a function of enzyme is inversely proportional to the toxin concentration in the testing sample. In the indirect ELISA, an aflatoxin protein conjugate is first prepared and then coated to the microtiter plate before assay. The plate is then incubated with specific rabbit antibody in the presence or absence of the homologous aflatoxin. The amount of antibody bound to the plate coated with aflatoxin conjugated protein is then determined by reaction with goat anti-rabbit IgG enzyme. Thus toxin in the sample and toxin in the solid-phase compete for the same binding site with the specific antibody in the solution. The indirect ELISA has been used for the analysis of a number of mycotoxins (Chu, 1991). Degan *et al.* (1989) developed method to analyse Aflatoxin involving the use of Europium ion (Eu) labeled antibody. Kolosovo *et al.* (2006), developed direct competitive ELISA for detection of Aflatoxin B_1 and an ELISA kit has been designed. This immunoassay was highly specific, sensitive, rapid, simple and suitable for aflatoxin monitoring. AFB_1 concentration determinable by ELISA ranged from 0.1 to 10 µg/L.

Immunochromatographic Technique

A rapid diagnostic method has been developed by Xiulan *et al.* (2005) in which they prepared an antibody – colloidal probe (conjugate) specific to aflatoxin B_1 (AFB_1). Nanogold – labeled probe was used to develop an immunochromatographic (IC) method for Aflatoxin B_1 analysis. With this method analysis can be completed within 10 min. reducing the detection time 6-10 times comparative with ELISA. Shim *et al.* (2007) developed an Immunochromatography (ICG) strip test using a nanocolloidal gold-antibody probe and optimized for the rapid detection of aflatoxin B_1. A monoclonal antibody specific to AFB_1 was produced from the cloned hybridoma cell (AF78), coupled with nanocolloidal gold and distributed on the conjugated pad of the ICG strip test. The visual detection limit of the ICG strip was 0.5 ng/ml. They analyzed 172 grain sample by ICG and compared with HPLC and showed good agreement with those obtained by HPLC. These strips are a potential screening tool for the detection of AFB_1 in samples and could be applied to the preliminary screening of mycotoxin in food and agricultural products, generating result within 15 minutes.

Polymerase Chain Reaction

Manonmani *et al.* (2006) developed rapid method for assessment of aflatoxigenic fungi in food using an indigenously designed primer pair for the aflatoxin regulatory gene *aflR* in Polymerase Chain Reaction. They extracted DNA from 28 different fungal strains on PCR template. Positive amplification was achieved only with DNA from Aflatoxigenic *A. flavus* and *A. parasiticus*. The detection limit for mycelium was determined as 0.05 g and ≥ 100 cfu respectively. Some of the genes coding for enzymes involved in aflatoxin biosynthesis have been cloned and sequenced which aided in the characterization of the aflatoxigenic molds. It is estimated that at least 16 enzymes are involved in the bioconversion of norsolorinic acid to aflatoxin (Chang *et al.*, 1995, Yu *et al.*, 1995). Among these genes, the nor-1 gene codes for reductase that convert norsolorinic acid to averantin (Trail *et al.*, 1994). The ver-1 gene is involved in the conversion of versicolorin A to Sterigmatocystin and omt-1 gene codes for O-methyltransferase that converts sterigmatocystin (Skory *et al.*, 1992). In addition to these structural genes, a regulatory gene, aflR, codes for regulatory protein that activates the pathway genes. On the basis of cloned and sequenced gene involved in the aflatoxin biosynthesis, specific primer had been designed for Polymerase Chain Reaction or Multiplex PCR method for detection of aflatoxigenic fungi

(Farber *et al.*, 1997, Geisen, 1996). Yang *et al.* (2004) applied multiplex PCR for the detection of potential aflatoxin-producing molds in Korean fermented foods and grains. Three genes, avfA, omtA, and ver-1, coding for key enzymes in aflatoxin biosynthesis, were used as aflatoxin-detecting target genes in multiplex PCR. DNA extracted from *Aspergillus flavus, Aspergillus parasiticus, Aspergillus oryzae, Aspergillus niger, Aspergillus terreus, Penicillium expansum*, and *Fusarium verticillioides* was used as PCR template to test specificity of the multiplex PCR assay. Positive results were achieved only with DNA that was extracted from the aflatoxigenic molds *A. flavus* and *A. parasiticus* in all three primer pairs. This result was supported by aflatoxin detection with direct competitive enzyme-linked immunosorbent assay (DC-ELISA).

Tandem Mass Spectrometry

The tandem MS method (Plattner, 1986) has been used extensively in the last few years. This method involves one stage of mass separation to select the compound of interest from the sample matrix (molecular ion, protonated molecular ion, or molecular anion) and a second stage of analysis after colloisionally activated dissociation (CAD) by collision with a target gas such as argon. Thus both the parent ion and the activated dissociated ion (daughter ions, or secondary ions) are analyzed. In contrast to selection ion mode (SIM), the monitoring system of tandem mass spectrometry is called "multiple reaction monitoring" (Plasencia *et al.*, 1990). This method has also been used in conjunction with Gas Chromatography (GC) (Plattner, 1986). Fast atom bombardment/tandem MS has been used for the detection of aflatoxin (Uyakual *et al.*, 1989). Lattanzio *et al.* (2007) developed liquid chromatography/tandem mass spectrometry method for simultaneous determination of aflatoxin B_1, B_2, G_1, G_2, Fumonisin, ochratoxin A, deoxynivalenol, zearalenone, T-2 and HT-2 toxins in maize.

Inactivation Method to Control Aflatoxin

Decontamination process must be technically and economically feasible. The FAO requirement for acceptable decontamination process is that the process must destroy, inactivate or remove aflatoxins, not produce, nor leave toxic (carcinogenic and mutagenic) residue in the food commodities and should destroy fungal spores and mycelium that could proliferate and produce new toxins under favorable conditions. Many detoxification methods of aflatoxin have been recommended and include: mechanical separation of contaminated seed, heat treatment, extraction using solvents, detoxification using solvent, detoxification using chemical agents and added sorbents. Some physical, chemical and biological methods to detoxify aflatoxin in feedstuffs are reviewed here.

Botanicals

Krishnamurthy and Shashikala (2006) studied the effect of *Withania somnifera, Hyptis suaveolens, Eucalyptus cimodora*, peel powder of *Citrus sinensis, Citrus medica* and *Punica granatum*, neem cake and Pongamia cake on Aflatoxin B_1 production by *A. flavus*. The all tested plant material was significantly effective in inhibiting aflatoxin B_1. Neem cake showed maximum inhibition (98.22 per cent at 10 per cent and 87.07 per cent at 5 per cent) followed by *H. suaveolens* (84.38 per cent at 5 per cent and 78.85 per cent at 10 per cent), Pongamia cake (81.35 per cent at 5 per cent and 75.17 per cent at 10 per cent), *Withania somnifera* (78.11 per cent at 10 per cent), *C. sinensis* ((70.84 per cent at 10 per cent) and *C. medica* (73.65 per cent at 5 per cent).

Leaf powder of *Ocimum* has been successfully used in inhibiting mould development and aflatoxin production on stored soyabean (Awuah, 1996). Neem leaf extracts were found to be very effective in controlling growth of mould and aflatoxin production (Bankole, 1997). Ghorbanian *et al.* (2007) studied the effect of neem (*Azadiracta indica* A. juss) leaf extract on the growth of *A. parasiticus* and production of aflatoxin. They found maximum inhibition *i.e.* 80-90 per cent at 50 per cent concentration. Similar finding was also reported by Bhatnagar and McCormic (1988) and Allameh *et al.* (2001). Bullerman *et al.* (1977) established that Cinnamon and clove oil and their principle components, cinnamic aldehyde and Eugenol, inhibited the growth and toxin of *A. parasiticus*.

Chandra *et al.* (2007) studied the synergistic effect of *Pongamia pinnata* (Bark) and *Tamarindus indica* (fruit) against *A. flavus* and found significant reduction in growth *i.e.* 62 per cent. Sanchez *et al.* (2005) studied the effect of ethanolic, methanolic and aqueous extract of *Agave asperrima* and *Agave striata* on growth and production of aflatoxin and observed remarkable reduction in aflatoxin synthesis. Hajare *et al.* (2005) reported that aqueous extract of Ajwain (*Trachyspermum ammi*) has aflatoxin Inactivation Factor (IF). Essential oil of *Thymus eriocalyx* and *Thymus x. porlock* showed inhibitory effect when exposed to *A. parasiticus*. The oil from above plants was found to be strongly fungicidal and inhibitory to aflatoxin production (Rasooli and Abyaneh, 2004). Chloroform extracts of *Garcinia indica* rinds was tested for the inhibition of *A. flavus* and Aflatoxin production using peanut powder as model food system. Aflatoxin was estimated by spectroflurophotometric and Thin Layer Chromatographic methods. At lower concentration *i.e.* 500-1000 ppm, the aflatoxin inhibition was much higher than the growth inhibition. At 3000 ppm the Aflatoxin production was completely inhibited (Tamil Selve *et al.*, 2003). As evident from the above account, the Botanicals have immense potential to control fungal growth and toxin and thus there is need of more research on purification of active fraction for commercialization as an alternate to Chemicals.

Heat

Normal home cooking such as boiling and frying (approx. 150°C) failed to destroy AFB_1 and AFG_1 in the solid state (Kamimura, 1989). Aflatoxins have high decomposition temperature ranging from 237°C to 306° C. Solid AFB_1 is quite stable to dry heating at temperature of 267°C (Behna, 1989). Degradation of aflatoxin by heat is also governed by the moisture content, ionic strength, and pH of the food. The moisture content is a critical factor; contaminated food that contains more moisture can more easily be inactivated by heat. Mann *et al.* (1967) observed that heating a cottonseed meal containing 30 per cent moisture at 100°C for 1h degraded 74.8 per cent of aflatoxins (B_1 + B_2) present in meal, whereas only 32.7 per cent of the toxins were destroyed after heating a similar meal containing 6.6 per cent moisture under the same condition.

Farah *et al.* (1983) cooked raw unshelled peanuts in 5 per cent NaCl solution in an autoclave at 116°C, 0.7 bar for 30 min. This treatment reduced the total content of aflatoxins (B_1 + G_1 + B_2 + G_2) by 80-100 per cent. Rustum *et al.* (1993) studied the effect of pH (5.0, 8.0, 10.2), temperature (121°C, 130°C, 140°C) and heating time (5s, 20s, 15 min). Heat treatment at pH 8.0 was not effective in reducing mutagenic activity. On the other hand, the treatment pH 10.2, 130°C, 20s and pH 10.2, 121°C, 15 min reduced the mutagenic activity by 78 per cent and 88 per cent respectively. Kumar *et al.* (2002) studied the effect of roasting at different moisture levels on the inactivation of Aflatoxin B_1 of food grains. The initial content of AFB_1 in inoculated groundnut kernel sample was 2836.4 ppb. Roasting at 150°C for 10 min at 10 per cent moisture level reduced AFB_1 by 48 per cent; whereas at the same temperature and time and 20 per cent and 30 per cent moisture AFB_1 was significantly reduced by 53.03 per cent and 56.02 per cent respectively.

Antagonistic Microorganism

Ciegler *et al.* (1966) have screened approximately 1000 microorganisms representing yeasts, molds, mold spores, actinomycetes, bacteria and algae for their ability either to destroy or to transform aflatoxin B_1 and G_1. Some molds and mold spores partially transformed aflatoxin B_1 to new fluorescing compounds. Only one of the bacteria *Flavobacterium aurantiacum*, NRRL B-184 removed aflatoxin from solution. Detoxification by cells of this microorganism was tested on milk, corn oil, peanut butter, corn, soyabeans and peanuts. The milk, corn oil, and peanut butter were artificially contaminated by adding 600, 700 and 700 $\mu g\ kg^{-1}$ of aflatoxin B_1 respectively, to 50 ml of milk and corn oil and 50 gm of peanut butter. When 2.0 x 10^{13} viable cells of *F. aurantiacum*, NRRL B-184 were added to each of these food products aflatoxin levels were reduced to 0 and a trace after 3 hours of incubation. Viable cells of *F. aurantiacum* were mixed with soyabeans, corn, and peanuts contaminated seeds which were then incubated for 12 hours at 28°C. Aflatoxins were completely removed from corn and peanuts but only 86 per cent were removed from soyabeans. Destruction of aflatoxins in solution by *F. aurantiacum* was confirmed by duckling tests. Aflatoxin contamination by the toxic strains of *A. flavus* inhibited by an atoxigenic strain of *A. flavus* in vivo and in liquid fermentation, and the atoxigenic strain was equally effective when applied at spore concentration either equal to or one-half those of the toxigenic strain (Cotty and Bayman, 1993). Displace toxigenic strains of *A. flavus* from agricultural fields with strains of *A. flavus* that do not produce aflatoxins; strategy is possible because of the great variability of phenotypes of *A. flavus* in agricultural fields and the common occurrence of atoxigenic strains (Cotty, 1989). Several atoxigenic strains of *A. flavus* isolated from agricultural fields in Arizona can reduce the aflatoxin contamination of developing cotton bolls caused by toxigenic strains (Cotty, 1990). *Lactobacillus casei pseudoplantarum* 371 isolated from a silage inoculant was found to inhibit aflatoxin B_1 and G_1 biosynthesis by *A. flavus* sub sp parasiticus NRRL 2999 in liquid medium (Gourama and Bullerman, 1997).

Radiation

Food irradiation is becoming a technique of potential application on commercial scale as it renders the food product sterile (Diehl, 1990).

Gamma Rays

The toxicity of Peanut meal contaminated with AFB_1 was reduced by 75 per cent and 100 per cent after irradiation with gamma rays at a dose of 1 and 10 kGy, respectively (Temcharoen and Thilly, 1982). Dose higher than 10kGy inhibited the seed germination and increased the peroxide value of the oil in gamma irradiated peanuts (Chiou *et al.*, 1990). A dose of 10 kGy completely inactivated AFB_1 and 95 per cent of AFG_1 in Dimethylsulphoxide-Water (1:9 v/v) solution (Mutluer and Erkoe, 1987). Addition of 1 ml of 5 per cent hydrogen peroxide to an aqueous AFB_1 solution (50 µg/ml) resulted in 37-100 per cent degradation of the toxin at a lower dose (2 kGy). The final degradation products showed no biological activity in Ames mutagenic test. The same treatment reduced the level of AFB_1 in peanut kernel by 73-80 per cent (Patel *et al.*, 1989). Bazza-Zernab *et al.* (2001) exposed the *A. flavus* isolate to gamma radiation dose level from 0.0 to 3.0 kGy. The gradual decrease in the growth occurred by increasing the irradiation dose upto 2.5 kGy. Low doses of gamma radiation did not affect its production upto 0.5 kGy, the mycelial weight markedly increased the total production reaching 3000 µg/L. Thereafter a decrease in its production was observed by increasing the dose. El-Bazza *et al.* (2001) found that increased dose of ionizing radiation for the spore of *A. flavus* led to an increase in the aflatoxin production reaching the maximum at a dose level of 1.0 kGy. No detectable aflatoxin was observed at 3.0 kGy. Aziz and Yousef (2002) showed that application of radiation at 10 kGy significantly detoxify aflatoxin B_1 by 82-88 per cent. Aziz *et al.* (2004) showed that at 4 kGy gamma rays in maize significantly destroyed 60.9 per cent-66.7 per cent of aflatoxin.There is number of reports which suggest that the molds are very sensitive to gamma radiation and the mycotoxin production decreased after irradiation of food (Refai *et al.*, 1996; Yousef *et al.*, 1999). It appears that the fungal strains, condition of storage, humidity, inoculum size and irradiation dose affect mold growth and toxin production (Mahrous *et al.*, 2002; Aziz *et al.*, 2004).

UV Rays

Aflatoxin B_1 absorbs UV light at 222, 265 and 362 nm with maximum absorption occurring at 362 nm, which lead to the formation of 12 photodegradation products (Samarjeeva *et al.*, 1995). The photodegradation products were less toxic to chick embryo than the parent toxins (Andrello *et al.*, 1967). Treatment of peanut oil with UV light for 2 h destroys 40-45 per cent aflatoxin (Shantha and Murthy, 1977). UV Radiation (30 min) treatment of dried figs artificially contaminated with AFB1 (250 µg/Kg) reduced the aflatoxin level by 45.7 per cent (Altug *et al.*, 1990).

Solar Radiation

Efficacy of solar energy has been studied in different countries. In India, Shantha and Murthy (1981) exposed the peanut cake in sunlight for 6h and found 50 per cent reduction in aflatoxin content. Naturally contaminated peanut flakes were exposed to sunlight for 14h and 50 per cent reduction in aflatoxin content was found (Shantha and Murthy, 1981).

In USA, Mahkoub and Bullermann (1988) exposed the olive oil for 10 and 40 min and found the 55 per cent and 95 per cent reduction of aflatoxin respectively.

Chemicals

Ghosh *et al.* (1996) studied the effect of propionic acid, sodiumbisulfite and sodium hydroxide on the biosynthesis of aflatoxin on groundcake. The aflatoxin free ground cake were treated with 0.25, 0.50, 0.75 and 1.0 per cent each of propionic acid, sodiumbisulfite or sodium hydroxide at three moisture levels of 10, 15 and 20 per cent. Out of three chemicals, the propionic acid was found to be most effective followed by sodium sulfite and sodium hydroxide. Acetic acid and propionic acid which are used in animal feeding are effective mold inhibitor (Sauer, 1997). Several studies have dealt with the use of propionates to control mold growth and toxin production by *A. flavus*, *A. parasiticus*, *A. ochraceus* and *Penicillium viridicatum* in artificially inoculated corn up to 29 weeks of storage (Vandergraf *et al.*, 1975). Propionic acid inhibits aflatoxin formation largely through inhibition of growth of *A. flavus* and *A. parasiticus* (Tsai *et al.*, 1984). Davis and Diener (1967) reported that sulfite inhibits a number of metabolic pathways in fungi. Therefore, it reduces the growth of fungus which leads to less aflatoxin production. Moerck *et al.* (1980) reported that sodium hydroxide at 2 per cent concentration was effective for reducing aflatoxin level. Aflatoxin in the maize grain with an initial concentration of 29 ng/g was completely degraded and 96.7 per cent degradation occurred in maize contaminated with 93 ng/g when treated with citric acid. Aflatoxin fluorescence strength of acidified samples was much weaker than untreated sample when observed in HPLC chromatogram (Mendez-Albores *et al.*, 2005).

Ammonia is one of the most effective reagents proposed for chemical inactivation of aflatoxin in contaminated peanut and cottonseed meals. Masri (1965) treated a toxic meal (Aflatoxin B1 content about 1 ppm) with ammonium hydroxide solution and biological test of treated meal indicated elimination toxicity. Dollear and Gardner (1966), reported inactivation of aflatoxin in cottonseed and peanut meals with anhydrous ammonia under pressure in the range of 20 to 43 psig. It has been observed that AFB_1 molecular structure is irreversibly altered if exposure to ammonia lasts long enough. The disadvantage of ammonia treatment is mainly related to the need to build special plant as ammonia corrodes metal and becomes explosive in the air at mixture over 15 per cent volume. Some effect on the chemical and qualitative characteristics of the feed *i.e.* undesirable brown color of the treated feed has also been observed.

Feuell (1966) reported that treatment of peanut meal with chlorine reduced its toxicity to duckling but did not prevent liver lesion. Aly *et al.* (2004) used the commercially hydrated sodium calcium alumino silicate (HSCAS) and Egyptian monmorillonite (EM) and found that it has capability of absorbing AFB_1 and Fumonisin B_1 in aqueous solution. HSCAS was reported to have high affinity to AFB_1. The HSCAS removed more than 80 per cent of the toxin from solution. *In vivo* studies demonstrated the role of HSCAS in preventing the mutagenicity and toxicity of AFB_1 (Phillips *et al.*, 1998).

Mckenzie *et al.* (1997) treated aqueous equimolar (32 µM) solution of Aflatoxin B_1, B_2, G_1, G_2, cyclopiazonic acid, Fumonisin B_1, Ochratoxin A, Patulin, Secalonic acid and Zearalenone with 2, 10 and/or 20 weight per cent Ozone (O_3) over a period of 5 min and analysed by HPLC. AFB_1 and AFG_1 were rapidly degraded using 2 per cent O_3, while AFB_2 and AFG_2 were more resistant to oxidation and required higher dose of O_3 (20 per cent) for rapid degradation. O_3 gas generated by corona discharge has been reported to degrade the aflatoxin in corn and cottonseed meals (Dollear *et al.*, 1968; Dwarkanath *et al.*, 1968) and in aqueous solution (Moeba *et al.*, 1988). Akacid®plus, a new member of guanidine based polymeric disinfectant was recently introduced for the first time as a potent inhibitor of *A. parasiticus* growth and its aflatoxin productivity (Razzaghi-Abyaneh *et al.*, 2006). Akacid®plus was developed by POC polymer Production GmbH, Vienna, as a new member with enhanced broad antimicrobial activity while significantly less toxicity compared to the former compound of this class. Akacid®plus is water soluble, nonflammable, nonexplosive, colorless and odorless formulation, which is composed of a mixture of two different polymeric guanidine compounds. It was accepted as biocide according to the new EU guidelines and it has *in vitro* antimicrobial activity against some important pathogenic bacteria and fungi (Kratzer *et al.*, 2006a, b; Buxbaum *et al.*, 2006).

Analysis of aflatoxin is difficult task because only trace amount of toxin are present in the animal feed and other food commodities. However, rapid progress in the development of new aflatoxin analytical methodology has been made during the last few years. Researchers have attempted to simplify and improve the existing analytical methods. New more versatile and sensitive methods have come to the market. After many years of laboratory research, immunoassay technique has gained wide acceptance as analytical method for aflatoxin and other toxins. The mycotoxin problem is serious in developing countries where conditions and agricultural practices are conductive to fungal growth and toxin production. The control of Aflatoxin can be achieved by good farm management and use of effective detoxification process which we have discussed in this review.

References

Allameh, A., Razzaghi-Abyaneh, M., Shams, M., Rezaee, M.B., Jaimand, K., 2001. Effects of neem extract on production of aflatoxin and activities of fatty acids synthetase, isocitrate dehydrogenase and glutathonase S-transferase in *Aspergillus parasiticus*. *Mycopathologia*, 154: 79–84.

Altug, T., Yousef, A.E. and Marth, E.H., 1990. Degradation of Aflatoxin B_1 in dried figs by sodium bisulfite with or without heat, Ultravoilet energy or hydrogen peroxide. *Journal of Food Protection*, 53: 581–582.

Aly, S.E., Abdel-Galil, M.M. and Abdel-Wahhab, M.A., 2004. Application of Adsorbent Agents Technology in the removal of Aflatoxin B_1 and Fumonisin B_1 from malt extract. *Food and Chemical Toxicology*, 42: 1825–1831.

Andrello, P.J., Beckwith, A.C., and Eppley, R.M., 1967. Photochemical changes of Aflatoxin B_1. *J. Assoc. Off. Anal. Chem.*, 50: 346–350.

Awuah, R.T., 1996. Possible utilization of plant products in grain storage. In: *Proceeding of the Workshop on Mycotoxin in Food in Africa*, (Ed.) K.F. Cardwell, 1995 at Cotonou; Benin, 6–10 November, pp. 32. Benin international Institute of Tropical Agriculture.

Aziz, N.H. and Youssef, B.M., 2002. Inactivation of naturally occurring of mycotoxins in some Egyptian foods and agricultural commodities by gamma radiation. *Egypt. J. Food Sci.*, 30: 167–177.

Aziz, N.H., Moussa, I.A.A., and Far, F.M.E., 2004. Reduction of fungi and mycotoxins formation in seed by gamma radiations. *Journal of Food Safety*, 24: 109–127.

Bankole, S.A., 1997. Effect of essential oil from two Nigerian medicinal plants (*Azadiracta indica* and *Murinda lucida*) on growth and Aflatoxin B_1 in maize grain by a toxigenic *Aspergillus flavus*. *Letter Applied Microbiology*, 24: 190–192.

Behna, V., 1989. Mycotoxin: Chemical, biological and environmental aspects. In: *Bioactive Molecules*, (Ed.) V. Behna. Elsevier Applied Science, London, pp. 114–150.

Bhat, R.V., Vasanthi, S., Rao, B.S., Rao, R.N., Rao, V.S., Nagarajan, K.V., Prasad, C.A.K., Vanchinathan, S., Ray, R., Rajat, Saha, S., Mukherjee, A., Ghosh, P.K., Toteja, G.S. and Saxena, B.N., 1997. Aflatoxin B_1 contamination in maize samples collected from different geographical regions of India: A multicentre study. *Food Additive and Contamination*, 14: 151–156.

Bhatnagar, D. and McCormic, S.P., 1988. The inhibitory effects of neem (*Azadiracta indica*) leaf extracts on aflatoxin synthesis in *Aspergillus parasiticus*. *Journal of American Oil chemist Society*, 65: 1166–1168.

Bilgrami, K.S., Prasad, T. and Sinha, K.K., 1983. *Aflatoxin in Food and Feed*. Allied Press, Bhagalpur, India, pp. 1–13.

Blesa, J., Soriano, J.M., Molto, J.C., Marin, R. and Manes, J., 2003. Determination of Aflatoxins in peanuts by matrix solid phase dispersion and Liquid chromatography. *Journal of Chromatography A.*, 1011: 49–54.

Broadbent, J.H., Cornelius, J.A. and Shone, G., 1963. The detection and estimation of Aflatoxin and groundnut material. *Analyst*, 88: 214–216.

Bullerman, L.B., Lieu, F.Y. and Seire, A.S., 1977. Inhibition of growth and aflatoxin production by cinnamic aldehyde and eugenol. *Journal of Food Science*, 42: 1107–1116.

Bullerman, L.B., 1986. Mycotoxin and food safety. *Food Technology*, 40: 59–66.

Buxbaum, A., Kratzer, C., Graninger, W. and Georgopoulos, A., 2006. Antimicrobial and toxicological profile of the new biocide Akacid®plus. *Journal of Antimicrobial Chemotheraphy*, 58: 193–197.

Calleri, E., Marrubini, G., Brurotti, G., Massoline, G. and Caccialanza, G., 2007. Development of an immunoaffinity monolithic disk for the online solid phase extraction and HPLC determination with fluorescence detection of aflatoxin B_1 in aqueous solution, 44(2): 396–403.

Chandra, Harish, Srivastava, Jatin, Sidhu, O.P., Rai, Nishant, Chauhan, Sachin and Nautiyal, A.R., 2007. Synergistic effect of *Pongamia* bark and *Tamarindus* fruit extract against aflatoxin producing fungi *i.e. Aspergillus flavus*. *Environment Conservation Journal*, 8(1–2): 101–107.

Chang, P.K., Ehrlich, K.C., Yu, J., Bhatnagar, D., and Cleveland, T.E., 1995. Increased expression of *Aspergillus parasiticus aflR*, encoding a sequence-specific DNA-binding protein, relieves nitrate inhibition of aflatoxin biosynthesis. *Applied Environmental Microbiology*, 61: 2372–2377.

Chiavaro, E., Dall, A.C., Galaverna, G., Biancardi, A., Gambarelli, E., Dossena, A. and Marchelli, R., 2001. New reversed phase liquid chromatographic method to detect aflatoxin in food and feed with cyclodextrins as fluorescence enhancer added to the elutent. *Journal of Chromatography A*, 937(1–2): 31–40.

Chiou, R.Y.Y., Lin, C.M., and Shyu, S.L., 1990. Property characterization of peanut kernels subjected to gamma irradiation and its effect on the outgrowth and aflatoxin production by *Aspergillus parasiticus*. *Journal of Food Science*, 55: 210–213.

Chu, F.S., 1991. *Development and use of immunoassay in detection of the ecologically important mycotoxins*. In: *Handbook of Applied Mycology, Vol 5: Mycotoxins*, (Eds.) D. Bhatnagar, E.B. Lillihoj and D.K. Arora. Marcel Dekker, New York.

Chu, F.S., 1986. *Immunoassay for mycotoxin*. In: *Modern Methods in the Analysis and Structural Elucidation of Mycotoxin*, (Ed.) R.J.Cole. Academic Press, New York, pp. 207–237.

Ciegler, A., Lillehoj, E.B., Peterson, R.E. and Hall, H.H., 1966. Microbial detoxification of aflatoxin. *Appl. Microbial.*, 14: 934–939.

Codex Alimentarius Commission, 1989. Report of the twentieth session of the codex committee on Food additive and contaminant. *Alinorm*, 89112: 16.

Coomes, T.J., and Sanders, J.C., 1963. The detection and estimation of aflatoxin in groundnuts and groundnut materials. *Analyst*, 88: 209–213.

Cotty, P.J., 1989.Virulence and cultural characteristics of two *Aspergillus flavus* strains pathogenic on cotton. *Phytopathology*, 79: 808–814.

Cotty, P.J., 1990. Effect of atoxigenic strains of *Aspergillus flavus* on aflatoxin contamination of developing cottonseed. *J. Plant Dis.*, 74: 233–235.

Cotty, P.J., and Bayman, P., 1993. Competitive exclusion of a toxigenic strain of *Aspergillus flavus* by an antioxigenic strain. *Phytopathology*, 83: 1283–1287.

Davis, N.D. and Diener, U.L., 1967. Inhibition of aflatoxin synthesis by p-aminobenzoic acid, potassium sulfite and potassium fluoride. *Applied Microbiology*, 23: 1580–1584.

Degan, P., Montagnoli, G. and Wild, C.P., 1989. Time resolved fluoroimmunoassay of aflatoxin. *Clinical Chemistry*, 35: 2308–2310.

Diehl, J.F., 1990. *Safety of Irradiated Foods*, (Ed.) J.F. Diehl. Marcel Dekker, New York, pp. 145–179.

Dollear, F.G. and Gardner, H.K., Jr., 1966. Inactivation and removal of aflatoxin. In: *Proc. 4th Nat. Peanut Res. Conf.*, Tifton, Georgia, July 14–15, pp. 72–81.

Dollear, F.G., Mann, G.E., Codifer, L.P., Jr, Gardner, H.K., Koltun, S.P. and Vix, H.L.E., 1968. Elimination of aflatoxin from peanut meal. *Journal of American Oil Chemist Society*, 45: 862–865.

Dwarakanath, C.T., Rayner, E.T., Mann, G.E., and Dollear, F.G., 1968. Reduction of aflatoxin level in cottonseed and peanut meals by ozonisation. *Journal of American Oil Chemist Society*, 45: 93–95.

El-Bazza Zernab, Farrag Hala, A., El-Fouly Mohie, E.D.Z., and El-Tablawy Scham, Y.M., 2001. Inhibitory effect of gamma radiation and *Nigella sativa* seeds oil on growth, spore germination and toxin production of fungi. *Radiation Physics and Chemistry*. 60:181–189.

Farah, Z., Martins, M.J.R. and Bachmann, M.R., 1983. Removal of aflatoxin in raw unshelled peanuts by traditional salt boiling process practiced in North East of Brazil. Labensm. *Wiss Technology*, 16: 122–124.

Farber, P., Geison, R. and Holzapfel, W.F., 1997. Detection of aflatoxigenic fungi in figs by a PCR reaction. *International Journal of Food Microbiology*, 36: 215–220.

Feuell, A.J., 1966. Aflatoxin in groundnuts. IX. Problems of detoxification. *Trop. Sci.*, 8: 61–70.

Gams, W., Christensen, M., Onions, A.H.S., and Pitt, J.I. and Samson, R.A., 1985. Intrageneric taxa of *Aspergillus*. In: *Advances in Penicillium and Aspergillus Systematics*, (Eds.) R.A. Samson and J.I. Pitt. Plenum press, New York, NY, pp. 55–62.

Geison, R., 1996. Multiplex polymerase chain reaction for the detection of potential aflatoxin and sterigmatocystin producing fungi. *Systematic Applied Microbiology*, 19: 388–392.

Ghorbanian, M., Razzaghi-Abyaneh, M., Allameh, A., Shams-Ghahfarokhi, M., and Qorbane, M., 2007. Study on the effect of neem (*Azadiracta indica* A. juss) leaf extract on the growth of *Aspergillus parasiticus* and production of aflatoxin by it at different incubation time, p. 1–5.

Ghosh, M.K., Chhabra, A., Atreja, P.P. and Chopra, R.C., 1996. Effect of heating with propionic acid, sodium bisulphate and sodium hydroxide on the biosynthesis of aflatoxin on ground cake, 60: 43–49.

Gourama, H. and Bullerma, L.B., 1997. Antiaflatoxigenic activity of *Lactobacillus casei pseudoplantarum*. *Int. J. Food Microbiol.*, 34: 131–143.

Hazare, S.S., Hajare, S.N. and Sharma, A., 2005. Aflatoxin inactivation using aqueous extract of Ajwain (*Trachyspermum ammi*) seeds. *Journal of Food Science*, 70(1): 271–279.

Hesseltine, C.W., 1986. Global significance of mycotoxins, In: *Mycotoxin and Phycotoxin*, (Eds.) P.S. Steygen and R. Vineggar. Elsevier Science Publishers, Amsterdam, pp. 1–18.

Hutchins, J.E. and Hagler, W.M., Jr., 1983. Rapid determination of aflatoxin in heavily contaminated Corn. *J. Assoc. of. Anal. Chem.*, 66: 1458–1465.

Kamimura, H., 1989. Removal of mycotoxin during food processing. In: *Mycotoxin and Phycotoxin*, (Eds.) S. Natori, K. Hashimoto and Y. Ueno. Elsevier Science Publishers. Amsterdam, pp. 169–176.

Kolosova, A.Y., Shim, W.B., Yang, Z.Y., Eremin, S.A. and Chung, D.H., 2006. Direct competitive ELISA based on a monoclonal antibody for detection of aflatoxin B_1, stabilization of ELISA kit components and application to grain samples. *Annal. Bioannal Chem.*, 384(1): 286–294.

Kratzer, C., Tobudie, S., Assadian, O., Buxbaum, A., Graninger, W., Georgopoulos, A., 2006a. Validation of Akacid plus as a room disinfectant in the hospital setting. *Applied Environmental Microbiology*, 72: 3826–3831.

Kratzer, C., Tobudie, S., Graninger, W., Buxbaum, A. and Georgopoulos, A., 2006b. *In vitro* antimicrobial activity of the novel polymeric guanidine Akacid plus. *Journal of Hospital Infection*, 63: 316–322.

Krishnamurthy, Y.L. and Shashikala, J., 2006. Inhibition of aflatoxin B_1 production of *Aspergillus flavus*, isolated from soyabean seeds by certain natural plant. *Letters in Applied Microbiology*, 43: 469–474.

Kumar, Harish, Jha, Y.K. and Chauhan, G.S., 2002. Detection and estimation of aflatoxin in food grains of Tarai region and effect of heat treatment and its inactivation, 39(5): 479–483.

Lattanzio, V.M., Solfrizzo, M., Powers, S. and Visconti, A., 2007. Simultaneous determination of aflatoxin, ochratoxin A and Fusarium toxins in maize by liquid chromatography/tandem mass spectrometry after multitoxin immunoaffinity cleanup. *Rapid Commun. Mass Spectrometry*, 21(20): 3253–3261.

Mahkoub, A. and Bullerman, L.B., 1988. Effect of storage time, sunlight, temperature and frying on stability of aflatoxin B_1 in olive oil. *Lebensm, Wiss. Technology*, 21: 29–32.

Mahrous, S.R., Youseff, B.M. and Aziz, N.H., 2002. Effects of gamma radiation in some microbial organoleptical and chemical aspects of fresh meat, ground beef and beef burger stored at 5°C, Egypt. *Journal of Radiation Science Application*, 13: 17–92.

Mann, G.E., Codifer, L.P. and Dollear, F.G., 1967. Effect of heat on aflatoxin in oil seed meals. *Journal of Agricultural and Food Science*, 15: 1090–1092.

Manonmani, H.K., Anand, S, Chandrashekhar, A. and Rati, E.R., 2006. Detection of aflatoxigenic fungi in selected food commodities by PCR. *Process Biochemistry*, 40: 2859–2864.

Masri, M.S., 1965. Biochemical evaluation of aflatoxin (Abstr.). In: *Western Experiment Station Collaborators Conference on the Importance of Mold Metabolites in Agricultural Products*, March 1–3. Program and Abstracts of papers, Albany, Cal., Western Utilization Res. And Dev. Division, Agricultural Research Service, U.S. Department of Agriculture, Processed, p. 9.

Mekenzie, K.S., Sarr, A.B., Mayura, K., Bailey, R.H., Miller, D.R., Roger, T.D., Norred, W.P., Voss, K.A., Plattner, R.D., Kubena, L.F. and Phillips, T.D., 1997. Oxidative degradation and detoxification of mycotoxins using a novel source of ozone, 35: 807–820.

Mendez-Albores, A., Arambula-Villa, G., Loarea-Pinna, M.G.P., Castano-Tostado, E. and Moreno-Maryinez, E., 2005. Safety and efficacy evaluation of aqueous citric acid to degrade B aflatoxin in maize. *Food and Chemical Toxicology*, 43: 233–238.

Moeba, H., Takamoto, Y., Kamimura, M., and Miura, T., 1988. Destruction and detoxification of aflatoxin with ozone. *Journal of Food Science*, 53: 667–668.

Moerck, K., MceLfresh, P., Wohlman, A. and Hilton, B.W., 1980. Aflatoxin destruction in corn using sodium bisulfite, sodium hydroxide and aqueous ammonia. *Journal of Food Protection*, 43: 571–574.

Mutluer, B. and Erkoe, F.U., 1987. Effects of gamma radiation on Aflatoxins. *Z.Lebensm. Unters. Forsch.*, 164: 398–401.

Nesheim, S., 1968. Conditions and techniques for thin layer chromatography of aflatoxins. *J. Am. Oil Chemists' Soc.*, 45(2), Program abstract No. 8; AACC–AOCS Joint meeting.

Otta, K.H., Papp, E. and Bagocsi, B., 2000. Determination of aflatoxin in food by over pressured layer chromatography. *J. Chrom. A.*, 882: 11–16.

Panalaks, T., and Scott, P.M., 1977. Sensitive silica gel packed flow cell for fluorometric detection of aflatoxins by high pressure liquid chromatography. *Journal of the Association of Official Analytical Chemists*, 60(3): 583–589.

Patel, U.D., Govindrajan, P. and Dave, P.J., 1989. Inactivation of aflatoxin B_1 by using the synergistic effect of hydrogen peroxide and gamma radiation. *Applied Environmental Microbiology*, 55: 465–467.

Pestka, J.J., Gaur, P.K. and Chu, F.S., 1980. Quantitation of Aflatoxin B_1 and Aflatoxin B_1 Antibody by an enzyme-linked immunosorbent microassay. *J. Appl. and Environ. Microbiol.*, 40: 1027–1031.

Peterson, R.E. and Ciegler, A., 1967. Separation of aflatoxin by two-dimensional thin-layer chromatography. *J. Chromatog.*, 31: 250–251.

PFA, 1954. *Prevention of Food Adulteration Act*, Ministry of Health, Govt. of India, Manak Bhavan, New Delhi.

Phillips, T.D., Kubena, J.F., Harvey, R.B., Taylor, D.S. and Heidelbaugh, N.D., 1998. Hydrated calcium: A high affinity sorbent for aflatoxin. *Poultry Science*, 67: 243–247.

Plasencia, J., Mirocha, C.J., Pawlosky, J. and Smith, J.F., 1990. Analysis of zearalenone and a-zearalenol in urine of ruminant using gas chromatography-tandem mass spectrometry. *Journal Association Off. Anal. Chem.*, 73: 973–980.

Plattner, R.D., 1986. Mass spectrometry–mass spectrometry as a tool for mycotoxin analysis. In: *Modern Method in the Analysis and Structural Elucidation of Mycotoxins*, (Ed.) R.J. Cole. Academic Press, New York, pp. 393–414.

Pons, W.A., Jr., 1979. High pressure liquid chromatographic determination of aflatoxins in corn. *J. Assoc. Off. Anal. Chem.*, 62: 586–594.

Rasooli, I. and Abyaneh, M.R., 2004. Inhibitory effect of Thymes oil on the growth and Aflatoxin production by *Aspergillus parasiticus*. *Food Control*, 15: 479–483.

Razzaghi-Ahyaneh, M., Shams-Ghahfarokhi, M., Eslamifar, A., Schmidt, O.J, Gharebaghi, R., Karimian, M., Nareri, A. and Sheikhi, M., 2006. Inhibitory effect of Akacid plus® growth and aflatoxin production by *Aspergillus parasiticus*. *Mycopathologia*, 61: 245–247.

Refai, M.K., Aziz, N.H., El-Far, F.M. and Hassan, A.A., 1996. Determination of Ochratoxin produced by *Aspergillus ochraceus* in feed stuffs and its control by gamma radiation. *Applied Rad. Iso.*, 47: 617–621.

Rustum, I.Y.S., Lopez-Leva, M.H. and Nair, B.M., 1993. Effect of pH and heat treatment on the mutagenic activity of peanut beverages contaminated with aflatoxin B_1. *Food Chemistry*, 46: 37–42.

Samarjeeeva, U., Sen., A.C., Cohen, M.D. and Wei, C.I., 1995. Detoxification of Aflatoxin in foods and feed by physical and chemical methods. *Journal of Food Protection*.53:489–501.

Sanchez, E., Heredia, N., Gareia, S. (2005). Inhibition of growth and mycotoxin production of *Aspergillus flavus* and *Aspergillus parasiticus* by extract of Agave species. *Int. J. Food Microbiology*, 98: 271–278.

Sauer, F., 1997. Control of yeast and molds with preservatives. *Food Technology*, 31: 66–67.

Seitz, L.M., 1975. Comparison of methods for aflatoxin analysis by high-pressure liquid chromatography. *J. Chromatography*, 104: 81–89.

Shantha, T. and Murthy, V.S., 1977. Photodestruction of aflatoxin in groundnut oil. *Indian Journal of Technology*, 15: 453–454.

Shantha, T. and Murthy, V.S., 1981. Use of sunlight to partially detoxify groundnut cake flour and casein contaminated with aflatoxin B_1. *Journal of Association of Analytical Chemist.*, 64: 291–293.

Shantha, T. and Murthy, V.S., Rai, E.R. and Prema, V., 1986. Detoxification of groundnut seeds by urea and sunlight. *Journal of Food Safety*, 7: 225–231.

Sharma, M. and Marquez, C., 2001. Determination of aflatoxins in domestic pet food (Dog and cat) using immunoaffinity column and HPLC. *Animal Feed Science and Technology*, 93: 109–114.

Shim, W.B., Yang, Z.Y., Kim, J.Y., Kang, S.J., Woo, G.J., Chung, Y.C., Sergei, A.E., and Chung, D.H., 2007. Development of Immunochromatography strip-test using nanocolloidal gold antibody probe for the rapid detection of aflatoxin B_1 in grain and feed samples. *Journal of Microbiology and Biotechnology*, 17(10): 1629–1637.

Skory, C.D., Chang, P.K., Cary, J. and Linz, J.E., 1992. Isolation and characterization of a gene from *Aspergillus parasiticus* associated with the conversion of versicolaria A to Sterigmatocystin in aflatoxin biosynthesis. *Applied Environmental Microbiology*, 58: 3537–3647.

Stroka, J. and Anklam, E., 2000. Development of a simplified densitometer for the determination of aflatoxins by thin layer chromatography. *Journal of Chromatography A*, 904: 263–268.

Stubblefield, R.D. and Shotwell, O.L., 1977. Reverse phase analytical and preparative high pressure liquid chromatography of aflatoxins. *J. Assoc. Off. Anal. Chem.*, 60: 784–790.

Tamil Selve, A., Joseph, G.S. and Jyaprakash, 2003. Inhibition of growth and Aflatoxin production in *Aspergillus flavus* by *Garcinia indica* extract and its antioxidant activity. *Food Microbiology*, 20: 455–460.

Tarter, E.J., Hanchay, J.P. and Scott, P.M., 1984. Improved liquid chromatographic method for determination of aflatoxins in peanut butter and other commodities. *J. Assoc. Off. Anal. Chemists.*, 67: 597–600.

Temcharoen, P. and Thilly, W.G., 1982. Removal of aflatoxin B_1 toxicity but not mutagenicity by 1 megarad gamma radiation of peanut meal. *Journal of Food Safety*, 4: 199–205.

Thean, J.E., Lornz, D.R., Wilson, D.M., Rodgers, K., and Gueldner, R.G., 1980. Extraction, cleanup, and quantitative determination of aflatoxins in corn. *J. Assoc. Off. Anal. Chem.*, 63: 631–633.

Trail, F., Chang, P.K., Cary, J. and Linz, J.E., 1994. Structural and functional analysis of the nor-1 gene involved in the biosynthesis of aflatoxin by *Aspergillus parasiticus. Applied Environmental Microbiology*, 60: 4078–4085.

Tsai, W.K.J., Shaoo, K.P.P. and Bullarman, L.B., 1984. Effect of sorbate and propionate on growth and aflatoxin production of sublethality injured *A. parasiticus*. *Journal of Food Science*, 49: 86–90.

Uyakual, D., Isobe, M. and Goto, T., 1989. Mycotoxin analysis by fast atom bombardment tandem mass spectrometry. *Journal Association Off. Anal. Chem.*, 72: 491–497.

Vandergraft, E.E., Hesseltime, C.W and shotell, O.L., 1975. Grain preservatives: Effects on aflatoxin and ochratoxin production. *Cereals Chemistry*, 52: 79–84.

Whitaker, T. B., 2000. Sampling techniques. In: *Methods in Molecular Biology Mycotoxin Protocols*, (Eds.) M.W. Trucksess, and A.E. Pohland. Humana Press, Totowa, NJ, 157: 11–24.

Whitaker, T.B. and Wiser, E.H., 1969. Theoretical investigations into the accuracy of sampling shelled peanuts for aflatoxin. *Journal of American Oil Chemists Society*, 46(7): 377–379.

Xiulan, S., Xiaolian, Z., Jian, T., Zhou, J. and Chu, F.S., 2005. Preparation of gold labeled antibody probe and its use in immunochromatography assay for detection of Aflatoxin B_1. *International Journal of Food Microbiology*, 99: 185–194.

Yang, Z.Y., Shim, W.B., Kim, J.H., Park, S.J., Kang, S.J., Nam., B.S. and Chung, D.H., 2004. Detection of aflatoxin producing molds in Korean fermented foods and grains by multiplex PCR. *Journal of Food Protection*, 67(11): 2622–2626.

Yousef, B.M., Mahrous, S.R. and Aziz, N.H., 2002. Effects of gamma radiation on aflatoxin B_1 production by *Aspergillus flavus* in ground beef stored at 5°C. *Journal of Food Safety*, 19: 231–239.

Yu, J., Chang, P.K., Cary, J.W., Wright, M., Bhatnagar, D., Cleveland, T.E., Payne, G.A. and Linz, J.E., 1995. Comparative mapping of aflatoxin pathway gene clusters in *Aspergillus parasiticus* and *Aspergillus flavus*. *Applied Environmental Microbiology*, 61: 2365–2371.

Ethnobotanical Wisdom and Microbial Studies on Medicinal Plants (2011) *Pages* **291–293**
Editors: **D.R. Khanna, Ashutosh Gautam, R. Bhutiani, Gagan Matta, Anil Dhiman**
Published by: **DAYA PUBLISHING HOUSE, NEW DELHI**

Chapter 41

Fungal Ecology of BLSB Infected Maize Fields

Nishat Asif and T.P. Mall
P.G. Department of Botany,
Kisan P.G. College, Bahraich, U.P.

ABSTRACT

Screening of fungal population was done from the rhizosphere/rhizoplane of BLSB infected maize fields from different locality of District Bahraich. Nineteen species were isolated from rhizoplane and twenty-three species from rhizosphere. *Rhizoctonia solani* and *Aspergillus flavus* were dominant in serial dilution whereas *Aspergillus flavus* was dominant by Warcup method.

Keywords: *Rhizosphere, Rhizoplane, Fungal population, Maize, BLSB, Rhizoctonia solani.*

Introduction

Maize (*Zea mays* L.) is a cereal grain commonly known as corn, originally referred to a granular particle. It occupies 3rd rank among principle staple of the world *i.e.* wheat and rice. India produces 12 million metric tons in total maize production of world. Inspite of all advance agricultural practices the crop suffers from severe losses every year. Poor seedling germination, attacks of pest and pathogen, microbial inoculum infested in the seeds also reduces the productivity.

Maize suffers from number of diseases *viz.*, Bacterial stalk rot, Pythium stalk rot, Rusts, Smuts, Charcoal rot, Maydis leaf blight, Banded leaf and Sheath Blight (BLSB) etc. Among which Bacterial stalk rot and Banded leaf and sheath blight are economically important. Banded leaf and sheath blight is caused by the most versatile and dreaded pathogen *Rhizoctonia solani* which causes 15-20 per cent yield loss annually (Saxena, 2002).

Rhizosphere is the well-recognized and specialized ecological niche for fungi, bacteria and actinomycetes. It is also a well-established region around root supporting beneficial and biotrophic microbes. Rhizosphere is the portion of the soil, which is around root supporting diversified microorganism, and is influenced by root exudates, soil factors, age and type of the host plant and molecular signals. Rhizospheric effect is indicated by the interaction of soil and rhizosphere microbes and their ratio. Significance of rhizosphere mycoflora in plant health has been reviewed. Present study has been undertaken to screen the mycobial population of maize plants infected with BLSB disease.

Materials and Methods

Five fields in three maize growing blocks of district Bahraich that falls in Northern Tarai belt of U.P. were selected. Soil samples were collected from the rhizosphere of maize plants, which were infected with BLSB disease in selected sites. The fungal population was recorded by Warcup method (Warcup, 1950). Fungal flora was isolated on PDA medium supplemented with streptomycin to supress the bacterial growth.

For isolation of rhizoplane mycoflora serial dilution method was followed (Kanaujia, 1972). Identification was done by mounting material in lactophenol and stained with cotton blue. Observations were done under microscope using standard manuals and monographs.

Results and Discussion

Twenty-three species belonging to sixteen genera were recorded from rhizosphere. *Aspergillus flavus* was recorded with 60 per cent frequency whereas *R.solani* and *Cercospora* sp.have 50 per cent; *Alternaria alternata* was recorded with 40 per cent; *A.tenuis, Aspergillus fumigatus, A.niger, A.terreus, Cladosporium herbarum, Monilia* sp., 30 per cent *Chaetomium globosum, Curvularia lunata, Fusarium oxysporum, Macrophomina phaseolina, Phoma* sp., *Trichoderma viride* and *Syncephalastrum* sp., 20 per cent; *A.ochraceous, Dreschelra* sp., *Fusarium roseum, Pythium* sp., *Penicillium citrinum* and *T.harzianum* frequency was 10 per cent only. *A.flavus* showed highest density (6) whereas *A.ochraceous, Pythium* sp., *P.citrinum, T.harzianum* was recorded in minimum density (1).

Isolation of rhizoplane mycoflora showed maximum frequency of *A.flavus* and *R.solani* 70 per cent whereas *A.tenuis, A.alternata* showed minimum frequency 10 per cent. *Fusarium oxysporum* and *A.niger* 60 per cent, *A.ochraceous* 50 per cent; *Fusarium roseum* and *Chaetomium globosum* 40 per cent; *T.viride, T. harzianum, P.citrinum, Dreschlera* sp., *Curvularia lunata, Cercospora* sp.30 per cent; *Phoma* sp., *Monilia* sp., *Cladosporium* sp.and *A.terreus* 20 per cent frequency was recorded. Highest density was recorded of *A.flavus* and *R.solani* (7); *F.oxysporum, A.niger* (6); *A.ochraceous* (5); *F.roseum* and *C.globosum* (4);*T.viride, T.harzianum,P.citrinum, Dreschlera* sp., *Curvularia lunata* and *Cercospora* sp. (3). *Phoma* sp., *Monilia* sp., *Cladosporium* sp. and *A. terreus* (2); *A.alternata* and *A.tenuis* showed minimum density (1) only. Several other workers have also reported the rhizosphere mycoflora. (Singh *et al.*, 2006; Sooting and Dhkar, 2003; Wani *et al.*, 2006; Roy, 2006).

Acknowledgement

Authors are thankful to Dr. S.P. Singh Principal, Kisan P.G. College, Bahraich for providing necessary facilities.

References

Kanaujia, R.S., 1972. Studies on phyllosphere fungi. VII. Foliar application of urea on certain ornamentals. *Acta. Mycologia*, 13: 281–288.

Roy, A.K., 2006. Documentation of maize rhizosphere microflora and their prospect in the management of toxigenic *Aspergillus flavus* strains and Aflatoxin production. In: *Proc. Nat. Symp. on Microbial Diversity and Plant Health Problems*, Held at D.D.U. Univ., Gorakhpur, Gorakhpur, December 18–19, pp. 11.

Saxena, S.C., 2002. Biointensive integrated disease management of banded leaf and sheath blight of maize. In: *Proc. 8th Asian Regional Maize Workshop*, Held at Bankok, Thailand, August 5–8, pp. 380–388.

Singh, P.R., Patel, K.G., Srivastava, K.A. and Arora, K.D., 2006. Microbial Diversity in the rhizosphere of the Parthenium weed. In: *Proc. Nat. Symp. on Microbial Diversity and Plant Health Problems*, Held at D.D.U. Univ., Gorakhpur, Gorakhpur, December 18–19, pp. 20.

Sooting, A.P. and Dhkar, M.S., 2003. Effect of two dominant rhizosphere fungal metabolites on germination of seeds of two varieties of chilli. *Indian Phytopath.*, 56(3): 348.

Wani, A.H., Khan, A.M., Boda, H.R. and Tasken–un–nisa, 2006. Studies on Rhizospheric soil fungi of brinjal in different localities of Baramulla, Kashmir. In: *Proc. Nat. Symp. on Microbial Diversity and Plant Health Problems*, Held at D.D.U. Univ., Gorakhpur, Gorakhpur, December 18–19, pp. 7.

Warcup, J.H., 1950. The soil plate method for the isolation of fungi from soil. *Nature*, 166: 117–118.

Ethnobotanical Wisdom and Microbial Studies on Medicinal Plants (2011) *Pages* **294–300**
Editors: **D.R. Khanna, Ashutosh Gautam, R. Bhutiani, Gagan Matta, Anil Dhiman**
Published by: **DAYA PUBLISHING HOUSE, NEW DELHI**

Chapter 42

The Anatomic Responses of *Lycopersicon esculentum* Mill. Due to the Tomato Leaf Curl Virus (TLCV)

Sarika Srivastava, G.P. Srivastava and J.P. Tewari
Plant Pathology Laboratory, M.L.K (P.G.) College, Balrampur, U.P.

ABSTRACT

An anatomical study of tomato plants infected by tomato leaf curl virus was conducted to elucidate the mode of infection of the causal virus. One to two months after whitefly transmission, the severe symptoms appear with thickening of the veins, curling of leaf and stunting of plant. Typically reorganization of leaf tissue consisted in replacement of the spongy parenchyma by a palisade parenchyma. Palisade parenchyma tissues were compact in comparison to healthy. Abnormal cambial activity was observed in conducting tissue. Weaker sclerenchyma rings were narrow and these were fewer narrow xylem vessels. Phloem necrosis was observed frequently in virus infected stem. Bronzing and discoloration sieve elements in phloem were also found in infected stem. The cortical parenchyma was wider and formation of mechanical and conducting tissues was reduced. In root of infected plant secondary thickening was less in comparison to healthy. Xylem vessels were narrow and with scanty phloem in diseased root. The number of stomata was also reduced in infected leaves.

Keywords: *Anatomical changes, Tomato, TLCV, Stomatal index.*

Introduction

Tomato leaf curl is a serious disease of tomato (*Lycopersicon esculentum* Mill.) that has been spreading in intensive tomato production area of eastern U.P. This disease can cause 50-90 per cent yield loss (Sastry and Singh, 1973; Saikia and Muniyappa, 1986, 1989). However, the incidence occasionally reaches up to 100 per cent. TLCV infected plants show a variety of symptoms including leaf curling, vein clearing yellowing and stunting etc. Tomato leaf curl disease is caused by Geminivirus (Genus-Begmovirus) and is transmitted by *Bemisia tabaci* Genn. (Vasudeva and Samraj, 1948; Cozosnek *et al.*, 1988; Chakraborty *et al.*, 2003).

Most of investigations carried on this disease are confined to symptology, transmission and physiological identification (Yassin and Nour, 1965). Generally virus infection is known to result in drastic histopathological changes. The effects of Tomato Leaf Curl Virus on the anatomic structure of plants have received less attention than comparable effects on plants infected with leaf curl virus. Therefore, the aim of present investigation is to analyse the anatomical changes of tomato roots, stems and leaf affected by TLCV.

Materials and Methods

In March 2006, tomato plants were raised under green house condition. For transmission of TLCV, the adult whiteflies (*Bemisia tabaci* Genn.) were collected from TLCV infected tomato plants. These whiteflies were then fed on infected tomato plants for 48 hrs in an insect proof cage to ensure the acquisition of virus. Later 20 whiteflies were removed and transferred to each healthy tomato plant kept in an insect proof cage in three replications. An equal number of plants were kept in green house as uninoculated controls. Leaves, stems and roots were sampled from infected and healthy plants and fixed in FAA after 24 hrs, washed in 70 per cent alcohol. The plant material were dehydrated with ethyl alcohol butanol grades and embedded in paraffin wax. Sections were cut on rotary microtome at 10µm thickness. These sections were stained with safranin and fast green and mounted in DPX (Johensen, 1940). The stomatal size, frequency and stomatal index were measured with a precalibrated stage and ocular micrometer. The stomatal index (SI) was calculated (Salisburg, 1928).

$$SI = \frac{\text{No. of stomata per unit area}}{\text{No. of stomata per unit area + No. of epidermal cell}} \times 100$$

Results and Discussion

Symptoms

Plants of tomato infected with TLCV were severely stunted, leaf curling and shrinking of leaves. Leaves were often bent downwards or upwards and were stiff. Figures 42.1 to 42.7 show T.S. of leaves and stems of infected and healthy plants.

Anatomical Observations in Root

The cells of epiblema were smaller in size and possessed lignified walls on its inner surface. The cortical and endodermal cells of diseased plants were relatively small with thicker walls. The pith region had large cells in middle region with less developed conducting tissue, *viz.*; phloem and xylem (Figure 42.1). Almost similar observations had also been made in tomato root infected with TMV. The epidermal cells were smaller in size and had large pith region (Dubey and Bhardwaj, 1982). The urdbean leaf crinkle virus caused reduction in cell size of phloem and xylem in urdbean. Epiblema, cortex endodermis, medullary rays and pith were found thickened (Sharma and Dubey, 1985).

Anatomical Observations in Stem

In diseased stem, the xylem vessels were fewer narrow and reduced in number. Phloem necrosis was observed in infected stem (Figure 42.2). The cortical parenchyma was wider and formation of mechanical and conducting tissues was reduced. The cell wall of the infected ground tissues was reduced in thickness.These results were found to be in conformity with the findings reported in groundnut infected with mosaic rosette virus (Singh, 1970). There was found to be marked reduction in the size of all the tissue of *Solanum khasianum* infected with green vein banding virus. Vascular bundles were reduced in sized and pith region greatly affected (Garg *et al.*, 1977). The cortical region was enlarged in tomato stem infected with TMV (Dubey and Bhradwaj, 1982).

Table 42.1: Thickness of Epidermal Cells of Healthy and Diseased Leaves (Mean of three replicate)

Sl.No.	*Healthy*		*Diseased*	
	Upper Epidermal Cell + Cuticle (μm)	*Lower Epidermal Cell + Cuticle (μm)*	*Upper Epidermal Cell + Cuticle (μm)*	*Lower Epidermal Cell + Cuticle (μm)*
1.	51.25	43.10	38.35	35.10
2.	50.50	41.32	36.20	32.26
3.	50.10	40.25	35.66	31.25
4.	49.41	40.20	35.43	30.45
5.	49.25	40.10	34.50	30.40
6.	46.50	39.80	34.12	30.10
C.D at 5%	4.35	7.28	2.51	4.8

Table 42.2: Thickness of Palisade Cells of Healthy and Diseased Leaves (Mean of three replicate)

Sl.No.	*Thickness in μm*	
	Healthy	*Diseased*
1.	143.50	136.68
2.	142.75	132.45
3.	141.75	132.32
4.	141.20	130.34
5.	140.61	128.25
6.	140.50	120.10
C.D at 5%	1.52	2.28

Anatomical Observations in Leaves

Epidermal cells and the tissues of the midrib region especially on dorsal surface of infected leaf were degenerate (Table 42.1). The development of midrib was irregular. In vascular region the xylem parenchyma, xylem vessels and phloem were greatly reduced. Leaf tissue consisted in replacement of the spongy parenchyma by a palisade parenchyma (Figure 42.3). The palisade cells of diseased leaves

Figure 42.1: Tomato Leaves Infected with TLCV

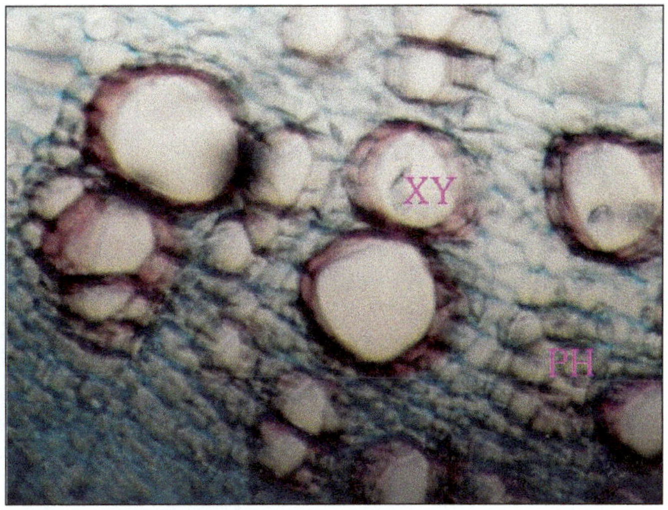

Figure 42.2: T.S. of Healthy Root

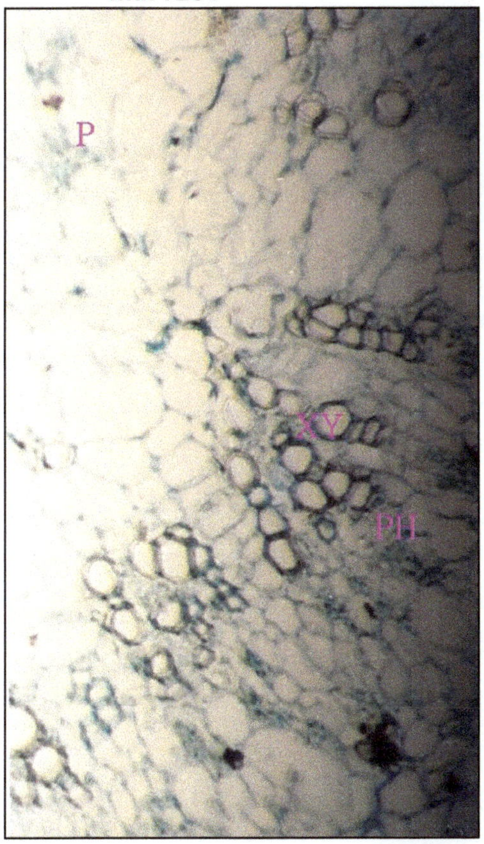

Figure 42.3: T.S. of Diseased Root Showing Large Pith Region (P) and Reduced Xylem (XY), Phloem (PH) Bar: 130 µm

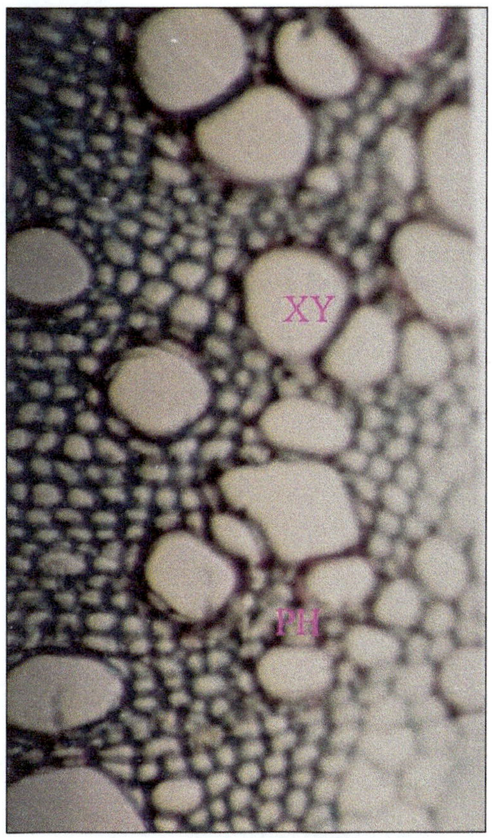

Figure 42.4: T.S. of Health Stem Bar: 110 µm

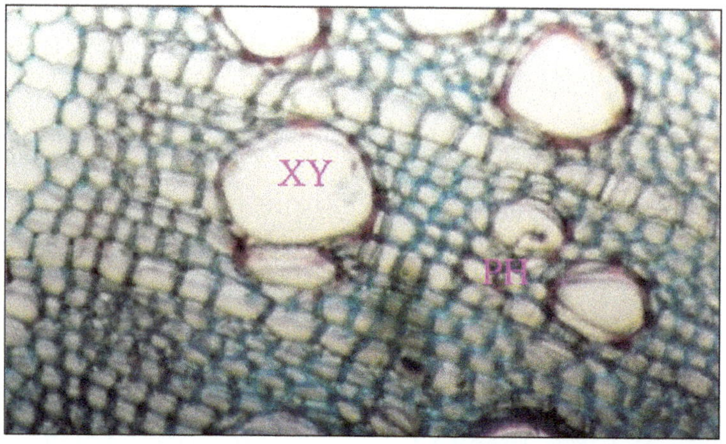

Figure 42.5: T.S. of Diseased Stem Showing Enlarge Reduced Xylem (XY), Necrosis of Phloem (PH)

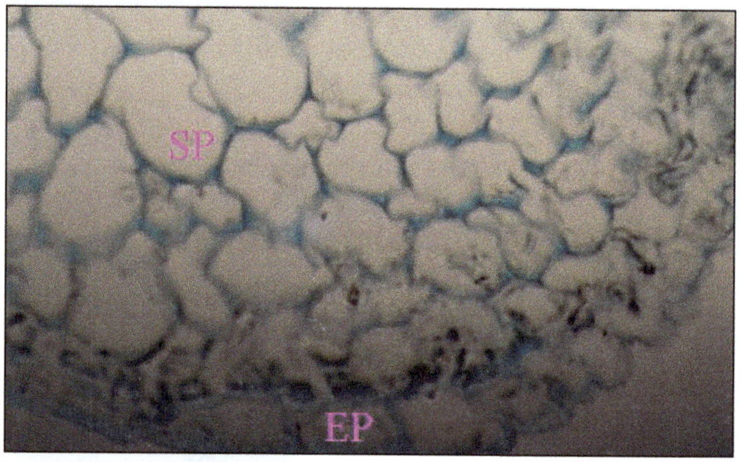

Figure 42.6: T.S. of Health Leaf Showing Spongy Parenchyma (SP)

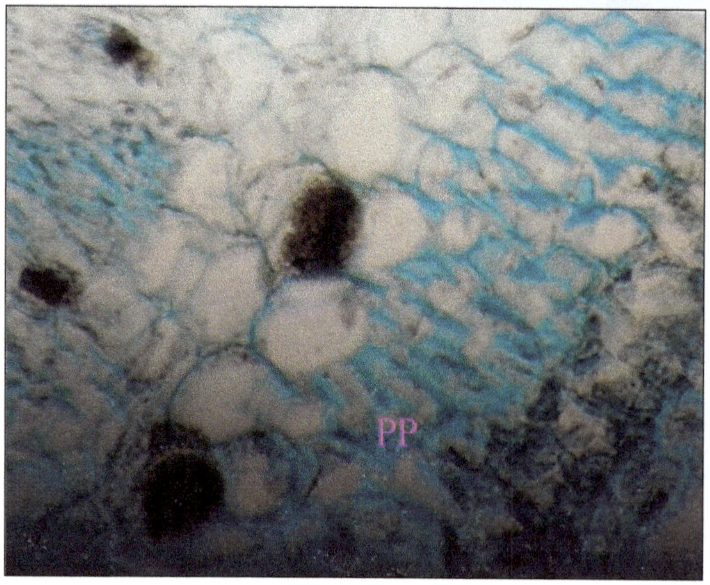

Figure 42.7: T.S. of Diseased Leaf Showing Degenerating Epidermal Cells (EP) and Thin and Less Densely Palisade Parenchyma (PP)

were thinner (Table 42.2) and less densely packed with chloroplast. The significant changes were shown in spongy tissues. The average stomatal frequency and stomatal index were significantly lower in diseased leaves, although their size was increased (Table 42.3). The chloroplasts were few and showed irregular shape. The epidermal, palisade and spongy cells reduced in *Vicia faba* infected by tomato ring spot virus (TRSV) (Smith and Mc Whorter, 1957). The leaf tissues consisted in replacement of the spongy parenchyma by palisade parenchyma and cambial activity was observed in the main phloem parenchyma of *Athea rosea* infected with Begmovirus (Bigare *et al.* 2001). In vascular region the phloem, xylem parenchyma and xylem vessels were greatly influenced in diseased tomato leaves with TMV (Dubey and Bhardwaj, 1982). The changes in colour as primary symptoms due to chlorosis seems to be due to the disintegration and disruption of chloroplasts in tobacco leaves infected by tobacco mosiac virus (Carrol and Kosuge, 1969).

Table 42.3: Thickness of Palisade Cells of Healthy and Diseased Leaves (Mean of three replicate)

Sl.No.	*Healthy Size of Stomata (μm)*				*Diseased Size of Stomata (μm)*			
	SF	*Length*	*Breath*	*SI*	*SF*	*Length*	*Breath*	*SI*
1.	617	17.95	9.18	26.07	483	23.63	14.45	31.4
2.	589	17.85	9.03	25.74	461	21.93	12.75	30.92
3.	578	17.34	7.65	25.54	440	21.42	12.11	29.86
4.	557	16.83	7.65	25.52	428	21.42	11.47	28.86
5.	515	15.30	7.65	25.15	373	20.90	10.83	27.93
6.	419	15.00	7.65	24.94	367	20.40	10.20	27.84
7.	431	14.00	7.30	24.28	317	20.20	9.60	21.53
C.D at 5%	42.37	1.38	0.92		30.18	2.56	1.84	

Acknowledgement

The authors thank to the Head, Department of Botany, M.L.K.P.G. College, Balrampur (U.P.) for providing library and laboratory facilities. Thanks are also due to Dr. D. S. Shukla and Miss. Anjana Shukla for their help during experimentation.

References

Bigare, L., Chazly, M., Salah, M., Ibrahim, M., Padidam, M., Nicole, M., Peterschmitt, Fauquet, C. and Thouvenel, J.C., 2001. Characterization of a new Begmovirus from Egypt infecting hollyhock (*Athea rosea*). *European J. Plant Pathology*, 107: 70–71.

Carrol, T.W. and Kosuge T., 1969. Changes in structure of chloroplast accompanying necrosis of tobacco leaves systemically infected with tobacco mosaic virus. *Phytopath.*, 59: 953–962.

Chakraborty, S., Pandey, P.K., Banerjee, M.K., Kaloo, G. and Fauquet C.M., 2003. Tomato leaf curl Gujarat virus, a new Begmovirus species causing a severe leaf curl disease of tomato in Varanasi, India. *Phytopath.*, 93(12): 1485–1495.

Cozosnek, H., Ber, R., Antignus, Y., Cohen, S., Navot, N. and Zamiri, D., 1988. Isolation of tomato leaf curl from India. *Plant Dis.*, 77: 168.

Dubey, G.S. and Bhardwaj S.V., 1982. Histopathological studies in tomato infected with tobacco mosaic virus. *Indian Phytopath.*, 35: 175–177.

Garg, R.C., Chowela, S.C. and Tyagi, S.N.S., 1977. Histopathological changes induced by green vein banding virus in *Solanum khasianum* Clarke. *Indian J. Plant Pathology*, 5(2): 129–132.

Johensen, D.A., 1940. *Plant Microtechnique*. McGraw Hill Book Co.

Saikia, A.K. and Muniyappa, V., 1989. Epidemiology and control of tomato leaf curl virus. In: *Abstract Nat. Seminar on Whitefly Transmitted Plant Virus Diseases*. IARI, New Delhi, p. 30.

Saikia, A.K. and Muniyappa, V., 1986. Epidemiology and control of tomato leaf curl virus in South India. *Trop. Agric.* (*Trinidad*), 66(4): 350–354.

Salisburg, E.J., 1928. On the causes and ecological significance of stomatal frequency with special reference to wood land flora. *Phil. Trans. Res. Soc.*, 216: 1–65.

Sastry, K.S.M. and Singh, S.J., 1973. Assessment of losses in tomato caused by tomato leaf curl virus. *Indian J. Mycol. Plant Pathol.*, 3: 50–54.

Sharma, I. and Dubey, G.S., 1985. Anatomy of urdbean leaf crinkle virus infected urdbean root. *Indian J. Plant Pathology*, 3(2): 236–237.

Singh, A., 1970. Studies on mosaic rosette of groundnut *Arachis hypogea* L. *Ph.D. Thesis*, Agra University, Agra, India.

Smith, F.H. and Mc Whorter F.P., 1957. Anatomical effects of tomato ring spot virus in *Vicia faba*. *Am. J. Bot.*, 44: 470–477.

Vasudeva, R.S. and Samraj, J., 1948. A leaf curl disease of tomato. *Phytopath.*, 38: 364–369.

Yassin, A.M. and Nour, M.A., 1965. Tomato leaf curl disease in Sudan and their relation to tobacco leaf curl. *Ann. App. Bio.*, 5: 207–217.

Ethnobotanical Wisdom and Microbial Studies on Medicinal Plants (2011) *Pages* **301–308**
Editors: **D.R. Khanna, Ashutosh Gautam, R. Bhutiani, Gagan Matta, Anil Dhiman**
Published by: **DAYA PUBLISHING HOUSE, NEW DELHI**

Chapter 43

Three Lignicolous Macrofungi from District Doda of Jammu Province (J&K), India

Sanjeev Kumar and Y.P. Sharma
Department of Botany, University of Jammu, J&K,

ABSTRACT

Three lignicolous macrofungal species, namely, *Fomes fomentarius* (L.ex Fr.) Fr., *Fomitopsis pinicola* (Fr.) Kar and *Fomes igniarius* (L.) Fr. belonging to aphyllophorales, collected from Kishtwar and Bhadarwah regions of district Doda of Jammu province (J and K), India, have been described in the present communication. Their taxonomic details, macro and micro-morphological characters have also been included. Critical microscopic observations and persual of literature revealed that *Fomes igniarius* is new addition to the macrofungal flora of Jammu and Kashmir while as *Fomes fomentarius* and *Fomitopsis pinicola* are the first authentic records for the Jammu province.

Keywords: *Aphyllophorales, Taxonomy, Fomes fomentarius, F. igniarius, Fomitopsis pinicola.*

Introduction

Doda, the largest district in Jammu Province of Jammu and Kashmir state of India, lies in the Pir Panjal range of North-West Himalaya (Kashmir Himalaya) between 74° 30'–76° 30′ E longitude and 32° 30'- 34° 15' N latitude covering an area of 11,691 sq km of which 5,555 sq km comes under forest cover (DOS, 2002). The region is mainly mountainous and vegetation is dominated by coniferous and mixed forests with sub-tropical to temperate climate with average annual rainfall of 107-150 cm. The predominant plant species are *Cedrus deodara* (Roxb.) G. Don, *Pinus wallichiana* A.B. Jackson, *Picea smithiana* (Wallich.) Boiss., *Abies pindrow* Royle, *Quercus* spp. L., *Juglans regia* L., *Alnus nepalensis* D. Don, *Ulmus wallichiana* Planch etc. Though the study area is endowed with vast phytodiversity including macrofungal resources wealth, hardly any report on the occurrence and distribution of lignicolous macrofungi from this region exists in the literature. Therefore, the present study was carried out with an objective to find out the lignicolous macrofungal species in the study area and thus provide more data on the lignicolous macrofungal flora of the region.

Materials and Methods

The collected specimens have been described and illustrated based on the field study of the fresh specimens. For the collection of these fungi, standard methods of collection, preservation, macro and microscopic studies were followed (Atri *et al.*, 2003; Kumar *et al.*, 1990; Major, 1974; Smith *et al.*, 1981) the shape, size, and colour of fresh specimens were recorded before preservation. All the measurements were taken and illustrations were made with the aid of Camera Lucida (Erma, Japan). Reagents used during microscopic analysis were 3 per cent KOH, lactophenol, cotton blue, 1 per cent phloxine and Melzer's Reagent. Crystals of 1,4-dichlorobenzene were used against insect infestation. The examined specimens have been deposited in the herbarium of Botany Department, University of Jammu, Jammu with accession numbers. For three-dimensional photography, the stereo Nikon camera (SMZ 800, Japan) was used.

Results and Discussion

1. *Fomes fomentarius* (L.ex Fr.) Fr.

Synonymy

Agaricus fomentarius (L.) Lam. (1783)

Boletus fomentarius L., Sp. *Plantarum*: 1176 (1753)

Polyporus fomentarius (L.) Fr., *Syst.mycol.* (Lundae) 1: 374 91821)

Ungulina fomentaria (L.) Pat., *Essai Tax. Hymenomyc.*: 102(1900)

Collection Examined

Jammu and Kashmir, Kandail forest area, Paddar, Kishtwar, lignicolous, growing on wood of *Cedrus deodara* (Roxb.) G. Don, scattered to gregarious, coniferous forest, Sanjeev Kumar and Y.P. Sharma, JUH 9637, July 18, 2006.

Carpophore 14-20 cm, bracket shaped or hoof shaped, dimidiate to ungulate, sessile, upper surface smooth, greyish to grey brown with faint semi-circular markings, margins cream coloured, hazel or light brown, perennial; Context antique brown to cinnamon brown; hyphal system trimitic, generative hyphae 1.6-3.2 μ, branched, clamped, rarely septate, skeletal hyphae 3.2-8.0 μ, thick walled, hyaline, branched, binding hyphae 4.8-6.4 μ, thick walled, branched, pores 150-240 x 120-255 μ, small, round, light brown, flesh brown, thick, suberose, soft, a hard thick crust, shiny below, blackish gray in cross section, basidiospores 4.8-12.8 x 3.2-8.0 μ, ellipsoidal, smooth, hyaline (Figure 43.1–43.7).

Habitat and Distribution

Fomes fomentarius is inedible, worldwide in distribution and has been reported from Africa, Asia, Europe and North America (Sinclair *et al.*, 1987). In India, it was recorded growing on twig of *Aesculus indica* Coleb. ex Cambess and *Juglans regia* from Dehradun (Bakshi *et al.*, 1971); on *Cedrus deodara* and

Figure 43.1: ***Fomes fomentarius*** **on Tree Trunk of** ***Cedrus deodara***

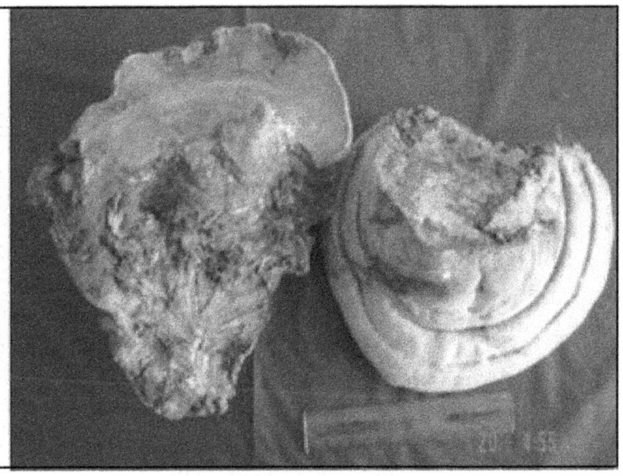

Figure 43.2: Carpophore of ***Fomes fomentarius***

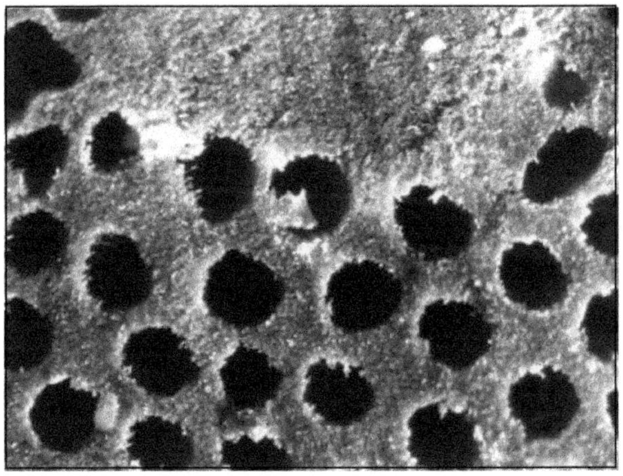

Figure 43.3: Pores of ***Fomes fomentarius***

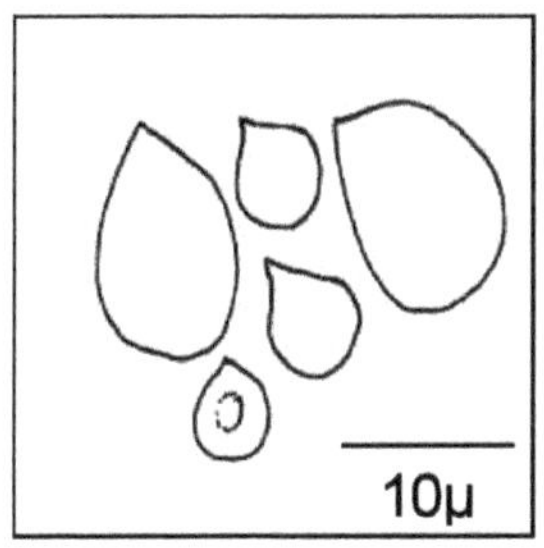

Figure 43.4: Basidiospores of ***Fomes fomentarius***

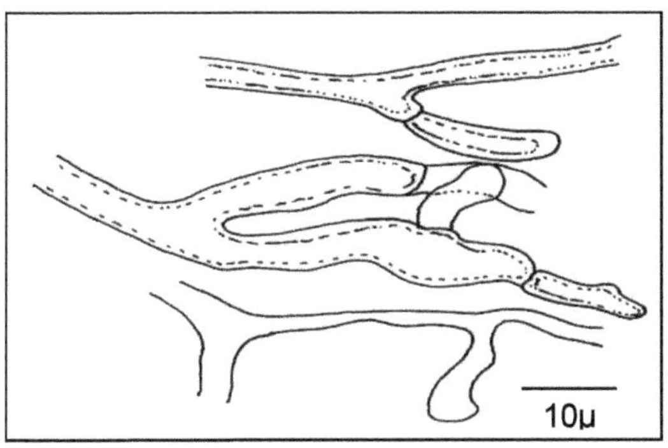

Figure 43.5: Generative hyphae of ***Fomes fomentarius***

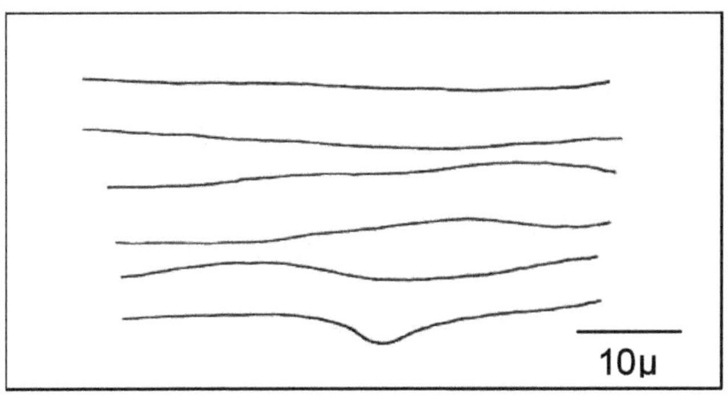

Figure 43.6: Skeletal hyphae of ***Fomes fomentarius***

Pyrus sp. L. from Himachal Pradesh (Thind and Rattan, 1971) and on *Betula* sp. Tourn. from Kashmir, Jammu and Kashmir (Abraham, 1991; Llyod, 1898-1925). This fungus is very well used by local population for fire production, decorative items etc. On comparing the taxonomic details of the present specimen with Teng (1988) and Schwarze (1994), it was close to their description except spore length.

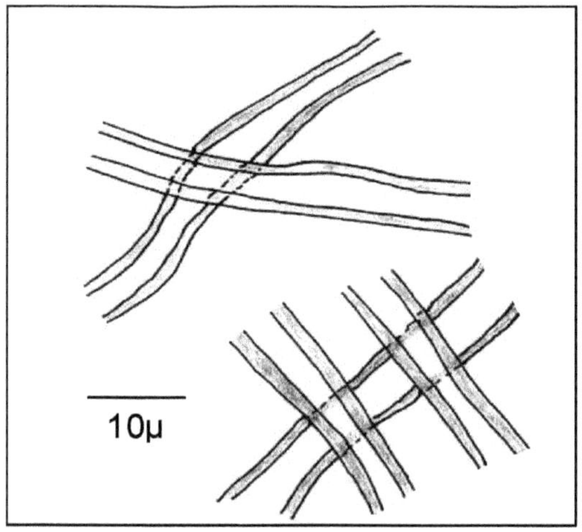

Figure 43.7: Binding hyphae of *Fomes fomentarius*

2. *Fomes igniarius* (L.) Fr.

Synonymy

Phellinus igniarius (L.) Quel. (1886)

Collection Examined

Jammu and Kashmir, Bhadarwah, Ramtund forest area, lignicolous, growing on wood of *Cedrus deodara*, scattered, Sanjeev Kumar and Y.P.Sharma, JUH 9641, January 10, 2006.

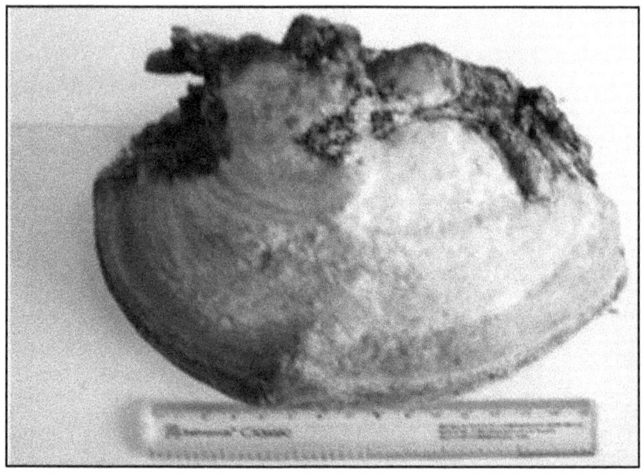

Figure 43.8: Carpophore of *Fomes igniarius*

Carpophore 15x16.5 cm, applanate, woody, hard, cinnamon brown, sessile, bracket like, perennial, pores 135-225 x 210-300 µ, oval to elongated; margins obtuse, sterile, ochraceous tawny to sudan brown; context argus brown, hard woody; hyphal system dimitic, generative hyphae 2.4- 8.0 µ, wide, thin to thick walled, septate, clamped, hyaline, skeletal hyphae 1.6-6.4 µ, wide, thick

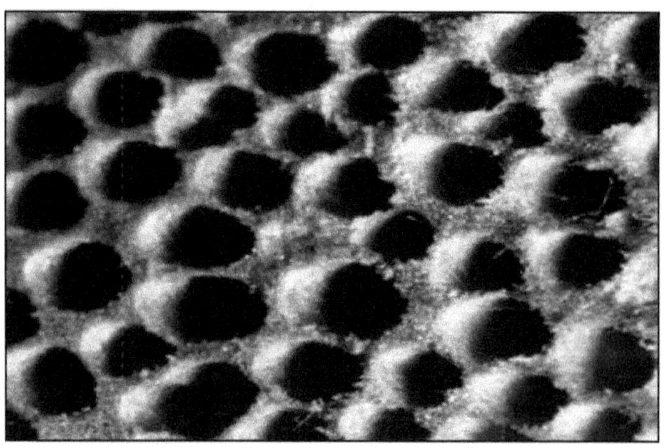

Figure 43.9: Pores of *Fomes igniarius*

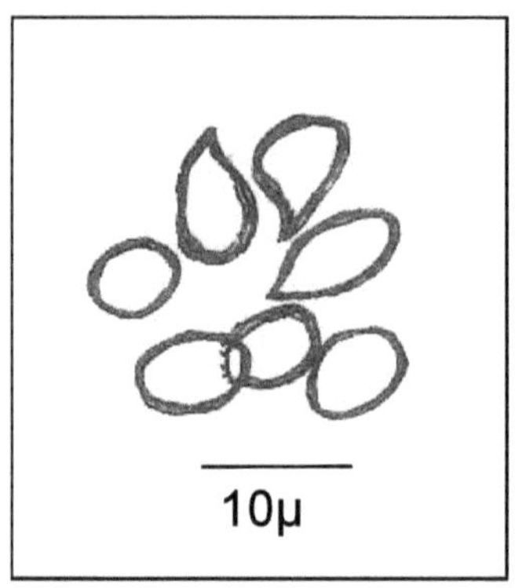

Figure 43.10: Basidiospores of *Fomes igniarius*

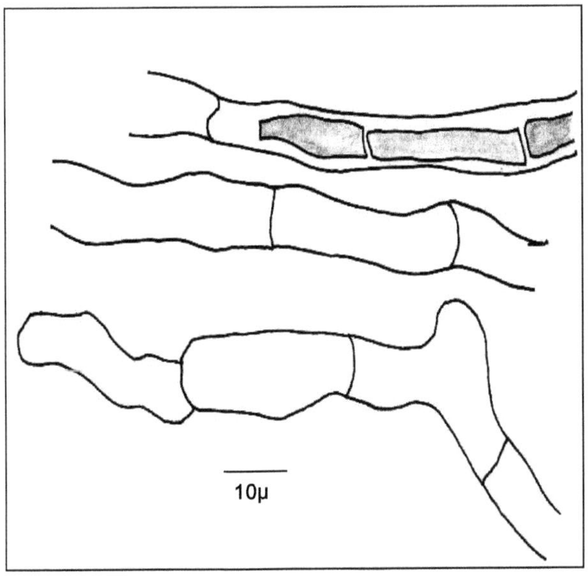

Figure 43.11: Generative Hyphae of *Fomes igniarius*

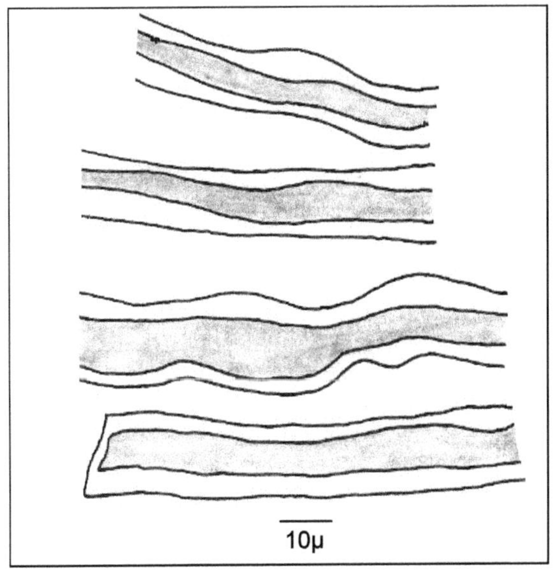

Figure 43.12: Skeletal Hyphae of *Fomes igniarius*

walled, aseptate, branched, basidiospores 4.8-6.4 x 1.6-3.2 µ, ellipsoidal to subglobose with thick walls (Figures 43.8–43.12).

Habitat and Distribution

Recorded on dead trees of *Picea morinda* Link. and *Abies pindrow* from Darjeeling, West Bengal Berkeley (1856). The above examined specimen resembles with the taxonomic details given by Natrajan and Kolandavelu (1998) and differs only with respect to basidiospore length.

3. *Fomitopsis pinicola* (Fr.) Kar.

Synonymy

Antrodia tuber (P.Karst.) P.Karst., *Finl. Basidsvamp.* (11): (1898)

Boletus fulvus Schaeff., *Fung. Bavar. Palat.* 4: 89 (1774)

Boletus pinicola Sw., *Sevenska Vet.Acad.hand.*, 1852: 88 (1810)

Fomes pinicola (Sw.) Fr. *Summa veg. Scand.*, Section Post. (Stockholm) (1849)

Polyporus pinicola (Sw.) Fr., *Syst.mycol.* (Lundae) 1: 372(1821)

Ungulina marginata (Fr.) Pat., *Essi Tax. Hymenomyc.*: 103 (1900)

Collection Examined

Jammu and Kashmir, Kishtwar, Chishoti village of Machail Padder, solitary, lignicolous, on wood of *Juglans regia* L. in mixed forest of *Juglans regia* L. and conifers, Sanjeev Kumar and Y.P. Sharma, JUH 9638, July19, 2006.

Carpophore 12 cm long and 11cm wide, hoof shaped, sessile, thick and flattened, yellowish turning slightly reddish brown and finally brownish black with semi-circular concentric markings, blackish crust shiny or pruinose, margins yellowed, perennial, pores 105 x 210 µ, irregular, cream or light brown sometimes reddish when rubbed, hyphal system dimitic, generative hyphae 4.0-20.0 μ wide, septate, thin walled to thick walled, branched with clamp connections, skeletal hyphae 1.6-6.4 μ wide, thick walled, branched, without clamps, slightly flexuous, binding hyphae 3.2-8.0 μ, wide, hyaline, aseptate, unbranched, basidiospores 4.8-11.2 x 2.4-6.4 μ, pale yellow to white, smooth, elliptical, chlamydospores 4.8 -14.4 x 3.2-11.2 μ, thick walled (Figures 43.13–43.19).

This species resembles with the taxonomic details given by Natrajan and Kolandavelu (1998) and differs only with respect to basidiospore length.

Habitat and Distribution

Fomitopsis pinicola is inedible and has been reported only from Kashmir in India by Abraham (1991). It is usually parasitic on broad–leaved and coniferous species but also grows as a saprophyte on dead tree trunks.

Figure 43.13: *Fomitopsis pinicola* on Tree Trunk of *Juglans regia*

Figure 43.14: Carpophore of *Fomitopsis pinicola*

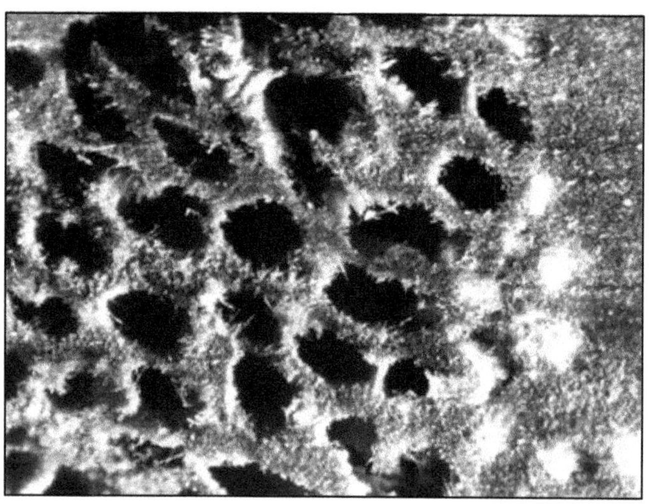

Figure 43.15: Pores of *Fomiitopsis pinicola*

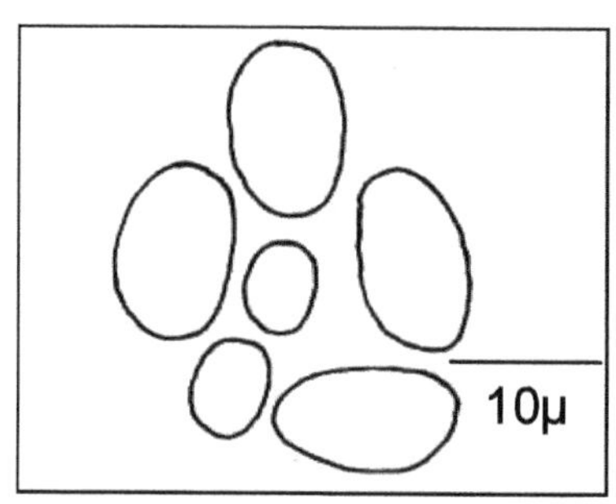

Figure 43.16: Basidiospores of *Fomitopsis pinicola*

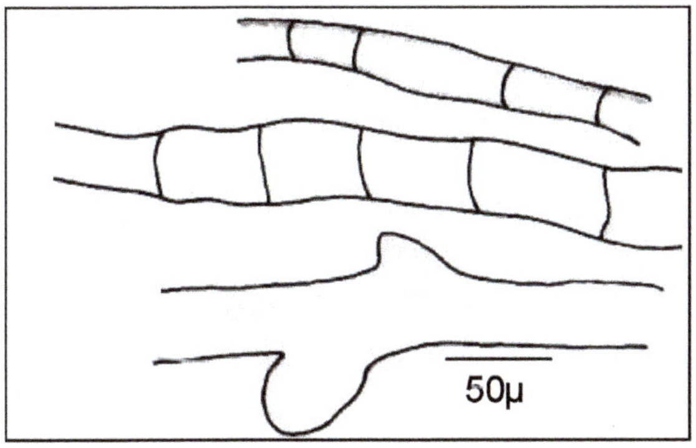

Figure 43.17: Generative Hyphae of *Fomitopsis pinicola*

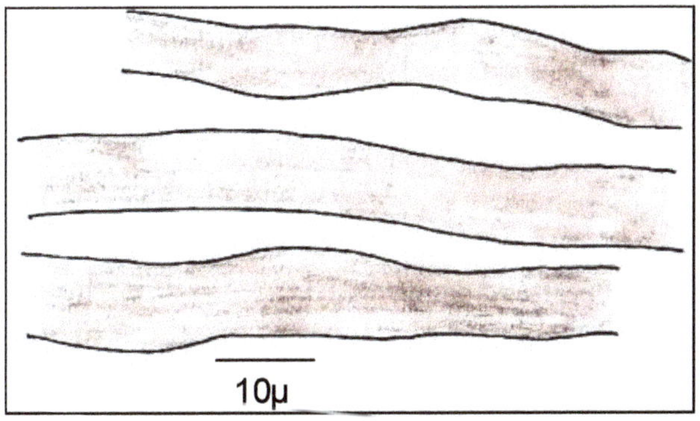

Figure 43.18: Skeletal Hyphae of *Fomitopsis pinicola*

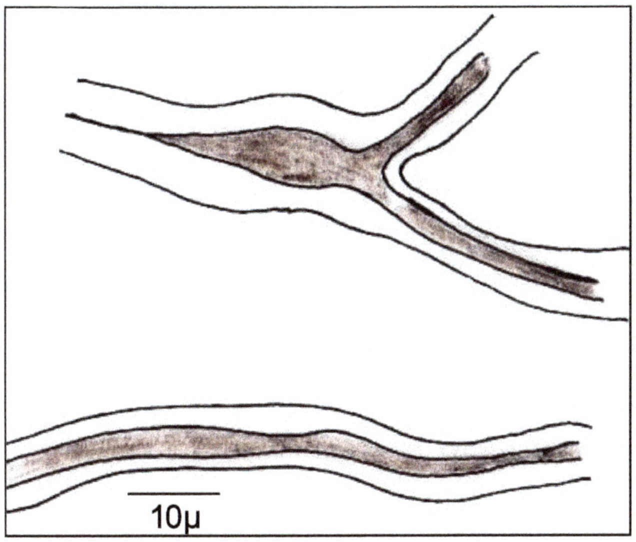

Figure 43.19: Binding Hyphae of *Fomitopsis pinicola*

Survey of literature (Bilgrami *et al.*, 1979, 1981, 1991; Jamaluddin *et al.*, 2004) shows that out of these three species, *Fomes igniarius* is new from Jammu and Kashmir state, whereas, *Fomes fomentarius* and *Fomitopsis pinicola* constitute the first authentic reports from the Jammu Province.

References

Abraham, S.P., 1991. Kashmir fungal flora: An overview. *Indian Mushrooms*, 13: 13–24.

Atri, N.S., Kaur, A. and Kaur, H., 2003. *Wild Mushrooms: Collection and Identification*. NRCM, Chambaghat, Solan, pp. 16.

Bakshi, B.K., Reddy, M.A.R., Puri, Y.N. and Singh, S., 1971. *Survey of the Disease of Important Native and Exotic Forest in India*, Pl. 480 Report, FRI, Dehradun.

Berkeley, M.J., 1856. Decades of Fungi, Decas 1–62 Nos. 1–620. In: Hooker London. *Journal of Botany*, 3–8: 1844–1856.

Bilgrami, K.S., Jamaluddin and Rizvi, M.A., 1979. *The Fungi of India,* Part I. Today and Tomorrow's Printers and Publishers, New Delhi, pp. 467.

Bilgrami, K.S., Jamaluddin and Rizvi, M.A., 1981. *The Fungi of India,* Part II. Today and Tomorrow's Printers and Publishers, New Delhi, pp. 128.

Bilgrami, K.S., Jamaluddin and Rizvi, M.A., 1991. *The Fungi of India.* Today and Tomorrow's Printers and Publishers, New Delhi, pp. 798.

DOS, 2002. *Digest of Statistics.* Government of Jammu and Kashmir, pp. 440.

Jamaluddin, Goswami, M.G. and Ojha, B.M., 2004. *Fungi of India 1989–2001.* Scientific Publishers, India, Jodhpur, pp. 326.

Kumar, A., Bhatt, R.P. and Lakhanpal, T.N., 1990. *The Amanitaceae of India.* Bishen Singh Mahendra Pal Singh. Dehradun. pp: 160.

Lloyd, C.G., 1898–1925. *Mycological Notes,* 1–75, 1–1364 Cincinnati, Ohio.

Major, A., 1974. *Mushrooms and Toadstool and Fungi.* John Bartholomew and Sons Limited, Edinburgh and London, pp. 263.

Natrajan, K. and Kolandavelu, K., 1998. *Resupinate Aphyllophorales of Tamil Nadu.*CAS in Botany, University of Madras. pp: 131.

Schwarze, F., 1994. Wood rotting fungi: *Fomes fomentarius* (L. ex Fr.) Fr. *Mycologist,* 8: 32–34.

Smith, A.H., Smith, H.V. and Weber, N.S., 1981. *How to Know Non-gilled Mushrooms.* Wm.C. Brown Company Publisher, USA. pp. 317.

Sinclair, W.A., Lyon, H.H. and Johanson, W.T., 1987. *Diseases of Trees and Shrubs.* Cornell University Press. pp: 346–347.

Teng, S.C., 1988. *A Contribution to Our Knowledge of the Higher Fungi of China.* Bishen Singh Mahendra Pal Singh, Dehradun, pp. 364.

Thind, K.S. and Rattan, S.S., 1971. The Polyporaceae of India–VIII. *Research. Bulletin of Punjab Univ.,* 22: 27–34.

Ethnobotanical Wisdom and Microbial Studies on Medicinal Plants (2011) *Pages* **309–311**
Editors: **D.R. Khanna, Ashutosh Gautam, R. Bhutiani, Gagan Matta, Anil Dhiman**
Published by: **DAYA PUBLISHING HOUSE, NEW DELHI**

Chapter 44

Population Dynamics of Mycoflora in BLSB Susceptible Maize Varieties

Nishat Asif and T.P. Mall
P.G. Department of Botany, Kisan P.G. College, Bahraich, U.P.

ABSTRACT

Seed mycoflora of three varieties susceptible to Banded leaf and Sheath blight disease of maize was isolated from fresh harvest of 2007 following agar plate method. The results indicated the presence of fifteen species belonging to eleven genera. Highest percentage of *Aspergillus niger* was recorded in all tested varieties. The effect of dominant fungi on seed germination, seedling growth, and their fresh and dry weight and vigour index was recorded. It showed that highest (94 per cent) per cent of seed germination was observed in control.

Keywords: *Maize, Seed mycoflora, Banded leaf and sheath blight, Susceptible, Rhizoctonia solani.*

Introduction

Maize (*Zea mays* L.) is an important cereal crop of the world. In India maize occupies fifth place in area and fourth in production among the major cereals grown. Poor seed germination, seedling growth and number of diseases are responsible for the low productivity. Seed borne mycoflora is a term indicating the association of fungi with seeds or parasitic, easily penetrating into the seed causes instant death or delayed systemic infestation of the emerging seedling. In India maize suffers from a number of diseases *viz.*, Rust and Smut of maize, Charcoal rot, Sclerotial blight (Banded leaf and sheath blight), Bacterial stalk rot, Turcicum leaf blight etc. Among these Banded leaf and sheath blight and bacterial stalk rot are economically important. Banded leaf and sheath blight disease is caused by the most versatile and dreaded pathogen *Rhizoctonia solani* (Kuhn). Banded leaf and sheath blight causes 15-20 per cent yield losses annually (Saxena, 2002).

The yield and emergence of seeds are known to be affected by a number of factors among which seed borne microbes are prominent. The longevity of seed is determined largely by fungi. Extensive work has been done on seed mycoflora of different crops in India (Singh and Chauhan, 1973; Borah *et al.*, 2003; Das *et al.*, 2003; Wattal and Puttoo, 2003).

Present study was carried out to check the seed mycoflora of some varieties popularly grown by maize farmers in district Bahraich, the percentage occurrence of isolated mycoflora and their effect on germination, seedling growth and vigour.

Materials and Methods

Seeds of three popular Cv. of maize *viz.*, Amar, Sweety and Suguna were collected from different seed lots in different agroecological regions of Bahraich district following ISTA (1993). These Cv. are susceptible to Banded leaf and sheath blight disease. The freshly harvested seeds of maize varieties 2007 were used as seed samples. For isolation of all possible mycoflora associated with 100 surface sterilized seeds by 2 per cent Sodium hypochlorite (NaOCl) and 100 unsterilized seeds were plated by following standard technique, agar plate method (APM). Observations were done after seven days of incubation at temperature 25°C. Number of colonies of each fungus were counted and percentage occurrence of each fungus was calculated.

For testing the pathogenic behaviour of dominant fungal forms towel paper method was followed (ISTA, 1993). 1×10^6 cfu of *Aspergillus niger*, *Aspergillus flavus* and *Aspergillus fumigatus* were inoculated with surface sterilized seeds. Observations were recorded after seven or eight days of incubation. Control was also maintained for comparison. Germination percentage, fresh and dry weight of root and shoot along with their length was observed. Vigour index was also calculated following Abdul-Baki and Anderson (1973).

Results and Discussion

The result reveals fifteen species associated to all tested varieties at varying level of incidence. Ten species were detected from Cv., Amar; nine from Sweety; ten from Suguna.

Cv.Amar

Aspergillus niger, *Aspergillus fumigatus*, *Aspergillus flavus*, *Aspergillus ochraceous*, *Alternaria tenuis*, *Curvularia lunata* were associated internally. *Rhizopus nigricans*, *Penicillium citrinum*, *Fusarium roseum* and *Phoma* sp. were associated externally.

Cv.Sweety

A.niger,*A.fumigatus*,*A.tenuis*,*C.lunata*,*R.nigricans* were associated externally while *A.flavus*,*F.roseum*,*Phoma* sp. were found associated internally.

Cv.Suguna

A.niger.*A.fumigatus*,*A.alternata*,*Dreschelra* sp.,*C.lunata*,*Mucor variens* were externally associated and *A.flavus*,*P.citrinum*,*F.roseum*,*Monilia* sp. were internally associated with seeds.Neither a single colony of *Rhizoctonia solani* was observed in all tested varieties.

The effect of seed borne fungi on germination showed the maximum (94 per cent) in control and minimum (74 per cent) in *A.flavus* inoculated seeds. Highest root length (10.0cm) was observed in

A.flavus and lowest (6.8cm) in *A.niger* inoculants. Highest shoot length (7.0cm) was recorded in *A.fumigatus* and lowest (3.2cm) in control.

Fresh and dry weight was maximum (0.269, 0.036) in *A.flavus* and minimum (0.128, 0.008) in control respectively. Fresh weight of root was recorded maximum (1.004) in *A.fumigatus* inoculants and minimum (0.070) in *A.niger* by causing seed rotting and seedling mortality. *Aspergillus niger* showed minimum effect on seed germination and seedling growth. *A. flavus* was pathogenic causing seed rotting and seedling mortality.

Loss in germinability have been reported by several workers (Kumhar *et al.*, 1987; Mehrotra *et al.*, 1992; Mondal *et al.*, 1981).

Acknowledgement

The authors are thankful to Dr. S.P. Singh Principal, Kisan P.G. College, Bahraich for providing necessary library and laboratory facilities.

References

Abdul-Baki, A. and Anderson, J.D., 1973. Vigour determination in soybean by multiple criteria. *Crop Sci.*, 13: 630–33.

Borah, H., Melhi, B.N. and Bora, K.N., 2003. Studies on Seed mycoflora of flood prone lowland rice genotypes. *Indian Phytopath.*, 56(3): 345

Das, K., Das, R., Barua, P. and Bora, K.N., 2003. Seed Mycoflora of Mezankari and its control. *Indian Phytopath.*, 56(3): 345.

ISTA, 1993. International rules for Seed Testing. *Seed Sci. Technol.*, 21: 296.

Kumhar, G.R., Agnihotri, J.P. and Gupta, A.K., 1987. Seed mycoflora of chickpea (*Cicer arietinum*) L. their effect on germination and vigour of seedlings and their control. *Indian Bot. Reptr.*, 6(2): 87–94.

Mehrotra, R.S., Rashmi, N. and Navneet, 1992. *Seed mycoflora of oilseed crop with special reference to mycotoxin production*.In: *Current Concept in Seed Biology* (Prof. S.N. Dixit Festschrift Volume), (Ed.) K.G. Mukherji. Naya Prakash Publishers, Kolkata, pp. 171–199.

Mondal, G.C., Nandi, D. and Nandi, B., 1981. Studies on deterioration of some oil seeds in storage. I. Variation in seed moisture infection and germinability. *Mycologia*, 83: 157–166.

Saxena, S.C., 2002. Bio-intensive integrated disease management of banded leaf and sheath blight of maize. In: *Proc 8th Asian Regional Maize Workshop*, held at Bankok, Thailand, August 5–8, pp. 380–388.

Singh, I. and Chauhan, J.S., 1973. Seed borne mycoflora of Methra (*Trigonella foenum-graecum* L.) and Kasuri-methi (*Trigonella carniculata* L.). *Indian Phytopath.*, 26: 749–751.

Wattal, S.K. and Puttoo, B.L., 2003. Fungal flora associated with Walnut (*Juglans regia*) and their effect on Kernel deterioration. *Indian Phytopath.*, 56(3): 310.

Ethnobotanical Wisdom and Microbial Studies on Medicinal Plants (2011) *Pages* **312–314**
Editors: **D.R. Khanna, Ashutosh Gautam, R. Bhutiani, Gagan Matta, Anil Dhiman**
Published by: **DAYA PUBLISHING HOUSE, NEW DELHI**

Chapter 45

Diversity in Phyllosphere Mycoflora Infected with Black Mildew

T.P. Mall and D.P. Singh
Post Graduate Department of Botany, Kisan P.G. College, Bahraich, U.P.

ABSTRACT

Correct identification of a fungus absolutely free from ambiguities is vital for its employment in applied discipline. The weeds and forest plants serve as reservoirs of leaf spot pathogens which on getting opportunity may spread to agricultural and horticulture plants. Keeping this view in mind the authors during January 9-12, 2008 while in Mahabaleshwar, surveyed the lush green vegetation for folicolous fungi on Jan 10, 2008. The authors collected eleven host plants being parasitized with nine fungi. Out of these two hosts *Syzygium* sp. Linn. (Myrtacear; HClo-48269) and *Flacourtia indica* Merrill (Flacourtiaceae; HClO- 48270) were found infected with novel species of *Meliola syzyginea* Singh and Mall sp. no. and *M. flacourticola* Singh and Mall sp. Nov. respectively. The fungal diversity of Phyllosphere mycoflora of the above said two hosts were worked out following standard methods. On *Syzygium* sp. Linn., *Meliola syzginea* Singh and Mall sp. Nov. only was representing frequency class D where as *Aspergillus niger* Van Tiegh, *A. flavus* L., *Curvularia* sp. (Waker) Boedifin was representing class C; *A. Terreus* Thom, *Scoratia* sp., *Mucor* sp. class B; *Fusarium* sp. Sheldon and *Penicillum* sp. class A. No fungus was found representing frequency class E. On *Flacourtia indica* Merrill, *Aspergillus niger* Van Tiegh only was found representing frequency class E where as *Meliola flacourticola* Singh and Mall sp. Nov. and *A. ustus* Thom and Church represented class D; *Trichoderma verdi* Pers. ex Fr., *Fusarium* sp. Sheldon. class C; *Rhizopus* sp. Ehrenb ex Corda, *Scoratia* sp. class B and *Trichoderma* sp., *Penicillium globarcum* class A.

Keywords: *Mycobial diversity, Phyllosphere, Black mildew, Species novum.*

Introduction

Correct identification of a fungus absolutely free from ambiguities is vital for its employment in applied discipline. The weeds and forest plants serve as reservoirs of leaf spot pathogens which on getting opportunity may spread to agricultural and horticulture plants. Keeping this view in mind the authors during January 09-12, 2008 while in Mahabaleshwar surveyed the lush green vegetation for follicolous fungi on January 10, 2008. The present study have been under taken to explore the diversity in Phyllosphere mycoflora Infected with Black Mildew.

Materials and Methods

The fungal diversity of Phyllosphere mycoflora of the two hosts *Syzygium* sp. and *Flacourtia indica* were worked out following standard methods as described by Mishra and Kanaujia, 1972a,b; Kanujia 1981, 1981a, 1982; Kanaujia and Singh, 1977, 1978; Singh and Mall, 2007, 2008. The mycoflora of phyllosphere was developed on PDA medium supplemented with antibiotic and incubated at 25±2°C after serial dilution of 1 ml washing of leaves surface as described by Ruinen (1961) and Kanaujia (1977). The samples were taken in 10 petriplates and observation were taken after the seven days. Morphotaxonomic determinations of taxa were done with the help of current literature and resident expertise available. New Holotypes have been deposited in HClO, IARL, New Delhi and all the isotypes retained in the departments herbarium for further reference.

Results and Discussion

The authors collected eleven host plants being parasitized with nine fungi. Out of these two hosts *Syzygium* sp. Linn. (Myrtaceae; HCIO- 48269) and *Flacourtia indica* Merrill (Flacourtiaceae; HCIO-48270) were found infected with novel species of *Meliola syzyginea* Singh and Mall sp. nov. and *M. flacourticola* Singh and Mall sp. nov. respectively. The fungal diversity of phyllosphere mycoflora of the above said two hosts were worked out following standard methods. On *Syzygium* sp. Linn., *Meliola syzginea* Singh and Mall sp. nov. only was representing frequency class D where as *Aspergillus niger* Van Tiegh, *A. flavus* L., *Curvularia* sp. (Waker) Boedifin was representing class C; *A. Terreus* Thom, *Scoratia* sp. *Mucor* sp. class B; *Fusarium* sp. Sheldon and *Penicillium* sp. class A. No fungus was found representing frequency class E. On *Flacourtia indica* Merrill, *Aspergillus niger* Van Tiegh only was found representing frequency class E where as *Meliola flacourticola* Singh and Mall sp. nov. and *A. ustus* Thom and Chruch represented class D; *Trichoderma verdi* Pers. ex Fr., *Fusarium* sp. Sheldon class C; *Rhizopus* sp. Ehrenb ex Corda, *Scoratia* sp. class B and *Trichoderma* sp., *Penicillium globarcum* class A. The moisture requirement, desiccation or waterlogging are important in determination of fungal population. Waid (1960) found that there is a great reduction in active growth of fungi whenever the moisture content is sufficiently high so as to reduce aeration.The different edaphic factors and cover vegetation influence mycoflora to an appreciable degree. At the same moment no single component may be spotted out as major microbial determinant. Environmental conditions and microbial composition interact in a very specific manner maintaining an equilibrium.

Acknowledgements

Authors are thankful to Dr. S.P. Singh Principal, Kisan P.G. College, Bahraich for providing facilities and to Prof. Kamal Emeritus Scientist, DST for helpful suggestions and encouragements.

References

Kanaujia, R.S., 1974. Studies on phyllosphere fungi. *Technology*, 2: 382–388.

Kanaujia, R.S., 1977. Observation on soil fungistasis *V. fungistasis* in relation to rhizosphere effect. *Acta Mycologia*, 13: 151–167.

Kanaujia, R.S., 1977. Studies on phyllosphere fungi. VII. Foliar application of urea on certain ornamentals. *Acta Mycologia*, 13: 281–288.

Kanaujia, R.S., 1978. Release of phytotoxic substances by decomposing roots of *Pennisetum typhoides* (Burm f.) stap of et Hubb. Their effect on soil fungi and crop seeds. *Plant and Soil*, 15: 75–85.

Kanaujia, R.S., 1981a. Studies on certain aspects of root surface fungi. I. Fungi on living roots of *Pennisetum typhoides* (Burum f.) stap of et Hubb. *Acta Mycologia*, 17: 5–25.

Kanaujia, R.S., 1981. Succession of fungi on decomposing *Pennisetum typhoides* (Burm f.) Stap of et Hubb. *Acta Mycologia*, 17: 27–40.

Kanaujia, R.S., 1982. Studies on certain aspects of root surface fungi. III. Effect of harvesting. *Acta Mycologia*, 18: 45–60.

Kanaujia, R.S. and Singh, C.S., 1978. Studies on certain ceological aspects of soil, fungi. Fungi in relation to locality type, cover vegetation and physico-chemical characters of the soils. *Sydowia, Annales Mycologici Ser II.*, 30: 112–121.

Kanaujia, R.S. and Singh, C.S., 1977. Distribution of fungi, bacteria and actinomycetes in relation to cropping pattern. *Environment India*, 1: 25–29.

Mishra, R.R. and Kanaujia, R.S., 1972. Studies on certain ecological aspect of soil fungi. *Trop. Ecol.*, 13: 5–11.

Mishra, R.R. and Kanaujia, R.S., 1972a. Studies on certain ecological aspect of soil fungi. Proceeding of the National Academy of Sciences, India, 42(B): 131–140.

Mishra, R.R. and Kanaujia, R.S., 1972b. Studies on soil fungistasis. II. *Abstract Indian J. Microbiol.*

Ruinen, J., 1961. The phyllosphere plant and soil, 15: 81–109.

Singh, D.P. and Mall, T.P., 2007. Mycobial diversity on certain aspects of sal forest. *Vegetos*, 20(2): 55–57.

Singh, D.P. and Mall, T.P., 2008. Mycobial diversity on certain aspects of teak forest internat. *J. Plant Sci.*, 3(1): 263–265.

Waid, J.S., 1960. The growth of fungi in soil. In: *Proc. Symp. Ecol., Soil Fungi*, Liverpool, 1958. Liverpool Univ. Press, pp. 55–75.

Ethnobotanical Wisdom and Microbial Studies on Medicinal Plants (2011) *Pages* **315–319**
Editors: **D.R. Khanna, Ashutosh Gautam, R. Bhutiani, Gagan Matta, Anil Dhiman**
Published by: **DAYA PUBLISHING HOUSE, NEW DELHI**

Chapter 46

In vitro: Antibacterial Activity of *Salsola kali* Linn.

Kumresh[1], Sanjeev[1], Navneet[2] and Prabhat[2]
[1]Department of Chemistry, Gurukula Kangri Vishwavidyalaya, Haridwar
[2]Department of Botany and Microbiology,
Gurukula Kangri Vishwavidyalaya, Haridwar

ABSTRACT

Plant based medicine were commonly used in India and all over world. Medicinal plants produce a variety of compounds of known therapeutic properties. The substances can either inhibit the growth of pathogens or kill them and have least toxicity to host cell are considered for developing new antimicrobial drugs. Plants are known to contain innumerable biologically active compounds, which possess antibacterial properties. Medicinal components from plants play an important role in conventional plant derived medicine. They have been a part of the evolution of human healthcare for thousands of years. Now a day, nearly eighty per cent of the global population turns to plant derive medicines as their first line of defence for maintaining health and combating diseases. Antibacterial activity of petroleum ether, ethanol and water extract of different parts of *Salsola kali* Linn. was studied against *E. coli*, *K. pneumoniae*, *S. aureus*, *B. subtilis* and *S. mutans*.

Keywords: *Antibacterial activity, Medicinal components, Medicinal plant, Salsola kali.*

Introduction

The ancient Ayurvedic literature describes *Salsola kali* Linn. ash as "Sarjjikakshara", used for the preparation of a large number of Ayurvedic compounds formulations (Gupta and Nath, 1984), which seems to correct the dysfunctions of liver (Kumar and Upadhyay, 1994). It has the properties of the fire. It is pungent, warm and acrid. It is efficacious in gulma, flatulence of stomach, diseases of belly, bile, worms, asthma, intestinal obstruction, enlargement of spleen and liver. The juice of the plant was said to be rich in soda and it shows excellent diuretic properties (Grieve, 1992). According to Hartwell the plant is used in folk remedies for the cancerous condition. Aerial parts of plant *Salsola kali* contained stigmasterol, campesterol, sitosterol and their glycosides (Prabhat *et al.*, 2005; Prabhat and Navneet, 2007). Medicinal herbs and their extract are widely used in the treatment of liver diseases like hepatitis, chlorosis and loss of appetite (Cupp, 1999). These exert physiological and therapeutic effect (Prabhat *et al.*, 2008). The compounds that are responsible for the medicinal properties of the drug are usually secondary metabolites. Antibacterial activity of petroleum ether, ethanol and water extract of aerial parts, stems, roots and seeds of *Salsola kali* Linn. was studied against *E. coli, K. pneumoniae, S. aureus, B. subtilis* and *S. mutans*.

Materials and Methods

The plant material was collected from the Vijay Nagar near Ganganagar in Rajasthan district. It was identified from Botanical Survey of India, Dehradun (Uttarakhand). The aerial parts, stems, roots and seeds of the plant were collected and dried at room temperature for one week, grounded and kept as dried powdered.

Powdered plant materials were extracted with three different solvents *i.e.* petroleum ether, ethanol and water by using soxhlet extractor for 24 hours. The polarity of the solvent would leach out compounds soluble in the particular solvent. The solvent was removed at a low temperature under reduced pressure to yield a thick syrup or semi solid mass. The fractions were further used to test for the antimicrobial activities.

Muller Hinton agar (Himedia No. 173) media was used to test the antimicrobial activity against *Escherichia coli* (MTCC-739), *Klebsiella pneumoniae* (MTCC-109), *Bacillus subtilis* (MTCC-441), *Staphylococcus aureus* (MTCC-96) and *Streptococcus mutans* (MTCC-890) by well diffusion method (Ahmad *et al.*, 1998; Ahmad *et al.*, 1999). 8 mm diameter wells were punched in the agar and filled with extracts and respective solvents for negative control and antibiotic (100 mg/ml) was used as positive control. The slants were incubated at 37 °C for 18-24 hours. The antimicrobial activity was evaluated by measuring the diameter of inhibition zone in mm and their percentage (Reddy, 2000).

$$\text{Percentage of Inhibition} = 100 - \frac{T \times 100}{C}$$

where,

C: Growth in standard

T: Growth in test

Results and Discussion

The plant parts showed broad spectrum antibacterial activity (Tables 46.1–46.4) *i.e.* the petroleum ether, ethanol and water extracts were active against both gram positive and gram negative bacteria. The *Salsola kali* extracts were found to be less effective as compared to ampicillin. Investigation into the folklore antimicrobial activity of *Salsola kali* against *Salmonella paratyphi* and *Serratia marcescens* were conducted (Prabhat *et al.*, 2005a). Here we report on the preliminary result from the antibacterial test of twelve crude extracts from plant species. The antibacterial activity of plant parts *i.e.* aerial parts, roots, stems, seeds and antibiotic ampicillin in all the three solvents against *Escherichia coli, Streptococcus mutans, Staphylococcus aureus, Bacillus subtilis* and *Klebsiella pneumoniae* at concentrations of 100 mg/ml is given in Tables 46.1–46.4.

Table 46.1: The Percentage of Inhibition of Stems Extract and Antibiotics (Ampicillin) Against Pathogens

Pathogens	*Zone of Inhibition (in mm) (Extract 100 mg/ml)*			*Zone of Inhibition (in mm) (Antibiotic 100 mg/ml)*			*Percentage of Inhibition*		
	Petroleum Ether	*Ethanol*	*Water*	*Petroleum Ether*	*Ethanol*	*Water*	*Petroleum Ether*	*Ethanol*	*Water*
E. coli	NI***	12	10	14	18	20	0.00	33.33	50.00
K.pneumoniae	10	20	20	22	27	30	54.54	25.92	33.33
S. aureus	20	25	20	25	29	30	20.00	13.79	33.33
B. subtilis	20	24	26	26	28	30	23.07	14.28	13.33
S. mutans	10	20	20	18	26	30	44.44	23.07	33.33

***: Represents no inhibition.

Table 46.2: The Percentage of Inhibition of Roots Extract and Antibiotics (Ampicillin) Against Pathogens

Pathogens	*Zone of Inhibition (in mm) (Extract 100 mg/ml)*			*Zone of Inhibition (in mm) (Antibiotic 100 mg/ml)*			*Percentage of Inhibition*		
	Petroleum Ether	*Ethanol*	*Water*	*Petroleum Ether*	*Ethanol*	*Water*	*Petroleum Ether*	*Ethanol*	*Water*
E. coli	NI***	14	10	14	18	20	0.00	22.22	50.00
K.pneumoniae	10	19	20	22	27	30	54.54	29.62	33.33
S. aureus	18	20	18	25	29	30	28.00	31.03	40.00
B. subtilis	20	18	20	26	28	30	23.07	35.71	33.33
S. mutans	12	20	20	18	26	30	33.33	23.07	33.33

***: Represents no inhibition.

The zone of inhibitions obtained in crude extracts of different parts of plants varied against pathogens while it was same in case of ampicillin. The results revealed that antibiotic is more effective as compared to crude extracts. The ethanolic extracts of seeds showed the highest activity against *Bacillus subtilis* followed by *Staphylococcus aureus, Klebsiella pneumoniae, Streptococcus mutans* and *Escherichia coli* as compared to other extracts.

The extracts of stems showed maximum activity followed by seeds, roots and aerial parts. The antibacterial therapy is going through crises due the rapidly increasing development of resistance to existing agents. Antibiotic resistance has increased substantially in the recent years and is posing an increasing therapeutic problem. The use of plants as primary health remedies is quite common due to their pharmacological properties. The plant produced a variety of phytoconstituents that have antibacterial activity. These compounds include stigmasterol, campesterol, b-sitosterol and their glycosides etc. The natural antimicrobial compounds in plants can inhibit the growth of bacteria by means of unknown mechanisms other than that of known antibiotics and for this reason the search for new antibiotics must continue. The present study has shown that the plant and its parts are potentially a rich source of antibacterial agents. The plant extracts inhibited the growth of tested pathogens. The crude extracts (100mg/ml) of each part was used for determination of their potency against pathogens and compared with antibiotic (100 mg/ml). The extent of antimicrobial activity of the extracts based on inhibition zone diameter has been described as low (10-14), moderate (15-20) and strong (21-26) by Ahmad *et al.* (1999). In our study, seeds and stems extracts showed strong activity against all pathogens and followed by roots and aerial part. The ethanolic and water extracts of each part is highly effective against all pathogens because more phytoconstituents were leached in these solvents. Our findings have validated the use of this medicinal plant for the treatment of microbial infections such as diarrhoea and fever. It seems important to recommend that further studies using isolated constituents instead of whole extracts must be done in this field. Health foundations have to increase their funding of these studies and research to help saving the lives of many peoples. This will also offer a great help in facing the emergence and spread of antimicrobial resistance.

Table 46.3: The Percentage of Inhibition of Seeds Extract and Antibiotics (Ampicillin) Against Pathogens

Pathogens	*Zone of Inhibition (in mm) (Extract 100 mg/ml)*			*Zone of Inhibition (in mm) (Antibiotic 100 mg/ml)*			*Percentage of Inhibition*		
	Petroleum Ether	*Ethanol*	*Water*	*Petroleum Ether*	*Ethanol*	*Water*	*Petroleum Ether*	*Ethanol*	*Water*
E. coli	NI***	15	10	14	18	20	0.00	16.16	50.00
K.pneumoniae	10	20	19	22	27	30	54.54	25.92	36.66
S. aureus	12	22	18	25	29	30	52.00	24.13	40.00
B. subtilis	10	25	20	26	28	30	61.53	10.71	33.33
S. mutans	NI***	20	21	18	26	30	0.00	23.07	30.00

***: Represents no inhibition.

Table 46.4: The Percentage of Inhibition of Aerial Parts Extract and Antibiotics (Ampicillin) Against Pathogens

Pathogens	*Zone of Inhibition (in mm) (Extract 100 mg/ml)*			*Zone of Inhibition (in mm) (Antibiotic 100 mg/ml)*			*Percentage of Inhibition*		
	Petroleum Ether	*Ethanol*	*Water*	*Petroleum Ether*	*Ethanol*	*Water*	*Petroleum Ether*	*Ethanol*	*Water*
E. coli	NI***	12	10	14	18	20	0.00	33.33	50.00
K.pneumoniae	10	20	18	22	27	30	54.54	25.92	40.00
S. aureus	10	15	16	25	29	30	60.00	48.27	46.66
B. subtilis	12	14	15	26	28	30	53.84	50.00	50.00
S. mutans	10	18	18	18	26	30	44.44	30.76	40.00

***: Represents no inhibition.

References

Ahmad, I. Mehmood, Z. and Mohammad, F., 1998. Screening of some Indian medicinal plants for their antimicrobial properties. *Journal of Ethnopharmacology*, 62: 183–193.

Ahmad, I., Beg, A.Z. and Mehmood, Z., 1999. Antimicrobial potency of selected medicinal plants with special interest in activity against phyto pathogenic fungi. *Indian Veterinary Medical Journal*, 23: 299–306.

Cupp, M.J., 1999. Herbal remedies: Adverse effects and drug Interaction. *Am. Fam. Physician*, 59: 1239–44.

Gupta, S. and Nath, N., 1984. *The Ayurvedic System of Medicine*, Vol. III. Neeraj Publishing House, Delhi, pp. 184–186.

Grieve, M. (F.R.H.S.), 1992. *A Modern Herbal Vol. 1*. Tiger Book International, London, pp. 357–358.

Kumar, K. and Upadhyay, V., 1994. Chemistry of traditional recipe used on hepatic disorders. *TOHO* (10): 136–144, Tokyo, Japan.

Prabhat and Navneet, 2007. *Antibacterial activity of medicinal plants against dental infections*. In: *Medicinal Plants: Conservation, Cultivation and Utilization*, (Eds.) D.R. Khanna *et al.* Daya Publishing House, New Delhi, pp. 139–146.

Prabhat, Navneet and Shri Krishna, 2005. Antimicrobial activity of Apamarga (*A. aspera*). *Indian Academy of Science Letters*, 28(11–12): 379–381.

Prabhat, Navneet and Shri Krishna, 2005a. Antibacterial activity *Mimuspselengi* (Bakul). *Env. Cons. J.*, 6(2): 59–61.

Prabhat, Navneet and Shri Krishna, 2008. Antibacterial activity of Harad (*Terminalia chebula*). *Journal of Botanical Society*, 87(3–4): 1–3.

Reddy, H.M., 2000. Indian medicinal plants. I. *Achyranthes Aspera* L. *J. Econ. Texon. Bot.*, 24(3): 627–630.

Ethnobotanical Wisdom and Microbial Studies on Medicinal Plants (2011) *Pages* **320–325**
Editors: **D.R. Khanna, Ashutosh Gautam, R. Bhutiani, Gagan Matta, Anil Dhiman**
Published by: **DAYA PUBLISHING HOUSE, NEW DELHI**

Chapter 47

In vitro Antimicrobial Activity of *Psidium guajava* (L.) Against Some Isolated Oral Pathogens

Ashok Kumar Singh[1], Gauri[1], Lalit Matela[1], Harish Chandra[2] and R.P. Bhatt[3]
[1]Department of Microbiology, Dolphin (P.G) Institute of Biomedical Science Manduwala, Dehradun, Uttarakhand
[2]Department of Microbiology, Gayatri College of Biomedical Science, G.M.S Road, Ballupur Chowk, Dehradun, Uttarakhand
[3]Department of Botany, H.N.B. Garhwal University, Srinagar, Garhwal, Uttarakhand

ABSTRACT

Methanolic extract of *Psidium guajava* stem was investigated for its antimicrobial activities against five isolated oral pathogens. Among the 50 sample collected from patients 26.25 per cent were identified as *Staphylococcus* spp., 25 per cent as *Micrococcus* spp., 23.75 per cent *Streptococcus* spp, 15 per cent *Corynebacterium* spp and 10 per cent *Proteus* spp.The strongest antimicrobial activity was observed against *Micrococcus* spp.

Keywords: *Psidium guajava, Antimicrobial.*

Introduction

In recent year antibiotic resistance to common antibiotic has been increasing day by day. The mechanism of resistance is either emergence of bacteria or mutation of one or more genes. In this situation, the undesirable side effect of certain antibiotic and the emergence of previously uncommon infections has forced scientist to look for new antimicrobial substance from medicinal plant and other sources (Marchese and Shito, 2001). The screening of plant extracts and plant products for antimicrobial activity has shown that plants represent a potential source of new antimicrobial agents (Amani *et al.*, 1998). Recent interest in chewing sticks and their extract has focused on their effects on organisms that are involved in oral infection. Dental caries and periodontal disease *i.e.* gingivitis and periodontitis are the most common chronic oral diseases worldwide. Although these diseases have affected human beings since prehistorical times, the prevalence of these diseases has greatly increased in modern times. The prevalence of dental diseases in India is approximately 60-65 per cent (Shouri, 1941) while the prevalence of periodontal disease is approximately 65-100 per cent (Kanal *et al.*, 1971). Dental caries is an infectious microbial disease that results in localized dissolution and destruction of the calcified tissues of teeth (Ross and Holbrook, 1994). Most of the investigators believe that development of caries of enamel is preceded by the formation of microbial plaque in the tooth (Gibbons *et al.*, 1963). The human oral cavity is habitat for about 500 cultivable and non cultivable bacterial species (Paster *et al.*, 2001) up to 100 species can be present in a particular oral cavity (Consensus, 1996) while the majority of these species are commensals, a subset is opportunistic pathogens. They have also been implicated in the etiology of a number of systemic diseases like infective endocarditis, (Barrau *et al.* 2004) respiratory infections, (Mojon and Bourbeau, 2003) cardiovascular diseases, (Okuda *et al.*, 2004) and brain abscess (Corson *et al.*, 2001). Oral bacterial isolates resistant to penicillin, metronidazole, tetracycline and macrolides have been reported by researchers from different countries (Sweeney *et al.*, 2004). Such resistant bacteria have also been isolated from infections at extra oral sites (Doern *et al.*, 1996). To cope with this problem of antibiotic resistance, a number of strategies such as reduced antibiotic use and antibiotic alternatives have been proposed (Hamilton, 2004). Among antibiotic alternatives are therapies derived from complementary and alternative medicine. In fact, there are an overwhelming number of studies on the antibacterial activities of plant and natural products derivatives. Recently some plant extracts have been shown to have activity against resistant bacterial strains (Gibbons *et al.*, 2003).

Psidium guajava (L.) (Guava) belong to the family Myrtaceae, grows nearly throughout the country up to 1500 m. it is cultivated commercially in almost all states. The total estimated area being about 50,000 hectares. The important guava growing states are: U.P. Bihar, Uttaranchal. Maharashtra, Assam, West Bengal, Andhra Pradesh and Chennai. About half of the total area is reported in Uttar Pradesh. The plant of guava is found almost in every garden of India. Guava is rich in tannins, phenols, triterpenes, flavonoids, essential oils, saponins, carotenoids, lectins, vitamins, fiber and fatty acids. Guava fruit is higher in vitamin C than citrus (80 mg of vitamin C in 100g of fruit) and contains appreciable amounts of vitamin A as well. Guava fruits are also a good source of pectin – a dietary fiber. The leaves of guava are rich in flavonoids, in particular, quercetin. Much of guava's therapeutic activity is attributed to these flavonoids. The flavonoids have demonstrated antibacterial activity. Quercetin is thought to contribute to the anti-diarrohea effect of guava; it is able to relax intestinal smooth muscle and inhibit bowel contractions. In addition, other flavonoids and triterpenes in guava leaves show antispasmodic activity. Guava also has antioxidant properties which are attributed to the polyphenols found in the leaves (Arima and Danno, 2002). Bark and leaf extracts have shown to have *in vitro* toxic action against numerous bacteria. In several studies guava showed significant antibacterial activity against such common diarrhoea causing bacteria as *Staphylococcus, Shigella, Salmonella, Bacillus, E. coli, Clostridium* and *Pseudomonas* (Abdelrahim *et al.*, 2002). The present study was undertaken to investigate the *in vitro* antibacterial activity of methanolic extract of twigs of *P. guajava* against oral pathogens.

Materials and Methods

Collection of Sample for Isolation of Oral Pathogen

Total 50 samples were collected from Shree "Mahant Indresh hospital" Patel Nagar, Dehradun. The samples from dentine of teeth were taken into a sterile cotton swab. The cotton swab was placed in each approximal site and then passed along the gingival margin into the next approximal site of both the upper and lower teeth.

Enrichment of the Sample

The cotton swabs were enriched in lactose broth (HiMedia) and Trypticase soy broth (HiMedia) for 12 hours at 37° C in the incubator. Then the enriched sample was streaked on both general Nutrient Agar and Selective Media Brain Heart Infusion Blood Agar (HiMedia), Mac Conkey Agar (HiMedia), Mannitol Salt Agar (HiMedia), and Vogel Johnson Agar (HiMedia).

Isolation and Identification of Pathogens

The pathogens associated with dental caries, gum and periodontal diseases were isolated from infected tissues and identified by standard morphological and biochemical tests on the surface of the growth of the isolates (Holt. *et al.*, 1994).

Collection of Plant Material

The Plant used were the air dried stem of *P. guajava* (Guava) collected from Forest Research Institute (FRI) Dehradun and Identified by Botanical Survey of India, Dehradun.

Extraction of Plant Materials

The air dried and powdered material was extracted, with methanol by soxhlet apparatus by removing the solvent with rotary evaporator (Butchi Type) crude extracts were obtained.

Antibacterial Assay

Preparation of Inoculum

The ideal inoculum after overnight incubation gives even semi confluent growth. Too heavy inoculum may reduce the size of inhibition zone by many antimicrobial agents from plant source. Using a straight wire touch 5-10 well isolated colonies of particular microorganism against which antimicrobial activity to be tested. Inoculate on the Nutrient Broth Medium. Incubate at 35-37°C for 4 – 6 hour. The density of the inoculums is adjusted to 108 cfu/ml by comparing with that of 0.5 Mc Farland Standard.

Agar Well Diffusion

0.1 ml of the original cultures (about106-107 cells) were added into sterile duplicate sets of Petri dishes and 25 ml of the molten (45 °C) Mueller Hinton Agar (HiMedia, Ltd) were poured into Petri dishes. The methanol extract (0.1 ml) were placed in wells (8 mm diameter) cut in the agar media and plates were incubated at 37 °C in the case of bacteria and at 30 °C in the case of yeasts (Kývanç and Akgul, 1986). The resulting inhibition zones obtained with bacteria were recorded after 24 hour.

Results and Discussion

The present study was conducted to investigate the antimicrobial activity methanolic extract of *P. guajava* against some isolated oral microorganism. Among the 50 sample collected from patients 26.25

per cent were identified as *Staphylococcus* spp., 25 per cent as *Micrococcus* spp., 23.75 per cent *Streptococcus* spp, 15 per cent *Corynebacterium* spp and 10 per cent *Proteus* spp. (Figure 47.1). Among the 40 microbial samples collected from adult patients, 10.63 per cent were identified as *Staph. aureus*, 4.32 per cent as *Strep. mutans*, 6.38 per cent *Strep. faecalis*, 2.12 per cent *Strep. pyogenis*, 5.25 per cent *Lactobacillus acidophilus*, 2.1 per cent *Ps. aeruginosa* and 4.3 per cent as *Candida albicans* (Al-Bayati and Sulaiman, 2008).

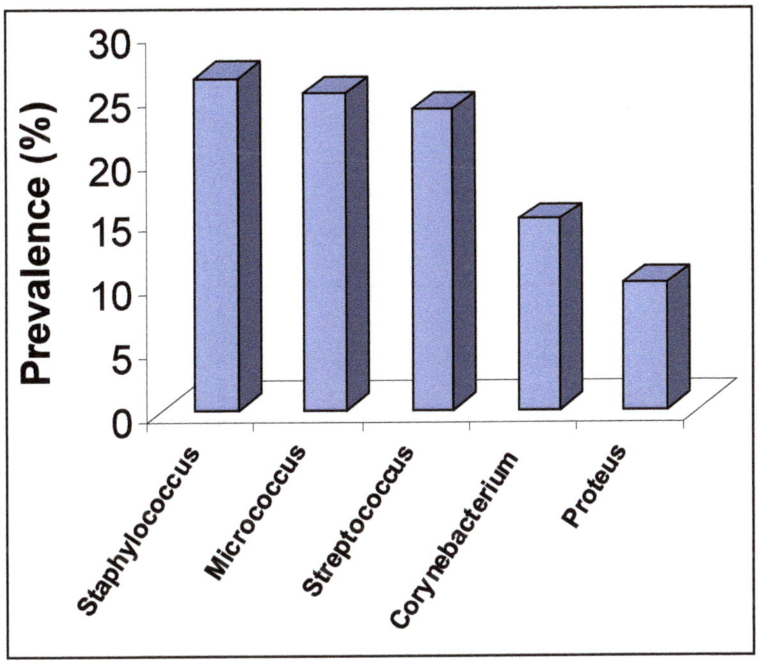

Figure 47.1: Prevalence of Isolated Microorganism

On the other hand, the methanolic extract of *P. guajava* showed activity against the entire isolated microorganism. Zone of inhibition ranged from 12.82±1.15–16.93±1.28 mm. Maximum activity was shown against *Micrococcus* sp. *i.e.* 16.93±1.28 mm and less active as compared to tested microorganism is against *Proteus* sp *i.e.* 12.82±1.15 mm. Standard drug Amoxicillin and Doxycycline was active against the entire test microorganism. The extracts of *P. guajava* compared with the standard antibiotics (Doxycycline and Amoxicillin). The plant extracts and antibiotics show almost equal antibacterial activity against oral pathogens. Previous studies reported that the methanolic extract of *Psidium guajava* was also shown to possess antibacterial effect on *Bacillus subtilis*, *Staphylococcus aureus*, *Escherichia coli* and *Pseudomonas aeruginosa* (Abdelrahim *et al.*, 2002). Gnan and Demello (1999) reported a complete inhibition of growth of *S. aureus*, *S. epidermidis* and *Salmonella typhimurium* caused by aqueous guava leaf extract at a concentration of 8 mg/ml. Veeira *et al.* (2001) reported the microbicidal effect of guava sprout extract (ethanol, acetone and water) upon toxigenic *S. aureus* and *E. coli*, performed using radial diffusion. Earlier pharmacological investigation indicated that its fruit and leaves possess antibacterial, hypoglycemic, anti-inflammatory, analgesic, antipyretic, spasmolytic and CNS depressant activities (Begum *et al.*, 2002). Jairaj *et al.* (1999) also reported the water, alcohol and chloroform extracts of leaves were effective against *Aeromonas hydrophila*, *Shigella* spp., *Vibrio* spp., *Staphylococcus* spp., *Sarcina lutea* and *Mycobacterium phlei*. It can be concluded that the methanolic extract are excellent oral hygiene agents and their use should be promoted based on scientific knowledge of their benefit and proper use. Because it is widely available in India and is inexpensive. *P. guajava* chewing sticks can be great help in developing countries with financial constrains and poor or limited oral health care facilities.

Table 47.1: Effect of Methanolic Extract Against Isolated Microorganism

Microorganism	*Zone of Inhibition (In mm)*			
	No. of Isolates	*MeOH**	*Amoxicillin*	*Doxycycline*
Micrococcus sp.	20	16.93±1.28	17.97±1.25	18.07±1.21
Corynebacterium sp.	12	15.81±1.22	16.67±1.54	16.05±1.62
Staphylococcus sp.	21	15.23±1.14	16.21±1.58	15.72±1.61
Streptococcus sp.	20	16.48±0.97	17.32±1.58	16.47±1.55
Proteus sp.	8	12.82±1.15	14.0±1.38	14.15±1.71

References

Abdelrahim, S.I., Almagboul, A.Z., Omer, M.E. and Elegami, A., 2002. Antimicrobial activity of *Psidium guajava*. *Fitoterapia*, 73: 713–715.

Al-Bayati, F.A. and Sulaiman, K.D., 2008. *In vitro* antimicrobial activity of *Salvadora persica* L. extracts against some isolated oral pathogens in Iraq. *Turkish Journal of Biotechnology*, 32: 57–62.

Amani, S., Isla, M.I., Vattuone, M.A., Poch, M.P., Cudmani, N.G. and Sampietro, A.R., 1998. Antimicrobial activities in some argentine medicinal plants. *Acta Horticulture*, 501: 115–122.

Arima, H. and Danno, G., 2002. Isolation of antimicrobial compounds from guava (*Psidium guajava*). *Biosci. Biotechnol Biochem.*, 66: 1727–1730.

Barrau, K., Boulamery, A., Imbert, G., Casalta, J.P., Habib, G., Messana T., Bonnet, J.L. Rubinstein, E. and Raoult, D., 2004. Causative organisms of infective endocarditis according to host status. *Clin. Microbiol. Infect.*, 10: 302–308.

Begum, S., Hassan, S.I. and Siddiqui, B.S., 2002. Two new triterpenoids from the fresh leaves of *Psidium guajava*. *Planta Medica*, 68: 1149–1152.

Consensus Report, 1996. Periodontal diseases: Pathogenesis and microbial factors. *An. Periodontal.*, 1: 926–932.

Corson, M.A., Postiethwaite, K.P. and Seymou, R.A., 2001. Are dental infections a cause of brain abscess? Case report and review of the literature. *Oral Dis.*, 7: 61–65.

Doern, G.V., Ferraro, M. J., Brueggmann, A.B. and Ruoff, K.L., 1996. Emergency of high rates of antimicrobial resistance among viridians group *Streptococci* in the United States. *J. Antimicrob. Agent. Chemotherapy*, 40: 891–894.

Gibbons, R.J. Socransky, S.S., Sawyer, S., Kapsinialis, B. and Macdonald, J.B., 1963. The microbiota of the gingival crevice area of man: II. The predominant cultivable organisms. *Arch. Oral. Biol.*, 8: 281–289.

Gibbons, S., Oluwatuyi, M., Veitch, N.C. and Gray, A.L., 2003. Bacterial resistance modifying agents from *Lycopus europaeus*. *Phytochemistry*, 62: 83–87.

Gnan, S.O. and Demello, M.T., 1999. Inhibition of *Staphylococcus aureus* by aqueous goiba extracts. *Journal of Ethanopharmacology*, 68: 103–108.

Hamilton, J.M., 2004. Antibiotic resistance from two perspectives: man and microbe. *Int. J. Antimicrob. Agents*, 23: 209–212.

Holt, J.G., Krieg, N.R., Sneath, P.H.A., Staley, J.T and Williams, S.T. (Eds.), 1994. *Bergey's Manual of Determinative Bacteriology*, 9th edn. Williams and Wilkin, Baltimore, Md.

Jairaj, P., Koohaswan, P., Wongkrajang, Y., Peungvicha, P., Suriyawong, P., Saraya, M.L.S. and Ruangsomboo, O., 1999. Anticough and antimicrobial activities of *Psidium guajava* Linn. Leaf extract. *Journal of Ethanopharmacology*, 67: 203–212.

Kanal, M., Roy, R.K. Chawla T. N., Kapoor, K. K., Kapoor, D.N., 1971. Gingivitis in the primary dentition among Lucknow urban and rural children. *Journal of Indian Dental Association*, 43: 709–713.

Kývanç, M. and Akgül, A., 1986. Antibacterial activities of essential oils from Turkish spices and citrus. *Flavour and Fragrance Journal*, p. 175–179.

Marchese, A. and Shito, G.C., 2001. Reisitance pattern of lower respiratory pathogen in Europe. *International Journal of Antimicrobial Agents*, 16: 25–29.

Mojon, P. and Bourbeau, J., 2003. Respiratory infection: How important is oral health? *Curr. Opin. Plum. Med.*, 9: 166–170.

Okuda, K., Kato, T. and Ishihara, K., 2004. Involvement of periodontopathic biofilm in vascular diseases. *Oral Dis.*, 10: 5–12.

Paster, B.J., Boches, S.K., Galvin, J.L., Ericson, R.E., Lau,C.N., Levanos, V.A., Sehasrabudhe, A. and Dewhirst, F.H., 2001. Bacterial diversity in human subgingival plaque. *J. Bacteriol.*, 183(12): 3770–3783.

Ross, P.W. and Holbrook, W.P., 1994. *Dental plaque*. In: *Clinical and Oral Microbiology*, Blackwell Scientific Publications, Glasgow, pp. 82.

Shouri, K.L., 1941. Dental caries in Indian children. *Indian Journal of Medical Research*, 29: 709–722.

Sweeney, L.C., Dave, J., Chambers, P.A. and Heritage, J., 2004. Antibiotic resistance in general dental practice: a cause for concern? *J. Antimicrob. Chemother.*, 53: 567–76.

Veeira, R.H.S.F., Rodrigues, D.P., Goncalves, F.A., Menezes, F.G.R., Aragao, J.S. and Sousa, O.V., 2001. Microbicidal effects of medicinal plant extracts (*Psidiun guajava* Linn. and *Carica papaya* Linn.) upon bacteria isolated from fish muscle and known to induce diarrhoea in children. *Rev. Inst. Med. Trop., S. Paulo*, 43(3): 145–148.

Ethnobotanical Wisdom and Microbial Studies on Medicinal Plants (2011) *Pages* **326–328**
Editors: **D.R. Khanna, Ashutosh Gautam, R. Bhutiani, Gagan Matta, Anil Dhiman**
Published by: **DAYA PUBLISHING HOUSE, NEW DELHI**

Chapter 48

Some Additional Records of Follicolous Fungi from North Central Tarai Forests of U.P.

D.P. Singh and T.P. Mall
Post Graduate Department of Botany, P.G. College, Bahraich, U.P.

ABSTRACT

The follicolous fungi was collected from Feb. 03-05, 2008 from Sirsa forest range of Shrawasti District. The authors collected twenty plant species, representing fourteen families, parasitized by sixteen fungi species. *Leptoxyphium buteae* was found on *Butea frondosa* (Fabaceae), where as *Pseudocercospora nigricans* on *Cassia occidentalis* (Fabaceae); *Pseudocercospora* sp. on *Hetrofragma* sp (Boraginaceae) *Alternaria* on *Achyranthes aspera* (Amranthaceae), *Syzygium* sp. (Myrtaceae) and *Corchorus olitoris* (Tiliaceae); *Corynespora* on *Lantana* in(Verbenaceae), *Croton roxburghii* (Euphorbiaceae), *Clerodendron* sp. (Verbenaceae); *Sirosporium lantanae* on *Lantana camera* (Verbenaceae); *Sirosporium* sp. on *Carica papaya* (Caricaceae); *Stenella tectonic* on *Tectona grandis* (Verbeaceae); *Stenella* sp. on *Eucalyptus lanculatous* (Myrtaceae); *Stenella grewiae* on *Grewia elastica* (Tiliaceae); *Cercospora* sp. on *Galycosmis pentaphyla* (Rutaceae), *Corchorus olitorius* (Tiliaceae); *Meliola* sp. on *Streblus asper* (Moraceae); *Astrostomella* on *Litsea chinensis* (Lauraceae); *Acrodytis* sp. on *Tinspora malaverica* (Menispermaceae); *Passalora* sp on *Eopatarium cannabinum* (Asteraceae); *Oidium* sp on *Syzygium* sp. (Myrtaceae) and *Coccinia indica* (Cucurbitaceae).

Keywords: *Follicolous fungi, Ethnomedicine.*

Introduction

The leaves provide a very suitable habitat for the growth and development of fungal pathogens by providing ample surface area and nutrient supply. Such leaf inhabiting fungi are known as follicolous fungi and invaded area of the leaf appear as leaf spot or leaf lesions. Taxonomic studies of such fungal forms have been generally considered as only of academic interest, taxonomic treatment of a fungal organism is the first requirement for any studies concerning its biology. Correct identification of a fungus absolutely free from ambiguities is vital for its employment in applied disciplines. Infact without being equipped for ascertaining the correct identity of a fungul pathogen all studies concerning its phytopathological aspects would be misleading. The weed and forest plants serve as reservoirs of leaf spot pathogens which on getting opportunity may spread to agriculture and horticulture plants keeping this in view the author surveyed the Sirsia forest range of Shrawasti District.

Materials and Methods

During collection, infected leaf samples were taken in separate polythene bags. Suitable amount of suface scrapping and free hand cut sections were prepared from infected portions of the leaf samples. Slides were prepared in cotton-blue lactophenol mixture, slides were examined and *Camera lucida* drawings were made which seems to be new. Morphotaxonomic determinations of taxa were done with the help of current literature and resident expertise available. All the fungal taxon were identified after making microscopic preparations and later confirmed by Prof. Kamal, Emeritus Scientist (DST), DDU Gorakhpur University, Gorakhpur.

Results and Discussion

The author surveyed the Sirsia forest range of Shrawasti District on Feb. 03-05, 2008 so as to collect the follicolous fungi. The authors collected twenty plant species, representing fourteen families, parasitized by sixteen fungi species. *Leptoxyphium buteae* was found on *Butea frondosa* (Fabaceae), where as *Pseudocercospora nigricans* on *Cassia occidentalis* (Fabaceae); *Pseudocercospora* sp. on *Hetrofragma* sp (Boraginaceae) *Alternaria* on *Achyranthes aspera* (Amranthaceae), *Syzygium* sp. (Myrtaceae) and *Corchorus olitoris* (Tiliaceae); *Corynespora* on *Lantana* (Verbenaceae), *Croton roxburghii* (Euphorbiaceae), *Clerodendron* sp. (Verbenaceae); *Sirosporium lantanae* on *Lantana camera* (Verbenaceae); *Sirosporium* sp. on *Carica papaya* (Caricaceae); *Stenella tectonic* on *Tectona grandis* (Verbeaceae); *Stenella* sp. on *Eucalyptus lanculatous* (Myrtaceae); *Stenella grewiae* on *Grewia elastica* (Tiliaceae); *Cercospora* sp. on *Galycosmis pentaphyla* (Rutaceae), *Corchorus olitorius* (Tiliaceae); *Meliola* sp. on *Streblus asper* (Moraceae); *Astrostomella* on *Litsea chinensis* (Lauraceae); *Acrodytis* sp. on *Tinospora malaverica* (Menispermaceae); *Passalora* sp. on *Eopatarium cannabinum* (Asteraceae); *Oidium* sp. on *Syzygium* sp. (Myrtaceae) and *Coccinia indica* (Cucurbitaceae).

The literature (Bilgrami *et al.*, 1981; 1991; Goos and Hosagoudar, 1998; Hosagoudar and Goos, 1990; Hosagoudar 1996; Hosagoudar *et al.*, 1997; Hosagoudar and Abrahum 1998; Hosagoudar *et al.*, 2007; Jamaluddin *et al.*, 2004; Jana *et al.*, 2005; and Singh and Mall, 2007) reveals that all fungal taxon has not been reported from north central tarai forest of Uttar Pradesh.

The follicolous fungal pathogens interfere with the manufacturing rate of food and other valuable substances by damaging the photosynthetic elements of living leaves, bringing about qualitative and quantitative damage in the living tissue of the host in various ways. Several leaf spot pathogens are known to produce toxins of the various kinds. When the leaf spots are numerous or large, there is a considerable reduction in the photosynthetic area of the leaf. Some times rapid defoliation occurs due to such infections. The productivity of the host plants is reduced. The weeds and forest plants serve as reservoir of leaf spot pathogens which may also spread to agriculture and horticultural plants. The destruction caused by these enemies of leaves is serious problem because they also cause degradation of quality of ethnomedicine present therein.

Acknowledgement

The authors are grateful to Sri Mohd. Ashan, Chief Wildlife Warden, Uttar Pradesh, Lucknow for giving his permission for field works, Dr. S.P. Singh Principal, Kisan P.G. College, Bahraich for providing necessary facilities and Prof. Kamal, Emeritus Scientist, DST for helpful suggestions and encouragements.

References

Bilgrami, K.S., Jamaluddin and Rizwi, M.A., 1981. *Fungi of India*, Part II. Today and Tomorrow's Printers and Publishers, New Delhi, pp. 140.

Bilgrami, K.S., Jamaluddin and Rizwi, M.A., 1991. *Fungi of India, Part I: List and References*. Today and Tomorrow's Printers and Publishers, New Delhi, pp. 778.

Goos, R.D. and Hosagoudar, V.B., 1998. *Meliola chennaiana* sp. Nov. and some additional records of fungi from India. *Mycotaxon*, 68: 41–46.

Hosagoudar, V.B. and Goos, R.D., 1990. Meliolaceous fungi from the state of Kerala, India–II: The genus Meliola. *Mycotaxon.*, 37: 217–271.

Hosagoudar, V.B., 1996. *Meliolales of India*. Botanical Survey of India, Calcutta, pp. 363.

Hosagoudar, V.B., Abraham, T.K. and Pushpangadam, P., 1997. *The Meliolineae: A Supplement*. Tropical Botanical Garden and Research Institute, Palode, Thiruvananthapuram, Kerala, pp. 201.

Hosagoudar, V.B. and Abraham, T.K., 1998. Some interesting meliolaceous fungi from Kerala. *India. J. Mycopath.*, 36: 95–102.

Hosagoudar, V.B., Archana, G.R. and Agarwal, D.K., 2007. Meliolates of Kerala of India–XXV. *India Phytopath.*, 60(2): 237–243.

Jamaluddin, Goswami, M.G. and Ojha, B.M., 2004. *Fungi of India 1989–2001*. Scientific Publishers (India), Jodhpur, pp. 326.

Jana, T.K., Ghose, S.N. and Das, A.K., 2005. Meliolaceae of Nagaland (India). *I. J. Mycopathol. Res.*, 43: 33–40.

Singh, D.P. and Mall, T.P., 2007. Follicolous fungi of medicinal plants in North Western Tarai region Uttar Pradesh. *Environmental Conservation Journal*, 8(3): 13–16.

Ethnobotanical Wisdom and Microbial Studies on Medicinal Plants (2011) *Pages* **329–334**
Editors: **D.R. Khanna, Ashutosh Gautam, R. Bhutiani, Gagan Matta, Anil Dhiman**
Published by: **DAYA PUBLISHING HOUSE, NEW DELHI**

Chapter 49

Bacteriological Characteristics of Raw Water of the River Tawi Near Sitlee Water Treatment Plant, Jammu

Meenakshi Khajuria and S.P.S. Dutta
Department.of Environmental Sciences, University of Jammu, Jammu

ABSTRACT

Bacteriological analysis of raw water of the river Tawi, near Sitlee water treatment plant, has shown the seasonal presence of *Klebsiella, Citrobacter, Escherichia* and *Pseudomonas*. Total bacterial count (MPN) recorded summer (April, May), monsoon and post-monsoon (June-October) increase and winter (November, December, February) decrease. It varied between 92 and >180 during both the years of study. MPN index above 10 /100 ml. indicates that raw water is not suitable for drinking purposes and comes under the category of unsatisfactory. It needs proper treatment before supplied to the consumers.

Keywords: *River Tawi water, Bacteriological, MPN.*

Introduction

Growing population, increasing living standards, rapid industrialization and wide sphere of human activities, have resulted in steady increase in the demand for water resources. This requires regular water monitoring to ensure supply of clean water to meet the consumers' demands. According to WHO, about 600 million cases of diarrhoea and 46,00,000 childhood deaths are reported per year because of contaminated water and lack of sanitation. Although some work on physico-chemical characteristics of water of the river Tawi has been attempted by Zutshi (1992), but there is no report on bacteriological characteristics of water of the river. The river Tawi, an important tributary of the river Chenab, having its origin in the middle Himalayas, below Seoj Dhar peak at Kalikund near Bhaderwah, is a major source of potable water supply prior to the inhabitants of Jammu city since 1916 when Dhaonthly was the only supply point. At present, water supply from this river for Jammu city is from Sitlee, Dhaonthly and Gorkha Nagar treatment plants. In order to assess the water quality of the river Tawi, two years study was undertaken and has been described.

Materials and Methods

Presumptive Coliform Count (MPN)

MPN count of coliform organisms was done by Multiple Tube method (Senior, 1989; APHA, 1998).

Confirmed Phase

Submitted all presumptive tubes showing growth, any amount of gas, or acidic reaction within 24 hrs. of incubation to the confirmed phase. Calculated the MPN value from the number of positive BGLB tubes (Senior, 1989; APHA, 1998).

Completed Phase

Using aseptic technique (Senior. 1989: APHA 1998), streaked one each MacConkey agar plate, from each tube of positive BGLB showing gas.

Faecal Coliform Count (MPN)

MPN count of faecal coliform was done by multiple tube method. Biochemical Tests: To confirm the presence of various coliform bacteria, the following biochemical tests (IMViC) (Senior, 1989: APHA, 1998) were conducted (Table 49.1).

Table 49.1: Biochemical Tests (IMViC) to Confirm the Presence of Various Coliform Bacteria

Organism	*Indole*	*Motility*	*TSI*	*MR*	*Urease*	*Citrate*	*PPA*	*H_2S*	*Dry Filter Paper/Oxidase*
Escherichia	+	+	+	+	–	–	–	–	–
Klebsiella	–	–	+	–	+	+	–	–	–
Citrobacter	+	+	+	+	–	+	–	+	–
Proteus	+	+	+	–	+	+	+	+	–
Pseudomonas	–	–	–	–	–	–	–	–	+

Results and Discussion

The results of various bacteriological characteristics of water have been tabulated in Tables 49.2 and 49.3 and depicted in Figure 49.1.

Qualitative Composition

In the raw water of the river Tawi. near Sitlee water treatment plant, *Klebsiella. Citrobacter, Escherichia* and *Pseudomonas* showed their seasonal qualitative presence, during both the years. During the first

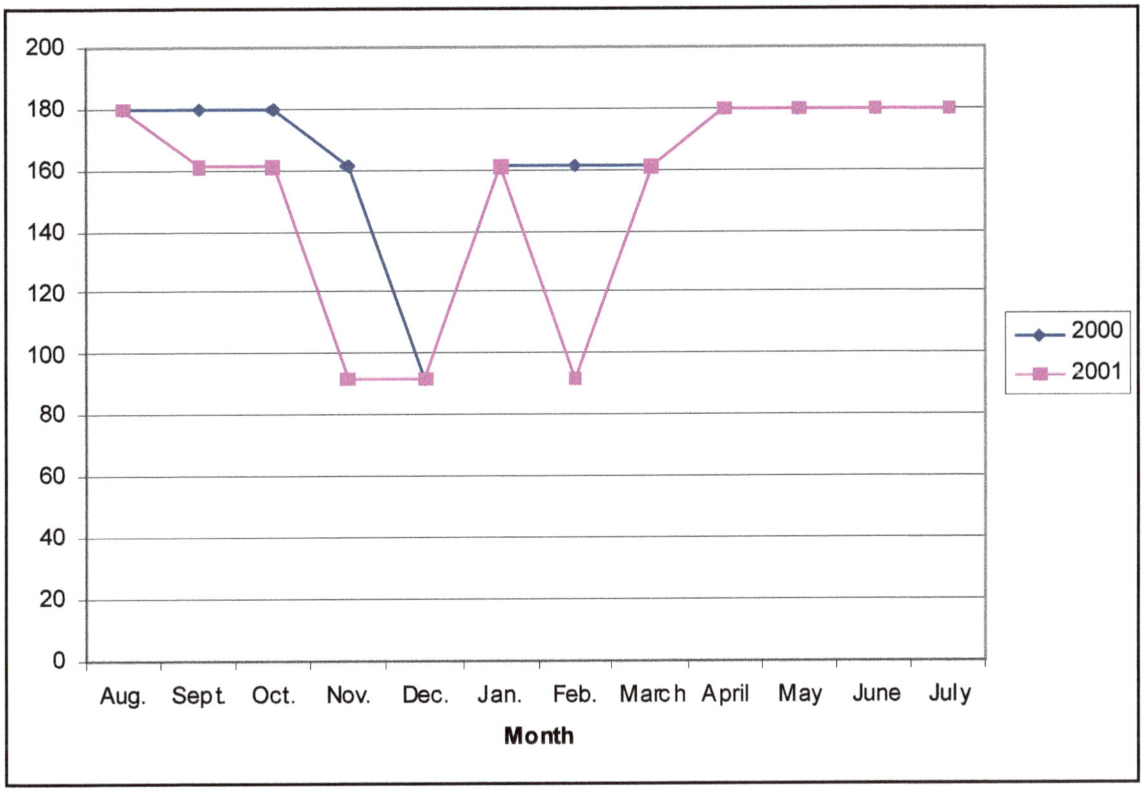

Figure 49.1: Monthly Quantitative Variations of Microbes (MPN/100 ml) in the River Tawi, Near Sitlee Water Treatment Complex, Jammu

year, *Klebsiella* was observed only once (June), *Pseudomonas* four times (September–November and July), *Escherichia* ten times (August, October–May and July) and *Citrobacter* eleven times (August–June). During the second year, *Citrobacter* showed its presence five times (November, December, April, May and July); *Klebsiella* six times (August, September, January and February, May and June); *Escherichia* seven times (August–October, December, March, April and July) and *Pseudomonas* eight times (October. November, January–March and May–July) (Table 49.1). Qualitative seasonal presence of these microbes in lotic waters has also been reported by Gupta and Gupta (1999), Gaur *et al.* (2000), Koshy and Nayar (2000), Fokmare and Musaddiq (2001), Singh *et al.* (2001), Srivastava (2002), Bhadra *et al.* (2003), Taylor (2003) and Sharma (2004). Entry of these microbes in river Tawi may result from inflow of sewage and human and animal excreta in the upstream region (Taylor, 2003) and surface runoff during rains (Taylor, 2003), Re-growth of some microbes present in sediments of river, derived from faecal pollution and enriched organic matter, may also account for the presence of these microbes in the river Tawi (Taylor, 2003).

Table 49.2: Monthly Variations in Qualitative Composition of Microbes in the River Tawi Near Sitlee Water Treatment Complex, Jammu

Month	*2000*	*2001*
August	*Escherichia, Citrobacter*	*Escherichia, Klebsiella*
September	*Citrobacter, Pseudomonas*	*Escherichia, Klebsiella*
October	*Citrobacter, Pseudomonas, Escherichia*	*Pseudomonas, Escherichia*
November	*Citrobacter, Pseudomonas, Escherichia*	*Pseudomonas, Citrobacter*
December	*Citrobacter Escherichia*	*Citrobacter, Escherichia*
January	*Citrobacter, Escherichia*	*Pseudomonas, Klebsiella*
February	*Citrobacter, Escherichia*	*Pseudomonas, Klebsiella*
March	*Citrobacter, Escherichia*	*Pseudomonas, Escherichia*
April	*Citrobacter, Escherichia*	*Citrobacter, Escherichia, Klebsiella*
May	*Citrobacter, Escherichia*	*Pseudomonas, Klebsiella, Citrobacter*
June	*Citrobacter, Klebsiella*	*Pseudomonas, Klebsiella*
July	*Escherichia, Pseudomonas*	*Citrobacter, Escherichia, Pseudomonas*

Table 49.3: Monthly Variations of Microbes (MPN/ 100 ml) in the River Tawi, Near Sitlee Water Treatment Complex, Jammu

Month	*2000*	*2001*
August	>180	>180
September	>180	161
October	>180	161
November	161	92
December	92	92
January	161	161
February	161	92
March	161	161
April	>180	>180
May	>180	>180
June	>180	>180
July	>180	>180

Quantitative Analysis

In the Tawi water, MPN index per 100 mL., during the first year, ranged between 92 (December) to >180 (August–October, April–July). During the second year, MPN range was between 92 (November, December, February) to >180 (August, April–July). A very high bacterial count, seen in the river Tawi raw water, has also been reported from lotic waters by Bhosle and Rao (2001), Fokmare and Musaddiq (2001), Singh *et al.* (2001), Begum *et al.* (2003), Rajurkar *et al.* (2003), Thakur *et al.* (2003), Bankar (2004) and Sharma (2004). Microbial enrichment in lotic waters is caused by discharge of sewage and entry of human (open defecation) and animal excreta and other decomposing organic matters (animal slaughtering and washing of stomach and other internal parts) at various places (Begum *et al.*, 2003). This indicates that bacteria, always and under all conditions, remain in water body and this signifies the organic pollution. Summer (April, May) microbial highest count may be attributed to bacterial survival and increased production under warm conditions and is in conformity with the findings of Gaur *et al.* (2000), Singh *et al.* (2001) and Bhadra *et al.* (2003). High temperature and increased surface runoff in catchment areas may account for monsoon (June–September) increase in MPN count in the river Tawi (Table 49.3). Comparison of MPN per 100ml with National and International Standards (BIS, 1991; WHO, 1992) reveals that the water quality exceeds the allowable limits of drinking water standards. The water quality remained unsatisfactory during both the years. as per British Ministry of Health, 1957 classification (Table 49.4).

Table 49.4: Comparison of Water Quality of the the River Tawi, Near Sitlee Water Treatment Complex, Jammu with National and International Standards

Parameter	*PCC/100 ml*	*WHO*	*Ac*	*Al*	*BIS*	*Ac*	*Al*
	1st Year	*2nd Year*	*0*	*10*		*0*	*10*
River Tawi 92 -> 180	92 -> 180						
British Ministry of Health	Excellent 0/100	Satisfactory 1–3/100					
	Suspicious 4–10/100	Unsatisfactory > 10/100					

FCC: Presumptive Coliform Count; Ac: Acceptable; Al: Allowable.

References

APHA, 1998. *Standard Methods for the Examination of Water and Wastewater*, 20th edn. American Public Health Association, 1015 Fifteenth Street, NW Washington, DC, pp. 2005-2605.

Bankar, J.M., 2004. Bacteriological characterstics of drinking water from public places in Satna District, Maharastra. *J. Aqua. Biol.*, 19(2): 1–6.

Begum, J., Ahmed, K. and Bora, K.N., 2003. Estimation of heterotrophic bacteria in differet source of drinking water in different seasons. *Nat. Environ. Poll. Tech.*, 2(4): 455– 457.

Bhadra, B., Mukherjee, S., Chakraborty, R. and Nanda, K., 2003. Physico-chemical and bacteriological investigation on river Torsa of North Bengal. *J. Environ. Biol.*, 24(2): 25–133.

Bhosle, A.B. and Rao, R., 2001. Comparative study of treated and untreated river water for potablity. *Poll. Res.*, 20(3): 475–479.

BIS: 10500, 1991. *Indian Standard Specification for Drinking Water* IS: 10500–91 (Bureau of Indian Standards), New Delhi, pp. 1–4.

British Ministry of Health, 1957. *The Bacteriological Examination of Water Supplies*. Report No. 71. Ministry of Health, London, U.K.

Fokmare, A.K. and Musaddiq, M., 2001. Comparative studies of physico-chemical and bacteriological quality of surface water and groundwater at Akola (Maharashtra). *Poll. Res.*, 20(4): 651–655.

Gaur, R.K., Khan, A.A., Alam, A. And Alam, M.A., 2000. Bacteriological quality of river Ganga from Narora to Kannauj: a comprehensive study. *Ind. J. Environ. Prot.*, 20(3): 165– 170.

Gupta, B.K. and Gupta, R.R., 1999. Physico-chemical and biological study of drinking water in Satna, Madhya Pradesh (India). *Poll. Res.*, 18(4): 523–525.

Koshy, M. and Nayar, T.V., 2000. Water quality of river Pamba at Kozhencherry. *Poll Res.*, 19(4): 665–668.

Rajurkar, N.S., Nongbri, B. and Patwardhan, A.M., 2003. Water quality status of river Umkhrah at Shillong. *Intl. J. Env. Prot.*, 23(9): 990–998.

Senior, B.W., 1989. Examination of water, milk, food and air. In: *Mackie and McCartney Practical Medical Microbiology*, 13th edn, Vol. II, (Eds.) J.G. Collee, J.P. Duguid, A.G. Fraser and B.P. Marmion. Chruchill Livingstone, Edinburgh, N.Y., p. 204–216.

Sharma, M.R., 2004. Water quality of river Beas in Hamirpur area of outer Himalayas. *Poll. Res.*, 23(1): 41–44.

Singh, R.K., Iqbal, S.A. and Seth, P.C., 2001. Bacteriological pollution in a stretch of river Narmada at Hoshangabad, Madhya Pradesh. *Poll. Res.*, 20(2): 211–213.

Srivastava, R.D., 2002. *It's Poison Everywhere.* In: Capital Story, Times of India Daily, 15th April.

Taylor, H., 2003. Surface waters. In: *The Handbook of Water and Wastewater Microbiology*, (Eds.) D. Mara and N. Horan. Academic Press, London, p. 611–626.

Thakur, R.S., Bhuyan, K. and Burna, R., 2003. Potability of water of Jorhat city in terms of bacteriological quality. *Ind. J. Environ. and Ecoplan.*, 7(1): 83–86.

WHO, 1992. *International Standard for Drinking Water.* World Health Organization, Geneva, Switzerland.

Zutshi, N., 1992. Effect of Jammu city sewage water on abiotic and biotic factors of the river Tawi, Jammu. *Ph.D. Thesis*, Jammu University, Jammu (J&K), India.

Ethnobotanical Wisdom and Microbial Studies on Medicinal Plants (2011) *Pages* **335–338**
Editors: **D.R. Khanna, Ashutosh Gautam, R. Bhutiani, Gagan Matta, Anil Dhiman**
Published by: **DAYA PUBLISHING HOUSE, NEW DELHI**

Chapter 50

Nyctanthes arbor-tristis Linn. (Harsinghar): A Potential Medicine

Prabhat[1], Navneet[1], Sanjeev[2] and Pramod[2]
[1]Department of Botany and Microbiology, [2]Department of Chemistry, Gurukula Kangri University, Haridwar

ABSTRACT

Nyctanthes arbor-tristis Linn. is widely used in the traditional medicinal systems of India. It possesses hepatoprotective, antileishmonial and antiallergic, antiviral and antifungal activities. The petroleum ether, ethanol and water extract of stem, root, flowers, seeds and leaves of the plant were screened for the antibacterial activity against *Escherichia coli, Bacillus subtilis, Staphylococcus aureus* and *Klebsiella pneumoniae* by using well diffusion method. The results were compared with reference antibiotic ampicillin. The water extract of root shows minimum inhibition zone (10 mm) against *S. aureus* and *B.subtilis* and petroleum ether extract of leaves against *E.coli* and ethanolic extract showed the maximum activity againnst *S. aureus* and *B. subtilis*.

Keywords: Nyctanthes arbor-tristis, Antibacterial activity, Pathogens.

Introduction

The development of bacterial resistance to presently available antibiotics has necessitated the search for new antibacterial agents (Prabhat and Navneet, 2007). The Gram-positive bacteria such as *S.aureus* are mainly responsible for post operative wound infection, toxic shock syndrome and food poisoning. The Gram-negative bacteria such as *E.coli* are present in human intestine and causes lower urinary tract infection, coenocytes or septicemia. Several plants are indicated in folk and other traditional systems of medicine that act as aseptic agents. *Nyctanthes arbor-tristis* commonly known as Harsinghar belongs to the family Oleaceae (Chitravanshi *et al.*, 1992). The tree is small in size and found in abundance in the forests of central India and sub-Himalayan region. Traditionally in India the plant is used in snake bite, animals bites, cancer, sores, ulcers, dysentery, menorrhaogia (Badam *et al.*, 1987) and obstinate sciatica. The leaf of *Nyctanthes arbor-tristis* is very active from the immunologic point of view. It strongly stimulates antigen specific and non-specific immunity as shown by increase in humoral and delayed type hypersensitivity response to sheep erythrocytes and in macrophage migration index (Puri *et al.*, 1994). In the present study the antibacterial activity of different parts *i.e.* seeds, root, stem, leaves and flower against *E. coli*, *B. subtilis*, *S. aureus* and *K. pneumoniae* were carried out.

Materials and Method

The plant material of *Nyctanthes arbor-tristis* was collected from Haridwar and the plant was identified by the Botanical Survey of India, Dehradun (Uttarakhand). The each part of the plant was dried separately in shade and powdered by using grinder. The 100 gm of powdered plant material was loaded in soxhlet assembly extracted successively with petroleum ether, ethanol and water (400ml of each solvent). By removing the solvents with vacuum evaporator crude extract was used for antibacterial activities. The Mueller Hinton Agar media (Hi media No. M 173) was poured (25ml.) into sterilized Petri plate and left for solidification at room temperature and used to test the antibacterial activity of the extracts prepared from the plant against *S.aureus*, *B.subtilis*, *E.coli* and *K.pneumoniae* by well diffusion or cup plate method (Ahmad *et.al.*, 1998 and Prabhat *et al.*, 2005a,b). 8mm diameter wells were punched in the agar media by borer and filled with extracts and antibiotic ampicillin (100 mg/ml) was used as positive control and solvent as negative control. The plates were incubated at 37°C for 24 hours to obtain inhibition zones.

Results and Discussion

Table 50.1 enumerates the effect of the different solvents and aqueous extract of the plant against 4 pathogens. All plant extracts showed significant activity against the bacteria. Similar results of biological activity of plant against fungi and bacteria were reported by Ahmad and Beg (2001).

The methanolic extracts showed the strong activity against both gram positive and gram negative bacteria. In general gram +ve organisms were more sensitive than gram-ve, similar differences in the sensitivity were also observed by Suresh and Chauhan (1992). The extracts are tested for their antibacterial activity against the Gram-positive and Gram-negative bacteria. The activity was compared with the antibiotic ampicillin. All the extracts show activity against *E.coli*, *K.pneumoniae*, *S.aureus* and *B.subtilis*. The inhibition zones of all parts of the plant extracts are found to be less as compared to ampicillin. The results indicate that the petroleum ether extracts exhibit the lower degree of antibacterial activity as compared to aqueous and ethanol. The maximum inhibition zone measured for antibiotic control is (25 mm) against *B. subtilis*. The minimum inhibition zone is (10 mm) of water extract of root and petroleum extract of leaves for *S.aureus*, *B.subtilis* and *E.coli* respectively. The ethanolic extract of leaves show maximum inhibition zone is (23 mm) against *S.aureus* and *B.subtilis*.

The inhibition zone by flowers extracts in ethanol and stems extract in water against *E. coli* was very effective nearly exhibition with ampicillin. Similarly water extracts of seeds, leaves, flowers and a stem inhibition is almost resembles inhibition in ampicillin against *K. pneumoniae*. The leaves extract with ethanol shows inhibition against *S.aureus* was found to be the same with ampicillin. The use of

Table 50.1: The Antibacterial Activity of *Nyctanthes arbor-tristis* Extracts in mm

Pathogen	*Root*			*Leaves*			*Seeds*			*Flowers*			*Stem*			*Ampicillin*
	P.E.	*EOH.*	*W*	*P.E.*	*EOH.*	*W*	*P.E.*	*EOH.*	*W*	*P.E.*	*EOH.*	*W*	*P.E.*	*EOH.*	*W*	
E.C.	12	17	16	10	18	14	13	20	20	16	22	20	14	20	21	23
K.P.	14	16	15	11	20	20	14	20	21	17	20	19	15	19	20	21
S.A.	13	15	10	12	23	20	14	21	20	18	21	20	14	22	20	24
B.S.	12	14	10	14	23	21	12	20	20	17	20	21	11	19	19	25

E.C.: *E.coli*; K.P.: *K. pneumoniae*; S.A.: *S. aureus*; B.S.: *B. subtilis*.

: The solvents did not showed any zone of inhibition. P.E.: Petroleum ether; EOH: Ethanol*; W: Water*.

leaves, seeds, and stem can be used in place of ampicillin antibiotic against some diseases caused by these bacteria.

References

Ahmad, I., Mehmood, Z., Mohammod, F., 1998 Screening of some Indian medicinal plants for their antimicrobial properties. *Journal of Ethnopharmacology*, 62: 183–193.

Ahmad, I. and Beg, A.Z., 2001. Antimicrobial and phytochemical studies on 45 Indian medicinal plants against multi-drug resistant human pathogens. *Journal of Ethnopharmacology*, 74: 113–123.

Badam, L., Rao, T.L.G., Andwaghu, V., 1987. *In vitro*: Antimalarial activity of fresh leaf juice of *Nyctanthes arbor-tristis* (Linn.). *Indian Journal of Parasitology*, 11(1): 13–14.

Chitravanshi,V.C.,Ghoshal, Singh, A.P., Prasad, B.N.K., Srivastava, V. and Tandan, J.S., 1992. Therapeutic action of *Nyctanthes arbor-tristis* L., Against caecal amoebiasis of rat. *International Journal of Pharmacognosy*, 30: 71–75.

Prabhat, Navneet, and Shri Kishna, 2005a. Antimicrobial activity of Apamarga (*A. aspera*). *Indian Academy of Science Letters*, 28(11–12): 379–381.

Prabhat, Navneet and Shri Krishna, 2005b. Antibacterial activity of *Mimusops elengi* (Bakul). *Environment Conservation Journal*, 6(2): 59–61.

Prabhat and Navneet, 2007. Antibacterial activity of medicinal plants against dental infections. In: *Medicinal plants: Conservation, Cultivation and Utilization*, (Eds.) D.R. Khanna *et al.* Daya Publishing House, New Delhi, pp. 139–146.

Puri, A., Saxena, R., Saxena, R.P, Saxena, K.C. Srivastava, V. and Tandan, J.S., 1994. Immunostimulant activity of *Nyctanthes arbor-tristis* (*L*). *Journal of Ethnopharmacology*, 42: 31–32.

Suresh, K. and Chauhan, K.K., 1992. A study of antimicrobial activity of *Calotropis procera* leaf extracts. *Geobios*, 19: 135–137.

Ethnobotanical Wisdom and Microbial Studies on Medicinal Plants (2011) *Pages* **339–343**
Editors: **D.R. Khanna, Ashutosh Gautam, R. Bhutiani, Gagan Matta, Anil Dhiman**
Published by: **DAYA PUBLISHING HOUSE, NEW DELHI**

Chapter 51

Antimicrobial Activity of *Vitex negundo* Leaf Extracts

Deepika[1], Padma Singh[1] and Vishal Upadhyay[2]
[1]Department of Microbiology, Kanya Gurukul Mahavidyalaya, (GKU), Haridwar
[2]Department of Botany and Microbiology, Gurukul Kangri University, Haridwar

ABSTRACT

Methanol, diethyl ether and acetone extracts of leaf of *Vitex negundo* were tested for their antibacterial activity against two human pathogenic bacteria *Staphylococcus aureus, Pseudomonas aeruginosa* and one fungus *Aspergillus niger* using the disc diffusion method. It was found that all the extracts produced inhibitory effect but the methanol extract of leaves exhibited a superior level of antimicrobial activity. The minimum inhibitory concentration (MIC) and minimum fungicidal concentration (MFC) was determined by broth dilution method. The results were compared with the reference antibiotics.

Keywords: *Antibacterial activity, Vitex negundo, Leaf extract, MIC.*

Introduction

Infectious diseases are a serious problem worldwide and account for high proportion of health problems in the developing countries (Sashikumar *et al.*, 2003). On the other hand some of the drugs currently in use result in adverse side effects (Covington, 1988). Therefore the search for new antimicrobial substances exhibiting minimal side effects is warranted (Kandil *et al.*, 1994). One of the most promising area in the search for new biologically active compounds are the plants used in traditional medicine (Alonso *et al.*, 1995).

Vitex negundo (verbenaceae), a large aromatic shrub upto 4-5 meters in height is found throughout the greater part of India growing upto an altitude of 1500m in the outer Himalayas. In addition to India, it is also found in Sri Lanka, Burma, China, Pakistan, Afghanistan, Malaysia, tropical Africa and the Philippines. It can be propagated readily by vegetative cutting. The leaves of *Vitex negundo* are reported to possess pesticidal, antifungal and antibacterial properties. The plant is also used as a commercial drug in the indigenous system of medicine (Anon, 1976).

Materials and Method

The present investigation was conducted to evaluate the antimicrobial potential of leaf extracts of *Vitex negundo* against three pathogenic microorganisms *Staphylococcus aureus*, *Pseudomonas aeruginosa* and *Aspergillus niger* which was obtained from Institute of Microbial Technology (IMTECH), Chandigarh. Local isolates were obtained from B.H.E.L Hospital, Haridwar.

The matured leaves of *Vitex negundo* were collected from Haridwar and its adjoining areas, dried in shade and crushed in mortar. The crushed leaves were extracted with methanol, diethyl ether and acetone for 24h using Soxhlet apparatus. The solvent was removed in rotary evaporator and the crude extract was used. Three different dilutions of the extracts will be prepared in dimethyl sulphoxide (DMSO).The working concentrations of the extracts were 200, 400 and 800 mg mL^{-1} respectively.

Table 51.1: Inhibition Zones of the Leaf Extracts of *Vitex negundo* against *Staphylococcus aureus*

Extract	*Concentration (mg/ml)*	**Effective Zone of Inhibition*	
		S. aureus (MTCC)	*S. aureus (Local)*
Methanol	800	17.6**±0.57	13.6**±0.57
	400	14.6**±0.57	12.0**±0
	200	11.3**±0.57	9.6**±0.57
Diethyl ether	800	17.3**±0.57	13.3**±0.57
	400	12.0**±1.0	11.3**±0.57
	200	9.6**±0.57	9.3**±0.57
Acetone	800	17.0**±0	12.0**±1.0
	400	14.3**±0.57	10.6**±0.57
	200	10.0**±0	9.3**±0.57
Ciprofloxacin	500 ppm	15**±1.0	12**±1.0
DMSO	100 per cent	—	—

Note: Values are the average of 3 replicates ± SD, *Effective zone of inhibition=Total zone of inhibition- diameter of the disc(5mm), **Significant at 0.05 per cent level

The antibacterial activity was tested *in vitro* by disc diffusion assay (Bauer *et al.*, 1966) on nutrient agar medium (NAM) and Sabouraud's dextrose agar medium(SDA) by taking Whattmann no 42 filter paper discs which was saturated with 10µl of different dilutions of different extracts. DMSO as negative control and ciprofloxacin, gentamycin and clotrimaxazole (500ppm) were used as positive control for *S. aureus*, *P. aeruginosa* and *A. niger* respectively. The plates were incubated at 37°C for 24 h in case of bacteria and at 28°C for 5 to 7 days in case of fungi. The activity was measured in terms of diameter of inhibition zone appearing around the filter paper discs saturated with leaf extracts. The tests were made in triplicates.

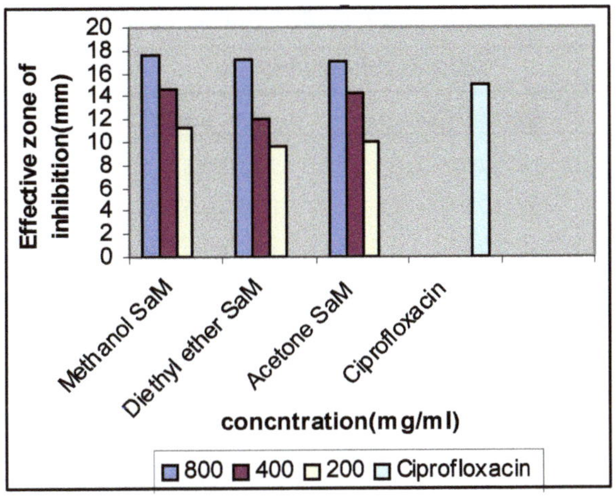

Figure 51.1: Comparative Analysis of the Activity of Leaf Extract of *V. negundo* and that of Antibiotic Control Against *S. aureus*

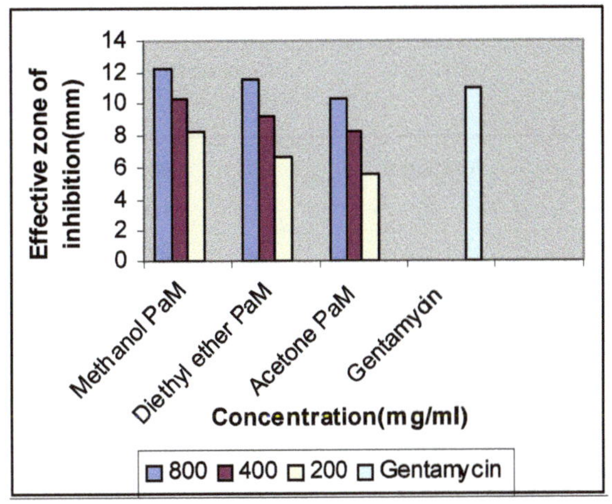

Figure 51.2: Comparative Analysis of the Activity of Leaf Extract of *V. negundo* and that of Antibiotic Control Against *P. aeruginosa*

The minimum inhibitory concentration (MIC) was determined in sterile microtiter plates each having 24 wells (Forbes *et al.*, 1998). 1µl of methanol extract was taken from the stock solution having a concentration of 800mg mL^{-1} and serially diluted using nutrient broth. 0.1 µl of the bacterial and fungal inoculum was added to each well and incubated. The tests were made in triplicates.The lowest concentration (highest dilution) of the plant extract preventing the turbidity is considered to be the MIC.

Table 51.2: Inhibition Zones of the Leaf Extracts of *Vitex negundo* Against *P. aeruginosa*

Extract	*Concentration (mg/ml)*	**Effective Zone of Inhibition*	
		P. aeruginosa (MTCC)	*P. aeruginosa (Local)*
Methanol	800	12.3**±0.57	10.6**±0.57
	400	10.3**±0.57	8.0**±1.0
	200	8.3**±0.57	5.3**±0.57
Diethyl ether	800	11.6**±0.57	9.0**±1.0
	400	9.3**±0.57	6.3**±0.57
	200	6.6**±1.15	3.3**±0.57
Acetone	800	10.3**±0.57	9.6**±0.57
	400	8.3**±0.57	7.6**±0.57
	200	5.6**±0.57	6.0**±0
Gentamycin	500ppm	11**±1.0	9**±1.0
DMSO	100 per cent	—	—

Note: Values are the average of 3 replicates ± SD, *Effective zone of inhibition=Total zone of inhibition- diameter of the disc(5mm), **Significant at 0.05 per cent level

Table 51.3: Inhibition Zones of the Leaf Extracts of *Vitex negundo* against *A. niger*

Extract	*Concentration (mg/ml)*	**Effective Zone of Inhibition*	
		A. niger (MTCC)	*A. niger (Local)*
Methanol	800	10.3**±0.57	7.6**±0.57
	400	7.6**±0.57	5.3**±0.57
	200	5.3**±0.57	3.6**±0.57
Acetone	800	9.3**±0.57	7.0**±1.0
	400	5.6**±0.57	5.3**±0.57
	200	4.0**±1.0	3.6**±1.15
Diethyl ether	800	6.6**±0.57	5.0**±1.0
	400	5.0**±0	3.6**±0.57
	200	3.6**±0.57	2.0**±1.0
Clotrimazaole	500ppm	10**±1.0	14±0.47
DMSO	100 per cent	—	—

Note: Values are the average of 3 replicates ± SD, *Effective zone of inhibition=Total zone of inhibition- diameter of the disc(5mm), **Significant at 0.05 per cent level

Table 51.4: MIC (Minimun Inhibitory Concentration) of Methanol Extract of *Vitex negundo* Against *S. aureus*

Replicates	*Minimum Inhibitory Concentration (MIC in mg/ml)*	
	S. aureus (MTCC)	*S. aureus (Local)*
1	12.5	25
2	12.5	25
3	6.25	12.5
Mean	10.4	20.8
SD	±2.9	± 5.8

Table 51.5: MIC (Minimum Inhibitory Concentration) of Methanol Extract of *Vitex negundo* Against *P. aeruginosa*

Replicates	*Minimum Inhibitory Concentration (MIC in mg/ml)*	
	P. aeruginosa (MTCC)	*P. aeruginosa (Local)*
1	25	50
2	25	50
3	12.5	25
Mean	20.8	41.6
SD	± 5.8	±7.2

Table 51.6: MIC (Minimum Inhibitory Concentration) of Methanol Extract of *Vitex negundo* Against *A. niger*

Replicates	*Minimum Inhibitory Concentration (MIC in mg/ml)*	
	A. niger (MTCC)	*A. niger (Local)*
1	200	200
2	200	400
3	400	400
Mean	41.6	333.3
SD	±94.2	±94.2

Results and Discussion

All the three dilutions of the leaf extract showed antimicrobial activity and the 800 mg/ml concentration was most effective and was comparable to the one unit solution of antibiotic (Tables 51.1–51.3 and Figures 51.1–51.3). The minimum inhibitory concentration (MIC) of the methanol extract of *V. negundo* against *S. aureus*, *P. aeruginosa* and *A. niger* was found to be 10.4, 20.8 and 41.6 mg/ml respectively and the minimum bactericidal concentration (MBC) and minimum fungicidal concentration (MFC) was found to be 41.6, 83.3 and 800 mg/ml respectively (Tables 51.4–51.6). The extract was found to be cidal in nature as there is no visible growth after subculturing the sample from the wells. Thus the results confirm the presence of antimicrobial components in the leaves of *Vitex negundo*.

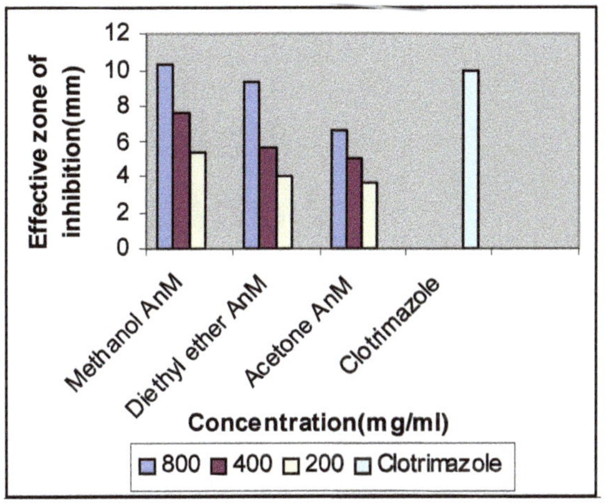

Figure 51.3: Comparative Analysis of the Activity of Leaf Extract of *V. negundo* and that of Antibiotic Control Against *A. niger*

References

Alonso, Paz E, Cerdeiras, M.P., Fernandez, J., Ferreira, F., Moyna, P., Soubes, M., Vazquez, A., Vero, S., and Zunino, L., 1995. Screening of Uruguayan medicinal plants for antimicrobial activity. *J. Ethnopharmacology*, 45: 67–70.

Anon, 1976. *Wealth of India: Raw Materials*, CSIR, New Delhi, 10: 522.

Bauer, A.W., Kirby, W.M.M., Sherris, J.C. and Truck, 1966. Antibiotic susceptibility testing by standardized single disc method. *American Journal of Clinical Pathology*, 45: 493.

Covington, R.T., 1988. Management of diarrhoea: Fact and comparison. *Drug News Letter*, 13: 1–2.

Forbes, B.A., Sahm, D.F., and Weissfeld, A.S., 1998. *Bailley and Scott's Diagnostic Microbiology*, 10th edn. Mosby Publication USA, pp. 252–258.

Kandil, O., Radwan, N.M., Hassan, A.B., Ames, A.M.M., Bana, H.A., and Ames W.M.M., 1994. Extracts and fractions of *Thymus capitatus* exhibit antimicrobial activities. *J. Ethnopharmacology*, 44: 19–24.

Sashikumar, J.M., Remya, M., Janardhanan, K., 2003. Antimicrobial activity of ethanomedical plants of Nilgiri Biosphere reserve and Western ghats. *Asian Journal of Microbial Tech. Env. Sci.*, 5:183–185.

Ethnobotanical Wisdom and Microbial Studies on Medicinal Plants (2011) *Pages* **344–351**
Editors: **D.R. Khanna, Ashutosh Gautam, R. Bhutiani, Gagan Matta, Anil Dhiman**
Published by: **DAYA PUBLISHING HOUSE, NEW DELHI**

Chapter 52

Antifungal Activity of *Lantana camara* L. and *Syzygium aromaticum* L. Against *Candida albicans*

Aditi Grover[1], B.S. Bhandari[1], Nishant Rai[2] and Harish Chandra[2]
[1]Department of Botany and Microbiology, H.N.B. Garhwal University
[2]Department of Microbiology, Gayatri College of Biomedical Science, Dehradun

ABSTRACT

Candida is a commensal organism found in 40-80 per cent of normal humans. Candidiasis is observed in immunocompromised individuals such as HIV-positive patients, it may also occur in blood and genital tract. The drug-resistant bacteria and fungal pathogens have further complicated the treatment of infectious diseases in immunocompromised, AIDS and cancer patients. The *in vitro* antimicrobial activity was performed by Agar disc diffusion and Agar well diffusion method. The crude aqueous fractions and Methanolic extracts of leaves of *Lantana camara* and buds of *Syzygium aromaticum* were investigated for their antifungal activity against yeast *Candida albicans*. Methanolic extract of *Syzygium* was found to be effective against *Candida* while aqueous extract showed no activity. Both aqueous and methanolic extract of *Lantana camara* showed no antifungal activity. These observations could be the basis for the usefulness of the buds of *Syzygium* in treatment remedies for candidiasis or microbial infections.

Keywords: *Candida albicans, Syzygium aromaticum, Lantana camara, Antifungal activity.*

Introduction

Infectious diseases are the world's leading cause of premature deaths, killing almost 50,000 people everyday. In recent years, drug resistance to human pathogenic bacteria has been commonly reported from all over the world (Piddock and Wise, 1989; Singh *et al.*, 1992; Mulligen *et al.*, 1993; Davis, 1994; Robin *et al.*, 1998). However, the situation is alarming in developing as well as developed countries due to indiscriminate use of antibiotics. The drug-resistant bacteria and fungal pathogens have further complicated the treatment of infectious diseases in immunocompromised, AIDS and cancer patients (Rinaldi, 1991; Diamond, 1993). In the present scenario, emergence of multiple drug resistance to human pathogenic organisms, has necessitated a search for new antimicrobial substances from other sources including plants. The increasing resistance to antifungal compounds and the reduced number of available drugs led us to search therapeutic alternatives among aromatic plants and their essential oils, empirically used by antifungal properties (Fontenelle *et al.*, 2007). The use of traditional medicine for the treatment of various diseases has been practiced for generation and a large number of populations in the country use traditional medicines for day to day healthcare needs. (Moshi *et al.*, 2006; Hamza *et al.*, 2006). Many efforts have been made to discover new antimicrobial compounds from various kinds of sources such as microbes, plants and animals. (Janovska *et al.*, 2003). Fungi are opportunistic organisms, which are ubiquitous in nature. Prolonged antibiotic therapy, invasive therapeutic procedures, immunosuppressive therapies and AIDS have all contributed to the rise in systemic fungal infections. (Advani, 1992). Ayurveda is the system of traditional medicine prevalent in India since 2000 B.C. Herbal plants from nature provide rational means for the treatment of many diseases, which are considered to be obstinate and incurable in other systems of medicine. It is a form of treatment by natural remedies, which makes use of the power of nature to restore human beings to a state of balance. Genuine and pure powdered medicinal herbs of Indian origin are used in herbal medicines, herbal extracts, herbal cosmetics and nutrition foods and other alternative medicines and herbal remedies for natural healings (Heinerman, 1996).

Plants that are traditionally used in treatment of bacterial and fungal infections or related ailments could be a good source for new, safe and biodegradable antimicrobial drugs (Hamza *et al.*, 2006) and could offer potential lead towards development of novel compounds that are active against pathogenic microbes (Runyoro *et al.*, 2006). In the present communication, an attempt has been made to explore antimycotic principles, which involves an investigation on the efficacy of essential oil against *Candida*. Ethnobotanical data is given in Table 52.1.

Table 52.1: Ethnobotanical Data of the Plants Used in the Present Study

Botanical Name	*Family*	*Common Name*	*Part Used*	*Traditional Uses (Chopra et al., 1992)*
Lantana camara L.	Verbenaceae	Ghaneri	Leaves	Decoction used for sore throat
Syzgium aromaticum L.	Myrtaceae	Laung	Bud	Stimulant, carminative used in dyspeps

Materials and Methods

Test Organisms

The strain of *Candida albicans ATCC 0231* was obtained from American type culture collection from IMTECH, Chandigarh and was biochemically and serologically characterized by standard methods. The culture was maintained at 4 ± 1°C on slants, plates of *Candida* Agar and corn meal agar (HiMedia) and overnight grown culture in sterile glycerol (15 per cent).

Plant Materials

Leaves of *Lantana camara* were collected from Garhwal region and bud of *Syzygium aromaticum* was procured from local market of Dehradun.

Preparation of Plant Extracts

Plant extracts were prepared by the methods of Alade and Irobi (1993) with minor modification. Both aqueous as well as alcoholic extracts were used to check the antifungal activity. The leaves of *L. camara* and bud of *Syzygium aromaticum* were air dried, ground and soaked in water for aqueous extract and in methanol for alcoholic extract for 72 hrs with continuous shaking and homogenized using high speed homogeniser (REMI RQ 127 A/D). It was filtered (Whatman No. 1) and extracts were obtained using soxhlet assembly. The extracts were concentrated in check at 30°C and stored at 4°C for further use.

Inoculum Preparation

For the preparation of inoculum, the tested fungus was cultured in mycological broth (HiMedia) at 37°C for 24hrs and standardized for the same absorbency, number 0.5 of the McFarland Standard, which corresponds to the order of 10^8 CFU/m (Barry and Thornsberry, 1985).

Antifungal Susceptibility Testing

The antifungal assay was performed by two methods *viz.* agar disc diffusion method (Bauer *et al.*, 1986) for aqueous extract and agar well diffusion method (Perez *et al.*, 1990) for solvent extract. The molten PDA (at 45°C) was inoculated with 100 µl of the inoculum (10^8cfu/ml) and poured into petri dishes. The plates were allowed to gel for an hour. 50 mg of both extracts were dissolved in 100 µl of sterile ethanol (15 per cent) separately and sterilized by passing through 0.45 µm millipore syringe filter. For agar disc diffusion method, the sterile disc (0.7 cm) Hi-Media, was saturated with 100 µl of both extract, allowed to dry and was introduced on the upper layer of the seeded agar plates separately. Sterile paper disc treated with 15 per cent of sterile ethanol was used as control. For agar well diffusion method, a well was prepared in the plates with the help of a cork-borer (0.85cm). 100 µl of the both extracts was introduced into the well in different plates. The plates were incubated overnight at 37 °C. Microbial growth was determined by measuring the diameter of zone of inhibition (ZOI). Pure solvents instead of extract were used as Control. The result was obtained by measuring zone diameter. The experiment was done three times and the mean values were taken.

Minimum Inhibitory Concentration

MIC method using broth dilution method was applied on extracts that proved their high efficacy against test microorganism by disc diffusion method and agar well diffusion method. The dried plant extract was dissolved in 15 per cent ethanol and serially diluted to fungal broth in order to observe their activities at low concentration. *Syzygium* extract was evaluated against *Candida* by dilution of stock to various concentrations from 100 mg/ml to 6.25 mg/ml. Equal volume of stock and fungal broth was mixed in a test tube. Specifically 0.1ml of standardized inoculum (10^8 CFU/ml) was added to each tube. The tubes were incubated aerobically at 37°C for 18-24 h. Two control tubes were maintained for each batch as antibiotic control (tube containing extract and the growth medium without inoculum) and organism control (the tube containing the growth medium, physiological saline and the inoculum). The lowest concentration (highest dilution) of the extract that produced no visible fungal growth (no

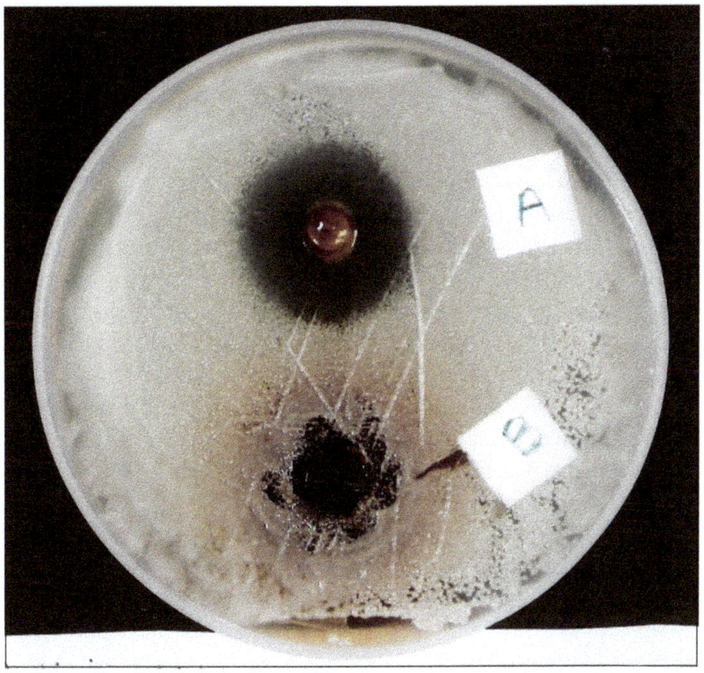

Figure 52.1: Antifungal Activity of *Syzygium aromaticum* Against *C. albicans*
A: Methanolic extract: B: Aqueous extract

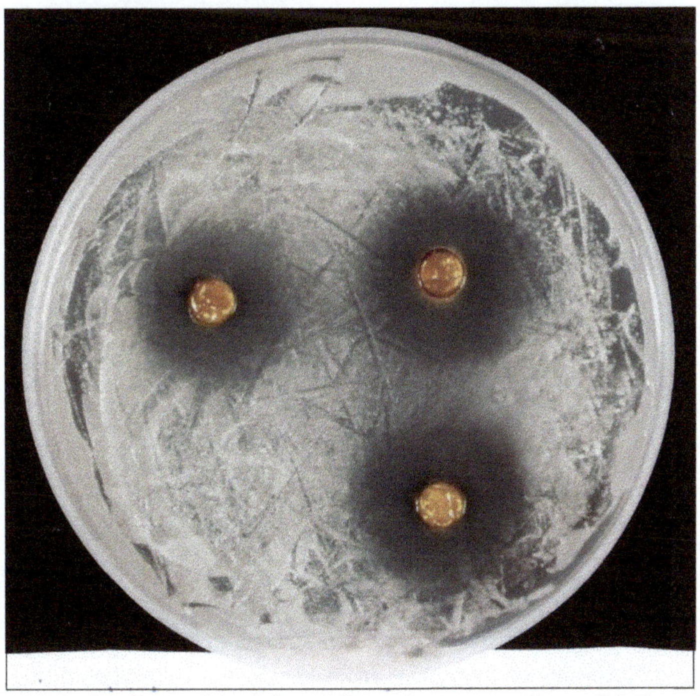

Figure 52.2: Antifungal Activity of *Syzygium aromaticum* (Methanolic) Against *C. albicans* Showing Zone of Inhibition

turbidity) when compared with the control tubes were regarded as MIC. However, MFC was determined by sub -culturing the test dilution on to a fresh drug-free medium and incubated further for 18-24 h. It is defined as the highest dilution that yielded no single fungal colony.

Results and Discussion

The antifungal activity of the extracts and their potency was quantitatively assessed by the presence or absence of inhibition zone and zone diameter respectively as shown in Table 52.2. Alcoholic extract was found to be a better solvent for extraction of antimicrobial active substances compared to water and hexane as also shown by Ahmad *et al.* (1998). The antifungal activity of the extracts and their potency was quantitatively assessed by the presence or absence of inhibition zone and zone diameter, respectively as shown in Table 52.3. The results showed that the growth of *C. albicans* was inhibited by *S. aromaticum*. Methanolic extract was found effective against *Candida* and displayed potent antifungal activity but *Candida* was not inhibited by aqueous extract (Figures 52.1 and 52.2). Zone of inhibition of *Syzygium* against *Candida* was 26 mm obtained after 24 hrs of incubation. MIC was observed at 6.25 mg/ml and MFC was observed at 12.50 mg/ml (Table 52.4). *Syzgium* extract evaluated against *Candida* by serial two fold dilution of stock concentration (100mg/ml) showed inhibition upto 6.25 mg/ml (Figure 52.3). *Syzygium* was more potent and maximum zone of inhibition was observed. Aqueous and methanolic extract of *L. camara* showed no activity, confirming that *Candida* is resistant to this extract. A drug resistant strain of *C. albicans* was found to be sensitive to the

Figure 52.3: Antifungal Activity of *Syzygium aromaticum* in Various Dilutions Against *C. albicans*

Well A: 100mg/ml; Well B: 50mg ml; Well C: 25mg/ml; Well D:12.5;mg/ml; Well E: 6.25mg/ml

tested plant extract. This indicated that antibiotic resistance does not interfere with antifungal action of plant extract. Similar reports on anticandidal activities of these plants were also reported by Ahmad and Beg, 2001 with varying degrees of potency. The difference in potency may be due to the stage of collection of the plant sample, different sensitivity of the test strains and method of extraction (Nimri *et al.*, 1999).

Table 52.2: Antifungal Activity of *Lantana camara* and *Syzygium aromaticum* Zone of Inhibition (mm)

	Lantana camara		*Syzygium aromaticum*	
	Aqueous	*Methanolic*	*Aqueous*	*Methanolic*
C. albicans	NI	NI	NI	2

NI: No Inhibition

Table 52.3: Antifungal Activity of *Syzygium aromaticum* Zone of Inhibition (mm)

Stock (100 mg/ml)	*50.00 mg/ml*	*25.00 mg/ml*	*12.50 mg/ml*	*6.25 mg/ml*
25.00	19.00	13.00	10.00	9.0

Note: ZOI in mm; < 10mm: Low activity; 10-20mm: Moderate activity; >20mm: High activity.

Table 52.4: Minimum Inhibitory Concentration (MIC) and Minimum Fungicidal Concentration (MFC) (*Syzygium aromaticum*)

Organism	*MIC (mg/ml)*	*MFC (mg/ml)*
C. albicans	6.25	12.5

Successful prediction of botanical compounds from plant material is largely dependent on the type of solvent used in the extraction procedure. The traditional healers or practitioners make use of water primarily as a solvent, but our studies showed that methanolic extracts of these plants were certainly much better and powerful. This may be due to the better solubility of the active components in the organic solvent (De Boer *et al.*, 2005). These observations can be rationalized in terms of polarity of the compounds being extracted by each solvent and in addition to their intrinsic bioactivity, by their ability to dissolve or diffuse in the different media used in the assay. The growth media also seem to play an important role in the determination of the antibacterial activity. (Lin *et al.*, 1999).

The results of the present study supports the traditional usage of the studied plants and suggests that some of the plant extracts possess compounds with antifungal properties that can be used as antifungal agents in new drugs for the therapy of infectious diseases caused by pathogens. The most active extracts can be subjected to isolation of the therapeutic antimicrobials and carry out further pharmacological evaluation.

References

Advani, S.H., 1992. Incidence of fungal disease increasing in cancer patients. *Fungal Challenge.*, 1:1–2.

Ahmad, I. and Beg, A.Z., 2001. Antimicrobial and phytochemical studies on 45 Indian medicinal plants against multi-drug resistant human pathogens. *Journal of Ethnopharmacology*, 74: 113–123.

Ahmad, I., Mehmood, Z. and Mohammad, F., 1998. Screening of some Indian medicinal plants for their antimicrobial properties. *Journal of Ethnopharmacology*, 62: 183–193.

Alade, P.I. and Irobi, O.N., 1993. Antimicrobial activities of crude leaf extracts of *Acalypha wilkensian*. *Journal of Ethnopharmacology*, 39: 171–174.

Barry, A.L. and Thornsberry, C., 1985. Susceptibility tests: Diffusion test procedures. In: *Manual of Clinical Microbiology*, (Eds.) E.H. Lennnette, A. Balows, W.J. Hausler and H.J. Shadomy. Am. Soc. for Microbiol., Washington, DC, p. 978–987.

Bauer, A.W., Kirby, W.M.M., Sherris, J.C.and Turck, M., 1996. Antibiotic susceptibility testing by standardized single disc method. *American Journal of Clinical Pathology*, 45:493–496.

Davis, J., 1994. Inactivation of antibiotic and the dissemination of resistance genes. *Science*, 264: 375–382.

De Boer, H.J., Kool, A., Broberg, A., Mziray, W.R., Hedberg, I., Levenfors, J.J., 2005. Antifungal and antibacterial activity of some herbal remedies from Tanzania. *J. Ethnopharmacol.*, 96: 461–469.

Diamond, R.D., 1993. The growing problem of mycoses in patients infested with human immunodeficiency virus. *Review of Infectious Diseases*, 13: 480–486.

Fontenelle, R.O.S., Morais, S.M., Brito, E.H.S., Brihante, R.S.N., Cordeiro, R.A., Nascimento, N.R.F., Kerntopf, M.R., Sidrim, J.J.C. and Rochna, M.F.G. 2007. Antifungal activity of essential oils of Croton species from the *Brazalian caatinga* biome. *Journal of Applied Microbiology* doi: 10.1111/j.1365–2672.2007.03707.x.

Hamza, O.J.M., Beukel, C., Matee, M. J., Ajam, V.V., Mikx, F. and Verweji, P.E., 2006. Antifungal activity of some Tanzanian plants used traditionally for the treatment of fungal infections. *J. Ethnopharmacol.*, 108(1): 124–132.

Heinerman, J. Heinerman, 1996. *Encyclopedia of Healing Herbs and Spices*. Parker Publishing Co-west Nyack, N.Y., pp. 255–257.

Janovska, D., Kubikova, K. and Kokoska, L., 2003. Screening for antimicrobial activity of some medicinal plant species of traditional Chinese medicine. *Czech. J. Food Sci.*, 21:107–111.

Moshi, M.J., Mbwambo, Z.H., Kapingu, M.C., Mhozya, V.H. and Marwa, C., 2006. Antimicrobial and brine shrimp lethality of extracts of *Terminalia mollis* Laws. *Afri. J. Trad. CAM.*, 3(3):1–10.

Mulligen, M.E., Murry-leisure, K.A., Ribner, B.S., Standiford, H.C., John, J.F., Karvick, J.A., Kauffman, C.A. and Yu, V.L., 1993. Methicillin resistant *Staphylococcus aureus*. *American Journal of Medicine*, 94: 313–328.

Nimri, L.F., Meqdam, M.M. and Alkofahi, A., 1999. Antibacterial activity of Jordanian medicinal plants. *Pharmaceutical Biology*, 37(3): 196–201.

Lin, J., Opoku, A.R., Geheeb-keller, M., Hutchings, A.D., Terblanche, S.E., Jager, A.K. and van Staden, J., 1999. Preliminary screening of some traditional Zulu medicinal plants for anti-inflammatory and anti-microbial activities. *J.Ethnopharmacol.*, 68: 267–274.

Perez, C., Pauli, M. and Bazerque, P., 1990. An antibiotic assay by the well agar method. *Acta Biologiae et Medicine Experimentalis*, 15: 113–115.

Piddock, K.J.V., Wise, R., 1989. Mechanisms of resistance to quinolones and clinical perspective. *Journal of Antimicrobial Chemotherapy*, 23: 475–483.

Rinaldi, M.G., 1991. Problems in the diagnosis of invasive fungal diseases. *Review of Infectious Diseases*, 13: 493–495.

Robin, E.H., Anril, W., Alexander, M., Leoto, M. and Keith, K., 1998.*Nasopharyngeal carriage* and antimicrobial resistance in isolates of *Streptococcus pneumoniae* and *Heamophilus influenzae* Type b in children under 5 years of agein Botswana. *International Journal of Infectious Diseases*, 3(1): 18–25.

Runyoro, D.K.B., Matee, M.I.N., Ngassapa, O.D., Joseph, C. C. and Mbwambo, Z.H., 2006. Screening of Tanzanian medicinal plants for anti-Candida activity. *BMC Complement Altern Med.*, 6: 11.

Singh, M., Chaudhary, M.A., Yadava, J.N.S. and Sanyal, S.C., 1992. The spectrum of antibiotic resistance in human and veterinary isolates of *Escherichia coli* collected from 1984–1986 in Northern India. *Journal of Antimicrobial Chemotherapy*, 29: 159–168.

Ethnobotanical Wisdom and Microbial Studies on Medicinal Plants (2011) *Pages* **352–354**
Editors: **D.R. Khanna, Ashutosh Gautam, R. Bhutiani, Gagan Matta, Anil Dhiman**
Published by: **DAYA PUBLISHING HOUSE, NEW DELHI**

Chapter 53

Bioevaluation of Antibacterial Potential of Sarpgandha (*Rauwolfia serpentina*)

Vishal Upadhyay[1], Deepika[2], Padma Singh[2] and P. Kaushik[1]
[1]Department of Botany and Microbiology, G.K.V., Haridwar
[2]Department of Microbiology, Kanya Gurukul Mahavidyalaya, Haridwar

ABSTRACT

The *in vitro* antibacterial activity of *Rauwolfia serpentina* plant extract has been investigated against *Escherichia coli, Staphylococcus aureus, Salmonella typhi* and *Klebsiella pneumoniae* using the disc diffusion method. The alcoholic extract was found effective against *Staphylococcus aureus* only. The antibacterial activity is attributed to the presence of alkaloids, which was confirmed by gas liquid chromatography and positive alkaloid test. The minimum inhibitory concentration (MIC) was determined by paper disc diffusion method. The results were compared with reference antibiotic tetracycline (one unit solution).

Keywords: *Antibacterial, Rauwolfia serpentina, Plant extract, Disc diffusion method, MIC.*

Introduction

The evaluation of plant extract for their antibacterial activity has been known for more than seventy years (Machat and Kankel, 1920). Various medicinal plants have been used for years in daily life to treat diseases all over the world. The medicinal herbs represent a rich source of antibacterial activity. Herbal medicines are still the mainstay of about 70 per cent of world population for health care (Kaushik and Dhiman, 1999). Evaluation of plant extract for their antimicrobial activity has been done by several workers. Ansari (1995) studied effect of plant extract against the pathogen of leaf sheath bright of rice. Chakroborty and Brantner (1999) studied antibacterial steroid alkaloids from the stem bark of *Holarrhena pubescens*. Earlier literature on antibacterial potential of plants has been reviewed by Kaushik and Dhiman (1999). Antibacterial activity of various plant parts of *Aerva persica* has been tested against human pathogenic bacterial strains and pathogenic fungal species (Gehlot and Bohra, 1998).

Materials and Methods

The plant material of *Rauwolfia serpentina* was collected from state Ayurvedic College, Gurukul, Haridwar, and bacterial strains *viz. Escherchia coli* (MTCC-739), *Staphylococcus aureus* (MTCC-537), *Salmonella typhi* (MTCC-531) and *Klebsiella pneumoniae* (MTCC-432) were self purchased from IMTECH, Chandigarh. For the preparation of plant extract, plant materials were first washed 2-3 times with tap water and then again with sterilized double distilled water. Finally the surface sterilization was done with 90 per cent ethyl alcohol. 100 grams of plant material were crushed in ware blender resulting in the formation of a paste which was mixed in 250 ml of absolute ethyl alcohol. Alcoholic extract so prepared was allowed to evaporate at room temperature until 80 ml of this was left. This extract was squeezed through double layer muslin cloth and filtered through Whatman's filter paper no-42 and was centrifuged at 5000 r.p.m. for 20 minutes and was then sterilized by passing through 0.2 micron disposable filters for primary screening of antibacterial testing procedure, 100 per cent, 50 per cent and 20 per cent dilution of extract were taken.

For antibacterial screening and minimum inhibitory concentration (MIC), agar and disc diffusion method was used (Bauer *et al.*, 1966). In this method nutrient agar medium was prepared and autoclave and then cooled up to 42–45°C. To each 100 ml nutrient agar medium 1.0 ml of 24 hrs old bacterial cultures was added from nutrient broth and then shaken properly to ensure complete distribution of microorganisms in the medium.

Table 53.1: The Antibacterial Effect of *Rauwolfia serpentina* Plant Extract

Test Organism	*Inhibition Zones in mm*								
	Antibiotic Zone in mm	*Extract Zone (A)*			*Control Alcohol Zone (B) in mm*	*D.W. Zone (C)*	*Effective Zone of Inhibition (A–B)*		
		100%	*50%*	*20%*			*100%*	*50%*	*20%*
Staphylococcus aureus	18	29.5	21	13.5	5	Nil	24.5	16	8.5
Salmonella typhi	19	–	–	–	5	Nil	–	–	–
Escherichia coli	17	–	–	–	5	Nil	–	–	–
Klebsiella pneumoniae	18	–	–	–	5	Nil	–	–	–

The culture medium which was already inoculated with bacterial suspension, was poured in Petri dishes when it was in solid phase, Whatman's filter paper no-42 discs, which were already dipped in different dilutions of the plant extract were placed on nutrient agar surface. D.W. and absolute alcohol served as negative control and the standard antibiotic tetracycline (one unit solution) as positive control. After inoculation plates were kept at 30° C for 24 hrs in incubator. The antibacterial activity was evaluated by measuring the diameter of the inhibition zone in mm. The alcoholic extract was tested for presence of alkaloid by Dragondraff's test also followed by Stahi (1969) and Gas liquid chromatography (GLC) which was carried under University Science instrumentation centre, Indian

Istitute of Technology (IIT) Roorkee. In GLC HP-5 column and FID detector 250°C were used for analysis. This column detected alkaloids from extract and solvent was ethyl alcohol.

Results and Discussion

Results of present investigations clearly indicate that the alcoholic extract of *Rauwolfia serpentina* is effective against *S.aureus* only, and was found non effective against *E.coli*, *S.typhi* and *K. pneumoniae.* The effective zone of inhibition was 24.5 mm against 100 per cent concentration, 16 mm against 50 per cent concentration and 8.5 mm against 20 per cent concentration (Table 53.1). The MIC was reported at 14 per cent extract concentration against *S.aureus.* The results indicated that the undiluted alcoholic extract more effective in comparison to the antibiotic.

References

Ansari, M.M., 1995. Control of sheath bright of rice by plant extract. *Indian Phytopathology*, 48(3): 268–270.

Bauer, A.W., Kirby, W.M.M., Sherris, J.C. and Turek, 1966. Antibiotic susceptibility testing by standardized single disc method. *American Journal of Clinical Pathology*, 45: 493.

Chakraborty, A. and Brantner, A.H., 1999. Antibacterial steroid alkaloids from the stem bark of *Holarrhena pubescens. Journal of Ethnaopharmacology*, 68: 339–344.

Gehlot, D. and Bohra, A., 1998. Antimicrobial activity of various plant part extracts of *Aerva persica. Advances in Plant Science*, 11(1): 109–111

Kaushik, P. and Dhiman, A.K., 1999. *Medicinal Plants and Raw Drugs of India.* Bishen Singh, Mahinder Pal Singh, Dehradun, p. xii+623.

Machat, D.I. and Kankel, W.M., 1920. *Proc. Soc. Exp. Bio. Med.*, p. 19–68.

Stahl, E., 1969. *Thin Layer Chromatography*, 2nd Edn. MRF Ashworth, Springer Verlag, Heidelberg.

Ethnobotanical Wisdom and Microbial Studies on Medicinal Plants (2011) *Pages* **355–366**
Editors: **D.R. Khanna, Ashutosh Gautam, R. Bhutiani, Gagan Matta, Anil Dhiman**
Published by: **DAYA PUBLISHING HOUSE, NEW DELHI**

Chapter 54

Stress Induced Alterations in the Outer Membrane of *Escherichia coli K-12* Strain

Bindu Arora
Department of Microbiology, Abeda Inamdar Sr. College, Pune

ABSTRACT

In the present study, gram-negative (*Escherichia coli K-12*) bacterial biomembrane involvement was studied in the presence of modulating factors such as EDTA, Mg^{+2} ions and EDTA and Mg^{+2} ions in combination. The release of proteins and their involvement during the transport of β-Lactams namely Ceftriaxone and Cefazolin were also studied. The broader applications of Ceftriaxone for pharmaceutical implications offer greater advantage as compared to pre-existing β-Lactams. Due to the availability of more signal molecules in the membranes there is enhanced toxicity at 5 mM EDTA concentration, and easy entrapment of antibiotics, thus enhanced sensitivity levels. A concentration of 15 mM Mg^{+2} ions was found to be toxic for *E.coli* whereas it exhibited luxuriant growth with decreasing Mg^{+2} ion concentration under antibiotic stress. On the contrary, when 5 mM EDTA is treated in combination with Mg^{+2}, it attributed reduced signals available on the membrane surface therefore, reduced drug sensitivity. To identify the involvement of specific proteins and to know the site of proteins released which are directly or indirectly involved in transport of antibiotics across the biological membrane, the protein release was monitored from intact cells, as well as, membrane vesicles derived from *E.coli* cells and studied upto a level of molecular weight determination and measured by using a high-pressure liquid chromatography (HPLC). The study confirms the induction of certain stress signal proteins from the outer membrane, thereby rendering the bacteria more susceptible to therapy.

Keywords: HPLC (High Pressure Liquid Chromatography), CEF (Ceftriaxone Sodium), CEZ (Cefazolin Sodium), OMPs (Outer Membrane Proteins).

Introduction

Outer membrane protein A (OmpA) is located in the membrane of *Escherichia coli* and other gram-negative bacteria and plays a multifunctional role in bacterial physiology and pathogenesis.(Alfredo *et al.*, 2006). Also, many new outer membrane proteins have recently been identified by proteomics techniques in *Escherichia coli*. In the outer membrane of gram-negative bacteria, the porins, are present in large amounts and form water-filled channels that permit the diffusion of small hydrophilic solutes across the membrane (Nikaido, 2003; Nikaido, 1994; Nikaido, 1985; Nikaido and Vaara, 1985) including bacterial nutrients and antimicrobials. The permeability of outer membranes of gram-negative bacteria to hydrophilic compounds is mostly due to presence of porin channels (Dela Vega and Delcour, 1995). Since β-Lactam antibiotics penetrate the outer membrane of gram-negative bacteria, resistance could also be caused by loss or deficiency of specific porins that reduce the outer membrane permeability to β-Lactam antibiotic. This might be an important factor in mediating β-Lactam resistance in multidrug *E.coli* (Nikaido, 2003). The Cephalosporins have been known to induce stress signal proteins, from the outer membrane, inhibiting cross-linking step of peptidoglycan biosynthesis in the cell wall of *E.coli* thereby rendering the bacteria more susceptible to therapy (Russel and Chopra, 1990). β-Lactams binding to these PBP's may be eliminated by the action of 3 enzymes, *viz.*, β-Lactamases, the acylases and the esterases. Bacteria may become more resistant to these β-Lactams by producing altered transpeptidases (PBP's) with greatly reduced affinity for the binding of these antibiotics (Nikaido *et al.*, 1990).

Materials and Methods

Escherichia coli K-12, wild type strain was procured from IMTECH,Chandigarh. Cultures were grown on Nutrient Agar at 37 °C±1 and stored at 4 °C. Ceftriaxone sodium (CEF) and Cefazolin Sodium (CEZ) was purchased from Aristo pharmaceutical India Ltd. (M.P.).

E.coli was grown for growth studies on Spencer and Guest (1973) media, pH 6.9. Incubation was carried out at 37°C±1 in a thermostatically controlled orbital shaker (Labline model No. 3521) under aerobic conditions with a platform rotation of 180 rev min^{-1}.

Growth was monitored under various conditions spectrophotometrically (Spectronics-20 Bausch and Lomb) at 550 nm.

For transport and protein release studies, membrane vesicles were prepared by Koning and Kaback (1973) with some modifications. Membrane protein analysis was carried out by Warburg and Christian (1941) method. The qualitative study of protein release was done by HPLC (Waters, USA) at 4,000 psi pressure, at absorbance of 280 nm. The solid phase was Protein-PAK column (Waters,USA).

Results and Discussion

A highly significant and negative correlation between concentration of CEF and CEZ with percent survival was found to be r = 0.646 (pd" 0.02) and r = -0.757 (pd" 0.02) respectively. *E.coli K-12* wild type strain was found to be ten times sensitive to CEF (LD_{50} of 0.005 ppm), than with CEZ (LD_{50} of 0.05 ppm) (Figure 54.1). This high sensitivity of *E.coli* to CEF as compared to CEZ was attributed to the high Na+ ion concentration outside *E.coli* (Na+ ion content of CEF-Na^+ ion, 83 mg/gm) cell wall as compared to its interior, leading to Na^+–H^+ ion antiporter activity and thus dissipation of pH gradient, ultimately interrupting the pmf and increase in the CEF uptake through its porins. Loss of porin-mediated resistance mechanism against cephalosporin has been observed among the multidrug resistant *E.coli* (Ananthan and Subha, 2005). In relation to sensitivity of *E.coli* cells to β-Lactams in the presence of

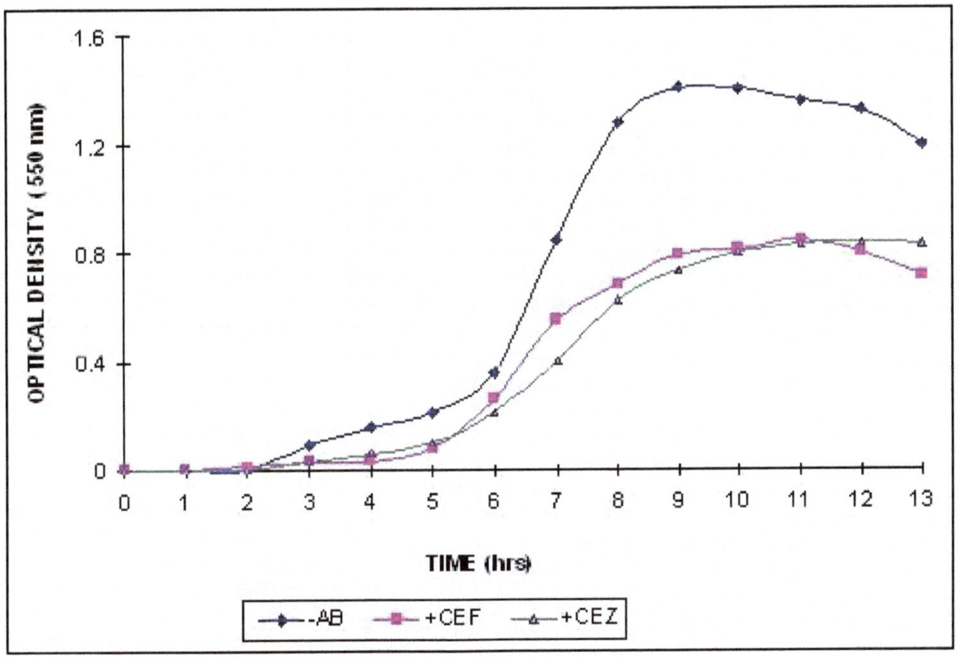

Figure 54.1: Effect of LD_{50} Concentrations of CEF and CEZ on the Growth of *E. coli*

varying H^+ ion concentrations, the protein thus, obtained in suspension medium shows variations with respect to time of incubation. The protein content present in this suspension medium (Tris-HCI Buffer), in the presence of β-Lactams as well as the release of proteins under antibiotic stress shows variations with respect to time of treatment (Figures 54.2a,b).

Bacterial adaptation to various environmental stresses has been extensively investigated (Yohannes *et al.*, 2004; Yohannes *et al.*, 2005). Chelation property of EDTA, thus, facilitates antibiotic transport, thereby resulting in antibiotic toxicity (Yamada *et al.*, 1978). It has also been demonstrated that the effect of Mg^{+2} ions at high concentrations include reduction in the growth rate, extension of lag phase and perturbation in morphology and physiology (Hughes and Poole, 1989). The cell extract of *E.coli* exhibited a significant degradation of EDTA only in the presence of Fe^{+3} ion. There have been some reports regarding the interaction of non-toxic metals with toxic metals or metal chelators diminishing or enhancing the toxicity of the latter by the former. The ion antagonism has been studied but not sufficiently. Leive (1968) found that the concentration of Na^+ ion and K^+ ion do not affect EDTA action but the divalent cation like Mg^{+2} ion prevent its action. Attempts to revive EDTA-treated cells with cation such as Mg^{+2} ion have met with slight success.

It was observed that with varying concentrations of EDTA (1.00, 2.50 and 5.00 mM), antibiotic-treated cells were inhibited more than normal cells. The protein release in CEF-treated cells along with 1.00, 2.50 and 5.00 mM EDTA concentrations was 0.245, 0.270 and 0.280 µg/ml; whereas with CEZ-treated cells was observed as 0.225, 0.260 and 0.275 µg/ml respectively, after 60 min of incubation, as illustrated in Figure 54.3a.

These observations were found consistent with the existing evidence (Matzushita *et al.*, 1978). The enhanced protein release with higher concentration of EDTA explained the pronounced activity

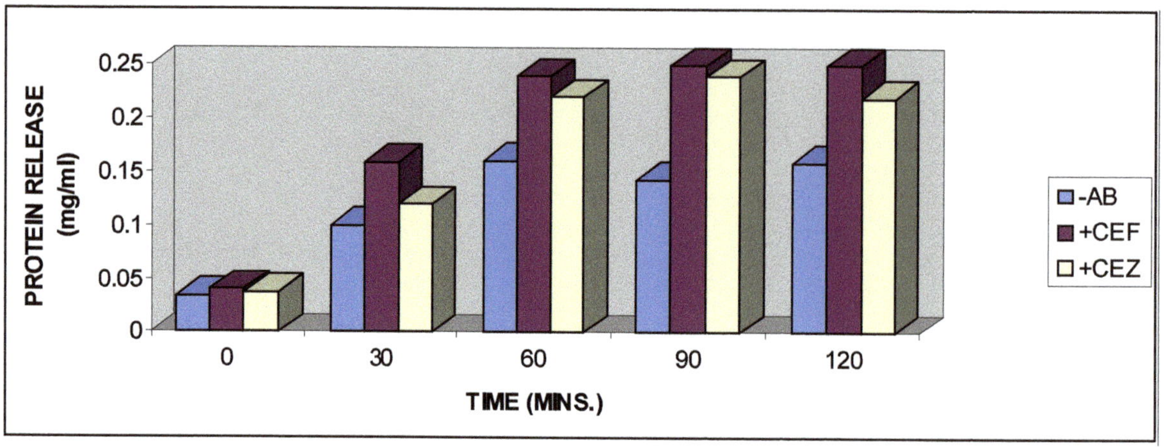

Figure 54.2a: Protein Release from Intact Cells of *E. coli* at Varying Time Intervals

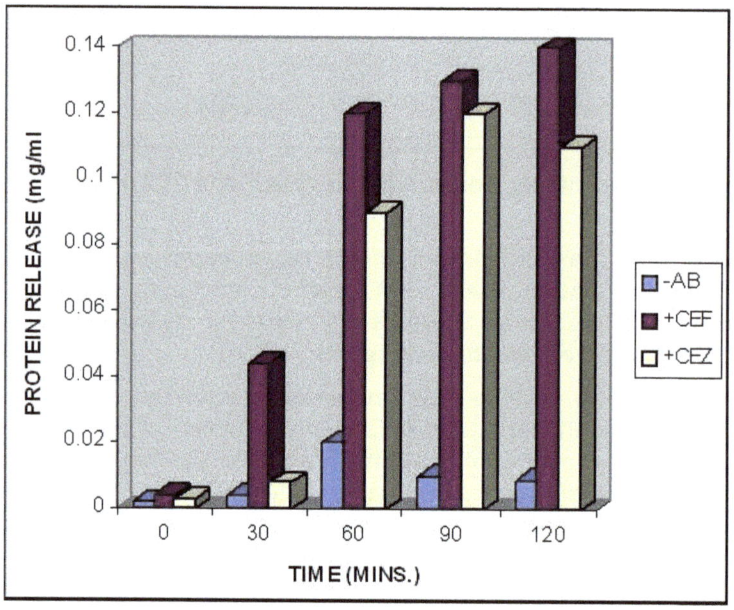

Figure 54.2b: Protein Release from Membrane Vesicles of *E. coli* at Varying Time Intervals

of antibiotics against the EDTA treated cells at higher concentration. The effect of varying concentrations of EDTA on the membrane vesicles of *E.coli* is illustrated in Figure 54.3 (a), (b).

In the presence of 1.00 mM EDTA, the release of protein was quite less *i.e.,* 0.13 and 0.10 µg/ml with CEF and CEZ, respectively in comparison to 2.50 mM and 5.00 mM EDTA (in combination with antibiotics) releasing 0.15 and 0.20 µg/ml in the presence of CEF; whereas, 0.14 and 0.19 µg/ml in the presence of CEZ respectively, in the medium after 60 mins of incubation. In contrast to that, the inhibition of 5.00 mM EDTA was more pronounced than 1.00 mM EDTA in membrane vesicles of *E.coli*. The less quantity of protein released with 1.00 mM EDTA supports the earlier observations that

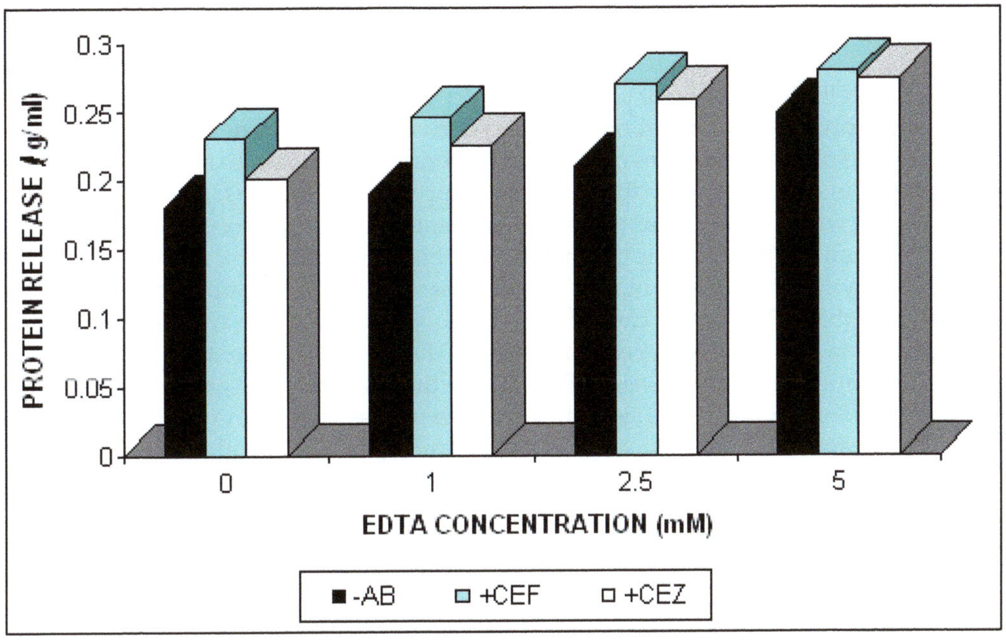

Figure 54.3a: Protein Release from Intact Cells of *E. coli* in Presence of Varying Conc. of EDTA

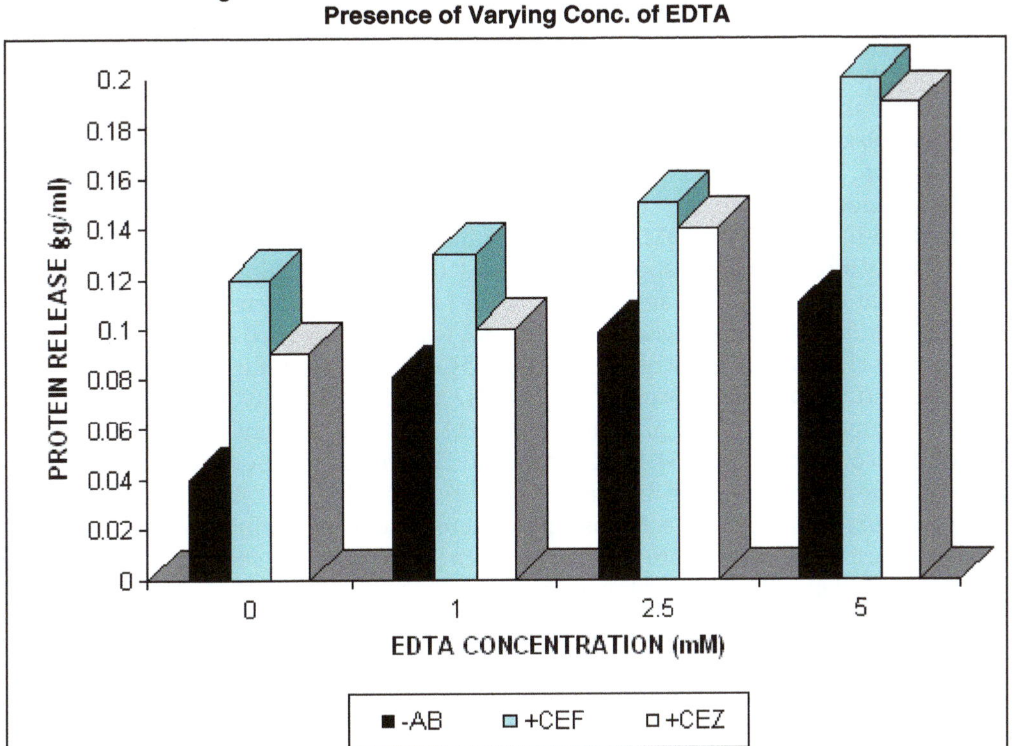

Figure 54.3b: Protein Release from Membrane Vesicles of *E. coli* in the Presence of Varying Conc. of EDTA

in growing cultures of *E.coli*, 5.00 mM EDTA along with cephalosporins showed maximum inhibition of growth.

In case of intact cells, the amount of protein released in the presence of 5.00 mM EDTA was four-fold greater than the protein from membrane vesicles derived from gram-negative bacteria. It is very obvious that outer membrane comprises of low protein content in contrast to intact cells which has all the cellular proteins in combination with structural component, as well as, functional components such as enzymes. But with intact cells, as well as, membrane vesicles the speciation of proteins differs with respect to their molecular size as apparent from the Figures 54.4 (a), (b). As far as molecular size is concerned, EDTA with cephalosporin resulted in release of similar group of proteins which had molecular size more than 67.00 kD. In case of EDTA also, peak C at 5 min was very prominent and differs in concentration reflecting high peak in case of EDTA and CEF, in contrast to EDTA and CEZ; after 60 min of incubation. In all the cases, peak height of the proteins were greater in the presence of CEF and CEZ. It is evident from growth, as well as, enzyme studies, that gram-negative cells were more sensitive to CEF treatment in contrast to CEZ as a counterpart. EDTA although potentiated the cephalosporin toxicity by many folds in case of CEF than CEZ. It is clear from the figure that the protein leaked out at 10 min run was at much higher conc. in case of CEF than CEZ. The protein release was found to be directly dependent upon the increasing concentration of EDTA from the cell. The amount of cellular protein, thus calculated by carrying a difference between protein release from intact cells and membrane vesicles shows a constant trend with a similar concentration at 1.00, 2.50 and 5.00 mM concentration, in case of both the cephalosporins. However, the difference in toxicity can be then considered as involvement of OM proteins exclusively in conferring sensitivity towards antibiotics.

In *E.coli*, when intact cells were treated with varying concentrations of Mg^{+2} ions, it was observed that the extent of protein release was quite higher as the concentration of Mg^{+2} ions was increased. The protein release from the cell wall of *E.coli* when treated with 1.00 mM, 5.00 mM, 10.00 mM and 15.00 mM Mg^{+2} ion was observed as 0.160, 0.175, 0.225 and 0.247 µg/ml; in the absence of antibiotics after 60 min of incubation. Thus, increasing concentration of Mg^{+2} ions had released more proteins. The inhibitory effects induced by high Mg^{+2} ion concentration which release more protein were explained in terms of stress-induced injury of OM. On the other hand in the presence of antibiotics, 15.00 mM Mg^{+2} ions drastically inhibits the transport of antibiotics, therefore resulting in fewer protein release. The protein release at 1.00, 5.00, 10.00 and 15.00 mM Mg^{+2} ions in the presence of CEF was found to be 0.229, 0.225, 0.215 and 0.190 µg/ml; whereas, with CEZ was observed as 0.202, 0.206, 0.20 and 0.185 µg/ml respectively after 60 min of incubation (Figure 54.4a). However, the protein release from the membrane vesicles (OM) of *E.coli* was observed as 0.042, 0.05, 0.072 and 0.097 µg/ml when the cells were suspended in the presence of 1,5,10 and 15.00 mM Mg^{+2} ion concentration respectively; after incubation period of 60 mins (Figure 54.4b). The variation in protein release at varying concentrations of Mg^{+2} ions could be attributed to the differential behaviour of Mg^{+2} ions at different concentrations *i.e.*, at lower concentration Mg^{+2} ions acted as a potentiator and a stabilizing agent of the membrane while at higher concentrations they are inhibitory. With β-Lactams, the contrast behaviour of Mg^{+2} ions has been observed in relation to protein release.

At MIC_{50} conc. of CEF and CEZ, the protein released in the medium when the cells were suspended with 1.00, 5.00, 10.00 and 15.00 mM Mg^{+2} ion concentrations was observed as 0.12, 0.11, 0.10 and 0.09 µg/ml with CEF, whereas with CEZ, the protein release was 0.11, 0.10, 0.08 and 0.07 µg/ml, when the aliquots were drawn after 60 min of incubation. This might be due to the fact that the antibiotic accessibility to the inside of the cell is reduced at higher concentration of Mg^{+2} ions which releases fewer proteins at high concentrations (15.00 mM Mg^{+2} ions).

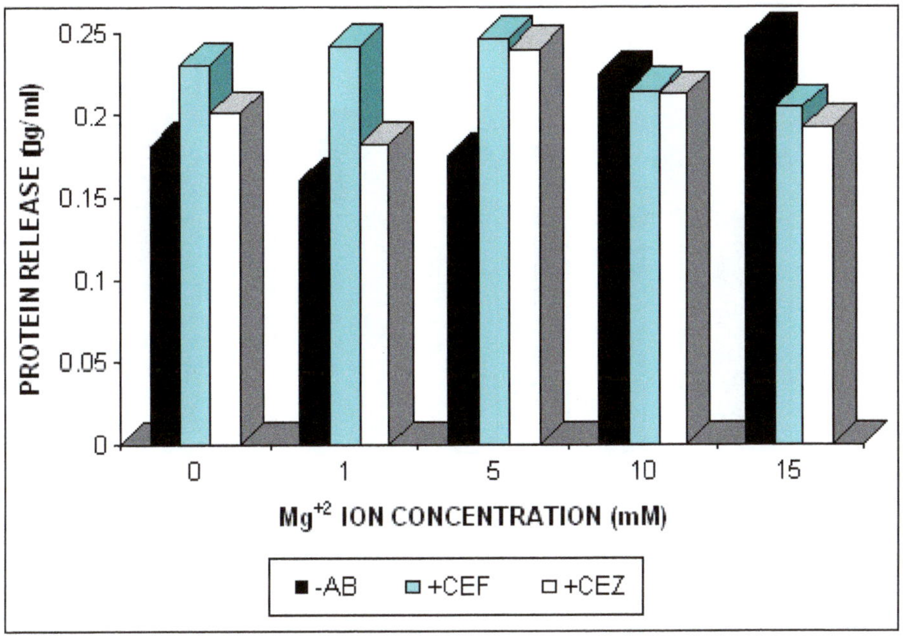

Figure 54.4a: Protein Release from Intact Cells of *E. coli* with and Without Antibiotics in Presence of Varying Concentrations of Mg^{+2} Ions

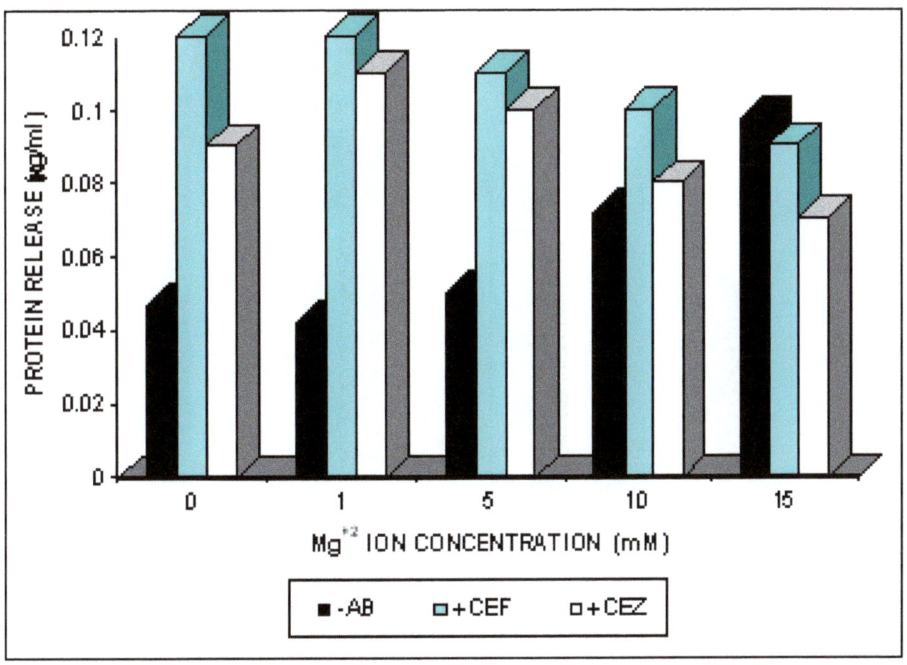

Figure 54.4b: Protein Release from Membrane Vesicles of *E. coli* in Presence of Varying Conc. of Mg^{+2}

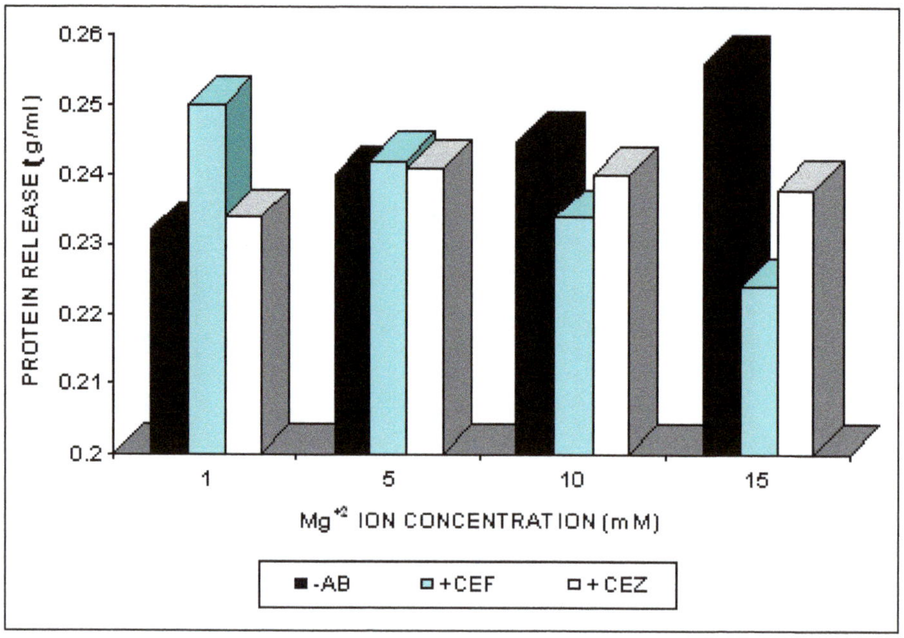

Figure 54.5a: Protein Release from Intact Cells of *E. coli* in the Presence of Varying Conc. of Mg^{+2} Each Followed by the Addition of 5mM EDTA

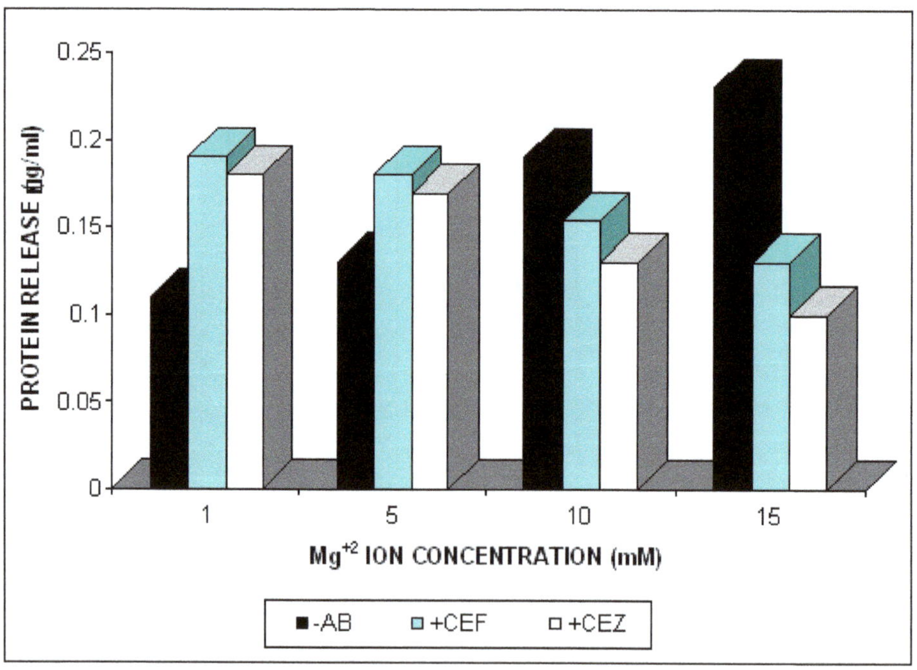

Figure 54.5b: Protein Release from Membrane Vesicles of *E. coli* in the Presence of Varying Conc. of Mg^{+2} Each in Combination with 5 mM EDTA

The effect of β-Lactams on the protein release of *E.coli* in the presence of EDTA as well as, Mg^{+2} ions have been studied. The extent of protein release was less when intact cells were suspended in 15 mM Mg^{+2} (in combination with 5 mM EDTA) than in 1.00 mM (in combination with 5.00 mM EDTA), after 60 min of incubation. The protein released in the medium when cells were suspended in 1.00, 5.00, 10.00 and 15.00 mM Mg^{+2} ion concentration (each in combination with 5.00 mM EDTA) was observed as 0.250,0.242, 0.234 and 0.224 µg/ml respectively, in the presence of CEF, whereas protein release was 0.234, 0.241, 0.240 and 0.238 µg/ml respectively, in the presence of CEZ; after 60 min of incubation (Figure 54.5a).

This must be attributed again to the differential behaviour of different concentrations of Mg^{+2} ions in relation to release of proteins. The enhanced protein release by antibiotics and EDTA had been limited by high concentrations of Mg^{+2} ions *i.e.*, 15.00 mM to a great extent. The protein release was observed as 0.19, 0.18, 0.154 and 0.13 µg/ml in the presence of CEF; and in the case of CEZ, the protein release from the membrane vesicles of *E.coli* was observed as 0.18, 0.17, 0.13 and 0.10 µg/ml, when the cells were suspended in 1.00 mM, 5.00 mM, 10.00 mM and 15.00 mM Mg+2 ion concentration (each supplemented with 5 mM EDTA) respectively, after 60 min of incubation (Figure 54.5b). The variation in protein release might be attributed to the fact that the inhibitory effect of antibiotics was potentiated with the help of EDTA, which has resulted in more protein release, however, Mg^{+2} ions would have retained the stability of OM thus protecting the synergistic behaviour of EDTA in combination with β-Lactam. Contrary to that Mg^{+2} ions at higher conc. when supplemented would have saturated the molecules of EDTA present into the medium thereby, protecting possible chelation of OM ions by EDTA; thus, maintaining integrity of cell wall.

As far as, speciation of protein leakage is concerned magnesium in combination with EDTA and cephalosporins resulted in release of possibly similar group of proteins that were greater than 67 kD which appeared after 11 min of run. In the absence of EDTA, Mg^{+2} ion causes effectively release of protein A and B in case of CEF; whereas with CEZ peak B was omitted (Figure 54.6c). A remarkable phenomenon observed with CEF when treated with EDTA and Mg^{+2} ion the potentiality of third generation cephalosporin can be possibly related by the appearance of peak G after 20 min of run resembling exactly with cytochrome "c" with a molecular weight of 12.5 kD (Figure 54.6d). The appearance of cytchrome "c" peak into the medium under the stress of CEF might have resulted in making gram-negative cells sensitive towards CEF by reducing cellular energy level required to carry out possible cellular metabolism. Omp's located in the outer membrane of *E.coli* plays a multifunctional role in bacterial physiology and pathogenesis (Alfredo *et al.*, 2006). Recently it has been identified by proteomics techniques that there are many new *E.coli* outer membrane proteins, out of eight predicted outer membrane proteins, the outer membrane localization for five- YftM, YaiO, YfaZ, CsgF, and YliI- are confirmed and also the sixth one-YfaL- may be an outer membrane autotransporter. Since β-Lactam antibiotics penetrate the outer membrane, resistance could also be caused by loss or deficiency of specific porins that reduce the outer membrane permeability to β-Lactam antibiotic. Also, the cause for antibacterial drug resistance has been known to be the active efflux.The study of protein release from intact cells, as well as, membrane vesicles of *E.coli* shows the direct involvement of outer membrane components in developing sensitivity towards β-Lactams. The observations, thus, confirms the greater effectiveness of CEF as compared to CEZ towards *E.coli*.

The extent of EDTA- induced outer membrane losses from cells of wild- type *Escherichia coli* K-12 were concentration dependent. An additional Mg^{+2} ions immediately following the EDTA treatment decreased the release of outer membrane proteins and reduced the leakage of periplasmic proteins, suggesting that the temporary increase in outer membrane "permeability" caused by EDTA treatment

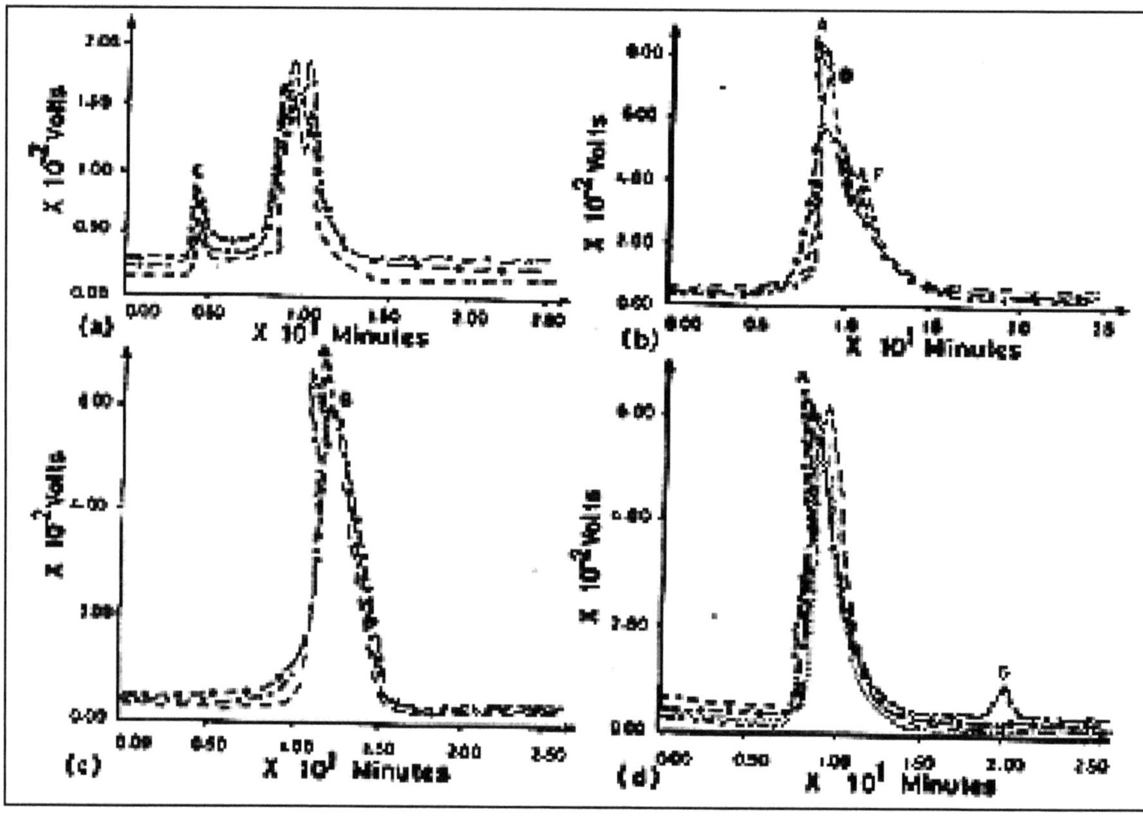

Figure 54.6: Molecular Weight Determination of Protein Released from *E. coli* Cells in the Presence of Antibiotic Stress as Monitored by HPLC, in the Presence of 5mM EDTA, 5mM Mg^{+2} Ions and 5 mM EDTA in Combination with 5 mM Mg^{+2} Ions, Separately After 60 mins of Incubation.

(a) Membrane vesicles- EDTA (5 mM) (b) Intact cells – EDTA (5 mM) (c) Intact cells – Mg^{+2} ions (5 mM) (d) Intact cells–EDTA (5 mM) + Mg^{+2} ions (5mM). Here — indicates –AB; -x-x indicates +CEF; -o-o- indicates + CEZ; ...... indicates control conditions.

was rapidly reversed by the redistribution of outer membrane components, a process which is favored by a low Mg^{+2} ion concentration. The envelope alterations caused by EDTA reveals that the biochemical, as well as, cellular disturbances were found to be more easily related to the toxic action of β-Lactams, CEF and CEZ. The susceptibility of bacterial culture for specific β-Lactams can be assessed by gradual exploration of membrane/cellular proteins responsible for making possible transport of antibiotics, across the cellular membrane against concentration gradient.

The study thus reveals that some of the membrane proteins as, well as some of the intracellular proteins get affected by the β-Lactam transport, revealing the mediation of direct antibiotic-protein interaction. The study confirms the orientation of the outer membrane and the energy status of the cell, determining the therapeutic value of cephalosporins.

Acknowledgement

The author is thankful to Dr. Kiran Singh (Barkatullah University, Bhopal), Dr. Rajendra Prasad (Jawahar Lal Nehru University, Delhi) and Mr. Pradeep Sahaya (National Institute of Immunology, Delhi) for all the support and help.

References

Alfredo, G., Torres, Yongguo, Li, Christopher, B. Tutt, Lijun Xin, Tonyia Eaves-Pyles and Lynn, Soong, 2006. Outer membrane protein A of *E. coli O157: H7* stimulates dendritic cell activation. *Infection and Immunity*, 74(5): 2676–2685.

Ananthan, S. and Subha, A., 2005. Cefoxitin resistance mediated by loss of a porin in clinical strains of *Klebsiella pneumoniae* and *Escherichia coli*, *Indian Journal of Medical Microbiology*, 23(1): 20–23.

Dela Vega, A.L. and Delcour, A.H., 1995. Cadaverine induces closing of *E.coli* porins. *EMBO J.*, 14: 6058–6065.

Hughes, M.N. and Poole, R.K., 1989. *Metals and Microorganisms*. Chapman and Hall, London, New York, pp. 412.

Koning, W.N. and Kaback, H.R., 1973. Anaerobic Transport in *E.coli* membrane vesicles. *Proc. Nat. Sci.*, 70: 3376–3381.

Leive, L., 1968. Studies on the permeability change produced in coliform bacteria by EDTA. *J. Biol. Chem.*, 243: 2373–2380.

Matzushita, K. Adachi, O. Shnagawa, E. and Ameyama, M., 1978. Isolation and characterization of outer and inner membranes from *Pseudomonas aeruginosa* and effect of EDTA on the membranes. *J. Biol. Chem.*, 83: 171–181.

Nikaido, H., 1985. Role of permeability barriers in resistance to β-Lactam antibiotics. *Pharmacol. Ther. J.*, 27(2): 197–231.

Nikaido, H., 1994. Porins and specific diffusion channels in bacterial outer membranes. *J. Biol. Chem.*, 269: 3905–3908.

Nikaido, H., 2003. Molecular basis of bacterial outer membrane permeability revisited. *Mol. Biol. Rev.*, 67(4): 593–656.

Nikaido, H., Liu, W. and Rosenberg, E.Y., 1990. Outer membrane permeability and β-Lactamase stability of dipolar ionic cephalosporins containing methoxyimino substituents. *Antimicrob. Agents Chemother.*, 34(2): 337–342.

Nikaido, H. and Vaara, M., 1985. Molecular basis of bacterial outer membrane permeability. *Microbial. Rev.*, 49(1): 1–32.

Russel, A.D. and Chopra, I., 1990. *Understanding Antibacterial Action and Resistance.* Ellis Harwood Series in Pharmaceutical Technology, England, U.K., pp. 19–227.

Spencer, M.E. and Guest, J.R., 1973. Isolation and properties of fumarate reductase mutants of *E.coli*. *J. Bacteriol.*, 114: 563–570.

Warburg, O. and Christian, W., 1941. Isolation and crystallization of enolase. *Biochem. J.*, 3(10): 384–421.

Yamada, J., Tatsuguchi, K. and Watanable, T., 1978. Effect of trialkyltin chlorides on microbial growth. *Agric. Biol. Chem.*, 42: 1167–1172.

Yohannes, E., Barnhart, D.M. and Slonczewski, J.L., 2004. pH-dependent catabolic protein expression during anaerobic growth of *Escherichia coli K–12. J. Bacteriol.*, 186:192–199.

Yohannes, E., Thurbur, A.E., Wilks, J.C., Tate, D.P. and Slonczewski, J.L., 2005. Polyamine stress at high pH in *E.coli K-1. BMC Microbiology*, 1471, 5: 59.

Ethnobotanical Wisdom and Microbial Studies on Medicinal Plants (2011) *Pages* **367–370**
Editors: **D.R. Khanna, Ashutosh Gautam, R. Bhutiani, Gagan Matta, Anil Dhiman**
Published by: **DAYA PUBLISHING HOUSE, NEW DELHI**

Chapter 55

Bioevaluation of *Cascabela thevatica* Plant Extract

Vishal Upadhyay[1], P. Kaushik[2] and Deepika Upadhyay[3]
[1]Ranbaxy Fine Chemicals Limited, SIDCUL, Haridwar
[2]Department of Botany and Microbiology, Gurukul Kangri University, Haridwar
[3]Department of Microbiology, Kanya Gurukul Mahavidyalaya, Haridwar

ABSTRACT

The *in vitro* antibacterial activity of *Cascabela thevatica* plant extract has been investigated against *Escherichia coli, Staphylococcus aureus, Salmonella typhi* and *Klebsiella pneumoniae* using the agar disc diffusion method. The alcoholic extract was found effective against *Escherichia coli, Staphylococcus aureus* and *Salmonella typhi* but doesn't show any activity against *Klebsiella pneumoniae.* The antibacterial activity is attributed to the presence of alkaloids, which was confirmed by gas liquid chromatography (GLC) and positive alkaloid test. The minimum inhibitory concentration (MIC) was determined by paper disc diffusion method and the results were compared with reference antibiotic tetracycline (one unit solution).

Keywords: *Antibacterial, Cascabela thevatica, Plant extract, Disc diffusion method, MIC.*

Introduction

The increasing prevalence of multi drug resistant strains with reduced susceptibility to antibiotics raises the specter of untreatable microbial infections and adds urgency to the search for new infection fighting strategies. The use of plant extracts with known antimicrobial properties can be of great significance in therapeutic treatments. The evaluation of plant extract for their antibacterial activity has known for more than seventy years (Machat and Kankel, 1920). Evaluation of plant extract for their antimicrobial activity has been done by several workers. Chakraborty and Brantnor (1999), studied antibacterial steroid alkaloids from the stem bark of *Holarrhena pubescens*. Kaushik and Kishore (1991) studied effect of alcoholic extract of *Pholidota articulata* Lind. against *Bacillus subtilis, Staphylococcus aureus* and *Klebsiella pneumoniae*. Antimicrobial activity of various plant parts of *Aerva persica* has been tested against human pathogenic bacterial strains and pathogenic fungal species (Gehlot and Bobra, 1998). Antifungal activity of *Cassia fistula* leaf extract has been tested against *Candida albicans* (Singh and Karnwal, 2006).

The microorganisms have developed resistance to many antibiotics. This has created immense clinical problems in the treatment of disease; therefore, there is a need to develop an alternative to these drugs for the treatment of disease. The medicinal, herbs represent a rich source of antibacterial activity (Dhar *et al.*, 1968).

Materials and Methods

The plant material of *Cascabela thevatica* was collected from B.H.E.L, Haridwar and the bacterial strains viz., *Escherichia coli* (MTCC-739), *Staphylococcus aureus* (MTCC-737), *Salmonella typhi* (MTCC-31), *Klebsiella pmiumoniae* (MTCC-432) were self purchased from IMTECH, Chandigarh.

For the preparation of plant extract, plant material were first washed with 2–3 times with tap water and then again with sterilized distilled water. Finally the surface sterilization was done with 90 per cent absolute ethyl alcohol. 100 gms of plant material were crushed in ware blender resulting in the formation of a paste which was mixed in 250 ml of absolute ethyl alcohol. Alcoholic extract so prepared was allowed to evaporate at room temperature until 80 ml of this was left. This extract was squeezed through double layer muslin cloth and filtered through Whatman filter paper no.42 and was centrifuged at 5000 rpm for 20 minutes and was then sterilized by passing through 0.2 micron disposable filters. For primary screening of antibacterial testing procedure 100 per cent, 50 per cent and 20 per cent dilutions of extract were taken. In 100 per cent dilution, no distilled water was added, in 50 per cent extract dilution 50 per cent part of distilled water and in 20 per cent dilution 80 per cent part of distilled water was mixed.

For antibacterial screening and minimum inhibitory concentration (MIC), agar disc diffusion method was used (Bauer and Kirby, 1966).In this method nutrient agar medium was prepared and autoclaved and then cooled up to 42–45°C. To each 100mI of nutrient agar medium 1.0 ml of 24h old bacterial culture of *E. coli, S. aureus, S. typhi* and *K. pneunioniae* was added from nutrient broth and then shaken properly to ensure complete distribution of microorganisms in the medium.

The culture medium which was already inoculated with bacterial suspension was poured in Petri dishes and when it was in solid phase, Whatman's filter paper no 42 discs, which were already dipped in different dilutions of the plant extract, were placed on nutrient agar surface. Distilled water and absolute alcohol served as negative control and the standard antibiotic tetracycline (one unit solution) as positive control. After inoculation plates were kept at 37° for 24 hrs in incubator. The antibacterial activity was evaluated by measuring the diameter of the inhibition zones in mm.

The alcoholic extract was tested for the presence of alkaloid by Dragondroff's test, also followed by Stahl (1969) and gas liquid chromatography (GLC), which was carried under university science instrumentation center, Indian Institute of Technology (IIT), Roorkee. In GLC HP-5 column and FID detector was used for analysis at 250°C temp. This column detected alkaloids from extract and solvent was ethyl alcohol.

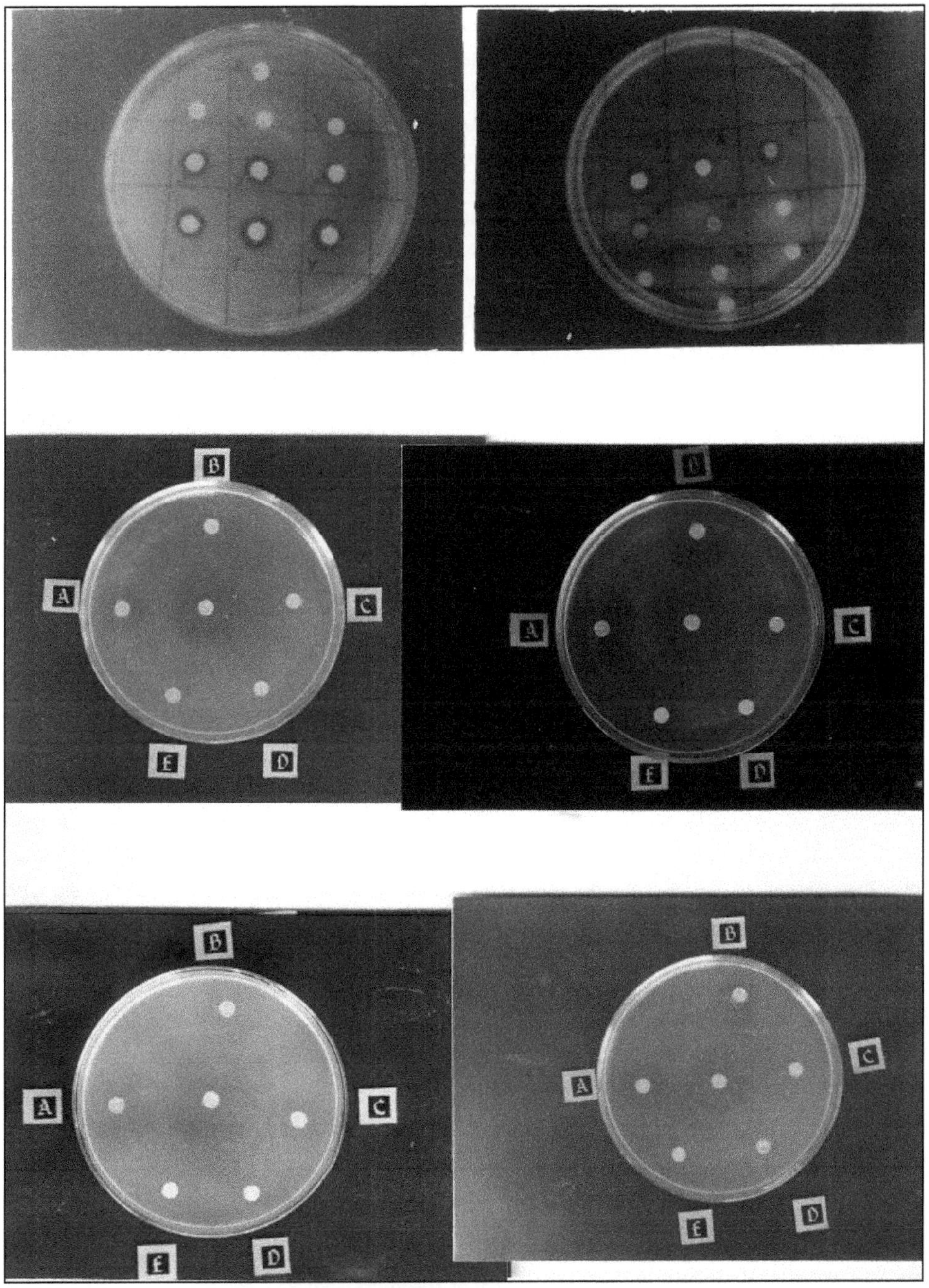

Figure 55.1: Showing MIC Against *E. coli* and *S. typhi* (Upper)
Showing Zone of Inhibition Against *E. coli* (Middle)
Showing Zone of Inhibition Against *S. typhi* (Lower)

Here, A: Showing 100 per cent Extract Potency; B: Showing 50 per cent Extract Potency; C: Showing 20 per cent Extract Potency; and Center Disc of Antibiotic Potency

Results and Discussion

Results of present investigation clearly indicates that the alcoholic extract of *Cascabela thevatica* is effective against *E. coli, S. aureus* and *S. typhi,* and was found non effective against *K. pneumoniae.* The size of effective zone of inhibition of undiluted (100 per cent) plant extract of *Cascabela thevalica* against *E. coli* measured 9.5 mm, *S. aureus* 12.7 mm and *S. typhi* 11.7 mm (Table 55.1). The MIC of *Cascabela thevatica* extract against *E. coli* was 33.3 per cent concentration, against *S. aureus* measured 11.3 per cent and *S. typhi* measured against 18.3 per cent concentration (Table 55.2) *S. aureus* was only bacterium, which was the most sensitive to undiluted (100 per cent) alcoholic extract of *C. thevatica.* In reference to antibiotic the zone of inhibition against *E. coli* were 15 mm, against *S. aureus* 16 mm and against *S. typhi* 13 mm which was nearly comparable to that at plant extract.

Table 55.1: Determination of the Antibacterial Activity of *C. thevatica* Plant Extract and Reference Antibiotic by Disc Diffusion Method

Test Organism	**Effective Inhibition Zone in mm*					
	Antibiotic Zone	*Extract Zone*			*Control Alcohol*	*Distilled Water Control*
		100 per cent	*50 per cent*	*20 per cent*		
Escherichia coli	15	9.5	6.7	–	Nil	Nil
Staphylococcus aureus	16	12.7	10.0	6.0	Nil	Nil
Salmonolla typhi	13	11.7	9.7	6.5	Nil	Nil
Klebsiella Pneumoniae	15	–	–	–	Nil	Nil

* Effective zone of inhibition = Total zone of inhibition – Diameter of the disc (5 mm)

Table 55.2: Minimum Inhibitory Concentration (MIC) of *C. thevatica*

Extract	*Name of Organisms*		
Alcoholic Plant Extract	*Escherchia coli*	*Staphylococcus aureus*	*Salmonella typhi*
MIC (in per cent)	33.3 per cent	11.3 per cent	18.3 per cent

References

Bauer, A.W., Kirby, W.M.M., Sherris, J.C. and Turck, 1966. Antibiotic susceptibility testing by standardized single disc method. *American Journal of Clinical Pathology,* 45: 493.

Chakraborty, A. and Brantner, A.H., 1999. *Journal of Ethnopharmacology,* 68: 339–344.

Dhar, M.N., Dhar, M.M., Dhawan, B.N., Mahrotra, B.N., and Ray, C., 1968. Screening of Indian plants for biological activity. Part-I. *Indian J. Expt. Biology,* 6: 232.

Gahlot, D. and Bohra, A., 1998. *Advances in Plant Sciences,* 11(1): 109–111.

Kaushik, P. and Kishor, N., 1991. *J. Orchid Society of India,* 5(1,2): 93–96.

Machat, D.I. and Kankel, W.H., 1920. *Proc. Sec. Exp. Bio. Med.,* 19: 68.

Singh, P. and Karnwal, P., 2006: Antifungal activity of *Cassia fistula* leaf extract against *Candida albicans. Indian Journal of Microbiology,* 46(2): 169–170.

Stahl, E., 1969. 2nd edn. MRF Ashworth Springer Verlag, Heidelberg.

Ethnobotanical Wisdom and Microbial Studies on Medicinal Plants (2011) *Pages* **371–381**
Editors: **D.R. Khanna, Ashutosh Gautam, R. Bhutiani, Gagan Matta, Anil Dhiman**
Published by: **DAYA PUBLISHING HOUSE, NEW DELHI**

Chapter 56

Establishing Relationship Between Ayurvedic System of Medicine and Ethnobotany

Anil Kumar Dhiman[1] and D.R. Khanna[2]
[1]Information Scientist, Gurukul Kangri University, Haridwar
[2]Professor, Department of Zoology and Environmental Science, Gurukul Kangri University, Haridwar

ABSTRACT

Ayurvedic system of medicine and Ethnobotany both are related to the searching of new medicines and improving the existing ones. Ethnobotany is the science of botany which deals with the folk uses of plants, especially for curing the diseases and ayurveda is also related to the formation and development of new drugs which cure the diseases in human beings. Thus, both ethnobotany and ayurvedic system of medicine can be correlated. Here such an attempt is made to establish a relationship between them.

Ayurvedic System of Medicine

Ayurveda is one of the great gifts of the sages of ancient India to mankind. It is not only a system of medicine in the conventional sense of curing disease but also a way of life that teaches us how to maintain and protect the health. It is a tradition with an antiquity comparable to that of life itself and it treats man as a *whole,* which is a combination of body, mind and soul. Therefore, it is a truly holistic and integral medical system.

The period from 800-1000 B.C. is considered as the golden era in the ayurvedic system of medicine, where the trinity of *Acharya Charak, Susruta and Vagbhatta* were born and written *Charak Samhita,*

Susruta Samhita and *Astanga Hridaya* respectively. These three persons are known as the *'Vriddha trai,'* in the history of ayurveda. Another triad or trinity of ayurveda is known as *'Laghu trai'*, which are the compilations by Madhavakar (12th century), who wrote *`Madhavanidana'*, concerned mainly with diagnosis; Sharngadhara (14th century), who wrote *`Sharngadhara Samhita'*, a systematic ayurvedic materia medica; and Acharya Bhava Mishra (15th century), a native of Benaras who wrote *`Bhavprakasha'* which contains more than 600 drugs including some foreign drugs too (Vidyalankar, 1981).

Ayurvedic system of medicine flourished well during above periods, but after the period of the *Tantras* and *Siddhas*, the glories of the medicine rapidly declined and the golden age of ayurveda ended when waves of *Muslim* invaders attacked the country. During invasion on India by *Mohammedan*, no original work could be done and ayurvedic medicine gradually began to decay. After the *Mughals*, Europeans, first the Portuguese then French and lastly English persons invaded the India. And later, British East India Company closed and banned ayurveda schools and opened allopathic colleges.

Though, the Britishers founded Botanical Survey of India in 1890 to protect and preserve plants of various types in India, but it was during the latter part of the British rule in India, when the movement towards the development of ayurveda was observed. In 1920's, Hakim Ajmal Khan started a regular teaching and training institution with a curriculum keeping in view the traditional methodology at Delhi with the founding of Ayurvedic and Unani Tibbia College and Pandit Madan Mohan Malaviya at Banaras Hindu University, Varanasi. This paved the way for starting of various ayurvedic institutions in many parts of the country.

Government of India appointed a committee under the chairmanship of Col. R.N. Chopra in 1946 and its recommendations in the form of a report were published in 1948 after the independence. Later, another committee under the chairmanship of Dr. C.G. Pandit was set up to follow up some of the recommendations of Chopra committee. As a result, Central Research Institute in Indigenous Systems of Medicines was established at Jamnagar (Gujarat) in 1952. Soon after a few years, Post Graduate Institute of Ayurveda was formed in the Banaras Hindu University, Varanasi. Now, AYUSH department is working for the development of ayurvedic system of medicine along with Unani, Siddha and Homeopathic system of medicines. National institute of Ayurveda was also established in 1976 at Jaipur (Rajasthan) in collaboration with Government of Rajashtan, which is working as a national centre for promoting Ayurveda. Institute of History of Medicine and Medical Research, Delhi and Central Council for Research in Unani Medicine (CCRUM) established in 1979 are working for the coordination and scientific research in Unani Medicine, which infact indirectly incorporate Ayurveda (Dhiman, 2006). Besides, there is a separate ayurvedic univeristy in Gujarat working for the development of ayurveda in India.

A separate unit called Amchi research unit was also set up in CCRAS in Ladakh region to carry out clinical researches and survey of local drug potentials in Tibetan System of Medicine, which is in vogue in Ladakh, Lahul and in some other regions. Amchi system is very alike the Ethnobotany field, where the experiences of local people are documented for further testing, likewise the local drug potential of Amchi people is documented in Siddha system and is further tested clinically, which led to the discovery of new medicines in the recent past.

What is Ethnobotany?

Ethnobotany is defined as the *study of the relationship between people and plants* and most commonly refers to the study of indigenous uses of plants. In other words, it is the marriage between cultural anthropology and botany, a study that investigates the roles of plants as medicine, nourishment, natural resources or gateways to the Gods. The term *Ethnobotany* was first applied by Harshberger (1895) to study the plants used by primitive and aboriginal people and the term became popular immediately and was widely accepted by Fewkes (1896), who was the first after Harshberger to adopt the term for the title of a paper.

It is a branch of botanical science, which is relatively much younger than ayurveda but it is now more than 100 years old and it has traveled through various stages of development. With the passage of time, its scope is changed and broadened, which may be seen through its various definitions. A developmental review on the definitions of Ethnobotany given by various learned scientists is reproduced as under:

Powers (1875) can be said the originator of Ethnobotany and he has given the term 'Aboriginal Botany' to Ethnobotany, which was introduced for the study of all forms of vegetation which aborigines used for commodities such as medicine, food, textiles and ornaments. To Rochebrune (1879), it is "the study of plants from archeological sites."

But it was Harshberger (1895), who coined the term in 1895 and defined it with the 'use of plants by the aboriginal people.' Jones (1941), defined it as "the study of the inter-relations of primitive man and plants." After a decade, he again revised his concept and treated Ethnobotany as the "unit of an ecological study specializing in the interaction of man and the plant world." Faulks (1958) considered the subject of Ethnobotany as "the total relationship between man and vegetation which means more than the scope of Economic Botany." Schultes (1962), told that "it is the study of relationships which exist between people of a primitive society and their plant environment."

Whereas, according to Jain (1967 & 1986), "Ethnobotany deals with the direct relationship of plants with man." Later the definition elaborated as "the total natural and traditional relationship and the interactions between men and his surrounding plant wealth" (Jain, 1987). While Vartak and Gadgil (1980) are of the opinion that "Ethnobotany is a branch of economic botany, a section of which deals with the role of plants in the life and culture of aborigines and tribal people." To Alcorn (1984), it is the "study of contextualized plant relationship and the interaction between man and his surrounding plant wealth."

Manilal's (1989) opinion is that "the term today has come to denote the entire realm of direct relationship between plants and man." Later Wickens (1990), has described it as "the study of useful plants prior to their commercial exploitation and eventual domestication. It includes the uses of plants both by tribal and non-tribal communities without any implication of primitive or developed societies." And very recently, according to Arora (1997), "Ethnobotany in wider context denotes the entire realm of useful relationship between plants and man."

So, it is seen that till few years back, interpretations varied greatly, now ethnobotany generally includes total natural relationship of plant kingdom with man and his animals. There is however, a misconception that prevails in the minds of many people that *Ethnobotany* is exclusively concerned with the relationships of only the primitive or aboriginal people with plants. But Jain and Mitra (1991) pointed out that the term ethnobotany should rightly be applied to natural and direct relationship with plants or any people, at any level of antiquity, primitiveness or acculturation and even to the most sophisticated gentlemen and women, if the latter associate any particular plant with festivals or offer flowers or fruits to their Gods or Deities. Yet, ethnobotany is generally taken as the science of relationship of only the primitive or aboriginal people with the plants.

Although the definitions of Ethnobotany seem to be very simple, it involves the holistic approach, where the entire reciprocal and dynamic aspect of human interaction with plants is anticipated. It is now also considered with microbes, lichens, mosses and to the lower plants not yet investigated or developed for practical use. It can be said that the total relationship of man with plants including lower as well as higher ones, is Ethnobotany. The relationships between man and plant kingdom may be fall under four broad categories – relationships useful to both man and plants, relationships useful to man but harmful to plants, relationships useful to plants but harmful to man and relationships harmful to both man and plants. However, the ethnobotanical studies appear to have been restricted among the tribal or rural people for recording their knowledge about plant wealth and for search of new resources of herbal drugs, edible plants and other aspects of plants. The other three aspects and the urban population have generally been left.

Whatever the case may be, broadly saying, it is restricted to the study of plants species and their medicinal uses in traditional ways (Ahmed and Borathakur, 2005). In addition, ethnobotanical data may be utilized by Economic Botanists for discovering new plant species, for fresh ideas to environment planner, a tool for basic selection of plant species for development of drugs by Pharmacologists, Phytochemists and Clinicians, as a new source of history of plant names for linguistics, a source for locating new germplasm for agriculturists etc. (Dhiman, 2001).

If we talk about the development of Ethnobotany in India, it is said to be originated by S.K. Jain around 1960s, when he started to make regional studies on plants and their uses by native people but the mission took speed with the establishment of Institute of Ethnobotany in NBRI, Lucknow by Jain (1992). And the prominent researchers who have done extensive research in this field include - S.K. Jain, S.K. Borthakur, N.C. Shah, J.K. Maheshwari, D.C. Pal, R.L.S. Sikarwar, Prabhakar Joshi etc. (Dhiman, 2010).

Relationship between Ayurveda and Ethnobotany

Although, we define Ayurveda as an alternative system of healing as subjects that are not taught in medical schools, it is worthwhile to mention that before the availability of synthetic drugs, plant-based remedies formed the basis of primary healthcare system. Herbal infusion, decoction and tinctures were house-hold remedies for common ailments, which are better known as *dadi ma ke nuskhe* in the Indian context. Ethnobotany is also dealt with the so-called house hold remedies, which are the experiences of the local people which are being used as generation to generation without testing them clinically, but they are providing positive results to overcome the various ailments.

The overall development of Ayurveda and traditional system of medicines can be depicted as given in Figure 56.1.

Ayurveda can be said as a well developed system of medicine based on philosophical ground and strong fundamental principals. And herbal medicines are the oldest remedies known to mankind (Dhiman, 1997). Herbs had been used by all cultures throughout history. India has one of the oldest, richest and most diverse cultural living traditions associated with the use of medicinal plants. Ethnobotany as seen through above figure leads to quality, efficacy and safety of health using plant parts or their products.

Now let us try to explain what is a drug? A drug, broadly speaking, is any chemical substance that, when absorbed into the body of a living organism, alters normal bodily function. There is no single, precise definition, as there are different meanings in medicine, government regulations, and colloquial usage. More elaborately, a drug is defined as "a chemical substance used in the treatment, cure, prevention, or diagnosis of disease or used to otherwise enhance physical or mental well-being. Drugs may be prescribed for a limited duration, or on a regular basis for chronic disorders. Drugs are

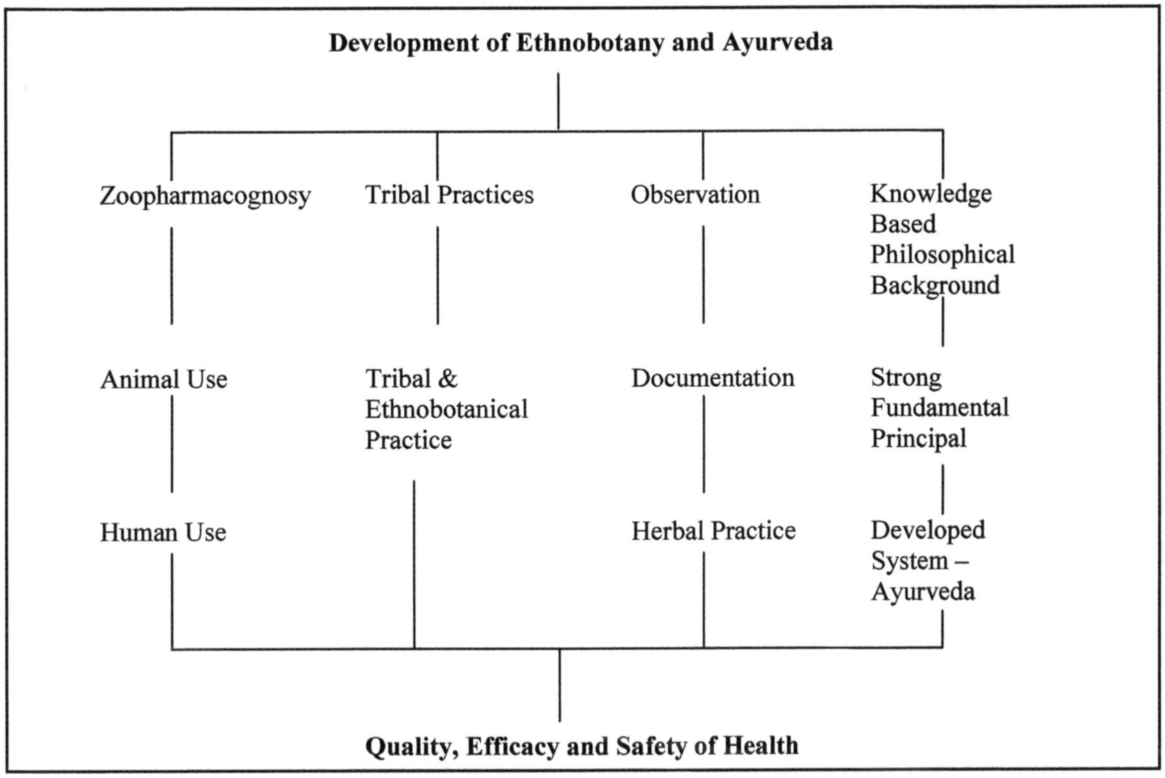

Figure 56.1: Development and Relationships of Ethnobotany and Ayurveda for Safety of Health

usually distinguished from endogenous bio-chemicals by being introduced from outside the organism. For example, insulin is a hormone that is synthesized in the body; it is called a hormone when it is synthesized by the pancreas inside the body, but if it is introduced into the body from outside, it is called a drug.

There are recreational drugs, which are the chemical substances that affect the central nervous system, such as narcotics or hallucinogens. They may be used for perceived beneficial effects on perception, consciousness, personality, and behaviour. Many natural substances such as beers, wines, and some mushrooms, blur the line between food and drugs, as when ingested they affect the functioning of both mind and body. However, some recreational drugs can cause addiction and habituation.

The research works in the field of medicines, are carried out world wide by people from various departments like Chemistry, Veterinary Science, Agriculture, Zoology, Biochemistry, Biotechnology, Siddha, Ayurveda and Unani system of medicines, Botany, Cosmetic, Forest, Homeopathy, Pharmacy and Medicine etc. But it is the *Ethnobotany* which is related with indigenous knowledge that is vital information for finding out and developing new drugs, but at present, indigenous knowledge is diminishing at an alarming rate. There is an urgent need to collect it before it is irretrievably lost. So, the ethno botanical surveys are carried out among tribals or people of particular community or rural areas of a particular district or place. The researchers collected information regarding their medical practices for the treatment of human ailments. Sometimes survey is conducted specifically such as practice for the treatment of hemorrhoids or veterinary purpose or increasing secretion of milk or women health disorders etc. The data are accrued through discussions and interviews with tribals, besides other rural people. Then they are confirmed by repeated enquiries and cross-questions in different seasons and in various villages/hamlets. The data are carefully recorded in audiotapes or chips in mobile phones etc. as well as field notebooks. Sometimes patients are contacted to confirm the diagnosis of the diseases. The collected medicinal plants are identified by local people of that area and authenticated by person working in any one of the departments such as - Botany, Pharmacognosy, Taxonomy, Horticulture, Botanical Survey of India, Siddha, Ayurveda, Forestry and Agriculture or by comparing the characteristics of the parts with herbarium specimens. Then, authenticated information about plants and their parts used as medicine are documented and published.

If we talk about the ayruveda, it is a fundamental truth that most of the drugs in ayurveda are herbal based. Besides, a majority of drugs that have been discovered in the past fifty or sixty years in the western countries are from such plants, as had been used in primitive societies or in indigenous medicines. These include the antihypertensive *reserpine* from *Rauvolfia serpentina,* anticancer *vinblastine* and *vincristine* from *Catharanthus roseus,* antimalarial quinine form *Cinchona ladgeriana,* antiglaucoma *pilocarpine* from *Pilocarpus jaborandi,* the topical *anesthetic* from *Erythroxylon coca,* anticough *codeine* from *Papver somniferum* and antiasthmatic and cold medicine *atropine* from *Atropa belladonna* and *Hyoscyamus niger* etc. (Dhiman, 2006). Thus majority of the drugs investigated or developed in the past few years are mostly plant-based or herbal drugs.

Ethnobotanically, some psychoactive plants have also been explored by Jain (2009), which are more or less similar in their properties that are described as sedative medicines and nervine tonic in ayurvedic system of medicine. These include – *Abutilon indicum, Alstonia scholaris, Cynodon dactylon, Holarrehena antidysenterica, Moringa olifera, Papaver somniferum, Pongammia pinnata, Rauvolfia serpentina, Ricuinus communis, Terminalia arjuna, T. bellirica,* and *T. chebula etc.* Thus, it is seen ethnobotany and ayurveda both are contributing collectively for the discoveries and development of herbal medicines.

A survey based on based on WHO Survey (2005) in 130 countries on Regulatory Requirements revealed that traditional system of medicine is used safely in 66 countries of the world. Table 56.1 revealed the actual data about the use of plants in the world.

Table 56.1: Safety Aspects of Traditional Medicine

Sl.No.	*Regulatory Requirements*	*Number of Countries*
1.	Special Requirements	82
2.	Traditional Use without harmful effects	66
3.	Reference on documented scientific research	53
4.	Others (Bibliographic data etc.)	21
5.	Same as Pharmaceuticals	57
6.	No requirement	28

Limitations of Herbal Drug Practice

It is a general truth that ayurvedic system of medicine and findings of ethnobotanical researches, which are herbal based cause no harmful effects in human beings, if taken properly. But there are various factors responsible for downing the use of herbal based practice. The following factors may be taken as major limitations for herbal drug practice (Singh, 2007):

Non-availability of Medicines

Herbal prescriptions range from single herb to poly-herbal formulations. In case of writing herbal prescriptions, availability of medicines is necessary. Non-availability of medicines at drug stores is one of the limitations of herbal or ayurvedic practice.

Variety of Medicinal Plants/Drugs

Another factor that limits the scope of herbal practice is the number of medicinal plants mentioned in respective pharmacopeias. There are about 850 medicinal plants in ayurveda, which are actively used for creating formulations. This causes difficulty in selecting the appropriate drug for use.

Adulterated/Spurious Medications

Several reports have been published regarding mixing of synthetic drugs like corticosteroids and pain-killers in herbal prescriptions. This has harmed the belief of many people in the alternative system of medicine.

Medicinal Plant Identification in Ayurveda

In ayurveda, several medicinal plants have controversial identities. For example, *Ashtavarga* is an important ingredient of various classical ayurvedic formulations like *Chavyanprasha*. *Ashtavarga* has been assigned various medicinal properties by Ancient Materia Medica dealing with ayurveda. The eight plants mentioned under *Ashtavarga* are considered to be medicinal plants of controversial origin, despite the fact that much work has been done on their botanical identity. This is depicted through Table 56.2.

Table 56.2: Medicinal Plants Used in Ashtavarga Drug Formulation

Sl.No.	*Ayurvedic Name*	*Botanical Name*	*Family*	*Part Used*
1.	Jivaka	*Malaxis muscifera* (Lindl.) Kuntze	Orchidaceae	Bulb
2.	Rishbhaka	*Malaxis acuminta* D.Don	Orchidaceae	Pseudo-bulb
3.	Meda	*Polygonatum verticillatum* (L.) All.	Polygonaceae	Rhizome
4.	Mahameda	*Polygonatum cirrhifolium* (Wall.) Royle	Polygonaceae	Rhizome
5.	Kakoli	*Roscoea procera* Wall.	Zingiberaceae	Root
6.	Kshira-Kakoli	*Fritillaria royeli* Hook.f. *Lilium polphyllum* D.Don	Liliaceae	Root
7.	Riddhi	*Habenaria intermedia* D.Don	Orchidaceae	Root
8.	Vriddhi	*Habenaria edgeworthii* Hook.f. ex Collett.	Orchidaceae	Root

Most of the ayurvedic drugs manufacturing companies using *Ashtavarga* in formulations are depicting botanical name mentioned in the Table 56.2. However, literature search reveals that certain medicinal plants of family Ascelpiadaceae are possible representatives of Kakoli, Kshira-kakoli and Meda. This is reproduced in Table 56.3.

Chavyanprasha is a hot-selling product in the Indian market and it is considered to be panacea for treatment of chronic respiratory ailments. Since *Ashtavarga* is a significant ingredient of *Chavyanprasha*, it is essential that exact plants be identified in the manufacture of the composite formulation.

Table 56.3: Alternative Representatives of Certain Ingredients of Ashtavarga

Sl.No.	*Ayurvedic Name*	*Botanical Name*	*Family*
1.	Kakoli	*Gymnema balsamicum*	Ascelpiadaceae
2.	Kshira-kakoli	*Gymnema lactiferum* (L.) R.Br. ex Schult.	Ascelpiadaceae
3.	Meda	*Gymnema aurantiacea*	Ascelpiadaceae

Pashanbheda is another example worth mentioning. It is a well-known lithontriptic (stone-dissolving) drug of ayurveda. Table 56.4 reveals the situation where, different plant species are used as *Pasahanbheda* in ayurvedic or traditional system of medicine in India.

Table 56.4: Medicinal Plants Used for Ancient Drug *Pasahanbheda*

Sl.No.	*Botanical Name*	*Family*
1.	*Aerva lanata* Juss.ex Schult.	Amaranthaceae
2.	*Bergenia ligulata* (Wall.) Engl.	Saxifragaceae
3.	*Bridelia crenulata* Roxb.	Euphorbiaceae
4.	*Bryophyllum calycinum* Salisb.	Crassulaceae
5.	*Coleus amboinicus* Lour.	Lamiaceae
6.	*Decalepis arayalpathra* Joseph & Chandrasekharan	*Periplocaceae*
7.	*Homonoia riporia* Lour.	Euphorbiaceae
8.	*Rotula aquatica* Lour.	Boraginaceae

So, various plants are being used under different names for composite formulation in different parts of the country. This needs proper identification of plants. Various adulterant and substitute plants have been described in details by Singh (1983), Kaushik and Dhiman (2000), Garg (2002), Dhiman (2004) and Dhiman (2006). As most of the Ethnobotanists come from botany field, they may help in correct identification of the plant species to ayurvedic scholars.

Lack of Awareness among Physicians

Recently, several clinical trials with Ayurvedic/herbal drugs have been published in reputed journals. As an instance, *Guggul* is widely used in ayurveda in the treatment of obesity. Animal and human trials have suggested that *Guggul* is useful for reducing high blood-cholesterol and triglycerides. This is perfect example of bridging the gap between traditional and modern medicine.

Lack of Awareness among Patients

There is a need to educate the patients seeking alternative medicine advice. It is needed to tell them categorically about the nature of herbal drugs. Herbal remedies are not magical bullets and have nothing to do with emergency conditions. However, chronic and degenerative diseases do respond to herbal remedies. While buying Ayurvedic products, the patients should always look for the label indicating 'metals with permissible limits'. They should also ensure the originality of the medication.

Bias against Herbal Medicine

False propaganda against herbal medicine should be discontinued immediately. Herbal system of medicine should be accepted with an open mind and its advantages and disadvantages should be kept in mind.

Inspite of above limitations, herbal system of medicine is flourishing well again. But standardization of ayurvedic drugs and plant materials are the need of the day. Many of them do not have standard identification tests or analytical procedures to maintain their consistent quality. Several pharmacopoeias containing monographs on plant materials describe only the physico-chemical parameters and are lacking in identification and quantification of phyto-constituents of therapeutic importance present in the plant. So research works need to be going on to estimate either single or simultaneously two constituents of same herb by using techniques for proper standardization of herb and its estimation in raw material or poly herbal formulations or its extracts or tissue culture of the plant. Likewise, those drugs or drug plants are investigated by the Ethnobotanists, their efficacy must be tested clinically and if these become really effective without any side effects, then they may be declared as the new drugs.

Conclusion

India has been a source of rich amount of the medicinal plant wealth since the antiquity. Now, as many as 16,000 plant species of flowering plants are known to be found in India and 3000 of them are recognized as of medicinal importance (Dhiman, 2006). It is seen that the ayurveda has developed as a result of experience, which does not cover a few years or decades rather it is a knowledge that accumulated over centuries and has a vast storehouse of hidden knowledge and a dive and swimming in it only will reveal its hidden secrets. Though progress is slow, the work is going on for the upliftment of ayurveda. Ethnobotanists are also working on the same line or pattern of research, where hidden knowledge of local people about the plant drugs is documented, but it must be put among the scientists for further clinical testing and to be a testified medicine.

Self medication with herbal remedies is increasing and is becoming popular, which is not advisable at all. Most of the herbal remedies like Garlic, Arjun, Ginkgo and Spirulina etc. are commonly available over the counter remedies. Medicinal plants used in ayurveda are also available in capsule or tablet form. This free sale should be controlled as there are increased reports of herbal-synthetic drug interactions in recent times. The patient should always try to consult a qualified physician before consuming any medication. It is also an inference that with increasing number of patients consulting herbal medicine practitioners, herbal clinical practice will be of more relevance in future, but the need of the hour is to take herbal medicine with an open mind. Herbal product claims, supported by scientific evidence should be used in clinical practice with confidence. This will enhance the physician-patient compliance and provide the much needed support to herbal based practice. Indian Council of Medical Research guidelines (2000) for safety on products of Indian System of Medicine has also given importance on testing of efficacy of drugs as :

"Since the substance to be tested is already in use in Indian Systems of Medicine or has been described in their texts, the need for testing its toxicity in animals has been considerably reduced. If there are reports suggesting toxicity or when the herbal preparation is to be used for more than 3 months it would be necessary to undertake 4-6 weeks toxicity study in 2 species of animals".

Nevertheless, the interest in screening plants for medicinal purpose grows because higher plants constitute a largely untapped source of novel compounds that might serve as leads for the development of new drugs. The march from plant extracts and molecules to the market would however be faster if the random testing of thousands of compounds against varied diseases were replaced by a selective approach based on clues from traditional knowledge. This is shown by not only the examples of *rouwolfia* and *guggul* but also by artemisinin. The plant *kwing hasavu* had been used in China for 2000 years as a febrifuge and its extract was tested for antimalarial activity because malaria was the most important among fevers. In all three instances, a plant extract was investigated for a specific pharmacological activity on the basis of traditional clues. The focus and intensity of this approach would be missing if thousands of compounds from numerous plants are tested for a variety of pharmacological activities. But there should be a will to accomplish the ayurvedic mission in the government, ayurvedic and scientific communities and also in the industry people (Valiathan, 2006) and enough efforts are also needed to promote the ethnobotanical researches to develop a safe, efficacious and healthy system of traditional medicine in the country.

References

Ahmed, A.A. and Borthakur, S.K., 2005. *Ethnobotanical Wisdom of Kahsis (Hynniew Treps) of Meghalaya*. Bishen Singh Mahendra Pal Singh, Dehradun, pp. 1-5.

Alcorn, J.B., 1984. *Huastec Mayan Ethnobotany*. University of Texas Press, Austin.

Arora, R.K., 1997. Ethnobotany and its Role in the Conservation and Use of Plant Genetic Resources in India. *Ethnobotany*. 9 : 6-15.

Dhiman, A.K., 1997. A Survey of Medicinal Plants of Haridwar and Adjoining Area vis-à-vis the Raw Plant Drugs being sold in Local Market. Unpublished *Ph.D. Thesis* in Botany. Gurukul Kangri University, Haridwar.

Dhiman, A.K., 2001. Ethnobotanists in New Millennium. *Advances in Plant Sciences*. 14 (II) (Supp.): 653-56.

Dhiman, A.K., 2004. *Common Drug Plants and Ayurvedic Remedies*. Reference Press, New Delhi.

Dhiman, A.K., 2006. *Ayurvedic Drug Plants*. Daya Publishing House, New Delhi.

Dhiman, A.K., 2010. Growth of Ethnbotanical Journal Literature and Citation Pattern of Ethnobotanists in India : A Biblimometric Study. Unpublished *Ph.D. Thesis* in Library Science. Jiwaji University, Gwalior.

Faulks, P.J., 1958. *An Introduction to Ethnobotany*. Moredale, London.

Fewkes, J. Walter., 1896. A Contribution to Ethnobotany. *American Anthropologists*. 9 : 14-21.

Garg, S., 1992. *Substitute and Adulterant Plants*. Periodical Experts Book Agency, Delhi.

Harshberger, J. W., 1895. Some New Ideas : the Plants Cultivated by Aboriginal People and How used in Primitive Commerce. *The [Daily] Evening Telegraph, Philadelphia*. 64 (134): 2.

Indian Council of Medical Research., 2000. *Ethical Guidelines for Biomedical Research on Human Subjects*. ICMR, New Delhi.

Jain, S.K., 1967. Ethnobotany – Its Scope and Study. *Indian Museum Bulletin*. 2 (1) : 39-43.

Jain, S.K., 1986. Ethnobotany. *Interdisciplinary Science Review*. 11 (3) : 285-92.

Jain, S.K., 1987. Ethnobotany – Its Concept and Relevance. *Presidential Address on Xth Botanical Conference*. 1-12.

Jain, S.K., 1992. Ethnobotany - What and Why. *Ethnobotany*. 4 : 75-79.

Jain, S.K., 2009. *Plants affecting Human Mind (Psychoactive Plants)*. Deep Publications, New Delhi.

Jain, S.K. and Mitra, R., 1991. Ethnobotany in India : Retrospect and Prospect. In S.K. Jain Edited '*Contribution to Indian Ethnobotany*.' Scientific Publishers, Jodhpur. Pp. 1-18.

Jones, V.H., 1941. The Nature and Scope of Ethnobotany. *Chronica Botanica.* 6 (10): 219-21.

Kaushik, P. and Dhiman, A.K., 2000. *Medicinal Plants and Raw Drugs of India.* Bishen Singh Mahendra Pal Singh, Dehradun.

Manilal, K.S., 1989. Linkages of Ethnobotany with other Sciences. *Ethnobotany.* 1 : 15-24.

Powers, Stephan, 1875. Aboriginal Botany. *California Academy of Sciences Proceedings.* 5 : 373-79.

Rochebrune, E.T., 1879. Recgercges d' ethnographie Botanique sur la Flore des Serpultures Peruviennes d'Ancon. *Actes Soc. Linn. Bordeaus.* 33 : 342-58.

Schultes, R.E., 1962. The Role of Ethnobotanists in the Search for New Medicinal Plants. *Lloydia.* 25 : 257-66.

Singh, A.P., 2007. Ethics in Herbal Medicine. *Ethnobotanical Leaflets.* 11: 206-211. Available at http://www.siu.edu/~ebl/leaflets/ethics.htm.

Singh, R. S., 1983. *Vanoushdhi Nirdeshika.* Uttar Pradesh Hindi Sansthan, Lucknow.

Vartak, V.D. and Gadgil, Madhav, 1980. Studies in Ethnobotany – A New Vistas in Botanical Science. *Biovigyanam.* 6 : 151-56.

World Health Organization, 2005. National Policy on Traditional Medicine and Regulation of Herbal Medicines: Report of a WHO Global Survey. WHO, Geneva.

Wickens, G.E., 1990. What is Economic Botany? *Economic Botany.* 44 : 12-28.

Valiathan. M. S., 2006. Ayurveda: Putting the House in Order. *Current Science.* 90 (1) : 5-6.

Vidyalankar, A.D., 1991. *Ayurved ka Itihas.* Uttar Pradesh Hindi Sansthan, Lucknow.

Index

F

www.ingramcontent.com/pod-product-compliance
Ingram Content Group UK Ltd.
Pitfield, Milton Keynes, MK11 3LW, UK
UKHW052006051026
15624UKWH00003B/590

9 789351 240396

OVERVIEW

The Endocrine Self-Assessment Program (ESAP™) is a self-study curriculum aimed at physicians wanting a self-assessment and a broad review of endocrinology. The ESAP Reference Edition consists of 120 brand-new multiple-choice questions in all areas of endocrinology, diabetes, and metabolism. There is extensive discussion of each correct answer, a comprehensive syllabus, and references. ESAP is updated annually with new questions.

The ESAP reference book is intended primarily for consultation and self-assessment of knowledge relating to endocrinology. As a reference book, educational credits are not available upon completion of the multiple-choice questions included. For information on educational products that include educational credit, please visit endocrine.org/store.

LEARNING OBJECTIVES

ESAP 2025 will allow learners to assess their knowledge of all aspects of endocrinology, diabetes, and metabolism.

Completion of this educational activity enables learners to accomplish key objectives:

- Recognize clinical manifestations of endocrine and metabolic disorders and select among current options for diagnosis, management, and therapy.
- Identify risk factors for endocrine and metabolic disorders and develop strategies for prevention.
- Evaluate endocrine and metabolic manifestations of systemic disorders.
- Use existing resources pertaining to clinical guidelines and treatment recommendations for endocrine and related metabolic disorders to guide diagnosis and treatment.

TARGET AUDIENCE

ESAP is a self-study curriculum aimed at physicians seeking initial certification or recertification in endocrinology, program directors interested in a testing and training instrument, and clinicians simply wanting a self-assessment and broad review of endocrinology.

STATEMENT OF INDEPENDENCE

The Endocrine Society has a policy of ensuring that the content and quality of this educational activity are balanced, independent, objective, and scientifically rigorous. The scientific content of this activity was developed under the supervision of the Endocrine Society's ESAP Faculty Working Group.

DISCLOSURE POLICY

The faculty, committee members, and staff who are in position to control the content of this activity are required to disclose to the Endocrine Society and to learners any relevant financial relationship(s) of the individual or spouse/partner that have occurred within the last 12 months with any commercial interest(s) whose products or services are related to the content. Financial relationships are defined by remuneration in any amount from the commercial interest(s) in the form of grants; research support; consulting fees; salary; ownership interest (eg, stocks, stock options, or ownership interest excluding diversified mutual funds); honoraria or other payments for participation in speakers' bureaus, advisory boards, or boards of directors; or other financial benefits. The intent of this disclosure is not to prevent planners with relevant financial relationships from planning or delivering content, but rather to provide learners with information that allows them to make their own judgments of whether these financial relationships may have influenced the educational activity with regard to exposition or conclusion. The Endocrine Society has reviewed all disclosures and resolved or managed all identified conflicts of interest, as applicable.

The following faculty reported relevant financial relationship(s): **Bradley D. Anawalt, MD** is a consultant for Biozen. **Adriana Ioachimescu, MD, PhD**, is a consultant for Crinetics, Xeris, and Camurus and has served as principal investigator for institution-directed research grants from Recordati. **Erik Imel, MD**, is a clinical trial investigator for Ultragenyx, Alexion, Amgen, Amolyt, Kyowa Kirin, Calciolytix, Spruce, Neurocrine, and Sanofi. He serves on advisory boards for Ultragenyx, Kyowa Kirin and Neurocrine. He is also a consultant for Novo Nordisk and serves on a Data and Safety Monitoring Board for Inozyme and Agios, and AshiBio. **Akshay Jain, MD**, has served on an advisory board for Abbott, Amgen, Novo Nordisk, AstraZeneca, Bausch Healthcare, Bayer, Boehringer Ingelheim, Dexcom, Eli Lilly, Embecta, GSK, HLS Therapeutics, Janssen, Medtronic, Novartis, Gilead Sciences, PocketPills, Insulet, Roche, Takeda, and Ypsomed. He has served as a speaker for Abbott, Amgen, Novo Nordisk, AstraZeneca, Bausch Healthcare, Bayer, Boehringer Ingelheim, Care to Know, CCRN, Dexcom, Diabetes Canada, Eli Lilly, Embecta, EOCI, GSK, HLS Therapeutics, Janssen, Liv, MDBriefcase, Medtronic, Moderna, Novartis, and Pfizer. He has also been a clinical researcher for Abbott, Amgen, and Novo Nordisk. **Kevin M. Pantalone, DO**, receives research support from Bayer, Novo Nordisk, and Twin Health. He is a speaker for AstraZeneca, Corcept Therapeutics, and Novo Nordisk. He is a consultant to Bayer, Boehringer Ingelheim, Corcept Therapeutics, Diasome, Eli Lilly, Merck, Novo Nordisk, and Sanofi. He is a principal investigator for Eli Lilly, Novo Nordisk, and Twin Health. **Aled Rees, PhD, MB BCh**, is a principal investigator for Ascendis, Diurnal Ltd, Neurocrine Biosciences, Sparrow Pharmaceuticals. He is also a speaker for Diurnal Ltd. **Vinaya Simha, MD**, serves on an advisory board for Pfizer. **Elias Spanakis, MD**, received research support from Dexcom and Tandem Diabetes, both of which were provided to Baltimore VA Medical Center and University of Maryland. **Adina Turcu, MD, MS**, receives research funds from the National Institutes of Health. She is an institutional clinical trial investigator for Astra Zeneca. **Thomas J. Weber, MD**, serves as a primary investigator and consultant for Kyowa Hakko Kirin Pharma, Inc. He is a member of the AgNovos Healthcare Data Safety and Monitoring Board. He also serves as a primary investigator and consultant for Ultragenyx. He is on the FDA Endocrine and Metabolic Drug Advisory Committee.

The following faculty reported no relevant financial relationships: **Danit Ariel, MD, MS; Palak Choksi, MD**; **Angela Leung, MD, MSc**; **Sarah Mayson, MD**; **Deepika Reddy, MD**; **Meera Shah, MB ChB**, and **Anu Sharma, MBBS.**

The medical editor for this program, **Abbie L. Young, MS, CGC, ELS(D)**, reported no relevant financial relationships.

The Endocrine Society staff associated with the development of content for this activity reported no relevant financial relationships.

DISCLAIMERS

The information presented in this activity represents the opinion of the faculty and is not necessarily the official position of the Endocrine Society.

USE OF PROFESSIONAL JUDGMENT:

The educational content in this self-assessment test relates to basic principles of diagnosis and therapy and does not substitute for individual patient assessment based on the health care provider's examination of the patient and consideration of laboratory data and other factors unique to the patient. Standards in medicine change as new data become available.

DRUGS AND DOSAGES:

When prescribing medications, the physician is advised to check the product information sheet accompanying each drug to verify conditions of use and to identify any changes in drug dosage schedule or contraindications.

POLICY ON UNLABELED/OFF-LABEL USE

The Endocrine Society has determined that disclosure of unlabeled/off-label or investigational use of commercial product(s) is informative for audiences and therefore requires this information to be disclosed to the learners at the beginning of the presentation. Uses of specific therapeutic agents, devices, and other products discussed in this educational activity may not be the same as those indicated in product labeling approved by the Food and Drug Administration (FDA). The Endocrine Society requires that any discussions of such "off-label" use be based on scientific research that conforms to generally accepted standards of experimental design, data collection, and data analysis. Before recommending or prescribing any therapeutic agent or device, learners should review the complete prescribing information, including indications, contraindications, warnings, precautions, and adverse events.

ACKNOWLEDGMENT OF COMMERCIAL SUPPORT

This activity is not supported by educational grant(s) or other funds from any commercial supporter.

PUBLICATION DATE: February 2025

Laboratory Reference Ranges

Reference ranges vary among laboratories. The listed reference ranges should be used when interpreting laboratory values presented in ESAP™. Conventional units are listed first with SI units in parentheses.

Lipid Values

High-density lipoprotein (HDL) cholesterol

- Optimal---------------------------- >60 mg/dL (SI: >1.55 mmol/L)
- Normal --------------------- 40-60 mg/dL (SI: 1.04-1.55 mmol/L)
- Low-------------------------------- <40 mg/dL (SI: <1.04 mmol/L)

Low-density lipoprotein (LDL) cholesterol

- Optimal-- <100 mg/dL (SI: <2.59 mmol/L) (for primary prevention); <70 mg/dL (SI: <1.81 mmol/L) (for secondary prevention)
- Low---------------------- 100-129 mg/dL (SI: 2.59-3.34 mmol/L)
- Borderline-high --------- 130-159 mg/dL (SI: 3.37-4.12 mmol/L)
- High --------------------- 160-189 mg/dL (SI: 4.14-4.90 mmol/L)
- Very high ------------------------ ≥190 mg/dL (SI: ≥4.92 mmol/L)

Non-HDL cholesterol

- Optimal-------------------------- <130 mg/dL (SI: <3.37 mmol/L)
- Borderline-high --------- 130-159 mg/dL (SI: 3.37-4.12 mmol/L)
- High ----------------------------- ≥240 mg/dL (SI: ≥6.22 mmol/L)

Total cholesterol

- Optimal-------------------------- <200 mg/dL (SI: <5.18 mmol/L)
- Borderline-high --------- 200-239 mg/dL (SI: 5.18-6.19 mmol/L)
- High ----------------------------- ≥240 mg/dL (SI: ≥6.22 mmol/L)

Triglycerides

- Optimal-------------------------- <150 mg/dL (SI: <1.70 mmol/L)
- Borderline-high --------- 150-199 mg/dL (SI: 1.70-2.25 mmol/L)
- High --------------------- 200-499 mg/dL (SI: 2.26-5.64 mmol/L)
- Very high ------------------------ ≥500 mg/dL (SI: ≥5.65 mmol/L)

Lipoprotein (a) --------------------------≤30 mg/dL (SI: ≤1.07 µmol/L)

Apolipoprotein B --------------------- 50-110 mg/dL (SI: 0.5-1.1 g/L)

Hematologic Values

Erythrocyte sedimentation rate ------------------------- 0-20 mm/h

Haptoglobin -------------------- 30-200 mg/dL (SI: 300-2000 mg/L)

Hematocrit------------------------- 41%-51% (SI: 0.41-0.51) (male); 35%-45% (SI: 0.35-0.45) (female)

Hemoglobin A_{1c}----------------------- 4.0%-5.6% (20-38 mmol/mol)

Hemoglobin --------------- 13.8-17.2 g/dL (SI: 138-172 g/L) (male); 12.1-15.1 g/dL (SI: 121-151 g/L) (female)

International normalized ratio --------------------------------0.8-1.2

Mean corpuscular volume (MCV)---------80-100 μm^3 (SI: 80-100 fL)

Platelet count----------------150-450 × 10^3/µL (SI: 150-450 × 10^9/L)

Protein (total) ----------------------------6.3-7.9 g/dL (SI: 63-79 g/L)

Reticulocyte count--- 0.5%-1.5% of red blood cells (SI: 0.005-0.015)

White blood cell count-------- 4500-11,000/µL (SI: 4.5-11.0 × 10^9/L)

Thyroid Values

Thyroglobulin --------- 3-42 ng/mL (SI: 3-42 µg/L) (after surgery and radioactive iodine treatment: <0.1 ng/mL [SI: <0.1 µg/L])

Thyroglobulin antibodies -----------------≤4.0 IU/mL (SI: ≤4.0 kIU/L)

Thyrotropin (TSH) -------------------------------------0.5-5.0 mIU/L

Thyrotropin-receptor antibodies (TRAb) ----------------- ≤1.75 IU/L

Thyroid-stimulating immunoglobulin-------- ≤120% of basal activity

Thyroperoxidase (TPO) antibodies --------<2.0 IU/mL (SI: <2.0 kIU/L)

Thyroxine (T_4) (free) ---------0.8-1.8 ng/dL (SI: 10.30-23.17 pmol/L)

Thyroxine (T_4) (total)------- 5.5-12.5 µg/dL (SI: 70.79-160.88 nmol/L)

Free thyroxine (T_4) index -------------------------------------- 4-12

Triiodothyronine (T_3) (free)------2.3-4.2 pg/mL (SI: 3.53-6.45 pmol/L)

Triiodothyronine (T_3) (total)----- 70-200 ng/dL (SI: 1.08-3.08 nmol/L)

Triiodothyronine (T_3), reverse ---- 10-24 ng/dL (SI: 0.15-0.37 nmol/L)

Triiodothyronine uptake, resin -----------------------------25%-38%

Radioactive iodine uptake -- 3%-16% (6 hours); 15%-30% (24 hours)

Endocrine Values

Serum

Aldosterone-------------------- 4-21 ng/dL (SI: 111.0-582.5 pmol/L)

Alkaline phosphatase ------------- 50-120 U/L (SI: 0.84-2.00 µkat/L)

Alkaline phosphatase (bone-specific) ---------≤20 µg/L (adult male); ≤14 µg/L (premenopausal female); ≤22 µg/L (postmenopausal female)

Androstenedione - 65-210 ng/dL (SI: 2.27-7.33 nmol/L) (adult male); 30-200 ng/dL (SI: 1.05-6.98 nmol/L) (adult female)

Antimullerian hormone ------- 0.7-19.0 ng/mL (SI: 5.0-135.7 pmol/L) (male, >12 years); 0.9-9.5 ng/mL (SI: 6.4-67.9 pmol/L) (female, 13-45 years); <1.0 ng/mL (SI: <7.1 pmol/L) (female, >45 years)

Calcitonin ----------------<16 pg/mL (SI: <4.67 pmol/L) (basal, male); <8 pg/mL (SI: <2.34 pmol/L) (basal, female); ≤130 pg/mL (SI: ≤37.96 pmol/L) (peak calcium infusion, male); ≤90 pg/mL (SI: ≤26.28 pmol/L) (peak calcium infusion, female)

Carcinoembryonic antigen ---------------- <2.5 ng/mL (SI: <2.5 µg/L)

Chromogranin A --------------------------- <93 ng/mL (SI: <93 µg/L)

Corticosterone ---53-1560 ng/dL (SI: 1.53-45.08 nmol/L) (>18 years)

Corticotropin (ACTH) ------------ 10-60 pg/mL (SI: 2.2-13.2 pmol/L)

Cortisol (8 AM) ---------------- 5-25 µg/dL (SI: 137.9-689.7 nmol/L)

Cortisol (4 PM) ------------------2-14 µg/dL (SI: 55.2-386.2 nmol/L)

C-peptide ----------------------- 0.5-2.0 ng/mL (SI: 0.17-0.66 nmol/L)

C-reactive protein -------------0.8-3.1 mg/L (SI: 7.62-29.52 nmol/L)

Cross-linked N-telopeptide of type 1 collagen ---------------------- 5.4-24.2 nmol BCE/mmol creat (male); 6.2-19.0 nmol BCE/mmol creat (female)

Dehydroepiandrosterone sulfate (DHEA-S)

Patient Age	Female	Male
18-29 years	44-332 µg/dL (SI: 1.19-9.00 µmol/L)	89-457 µg/dL (SI: 2.41-12.38 µmol/L)
30-39 years	31-228 µg/dL (SI: 0.84-6.78 µmol/L)	65-334 µg/dL (SI: 1.76-9.05 µmol/L)
40-49 years	18-244 µg/dL (SI: 0.49-6.61 µmol/L)	48-244 µg/dL (SI: 1.30-6.61 µmol/L)
50-59 years	15-200 µg/dL (SI: 0.41-5.42 µmol/L)	35-179 µg/dL (SI: 0.95-4.85 µmol/L)
≥60 years	15-157 µg/dL (SI: 0.41-4.25 µmol/L)	25-131 µg/dL (SI: 0.68-3.55 µmol/L)

Deoxycorticosterone ------ <10 ng/dL (SI: <0.30 nmol/L) (>18 years)
1,25-Dihydroxyvitamin D_3 ----- 16-65 pg/mL (SI: 41.6-169.0 pmol/L)
Estradiol --------------- 10-40 pg/mL (SI: 36.7-146.8 pmol/L) (male);
  10-180 pg/mL (SI: 36.7-660.8 pmol/L) (follicular, female);
  100-300 pg/mL (SI: 367.1-1101.3 pmol/L) (midcycle, female);
  40-200 pg/mL (SI: 146.8-734.2 pmol/L) (luteal, female);
  <20 pg/mL (SI: <73.4 pmol/L) (postmenopausal, female)
Estrone ---------------- 10-60 pg/mL (SI: 37.0-221.9 pmol/L) (male);
  17-200 pg/mL (SI: 62.9-739.6 pmol/L) (premenopausal female);
  7-40 pg/mL (SI: 25.9-147.9 pmol/L) (postmenopausal female)
α-Fetoprotein --------------------------------- <6 ng/mL (SI: <6 µg/L)
Follicle-stimulating hormone (FSH) ---------------------------------
  1.0-13.0 mIU/mL (SI: 1.0-13.0 IU/L) (male);
  <3.0 mIU/mL (SI: <3.0 IU/L) (prepuberty, female);
  2.0-12.0 mIU/mL (SI: 2.0-12.0 IU/L) (follicular, female);
  4.0-36.0 mIU/mL (SI: 4.0-36.0 IU/L) (midcycle, female);
  1.0-9.0 mIU/mL (SI: 1.0-9.0 IU/L) (luteal, female);
  >30.0 mIU/mL (SI: >30.0 IU/L) (postmenopausal, female)
Free fatty acids ----------------- 10.6-18.0 mg/dL (SI: 0.4-0.7 nmol/L)
Gastrin ---------------------------------- <100 pg/mL (SI: <100 ng/L)
Growth hormone (GH) -- 0.01-0.97 ng/mL (SI: 0.01-0.97 µg/L) (male);
  0.01-3.61 ng/mL (SI: 0.01-3.61 µg/L) (female)
Homocysteine -------------------------- ≤1.76 mg/L (SI: ≤13 µmol/L)
β-Human chorionic gonadotropin (β-hCG) --------------------------
  <3.0 mIU/mL (SI: <3.0 IU/L) (nonpregnant female);
  >25 mIU/mL (SI: >25 IU/L) indicates a positive pregnancy test
β-Hydroxybutyrate -------------------- <6.3 mg/dL (SI: <0.6 mmol/L)
17-Hydroxypregnenolone ------ 29-189 ng/dL (SI: 0.87-5.69 nmol/L)
17α-Hydroxyprogesterone <220 ng/dL (SI: <6.67 nmol/L) (adult male);
  <80 ng/dL (SI: <2.42 nmol/L) (follicular, female);
  <285 ng/dL (SI: <8.64 nmol/L) (luteal, female);
  <51 ng/dL (SI: <1.55 nmol/L) (postmenopausal, female)
25-Hydroxyvitamin D ---- <20 ng/mL (SI: <49.9 nmol/L) (deficiency);
  21-29 ng/mL (SI: 52.4-72.4 nmol/L) (insufficiency);
  30-80 ng/mL (SI: 74.9-199.7 nmol/L) (optimal levels);
  >80 ng/mL (SI: >199.7 nmol/L) (toxicity possible)
Inhibin B --------------------------- 15-300 pg/mL (SI: 15-300 ng/L)

Insulinlike growth factor 1 (IGF-1)

Patient Age	Female	Male
18 years	162-541 ng/mL (SI: 21.2-70.9 nmol/L)	170-640 ng/mL (SI: 22.3-83.8 nmol/L)
19 years	138-442 ng/mL (SI: 18.1-57.9 nmol/L)	147-527 ng/mL (SI: 19.3-69.0 nmol/L)
20 years	122-384 ng/mL (SI: 16.0-50.3 nmol/L)	132-457 ng/mL (SI: 17.3-59.9 nmol/L)
21-25 years	116-341 ng/mL (SI: 15.2-44.7 nmol/L)	116-341 ng/mL (SI: 15.2-44.7 nmol/L)
26-30 years	117-321 ng/mL (SI: 15.3-42.1 nmol/L)	117-321 ng/mL (SI: 15.3-42.1 nmol/L)
31-35 years	113-297 ng/mL (SI: 14.8-38.9 nmol/L)	113-297 ng/mL (SI: 14.8-38.9 nmol/L)
36-40 years	106-277 ng/mL (SI: 13.9-36.3 nmol/L)	106-277 ng/mL (SI: 13.9-36.3 nmol/L)
41-45 years	98-261 ng/mL (SI: 12.8-34.2 nmol/L)	98-261 ng/mL (SI: 12.8-34.2 nmol/L)
46-50 years	91-246 ng/mL (SI: 11.9-32.2 nmol/L)	91-246 ng/mL (SI: 11.9-32.2 nmol/L)
51-55 years	84-233 ng/mL (SI: 11.0-30.5 nmol/L)	84-233 ng/mL (SI: 11.0-30.5 nmol/L)
56-60 years	78-220 ng/mL (SI: 10.2-28.8 nmol/L)	78-220 ng/mL (SI: 10.2-28.8 nmol/L)
61-65 years	72-207 ng/mL (SI: 9.4-27.1 nmol/L)	72-207 ng/mL (SI: 9.4-27.1 nmol/L)
66-70 years	67-195 ng/mL (SI: 8.8-25.5 nmol/L)	67-195 ng/mL (SI: 8.8-25.5 nmol/L)
71-75 years	62-184 ng/mL (SI: 8.1-24.1 nmol/L)	62-184 ng/mL (SI: 8.1-24.1 nmol/L)
76-80 years	57-172 ng/mL (SI: 7.5-22.5 nmol/L)	57-172 ng/mL (SI: 7.5-22.5 nmol/L)
>80 years	53-162 ng/mL (SI: 6.9-21.2 nmol/L)	53-162 ng/mL (SI: 6.9-21.2 nmol/L)

Insulinlike growth factor binding protein 3 ------------- 2.5-4.8 mg/L
Insulin ------------------------- 1.4-14.0 µIU/mL (SI: 9.7-97.2 pmol/L)
Islet-cell antibody assay ------- 0 Juvenile Diabetes Foundation units
Luteinizing hormone (LH) --- 1.0-9.0 mIU/mL (SI: 1.0-9.0 IU/L) (male);
  <1.0 mIU/mL (SI: <1.0 IU/L) (prepuberty, female);
  1.0-18.0 mIU/mL (SI: 1.0-18.0 IU/L) (follicular, female);
  20.0-80.0 mIU/mL (SI: 20.0-80.0 IU/L) (midcycle, female);
  0.5-18.0 mIU/mL (SI: 0.5-18.0 IU/L) (luteal, female);
  >30.0 mIU/mL (SI: >30.0 IU/L) (postmenopausal, female)
Metanephrines (plasma fractionated)
  Metanephrine ----------------------- <99 pg/mL (SI: <0.50 nmol/L)
  Normetanephrine ----------------- <165 pg/mL (SI: <0.90 nmol/L)
75-g oral glucose tolerance test blood glucose values --------------
  60-100 mg/dL (SI: 3.3-5.6 mmol/L) (fasting);
  <200 mg/dL (SI: <11.1 mmol/L) (1 hour);
  <140 mg/dL (SI: <7.8 mmol/L) (2 hour); between 140-200 mg/dL
  (SI: 7.8-11.1 mmol/L) is considered impaired glucose tolerance or

prediabetes; greater than 200 mg/dL (SI: >11.1 mmol/L) is a sign of diabetes mellitus

50-g oral glucose tolerance test for gestational diabetes ---------- <140 mg/dL (SI: <7.8 mmol/L) (1 hour)

100-g oral glucose tolerance test for gestational diabetes --------- <95 mg/dL (SI: <5.3 mmol/L) (fasting); <180 mg/dL (SI: <10.0 mmol/L) (1 hour); <155 mg/dL (SI: <8.6 mmol/L) (2 hour); <140 mg/dL (SI: <7.8 mmol/L) (3 hour)

Osteocalcin ---------------------- 9.0-42.0 ng/mL (SI: 9.0-42.0 μg/L)

Parathyroid hormone, intact (PTH) - 10-65 pg/mL (SI: 1.1-6.9 pmol/L)

Parathyroid hormone–related protein (PTHrP) ----------- <2.0 pmol/L

Progesterone -------------------- ≤1.2 ng/mL (SI: ≤3.8 nmol/L) (male); ≤1.0 ng/mL (SI: ≤3.2 nmol/L) (follicular, female); 2.0-20.0 ng/mL (SI: 6.4-63.6 nmol/L) (luteal, female); ≤1.1 ng/mL (SI: ≤3.5 nmol/L) (postmenopausal, female); >10.0 ng/mL (SI: >31.8 nmol/L) (evidence of ovulatory adequacy)

Proinsulin ------------------- 26.5-176.4 pg/mL (SI: 3.0-20.0 pmol/L)

Prolactin ----------------- 4-23 ng/mL (SI: 0.17-1.00 nmol/L) (male); 4-30 ng/mL (SI: 0.17-1.30 nmol/L) (nonlactating female); 10-200 ng/mL (SI: 0.43-8.70 nmol/L) (lactating female)

Prostate-specific antigen (PSA) -------------------------------------- <2.0 ng/mL (SI: <2.0 μg/L) (≤40 years); <2.8 ng/mL (SI: <2.8 μg/L) (≤50 years); <3.8 ng/mL (SI: <3.8 μg/L) (≤60 years); <5.3 ng/mL (SI: <5.3 μg/L) (≤70 years); <7.0 ng/mL (SI: <7.0 μg/L) (≤79 years); <7.2 ng/mL (SI: <7.2 μg/L) (≥80 years)

Renin activity, plasma, sodium replete, ambulatory ---------------- 0.6-4.3 ng/mL per h

Renin, direct concentration ---------- 4-44 pg/mL (SI: 0.1-1.0 pmol/L)

Sex hormone–binding globulin (SHBG) -------------- 1.1-6.7 μg/mL (SI: 10-60 nmol/L) (male); 2.2-14.6 μg/mL (SI: 20-130 nmol/L) (female)

α-Subunit of pituitary glycoprotein hormones ---------------------- <1.2 ng/mL (SI: <1.2 μg/L)

Testosterone (bioavailable) ----- 0.8-4.0 ng/dL (SI: 0.03-0.14 nmol/L) (20-50 years, female on oral estrogen); 0.8-10.0 ng/dL (SI: 0.03-0.35 nmol/L) (20-50 years, female not on oral estrogen); 83.0-257.0 ng/dL (SI: 2.88-8.92 nmol/L) (male 20-29 years); 72.0-235.0 ng/dL (SI: 2.50-8.15 nmol/L) (male 30-39 years); 61.0-213.0 ng/dL (SI: 2.12-7.39 nmol/L) (male 40-49 years); 50.0-190.0 ng/dL (SI: 1.74-6.59 nmol/L) (male 50-59 years); 40.0-168.0 ng/dL (SI: 1.39-5.83 nmol/L) (male 60-69 years)

Testosterone (free) ----- 9.0-30.0 ng/dL (SI: 0.31-1.04 nmol/L) (male); 0.3-1.9 ng/dL (SI: 0.01-0.07 nmol/L) (female)

Testosterone (total) ---- 300-900 ng/dL (SI: 10.4-31.2 nmol/L) (male); 8-60 ng/dL (SI: 0.3-2.1 nmol/L) (female)

Vitamin B_{12} ------------------- 180-914 pg/mL (SI: 133-674 pmol/L)

Chemistry Values

Alanine aminotransferase ---------- 10-40 U/L (SI: 0.17-0.67 μkat/L)

Albumin ---------------------------------- 3.5-5.0 g/dL (SI: 35-50 g/L)

Amylase -------------------------- 26-102 U/L (SI: 0.43-1.70 μkat/L)

Anion gap ---------------------------- 3-11 mEq/L (SI: 3-11 mmol/L)

Aspartate aminotransferase ------- 20-48 U/L (SI: 0.33-0.80 μkat/L)

Bicarbonate ----------------------- 21-28 mEq/L (SI: 21-28 mmol/L)

Bilirubin (total) ------------------ 0.3-1.2 mg/dL (SI: 5.1-20.5 μmol/L)

Blood gases

Po_2, arterial blood ------------ 80-100 mm Hg (SI: 10.6-13.3 kPa)

Pco_2, arterial blood ---------------- 35-45 mm Hg (SI: 4.7-6.0 kPa)

Blood pH -- 7.35-7.45

Calcium ------------------------ 8.2-10.2 mg/dL (SI: 2.1-2.6 mmol/L)

Calcium (ionized) ------------- 4.60-5.08 mg/dL (SI: 1.2-1.3 mmol/L)

Carbon dioxide --------------------- 22-28 mEq/L (SI: 22-28 mmol/L)

CD_4 cell count --------------------- 500-1400/μL (SI: 0.5-1.4 × 10^9/L)

Chloride ------------------------- 96-106 mEq/L (SI: 96-106 mmol/L)

Creatine kinase ------------------- 50-200 U/L (SI: 0.84-3.34 μkat/L)

Creatinine ------------- 0.7-1.3 mg/dL (SI: 61.9-114.9 μmol/L) (male); 0.6-1.1 mg/dL (SI: 53.0-97.2 μmol/L) (female)

Ferritin ----------------------- 15-200 ng/mL (SI: 33.7-449.4 pmol/L)

Folate ------------------------------------- ≥4.0 ng/mL (SI: ≥4.0 μg/L)

Glucose --------------------------- 70-99 mg/dL (SI: 3.9-5.5 mmol/L)

γ-Glutamyltransferase -------------- 2-30 U/L (SI: 0.03-0.50 μkat/L)

Iron ---------------------- 50-150 μg/dL (SI: 9.0-26.8 μmol/L) (male); 35-145 μg/dL (SI: 6.3-26.0 μmol/L) (female)

Lactate dehydrogenase ----------- 100-200 U/L (SI: 1.7-3.3 μkat/L)

Lactic acid --------------------- 5.4-20.7 mg/dL (SI: 0.6-2.3 mmol/L)

Lipase ---------------------------- 10-73 U/L (SI: 0.17-1.22 μkat/L)

Magnesium ---------------------- 1.5-2.3 mg/dL (SI: 0.6-0.9 mmol/L)

Osmolality -------------- 275-295 mOsm/kg (SI: 275-295 mmol/kg)

Phosphate ---------------------- 2.3-4.7 mg/dL (SI: 0.7-1.5 mmol/L)

Potassium ---------------------- 3.5-5.0 mEq/L (SI: 3.5-5.0 mmol/L)

Prothrombin time -- 8.3-10.8 s

Serum urea nitrogen --------------- 8-23 mg/dL (SI: 2.9-8.2 mmol/L)

Sodium ---------------------- 136-142 mEq/L (SI: 136-142 mmol/L)

Transferrin saturation ------------------------------------ 14%-50%

Troponin I -------------------------------- <0.6 ng/mL (SI: <0.6 μg/L)

Tryptase ------------------------------ <11.5 ng/mL (SI: <11.5 μg/L)

Uric acid -------------------- 3.5-7.0 mg/dL (SI: 208.2-416.4 μmol/L)

Urine

Albumin ------------ 30-300 μg/mg creat (SI: 3.4-33.9 μg/mol creat)

Albumin-to-creatinine ratio ------------------------- <30 mg/g creat

Aldosterone --------------------- 3-20 μg/24 h (SI: 8.3-55.4 nmol/d) (should be <12 μg/24 h [SI: <33.2 nmol/d] with oral sodium loading—confirmed with 24-hour urinary sodium >200 mEq)

Calcium ---------------------- 100-300 mg/24 h (SI: 2.5-7.5 mmol/d)

Catecholamine fractionation

Normotensive normal ranges:

Dopamine ------------------- <400 μg/24 h (SI: <2610 nmol/d)

Epinephrine -------------------- <21 μg/24 h (SI: <115 nmol/d)

Norepinephrine -----------------<80 μg/24 h (SI: <473 nmol/d)
Citrate -------------------320-1240 mg/24 h (SI: 16.7-64.5 mmol/d)
Cortisol --------------------------- 4-50 μg/24 h (SI: 11-138 nmol/d)
Cortisol following dexamethasone-suppression test (low-dose:
2 day, 2 mg daily) ---------------- <10 μg/24 h (SI: <27.6 nmol/d)
Creatinine---------------------- 1.0-2.0 g/24 h (SI: 8.8-17.7 mmol/d)
Glomerular filtration rate (estimated) ------->60 mL/min per 1.73 m^2
5-Hydroxyindole acetic acid----- 2-9 mg/24 h (SI: 10.5-47.1 μmol/d)
Iodine (random)-- >100 μg/L
17-Ketosteroids ---- 6.0-21.0 mg/24 h (SI: 20.8-72.9 μmol/d) (male); 4.0-17.0 mg/24 h (SI: 13.9-59.0 μmol/d) (female)
Metanephrine fractionation
Normotensive normal ranges:
Metanephrine -------- <261 μg/24 h (SI: <1323 nmol/d) (male); <180 μg/24 h (SI: <913 nmol/d) (female)
Normetanephrine -------------------- age and sex dependent
Total metanephrine ------------------ age and sex dependent
Osmolality ------------150-1150 mOsm/kg (SI: 150-1150 mmol/kg)
Oxalate ----------------------------- <40 mg/24 h (SI: <456 mmol/d)
Phosphate --------------------0.9-1.3 g/24 h (SI: 29.1-42.0 mmol/d)
Potassium --------------------- 17-77 mEq/24 h (SI: 17-77 mmol/d)
Sodium ----------------------40-217 mEq/24 h (SI: 40-217 mmol/d)
Uric acid ----------------------------<800 mg/24 h (SI: <4.7 mmol/d)

Saliva

Cortisol (salivary), midnight -----------<0.13 μg/dL (SI: <3.6 nmol/L)

Semen

Semen analysis--------------- >20 million sperm/mL; >50% motility

Abbreviations

ACTH --corticotropin
ACE inhibitor-------------- angiotensin-converting enzyme inhibitor
ALT --------------------------------------- alanine aminotransferase
AST ------------------------------------ aspartate aminotransferase
BMI -- body mass index
CDC ------------------- Centers for Disease Control and Prevention
CNS --------------------------------------- central nervous system
CT-- computed tomography
DHEA -------------------------------------- dehydroepiandrosterone
DHEA-S---------------------------- dehydroepiandrosterone sulfate
DNA --- deoxyribonucleic acid
DPP-4 inhibitor --------------------dipeptidyl-peptidase 4 inhibitor
DXA------------------------------dual-energy x-ray absorptiometry
FDA----------------------------------Food and Drug Administration
FGF-23 -------------------------------- fibroblast growth factor 23
FNA --- fine-needle aspiration
FSH ----------------------------------- follicle-stimulating hormone
GH --- growth hormone
GHRH------------------------- growth hormone–releasing hormone
GLP-1 receptor agonist----- glucagonlike peptide 1 receptor agonist
GnRH ----------------------------gonadotropin-releasing hormone
hCG -------------------------------- human chorionic gonadotropin
HDL---high-density lipoprotein
HIV--------------------------------- human immunodeficiency virus
HMG-CoA reductase inhibitor---------- 3-hydroxy-3-methylglutaryl coenzyme A reductase inhibitor
IGF-1------------------------------------ insulinlike growth factor 1
LDL --low-density lipoprotein
LH -- luteinizing hormone
MCV -------------------------------------mean corpuscular volume
MIBG----------------------------------- *meta*-iodobenzylguanidine
MRI ---------------------------------- magnetic resonance imaging
NPH insulin -------------------- neutral protamine Hagedorn insulin
PCSK9 inhibitor---- proprotein convertase subtilisin/kexin 9 inhibitor
PET ---------------------------------positron emission tomography
PSA -------------------------------------- prostate-specific antigen
PTH ---parathyroid hormone
PTHrP------------------------parathyroid hormone–related protein
SGLT-2 inhibitor -----------sodium-glucose cotransporter 2 inhibitor
SHBG ------------------------------- sex hormone–binding globulin
T_3 -- triiodothyronine
T_4 -- thyroxine
TPO antibodies ------------------------- thyroperoxidase antibodies
TRH -------------------------------- thyrotropin-releasing hormone
TRAb ------------------------------------ TSH-receptor antibodies
TSH -- thyrotropin
VLDL---------------------------------- very low-density lipoprotein

QUESTIONS

ENDOCRINE SELF-ASSESSMENT PROGRAM 2025

Part I

1 A 27-year-old woman presents to the emergency department with abdominal pain, nausea, and vomiting. The pain radiates to her right flank. She has no fever, chills, or changes in bowel or bladder function. She delivered a healthy baby via cesarean delivery 1 year ago, and she is still breastfeeding. Her medical history is otherwise unremarkable. She takes no medications. Abdominal CT with contrast is performed, which identifies a 6.4-cm heterogeneous, right adrenal mass with central calcifications and a prominent capsule (*see images, arrows*).

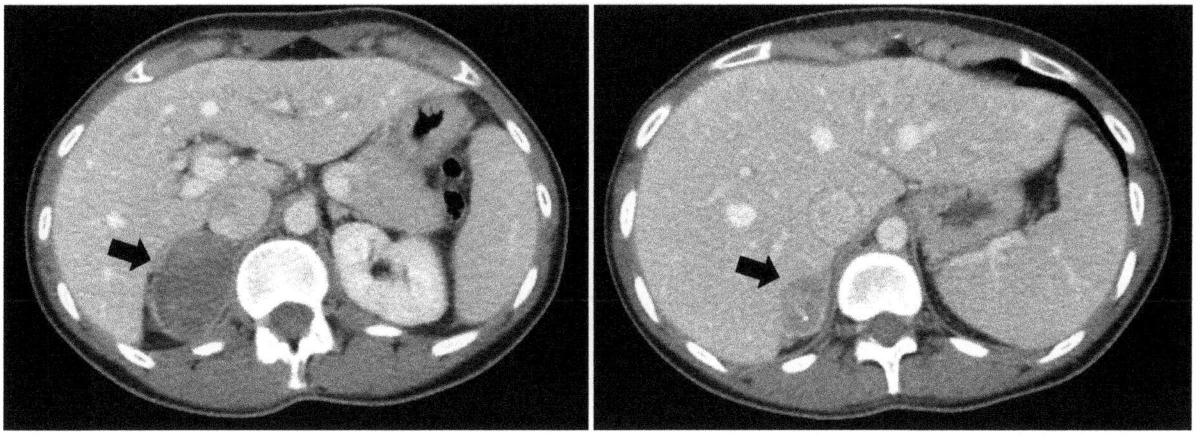

On physical examination, her blood pressure is 104/75 mm Hg and pulse rate is 72 beats/min. There is no orthostasis. Her height is 65 in (165 cm), and weight is 136 lb (61.7 kg) (BMI = 22.6 kg/m^2). She has no plethora, bruises, striae, acne, or hirsutism. Muscle strength is within normal limits. There is no spine or flank tenderness and no peripheral edema.

She is discharged home after intravenous hydration and is referred to endocrinology.

Laboratory test results after her endocrinology evaluation:

Serum cortisol (8 AM) = 22 µg/dL (5-25 µg/dL) (SI: 606.9 nmol/L [137.9-689.7 nmol/L])
ACTH (8 AM) = 24 pg/mL (10-60 pg/mL) (SI: 5.3 pmol/L [2.2-13.2 pmol/L])
Serum aldosterone = 6.2 ng/dL (4-21 ng/dL) (SI: 172.0 pmol/L [111.0-582.5 pmol/L])
Plasma renin activity = 3.3 ng/mL per h (0.6-4.3 ng/mL per h)
DHEA-S = 157 µg/dL (44-332 µg/dL) (SI: 4.25 µmol/L [1.19-9.00 µmol/L])
Plasma metanephrine = 98.6 pg/mL (<99 pg/mL) (SI: 0.5 nmol/L [<0.50 nmol/L])
Plasma normetanephrine = 274.7 pg/mL (<165 pg/mL) (SI: 1.5 nmol/L [<0.90 nmol/L])

An ^{18}F-FDG PET-CT shows uptake at the background level.

Which of the following is the best management approach for this patient?

A. Perform adrenal protocol CT, with and without contrast
B. Follow-up with abdominal CT in 3 months
C. Perform MIBG scan
D. Refer to surgery for right adrenalectomy
E. Start phenoxybenzamine and aggressive hydration

2 A 62-year-old man with type 2 diabetes of 8 years duration is seen for follow-up. He is currently on metformin, 1000 mg twice daily, and dulaglutide, 3 mg once weekly. He reports mild symptoms of bloating and intermittent diarrhea but feels well overall. A recent eye examination did not show retinopathy, and he does not have any other known chronic complications of diabetes. Hypertension is well controlled on lisinopril, 20 mg daily, and he also takes atorvastatin, 80 mg daily. He lost 15 lb (6.8 kg) after starting dulaglutide 3 years ago, but his weight has been stable over the last year. He has been less active over the last few months because of gout flares, but he has since resumed allopurinol and anticipates restarting regular walking this summer.

On physical examination, his blood pressure is 128/80 mm Hg and pulse rate is 76 beats/min. His height is 65.7 in (167 cm), and weight is 198.4 lb (90 kg) (BMI = 32.3 kg/m^2).

Laboratory test results:

Fasting plasma glucose = 124 mg/dL (70-99 mg/dL) (SI: 6.9 mmol/L [3.9-5.5 mmol/L])
Hemoglobin A_{1c} = 6.9% (4.0%-5.6%) (52 mmol/mol [20-38 mmol/mol])
Serum creatinine = 1.1 mg/dL (0.7-1.3 mg/dL) (SI: 97.2 μmol/L [61.9-114.9 μmol/L])
Total cholesterol = 188 mg/dL (<200 mg/dL [optimal]) (SI: 4.87 mmol/L [<5.18 mmol/L])
Triglycerides = 205 mg/dL (<150 mg/dL [optimal]) (SI: 2.32 mmol/L [<1.70 mmol/L])
HDL cholesterol = 48 mg/dL (>60 mg/dL [optimal]) (SI: 1.24 mmol/L [>1.55 mmol/L])
LDL cholesterol = 89 mg/dL (<100 mg/dL [optimal]) (SI: 2.31 mmol/L [<2.59 mmol/L])

The lipid panel results are similar to those obtained 6 months ago.

In addition to advice on intensifying lifestyle intervention for weight loss, which of the following is the best next step?

A. Add bempedoic acid, 180 mg daily
B. Add colesevelam, 6 tablets of 625 mg daily
C. Add evolocumab, 140 mg every 2 weeks
D. Add ezetimibe, 10 mg daily
E. Recommend no changes to current lipid-lowering therapy

3 A 20-year-old woman with questions about her risk of developing type 1 diabetes presents for evaluation. She reports that her 24-year-old sister was diagnosed with type 1 diabetes 7 years ago. Her sister informed her about a recently approved medication that might reduce the likelihood of at-risk patients, such as siblings of persons with type 1 diabetes, from developing diabetes. The patient's primary care provider checked antibodies for type 1 diabetes, and she tested positive. She has no polyuria, polydipsia, or recent weight loss. She is very active and runs 6 miles 3 days per week. She also reports eating small amounts of carbohydrates (typically consuming less than 60 g daily).

Laboratory test results:

Glutamic acid decarboxylase 65 antibodies, detectable
Insulinoma-associated protein (IA-2) antibodies, detectable
Zinc transporter (ZnT8) antibodies, undetectable
Hemoglobin A_{1c} = 6.3% (4.0%-5.6%) (45 mmol/mol [20-38 mmol/mol])

Two-hour oral glucose tolerance test results:

Fasting blood glucose = 108 mg/dL (SI: 6.0 mmol/L)
1-hour glucose value = 167 mg/dL (SI: 9.3 mmol/L)
2-hour glucose value = 208 mg/dL (SI: 11.5 mmol/L)

Second 2-hour oral glucose tolerance test results:

Fasting blood glucose = 112 mg/dL (SI: 6.2 mmol/L)
1-hour glucose value = 162 mg/dL (SI: 9.0 mmol/L)
2-hour glucose value = 210 mg/dL (SI: 11.7 mmol/L)

On physical examination, her blood pressure is 122/80 mm Hg and pulse rate is 64 beats/min. Her height is 66 in (167.6 cm), and weight is 132 lb (59.9 kg) (BMI = 21.3 kg/m^2).

Which of the following is the best next step in this patient's management?

A. Initiate teplizumab therapy
B. Measure hemoglobin A_{1c} again
C. Perform insulin antibody titer
D. Perform islet-cell antibody titer
E. Refer for diabetes education

4 A 68-year-old woman presents with a new diagnosis of osteoporosis. She has a history of fibromyalgia, Graves disease status post radioactive iodine treatment many years ago, HR+ breast cancer treated with chemotherapy and radiation (currently on year 4 of aromatase inhibitor therapy), and gastroesophageal reflux disease. She has no history of fractures or kidney stones. She recently underwent placement of a dental implant that has now healed. Her family history is relevant for hip fracture in her father at age 70 years, breast cancer in her mother, and osteoporosis and breast cancer in her older sister. Current medications include an albuterol inhaler as needed, ranitidine, calcium citrate, cholecalciferol, anastrozole, and levothyroxine.

Laboratory test results:

Serum 25-hydroxyvitamin D = 22 ng/mL (30-80 ng/mL [optimal]) (SI: 44.9 nmol/L [7.4.9-199.7 nmol/L])
TSH = 3.2 mIU/L (0.5-5.0 mIU/L)
Free T_4 = 1.20 ng/dL (0.8-1.8 ng/dL) (SI: 1.3 pmol/L [10.30-23.17 pmol/L])
Complete blood cell count, normal
Erythrocyte sedimentation rate = 28 mm/h (0-20 mm/h)
Complete metabolic panel, normal
24-hour urinary calcium, normal

DXA documents the following values:

Lumbar spine T-score = −2.5
Femoral neck T-score = −2.9

After discussion of osteoporosis treatment options, she decides to proceed with intravenous zoledronic acid. The day after infusion, she reports flulike symptoms, night sweats, myalgias, and intense pain and swelling of the right eye accompanied by right-sided headache. Periorbital edema, erythema, and conjunctival hyperemia are noted. Visual acuity is normal with no retinal involvement. Contrast-enhanced MRI shows proptosis of the right globe, orbital soft-tissue thickening, and intraorbital enhancement without diffusion-weighted imaging restriction.

Which of the following is the most likely explanation of this patient's visual symptoms?

A. Bisphosphonate-induced orbital inflammation
B. Giant-cell arteritis
C. Metastatic breast cancer
D. Orbital cellulitis
E. Thyroid eye disease

5 A 61-year-old man is referred for evaluation of possible hypogonadism and hyperprolactinemia. He was initially evaluated for erectile dysfunction and was told that his "testosterone is low, and prolactin is high; a hormone specialist would be helpful." He reports a decline in libido over the past 18 months. He has a good relationship with his girlfriend, and they have had an active and satisfying sex life. He reports decreased tumescence with erections during sexual activity and fewer spontaneous morning erections. His only notable medical history is osteoarthritis. He takes no medications.

On physical examination, his blood pressure is 128/80 mm Hg, pulse rate is 68 beats/min, and respiratory rate is 12 breaths/min. His height is 71 in (180.3 cm), and weight is 158 lb (71.7 kg) (BMI = 22.0 kg/m^2). He has no gynecomastia. Jugular venous pressure is normal, and heart rate is regular with no gallop. He has no edema. His penis is normal, and testes are 35 cc bilaterally. Skin has normal thickness and no hyperpigmentation or ecchymoses. He has normal sensation to light touch in the perineal region. Cremasteric reflex is normal.

Laboratory test results:

Serum total testosterone = 260 ng/dL (264-916 ng/dL) (SI: 9.0 nmol/L [9.2-31.8 nmol/L]) (assay validated and harmonized by the CDC)
Repeat serum testosterone (sample drawn at 7:30 AM) = 270 ng/dL (SI: 9.4 nmol/L)
(Calculated free testosterone, low-normal)
SHBG = 3.60 μg/mL (1.1-6.7 μg/mL) (SI: 32 nmol/L [10-60 nmol/L])
FSH = 30.0 mIU/mL (1.0-13.0 mIU/mL) (SI: 30.0 IU/L [1.0-13.0 IU/L])
LH = 2.0 mIU/mL (1.0-9.0 mIU/mL) (SI: 2.0 IU/L [1.0-9.0 IU/L])
Free T_4 = 1.0 ng/dL (0.8-1.8 ng/dL) (SI:12.87 pmol/L [10.30-23.17 pmol/L])
TSH = 1.8 mIU/L (0.5-5.0 mIU/L)
Prolactin = 24 ng/mL (4-23 ng/mL) (SI: 1.04 nmol/L [0.17-1.00 nmol/L])

Which of the following is the best next step in this patient's evaluation?

A. Karyotype analysis
B. Sellar MRI
C. Seminal fluid analysis
D. Serum estradiol measurement
E. Serum inhibin B measurement

6 A 63-year-old man reports gradual bilateral loss of vision over 2 years, which did not improve after cataract surgery. Ophthalmologic evaluation indicates optic neuropathy. His medical history is notable for prediabetes, hypertension, hyperlipidemia, remote transient ischemic stroke, and nephrolithiasis. Medications are rosuvastatin, 20 mg daily; lisinopril, 20 mg daily; and aspirin, 325 mg daily.

On physical examination, he has noninflammatory arthritic nodules on distal interphalangeal joints and bitemporal superior quadrantanopia.

Brain MRI reveals a 4.3 × 3.6 × 2.5-cm pituitary mass with suprasellar extension and bilateral cavernous sinus invasion (*see images*).

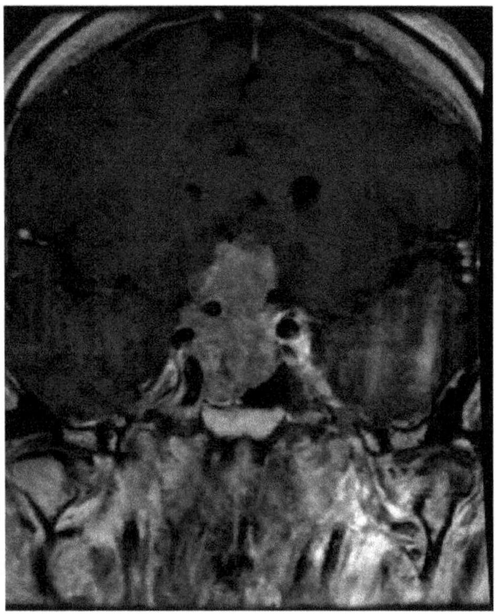

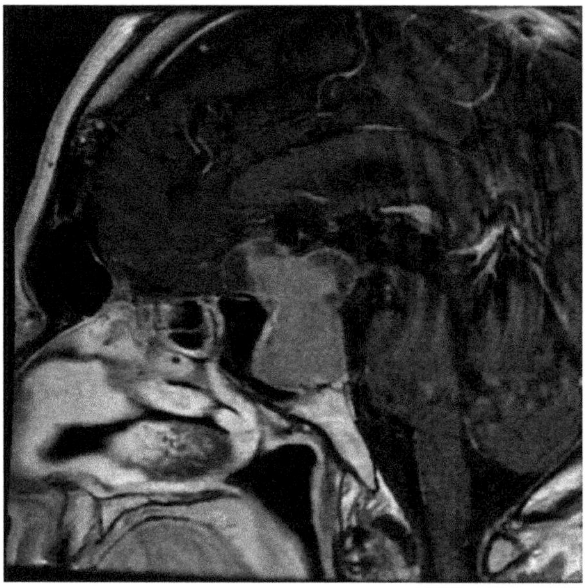

Laboratory test results (sample drawn at 8 AM while fasting):

Prolactin = 1609 ng/mL (4-23 ng/mL) (SI: 69.96 nmol/L [0.17-1.00 nmol/L])
TSH = 1.04 mIU/L (0.5-5.0 mIU/L)
Free T_4 = 1.00 ng/dL (0.8-1.8 ng/dL) (SI: 12.87 pmol/L [10.30-23.17 pmol/L])
FSH = 4.0 mIU/mL (1.0-13.0 mIU/mL) (SI: 4.0 IU/L [1.0-13.0 IU/L])

LH = 3.7 mIU/mL (1.0-9.0 mIU/mL) (SI: 3.7 IU/L [1.0-9.0 IU/L])
Total testosterone = 135 ng/dL (300-900 ng/dL) (SI: 4.7 nmol/L [10.4-31.2 nmol/L])
Cortisol = 9.7 μg/dL (5-25 μg/dL) (SI: 267.6 nmol/L [137.9-689.7 nmol/L])
ACTH = 69.2 pg/mL (10-60 pg/mL) (SI: 15.2 pmol/L [2.2-13.2 pmol/L])
GH = 1.99 ng/mL (0.01-0.97 ng/mL) (SI: 1.99 μg/L [0.01-0.97 μg/L])
Hemoglobin A_{1c} = 5.9% (4.0%-5.6%) (41% [20-38 mmol/mol])
Chemistry panel, normal

Which of the following is the best next step in managing this patient's pituitary tumor?

A. Measure 24-hour urinary cortisol
B. Measure GH 120 minutes after oral administration of 75 g glucose
C. Measure IGF-1
D. Refer for transsphenoidal tumor removal
E. Start cabergoline, 0.5 mg twice weekly

7 A 28-year-old woman (G2, P2) with a history of gestational diabetes in her 2 previous pregnancies is referred for evaluation. Her younger sister was recently diagnosed with diabetes, which prompted the patient to seek care. Her mother and maternal grandfather have diabetes. The patient was treated with insulin during both of her pregnancies. She had a normal delivery 6 months ago with no complications, and she is currently breastfeeding and takes no medications. She has never checked her blood glucose values outside of pregnancy. She has not been hospitalized for anything other than childbirth.

On physical examination, her blood pressure is 110/62 mm Hg and pulse rate is 68 beats/min. Her height is 65 in (165.1 cm), and weight is 144 lb (65.3 kg) (BMI = 24.0 kg/m^2).

Results of point-of-care testing at this visit and her most recent 100-g oral glucose tolerance test during pregnancy are shown (*see table*).

Measurement	At 20 weeks' gestation	Present
Hemoglobin A_{1c}	...	6.4% (46 mmol/mol)
Fasting glucose	92 mg/dL (SI: 5.1 mmol/L)	101 mg/dL (SI: 5.6 mmol/L)
1-hour glucose	195 mg/dL (SI: 10.8 mmol/L)	...
2-hour glucose	189 mg/dL (SI: 10.5 mmol/L)	...
3-hour glucose	155 mg/dL (SI: 8.6 mmol/L)	...

Assessing for pathogenic variants in which of the following genes will most likely establish the cause of this patient's abnormal test results?

A. *CFTR*
B. *FOXP3*
C. *HLA-DQA1*
D. *HNF1A*
E. *TCF7L2*

8 A 37-year-old woman was diagnosed with primary ovarian insufficiency at age 36 after prolonged amenorrhea and multiple elevated FSH measurements documented by her referring provider. Extensive evaluation, including karyotype analysis and genetic testing, has not identified an etiology. She reports occasional hot flashes, vaginal dryness, dyspareunia, and increase in baseline anxiety. Her libido is low, and she has had both personal and professional stressors. She does not plan to pursue fertility treatment and is comfortable with the idea of not having biological children, but an unintended pregnancy would be fine. She prefers to avoid unpredictable bleeding. Family history is notable for osteoporosis in her mother.

On physical examination, her blood pressure is 110/69 mm Hg and pulse rate is 71 beats/min. Her height is 65 in (165.1 cm), and weight is 127 lb (57.6 kg) (BMI = 21.1 kg/m^2). She appears well nourished and fully feminized with mature secondary female characteristics. There are no acute findings on examination.

Which of the following is the best next step in this patient's treatment?

A. 1-20 mcg ethinyl estradiol oral combined hormonal contraception, used cyclically
B. 1-20 mcg ethinyl estradiol oral combined hormonal contraception, used continuously
C. 50 mcg estradiol patch and micronized progesterone, 100 mg nightly
D. 100 mcg estradiol patch and medroxyprogesterone acetate, 10 mg cycled days 1 to 12 of each month
E. 150 mcg estradiol patch and micronized progesterone, 200 mg cycled days 1 to 12 of each month

9 A 43-year-old woman with type 2 diabetes undergoes Roux-en-Y gastric bypass and loses 88 lb (45 kg) over the next year. Her hemoglobin A_{1c} level normalizes, and she is able to cease taking diabetes medications. She subsequently stops following up with the bariatric clinic. Two years later, she presents with proximal muscle weakness and pain in her upper legs with walking. Radiographs show bilateral medial femoral stress fractures. She has had no prior fractures. She takes elemental calcium (as calcium citrate), 200 mg twice daily. She has been taking a multivitamin, and cholecalciferol, 2000 international units daily (50 mcg daily).

DXA documents a spine T-score of –3.1 and a left femoral neck T-score of –2.7.

Laboratory test results:

Calcium = 7.7 mg/dL (8.2-10.2 mg/dL) (SI: 1.9 mmol/L [2.1-2.6 mmol/L])
Albumin = 3.8 g/dL (3.5-5.0 g/dL) (SI: 38 g/L [35-50 g/L])
Phosphate = 2.2 mg/dL (2.3-4.7 mg/dL) (SI: 0.7 mmol/L [0.7-1.5 mmol/L])
Magnesium = 1.6 mg/dL (1.5-2.3 mg/dL) (SI: 0.66 mmol/L [0.6-0.9 mmol/L])
Alkaline phosphatase = 187 U/L (50-120 U/L) (SI: 3.12 µkat/L [0.84-2.00 µkat/L])
Creatinine = 0.8 mg/dL (0.6-1.1 mg/dL) (SI: 70.7 µmol/L [53.0-97.2 µmol/L])
Intact PTH = 127 pg/mL (10-65 pg/mL) (SI: 13.5 pmol/L [1.1-6.9 pmol/L])
25-Hydroxyvitamin D = 28 ng/mL (30-80 ng/mL [optimal]) (SI: 69.9 nmol/L [74.9-199.7 nmol/L])
1,25-Dihydroxyvitamin D_3 = 68 pg/mL (16-65 pg/mL) (SI: 176.8 pmol/L [41.6-169.0 pmol/L])
Hematocrit = 31% (35%-45%) (SI: 0.31 [0.35-0.45)])
Urinary calcium = 32 mg/24 h (100-300 mg/24 h) (SI: 0.8 mmol/d [2.5-7.5 mmol/d])

Which of the following is the best next step in this patient's management?

A. Increase the calcium citrate dosage to 400 mg 3 times daily
B. Increase the cholecalciferol dosage to 50,000 international units daily (1250 mcg daily)
C. Start anabolic therapy with teriparatide
D. Start antiresorptive therapy with alendronate
E. Start treatment with phosphate salts

10 A 35-year-old man with a 4-year history of type 1 diabetes presents as a new patient. He has been very invested in diabetes management, which initially consisted of multiple daily insulin injections, followed by a switch to a commercially available sensor-augmented insulin pump.

Data while on insulin pump therapy:

Hemoglobin A_{1c} = 6.8% (4.0%-5.6%) (51 mmol/mol [20-38 mmol/mol])
Time-in-range (70-180 mg/dL [SI: 3.9-10.0 mmol/L]) = 80%
Time-above-range (>180-250 mg/dL [SI: >10.0-13.9 mmol/L]) = 10%
Time-above-range (>250 mg/dL [SI: >13.9 mmol/L]) = 1%

Time-below-range (<70 mg/dL [SI: <3.9 mmol/L]) = 4%
Time-below-range (<54 mg/dL [SI: <3.0 mmol/L]) = 5%

Four months ago, he switched to a do-it-yourself automated insulin delivery (DIY AID) pump system, as he wanted to further tighten his glycemic control and avoid episodes of hypoglycemia. He is motivated to continue with this system and feels it is the right fit for him. In the few months that he has used it, he thinks his glycemic control has improved.

Which of the following is the best next step in this patient's diabetes management?

A. Continue DIY AID and advise him to use online DIY forums for help titrating insulin
B. Counsel on the potential drawbacks of using a non–FDA-approved device (DIY AID) and institute a back-up plan in case of device failure
C. Discharge the patient from the clinical practice if he plans to continue DIY AID, as these systems are not FDA approved
D. Stop DIY AID and resume multiple daily injections
E. Stop DIY AID and resume previous sensor-augmented insulin pump

11 A 55-year-old woman presents for the management of primary hypothyroidism. Hypothyroidism secondary to Hashimoto thyroiditis was diagnosed at age 28 years, and she has since been on thyroid hormone replacement therapy. She has hyperlipidemia that is well controlled with diet and exercise and no other notable medical history.

On physical examination, her height is 66 in (167.6 cm), and weight is 142 lb (64.4 kg) (BMI = 22.9 kg/m^2).

She takes levothyroxine, 100 mcg once daily. She reports not having missed any doses and ingests the medication with only water upon waking and at least 1 hour before eating breakfast and having her morning coffee. She takes no other medications and only a women's multivitamin at bedtime nightly.

She is asymptomatic and does not report any hypothyroid or hyperthyroid symptoms or abnormal bowel habits. However, her serum TSH concentrations have been widely erratic, ranging from 5.9 to 15.3 mIU/L over the past 3 years, whereas previously they were stable and within the reference range. Review of pharmacy records shows that she has been refilling her levothyroxine at the prescribed intervals. Celiac disease screening is negative (serum tissue transglutaminase [TTG] IgA concentration).

An oral levothyroxine absorption test is recommended to delineate possible reasons for the varying serum TSH levels. After an overnight fast, she arrives at clinic and is administered 10 tablets of 100 mcg levothyroxine (total of 1000 mcg) from her own supply by mouth and ingests it with water only. Serum free T_4 is measured at baseline and hourly thereafter for 6 hours. The baseline free T_4 concentration is 1.0 ng/dL (12.9 pmol/L), and the peak free T_4 concentration is 1.3 ng/dL (16.7 pmol/L), obtained at the fourth hour of the test.

Which of the following is the best interpretation of these data?

Answer	Was the absorption test appropriately performed?	Interpretation
A.	No	Her usual 100 mcg of daily levothyroxine should have been administered instead of the 1000 mcg dose
B.	No	Serum free T_3, not serum free T_4, should have been serially measured
C.	No	Serum TSH, not serum free T_4, should have been serially measured
D.	Yes	The rise in free T_4 is appropriate; potential poor adherence should be investigated
E.	Yes	She has decreased ability to absorb oral levothyroxine and should be referred to a gastroenterologist for further evaluation

12 A 57-year-old woman presents for a weight management follow-up appointment. She has systemic lupus for which she takes prednisone intermittently. She reports difficulty with weight management since undergoing menopause in her early 50s. In the last 2 years, her physical activity has been restricted by symptoms of cervical radiculopathy.

Approximately 6 months ago, she was prescribed phentermine/topiramate ER, 7.5/46 mg daily, which she is tolerating without adverse effects. She reports staying satiated between meals and is not tempted to snack. She is, however, concerned about the lack of weight loss in the last month. She is engaged in moderate-intensity exercise for about 60 minutes each week.

On physical examination, her blood pressure is 128/72 mm Hg and pulse rate is 82 beats/min. Her height is 66.1 in (167.9 cm), and weight is 206 lb (93.4 kg) (BMI = 33.1 kg/m^2). She has symmetric weight distribution between her upper and lower body.

Her weight trend is as follows:

Weight at initiation of phentermine/topiramate: 226 lb (102.5 kg)
Weight 3 months ago: 208 lb (94.3 kg)
Weight 1 month ago: 205 lb (93.0 kg)

Laboratory test result:

Hemoglobin A_{1c} = 5.9% (4.0%-5.6%) (41 mmol/mol [20-38 mmol/mol])

In addition to arranging a dietician visit, which of the following should be recommended next?

A. Add metformin to current regimen
B. Add orlistat to current regimen
C. Continue current regimen
D. Stop phentermine/topiramate and counsel patient on strategies to increase physical activity
E. Switch from phentermine/topiramate to naltrexone/bupropion

13 A 25-year-old woman presents for a second opinion regarding an adrenal nodule. She has multiple endocrine neoplasia type 2A, inherited from her mother, that was confirmed with genetic testing in infancy (C634W pathogenic variant in the *RET* gene). At age 5 years, she underwent prophylactic thyroidectomy, and pathology demonstrated C-cell hyperplasia. Her only current medication is levothyroxine. Calcitonin, carcinoembryonic antigen, serum calcium, and plasma metanephrine concentrations have been monitored, and all results have been normal. She recently presented to the emergency department with acute flank pain. Abdominal CT without contrast showed a heterogenous, 1.0-cm right adrenal nodule (*see image, arrow*).

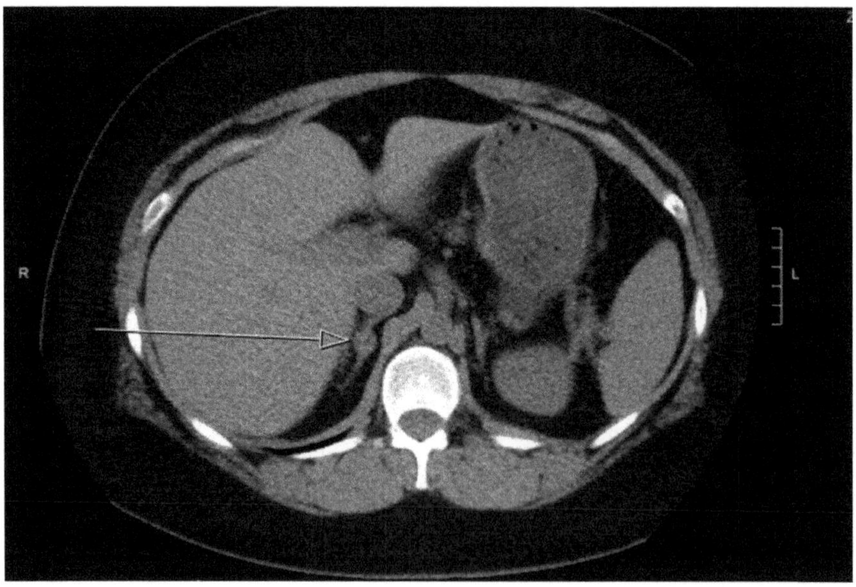

On review of systems, she reports daily headaches, panic attacks, sweating, and anxiety for the last several years. She has had migraines since high school.

On physical examination, her blood pressure is 109/65 mm Hg and pulse rate is 70 beats/min. Her height is 62.2 in (158 cm), and weight is 239 lb (108.4 kg) (BMI = 43.4 kg/m^2). She has no plethora, bruises, striae, acne, or hirsutism. Her muscle strength is normal. There is no spine tenderness and no edema.

Laboratory test results:

Serum cortisol (8 AM) = 20 µg/dL (5-25 µg/dL) (SI: 551.8.9 nmol/L [137.9-689.7 nmol/L])
ACTH (8 AM) = 18 pg/mL (10-60 pg/mL) (SI: 4.0 pmol/L [2.2-13.2 pmol/L])
Serum aldosterone = 5.5 ng/dL (4-21 ng/dL) (SI: 152.6 pmol/L [111.0-582.5 pmol/L])
Plasma renin activity = 3.0 ng/mL per h (0.6-4.3 ng/mL per h)
DHEA-S = 190 µg/dL (44-332 µg/dL) (SI: 5.15 µmol/L [1.19-9.00 µmol/L])
Plasma metanephrine = 118.3 pg/mL (<99 pg/mL) (SI: 0.6 nmol/L [<0.50 nmol/L])
Plasma normetanephrine = 73.3 pg/mL (<165 pg/mL) (SI: 0.4 nmol/L [<0.90 nmol/L])
TSH = 0.6 mIU/L (0.5-5.0 mIU/L)
Basic metabolic panel (including calcium), normal
Calcitonin = <2 pg/mL (<8 pg/mL) (SI: <0.58 pmol/L [<2.34 pmol/L])
Carcinoembryonic antigen = <0.6 ng/mL (<2.5 ng/mL) (SI <0.6 µg/L [<2.5 µg/L])

Which of the following is the best management approach for this patient?

A. Decrease the levothyroxine dosage
B. Follow-up with plasma metanephrine measurement in 3 to 6 months
C. Perform cortical-sparing right adrenalectomy
D. Perform ^{68}Ga-DOTATATE scan
E. Perform total right adrenalectomy

14 A 64-year-old man with a 10-year history of type 1 diabetes presents to establish care. He plans to go on a cruise in 3 months. He reports good glycemic control, and his most recent hemoglobin A_{1c} value is 7.3% (56 mmol/mol). Current medications are metformin, pioglitazone, glipizide, and atorvastatin. He has been vaccinated for COVID-19 but has not received a booster. He is worried about risk of infection on the cruise. He is unsure of other vaccination history, although he has had all childhood vaccinations. He was infected with chickenpox as a child and had pneumonia 4 years ago.

On physical examination, his blood pressure is 134/88 mm Hg and pulse rate is 92 beats/min. His height is 71 in (180.3 cm), and weight is 237 lb (107.5 kg) (BMI = 33.1 kg/m^2).

In addition to a COVID-19 vaccine booster and Tdap (tetanus, diphtheria, pertussis), which of the following vaccinations should this patient receive now?

A. Hepatitis B, pneumococcal, inactive influenza, zoster
B. Hepatitis B, pneumococcal, inactive influenza, varicella
C. Hepatitis B, pneumococcal, active influenza, zoster
D. Hepatitis A, pneumococcal, active influenza, zoster
E. Hepatitis A, pneumococcal, inactive influenza, varicella

15 A 41-year-old previously healthy woman on no medications has abdominal ultrasonography for evaluation of nonspecific abdominal pain. Fatty liver is identified. Her family history is notable for cirrhosis of unknown cause in her father.

On physical examination, her height is 66.5 in (169 cm) and weight is 161 lb (73 kg) (BMI = 25.6 kg/m^2). There is no acanthosis nigricans, and body fat distribution is normal.

Laboratory test results:

Fasting plasma glucose = 84 mg/dL (70-99 mg/dL) (SI: 4.7 mmol/L [3.9-5.5 mmol/L])
Total bilirubin = 0.8 mg/dL (0.3-1.2 mg/dL) (SI: 13.7 µmol/L [5.1-20.5 µmol/L])
ALT = 70 U/L (10-40 U/L) (SI: 1.17 µkat/L [0.17-0.67 µkat/L])
AST = 39 U/L (20-48 U/L) (SI: 0.65 µkat/L [0.33-0.80 µkat/L])
Alkaline phosphatase = 93 U/L (50-120 U/L) (SI: 1.55 µkat/L [0.84-2.00 µkat/L])
Serum creatinine = 1.0 mg/dL (0.6-1.1 mg/dL) (SI: 88.4 µmol/L [53.0-97.2 µmol/L])
Total cholesterol = 134 mg/dL (<200 mg/dL [optimal]) (SI: 3.47 mmol/L [<5.18 mmol/L])
Triglycerides = 46 mg/dL (<150 mg/dL [optimal]) (SI: 0.52 mmol/L [<1.70 mmol/L])
HDL cholesterol = 82 mg/dL (>60 mg/dL [optimal]) (SI: 2.12 mmol/L [>1.55 mmol/L])
LDL cholesterol = 43 mg/dL (<100 mg/dL [optimal]) (SI: 1.11 mmol/L [<2.59 mmol/L])
Apolipoprotein B = 36 mg/dL (50-110 mg/dL) (SI: 0.36 g/L [0.5-1.1 g/L])
TSH = 1.5 mIU/L (0.5-5.0 mIU/L)
25-Hydroxyvitamin D = 38 ng/mL (30-80 ng/mL [optimal]) (SI: 94.8 nmol/L [74.9-199.7 nmol/L])
Vitamin B_{12} = 672 pg/mL (180-914 pg/mL) (SI: 496 pmol/L [133-674 pmol/L])
Vitamin A = 55 µg/dL (32.5-78 µg/dL)
Vitamin E = 15 mg/L (5.5-17 mg/L)

She subsequently undergoes liver magnetic resonance elastography, and findings are normal.

Which of the following is the best next step in this patient's management?

A. Avoid high-fat diet and weight gain
B. Perform screening colonoscopy
C. Start a GLP-1 receptor agonist
D. Start moderate-intensity statin therapy for primary prevention
E. Start vitamin supplements

16 A 75-year-old woman with a history of hypertension, gastroesophageal reflux disease, and osteopenia is seen during an initial telehealth visit for a recent diagnosis of hyperthyroidism. Symptoms preceding her diagnosis included reduced appetite and weight loss over nearly 2 years. She has no fatigue, palpitations, tremors, heat intolerance, diaphoresis, changes in bowel movements, anxiety, or difficulty sleeping. She reports no difficulty breathing, voice changes, or anterior neck pain. She describes a sensation of food "sticking in her throat." She specifically has no current eye symptoms (ie, pain, irritation, or diplopia). There is no history of fracture. Her 45-year-old son has a thyroid disorder for which he takes medication.

The patient's prescribed medications are omeprazole, lisinopril, and cholecalciferol.

Her reported vital signs are blood pressure of 131/81 mm Hg and pulse rate of 97 beats/min. Her height is 63 in (160 cm), and weight is 134 lb (60.8 kg) (BMI = 23.74 kg/m²).

Initial evaluation of hyperthyroidism was completed under the care of her primary care provider.

Laboratory test results:

TSH = 0.01 mIU/L (0.5-5.0 mIU/L)
Free T_4 = 3.44 ng/dL (0.8-1.8 ng/dL) (SI: 44.28 pmol/L [10.30-23.17 pmol/L])
Total T_3 = 313 ng/dL (70-200 ng/dL) (SI: 4.82 nmol/L [1.08-3.08 nmol/L])
Complete blood cell count, normal
Alkaline phosphatase = 234 U/L (50-120 U/L) (SI: 3.91 µkat/L [0.84-2.00 µkat/L])
Complete metabolic panel, otherwise normal

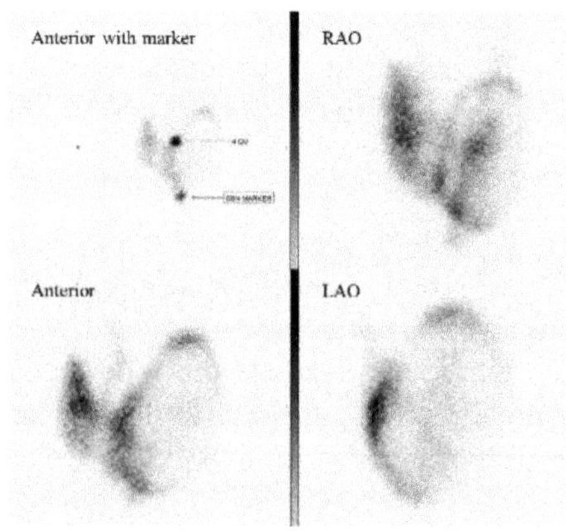

Thyroid uptake and scan is performed (*see image*). Uptake at 6 hours is 43% (normal, 7%-20%). Uptake at 24 hours is 58% (normal, 14%-30%).

In addition to prescribing a β-adrenergic blocker, which of the following is the most appropriate next step in this patient's management?

A. Neck CT
B. Neck ultrasonography
C. Radioactive iodine therapy
D. Serum calcitonin measurement
E. Thyroidectomy

17

A 19-year-old man with no notable medical history presents with an 8-month history of mid and lower back pain and recently documented high bone mass. He describes the pain as gradual in onset and progressive when he started biking about 10 miles per day. He stopped biking and was eventually referred to physical therapy; however, stretching made the pain worse.

He has not experienced trauma-related or fragility fractures, falls, or injuries. He is sexually active and has no symptoms of erectile dysfunction. There is no history of early loss of teeth, vision problems, or hearing loss. He drinks city water and has a water dispenser at home with a reverse osmosis system. He enjoys swimming in the summer. He does not use intravenous drugs, drink alcohol, or smoke cigarettes. Medications include a multivitamin started 2 to 3 weeks ago and occasional ibuprofen for pain relief (less than 2 to 3 times per week, 400 mg). He recently started using a medicated cream for skin lightening. His maternal grandmother has osteoporosis, but there is otherwise no relevant family history of kidney insufficiency or kidney stones.

On physical examination, his height is 70 in (177.8 cm) and weight is 165 lb (74.8 kg) (BMI = 23.7 kg/m^2). He has a normal body habitus and facies, no limb-length discrepancy, and no finger or chest deformities. There is no cleft palate, and dental hygiene is good. His spine is normal without kyphosis, scoliosis, or tenderness to palpation. There is no paraspinal tenderness. The rest of the examination findings are normal.

Laboratory test results:

- Calcium = 9.5 mg/dL (8.2-10.2 mg/dL) (SI: 2.4 mmol/L [2.1-2.6 mmol/L])
- Phosphate = 3.8 mg/dL (2.3-4.7 mg/dL) (SI: 1.2 mmol/L [0.7-1.5 mmol/L])
- Serum creatinine = 0.95 mg/dL (0.7-1.3 mg/dL) (SI: 84.0 μmol/L (61.9-114.9 μmol/L)
- Lactate dehydrogenase = 162 U/L (100-200 U/L) (SI: 2.71 μkat/L [1.7-3.3 μkat/L])
- Liver function tests (including γ-glutamyltransferase), normal
- Complete blood cell count, normal
- Serum 25-hydroxyvitamin D = 36 ng/mL (30-80 ng/mL [optimal]) (SI: 89.9 nmol/L [74.9-199.7 nmol/L])
- Serum albumin = 4.5 g/dL (3.5-5.0 g/dL) (SI: 45 g/L [35-50 g/L])
- TSH = 3.14 mIU/L (0.5-5.0 mIU/L)
- Free T_4 = 1.2 ng/dL (0.8-1.8 ng/dL) (SI: 15.44 pmol/L [10.30-23.17 pmol/L])
- Total alkaline phosphatase = 93 U/L (50-120 U/L) (SI: 1.55 μkat/L [0.84-2.00 μkat/L])
- Bone-specific alkaline phosphatase = 19.2 μg/L (≤20 μg/L)
- Ionized calcium = 1.32 mg/dL (4.60-5.08 mg/dL) (SI: 0.33 mmol/L [1.15-1.27 mmol/L])
- PTH = 55 pg/mL (10-65 pg/mL) (SI: 5.8 pmol/L [1.1-6.9 pmol/L])
- Serum tryptase, normal
- Creatine kinase = 305 U/L (50-200 U/L) (SI: 5.09 μkat/L [0.84-3.34 μkat/L])

DXA documents the following values:

- Lumbar spine Z-score = +4.6
- Total hip Z-score = +3.5
- Distal one-third forearm Z-score = +0.7

Radiographic images are shown (*see images*).

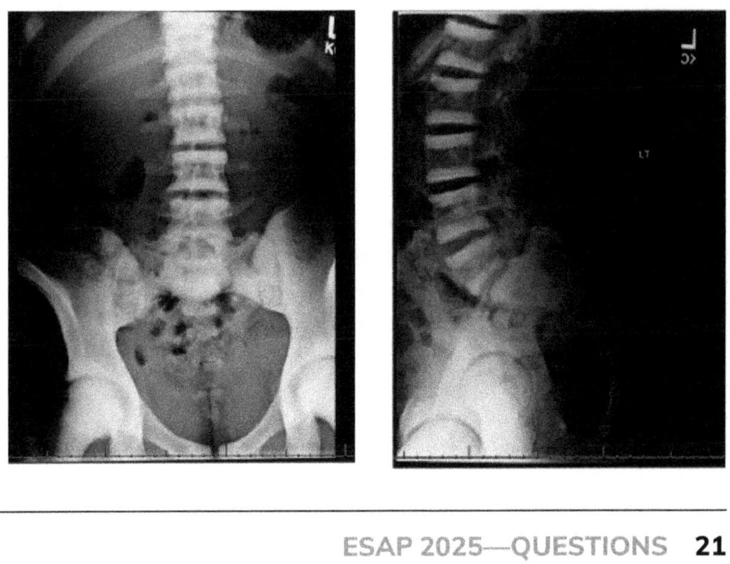

Bone scan shows diffusely increased uptake in the axial and appendicular skeleton consistent with a metabolic superscan.

Which of the following is the most likely cause of this patient's radiographic findings?

A. Endogenous ochronosis
B. Fluorosis
C. Hepatitis C–associated osteosclerosis
D. Osteopetrosis
E. Sclerosteosis

18 A 49-year-old woman with a history of type 1 diabetes presents for evaluation and management. Her medical history is notable for hypertension and dyslipidemia. She also uses a continuous glucose monitoring device. She is frustrated by frequent hypoglycemia alerts, which sometimes occur twice a week in the overnight hours. Hypoglycemia alerts often wake her and negatively affect her sleep. She usually takes insulin glargine at 10 PM. Her last short-acting insulin dose is administered with her evening meal, usually at 8 PM. She typically sleeps until 9 AM. Her current hemoglobin A_{1c} value is 6.9% (52 mmol/mol).

Her treatment consists of insulin glargine, 31 units daily; short-acting insulin aspart, 11 units with each meal; and correctional insulin, 1 unit for every 30 mg/dL (1.7 mmol/L) glucose above 130 mg/dL (>7.2 mmol/L).

Continuous glucose monitoring download shows the following values:

Glucose management indicator (GMI) = 6.9% (52 mmol/mol)
Average glucose concentration = 160 mg/dL (SI: 8.9 mmol/L)
Time-in-range (70-180 mg/dL [SI: 3.9-10.0 mmol/L]) = 73%
Time-above-range (180-250 mg/dL [SI: >10.0-13.9 mmol/L]) = 20%
Time-above-range (>250 mg/dL [SI: >13.9 mmol/L]) = 5.0%
Time-below-range (<70 mg/dL [SI: <3.9 mmol/L]) = 2.0%
Time-below-range (<54 mg/dL [SI: <3.0 mmol/L]) = 0.5%
Coefficient of variation = 34.1%

A continuous glucose monitoring report from one of the days that she experiences hypoglycemia is shown (*see image*).

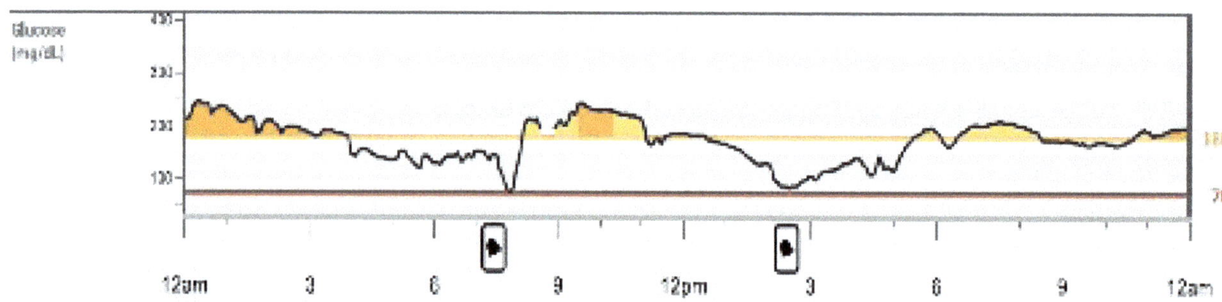

Which of the following is the most appropriate management plan for this patient?

A. Change the nocturnal low-glucose threshold alarm to reduce the number of hypoglycemia alarms
B. Decrease the dosage of basal insulin glargine
C. Decrease the short-acting basal insulin dose with dinner
D. Discuss proper continuous glucose monitor placement techniques with the patient
E. Recommend insulin pump initiation

19 A 70-year-old man with a medical history of hypercholesterolemia presents with a 4-month history of fatigue, poor appetite, nausea, and hot flashes. One month before presentation, he was hospitalized for acute pancreatitis. He has no history of gallstones, alcohol use, hypertriglyceridemia, or headaches. His only current medication is atorvastatin, 10 mg daily.

On physical examination, his blood pressure is 102/61 mm Hg and pulse rate is 77 beats/min. His height is 70.8 in (180 cm), and weight is 159 lb (72 kg) (BMI = 22.3 kg/m^2). He appears tired and pale and has mild tenderness in the epigastrium. Examination findings are otherwise unremarkable. Visual fields are normal.

Laboratory test results:

Hemoglobin = 12.0 g/dL (13.8-17.2 g/dL) (SI: 120 g/L [138-172 g/L])
Calcium = 9.1 mg/dL (8.2-10.2 mg/dL) (SI: 2.3 mmol/L [2.1-2.6 mmol/L])
Serum cortisol (8 AM) = 4 µg/dL (5-25 µg/dL) (SI: 110.4 nmol/L [137.9-689.7 nmol/L])
TSH = 0.34 mIU/L (0.5-5.0 mIU/L)
Free T_4 = 0.63 ng/dL (0.8-1.8 ng/dL) (SI: 8.11 pmol/L [10.30-23.17 pmol/L])
Testosterone = 88 ng/dL (300-900 ng/dL) (SI: 3.1 nmol/L [10.4-31.2 nmol/L])
FSH = 0.2 mIU/mL (1.0-13.0 mIU/mL) (SI: 0.2 IU/L [1.0-13.0 IU/L])
hCG, normal
ACE, normal

Serum cortisol concentrations after ACTH stimulation are 10 µg/dL (276 nmol/L) at 30 minutes and 12 µg/dL (331 nmol/L) at 60 minutes. Pituitary-directed MRI is shown (*see images*). Findings on chest x-ray are normal.

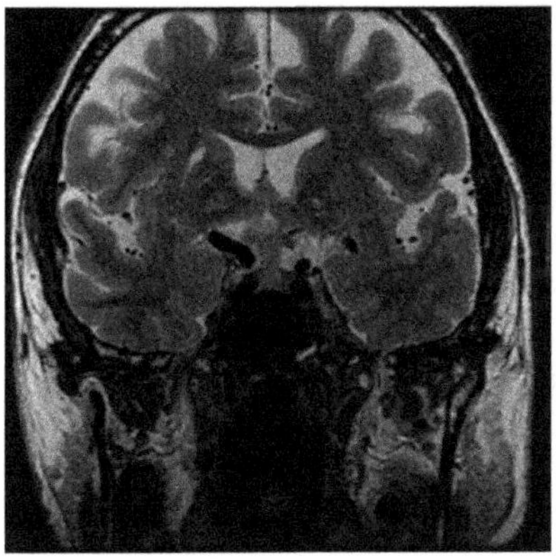
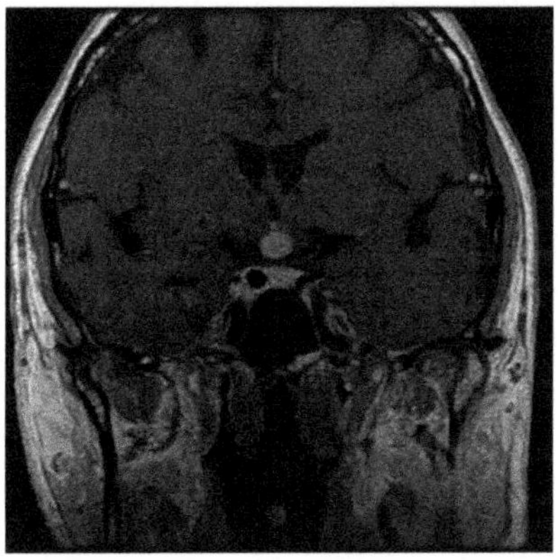
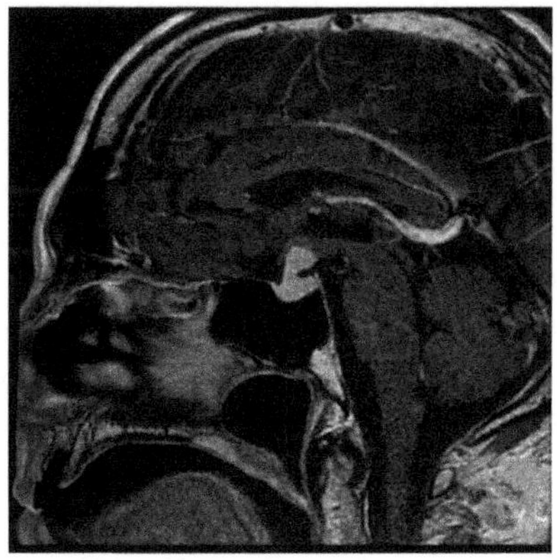

The following medications are initiated: hydrocortisone, 15 mg every morning; levothyroxine, 50 mcg daily; and testosterone gel, 50 mg daily. After 3 months, imaging indicates enlargement of the mass. Transsphenoidal biopsy is performed.

Which of the following is the most likely finding on neuropathological examination?

A. Eosinophilic infiltrate with Langerhans cells
B. Lymphoplasmacytic infiltrate with positive IgG-4 staining
C. Necrotizing granuloma
D. Noncaseating epithelioid-cell granuloma
E. Papillary craniopharyngioma

20 A 67-year-old man has developed erythrocytosis after initiation of testosterone therapy. His pretreatment hematocrit was 42% (0.42). Hypogonadism was diagnosed after an incidental finding of a nonfunctional macroadenoma 6 months ago. Because the macroadenoma abutted the optic chiasm on sellar imaging, he underwent transsphenoidal surgery. He has had normal thyroid hormone and prolactin concentrations and a normal result on cosyntropin-stimulation testing performed after pituitary surgery.

He reports no daytime sleepiness, and his wife says that he snores intermittently but not particularly loudly. She reports no episodes of apnea. He takes care of his young grandchildren 3 days a week. He is very concerned about secondary transfer of testosterone to his grandsons and wife, so he has opted to use intramuscular testosterone administered on a weekly basis. Aside from the nonfunctional macroadenoma, his medical history is remarkable only for migraine headaches. Medications are testosterone cypionate, 100 mg intramuscularly weekly, and ibuprofen and sumatriptan as needed for migraines.

On physical examination, his blood pressure is 125/80 mm Hg, pulse rate is 80 beats/min, and respiratory rate is 12 breaths/min. His height is 69 in (175.3 cm), and weight is 190 lb (86.2 kg) (BMI = 28.1 kg/m^2). Neck thickness is normal. Lungs are clear to auscultation with tidal volume breaths. He has no gynecomastia. His penis is normal, and testes are 15 cc bilaterally.

Laboratory test results (sample drawn 3.5 days after his last testosterone injection):

Hematocrit = 52% (41%-51%) (SI: 0.51 [0.41-0.51])
Serum total testosterone = 670 ng/dL (264-916 ng/dL) (SI: 23.2 nmol/L [9.2-31.8 nmol/L]) (assay validated and harmonized by the CDC)

Which of the following is the best next step in this patient's management?

A. Arrange for outpatient polysomnography
B. Assess for pathogenic variants in the *JAK2* gene
C. Change the route of testosterone therapy from intramuscular to transdermal
D. Decrease the dosage of testosterone cypionate to 75 mg weekly
E. Recommend phlebotomy or blood donation every 3 months

21 A 45-year-old woman is referred for evaluation of skin thinning, fatigue, dizziness, and easy bruising. Her medical history includes primary hypothyroidism, eczema, bronchiectasis, allergic bronchopulmonary aspergillosis, and asthma. Her current medications are inhaled budesonide/formoterol, inhaled salbutamol, itraconazole, thyroxine, nebulized hypertonic saline, and topical 1% hydrocortisone cream. She has no history of cigarette smoking and consumes no alcohol.

On physical examination, her blood pressure is 123/78 mm Hg and pulse rate is 52 beats/min. Her height is 65 in (165 cm), and weight is 124 lb (56.4 kg) (BMI = 20.7 kg/m^2). She has facial plethora, thin skin, and bruising.

Laboratory test results:

Sodium = 140 mEq/L (136-142 mEq/L) (SI: 140 mmol/L [136-142 mmol/L])
Potassium = 3.8 mEq/L (3.5-5.0 mEq/L) (SI: 3.8 mmol/L (3.5-5.0 mmol/L)
Urea = 4.2 mmol/L (2.5-7.8 mmol/L)
Creatinine = 0.67 mg/dL (0.6-1.1 mg/dL) (SI: 59.0 μmol/L [53.0-97.2 μmol/L])
TSH = 1.62 mIU/L (0.30-4.40 mIU/L)

Which of the following laboratory studies should be ordered next?

A. 8-AM cortisol and ACTH measurement
B. 24-Hour urinary cortisol measurement
C. Adrenal antibody assessment
D. Late-night salivary cortisol measurement
E. Overnight 1-mg dexamethasone-suppression test

22 A 62-year-old woman without osteoporosis has breast cancer with bone metastases. She is treated with left mastectomy followed by external beam irradiation. She then is treated with an aromatase inhibitor and denosumab, 120 mg monthly. After 5 years of therapy, she stops denosumab but continues the aromatase inhibitor. She has well-controlled hypertension managed with hydrochlorothiazide for more than 10 years, during which time her serum calcium concentration has mainly fluctuated between 9.4 and 9.9 mg/dL (2.35 and 2.47 mmol/L). Fifteen months after discontinuing monthly denosumab, she presents with polydipsia, polyuria, an acute increase in serum creatinine from 0.7 to 1.4 mg/dL (61.9 to 123.8 µmol/L), and elevated serum calcium. Imaging documents stability of bone lesions over the past 5 years and no new masses or evidence of recurrence. Mammography of the right breast is normal. She has no history of fracture. She is typically physically active and exercises regularly. She is not taking supplements other than a multivitamin containing 800 international units (20 mcg daily) of vitamin D.

On physical examination, her height is 65.4 in (166 cm) and weight is 123.5 lb (56 kg) (BMI = 20.3 kg/m^2). Dentition is normal. There are no neck masses. She has a surgical scar from left mastectomy and no masses in the right breast or either axilla. Findings on cardiovascular, lung, and abdomen examinations are normal. There is no bony deformity or tenderness.

Laboratory test results:

Calcium = 11.5 mg/dL (8.2-10.2 mg/dL) (SI: 2.9 mmol/L [2.1-2.6 mmol/L])
Albumin = 3.2 g/dL (3.5-5.0 g/dL) (SI: 32 g/L [35-50 g/L])
Phosphate = 4.3 mg/dL (2.3-4.7 mg/dL) (SI: 1.4 mmol/L [0.7-1.5 mmol/L])
Alkaline phosphatase = 88 U/L (50-120 U/L) (SI: 1.47 µkat/L [0.84-2.00 µkat/L])
Creatinine = 1.4 mg/dL (0.6-1.1 mg/dL) (SI: 123.8 µmol/L [53.0-97.2 µmol/L])
Estimated glomerular filtration rate = 43 mL/min per 1.73 m^2 (>60 mL/min per 1.73 m^2)
Intact PTH = <6 pg/mL (10-65 pg/mL) (SI: <0.6 pmol/L [1.1-6.9 pmol/L])
PTH-related peptide = 0.8 pmol/L (<2.0 pmol/L)
25-Hydroxyvitamin D = 36 ng/mL (30-80 ng/mL [optimal]) (SI: 89.9 nmol/L [74.9-199.7 nmol/L])
1,25-Dihydroxyvitamin D_3 = 23 pg/mL (16-65 pg/mL) (SI: 59.8 pmol/L [41.6-169.0 pmol/L])
Cross-linked N-telopeptide of type 1 collagen (urine NTX) = 82 nmol BCE/mmol creat (19-63 nmol BCE/mmol creat)
Urinary calcium-to-creatinine ratio = 0.57 mg/mg (<0.22 mg/mg) (SI: 1.6 mmol/mmol [<0.61 mmol/mmol])

Findings on serum protein electrophoresis and urine protein electrophoresis are normal.

DXA documents the lowest T-score to be –0.3.

Which of the following is the most likely mechanism of this patient's hypercalcemia?

A. Acute kidney insufficiency
B. Effect of denosumab cessation
C. Granulomatous disease
D. Hydrochlorothiazide effect
E. Recurrence of breast cancer

23 A 28-year-old man seeks evaluation for 2 to 3 months of moderate fatigue, constipation, and progressive abdominal distention. He has no notable medical history and is not taking any medications.

On physical examination, his blood pressure is 126/78 mm Hg and resting pulse rate is 58 beats/min. His height is 70 in (177.8 cm), and weight is 169 lb (76.7 kg) (BMI = 24.2 kg/m^2). Findings on thyroid examination show mild symmetric thyromegaly and no palpable thyroid nodules. He has a distended abdomen, and hepatomegaly is appreciated with an estimated liver span of 23 cm at the midclavicular line.

Laboratory test results:

Serum TSH = 62.3 mIU/L (0.5-5.0 mIU/L)
Free T_4 = 0.6 ng/dL (0.8-1.8 ng/dL) (SI: 7.7 pmol/L [10.30-23.17 pmol/L])
Free T_3 = 1.2 pg/mL (2.3-4.2 pg/mL) (SI: 1.8 pmol/L [3.53-6.45 pmol/L])
TPO antibodies, negative
Thyroglobulin antibodies, negative
Liver function panel, normal

Primary hypothyroidism is diagnosed, and he is prescribed levothyroxine, 125 mcg daily, with instructions on proper administration. Repeat TSH measurement 4 weeks later is 65.0 mIU/L. Despite escalations of the levothyroxine dosage to 137 mcg and then 150 mcg daily in the ensuing months, his TSH concentration remains similarly elevated.

Which of the following assessments would establish the cause of this patient's symptoms?

A. Abdominal MRI
B. Genetic testing
C. Measurement of serum thyroglobulin
D. Pituitary MRI
E. Thyroid radioiodine uptake

24 A 28-year-old man with an 18-year history of type 1 diabetes presents as a new patient. He is very comfortable managing his diabetes, including carbohydrate counting and diligently monitoring blood glucose using a real-time continuous glucose monitoring device. He was recently seen by a dietitian at the diabetes center, who confirmed he has appropriate knowledge of carbohydrate counting. In the past, he tried using continuous subcutaneous insulin infusion therapy, but he was not comfortable wearing an insulin pump and switched back to a basal-bolus insulin regimen. He noted variability in his blood glucose values that improved slightly after he switched from insulin glargine, U100 twice daily, to insulin glargine, U300 once daily. Before switching, his coefficient of variation on his continuous glucose monitoring reports was between 45% and 48%; after switching to U300 insulin glargine, his coefficient of variation is now between 40% and 42%.

Current medications are insulin glargine U300, 26 units once daily, and insulin aspart with an insulin-to-carbohydrate ratio of 1:6, insulin sensitivity factor of 1:50, and target blood glucose range of 100 to 130 mg/dL (5.6-7.2 mmol/L). He uses 4-mm needles for insulin administration and reports lifting skin before injecting insulin.

On physical examination, his blood pressure is 112/74 mm Hg and pulse rate is 72 beats/min. His height is 70 in (177.8 cm), and weight is 171 lb (77.6 kg) (BMI = 24.5 kg/m^2). He has no acanthosis nigricans. Chest is clear to auscultation, and there are normal heart sounds. He has 4 small subcutaneous lumps palpable on his abdomen that a previous physician said were lipomas.

Laboratory test results:

Hemoglobin A_{1c} = 7.4% (4.0%-5.6%) (57 mmol/mol [20-38 mmol/mol])
Serum creatinine = 0.8 mg/dL (0.7-1.3 mg/dL) (SI: 70.7 µmol/L [61.9-114.9 µmol/L])
Estimated glomerular filtration rate = 88 mL/min per 1.73 m^2 (>60 mL/min per 1.73 m^2)
Urinary albumin-to-creatinine ratio = 15 mg/g creat (<30 mg/g creat)

Which of the following is the most likely cause of this patient's glycemic variability?

A. Inaccurate carbohydrate counting
B. Lifting skin before administering insulin
C. Presence of lipohypertrophy
D. Unintentional intramuscular injection
E. Use of 4-mm needles for insulin administration

25 For the last week, a 48-year-old woman has been feeling intermittently feverish and more fatigued than usual. She underwent laparoscopic Roux-en-Y gastric bypass 20 years ago for medically complicated obesity. She reports adherence to her vitamin supplementation as prescribed.

Her current medications are vitamin D_3, 2000 international units daily (50 mcg daily); vitamin B_{12}, 1000 mcg monthly injection; zinc sulfate, 220 mg twice daily; prenatal chewable multivitamin daily; and calcium carbonate, 500 mg twice daily. She received an intravenous iron infusion 1 month ago.

On physical examination, she appears unwell with poor skin turgor. Her blood pressure 102/62 mm Hg, pulse rate is 95 beats/min, and temperature is 100.4°F (38°C). Her height is 66.1 in (167.9 cm), and weight is 158 lb (71.6 kg) (BMI = 25.4 kg/m^2). She has normal breath sounds, and her abdomen is soft and nontender to palpation. There are no rashes. Neurological examination is notable for an unsteady gait.

Laboratory test results:

Hemoglobin = 11.1 g/dL (12.1-15.1 g/dL) (SI: 111 g/L [121-151 g/L])
Platelet count = 285 × 10^3/µL (150-450 × 10^3/µL) (SI: 285 × 10^9/L [150-450 × 10^9/L])

White blood cell count = 2600/μL (4500-11,000/μL) (SI: 2.6 × 10^9/L [4.5-11.0 × 10^9/L])
Neutrophil count = 200/μL (2500-7000/μL) (SI: 0.2 × 10^9/L [2.5-7.0 × 10^9/L])

In addition to treating neutropenic fever, which of the following should be measured in this patient's serum to evaluate her presentation?

A. Copper
B. Ferritin
C. Vitamin B_{12}
D. Vitamin E
E. Zinc

26 A 31-year-old woman with a history of B-cell acute lymphoblastic leukemia at age 23 years (status post C10403 induction/consolidation and 1 cycle of blinatumomab and allogenic peripheral blood stem-cell transplant) now presents with primary ovarian insufficiency. She is treated with a 100-mcg estradiol patch and a progestin intrauterine device for endometrial protection. Most of her initial symptoms are well controlled except for some vasomotor symptoms, low libido, and persistent vaginal dryness. Her medical history is notable for venous thromboembolism while on combined hormonal contraception.

On physical examination, her blood pressure is 112/75 mm Hg and pulse rate is 88 beats/min.

Her height is 67 in (170.2 cm), and weight is 192 lb (87.1 kg) (BMI = 30.1 kg/m²). On speculum examination, she has pale and dry vulvovaginal mucosa with petechiae and diminished vaginal rugae. Examination findings are otherwise unremarkable.

Which of the following is the best next step in this patient's treatment?

A. Add vaginal estrogen
B. Measure estradiol
C. Measure testosterone
D. Recommend consultation with a sex therapist
E. Recommend vaginal laser treatment

27 A 22-year-old woman is concerned about her inability to lose weight. Since moving away from home for college 4 years ago, she has steadily gained weight (total of 25 lb [11.3 kg]). She does not cook; she typically orders in or eats out. She reports a balanced diet, in which she incorporates salads and fruits. She runs 3 times a week, but she has noticed declining endurance. She also reports hair loss and irregular periods over the past 1 to 2 years. Evaluation by a gynecologist included transabdominal ovarian ultrasonography, which revealed bilateral ovarian cysts. Prediabetes was diagnosed 1 year ago, and she takes metformin, 500 mg twice daily. She takes no other health supplements or over-the-counter medications. She does not use glucocorticoid injections or inhalers.

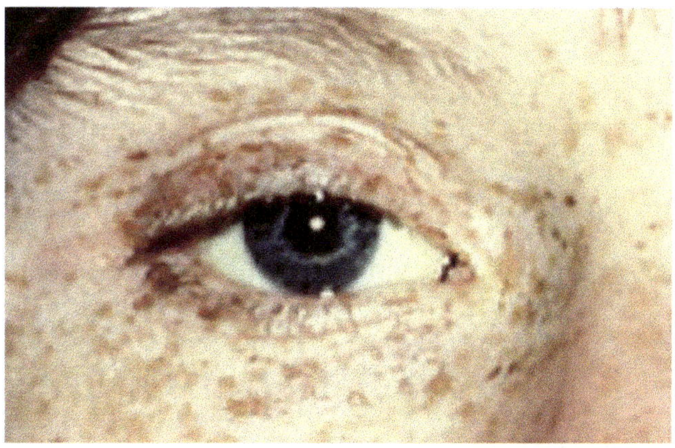

Image courtesy of William Young.

On physical examination, her blood pressure is 130/80 mm Hg and pulse rate is 72 beats/min. Her height is 66 in (167.6 cm), and weight is 190 lb (86.2 kg) (BMI = 30.7 kg/m²). Excess adiposity is centrally distributed, but without sparing the limbs. A mild dorsocervical hump is present. There is no muscle wasting, and muscle strength is normal. She has no plethora, bruises, acne, or hirsutism. Multiple lentiginous macules are present around both eyes (*see image*). She has a few 5- to 7-cm wide pink striae on her abdomen. No alopecia is observed.

Laboratory test results:

- Serum cortisol (8 AM) = 17 µg/dL (5-25 µg/dL) (SI: 469.0 nmol/L [137.9-689.7 nmol/L])
- ACTH (8 AM) = 9 pg/mL (10-60 pg/mL) (SI: 2.0 pmol/L [2.2-13.2 pmol/L])
- DHEA-S = 45 µg/dL (44-332 µg/dL) (SI: 1.22 µmol/L [1.19-9.00 µmol/L])
- Testosterone = 45 ng/dL (8-60 ng/dL) (SI: 1.6 nmol/L [0.3-2.1 nmol/L])
- TSH = 2.7 mIU/L (0.5-5.0 mIU/L)
- Basic metabolic panel, normal
- Hemoglobin A_{1c} = 6.3% (4.0%-5.6%) (45 mmol/mol [20-38 mmol/mol])
- Urinary cortisol = 45 µg/24 h (4-50 µg/24 h) (SI: 124 nmol/d [11-138 nmol/d])

Laboratory values following a 1-mg dexamethasone-suppression test (8 AM):

- Serum cortisol = 3.0 µg/dL (SI: 82.8 nmol/L)
- ACTH = <5 pg/mL (SI: <1.1 pmol/L)

Abdominal CT without contrast shows normal-appearing adrenal glands.

Which of the following is the best next step in this patient's management?

A. ^{68}Ga-DOTATATE scan
B. Oral contraceptive pill with third-generation progestin
C. Pituitary-directed MRI
D. Sequential low-dose/high-dose dexamethasone-suppression test
E. Synthetic glucocorticoid screen

28

A 45-year-old woman is referred following recent confirmation of pregnancy. She is 14 weeks' gestation and has a history of recurrent miscarriage. Two years ago, she underwent transsphenoidal surgery for a lactotroph-somatotroph pituitary macroadenoma. Postoperative MRI showed residual tumor extending into the left cavernous sinus without suprasellar extension. She takes aspirin and octreotide LAR. She does not smoke cigarettes and consumes no alcohol.

On physical examination, her blood pressure is 121/65 mm Hg and pulse rate is 76 beats/min. Her height is 61 in (155 cm), and weight is 152 lb (69 kg) (BMI = 28.7 kg/m^2). There is no visual field deficit on confrontation testing.

Laboratory test results:

- IGF-1 = 292 ng/mL (98-261 ng/mL) (SI: 38.3 nmol/L [12.8-34.2 nmol/L])
- Prolactin = 107 ng/mL (4-30 ng/mL) (SI: 4.65 nmol/L [0.17-1.30 nmol/L])
- TSH = 2.02 mIU/L (0.5-5.0 mIU/L)
- Free T_4 = 0.94 ng/dL (0.8-1.8 ng/dL) (SI: 12.1 pmol/L [10.30-23.17 pmol/L])

Which of the following is the best management plan for this patient?

A. Continue octreotide LAR throughout pregnancy
B. Recommend transsphenoidal surgery
C. Replace octreotide LAR with bromocriptine
D. Replace octreotide LAR with cabergoline
E. Stop octreotide LAR

29

A 58-year-old man with a history of coronary artery disease, hypertension, and mild cognitive impairment presents to establish care. He is asymptomatic. He has had no neck operations and no radiation exposure. Family history is relevant for hip fracture in his mother at age 68 years; his father is healthy and has no history of osteoporosis or hip fracture. There is no family history of thyroid disease.

On physical examination, his blood pressure is 130/82 mm Hg and pulse rate 93 beats/min and regular. His height is 63.5 in (161.3 cm), and weight is 156 lb (70.8 kg) (BMI = 27.2 kg/m^2). Neck examination reveals a palpable right thyroid nodule. Reflexes are slightly brisk. He has no tremors of his outstretched hands and no spinal tenderness. His gait is normal.

Laboratory test results:

TSH = 0.02 mIU/L (0.5-5.0 mIU/L)
Free T_4 = 1.7 ng/dL (0.8-1.8 ng/dL) (SI: 21.88 pmol/L [10.30-23.17 pmol/L])
Total T_3 = 150 ng/mL (70-200 ng/mL) (SI: 2.31 nmol/L [1.08-3.08 nmol/L])
Calcium = 9.5 mg/dL (8.2-10.2 mg/dL) (SI: 2.4 mmol/L [2.1-2.6 mmol/L])
Alkaline phosphatase = 82 U/L (50-120 U/L) (SI: 1.37 μkat/L [0.84-2.00 μkat/L])
Creatinine = 1.37 mg/dL (0.7-1.3 mg/dL) (SI: 121.11 μmol/L [61.9-114.9 μmol/L])
Albumin = 4.3 g/dL (3.5-5.0 g/dL) (SI: 43 g/L [35-50 g/L])
Testosterone = 280 ng/dL (300-900 ng/dL) (SI: 9.7 nmol/L [10.4-32.1 nmol/L])
Complete blood cell count, normal

The TSH concentration measured in 8 weeks is 0.04 mIU/L.

In additional to evaluating the subclinical hyperthyroidism, which of the following is the best next step?

A. Order a loop recorder
B. Perform DXA
C. Recommend no additional monitoring
D. Start alendronate
E. Start testosterone replacement

30 A 20-year-old woman with a medical history of a functional ovarian cyst and an anxiety disorder is seen in consultation for a recent diagnosis of hyperthyroidism. Symptoms include palpitations, loose stools, anxiety, and difficulty concentrating. The patient has had no dysphagia, difficulty breathing, voice changes, or anterior neck pain. There are no ophthalmologic concerns. Her mother has a history of hypothyroidism. The patient's prescribed medications are escitalopram and a combined oral contraceptive. She takes vitamins and supplements as recommended by her naturopathic provider.

On physical examination, her blood pressure is 126/72 mm Hg and pulse rate is 101 beats/min. Her height is 65 in (165 cm), and weight is 112 lb (50.9 kg) (BMI = 18.6 kg/m^2). She appears restless. Findings on eye examination demonstrate lid lag but no chemosis or proptosis. Oral examination reveals erosion of dental enamel. Her thyroid gland is normal in size and without nodules. There is a fine tremor of the upper extremities. Deep tendon reflexes are brisk.

Laboratory test results:

Measurement	4 weeks ago	Today	Reference range
TSH	0.01 mIU/L	0.01 mIU/L	0.5-5.0 mIU/L
Free T_4	3.2 ng/dL (SI: 41.19 pmol/L)	2.9 ng/dL (SI: 37.33 pmol/L)	0.8-1.8 ng/dL (SI: 10.30-23.17 pmol/L)
Total T_3	172 ng/dL (SI: 2.65 nmol/L)	165 ng/dL (SI: 2.54 nmol/L)	70-200 ng/dL (SI: 1.08-3.08 nmol/L)
Thyroglobulin	<3 ng/mL (SI: <3 μg/L)	...	3-42 ng/mL (SI: 3-42 μg/L)
Thyroglobulin antibodies	<4.0 IU/mL (SI: <4.0 kIU/L)	...	≤4.0 IU/mL (SI: ≤4.0 kIU/L)
TPO antibodies	<2.0 IU/mL	...	<2.0 IU/mL

Thyroid uptake and scan is performed (*see image*). Uptake at 24 hours is 0.93% (normal, 14%-30%).

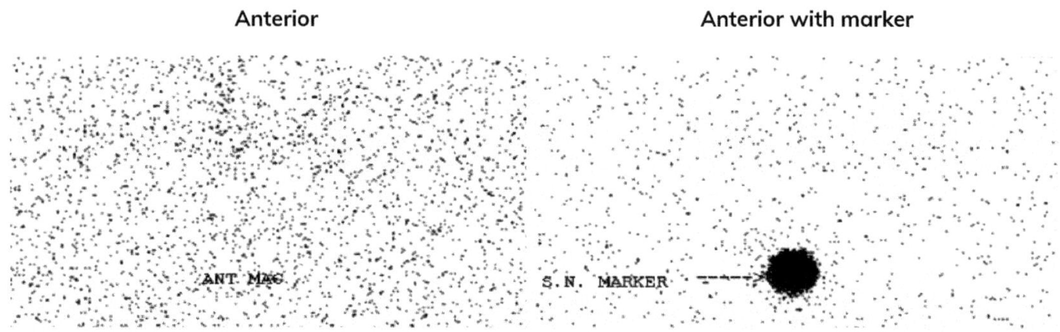

Which of the following is the most appropriate next step in this patient's management?

A. Fecal T_4 measurement
B. Initiation of methimazole
C. Pelvic radioimaging using ^{123}I
D. Thyroid function tests in 4 to 6 weeks
E. TRAb measurement

31 A 66-year-old woman is concerned about ongoing, worsening fatigue. Since menopause at age 55 years, she has had decreased energy. She must take quick, intermittent walks at work to avoid falling asleep at her desk despite sleeping more than 7 hours at night. She has no trouble falling asleep. She does snore at night. She has a history of type 2 diabetes, hypertension, dyslipidemia, and metabolic dysfunction-associated steatohepatitis. Current medications are semaglutide, insulin glargine, lisinopril, and atorvastatin.

On physical examination, her blood pressure is 142/95 mm Hg and pulse rate is 96 beats/min. Her height is 63 in (160 cm), and weight is 215 lb (97.5 kg) (BMI = 38.1 kg/m^2). She has truncal obesity, and her neck circumference is 17 in (43.2 cm).

Which of the following is the best next step in this patient's care?

A. Complete blood cell count with ferritin measurement
B. Nocturnal oximetry
C. Polysomnography
D. Serum cortisol measurement (8 AM)
E. STOP-Bang questionnaire

32 A 75-year-old man presents with a right hip fracture after a tripping over carpet. He is admitted to the hospital after surgery. Two years ago, he had a vertebral fracture treated with kyphoplasty but has had no previous osteoporosis treatment. He has a history of stroke and has mild cognitive impairment.

On physical examination, his height is 71 in (180.3 cm) and weight is 148 lb (67.1 kg) (BMI = 20.6 kg/m^2). He is edentulous and has moderate kyphosis with decreased space between his ribs and pelvis.

Spine radiograph documents kyphoplasty at T12 and additional vertebral compression fractures at T5 and T6. DXA documents the lowest T-score to be −2.9 at the spine.

He is going to a rehabilitation center for recovery, but he does not have a primary care physician. His family lives in another state.

Laboratory test results:

Serum protein electrophoresis, normal
Urine protein electrophoresis, normal
25-Hydroxyvitamin D = 26 ng/mL (30-80 ng/mL [optimal]) (SI: 64.9 nmol/L [74.9-199.7 nmol/L])
Calcium = 9.6 mg/dL (8.2-10.2 mg/dL) (SI: 2.4 mmol/L [2.1-2.6 mmol/L])
Albumin = 3.6 g/dL (3.5-5.0 g/dL) (SI: 36 g/L [35-50 g/L])
Phosphate = 4.3 mg/dL (2.3-4.7 mg/dL) (SI: 1.4 mmol/L [0.7-1.5 mmol/L])

Creatinine = 1.4 mg/dL (0.7-1.3 mg/dL) (SI: 106.1 μmol/L [61.9-114.9 μmol/L])
Estimate glomerular filtration rate = 52 mL/min per 1.73 m^2 (>60 mL/min per 1.73 m^2)
Creatinine clearance (Cockcroft-Gault equation) = 43 mL/min
Intact PTH = 61 pg/mL (10-65 pg/mL) (SI: 6.4 pmol/L [1.1-6.9 pmol/L])
Alkaline phosphatase = 83 U/L (50-120 U/L) (SI: 1.39 μkat/L [0.84-2.00 μkat/L])
Total testosterone = 235 ng/dL (300-900 ng/dL) (SI: 8.2 nmol/L [10.4-31.2 nmol/L])

In addition to recommending adequate daily calcium intake, which of the following is the best treatment of this patient's osteoporosis?

A. Abaloparatide during this hospitalization
B. Alendronate after the fracture heals
C. Cholecalciferol, 50,000 international units weekly (1250 mcg weekly)
D. Intravenous zoledronate during this hospitalization
E. Transdermal testosterone

33 A 19-year-old woman is seen for management of severe hypertriglyceridemia. She was first noted to have a serum triglyceride concentration greater than 2000 mg/dL (>22.60 mmol/L) at 2 years of age. Diabetes was diagnosed at age 14 years, and she had one episode of pancreatitis at age 16 years. Current lipid-lowering therapy consists of fenofibrate, 145 mg daily, and omega-3 fatty acids, 4 g daily. In addition, she takes insulin detemir, 40 units twice daily; insulin lispro, 40 units before each meal; metformin; and pioglitazone. She is of consanguineous parentage, but there is no family history of diabetes or dyslipidemia.

On physical examination, her blood pressure is 112/76 mm Hg and pulse rate is 84 beats/min and regular. Her height is 64.2 in (163 cm), and weight is 112.4 lb (51 kg) (BMI = 19.2 kg/m^2). She has a muscular appearance with generalized loss of subcutaneous fat and multiple eruptive xanthomas over the extremities and trunk (*see image*). The liver is palpable 4 cm below the right costal margin.

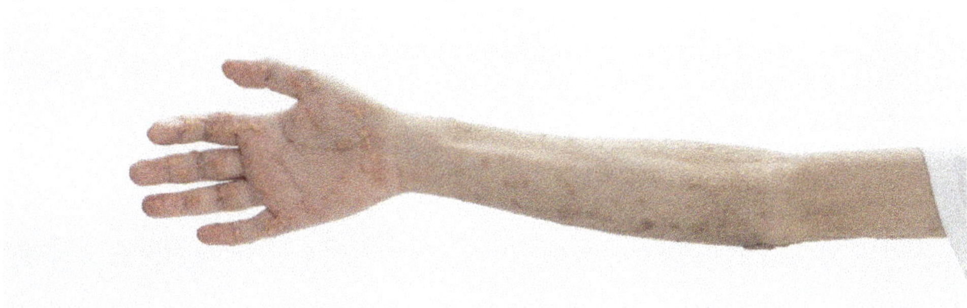

Laboratory test results:

Fasting plasma glucose = 259 mg/dL (70-99 mg/dL) (SI: 14.4 mmol/L [3.9-5.5 mmol/L])
Hemoglobin A_{1c} = 11.8% (4.0%-5.6%) (105 mmol/mol [20-38 mmol/mol])
Total bilirubin = 1.8 mg/dL (0.3-1.2 mg/dL) (SI: 30.9 μmol/L [5.1-20.5 μmol/L])
ALT = 47 U/L (10-40 U/L) (SI: 0.78 μkat/L [0.17-0.67 μkat/L])
AST = 25 U/L (20-48 U/L) (SI: 0.42 μkat/L [0.33-0.80 μkat/L])
Total cholesterol = 577 mg/dL (<200 mg/dL [optimal]) (SI: 14.94 mmol/L [<5.18 mmol/L])
Triglycerides = 3675 mg/dL (<150 mg/dL [optimal]) (SI: 41.53 mmol/L [<1.70 mmol/L])
HDL cholesterol = 10 mg/dL (>60 mg/dL [optimal]) (SI: 0.26 mmol/L [>1.55 mmol/L])

Which of the following would be the most effective treatment for this patient's hypertriglyceridemia?

A. Intensification of insulin therapy to target hemoglobin A_{1c} <7.0% (<53 mmol/mol)
B. Lipoprotein lipase gene replacement therapy
C. Recombinant human leptin replacement therapy
D. Semaglutide
E. Volanesorsen (apolipoprotein C-III antisense oligonucleotide)

34 A previously healthy 44-year-old man presents with low libido, breast tenderness, headaches, and fatigue and is diagnosed with a 2.5-cm pituitary tumor. Family history is negative for endocrine disorders.

Laboratory test results at initial presentation:

Prolactin = 450 ng/mL (4-23 ng/mL) (SI: 19.6 nmol/L [0.17-1.00 nmol/L])
Total testosterone = 99 ng/dL (300-900 ng/dL) (SI: 3.4 nmol/L [10.4-31.2 nmol/L])
Cortisol (8 AM) = 3 µg/dL (5-25 µg/dL) (SI: 82.8 nmol/L [137.9-689.7 nmol/L])

In the past year, he has been taking cabergoline, 1 mg twice weekly, and hydrocortisone, 20 mg in the morning and 10 mg in evening, and he reports good adherence and tolerance. He continues to have decreased libido but reports that breast tenderness and headaches have resolved. He has 2 biological children aged 12 and 10 years and is not interested in fertility.

On physical examination, his blood pressure is 110/71 mm Hg and pulse rate is 85 beats/min. His height is 71 in (180.3 cm), and weight is 195 lb (88.5 kg) (BMI = 27.3 kg/m^2). He has generalized fat distribution, bilateral adipomastia, normal visual fields on confrontation, and normal heart sounds on auscultation.

Laboratory test results (sample drawn at 8 AM while fasting):

Prolactin = 17 ng/mL (4-23 ng/mL) (SI: 0.74 nmol/L [0.17-1.00 nmol/L])
Total testosterone = 125 ng/dL (300-900 ng/dL) (SI: 4.3 nmol/L [10.4-31.2 nmol/L])
FSH = 3.0 mIU/mL (1.0-13.0 mIU/mL) (SI: 3.0 IU/L [1.0-13.0 IU/L])
LH = 2.7 mIU/mL (1.0-9.0 mIU/mL) (SI: 2.7 IU/L [1.0-9.0 IU/L])
Chemistry panel, normal
Hemogram, normal

MRI of the pituitary gland indicates a decrease in the maximal tumor size diameter (now 1.5 cm) and no contact with the optic chiasm. Echocardiography done 2 years ago showed trace tricuspid regurgitation.

Which of the following is the best next step in this patient's care?

A. Increase the cabergoline dosage to 1.5 mg twice weekly
B. Measure bone mineral density
C. Measure free testosterone now
D. Measure total testosterone in 12 months
E. Perform echocardiography

35 A 54-year-old woman seeks evaluation for palpitations, heat intolerance, fatigue, and worsening anxiety that have been intermittently present for the past 3 years. She has unintentionally lost 8 lb (3.6 kg) during this timeframe. She is postmenopausal (last menstrual period was at age 51 years).

On physical examination, her blood pressure is 138/92 mm Hg and resting pulse rate is 98 beats/min. Her height is 61 in (154.9 cm), and weight is 120 lb (54.4 kg) (BMI = 22.7 kg/m^2). She appears well, although she is slightly restless. She has a normal-sized thyroid gland with no palpable nodules and a bilateral, mild, fine tremor of her outstretched hands. Her eye examination shows no exophthalmos or conjunctival redness, and her extraocular eye muscles display normal range of motion in all directions of gaze. Her skin is warm. She has no tenderness during physical examination of her neck, chest, or abdomen and has no palpable masses.

Laboratory test results:

Hematocrit, normal
Basic metabolic panel, normal
FSH = 52.0 mIU/mL (>30.0 mIU/mL) (SI: 52.0 IU/L [>30.0 IU/L])
TSH = <0.09 mIU/L (0.5-5.0 mIU/L) (similar to a TSH value obtained 9 months ago)
Free T_4 = 2.0 ng/dL (0.8-1.8 ng/dL) (SI: 25.7 pmol/L [10.30-23.17 pmol/L])
Free T_3 = 4.5 pg/mL (2.3-4.2 pg/mL) (SI: 6.9 pmol/L [3.53-6.45 pmol/L])
TPO antibodies = <2.0 IU/mL (<2.0 IU/mL) (SI: <2.0 kIU/L [<2.0 kIU/L])

TRAb = ≤1.75 IU/L (≤1.75 IU/L)
Serum thyroglobulin = 298 ng/mL (3-42 ng/mL) (SI: 298 μg/L [3-42 μg/L])

An ^{123}I thyroid nuclear uptake and scan shows less than 1% uptake in the neck.

Which of the following is this patient's most likely diagnosis?

A. Antibody-negative Graves disease
B. hCG-producing tumor
C. Struma ovarii
D. Subacute thyroiditis
E. Thyrotoxicosis factitia

36

A 63-year-old man with hypogonadism and a history of low-trauma fracture presents for a follow-up appointment. For the past 3 years, he has experienced decreased libido and fewer early-morning erections. Six months ago, he had acute onset of severe back pain when he lifted a bowling ball. X-ray of the thoracic spine demonstrated a compression fracture at the 11th thoracic vertebra. The pain remitted after 8 weeks. His medical history includes squamous cell carcinoma of the neck (treated with resection and external-beam radiation 8 years ago), hyperlipidemia, hypertension, and benign prostatic hyperplasia. Medications are simvastatin, lisinopril, amlodipine, and tamsulosin. He reports intermittent nocturia while on tamsulosin.

On physical examination, his blood pressure is 128/80 mm Hg, pulse rate is 68 beats/min, and respiratory rate is 12 breaths/min. His height is 73 in (185.4 cm), and weight is 197 lb (89.4 kg) (BMI = 26.0 kg/m^2). Lungs are clear to auscultation with tidal volume breaths. He has 3-cm, slightly tender gynecomastia. Jugular venous pressure is normal, and heart rate is regular with no gallop. There is no edema. His penis is normal, and testes are 15 cc bilaterally. He has no hyperpigmentation.

Laboratory test results:

Hematocrit = 38% (41%-51%) (SI: 0.38 [0.41-0.51])
Serum total testosterone = 250 ng/dL (264-916 ng/dL) (SI: 8.7 nmol/L [9.2-31.8 nmol/L]) (assay validated and harmonized by the CDC)
Repeat serum testosterone (fasting sample drawn at 7:30 AM) = 260 ng/dL (SI: 9.0 nmol/L)
(Calculated free testosterone, low)
SHBG = 3.9 μg/mL (1.1-6.7 μg/mL) (SI: 35 nmol/L [10-60 nmol/L])
FSH = 2.0 mIU/mL (1.0-13.0 mIU/mL) (SI: 2.0 IU/L [1.0-13.0 IU/L])
LH = 1.0 mIU/mL (1.0-9.0 mIU/mL) (SI: 1.0 IU/L [1.0-9.0 IU/L])
Free T_4 = 1.1 ng/dL (0.8-1.8 ng/dL) (SI: 14.2 pmol/L [10.30-23.17 pmol/L])
TSH = 2.3 mIU/L (0.5-5.0 mIU/L)
Prolactin = 12 ng/mL (4-23 ng/mL) (SI: 0.52 nmol/L [0.17-1.00 nmol/L])
PSA = 1.8 ng/mL

Brain MRI shows a normal pituitary gland and no masses. DXA documents minimal degenerative changes in the lumbar spine with a few foci of calcinosis. Vertebrae appear normal, except for T11 (compression fracture). The L1-L4 T-score is −2.3. The right hip (the hip with lowest T-score [−2.1]) has good positioning and appropriate marking of the region of interest, with no bony abnormalities.

Which of the following is the best initial treatment for this patient?

A. Zoledronic acid
B. Intramuscular testosterone undecanoate
C. Monthly subcutaneous injection of romosozumab
D. Transdermal testosterone gel
E. Transdermal testosterone gel plus zoledronic acid

37 A 21-year-old man presents to discuss his risk of developing type 1 diabetes. He reports that his 18-year-old sister was diagnosed with type 1 diabetes 5 years ago. His primary care provider recently encouraged him to see an endocrinologist because of a recently approved medication to reduce the risk of developing diabetes in at-risk siblings of patients with type 1 diabetes. He reports that his primary care provider assessed him for autoimmune markers of type 1 diabetes, and that he tested positive. He has no polyuria, polydipsia, or weight loss.

On physical examination, his blood pressure is 110/76 mm Hg and pulse rate is 72 beats/min. His height is 70 in (177.8 cm), and weight is 156 lb (70.7 kg) (BMI = 22.4 kg/m^2).

Laboratory test results:

Glutamic acid decarboxylase 65 antibodies, detectable
Zinc transporter (ZnT8) antibodies, detectable
Hemoglobin A_{1c} = 5.6% (4.0%-5.6%) (SI: 38 mmol/mol [20-38 mmol/mol])
Fasting blood glucose = 88 mg/dL (70-99 mg/dL) (SI: 4.9 mmol/L [3.9-5.5 mmol/L])

Which of the following is the best next step in this patient's management?

A. Measure serum C-peptide
B. Perform 2-hour oral glucose tolerance test
C. Perform islet-cell antibody titer
D. Repeat hemoglobin A_{1c} measurement
E. Start teplizumab therapy

38 A 72-year-old woman with a medical history of nephrolithiasis, osteopenia, and hypertension is referred for hypothyroidism. Three months ago, the patient's primary care provider noted that she had an elevated serum TSH value on routine laboratory testing, which was similar on repeat testing. She was prescribed a generic levothyroxine tablet at a dosage of 50 mcg daily. She took the medication for 2 weeks but subsequently discontinued it when she experienced new insomnia, which she attributed to levothyroxine. At a subsequent visit with her primary care provider, she was prescribed a branded levothyroxine tablet at a dosage of 25 mcg daily, but she has not started taking this yet. She presents today for a second opinion regarding management.

The patient is in her usual state of health. She reports some daytime fatigue, similar over several years, requiring a midday nap. She describes a depressed mood and reduced interest in social activities since the death of her husband 7 months ago. She also endorses constipation, which she has been managing with a high-fiber diet. There is no history of thyroid surgery or radioactive iodine therapy. In addition to levothyroxine, her currently prescribed medications include losartan, hydrochlorothiazide, and cholecalciferol.

On physical examination, her blood pressure is 119/72 mm Hg and pulse rate is 60 beats/min. Her height is 62 in (157.5 cm), and weight is 110 lb (49.9 kg) (BMI = 20.1 kg/m^2). Her thyroid gland is normal in size and without nodules. Her deep tendon reflexes are 2+. The rest of the examination findings are normal.

Laboratory test results:

Measurement	Initial laboratory test results 3 months ago	Repeat laboratory test results (before levothyroxine) at 8 AM	Reference range
TSH	6.3 mIU/L	6.8 mIU/L	0.5-5.0 mIU/L
Free T_4	1.1 ng/dL (SI: 14.16 pmol/L)	1.0 ng/dL (SI: 12.87 pmol/L)	0.8-1.8 ng/dL (SI: 10.30-23.17 pmol/L)
TPO antibodies	<2.0 IU/mL	...	<2.0 IU/mL
LDL cholesterol	93 mg/dL (SI: 2.41 mmol/L)	...	<100 mg/dL (SI: <2.59 mmol/L)
Triglycerides	180 mg/dL (SI: 2.03 mmol/L)	...	<150 mg/dL (SI: <1.70 mmol/L)

Which of the following is the most appropriate next step in this patient's management?

A. Measure serum thyroglobulin antibodies
B. Measure serum total T_3
C. Prescribe branded levothyroxine gel cap
D. Provide reassurance that treatment with thyroid hormone is not required now
E. Recommend the branded levothyroxine tablet as prescribed by her primary care provider

39 A 31-year-old woman presents for evaluation of infertility. She and her husband have been trying to conceive for the past year. Menarche was at age 12 years. She reports irregular periods, and she menstruates every 30 to 60 days. She has struggled to lose weight most of her life. She has type 2 diabetes and is treated with metformin, 1000 mg twice daily. She also takes a prenatal multivitamin. She has occasional acne, particularly prior to menses, and has cyclical breast tenderness before menses. She has no galactorrhea or visual field deficits. She is physically active. She is of Mediterranean heritage.

On physical examination, her blood pressure is 109/75 mm Hg and pulse rate is 78 beats/min. Her height is 67 in (170.2 cm), and weight is 192 lb (87.1 kg) (BMI = 30.1 kg/m^2). She has an olive skin complexion and dark hair. Hirsutism is not present on examination. However, she bleaches her upper lip hair and reports plucking minimal dark hair on her chest and periareolar area. She has uniform distribution of excess adiposity. Muscle strength is normal. Her skin is warm, without plethora or bruises. She has thin, pale abdominal striae and 2 acne lesions on her chin. There are no features of acromegaly.

Laboratory test results (sample collected at 8 AM, first week after last menstrual period):

Complete blood cell count, normal
Metabolic panel, normal
Hemoglobin A_{1c} = 6.3% (4.0%-5.6%) (45 mmol/mol [20-38 mmol/mol])
TSH = 2.6 mIU/L (0.5-5.0 mIU/L)
Free T_4 = 1.1 ng/dL (0.8-1.8 ng/dL) (SI: 14.2 pmol/L [10.30-23.17 pmol/L])
Prolactin = 10 ng/mL (4-30 ng/mL) (SI: 0.43 nmol/L [0.17-1.30 nmol/L])
Total testosterone = 47 ng/dL (8-60 ng/dL) (SI: 1.6 nmol/L [0.3-2.1 nmol/L])
DHEA-S = 580 µg/dL (31-228 µg/dL) (SI: 15.7 µmol/L [0.84-6.78 µmol/L])
DHEA = 35 ng/mL (<13 ng/mL) (SI: 121.5 nmol/L [<45.1 nmol/L])
Androstenedione = 60 ng/dL (30-200 ng/dL) (SI: 2.10 nmol/L [1.05-6.98 nmol/L])
17-Hydroxyprogesterone = 238 ng/dL (<285 ng/dL) (SI: 7.21 nmol/L [<8.64 nmol/L])
Estradiol = 19 pg/mL (10-180 pg/mL) (SI: 69.7 pmol/L [36.7-660.8 pmol/L])
LH = 8.0 mIU/mL (0.5-18.0 mIU/mL) (SI: 8.0 IU/L [0.5-18.0 IU/L])
FSH = 4.1 mIU/mL (1.0-9.0 mIU/mL) (SI: 4.1 IU/L [1.0-9.0 IU/L])
Cortisol = 16 µg/dL (5-25 µg/dL) (SI: 441.4 nmol/L [137.9-689.7 nmol/L])
ACTH = 77 pg/mL (10-60 pg/mL) (SI: 16.9 pmol/L [2.2-13.2 pmol/L])

Which of the following is the best next step in this patient's management?

A. Adrenal CT
B. Cosyntropin-stimulation test
C. Genetic testing for pathogenic variants in the *HSD3B2* gene
D. Ovarian ultrasonography
E. Pituitary-directed MRI

40 A 33-year-old woman with a history of type 1 presents for follow-up evaluation and management. This is the first follow-up visit after her treatment regimen was switched from a traditional nonhybrid insulin pump to a hybrid closed-loop insulin pump system (automated insulin delivery). She is using automated insulin delivery that adjusts basal rates automatically and administers autocorrection insulin doses as needed to achieve normoglycemia. She is frustrated with her glycemic control, and she experiences frequent hypoglycemic episodes.

She is considering switching back to her previous nonhybrid insulin pump. Her current hemoglobin A_{1c} value is 7.4% (57 mmol/mol).

Continuous glucose monitoring download shows the following values:

Average glucose concentration = 163 mg/dL (SI: 9.0 mmol/L)
Time-in-range (70-180 mg/dL [SI: 3.9-10.0 mmol/L]) = 63%
Time-above-range (>180-250 mg/dL [SI: >10.0-13.9 mmol/L]) = 20%
Time-above-range (>250 mg/dL [SI: >13.9 mmol/L]) = 11%
Time-below-range (<70 mg/dL [SI: <3.9 mmol/L]) = 6%
Time-below-range (<54 mg/dL [SI: <3.0 mmol/L]) = 3%
Coefficient of variation = 34.1%

Auto mode is used 96% of the time. Basal rate, insulin-to-carbohydrate ratio, and insulin sensitivity factor ratio are shown (*see table*).

Start time	Basal rate (manual mode)	Insulin-to-carbohydrate ratio	Insulin sensitivity factor
Midnight	0.7 units/h	1:15	1:20 mg/dL (1:1.7 mmol/L)
6 AM	0.8 units/h	1:15	1:20 mg/dL (1:1.7 mmol/L)
9 AM	0.9 units/h	1:15	1:20 mg/dL (1:1.7 mmol/L)
2 PM	0.8 units/h	1:15	1:20 mg/dL (1:1.7 mmol/L)

A continuous glucose monitoring report is shown (*see image*). The patient says the day shown was a difficult one for her. At 1:30 AM, she experienced hypoglycemia, which she treated. She then had 2 additional hypoglycemic episodes at 6:30 PM and 11:30 PM, which she also treated. She ate 2 meals that day: an early breakfast at 5 AM and lunch just before 1 PM. She skipped dinner but had a snack at 4 PM.

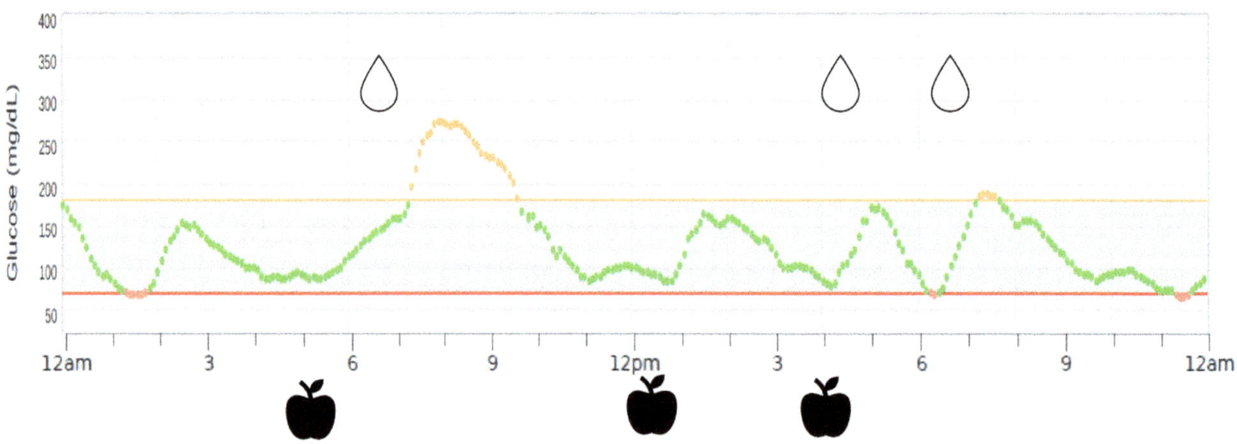

Apples indicate food intake and drops indicate autocorrections.

Her treatment consists of insulin glargine, 31 units daily; short-acting insulin aspart, 11 units with each meal; and correctional insulin, 1 unit for every 30 mg/dL (1.7 mmol/L) glucose is above 130 mg/dL (>7.2 mmol/L).

In addition to providing education about the appropriate treatment plans/actions for hypoglycemia, which of the following is the best management plan for this patient to reduce the number of hypoglycemic episodes?

Answer	Insulin pump therapy	Insulin adjustments
A.	Continue hybrid closed-loop insulin pump therapy on auto mode	Change insulin-to-carbohydrate ratio settings to decrease insulin delivered at meals
B.	Continue hybrid closed-loop insulin pump therapy on auto mode	Change sensitivity factor settings to decrease insulin delivered through correction boluses
C.	Continue insulin pump therapy, but switch to manual mode	Decrease the basal rate settings to decrease insulin delivery throughout the day
D.	Continue insulin pump therapy, but switch to manual mode	Change insulin-to-carbohydrate ratio settings to decrease insulin delivered at meals
E.	Discontinue insulin pump therapy	Initiate a basal-bolus multiple daily injection regimen at 10% lower total insulin dosing

41 A 28-year-old woman (G0, P0) is seen for prenatal counseling. Primary hypothyroidism secondary to Hashimoto thyroiditis was diagnosed at age 14 years, and she has since been on a stable dosage of levothyroxine. Her current regimen is levothyroxine, 125 mcg by mouth once daily, and she takes no other medications. On this dosage, her serum TSH concentration has been consistently 2.5 mIU/L. Her serum TPO antibody titer is 590 mIU/L (590 kIU/L). She is asymptomatic and reports no hypothyroid or hyperthyroid symptoms. She would like to plan for her first pregnancy within the next 12 months.

Which of the following is the best management plan?

A. Advise a gluten-free diet until she conceives
B. Begin liothyronine, 10 mcg daily, in addition to the levothyroxine that she is taking
C. Recommend that she contact the clinic immediately upon confirmed pregnancy
D. Start a daily supplement containing 150 mcg of potassium iodide
E. Start selenium, 200 mcg once daily

42 A 50-year-old woman presents with a long history of a very slowly enlarging right cranial bony mass (over the course of years) and associated hearing loss. This mass was first noted in childhood. At age 2 years, an ovarian cyst presented with vaginal bleeding. In recent years, she has been treated with a cochlear implant. She has had multiple fractures in her right femur, chronic bone pain in her right upper leg, and headaches. She notes some intermittent pain in the area of the cranial mass, but it is not tender when pressing on it. She has hypertension treated with lisinopril and type 2 diabetes managed with metformin.

On physical examination, she has obvious bony enlargement of the right side of the skull covering several centimeters, beginning in the frontotemporal region and extending to the parietal bone. This is nontender and without overlying redness. She has a cochlear implant on the right side. There is mild right proptosis, but eye movements, pupillary function, optic discs, and visual fields are normal. Visual acuity is normal. Findings on dental examination are normal. The thyroid gland is normal in size without palpable nodules. Findings on cardiovascular, lung, breast, and abdominal examination are normal. She has hyperpigmented macules with irregular borders on her right flank, arm, and leg. The right upper leg is nontender to palpation. The right leg is shorter than the left, and she walks with an antalgic gait. There is no tenderness over the ribs.

DXA documents a spine T-score of +1.5 and a left femoral neck T-score of –0.1. Technetium bone scan is shown (*see image*).

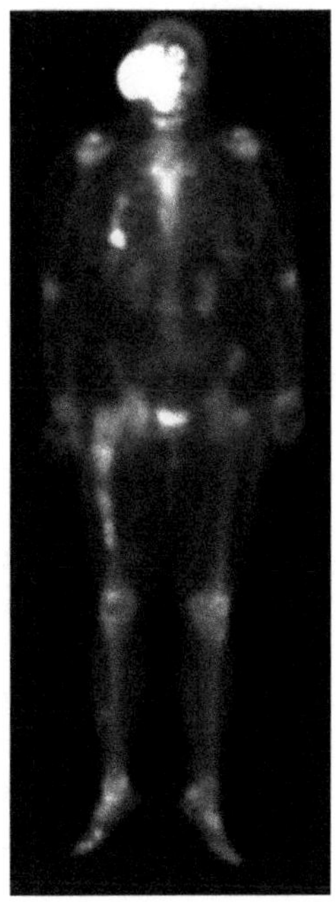

Laboratory test results:

Calcium = 9.5 mg/dL (8.2-10.2 mg/dL) (SI: 2.4 mmol/L [2.1-2.6 mmol/L])
Albumin = 3.9 g/dL (3.5-5.0 g/dL) (SI: 39 g/L [35-50 g/L])
Phosphate = 2.7 mg/dL (2.3-4.7 mg/dL) (SI: 0.9 mmol/L [0.7-1.5 mmol/L])
Creatinine = 0.9 mg/dL (0.6-1.1 mg/dL) (SI: 79.6 μmol/L [53.0-97.2 μmol/L])
Alkaline phosphatase = 214 U/L (50-120 U/L) (SI: 3.57 μkat/L [0.84-2.00 μkat/L])
Bone-specific alkaline phosphatase = 69.1 μg/L (≤22 μg/L)
Intact PTH = 31 pg/mL (10-65 pg/mL) (SI: 3.3 pmol/L [1.1-6.9 pmol/L])
25-Hydroxyvitamin D = 33 ng/mL (30-80 ng/mL [optimal]) (SI: 82.4 nmol/L [74.9-199.7 nmol/L])
TSH = 2.7 mIU/L (0.5-5.0 mIU/L)

Which of the following is the best treatment approach for this patient?

A. Burosumab
B. External-beam radiation
C. Intravenous bisphosphonate
D. Oral bisphosphonate
E. Surgical resection of the skull lesion

43 A 68-year-old man is seen for follow-up of prediabetes. He was noted to have a fasting plasma glucose concentration of 116 mg/dL (6.4 mmol/L) and a hemoglobin A_{1c} value of 6.4% (46 mmol/mol) 2 years ago when he first presented with atypical chest pain. He had an abnormal stress test and underwent percutaneous coronary intervention with stent placement. He has done very well since then, regularly exercising 45 to 60 minutes most days of the week. He decreased his intake of refined sugars and energy-dense foods and has lost about 25 lb (11.3 kg). He could not tolerate metformin because of diarrhea. Similarly, he tried atorvastatin and simvastatin, both of which had to be stopped because of severe muscle pain. He is currently taking rosuvastatin, 5 mg twice weekly, and ezetimibe, 10 mg daily. He tried increasing the frequency of rosuvastatin doses, but he experienced significant muscle cramps.

On physical examination, his blood pressure is 118/76 mm Hg and pulse rate is 70 beats/min. His height is 70.9 in (180 cm), and weight is 176.4 lb (80 kg) (BMI = 24.7 kg/m^2).

Laboratory test results:

Fasting plasma glucose = 102 mg/dL (70-99 mg/dL) (SI: 5.7 mmol/L [3.9-5.5 mmol/L])
Hemoglobin A_{1c} = 6.1% (4.0%-5.6%) (43 mmol/mol [20-38 mmol/mol])
ALT = 27 U/L (10-40 U/L) (SI: 0.45 μkat/L [0.17-0.67 μkat/L])
AST = 29 U/L (20-48 U/L) (SI: 0.48 μkat/L [0.33-0.80 μkat/L])
Creatine kinase = 93 U/L (50-200 U/L) (SI: 1.55 μkat/L [0.84-3.34 μkat/L])
Serum creatinine = 1.0 mg/dL (0.7-1.3 mg/dL) (SI: 88.4 μmol/L [61.9-114.9 μmol/L])
Total cholesterol = 184 mg/dL (<200 mg/dL [optimal]) (SI: 4.77 mmol/L [<5.18 mmol/L])
Triglycerides = 146 mg/dL (<150 mg/dL [optimal]) (SI: 1.65 mmol/L [<1.70 mmol/L])
HDL cholesterol = 42 mg/dL (>60 mg/dL [optimal]) (SI: 1.09 mmol/L [>1.55 mmol/L])
LDL cholesterol = 112 mg/dL (<100 mg/dL [optimal]) (SI: 2.90 mmol/L [<2.59 mmol/L])
High-sensitivity C-reactive protein = 3.6 mg/L (0.8-3.1 mg/L) (SI: 34.3 nmol/L [7.62-29.52 nmol/L])

In addition to encouraging him to persist with healthy lifestyle choices, which of the following is the best recommendation?

A. Change rosuvastatin to daily pravastatin
B. Initiate a retrial of metformin
C. Start bempedoic acid
D. Start semaglutide
E. Start stanol ester supplement

44 A 65-year-old man is referred for evaluation of a 3.7-cm right adrenal nodule that was discovered incidentally on abdominal CT when he presented with abdominal distention (*see image*). The nodule is homogeneous and has a density of −10 Hounsfield units. Incidental note is made of loss of vertebral body height at multiple levels in the thoracic and lumbar spine. His medical history includes a recent full-thickness tear of his left Achilles tendon, asthma, deep vein thrombosis, and hypertension. His personal and family history is negative for malignancies. There is no family history of diabetes. His medications include amlodipine, ramipril, doxazosin, simvastatin, spironolactone, and bumetanide. He has no history of cigarette smoking and consumes no alcohol.

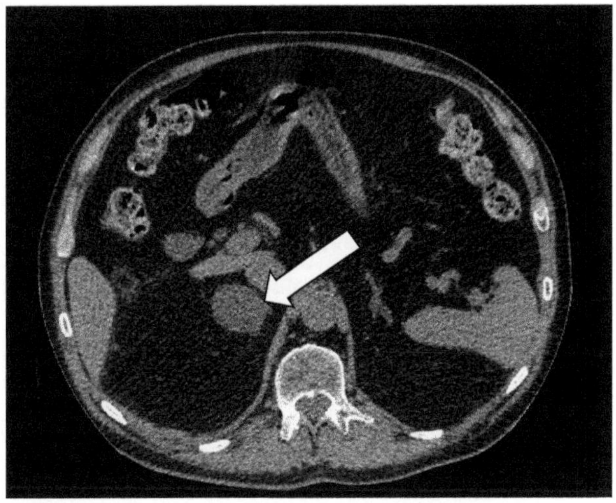

On physical examination, his blood pressure is 172/107 mm Hg and pulse rate is 80 beats/min. His height is 71 in (180 cm), and weight is 243 lb (110.2 kg) (BMI = 33.9 kg/m^2). He has peripheral edema and a distended abdomen. He has bruises on both arms but no striae on his abdomen. He has proximal muscle weakness.

Laboratory test results:

DHEA-S = 74 µg/dL (25-131 µg/dL) (SI: 2.00 µmol/L [0.68-3.55 µmol/L])
Potassium = 3.8 mEq/L (3.5-5.0 mEq/L) (SI: 3.8 mmol/L [3.5-5.0 mmol/L])
Aldosterone = 2.5 ng/dL (4-21 ng/dL) (SI: 70.0 pmol/L (111.0-582.5 pmol/L)
Plasma renin activity = 1.4 ng/mL per h (0.6-4.3 ng/mL per h)
Plasma metanephrines, normal
Hemoglobin A_{1c} = 6.6% (4.0%-5.6%) (49 mmol/mol [20-38 mmol/mol])

Laboratory values following a 1-mg dexamethasone-suppression test (8 AM):

Serum cortisol = 10.0 µg/dL (SI: 277 nmol/L)
Dexamethasone = 6.5 nmol/L (>3.3 nmol/L)
Midnight salivary cortisol = 0.16 µg/dL and 0.18 µg/dL (<0.13 µg/dL) (SI: 4.4 nmol/L and 5.0 nmol/L [<3.6 nmol/L])

Which of the following is the best next step in this patient's management?

A. Measure 24-hour urinary free cortisol
B. Measure ACTH
C. Reassess the aldosterone-to-renin ratio after discontinuation of spironolactone
D. Recommend right adrenalectomy
E. Start metyrapone therapy

45 A 60-year-old woman with hypothyroidism (treated with levothyroxine), obesity, and obstructive sleep apnea presents for follow-up of a right thyroid nodule that was initially identified 2 years ago. The nodule was detected on PET/CT completed during an evaluation for a pulmonary nodule. It did not demonstrate elevated fluorodeoxyglucose (FDG) uptake. Subsequent neck ultrasonography revealed a 3.2 × 2.2 × 2.1-cm nodule in the right midpole of the thyroid gland, no left thyroid nodules, and benign-appearing cervical lymph nodes. Ultrasound-guided FNA of the right thyroid nodule at that time demonstrated benign cytopathology (Bethesda II).

Repeat ultrasonography performed 2 weeks ago demonstrated increased size of the right thyroid nodule to 3.7 × 2.8 × 2.6 cm. Repeat ultrasound-guided FNA was performed, again demonstrating benign cytopathology. The patient presents today to review these results.

She has no current symptoms. There is no history of head or neck ionizing radiation exposure or family history of thyroid malignancy.

On physical examination, her blood pressure is 120/77 mm Hg and pulse rate is 80 beats/min. Her height is 63 in (160 cm), and weight is 239 lb (108.4 kg) (BMI = 42.3 kg/m^2). There is a firm, mobile, 3-cm nodule in the right thyroid lobe. There is no cervical lymphadenopathy. The rest of the examination findings are normal.

The most recent thyroid ultrasound images are shown (*see images*).

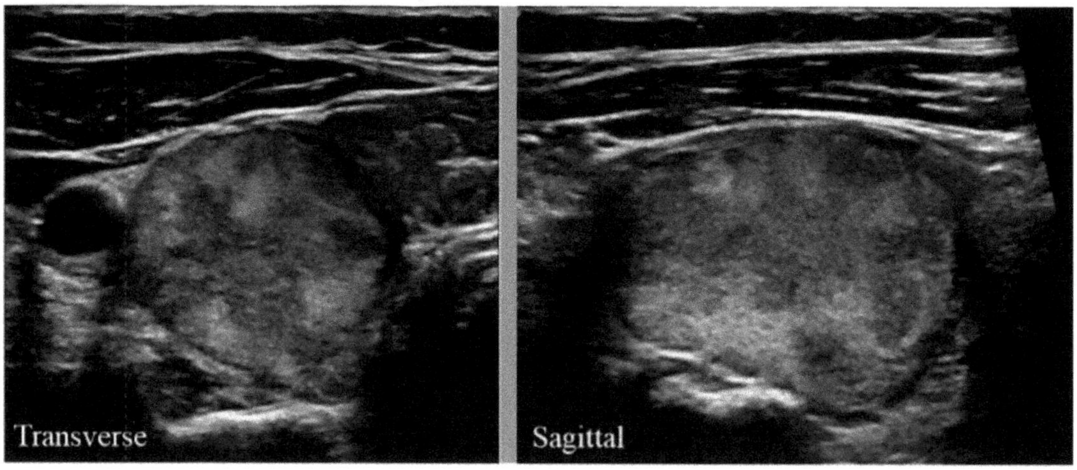

Laboratory test result:

TSH = 2.98 mIU/L (0.5-5.0 mIU/L)

Which of the following is the most appropriate next step in this patient's management?

A. Perform clinical assessment, including neck exam, in 12 months
B. Perform neck CT with contrast now
C. Perform neck ultrasonography in 12 months
D. Perform neck ultrasonography in 24 months
E. Refer for right thyroid lobectomy

46 A 52-year-old man with type 2 presents for initial evaluation and management. His medical history is notable for congestive heart failure, hypertension, and mild diabetic retinopathy. His outpatient regimen consists of insulin glargine, 22 units daily, and metformin, 1000 mg twice daily. His BMI is 30.5 kg/m^2. His family history is unremarkable, and he has no personal history of cigarette smoking, alcohol, or illicit drug use. His current blood pressure is 158/91 mm Hg, and pulse rate is 76 beats/min.

Laboratory test results:

Hematocrit = 39.8% (41%-51%) (SI: 0.398 [0.41-0.51])
Hemoglobin = 11.1 g/dL (13.8-17.2 g/dL) (SI: 111 g/L [138-172 g/L]])
Creatinine = 1.0 mg/dL (0.7-1.3 mg/dL) (SI: 88.4 μmol/L [61.9-114.9 μmol/L])
Urinary albumin-to-creatinine ratio = 323 mg/g and 311 mg/g (checked twice) (<30 mg/g creat)
Hemoglobin A_{1c} = 8.2% (4.0%-5.6%) (66 mmol/mol [20-38 mmol/mol])

Because of the elevated hemoglobin A_{1c} value, adjustments to his diabetes regimen are discussed. Several options are presented, and after a thorough discussion about risks and benefits, an SGLT-2 inhibitor is prescribed.

Four months later, the patient presents for follow-up evaluation and management. His hemoglobin A_{1c} measurement is now 8.5% (69 mmol/mol). The patient states he is adherent to his medication regimen. Fructosamine measurement is ordered as part of the evaluation, and the concentration is 332 μmol/L (equivalent to hemoglobin A_{1c} value of 8.2% [66 mmol/mol]). After discussing different treatment options, he is prescribed a GLP-1 receptor agonist. His current blood pressure is 155/90 mm Hg, and pulse rate is 81 beats/min.

Three months later, his hemoglobin A_{1c} value is still 8.2% (66 mmol/mol).

Which of the following is the most appropriate next step in the management of this patient's uncontrolled hyperglycemia?

A. Add a sulfonylurea
B. Add a thiazolidinedione
C. Evaluate for causes of medication nonadherence
D. Increase the basal insulin glargine dosage
E. Initiate prandial insulin

47

A 53-year-old woman is referred for evaluation of a 4.9-cm right adrenal nodule that was discovered incidentally on abdominal CT when she presented with right upper-quadrant abdominal pain (*see image, arrow*). The nodule is homogeneous and has negative Hounsfield unit density. Her medical history includes depression, Ehlers-Danlos syndrome, and monoclonal gammopathy of undetermined significance. Her personal and family history is negative for malignancies. Her father died after a myocardial infarction in his early 50s. Her medications include cholecalciferol, amitriptyline, and cyanocobalamin. She has a longstanding history of cigarette smoking and consumes 1 glass of wine per week.

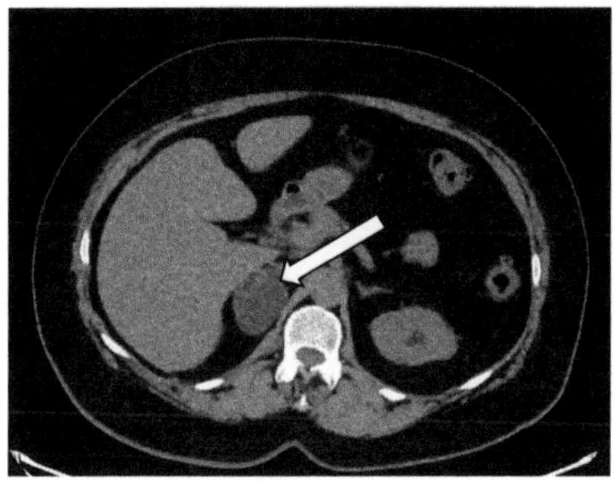

On physical examination, her blood pressure is 115/70 mm Hg and pulse rate is 76 beats/min. Her height is 65 in (165.1 cm), and weight is 216 lb (98 kg) (BMI = 35.9 kg/m^2). She has uniformly distributed excess adiposity. She has no plethora or bruises but has some pale striae on her abdomen. Muscle strength is within normal limits.

Laboratory test results (sample drawn at 8 AM):

Serum aldosterone = 9.4 ng/dL (4-21 ng/dL) (SI: 260 pmol/L [111.0-582.5 pmol/L])
Plasma renin activity = 0.6 ng/mL per h (0.6-4.3 ng/mL per h)
Urinary metanephrines, normal
Midnight salivary cortisol = 0.05 μg/dL (<0.13 μg/dL) (SI: 1.3 nmol/L [<3.6 nmol/L])
Urinary free cortisol = 25.7 μg/24 h (4-50 μg/24 h) (SI: 71 nmol/d [11-138 nmol/d])
DHEA-S = 7.4 μg/dL (15-200 μg/dL) (SI: 0.20 μmol/L [0.41-5.42 μmol/L])
ACTH = 72.5 pg/mL (10-60 pg/mL) (SI: 16.0 pmol/L [2.2-13.2 pmol/L])

Laboratory test results after a 1-mg dexamethasone-suppression test (8 AM):

Serum cortisol = 3.4 μg/dL (SI: 95 nmol/L)
Dexamethasone = 5.4 nmol/L (>3.3 nmol/L)

Which of the following is the best next step in this patient's management?

A. Perform CT of the thorax, abdomen, and pelvis
B. Perform bilateral inferior petrosal sinus sampling
C. Perform pituitary-directed MRI
D. Recommend right adrenalectomy
E. Repeat ACTH measurement using a separate immunoassay method

48

A 20-year-old woman has had 2 tibial stress fractures and 1 metatarsal stress fracture while running during the past 2 years. Menarche was at age 12 years, but she has been amenorrheic for the last 3 years.

On physical examination, she appears cachectic. Her blood pressure is 88/42 mm Hg, and pulse rate is 49 beats/min. Her height is 61 in (155 cm) (10th percentile), and weight is 75 lb (34 kg) (BMI = 14.2 kg/m^2). Her BMI has decreased since last year. Sclerae are white, and pupils are equal and reactive. The parotid glands are enlarged but nontender. She has poor dental enamel with evident erosion. Other than bradycardia, findings on heart and lung examinations are normal. There is no hepatomegaly. Breast development and pubic hair are Tanner stage 5. There is no kyphosis, scoliosis, or peripheral skeletal deformities. She has no tenderness in the lower extremities, and her gait is normal.

On DXA, her lumbar spine Z-score is −3.0, and right femoral neck Z-score is −2.8.

She is taking a multivitamin; cholecalciferol, 1000 international units daily (25 mcg daily); and elemental calcium, 1600 mg daily. She is interested in hormone-based therapies to improve her bone health.

Laboratory test results:

Sodium = 131 mEq/L (136-142 mEq/L) (SI: 131 mmol/L [136-142 mmol/L])
Creatinine = 0.4 mg/dL (0.6-1.1 mg/dL) (SI: 35.4 μmol/L [53.0-97.2 μmol/L])
Calcium = 8.5 mg/dL (8.2-10.2 mg/dL) (SI: 2.1 mmol/L [2.1-2.6 mmol/L])
Albumin = 2.7 g/dL (3.5-5.0 g/dL) (SI: 27 g/L [35-50 g/L])
Phosphate = 2.4 mg/dL (2.3-4.7 mg/dL) (SI: 0.8 mmol/L [0.7-1.5 mmol/L])
Alkaline phosphatase = 110 U/L (50-120 U/L) (SI: 1.84 μkat/L [0.84-2.00 μkat/L])
Intact PTH = 33 pg/mL (10-65 pg/mL) (SI: 3.5 pmol/L [1.1-6.9 pmol/L])
25-Hydroxyvitamin D = 58 ng/mL (30-80 ng/mL [optimal]) (SI: 144.8 nmol/L [74.9-199.7 nmol/L])
Urinary calcium = 237 mg/24 h (100-300 mg/24 h) (SI: 5.9 mmol/d [2.5-7.5 mmol/d])

In addition to efforts to attain a healthy weight, which of the following is the most appropriate next step in the management of this patient's bone health?

A. Increase elemental calcium intake to 2000 mg daily
B. Start oral combined hormonal contraceptive pills
C. Start oral DHEA
D. Start transdermal estrogen with cyclic oral progesterone
E. Start transdermal testosterone

49 A 48-year-old woman has experienced extended durations of amenorrhea in the last 2 years, followed by lighter bleeding. The longest duration without a menstrual cycle was 5.5 months. She has significant vasomotor symptoms, sleep disturbances, irritability, and anxiety. She has no personal history of venous thromboembolism, but her mother had a lower-extremity deep vein thrombosis after surgery. Years ago, she had a functional medicine evaluation that identified an *MTHFR* variant (Ala1298Cys), although this result is not available in her electronic medical record. She is sexually active, with ovum meeting sperm. She used an oral combined hormonal contraceptive pill between the ages of 18 and 20 years, but she is hesitant to use this medication again because of her family history.

Which of the following is best next step in this patient's management?

A. Measure estradiol and FSH
B. Prescribe a low-dosage oral combined hormonal contraceptive pill
C. Prescribe a methylated B-complex vitamin
D. Prescribe transdermal estradiol and micronized progesterone
E. Repeat thrombophilia screening

50 A 48-year-old man was diagnosed with type 2 diabetes 5 years ago. He has diligently worked on lifestyle modifications and incorporated dietician recommendations. Through these efforts, he has lost 12 lb (5.4 kg), and his weight has been stable for the past 6 months. He has bilateral knee osteoarthritis (right > left) that is causing considerable pain and is also limiting his ability to exercise on a regular basis. He also has obstructive sleep apnea (treated with continuous positive airway pressure), hypertension, dyslipidemia, and depression. He self-monitors blood glucose 3 to 4 times weekly, and his fasting values are typically in the range of 90 to 120 mg/dL (5.0-6.7 mmol/L) and 2-hour postprandial values are in the range of 140 to 200 mg/dL (7.8-11.1 mmol/L). He has not had any episodes of hypoglycemia.

Current medications include dapagliflozin, 10 mg daily; metformin, 500 mg twice daily; dulaglutide, 1.5 mg weekly; candesartan, 16 mg once daily; amlodipine, 5 mg once daily; atorvastatin, 40 mg once daily; ibuprofen, 200 mg 3 times daily; and oxycodone/acetaminophen, 5/500 mg as needed for pain (currently once or twice daily).

On physical examination, his blood pressure is 152/91 mm Hg and pulse rate is 78 beats/min. His height is 69 in (175.3 cm), and weight is 239 lb (108.4 kg) (BMI = 35.2 kg/m^2).

Laboratory test results:

Hemoglobin A_{1c} = 6.4% (4.0%-5.6%) (46 mmol/mol [20-38 mmol/mol])
Serum creatinine = 0.8 mg/dL (0.7-1.3 mg/dL) (SI: 70.7 μmol/L [61.9-114.9 μmol/L])
Estimated glomerular filtration rate = 78 mL/min per 1.73 m^2 (>60 mL/min per 1.73 m^2)
LDL cholesterol = 65 mg/dL (<100 mg/dL [optimal]) (SI: 1.68 mmol/L [<2.59 mmol/L])
HDL cholesterol = 42 mg/dL (>60 mg/dL [optimal]) (SI: 1.09 mmol/L [>1.55 mmol/L])
Triglycerides = 95 mg/dL (<150 mg/dL [optimal]) (SI: 1.07 mmol/L [<1.70 mmol/L])
Urinary albumin-to-creatinine ratio = 15 mg/g creat (<30 mg/g creat)

Which of the following is the best next step in this patient's management?

A. Continue current management
B. Increase the dulaglutide dosage to 3.0 mg once weekly
C. Increase the metformin dosage to 1000 mg twice daily
D. Switch from dulaglutide to tirzepatide at the highest tolerated dosage
E. Start naltrexone/bupropion to treat obesity

51 A 35-year-old competitive body builder is referred for management of infertility. He reports that he has used androgen steroids since his early 20s. He typically takes multiple androgens for 12 to 16 weeks followed by intramuscular testosterone cypionate at a dosage of 125 to 150 mg weekly for 8 to 12 weeks ("blast and cruise"). He and his wife of 2 years would like to conceive, so he stopped all androgenic steroid administration 1 year ago. He reports low libido, but he and his wife have had unprotected vaginal intercourse 1 to 3 times weekly for the past year. He has no history of medical illnesses and takes no medications.

His 34-year-old wife is healthy and has normal menses every 28 days. She uses a smartphone app with a urinary kit that indicates that she has likely been ovulating regularly during each menstrual cycle.

On physical examination, his blood pressure is 128/80 mm Hg, pulse rate is 68 beats/min, and respiratory rate is 12 breaths/min. His height is 71 in (180.3 cm), and weight is 144 lb (65.3 kg) (BMI = 20.1 kg/m^2). He is lean and muscular. He has 2.5-cm, nontender gynecomastia. Jugular venous pressure is normal, and heart rate is regular with no gallop. He has no edema. His penis is normal, and testes are 12 cc bilaterally. Bilateral vasa deferentia are palpable. He has no hyperpigmentation.

Laboratory test results:

Hematocrit = 39% (41%-51%) (SI: 0.39 [0.41-0.51])
Serum total testosterone = 220 ng/dL (264-916 ng/dL) (SI: 7.6 nmol/L [9.2-31.8 nmol/L]) (assay validated and harmonized by the CDC)
Repeat serum testosterone (sample drawn at 7:30 AM) = 206 ng/dL (SI: 7.1 nmol/L)
Calculated free testosterone, low
SHBG = 3.4 μg/mL (1.1-6.7 μg/mL) (SI: 30 nmol/L [10-60 nmol/L])
FSH = 2.0 mIU/mL (1.0-13.0 mIU/mL) (SI: 2.0 IU/L [1.0-13.0 IU/L])
LH = 1.0 mIU/mL (1.0-9.0 mIU/mL) (SI: 1.0 IU/L [1.0-9.0 IU/L])
Free T_4 = 1.0 ng/dL (0.8-1.8 ng/dL) (SI: 12.87 pmol/L [10.30-23.17 pmol/L])
TSH = 1.8 mIU/L (0.5-5.0 mIU/L)
Prolactin = 12 ng/mL (4-23 ng/mL) (SI: 0.52 nmol/L [0.17-1.00 nmol/L])
Seminal fluid analysis = 50,000 sperm/mL (>15 million/mL)

Brain MRI shows a normal pituitary gland and no masses.

Which of the following is the best recommendation for this patient now?

A. Daily oral anastrozole
B. Daily oral clomiphene
C. hCG subcutaneously 3 times weekly
D. Reassurance and watchful waiting; serum hormone measurements and seminal fluid analysis every 3 months
E. Referral to a specialist who performs microscopic testicular sperm extraction

52 A 17-year-old man is referred for evaluation of a 2-month history of progressive lower limb weakness, leading to difficulty in climbing stairs. He describes simultaneous weight gain of 30 lb (13.6 kg), back pain, and loss of height. He has a history of asthma.

On physical examination, his blood pressure is 165/90 mm Hg and pulse rate is 90 beats/min. His height is 68 in (172.7 cm), and weight is 267 lb (121 kg) (BMI = 40.6 kg/m^2). He has peripheral edema and a distended abdomen. Bruises are observed on both arms, but no abdominal striae are present. He has proximal muscle weakness.

Evaluation confirms a diagnosis of ACTH-dependent hypercortisolism secondary to a pituitary macroadenoma.

On day 1 after pituitary surgery, he develops paraparesis with hypertonicity. Power is reduced in his legs (2/5), and he has sensory loss below the midthoracic region.

Laboratory test results:

25-Hydroxyvitamin D = 5.6 ng/mL (30-80 ng/mL [optimal]) (SI: 14 nmol/L [74.9-199.7 nmol/L])
Creatine kinase = 53 U/L (50-200 U/L) (SI: 0.89 µkat/L [0.84-3.34 µkat/L])
Electrolytes, normal
Serum cortisol day 5 after surgery (9 AM) = 2.5 µg/dL (5-25 µg/dL) (SI: 70.0 nmol/L [137.9-689.7 nmol/L])

MRI of the spine is shown (*Figure 1*).

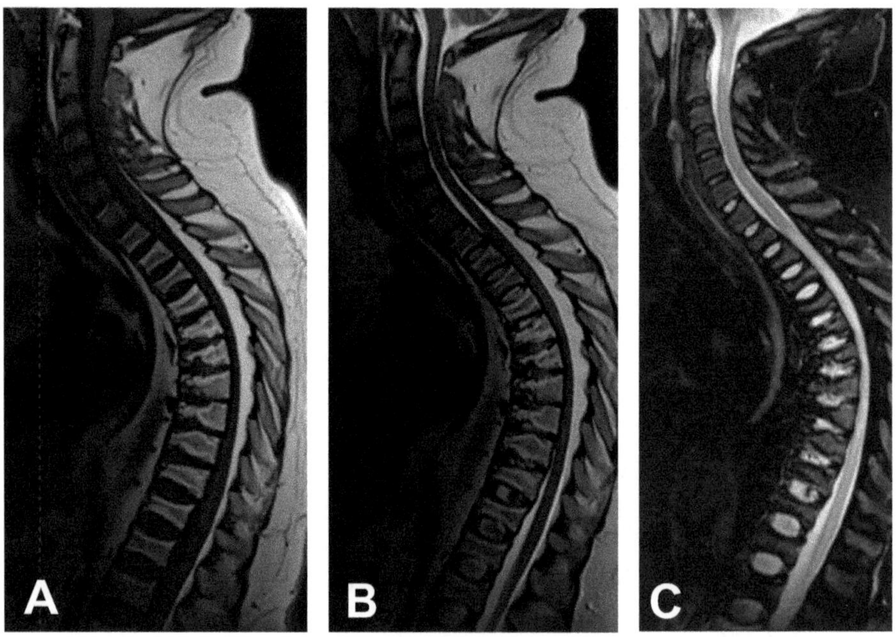

Panel A, sagittal T1-weighted; Panel B, sagittal T2-weighted; and Panel C, STIR sequence (short Tau inversion recovery). Adapted with permission of Bioscientifica Ltd. from Ahmad S et al. *Endocrinol Diabetes Case Rep*, 2020; 2020:20-0111.

Which of the following best explains this patient's presentation?

A. Osteoporotic fractures of the spine
B. Proximal myopathy secondary to hypercortisolism
C. Spinal cord infarction
D. Spinal epidural lipomatosis
E. Vitamin D deficiency

53 A 63-year-old man returns for continued management of type 2 diabetes that was diagnosed 12 years ago. He has a medical history of hypertension, hyperlipidemia, and chronic kidney disease with severely increased albuminuria. Peripheral neuropathy is well controlled with duloxetine.

Current medications:

Sitagliptin, 100 mg once daily
Hydrochlorothiazide, 25 mg once daily
Lisinopril, 10 mg once daily
Rosuvastatin, 10 mg once daily
Duloxetine, 60 mg once daily

On physical examination, his blood pressure is 134/84 mm Hg and pulse rate is 80 beats/min. His height is 66 in (167.6 cm), and weight is 218 lb (98.9 kg) (BMI = 35.2 kg/m^2). He has 1+ pitting edema and 2+ dorsalis pedis pulse in the bilateral lower extremities. His lungs are clear to auscultation, and his heart has a regular rate and rhythm with no audible murmur. There is decreased sensation to 10-g monofilament testing on the distal plantar aspect of his feet bilaterally.

Laboratory test results:

Hemoglobin A_{1c} = 6.3% (4.0%-5.6%) (45 mmol/mol [3.9-5.5 mmol/L])
Potassium = 5.2 mEq/L (3.5-5.0 mEq/L) (SI: 5.2 mmol/L [SI: 3.5-5.0 mmol/L])
Creatinine = 2.5 mg/dL (0.7-1.3 mg/dL) (SI: 221.1 µmol/L [61.9-114.9 µmol/L])
Estimated glomerular filtration rate = 28 mL/min per 1.73 m^2 (>60 mL/min per 1.73 m^2)
Albumin-to-creatinine ratio = 630 mg/g creat (<30 mg/g creat)

Which of the following medications should be started as the best next step in this patient's treatment?

A. Acarbose
B. Dapagliflozin
C. Finerenone
D. Insulin glargine
E. Pioglitazone

54 A 49-year-old man is referred to discuss weight management. He has always considered himself "heavy" but had no difficulty participating in sports until he developed chronic pain secondary to a slipped back disk 10 years ago. He takes buprenorphine for the management of chronic pain. Several years ago, he was prescribed clomiphene for a period to treat low testosterone and he wonders whether this should be resumed. He has longstanding low libido and poor erectile function. He does not drink alcohol. He has well-treated sleep apnea.

His current medications are atorvastatin, 20 mg daily; sublingual buprenorphine, 2 mg twice daily; furosemide, 20 mg twice daily; and venlafaxine, 75 mg daily.

On physical examination, his blood pressure is 138/76 mm Hg and pulse rate is 78 beats/min and regular. His height is 72.4 in (184 cm), and weight is 407 lb (185 kg) (BMI = 54.6 kg/m^2). He has generalized body weight distribution between the upper and lower body and has a dorsocervical fat bulge.

Laboratory test results:

Total testosterone by mass spectroscopy (8 AM) = 84 ng/dL (300-900 ng/dL) (SI: 2.91 nmol/L [0.31-1.04 nmol/L])
Total testosterone by mass spectroscopy (8 AM) 2 weeks later = 94 ng/dL (SI: 3.26 nmol/L)
Serum prolactin = 18 ng/mL (4-23 ng/mL) (SI: 0.78 nmol/L [0.17-1.00 nmol/L])
TSH = 1.6 mIU/L (0.5-5.0 mIU/L)
Free T_4 = 1.1 ng/dL (0.8-1.8 ng/dL) (SI: 14.16 pmol/L [10.30-23.17 pmol/L])
LH = 1.0 mIU/mL (1.0-9.0 mIU/mL) (SI: 1.0 IU/L [1.0-9.0 IU/L])
FSH = 1.5 mIU/mL (1.0-13.0 mIU/mL) (SI: 1.5 IU/L [1.0-13.0 IU/L])
SHBG = 6.0 μg/mL (1.1-6.7 μg/mL) (SI: 53.4 nmol/L [10-60 nmol/L])

Which of the following is the best next step in this patient's management?

A. Perform 1-mg overnight dexamethasone-suppression test
B. Perform pituitary-directed imaging
C. Prescribe high-dosage liraglutide
D. Refer to pain management clinic to wean off buprenorphine
E. Restart clomiphene

55 A 59-year-old man with a history of type 2 diabetes, obesity, and coronary heart disease presents for routine follow-up. Current medications are metformin, insulin aspart, insulin glargine, rosuvastatin, aspirin, and olmesartan. He is up-to-date on routine eye examination, and he has longstanding peripheral neuropathy. He drinks 4 to 5 glasses of wine per week and has no history of cigarette smoking.

On physical examination, his blood pressure is 138/88 mm Hg and pulse rate is 88 beats/min. His height is 68 in (172.7 cm), and weight is 205 lb (93.0 kg) (BMI = 31.2 kg/m^2). He has mild bilateral pitting edema. Lung and heart sounds are normal. There is decreased sensation to monofilament up to his ankles.

Laboratory test results:

Estimated glomerular filtration rate = 87 mL/min per 1.73 m^2 (>60 mL/min per 1.73 m^2)
Creatinine = 1.0 mg/dL (0.7-1.3 mg/dL) (SI: 88.4 μmol/L [61.9-114.9 μmol/L])
Hemoglobin = 15.2 g/dL (13.8-17.2 g/dL) (SI: 152 g/L [138-172 g/L])
Platelet count = 272 × 10^3/μL (150-450 × 10^3/μL) (SI: 272 × 10^9/L [150-450 × 10^9/L])
Hemoglobin A_{1c} = 7.4% (4.0%-5.6%) (SI: 57 mmol/mol [20-38 mmol/mol])
AST = 63 U/L (20-48 U/L) (SI: 1.05 μkat/L [0.33-0.80 μkat/L])
ALT = 40 U/L (10-40 U/L) (SI: 0.67 μkat/L [0.17-0.67 μkat/L])

His calculated Fibrosis-4 score is 2.16.

Vibration-controlled transient elastography results:

Measurement	Result	Normal
CAP (controlled attenuation parameter)	369 dB/m	<288 dB/m
LSM (liver stiffness measurement)	8.9 kPa	<8.0 kPa

Which of the following is the best next step in this patient's management?

A. Refer to hepatology
B. Refer to nutritionist to discuss the Mediterranean diet
C. Start dapagliflozin, 5 mg daily
D. Start pioglitazone, 15 mg daily
E. Stop rosuvastatin

56 A 67-year-old woman is referred for evaluation of fatigue and weight loss. Her medical history includes hypertension, chronic obstructive pulmonary disease, dyslipidemia, and chronic back pain. She has smoked 1 pack of cigarettes daily for more than 40 years. She does not drink alcohol or consume illicit substances. Over the past several months, she has become very tired, finds it difficult to function, and takes naps throughout the day. She reports diffuse muscle weakness and pain. She has occasional episodes of nausea and vomiting.

Her medications include lisinopril, 25 mg daily; atorvastatin, 40 mg daily; fluticasone propionate-salmeterol inhaler once daily; and oxycodone, 10 mg 3 times daily as needed. She reports regularly needing all 3 doses each day for the past 4 months.

On physical examination, her blood pressure is 108/70 mm Hg and pulse rate is 85 beats/min. Her height is 65.5 in (166.4 cm), and weight is 198.5 lb (90.0 kg) (BMI = 32.5 kg/m^2). On auscultation of her lungs, there are some opening rhonchi and no wheezes. Her heart rhythm is regular. Her abdomen is nontender to palpation. Her muscle strength is 4/5 symmetrically, in both proximal and distal muscle groups. She has no edema. Her skin is dry, and she has no hirsutism, plethora, hyperpigmentation, or vitiligo.

Laboratory test results (sample collected at 8 AM):

Complete blood cell count, normal
Metabolic panel, normal
TSH = 4.6 mIU/L (0.5-5.0 mIU/L)
Free T_4 = 1.1 ng/dL (0.8-1.8 ng/dL) (SI: 14.2 pmol/L [10.30-23.17 pmol/L])
Cortisol = 3.6 µg/dL (5-25 µg/dL) (SI: 99.3 nmol/L [137.9-689.7 nmol/L])
ACTH = 5 pg/mL (10-60 pg/mL) (SI: 1.1 pmol/L [2.2-13.2 pmol/L])
DHEA-S = 25 µg/dL (15-157 µg/dL) (SI: 0.68 µmol/L [0.41-4.25 µmol/L])

Pituitary-directed MRI shows a subtle 2-mm hypoenhancing area in the midportion of the pituitary gland, located to the right of midline. The pituitary stalk is normal. The suprasellar region appears intact, and the optic chiasm is normal.

Which of the following is the best next step in this patient's care?

A. Discuss switching to a nonglucocorticoid inhaler with her treating physician
B. Initiate oral hydrocortisone twice daily
C. Measure prolactin and IGF-1
D. Order a sleep study
E. Stop oxycodone and measure 8-AM cortisol and ACTH in 2 months

57 A 29-year-old woman (G1, P1) presents with occasional palpitations and mild heat intolerance that began approximately 1 month ago. She gave birth to a healthy baby girl 2 months ago and has been breastfeeding since.

On physical examination, she is afebrile, blood pressure is 128/74 mm Hg, and pulse rate is 105 beats/min. Her height is 62 in (157.5 cm), and weight is 128 lb (58.1 kg) (BMI = 23.4 kg/m^2). She has a fine, bilateral tremor of her outstretched hands. There is no exophthalmos, goiter, or thyroid bruit.

Laboratory test results:

TSH = <0.01 mIU/L (0.5-5.0 mIU/L)
Free T_4 = 2.9 ng/dL (0.8-1.8 ng/dL) (SI: 36.1 pmol/L [10.30-23.17 pmol/L])
Free T_3 = 6.0 pg/mL (2.3-4.2 pg/mL) (SI: 9.2 pmol/L [3.53-6.45 pmol/L])
Serum TPO antibodies = 8.0 IU/mL (<2.0 IU/mL) (SI: 8.0 kIU/L [<2.0 kIU/L])

Given her condition, this patient is most at risk for which of the following?

A. Future infertility
B. Neck pain and tenderness in the next week
C. Persistence of symptoms as long as she is breastfeeding
D. Recurrence of the same condition with future pregnancies
E. Thyroid eye disease

58 A 28-year-old transgender woman is seen for feminizing gender-affirming hormone therapy. Estradiol and spironolactone have helped her achieve target estradiol concentrations in the premenopausal range for cisgender women and a total testosterone concentration less than 50 ng/dL (<1.7 nmol/L) for 13 months. She has undergone facial feminization surgery with an excellent outcome and is pleased with the redistribution of her body fat, decreased muscle mass, decreased terminal hair growth, and decreased spontaneous erections and libido. However, she continues to have dysphoria related to the small size of her breasts. She is interested in pursuing a strategy to augment breast development. She has no personal history of depression and no family history of breast cancer, venous thromboembolism, stroke, or cardiovascular disease.

Which of the following is the best next step in this patient's treatment?

A. Increase the estradiol dosage to induce further ductal and lobular formation
B. Refer for affirming breast augmentation surgery
C. Refer for gender expression care
D. Refer for mental health services to address persisting dysphoria
E. Start micronized progesterone to induce ductal side-branching and acinar development

59 A 67-year-old woman presents for evaluation of postmenopausal osteoporosis. She had 2 recent mid-thoracic vertebral compression fractures, and she has a history of ankle fractures. She notes bone pain in the right hip, thigh, and knee, as well as chronic back pain. She walks with an antalgic gait. She has frequent headaches and recent-onset hearing loss. She does not take glucocorticoids. She has never been treated with osteoporosis medications. She takes a proton-pump inhibitor for Barrett esophagus. For the last 2 years, she has taken cholecalciferol, 2000 international units daily (50 mcg daily), and elemental calcium, 600 mg twice daily.

On physical examination, her blood pressure is 138/81 mm Hg and pulse rate is 76 beats/min. Her height is 64 in (162.6 cm) (previous self-reported maximal height = 66 in [167.6 cm]), and current weight is 123 lb (55.8 kg) (BMI = 21.0 kg/m^2). She does not appear cushingoid. She has good dentition. She has kyphosis, and when standing with her back against the wall, the occiput is 3 cm away from the wall. Her rib-to-pelvis distance is 3.4 cm, or slightly less than 2 fingerbreadths. She has pain in her right hip with walking, but no tenderness to palpation. Her right upper thigh is warmer than her left upper thigh. She has no edema, erythema, or bruising. Pigmentation is normal.

Laboratory test results:

- 25-Hydroxyvitamin D = 53 ng/mL (30-80 ng/mL [optimal]) (SI: 132.3 nmol/L [74.9-199.7 nmol/L])
- Calcium = 9.9 mg/dL (8.2-10.2 mg/dL) (SI: 2.5 mmol/L [2.1-2.6 mmol/L])
- Albumin = 3.6 g/dL (3.5-5.0 g/dL) (SI: 36 g/L [35-50 g/L])
- Phosphate = 4.3 mg/dL (2.3-4.7 mg/dL) (SI: 1.4 mmol/L [0.7-1.5 mmol/L])
- Alkaline phosphatase = 165 U/L (50-120 U/L) (SI: 2.76 µkat/L [0.84-2.00 µkat/L])
- Creatinine = 0.7 mg/dL (0.6-1.1 mg/dL) (SI: 61.9 µmol/L [53.0-97.2 µmol/L])
- Intact PTH = 58 pg/mL (10-65 pg/mL) (SI: 6.1 pmol/L [1.1-6.9 pmol/L])
- Serum protein electrophoresis, normal
- Urine protein electrophoresis, normal

DXA documents a lumbar spine T-score of −3.3 and a left femoral neck T-score of −2.1. Radiograph and technetium scan of the right pelvis and femur are shown (*see images*).

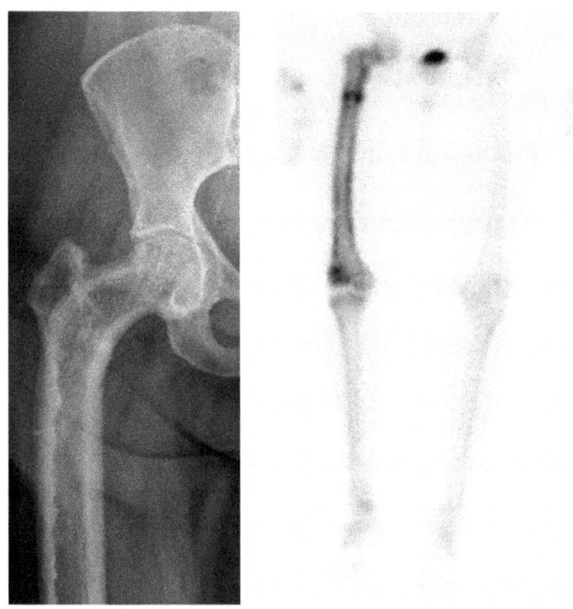

Images courtesy of Hagop Ghareebian and Erik A. Imel.

Which of the following is the best medication to manage this patient's fracture risk?

A. Asfotase alfa
B. Burosumab
C. Raloxifene
D. Teriparatide
E. Zoledronate

60 A 71-year-old woman with type 2 diabetes undergoes elective cholecystectomy for chronic right upper-quadrant pain secondary to cholelithiasis. The patient has a known allergy to propofol, so etomidate is used to assist with anesthesia induction. The operation is successfully completed, and she is transferred to the regular nursing floor. However, later that afternoon, she becomes confused and develops a deep and rapid breathing pattern.

She has a 10-year history of type 2 diabetes complicated by peripheral neuropathy controlled with gabapentin. She takes metformin XR, 500 mg twice daily; dapagliflozin, 10 mg daily; dulaglutide, 1.5 mg weekly; and insulin glargine, 20 units once daily. Her regimen of diabetes medications has been stable for the past 3 years. Her preoperative hemoglobin A_{1c} measurement was 7.2% (55 mmol/mol). She has a 20 pack-year history of cigarette smoking, hypertension treated with lisinopril/hydrochlorothiazide, and hyperlipidemia treated with atorvastatin. Her husband reports that she had been on nothing-by-mouth status since midnight on the day of surgery but that she has been consuming a limited diet over the past few weeks because of ongoing right upper-quadrant pain. He reports the patient took her regularly scheduled dose of insulin glargine last evening at bedtime.

On physical examination, her blood pressure is 122/64 mm Hg and pulse rate is 96 beats/min. Her height is 68 in (173 cm), and weight is 194 lb (88 kg) (BMI = 29.5 kg/m^2).

Laboratory test results:

Hemoglobin A_{1c} = 7.2% (4.0%-5.6%) (SI: 55 mmol/mol [20-38 mmol/mol])
Glucose = 164 mg/dL (70-99 mg/dL) (SI: 9.10 mmol/L [3.9-5.5 mmol/L])
Creatinine = 1.6 mg/dL (0.6-1.1 mg/dL) (SI: 141.5 µmol/L [53.0-97.2 µmol/L])
Serum urea nitrogen = 28 mg/dL (8-23 mg/dL) (SI: 10.0 mmol/L [2.9-8.2 mmol/L])
Sodium = 138 mEq/L (136-142 mEq/L) (SI: 138 mmol/L [136-142 mmol/L])
Potassium = 3.6 mEq/L (3.5-5.0 mEq/L) (SI: 3.6 mmol/L [3.5-5.0 mmol/L])
Chloride = 99 mEq/L (96-106 mEq/L) (SI: 99 mmol/L [96-106 mmol/L])
Bicarbonate = 15.0 mEq/L (21-28 mEq/L) (SI: 15 mmol/L [21-28 mmol/L])
Hemoglobin = 11.9 g/dL (12.1-15.1 g/dL) (SI: 119 g/L [121-151 g/L)
White blood cell count = 10,900/µL (4500-11,000/µL) (SI: 10.9 × 10^9/L [4.5-11.0 × 10^9/L])
pH = 7.2 (7.35-7.45)
Po_2 arterial blood = 92.0 mm Hg (80-100 mm Hg) (SI: 12.3 kPa [10.6-13.3 kPa])
Pco_2, arterial blood = 40.0 mm Hg (35-45 mm Hg) (SI: 5.3 kPa [4.7-6.0 kPa])
Lactic acid = 6.4 mg/dL (5.4-20.7 mg/dL) (SI: 0.71 mmol/L [0.6-2.3 mmol/L])
β-Hydroxybutyrate = 7.9 mg/dL (<3.0 mg/dL) (SI: 758.8 µmol/L [<288.2 µmol/L])

Which of the following medications most likely contributed to this patient's decompensation?

A. Dapagliflozin
B. Dulaglutide
C. Etomidate
D. Lisinopril/hydrochlorothiazide
E. Metformin

61 A 68-year-old man is seen for follow-up of hypercholesterolemia. He has heterozygous familial hypercholesterolemia and has previously been on LDL apheresis. He was able to discontinue the procedure 6 years ago after starting evolocumab, 140 mg every 2 weeks. He also takes atorvastatin, 80 mg daily, and ezetimibe, 10 mg daily. He previously underwent percutaneous coronary intervention with stent placement but is doing well now without symptoms of angina or exertional dyspnea. He has multiple comorbidities, including type 2 diabetes, hypertension, osteoarthritis, and chronic obstructive pulmonary disease. Additional medications include metformin, weekly semaglutide injection, lisinopril, hydrochlorothiazide, metoprolol, amlodipine, multiple nutritional supplements, and an inhaled long-acting bronchodilator with glucocorticoid. In addition, he recently started donepezil for memory issues.

Laboratory test results:

Total cholesterol = 220 mg/dL (<200 mg/dL [optimal]) (SI: 5.70 mmol/L [<5.18 mmol/L])
Triglycerides = 195 mg/dL (<150 mg/dL [optimal]) (SI: 2.20 mmol/L [<1.70 mmol/L])
HDL cholesterol = 45 mg/dL (>60 mg/dL [optimal]) (SI: 1.17 mmol/L [>1.55 mmol/L])
LDL cholesterol = 144 mg/dL (<100 mg/dL [optimal]) (SI: 3.73 mmol/L [<2.59 mmol/L])

He appears well and has had no significant weight changes.

Over the last 3 to 4 years, his LDL-cholesterol concentration had ranged from 80 to 100 mg/dL (2.07-2.59 mmol/L).

Which of the following is the best next step in this patient's management?

A. Change atorvastatin to rosuvastatin
B. Change evolocumab to alirocumab
C. Change evolocumab to inclisiran
D. Increase the evolocumab dosage to 420 mg every 4 weeks
E. Restart LDL apheresis every 2 weeks

62 A 33-year-old woman with Cushing disease presents for evaluation 3 months after transsphenoidal removal of a 1.6-cm pituitary adenoma, confirmed by pathology with positive ACTH immunostaining. She has had persistent weight gain, muscle weakness, and difficulty concentrating. She has an intrauterine device and reports she is not interested in pregnancy. She takes amlodipine, 5 mg daily, and hydrochlorothiazide, 25 mg daily.

On physical examination, her blood pressure is 145/91 mm Hg and pulse rate is 99 beats/min. Her height is 71.5 in (181.6 cm), and weight is 220.5 lb (100 kg) (BMI = 30.3 kg/m^2). She has a round face, centripetal fat distribution, and pink abdominal striae.

Laboratory test results:

Serum cortisol after 1 mg dexamethasone overnight = 8.0 μg/dL (SI: 220.7 nmol/L)
Urinary free cortisol = 80 μg/day (4-50 μg/24 h) (SI: 221 nmol/d [11-138 nmol/d])
Fasting glucose = 115 mg/dL (70-99 mg/dL) (SI: 6.4 mmol/L [3.9-5.5 mmol/L])

Postoperative MRI indicates a 0.5-cm residual tumor in the sella and left cavernous sinus. After discussing treatment options with the patient, pasireotide LAR, 10 mg monthly, is prescribed.

Which of the following is the most appropriate next step in glucose management for this patient?

A. Discuss how to recognize symptoms of hyperglycemia and ketoacidosis
B. Measure hemoglobin A_{1c} after 3 injections of pasireotide LAR
C. Start liraglutide, 0.6 mg daily
D. Start metformin, 500 mg twice daily
E. Stop hydrochlorothiazide

63 A 57-year-old man with type 2 diabetes diagnosed 5 years ago reports erectile dysfunction for which his primary care provider prescribed sildenafil. He tried has sildenafil on 5 different occasions, and while this drug resulted in modest improvement of penile tumescence with sexual stimulation, his erections remained inadequate for sexual intercourse. His primary care provider then prescribed tadalafil. The patient tried tadalafil on 6 different days and had a similar modest improvement in erections, but they remained inadequate for sexual activity. He reports normal libido. His partner is interested in having sexual activity, and their relationship is strong. He feels discouraged about his lack of response to sildenafil and tadalafil.

The patient plays tennis 3 times weekly and walks 3 to 4 miles on the other days. He reports no angina or dyspnea out of proportion to activity. He has dyslipidemia and hypertension. There is no diabetic retinopathy.

Medications are metformin, empagliflozin, and losartan.

On physical examination, his blood pressure is 128/80 mm Hg, pulse rate is 68 beats/min, and respiratory rate is 12 breaths/min. His height is 73 in (184.4 cm), and weight is 220 lb (99.8 kg) (BMI = 29.0 kg/m^2). He has no gynecomastia. Jugular venous pressure is normal, and heart rate is regular with no gallop. Peripheral pulses are normal. He has no edema. His penis is normal, and testes are 25 cc bilaterally. He has normal sensation to light touch in the perineal region and normal cremasteric and bulbocavernosal reflexes.

Laboratory test results:

Serum total testosterone = 375 ng/dL (264-916 ng/dL) (SI: nmol/L [9.2-31.8 nmol/L]) (assay validated and harmonized by the CDC)

Calculated free testosterone, normal

SHBG = 2.5 µg/mL (1.1-6.7 µg/mL) (SI: 22 nmol/L [10-60 mmol/L])

In addition to discontinuing tadalafil, which of the following should be recommended as the best next step in this patient's treatment?

A. Alprostadil (intracavernous injection)
B. Alprostadil (intraurethral suppository)
C. Oral vardenafil
D. Penile vacuum pump therapy
E. Referral to a urologist who is experienced with implantation of penile prostheses

64 A 64-year-old woman with a history of type 2 is admitted to the hospital for evaluation and management of pneumonia. Her primary care provider had initially prescribed oral antibiotic treatment, but because of persistent symptoms she was hospitalized and is being treated with intravenous antibiotics. Her outpatient diabetes regimen consists of U500 insulin comprising 60 units at breakfast, 40 units with lunch, and 60 units before dinner. She also takes semaglutide, 1 mg subcutaneously weekly, and metformin, 1000 mg per dose twice daily. Her most recent hemoglobin A_{1c} value, just before admission, was 8.1% (65 mmol/mol). The primary team has ordered a carbohydrate-controlled diet and requested a consultation for inpatient diabetes management.

Which of the following represents the most appropriate changes to make to this patient's regimen while she is hospitalized?

A. Continue U500 insulin at the current outpatient dosage but discontinue metformin and semaglutide
B. Continue U500 insulin (but decrease the dosage slightly compared with the outpatient regimen) and continue semaglutide and metformin
C. Discontinue U500 insulin and semaglutide but continue metformin
D. Discontinue U500 insulin, metformin, and semaglutide and start basal-bolus U100 insulin at equivalent doses compared with the outpatient regimen
E. Discontinue U500 insulin, metformin, and semaglutide and start basal-bolus U100 insulin at significantly decreased doses compared with the outpatient regimen

65 A 63-year-old woman presents with fatigue and weakness in her thighs. Although she has recently lost some weight following dietary restriction to 1200 calories daily, she has noticed more abdominal obesity and accumulation of fat around her neck. She reports recent hair loss from her head and facial hirsutism that requires regular hair removal. She endorses easy bruising and skin thinning but no striae. She has had hypertension for many years, with worsening control over time. She was recently hospitalized with epistaxis and hypertensive urgency. During this hospitalization, her serum potassium concentration was as low as 1.9 mEq/L (1.9 mmol/L). Lisinopril, spironolactone, and oral potassium chloride are started.

On physical examination, her blood pressure is 148/88 mm Hg and pulse rate is 68 beats/min. Her height is 65.3 in (166 cm), and weight is 218 lb (99 kg) (BMI = 35.9 kg/m^2). She has central obesity. Supraclavicular and dorsocervical fat pads are present. She has facial fullness without plethora. There is no frontal bossing or prognathism. Lungs are clear to auscultation bilaterally. Heart rhythm is regular. No acanthosis nigricans is noted. She has slight skin thinning and a few bruises on her right forearm, which are attributed to the recent hospitalization. She has no abdominal striae. She exhibits normal muscle strength in her upper extremities, with 4/5 strength of bilateral hip flexors. Her mood and affect are appropriate.

Laboratory test results obtained during the hospitalization:

Complete blood cell count, normal
Cortisol (8 AM) = 36 µg/dL (5-25 µg/dL)
(SI: 993 nmol/L [137.9-689.7 nmol/L])
ACTH (8 AM) = 169 pg/mL (10-60 pg/mL)
(SI: 37.2 pmol/L [2.2-13.2 pmol/L])
Plasma aldosterone = 4 ng/dL (4-21 ng/dL)
(SI: 111.0 pmol/L [111.0-582.5 pmol/L])
Plasma renin activity = 0.2 ng/mL per h
(0.6-4.3 ng/mL per h)
Urinary free cortisol =
650 µg/24 h (4-50 µg/24 h)
(SI: 1794 nmol/d [11-138 nmol/d])
Urinary creatinine =
1.6 g/24 h (1.0-2.0 g/24 h)
(SI: 14.1 mmol/d [8.8-17.7 mmol/d])
Urine volume = 2.1 L

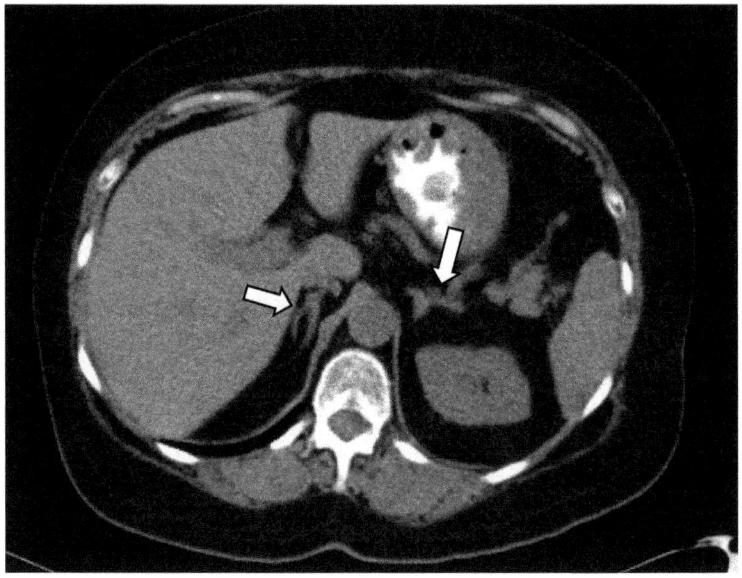

Abdominal CT shows subtle bilateral adrenal nodularity (*see image*).

Which of the following is the most likely cause of this patient's hypokalemia?

A. Cushing disease
B. Ectopic Cushing syndrome
C. Glucocorticoid-remediable aldosteronism
D. Primary pigmented nodular adrenal disease
E. Stress-induced hypercortisolism

66 A 43-year-old woman presents for follow-up of a 3.7-cm right thyroid nodule. She has undergone ultrasound-guided FNA biopsy twice with benign cytopathology on both occasions. She is concerned about the cosmetic appearance of her thyroid nodule and reports anterior neck pressure when she sleeps on her right side. She states that she is interested in radiofrequency ablation as a potential treatment option.

On physical examination, her blood pressure is 122/74 mm Hg and pulse rate is 74 beats/min. Her height is 67 in (170 cm), and weight is 146 lb (66.2 kg) (BMI = 22.9 kg/m^2). Her voice is normal. There is a palpable right thyroid nodule measuring approximately 4 cm without substernal extension. The rest of the examination findings are normal.

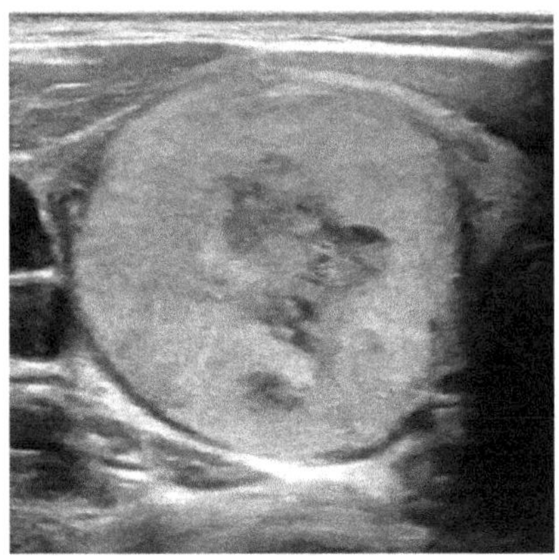

Laboratory test result:

TSH = 2.7 mIU/L (0.5-5.0 mIU/L)

Neck ultrasonography demonstrates a 3.7-cm, isoechoic, predominantly solid thyroid nodule in the right thyroid lobe with central cystic area and no contralateral thyroid nodules (*see image*). The visualized cervical lymph nodes appear normal.

Regarding management, this patient should be counseled that:

A. Radiofrequency ablation is not recommended; refer for right thyroid lobectomy
B. Radiofrequency ablation is not recommended; refer for radioactive iodine therapy
C. Radiofrequency ablation is not recommended; refer for ethanol ablation
D. Radiofrequency ablation is recommended; refer for a procedural consultation
E. She may be an appropriate candidate for radiofrequency ablation but first must undergo flexible laryngoscopy

67 A 36-year-old woman with a history of type 1 diabetes presents to the diabetes clinic for follow-up evaluation and management. Her medical history is notable for depression, sickle cell disease, and hypertension. She is treated with insulin glargine, 18 units daily at 12 PM (at lunchtime). She is also receiving short-acting insulin aspart, 6 units with each meal, and correctional insulin, 1 unit for every 50 mg/dL (2.8 mmol/L) above 150 mg/dL (>8.3 mmol/L). She usually has breakfast at 8 AM, lunch at 12 PM, and dinner at 6:30 PM. Her only other medication is hydroxyurea that she started almost 1 month ago, which she takes in the morning just before breakfast (8 AM).

The patient has recently started to experience marked hyperglycemia. She explains that her glucose concentration (as monitored by her real-time continuous glucose monitoring) starts rising from approximately 140 mg/dL (8.3 mmol/L) around 1:30 to 2 PM, and it becomes significantly elevated (300 mg/dL [16.7 mmol/L]). Her glucose management indicator is now 7.5% (58 mmol/mol), which is higher than the value that the patient had 3 months ago (6.8% [51 mmol/mol]). The continuous glucose monitoring report shows a pattern of hyperglycemia around 2 PM that usually persists until 6 or 6:30 PM.

Which of the following is the best next step in this patient's management?

A. Continue current management
B. Direct the patient to measure point-of-care fingerstick glucose values several times between 2 and 6 PM
C. Increase her prelunch short-acting insulin aspart dose
D. Recommend an automated insulin delivery system (insulin pump initiation)
E. Switch insulin glargine administration from 12 PM to bedtime

68 A 28-year-old man is referred to discuss weight management. He underwent resection of a suprasellar mass 4 months ago, followed by proton-beam radiation treatment that was completed 6 weeks before the appointment. Pathology results confirm the mass to be a mixed germ-cell tumor.

The patient is on hormone replacement for panhypopituitarism. His main concern is rapid weight gain as a result of appetite dysregulation. He reports being "always hungry" and has developed a preference for simple carbohydrates. His baseline weight before surgery was 220 lb (100 kg).

His weight trend over the last 4 months is shown:

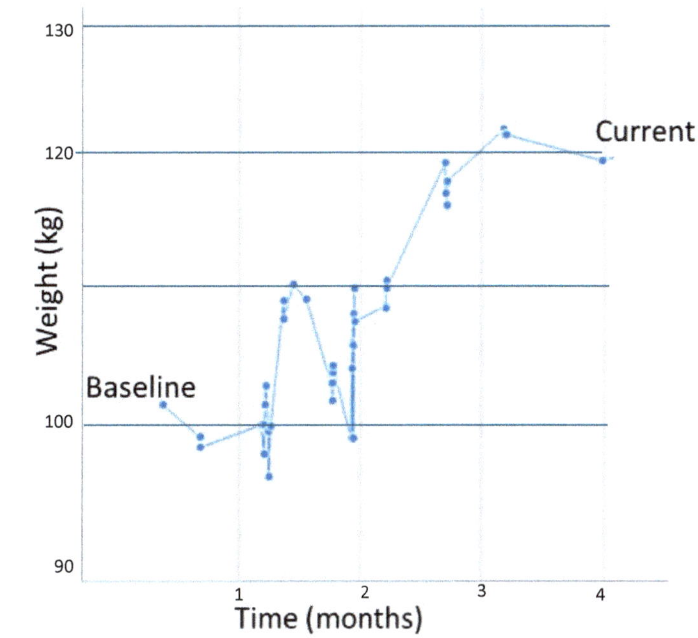

On physical examination, his blood pressure is 122/74 mm Hg and pulse rate is 88 beats/min. His height is 65.7 in (167 cm), and weight is 264 lb (120 kg) (BMI = 43 kg/m^2). He is mildly cushingoid in appearance with scattered acne on his face. He has abdominal obesity.

His current medications are hydrocortisone, 20 mg in the morning and 10 mg in the afternoon; levothyroxine, 150 mcg daily; testosterone gel, 30 mg daily; and pantoprazole, 40 mg daily.

Laboratory test results:

Free T_4 = 1.0 ng/dL (0.8-1.8 ng/dL) (SI: 12.87 pmol/L [10.30-23.17 pmol/L])
Serum sodium = 139 mEq/L (136-142 mEq/L) (SI: 139 mmol/L [136-142 mmol/L])
Total testosterone = 282 ng/dL (300-900 ng/dL) (SI: 9.8 nmol/L [10.4-31.2 nmol/L])
IGF-1 = 80 ng/mL (117-321 ng/mL) (SI: 10.48 nmol/L [15.3-42.1 nmol/L]) (Z-score = −1.65 SD)
Fasting glucose = 116 mg/dL (70-99 mg/dL) (SI: 6.4 mmol/L [3.9-5.5 mmol/L])

Which of the following is most likely to benefit this patient's weight management efforts now?

A. Increase physical activity, aiming for 200 minutes per week of moderate-intensity exercise
B. Provide dietary counseling on implementing caloric restriction
C. Reduce the hydrocortisone dosage to 15 mg in the morning and 5 mg in the afternoon
D. Start GH replacement
E. Start phentermine/topiramate ER

69 A 62-year-old man presents for continued management of type 2 diabetes diagnosed 12 years ago. He takes dapagliflozin, 10 mg daily, and insulin degludec, 40 units daily at bedtime. He has a history of coronary artery disease and congestive heart failure with reduced ejection fraction. Echocardiography recently documented an ejection fraction of 30%. He reports some shortness of breath, particularly with prolonged exertion or walking up stairs. He also has swelling in his lower extremities. While his level of albuminuria has remained relatively stable, his estimated glomerular filtration rate has continued to slowly decline (most recent measurement = 22 mL/min per 1.73 m^2). He reports adherence to a low-potassium diet. A recent 12-lead electrocardiogram was found to be consistent with normal sinus rhythm and with no abnormalities noted.

He has a 40 pack-year history of cigarette smoking, but he quit 12 years ago. He has hypertension treated with metoprolol, losartan, and furosemide and hyperlipidemia treated with rosuvastatin.

On physical examination, his blood pressure is 128/72 mm Hg and pulse rate is 66 beats/min. His height is 73 in (185.4 cm), and weight is 239 lb (108.4 kg) (BMI = 31.5 kg/m^2).

Laboratory test results:

Hemoglobin A_{1c} = 7.7% (4.0%-5.6%) (SI: 61 mmol/mol [20-38 mmol/mol])
Serum creatinine = 3.1 mg/dL (0.7-1.3 mg/dL) (SI: 274.1 μmol/L [61.9-114.9 μmol/L])
Estimated glomerular filtration rate = 22 mL/min per 1.73 m^2 (>60 mL/min per 1.73 m^2)
Potassium = 5.2 mEq/L (3.5-5.0 mEq/L) (SI: 5.2 mmol/L [3.5-5.0 mmol/L])
Potassium 6 months ago = 5.1 mEq/L (SI: 5.1 mmol/L)

Which of the following is the best next step in this patient's management?

A. Continue current management
B. Reduce the dapagliflozin dosage to 5 mg daily
C. Start finerenone
D. Stop dapagliflozin
E. Stop losartan

70 A 61-year-old woman presents to the endocrine clinic for mild hypercalcemia noted during evaluation after a left Colles fracture. In addition to hypertension, her medical history is notable for stage 1 unilateral breast cancer diagnosed at age 50 years, status post lumpectomy, radiation, and anastrozole treatment for 5 years. Uterine leiomyosarcoma was diagnosed at age 55 years, status post total hysterectomy with an incidental finding of a large cecal mass with pathology revealing spindle cells consistent with pelvic leiomyosarcoma. She has had no exposure to long-term steroid treatment. She is up to date with dental exams without any concerning features. She does not smoke cigarettes, drink alcohol, or use recreational drugs. Her 31-year-old daughter was recently diagnosed with mild hypercalcemia, and her maternal aunt was diagnosed with kidney cysts at age 58 years.

On physical examination, her blood pressure is 124/74 mm Hg and pulse rate is 86 beats/min. Her height is 63 in (160 cm), and weight is 157 lb (71.2 kg) (BMI = 27.8 kg/m^2). There are no palpable neck masses. Her left forearm is in a cast, and the rest of the examination findings are unremarkable.

Laboratory test results:

Complete blood cell count, normal
Serum 25-hydroxyvitamin D = 32 ng/mL (30-80 ng/mL [optimal]) (SI: 79.9 nmol/L [7.4.9-199.7 nmol/L])
Calcium = 11.2 mg/dL (8.2-10.2 mg/dL) (SI: 2.8 mmol/L [2.1-2.6 mmol/L])
Phosphate = 3.8 mg/dL (2.3-4.7 mg/dL) (SI: 1.2 mmol/L [0.7-1.5 mmol/L])
Serum creatinine = 0.95 mg/dL (0.6-1.1 mg/dL) (SI: 84 µmol/L (53.0-97.2 µmol/L)
Intact PTH = 66 pg/mL (10-65 pg/mL) (SI: 7.0 pmol/L [1.1-6.9 pmol/L])
TSH = 3.14 mIU/L (0.5-5.0 mIU/L)
Alkaline phosphatase = 95 U/L (50-120 U/L) (SI: 1.59 µkat/L [0.84-2.00 µkat/L])
Urinary calcium = 116 mg/24 h (100-300 mg/24 h) (SI: 2.9 mmol/d [2.5-7.5 mmol/d])

Testing for which of the following genetic conditions would be most appropriate?

A. Familial hypocalciuric hypercalcemia
B. Familial isolated hyperparathyroidism
C. Hyperparathyroidism–jaw tumor syndrome
D. Multiple endocrine neoplasia type 1
E. Multiple endocrine neoplasia type 2A

71

A 26-year-old woman was diagnosed with polycystic ovary syndrome after she presented to her primary care physician with a history of irregular menstrual cycles, cystic acne, and hirsutism. She has been referred to endocrinology for initial evaluation. She hopes to conceive in 2 to 3 years and wants to understand how she can improve her overall health.

On physical examination, her height is 69 in (175.3 cm) and weight is 148 lb (67.13 kg) (BMI = 21.9 kg/m^2). She has generalized fat distribution and acanthosis nigricans in the bilateral axillae.

Baseline laboratory test results (sample collected while fasting):

TSH = 2.9 mIU/L (0.5-5.0 mIU/L)
Prolactin = 12 ng/mL (4-30 ng/mL) (SI: 0.52 nmol/L [0.17-1.30 nmol/L])
17-Hydroxyprogesterone, normal
Hemoglobin A_{1c} = 5.5% (4.0%-5.6%) (37 mmol/mol [20-38 mmol/mol])
Total cholesterol = 145 mg/dL (<200 mg/dL [optimal]) (SI: 3.76 mmol/L [<5.18 mmol/L])
Triglycerides = 186 mg/dL (<150 mg/dL [optimal]) (SI: 2.10 mmol/L [<1.70 mmol/L])
HDL cholesterol = 59 mg/dL (>60 mg/dL [optimal]) (SI: 1.53 mmol/L [>1.55 mmol/L])
LDL cholesterol (calculated) = 86 mg/dL (<100 mg/dL [optimal]) (SI: 2.23 mmol/L [<2.59 mmol/L])
ALT = 53 U/L (10-40 U/L) (SI: 0.72 µkat/L [0.17-0.67 µkat/L])
AST = 72 U/L (20-48 U/L) (SI: 1.20 µkat/L [0.33-0.80 µkat/L])
Hepatitis profile, normal

Which of the following is best next step in this patient's overall health management?

A. Perform liver ultrasonography
B. Perform oral glucose tolerance test
C. Perform vibration-controlled transient elastography
D. Prescribe inositol
E. Prescribe metformin

72 A 38-year-old woman presents for evaluation of symptoms of hypoglycemia. She underwent Roux-en-Y gastric bypass 10 years ago and did well until about 6 months ago. She describes episodes of shakiness and sweats that occur 2 to 3 hours after eating about twice a month. If she does not treat promptly with glucose tablets, symptoms can progress to confusion and difficulty remembering how to administer glucose tablets. She met with a bariatric dietician and now eats several small meals each day, although this has not helped reduce the symptoms. She has not experienced symptoms while asleep or during periods of prolonged fasting.

The patient undergoes a mixed-meal study with a standard carbohydrate and protein sandwich but does not develop typical symptoms during the 4-hour meal protocol. Later in the day, about 2 hours after eating some ice cream, she develops typical symptoms.

Laboratory test results (sample drawn at the time of symptoms):

Plasma glucose = 24 mg/dL (70-99 mg/dL) (SI: 1.3 mmol/L [3.9-5.5 mmol/L])
Insulin = 34.7 µIU/mL (1.4-14.0 µIU/mL) (SI: 241.0 pmol/L [9.7-97.2 pmol/L])
C-peptide = 8.0 ng/mL (0.5-2.0 ng/mL) (SI: 2.65 nmol/L [0.17-0.66 nmol/L])
β-Hydroxybutyrate = <1.0 mg/dL (<3.0 mg/dL) (SI: <100 µmol/L [<288.2 µmol/L])
Serum creatinine = 0.7 mg/dL (0.6-1.1 mg/dL) (SI: 61.9 µmol/L [53.0-97.2 µmol/L])
Sulfonylurea screen, negative

On physical examination, her blood pressure is 102/66 mm Hg and pulse rate is 92 beats/min. Her height is 65.7 in (167 cm), and weight is 145 lb (65.8 kg) (BMI = 23.6 kg/m^2). She has global loss of subcutaneous fat stores.

Her current medications are bupropion XL, 150 mg daily; vitamin B_{12}, 1000 mcg monthly; multivitamin with minerals twice daily; and a combined oral contraceptive pill.

Which of the following is the best next step in determining the etiology of this patient's symptoms?

A. CT of the abdomen/pelvis with pancreas protocol (triple-phase)
B. Instructing the patient to keep a detailed food record
C. Oral glucose tolerance test
D. Selective arterial calcium-stimulation with hepatic vein sampling
E. Supervised 72-hour fast

73 A 42-year-old woman (G4, P3) with type 2 diabetes and hypertension presents for follow-up. She is currently 8 weeks pregnant although she has not yet had initial ultrasonography. Medications are metformin, insulin glargine, and valsartan. Her previous pregnancies were complicated by preeclampsia and preterm birth, and both babies were small-for-gestational-age. She is anxious regarding this pregnancy's outcome.

On physical examination, her blood pressure is 120/85 mm Hg and pulse rate is 96 beats/min. Her height is 65 in (165.1 cm), and weight is 217 lb (98.4 kg) (BMI = 36.1 kg/m^2).

Her most recent hemoglobin A_{1c} measurement is 7.0% (53 mmol/mol).

Which of the following is the best management plan for this patient?

A. Stop insulin glargine and valsartan, start atenolol and low-dosage aspirin
B. Stop insulin glargine and valsartan, start daily exercise and insulin NPH
C. Stop metformin and valsartan, start atenolol and low-dosage aspirin
D. Stop metformin and valsartan, start daily exercise and labetalol
E. Stop metformin and valsartan, start hydralazine and low-dosage aspirin

74 A 57-year-old man with end-stage kidney disease receives hemodialysis 3 days per week. He develops tertiary hyperparathyroidism and undergoes 3- and one-half gland parathyroidectomy. Before surgery, his serum calcium concentration was 11.9 mg/dL (2.3 mmol/L) and PTH concentration was 837 pg/mL (88.8 pmol/L). After surgery he has severe hypocalcemia for several days and requires repeated doses of intravenous calcium. He is eventually discharged on high dosages of calcium and calcitriol. He returns a few weeks later with severe hypocalcemia, muscle cramps, and parasthesias.

He accurately describes how he is supposed to take his medications, and he states he has not missed any doses (family members live with him and confirm accurate dosing). He does not have any gastrointestinal concerns or symptoms of malabsorption.

His medication list includes calcitriol, 2 mcg 3 times daily; calcium citrate, 2400 mg 3 times daily; cholecalciferol, 4000 international units daily (100 mcg daily); lisinopril, 20 mg daily; sevelamer, 1600 mg 3 times daily; carvedilol, 6.25 mg twice daily; and amlodipine, 5 mg daily.

On physical examination, he is comfortable and in no distress, but he has positive Chvostek sign. He demonstrates forearm and hand muscle spasms during blood pressure measurement. Findings on heart, lung, and abdominal examinations are normal. He has mild pedal edema. Electrocardiography shows a prolonged QT interval.

Despite further increasing the calcitriol dosage and providing extra doses of calcium and additional intravenous calcium gluconate, his laboratory test results 1 day after dialysis (third day of hospital admission) are as follows:

Intact PTH = <6 pg/mL (10-65 pg/mL) (SI: <0.6 pmol/L [1.1-6.9 pmol/L])
Calcium (total) = 6.2 mg/dL (8.2-10.2 mg/dL) (SI: 1.6 mmol/L [2.1-2.6 mmol/L])
Albumin = 3.2 g/dL (3.5-5.0 g/dL) (SI: 32 g/L [35-50 g/L])
Phosphate = 6.2 mg/dL (2.3-4.7 mg/dL) (SI: 2.0 mmol/L [0.7-1.5 mmol/L])
Alkaline phosphatase = 88 U/L (50-120 U/L) (SI: 1.47 µkat/L [0.84-2.00 µkat/L])
25-Hydroxyvitamin D = 36 ng/mL (30-80 ng/mL [optimal]) (SI: 89.9 nmol/L [74.9-199.7 nmol/L])
Tissue transglutaminase antibodies, not detectable

Which of the following is the best next step in improving this patient's calcium balance before hospital discharge?

A. Double the calcitriol dosage
B. Remind the patient not to miss calcium and calcitriol doses
C. Switch calcium citrate to calcium carbonate
D. Switch sevelamer to lanthanum carbonate
E. Switch to intravenous calcium only

75 A 31-year-old woman presents with worsening galactorrhea that first appeared 6 months ago. Her only notable medical history is Graves disease. Because she was unable to tolerate methimazole and propylthiouracil (both were associated with severe rash), she underwent total thyroidectomy 1 year ago. Her medications are levothyroxine, 100 mcg daily, and an oral contraceptive pill. Other symptoms that she has noticed over the past 6 months include a bifrontal headache, irregular menses, 20-lb (9.0-kg) weight gain, and generalized fatigue.

On physical examination, her blood pressure is 126/70 mm Hg and pulse rate is 58 beats/min. Her height is 65 in (165.1 cm), and weight is 157 lb (71.2 kg) (BMI = 26.1 kg/m^2). She appears well, is alert, and has a healed thyroidectomy scar. There is expressible milky discharge bilaterally but no discrete breast masses. She has no other notable examination findings.

Laboratory test results:

Serum prolactin = 90 ng/mL (4-30 ng/mL) (SI: 3.9 nmol/L [0.17-1.30 nmol/L])
TSH = 494.0 mIU/L (0.5-5.0 mIU/L)
Free T_4 = 0.2 ng/dL (0.8-1.8 ng/dL) (SI: 2.57 pmol/L [10.30-23.17 pmol/L])
Free T_3 = 0.8 pg/mL (2.3-4.2 pg/mL) (SI: 1.23 pmol/L [3.53-6.45 pmol/L])
TPO antibodies = 8.0 IU/L (<2.0 IU/mL) (SI: 8.0 kIU/L [<2.0 kIU/L])
Cortisol (8 AM) = 29.0 µg/dL (5-25 µg/dL) (SI: 717.3 nmol/L [137.9-689.7 nmol/L])
Concurrent ACTH (8 AM) = 30 pg/mL (10-60 pg/mL) (SI: 6.6 pmol/L [2.2-13.2 pmol/L])
IGF-1 = 199 ng/mL (113-297 ng/mL) (SI: 26.1 nmol/L [14.8-38.9 nmol/L])
β-hCG = <3.0 mIU/mL (<3.0 mIU/mL) (SI: <3.0 IU/L [<3.0 IU/L])

Which of the following best explains this patient's presentation?

A. Cushing disease
B. Early pregnancy
C. Mixed GH- and prolactin-secreting pituitary adenoma
D. Primary hypothyroidism
E. TSH-secreting pituitary adenoma

76 A 73-year-old man undergoes evaluation for headaches and vision changes. Brain MRI indicates a 2.3-cm pituitary macroadenoma with optic chiasm compression.

Preoperative laboratory test results:

Prolactin = 16.9 ng/mL (4-23 ng/mL) (SI: 0.73 nmol/L [0.17-1.00 nmol/L])
TSH = 1.55 mIU/L (0.5-5.0 mIU/L)
Free T_4 = 1.10 ng/dL (0.8-1.8 ng/dL) (SI: 14.2 pmol/L [10.30-23.17 pmol/L])
FSH = 24.0 mIU/mL (1.0-13.0 mIU/mL) (SI: 24.0 IU/L [1.0-13.0 IU/L])
LH = 9.7 mIU/mL (1.0-9.0 mIU/mL) (SI: 9.7 IU/L [1.0-9.0 IU/L])
α-Subunit = 0.6 ng/mL (0.1-0.5 ng/mL)
Total testosterone = 115 ng/dL (300-900 ng/dL) (SI: 4.0 nmol/L [10.4-31.2 nmol/L])
Cortisol (8 AM) = 8.7 μg/dL (5-25 μg/dL) (SI: 240.0 nmol/L [137.9-689.7 nmol/L])
ACTH = 49.2 pg/mL (10-60 pg/mL) (SI: 10.8 pmol/L [2.2-13.2 pmol/L])
GH = 1.99 ng/mL (0.01-0.97 ng/mL) (SI: 1.99 μg/L [0.01-0.97 μg/L])
IGF-1 = 100 ng/mL (62-184 ng/mL) (SI: 13.1 nmol/L [8.1-24.1 nmol/L])
Basic metabolic panel, normal

The patient undergoes transsphenoidal surgery. Pathologic evaluation of the tumor specimen indicates positive steroidogenic factor 1 and LH immunohistochemistry expression. The MIB-1 proliferation index is 2.5%.

The patient relocates and is lost to follow-up. Three years later, he presents with osteoporosis after diagnosis of a painful thoracic compression fracture. He has no headaches or vision changes. Medications include oxycodone, 10 mg every 6 hours, and alendronate, 70 mg weekly.

Laboratory test results:

Testosterone = 79 ng/dL (SI: 2.7 nmol/L)
FSH = 27.0 mIU/mL (SI: 27.0 IU/L)
LH = 10.0 mIU/mL (SI: 10.0 IU/L)
Prolactin = 23 ng/mL (SI: 1.0 nmol/L)
Hemoglobin = 10.1 g/dL (13.8-17.2 g/dL) (SI: 101 g/L [138-172 g/L])
Hematocrit = 31% (41%-51%) (SI: 0.31 [0.41-0.51])

Pituitary MRI identifies a 0.4-cm residual tumor in the sella and right cavernous sinus.

DXA scan reveals osteoporosis in the lumbar spine with a T-score of −3.5.

He expresses concern about tumor blood markers in elevated ranges.

Which of the following is the best next step in this patient's management?

A. Recommend radiotherapy for the residual adenoma
B. Recommend transsphenoidal reoperation
C. Start dopamine agonist therapy
D. Start testosterone replacement
E. Wait to initiate therapy until opioids are stopped and hormones are reevaluated

77 A 55-year-old woman with a history of type 2 diabetes presents to the emergency department where she is diagnosed with pneumonia and acute kidney failure. She is admitted to the hospital for further evaluation and management. She has had uncontrolled diabetes for many years, and her most recent hemoglobin A_{1c} value, just before admission, was 10.2% (88 mmol/mol). Her medical history is notable for congestive heart failure, hypertension, dyslipidemia, diabetic neuropathy, and diabetic retinopathy. Her outpatient regimen consists of insulin glargine, 32 units daily; short-acting insulin aspart, 10 units with each meal; and correctional insulin, 1 unit for every 25 mg/dL (1.4 mmol/L) above 150 mg/dL (>8.3 mmol/L). Two months ago, her primary care physician prescribed metformin, 500 mg twice daily, and empagliflozin, 12.5 mg daily. She has no history of cigarette smoking or drug use. She drinks alcohol only periodically.

Laboratory test results:

- Hematocrit = 37.2% (35%-45%) (SI: 0.372 [0.35-0.45])
- Hemoglobin = 10.6 g/dL (12.1-15.1 g/dL) (SI: 106 g/L [121-151 g/L])
- Creatinine = 1.7 mg/dL (0.6-1.1 mg/dL) (SI: 150.3 µmol/L [53.0-97.2 µmol/L])
- Estimated glomerular filtration rate = 31 mL/min per 1.73 m^2 (>60 mL/min per 1.73 m^2)
- Potassium = 4.1 mEq/L (3.5-5.0 mEq/L) (SI: 4.1 mmol/L [3.5-5.0 mmol/L])
- Bicarbonate = 15 mEq/L (21-28 mEq/L) (SI: 15 mmol/L [21-28 mmol/L])
- Anion gap = 10 mEq/L (3-11 mEq/L) (SI: 10 mmol/L [3-11 mmol/L])
- Arterial blood gases
  - pH = 7.28 (7.35-7.45)
  - Po_2 = 76.0 mm Hg (80-100 mm Hg) (SI: 10.1 kPa [10.6-13.3 kPa])
  - Pco_2 = 36 mm Hg (35-45 mm Hg) (SI: 4.8 kPa [4.7-6.0 kPa])

During hospitalization, there have been very few point-of-care glucose measurements: 210 mg/dL (11.7 mmol/L), 185 mg/dL (10.3 mmol/L), and 222 mg/dL (12.3 mmol/L). Because the hospitalist team is concerned about metabolic acidosis, the patient is transferred to the intensive care unit and an endocrine consultation is requested.

Which of the following is the most likely explanation of the patient's acid-base disorder?

A. Acute kidney failure
B. Diabetes-related ketoacidosis
C. Euglycemic diabetic ketoacidosis
D. Metformin-induced acidosis
E. Type 4 renal tubular acidosis

78 A 32-year-old man with type 1 diabetes reports that he and his wife have been unable to conceive after having vaginal intercourse 2 to 3 times weekly without contraception for the past year. Diabetes was diagnosed at age 17 years. He began using an insulin pump 5 years ago, and since that time his hemoglobin A_{1c} values have ranged from 6.5% to 7.1% (48-54 mmol/mol). His continuous glucose monitoring data demonstrate that his glucose management index has been 6.3% to 6.8% with greater than 90% time-in-range. He has had no recorded glucose values less than 54 mg/dL (<3.0 mmol/L) in the past 3 years and no severe hypoglycemic episodes in the past 10 years. He reports no cloudiness to his urine when he has postcoital micturition. He exercises 45 to 60 minutes 5 days weekly and has no notable medical history aside from diabetes. His wife is 29 years old and healthy and has regular menses.

On physical examination, he is lean and muscular. His blood pressure is 128/80 mm Hg, pulse rate is 68 beats/min, and respiratory rate is 12 breaths/min. His height is 68 in (172.7 cm), and weight is 145 lb (65.8 kg) (BMI = 22.0 kg/m^2). He has no gynecomastia. Jugular venous pressure is normal, and heart rate is regular with no gallop. He has no edema. Penis is normal, and testes are 20 cc bilaterally. Bilateral vasa deferentia are palpable. He has no hyperpigmentation or ecchymoses.

Laboratory test results:

- Hematocrit = 44 % (41%-51%) (SI: 0.44 [0.41-0.51])
- Serum total testosterone = 570 ng/dL (264-916 ng/dL) (SI: 19.8 nmol/L [9.2-31.8 nmol/L]) (assay validated and harmonized by the CDC)

Calculated free testosterone, normal
SHBG = 5.1 μg/mL (1.1-6.7 μg/mL) (SI: 45 nmol/L [10-60 nmol/L])
FSH = 2.0 mIU/mL (1.0-13.0 mIU/mL) (SI: 2.0 IU/L [1.0-13.0 IU/L])
LH = 4.0 mIU/mL (1.0-9.0 mIU/mL) (SI: 4.0 IU/L [1.0-9.0 IU/L])
TSH = 1.8 mIU/L (0.5-5.0 mIU/L)
Prolactin = 26 ng/mL (4-23 ng/mL) (SI: 1.1 nmol/L [0.17-1.00 nmol/L])
Seminal fluid analysis = 0 sperm/mL (>15 million/mL)

Findings on transcrotal ultrasonography are normal.

Which of the following is the best initial treatment for this man considering the goal to conceive?

A. Cabergoline
B. Clomiphene
C. L-carnitine and vitamin E
D. Referral to specialist for testicular sperm extraction
E. rhFSH

79 A 71-year-old woman presents for management of type 2 diabetes diagnosed 15 years ago. Her hemoglobin A_{1c} level has hovered around 8.5% (69 mmol/mol) for the past few years. She takes metformin, 500 mg twice daily; empagliflozin, 10 mg daily; and insulin glargine, 20 units daily at bedtime. Her most recent hemoglobin A_{1c} value was 8.2% (61 mmol/mol). She has hypertension, hyperlipidemia, and stage 3 chronic kidney disease. In addition to her diabetes medications, she takes enalapril and rosuvastatin.

To address her elevated hemoglobin A_{1c}, semaglutide, 0.25 mg once weekly, is initiated and subsequently titrated to 1.0 mg weekly. After 3 months of therapy, she reports feeling very good, but she has early satiety and reduced appetite. She has lost 18 lb (8.2 kg) and recognizes she is eating much less than before. She notes that these symptoms have been more pronounced since her last dosage increase to 1.0 mg weekly.

Six weeks after starting the dosage of 1 mg weekly, she contacts the clinic to report an episode of feeling weak and dizzy that occurred while she was walking at the mall. She describes her symptoms as the typical symptoms she notices when her blood glucose is low. However, at that moment, her continuous glucose monitor reading was 115 mg/dL (6.4 mmol/L), and she confirmed her blood glucose was not low with a fingerstick reading (122 mg/dL [6.8 mmol/L]). She sat down and drank a sugary beverage, which she reports made her feel better.

On physical examination, her blood pressure is 108/68 mm Hg, and pulse rate is 82 beats/min. Her height is 66 in (167.6 cm), and weight is 143 lb (64.9 kg) (BMI = 23.1 kg/m^2).

Laboratory test results:

Hemoglobin A_{1c} = 6.9% (4.0%-5.6%) (SI: 52 mmol/mol [20-38 mmol/mol])
Glucose = 145 mg/dL (70-99 mg/dL) (SI: 8.0 mmol/L [3.9-5.5 mmol/L])
Serum creatinine = 2.2 mg/dL (0.6-1.1 mg/dL) (SI: 194.5 μmol/L [53.0-97.2 μmol/L])
Estimated glomerular filtration rate = 23 mL/min per 1.73 m^2 (>60 mL/min per 1.73 m^2)
  Baseline serum creatinine = 1.1 mg/dL (0.6-1.1 mg/dL) (SI: 97.3 μmol/L [53.0-97.2 μmol/L])
  Baseline estimated glomerular filtration rate = 55 mL/min per 1.73 m^2 (>60 mL/min per 1.73 m^2)
Serum urea nitrogen = 52 mg/dL (8-23 mg/dL) (SI: 18.6 mmol/L [2.9-8.2 mmol/L])
Sodium = 145 mEq/L (136-142 mEq/L) (SI: 145 mmol/L [136-142 mmol/L])
Potassium = 5.7 mEq/L (3.5-5.0 mEq/L) (SI: 5.7 mmol/L [3.5-5.0 mmol/L])
Chloride = 109 mEq/L (96-106 mEq/L) (SI: 109 mmol/L [96-106 mmol/L])
Anion gap = 8 mEq/L (3-11 mEq/L) (SI: 8 mmol/L [3-11 mmol/L])
Bicarbonate = 28 mEq/L (21-28 mEq/L) (SI: 28 mmol/L [21-28 mmol/L])
β-Hydroxybutyrate = 3.4 mg/dL (<3.0 mg/dL) (SI: 326.6 μmol/L [<288.2 μmol/L])
Lactic acid = 8.4 mg/dL (5.4-20.7 mg/dL) (SI: 0.93 mmol/L [0.6-2.3 mmol/L])

Which of the following is most likely responsible for this patient's current clinical condition?

A. Empagliflozin
B. Enalapril
C. Insulin glargine
D. Metformin
E. Semaglutide

80 A 60-year-old woman with a 40-year history of type 1 diabetes is seen for follow-up. Diabetes control has improved after being on automated insulin delivery for the last 2 years. She has stable nonproliferative retinopathy and microalbuminuria. She does not have known atherosclerotic cardiovascular disease. However, cardiac CT obtained by her primary care physician last year showed evidence of subclinical atherosclerosis with a coronary artery calcium score of 85 (88th percentile for age-, sex-, and race-matched individuals without cardiovascular disease or diabetes).

Laboratory test results:

Total cholesterol = 190 mg/dL (<200 mg/dL [optimal]) (SI: 4.92 mmol/L [<5.18 mmol/L])
Triglycerides = 105 mg/dL (<150 mg/dL [optimal]) (SI: mmol/L [<1.70 mmol/L])
HDL cholesterol = 85 mg/dL (>60 mg/dL [optimal]) (SI: 2.20 mmol/L [>1.55 mmol/L])
LDL cholesterol = 84 mg/dL (<100 mg/dL [optimal]) (SI: 2.18 mmol/L [<2.59 mmol/L])

She is advised to start moderate-intensity statin therapy for primary prevention, but she thinks this is unnecessary, especially in view of her high "good cholesterol."

Which of the following tests might better reflect this patient's increased risk of atherosclerotic cardiovascular disease?

A. Apolipoprotein B measurement
B. Cholesterol efflux capacity
C. LDL particle number and size by nuclear magnetic resonance spectroscopy
D. Lipoprotein (a) measurement
E. Remnant lipoprotein cholesterol measurement

81 Six months ago, a 68-year-old woman underwent total thyroidectomy for toxic multinodular goiter. She developed postsurgical hypoparathyroidism after surgery. She has no history of kidney stones. Over the last month, she has experienced tingling in her hands and feet and around her mouth. She has been treated with calcitriol, 0.50 mcg twice daily; elemental calcium carbonate, 1200 mg twice daily; and vitamin D, 1000 international units daily (25 mcg daily). She also eats 1 serving each of yogurt and cheese and green leafy vegetables per day. She is frustrated by the amount of calcium she needs to take.

On physical examination, her blood pressure is 148/86 mm Hg and pulse rate is 73 beats/min. Her height is 64.9 in (165.0 cm), and weight is 142 lb (64.4 kg) (BMI = 23.7 kg/m^2). Chvostek and Trousseau signs are absent. There are no palpable neck masses. Costovertebral angle tenderness is absent. The rest of the examination findings are unremarkable.

Laboratory test results:

Calcium = 8.3 mg/dL (8.2-10.2 mg/dL) (SI: 2.1 mmol/L [2.1-2.6 mmol/L])
Phosphate = 4.3 mg/dL (2.3-4.7 mg/dL) (SI: 1.4 mmol/L [0.7-1.5 mmol/L])
Creatinine = 1.0 g/dL (0.6-1.1 mg/dL) (SI: 88.4 μmol/L [53.0-97.2 μmol/L])
Albumin = 4.0 g/dL (3.5-5.0 g/dL) (SI: 40 g/L [35-50 g/L])
Ionized calcium = 4.73 mg/dL (4.60-5.08 mg/dL) (SI: 1.18 mmol/L [1.16-1.31 mmol/L])
Intact PTH = <5 pg/mL (10-65 pg/mL) (SI: <0.5 pmol/L [1.1-6.9 pmol/L])
25-Hydroxyvitamin D = 22 ng/mL (30-80 ng/mL [optimal]) (SI: 54.9 nmol/L [74.9-199.7 nmol/L])
Magnesium = 1.8 mg/dL (1.5-2.3 mEq/L) (SI: 0.74 mmol/L [0.6-0.9 mmol/L])

Alkaline phosphatase = 75 U/L (50-120 U/L) (SI: 1.25 μkat/L [0.84-2.00 μkat/L])
24-hour urinary calcium = 388 mg/24 h (100-300 mg/24 h) (SI: 9.7 mmol/d [2.5-7.5 mmol/d])

Kidney ultrasonography does not show nephrocalcinosis.

Which of the following is the best next step in treating this patient's hypoparathyroidism?

A. Add hydrochlorothiazide
B. Decrease the calcitriol dosage
C. Increase the calcium carbonate dosage
D. Start magnesium
E. Switch from calcium carbonate to calcium citrate

82 A 38-year-old woman presents for follow-up of stage I papillary thyroid cancer (diffuse sclerosing variant that was first diagnosed 2 years ago) and postoperative hypothyroidism. Her initial treatment included total thyroidectomy with central and bilateral lateral neck dissections. Pathology demonstrated multifocal papillary thyroid carcinoma, the largest focus 0.2 cm, with tumor involving both lobes and the isthmus, and extensive Hashimoto thyroiditis. No extrathyroidal extension was noted, and all surgical margins were negative. A total of 28 of 98 lymph nodes were positive for metastasis without extranodal extension. The largest lymph node measured 2 cm in size. Two months after surgery, she was treated with 102 mCi ^{131}I. Her posttherapy whole-body scan demonstrated scant residual normal and/or malignant thyroid tissue in the left thyroid bed, and no evidence of metastatic disease.

Today the patient reports that she has been feeling well and has had no interval events. She is currently taking generic levothyroxine, 175 mcg daily, on an empty stomach in the morning.

On physical examination, her blood pressure is 100/74 mm Hg and pulse rate is 66 beats/min. Her height is 69 in (175 cm), and weight is 240 lb (108.9 kg) (BMI = 35.4 kg/m^2). Anterior and bilateral lateral neck scars are visible. There is no mass in the thyroid bed and no cervical lymphadenopathy.

Laboratory test results:

Measurement	2 Years ago	1 Year ago	Today	Reference range
TSH	0.29 mIU/L	0.18 mIU/L	0.5 mIU/L	0.5-5.0 mIU/L
Free T_4	1.6 ng/dL (SI: 20.59 pmol/L)	1.5 ng/dL (SI: 19.31 pmol/L)	1.4 ng/dL (SI: 18.02 pmol/L)	0.8-1.8 ng/dL (SI: 10.30-23.17 pmol/L)
Thyroglobulin	<0.1 ng/mL (SI: <0.1 μg/L)	<0.1 ng/mL (SI: <0.1 μg/L)	<0.1 ng/mL (SI: <0.1 μg/L)	3-42 ng/mL (SI: 3-42 μg/L)
Thyroglobulin antibodies	49.0 IU/mL (SI: 49.0 kIU/L)	13.0 IU/mL (SI: 13.0 kIU/L)	10.0 IU/mL (SI: 10.0 kIU/L)	≤4.0 IU/mL (SI: ≤4.0 kIU/L)

Neck ultrasonography performed today demonstrates no suspicious nodules, lymph nodes, or lymphadenopathy.

Which of the following is the most appropriate next step in this patient's evaluation?

A. Change to branded levothyroxine preparation
B. Continue generic levothyroxine but increase dosage to target serum TSH <0.1 mIU/L
C. Perform rhTSH-stimulated radioactive iodine (^{123}I) whole-body scanning
D. Repeat measurement of serum thyroglobulin after pretreatment with a heterophile-blocking agent
E. Repeat neck ultrasonography and measure serum thyroglobulin and thyroglobulin antibodies in 12 months

83 A 36-year-old man is referred following evaluation in the neurology clinic for possible temporal lobe seizures. Head MRI shows a 1.5-cm pituitary macroadenoma extending into the right cavernous sinus but without compression of the optic chiasm. He has no notable medical history and takes no regular medication. There is no family history of pituitary disease. He has not had headaches, sweating, change in appearance, change in weight, weakness, or bruising.

On physical examination, his blood pressure is 136/72 mm Hg and pulse rate is 68 beats/min. His height is 69 in (176 cm), and weight is 181 lb (82 kg) (BMI = 26.4 kg/m^2). There are no clinical signs of cortisol or GH excess. Visual fields are full to confrontation testing.

Laboratory test results (sample drawn at 8 AM):

Cortisol = 18.2 µg/dL (5-25 µg/dL) (SI: 503 nmol/L [137.9-689.7 nmol/L])
Free T_4 = 0.95 ng/dL (0.8-1.8 ng/dL) (SI: 12.2 pmol/L [10.30-23.17 pmol/L])
TSH = 2.26 mIU/L (0.5-5.0 mIU/L)
Prolactin = 21 ng/mL (4-23 ng/mL) (SI: 0.91 nmol/L [0.17-1.00 nmol/L])
LH = 2.4 mIU/mL (1.0-9.0 mIU/mL) (SI: 2.4 IU/L [1.0-9.0 IU/L])
FSH = 3.4 mIU/mL (1.0-13.0 mIU/mL) (SI: 3.4 IU/L [1.0-13.0 IU/L])
Testosterone = 466.9 ng/dL (300-900 ng/dL) (SI: 16.2 nmol/L [10.4-31.2 nmol/L])

Which of the following is the best next step in this patient's management?

A. Measure serum IGF-1
B. Perform a 1-mg overnight dexamethasone-suppression test
C. Perform visual field testing
D. Recommend transsphenoidal surgery
E. Repeat MRI in 6 to 12 months

84 A 67-year-old woman with type 2 diabetes presents to the diabetes clinic for evaluation and management of hyperglycemia. Her medical history is notable for papillary thyroid cancer (treated successfully 25 years ago), dyslipidemia, and celiac disease. Her current medication regimen is metformin, 1000 mg twice daily; empagliflozin, 25 mg daily; and levothyroxine, 100 mcg daily. Type 2 diabetes was diagnosed 4 years ago, and her glycemic control has always been excellent. Her point-of-care hemoglobin A_{1c} measurement at today's visit is 7.5% (58 mmol/mol), despite her efforts of modifying her diet and regularly exercising almost 45 minutes per day. The patient states that she checks glucose values in the morning, and they range between 90 and 120 mg/dL (5.0-6.7 mmol/L).

On physical examination, her height is 64 in (162.6 cm) and weight is 140 lb (63.5 kg) (BMI = 24.0 kg/m^2). Examination findings are normal.

Which of the following medications should be added to this patient's current regimen as the best next step in her management?

A. Basal insulin
B. DPP-4 inhibitor
C. Prandial insulin
D. Sulfonylurea
E. No additional medication

85 A 32-year-old woman with no notable medical history presents with low back and left hip pain that has been worsening for the last 2 years. She spends long hours at a desk job. She initially tried over-the-counter pain medications with temporary relief. Several weeks ago, she developed right wrist pain and started using a wrist splint with minimal improvement. She has no family history of rheumatoid arthritis. The patient's roommate recently moved out because of her mood swings and erratic behavior. Her only medication is ibuprofen, 400 to 800 mg every 8 hours as needed. She takes no herbal supplements or multivitamins. She used to drink 10 alcoholic beverages per week, but she stopped drinking alcohol 3 years ago. She does not use intravenous drugs.

On physical examination, her blood pressure is 120/70 mm Hg, and pulse rate is 96 beats/min. Her height is 65 in (165.1 cm), and weight is 107 lb (48.5 kg) (BMI = 17.8 kg/m^2). She appears her stated age and is slightly disheveled. She is wearing a wrist splint. Her lips appear chapped. She has no palpable neck masses. Very mild tremors of her outstretched hands are noted, and digits of both hands appear enlarged with bony prominences. There is no spine or hip tenderness.

Laboratory test results:

Total calcium = 9.1 mg/dL (8.2-10.2 mg/dL) (SI: 2.3 mmol/L [2.1-2.6 mmol/L])
25-Hydroxyvitamin D = 32 ng/mL (30-80 ng/mL [optimal]) (SI: 79.9 nmol/L [74.9-199.7 nmol/L])
PTH = 72 pg/mL (10-65 pg/mL) (SI: 7.6 pmol/L [1.1-6.9 pmol/L])
Phosphate = 2.5 mg/dL (2.3-4.7 mg/dL) (SI: 0.8 mmol/L [0.7-1.5 mmol/L])
Albumin = 4.0 g/dL (3.5-5.0 g/dL) (SI: 40 g/L [35-50 g/L])
Creatinine = 0.8 mg/dL (0.6-1.1 mg/dL) (SI: 70.7 μmol/L [53.0-97.2 μmol/L])
Alkaline phosphatase = 240 U/L (50-120 U/L) (SI: 4.00 μkat/L [0.84-2.00 μkat/L])
TSH = 2.05 mIU/L (0.5-5.0 mIU/L)
Osteocalcin = 64 ng/mL (9.0-42.0 ng/mL) (SI: 64 μg/L [9.0-42.0 μg/L])
Liver function, normal

Radiographs of the spine and pelvis show diffuse osteosclerosis. Hand x-rays reveal periosteal new bone at the proximal and middle phalanges. DXA documents the following values:

Lumbar spine Z score = +5.5
Femoral neck Z score = +4.2
Total hip Z score = +3.8

Bone scans show diffusely increased uptake at the spine, hips, shoulders, knees, and hands.

Based on the clinical findings, measuring which of the following would be the best next step?

A. IGF-1
B. Plasma fluoride
C. Reactive hepatitis C antibodies
D. Rheumatoid factor
E. Serum tryptase

86 A 32-year-old woman has a history of panhypopituitarism after transsphenoidal resection of a pituitary macroadenoma with pituitary apoplexy. She reports regular menstrual cycles before pituitary resection. Her primary gynecologist prescribed an oral combined hormonal contraceptive pill after pituitary surgery. She has no pelvic pain and no history of pelvic infection or pelvic surgery. The patient and her female partner would like to build their family using donor sperm, and the patient would like to be both the egg source and carry the pregnancy. Ovarian reserve test results are normal, with an antimullerian hormone concentration of 5.02 ng/mL (35.9 pmol/L) and antral follicle count of 17. Findings on sonohysterography with saline are normal, with bilaterally patent fallopian tubes noted.

Which of the following is the best next step in this patient's treatment?

A. Clomiphene citrate
B. Human menopausal gonadotropins (hMG)
C. Human menopausal gonadotropins (hMG) followed by hCG
D. Letrozole
E. Oocyte donation and in vitro fertilization

87 A 49-year-old woman with a medical history of major depressive disorder is referred to discuss weight management.

She underwent a laparoscopic sleeve gastrectomy for weight loss 5 years ago and lost approximately 25% of her body weight. She was able to maintain this weight until about a year ago when she slowly started regaining weight secondary to changes in her regimen of mood-modulating medications and difficulty adhering to a physical activity routine.

Weekly semaglutide injections were started 6 months ago, which she uptitrated per protocol. She reports good appetite suppression on the medication, fullness when she eats, and no adverse effects. She is concerned about her appearance and wishes to lose another 30 lb (13.6 kg). When semaglutide was initiated, her weight was 235 lb (106.6 kg [BMI = 42.8 kg/m^2]). Three months ago, she weighed 207 lb (93.9 kg [BMI = 37.7 kg/m^2]), and today she weighs 168 lb (76.2 kg [BMI = 30.6 kg/m^2]). In total, she has lost 67 lb (30.4 kg), or 28.5% of her body weight, since initiating semaglutide. She also reports that her menstrual periods are more irregular and that she has only had a menstrual period for 3 of the last 5 months.

Current oral medications are a multivitamin with minerals twice daily; vitamin D_3, 2000 international units daily (50 mcg daily); vitamin B_{12}, 1000 mcg daily; modafinil, 200 mg daily; aripiprazole, 10 mg daily; and prazosin, 2 mg daily.

On physical examination, she has dry skin and coarse hair. She has loose skin overlying the abdomen. She is able to get on and off the examination table without difficulty. She has no peripheral edema.

Laboratory test results:

Fasting glucose = 91 mg/dL (70-99 mg/dL) (SI: 5.1 mmol/L [3.9-5.5 mmol/L])
TSH = 3.0 mIU/L (0.5-5.0 mIU/L)

Given this patient's rapid and ongoing weight loss, which of the following is the best next step?

A. Discontinue semaglutide
B. Measure 24-hour urinary protein excretion
C. Measure serum FSH, LH, and estradiol
D. Perform a double contrast upper gastrointestinal fluoroscopic study
E. Refer to behavioral psychology

88 A 29-year-old woman with history of polycystic ovary syndrome is admitted to the hospital with acute shortness of breath. CT angiography detects a pulmonary embolism, and a large subdiaphragmatic mass is partially visualized. Subsequent abdominal MRI reveals a right suprarenal mass measuring up to 14 cm (*see image, arrow*). There is near-occlusive tumor extension into the inferior vena cava, invasion of hepatic segment 8, and possible involvement of the upper pole of the right kidney. No metastases or abnormal lymph nodes are identified on abdominal imaging.

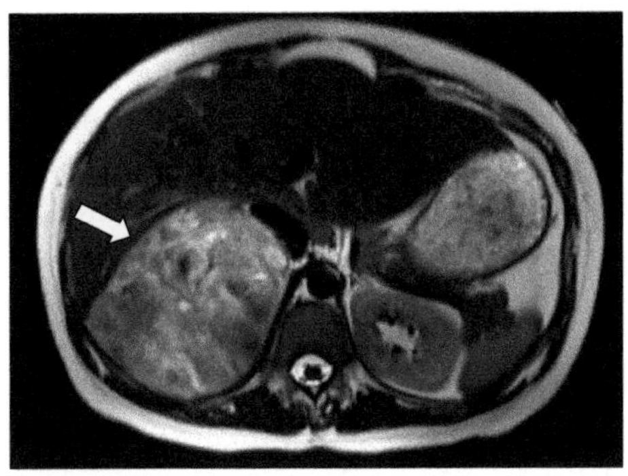

On physical examination, her blood pressure is 134/75 mm Hg and pulse rate is 88 beats/min. Her height is 65 in (165.1 cm), and weight is 172 lb (78.0 kg) (BMI = 28.6 kg/m^2). She is awake and breathing comfortably on oxygen delivered by nasal canula. Her oxygen saturation is 98%. She has mild plethora and a dorsocervical fat pad. There are no bruises or hirsutism. She has facial acne, and dark purple striae are present on her abdomen.

Laboratory test results obtained during the hospitalization:

Serum cortisol (8 AM) = 34.3 μg/dL (5-25 μg/dL) (SI: 946.3 nmol/L [137.9-689.7 nmol/L])
ACTH (8 AM) = 2 pg/mL (10-60 pg/mL) (SI: 0.4 pmol/L [2.2-13.2 pmol/L])
Serum aldosterone = <4 ng/dL (4-21 ng/dL) (SI: <111.0 pmol/L [111.0-582.5 pmol/L])
Plasma renin activity = 0.2 ng/mL per h (0.6-4.3 ng/mL per h)
DHEA-S = 2885 μg/dL (44-332 μg/dL) (SI: 78.18 μmol/L [1.19-9.00 μmol/L])
Total testosterone = 197 ng/dL (8-60 ng/dL) (SI: 6.8 nmol/L [0.3-2.1 nmol/L])
Urinary cortisol = 718 μg/24 h (4-50 μg/24 h) (SI: 1982 nmol/d [11-138 nmol/d])

Following appropriate therapy for pulmonary embolism, which of the following is the best next step in this patient's management?

A. Admit to the intensive care unit and initiate etomidate
B. Perform FDG PET-CT
C. Refer for right adrenalectomy
D. Start metyrapone
E. Start mitotane

89

A 51-year-old woman is evaluated for episodic tachycardia, dyspnea, and dizziness for which she has presented multiple times to the emergency department over the past 5 years. During one of the episodes, she was documented to be in atrial fibrillation with rapid ventricular response. Her medical history includes hearing deficit, hyperactive behavior, depression, and mild developmental delay. Her only medications are diltiazem and atenolol, which were prescribed following her episode of atrial fibrillation. She was adopted and does not know her family history.

On physical examination, her blood pressure is 128/76 mm Hg and she has an irregularly irregular heart rate of 82 beats/min. Her height is 62 in (157.48 cm), and weight is 124 lb (56.2 kg) (BMI = 22.7 kg/m^2). She appears fidgety and anxious. There is no exophthalmos, and visual fields are normal to confrontation. She has a bilateral diffusely symmetric goiter that is approximately 2 times the normal size with no palpable nodules. Cardiac auscultation is notable for tachycardia. Her pulmonary and abdominal examinations are unrevealing, and there is no peripheral edema. She has a fine tremor of her outstretched hands bilaterally, and her deep tendon reflexes are brisk.

Laboratory test results:

Serum TSH = 1.89 mIU/L (0.5-5.0 mIU/L)
Free T_4 = 3.8 ng/dL (0.8-1.8 ng/dL) (SI: 48.9 pmol/L [10.30-23.17 pmol/L])
Total T_3 = 258 ng/dL (70-200 ng/dL) (SI: 3.97 nmol/L [1.08-3.08 nmol/L])
TRAb = ≤1.75 IU/L (≤1.75 IU/L)
Thyroid-stimulating immunoglobulin = ≤120% of basal activity
TPO antibodies = <2.0 IU/mL (<2.0 IU/mL) (SI: <2.0 kIU/L [<2.0 kIU/L])

Which of the following is this patient's most likely diagnosis?

A. Familial dysalbuminemic hyperthyroxinemia
B. Presence of heterophilic antibodies against T_3 and T_4
C. Presence of heterophilic antibodies against TSH
D. Resistance to thyroid hormone (β subtype)
E. TSH-secreting pituitary adenoma

90

A 32-year-old woman (G4, P3, L3) is currently in the 14th week of gestation. She had gestational diabetes in 2 of her 3 previous pregnancies. Despite use of basal insulin during her last pregnancy, her glycemic control was still suboptimal. Subsequently, the pregnancy was complicated by preeclampsia, necessitating cesarean delivery at 34 weeks' gestation, and the baby was admitted to the neonatal intensive care unit.

She recently completed a 100-g oral glucose tolerance test and has been referred to endocrinology:

Fasting glucose = 109 mg/dL (<95 mg/dL) (SI: 6.0 mmol/L [<5.3 mmol/L])
1-Hour glucose = 175 mg/dL (<180 mg/dL) (SI: 9.7 mmol/L [<10.0 mmol/L])
2-Hour glucose = 160 mg/dL (<155 mg/dL) (SI: 8.9 mmol/L [<8.6 mmol/L])
3-Hour glucose = 131 mg/dL (<140 mg/dL) (SI: 7.3 mmol/L [<7.8 mmol/L])

During the office visit, she shares that during her last pregnancy, she often skipped insulin injections because she has a fear of needles. She has had no nausea, vomiting, or bleeding with the current pregnancy. Her blood pressure is 110/78 mm Hg.

Various dietary modifications and physical activities for the management of gestational diabetes are discussed. She inquires about the role of oral pharmacotherapy instead of insulin therapy in the management of diabetes.

Which of the following is the appropriate management strategy if she needs pharmacotherapy during this pregnancy?

A. Add glipizide
B. Add metformin
C. Add semaglutide
D. Start insulin pump therapy
E. Start subcutaneous insulin therapy and teach her partner how to administer it

91 A 75-year-old man being treated for lung cancer presents with severe hypercalcemia (15.2 mg/dL [3.8 mmol/L]). He also has congestive heart failure and baseline stage 3b chronic kidney disease. He receives 2 L of normal saline over the course of a day and then zoledronate, 4 mg intravenously. Intravenous fluids are continued, but he requires oxygen, 4 L by nasal cannula.

On physical examination on hospital day 5, his blood pressure is 146/84 mm Hg and pulse rate is 78 beats/min (standing), and blood pressure is 141/82 mm Hg and pulse rate is 86 beats/min (sitting). Oxygen saturation is 92% while receiving oxygen (4 L by nasal cannula). He is awake but mildly dyspneic. Oral mucosa is moist. His chest is barrel-shaped, and he has basilar rales but no dullness to percussion. His heart has a regular rate and rhythm and a systolic murmur. He has mild jugular venous distention. Bedside ultrasonography shows elevated neck veins. He has 2+ peripheral edema.

Hypercalcemia persists and another dose of zoledronate, 4 mg, is administered 10 days later.

Laboratory test results 1 week after the second dose of zoledronate:

Calcium = 12.1 mg/dL (8.2-10.2 mg/dL) (SI: 3.0 mmol/L [2.1-2.6 mmol/L])
Albumin = 2.8 g/dL (3.5-5.0 g/dL) (SI: 28 g/L [35-50 g/L])
Phosphate = 4.3 mg/dL (2.3-4.7 mg/dL) (SI: 1.4 mmol/L [0.7-1.5 mmol/L])
Alkaline phosphatase = 88 U/L (50-120 U/L) (SI: 1.47 μkat/L [0.84-2.00 μkat/L])
Creatinine = 2.1 mg/dL (0.6-1.1 mg/dL) (SI: 185.6 μmol/L [53.0-97.2 μmol/L])
Creatinine clearance = 20 mL/min
Intact PTH = <6 pg/mL (10-65 pg/mL) (SI: <0.6 pmol/L [1.1-6.9 pmol/L])
PTHrP = 3.5 pmol/L (<2.0 pmol/L)
25-Hydroxyvitamin D = 36 ng/mL (30-80 ng/mL [optimal]) (SI: 89.9 nmol/L [74.9-199.7 nmol/L])
1,25-Dihydroxyvitamin D_3 = 10 pg/mL (16-65 pg/mL) (SI: 26.0 pmol/L [41.6-169.0 pmol/L])

Which of the following is the best next step in managing this patient's hypercalcemia?

A. Double the intravenous fluid rate
B. Repeat zoledronate dosing
C. Treat with calcitonin
D. Treat with cinacalcet
E. Treat with denosumab

92 A 24-year-old transman (preferred pronouns he/him) has recently moved from another part of the country and has been referred for continuation of gender-affirming hormone therapy. He has been living as a man since age 16 years and has been taking testosterone since age 17. He lives in a steady polyamorous relationship with a transwoman and a ciswoman. All 3 members of this polyamorous relationship occasionally have sex with cismen and ciswomen outside of the core polyamorous group. When asked, he reports that he might like to conceive a child in the next 2 to 3 years.

He has been amenorrheic with no vaginal bleeding since age 19 years. He has regular gynecological examinations and had a normal result on a Papanicolaou test 6 months ago. He underwent bilateral mastectomy (top surgery) 3 years ago. His only medication is testosterone, 75 mg intramuscularly weekly.

On physical examination, his blood pressure is 128/80 mm Hg, pulse rate is 68 beats/min, and respiratory rate is 12 breaths/min. His height is 66 in (167.6 cm), and weight is 143 lb (64.9 kg) (BMI = 23.1 kg/m^2). Lungs are clear to auscultation with tidal volume breaths. No breast tissue is palpable, and he has well-healed mastectomy scars. He has no edema. Per recent gynecology note, he has normal introitus, vaginal mucosa, and cervical os.

Laboratory test results (sample drawn 3.5 days after injection):
Hematocrit = 45% (35-45%) (SI: 0.45 [0.35-0.45])
Serum total testosterone = 475 ng/dL (264-916 ng/dL) (SI: 16.5 nmol/L [9.2-31.8 nmol/L]) (assay validated and harmonized by the CDC)

Which of the following is the best contraceptive for this patient?

A. Condoms for vaginal intercourse with a sex partner with a penis
B. Increased dosage of testosterone
C. Oral micronized progesterone
D. Subcutaneous etonogestrel implant
E. No additional contraceptive measures are necessary

93 A 49-year-old man is evaluated in the emergency department for palpitations associated with lightheadedness. His wearable smartwatch had alerted him of abnormal pulse morphology compatible with atrial fibrillation. Electrocardiography documents sinus tachycardia. Bloodwork indicates normal complete blood cell count and chemistry panel results and a TSH value of 5.9 mIU/L (0.5-5.0 mIU/L). His symptoms spontaneously resolve. He is advised to start levothyroxine, 25 mcg daily, and metoprolol, 50 mg twice daily, and is evaluated in the endocrine clinic the following week.

On physical examination, he appears anxious. His blood pressure is 150/83 mm Hg and pulse rate is 91 beats/min. His height is 68 in (172.7 cm), and weight is 155 lb (70.3 kg) (BMI = 23.6 kg/m^2). Examination findings are remarkable for fine hand tremor and mobile nontender thyromegaly without nodules or neck lymphadenopathy.

He reports being in his usual state of health until a few months ago when he started to experience tremor and occasional palpitations. He had lost 4.4 lb (2 kg), which he attributes to eating better and engaging in regular exercise. He takes a multivitamin daily and the medications prescribed in the emergency department. He has no family history of thyroid disease.

Laboratory test results (drawn at 8 AM):
TSH = 6.01 mIU/L (0.5-5.0 mIU/L)
Free T_4 = 2.5 ng/dL (0.8-1.8 ng/dL) (SI: 32.2 pmol/L [10.30-23.17 pmol/L])
Free T_3 = 5.1 pg/mL (2.3-4.2 pg/mL) (SI: 7.83 pmol/L [3.53-6.45 pmol/L])
α-Subunit = 1.5 ng/mL (<0.86 µg/L) (SI: 1.5 µg/L [<0.86 µg/L])
Prolactin = 10 ng/mL (4-23 ng/mL) (SI: 0.43 pmol/L [3.0-20.0 pmol/L])
IGF-1 = 228 ng/mL (91-246 ng/mL) (SI: 29.9 nmol/L [11.9-32.2 nmol/L])
Cortisol = 12 µg/dL (5-25 µg/dL) (SI: 331.1 nmol/L [137.9-689.7 nmol/L])
ACTH = 35 pg/mL (10-60 pg/mL) (SI: 7.7 pmol/L [2.2-13.2 pmol/L])

Pituitary-directed MRI identifies a 1.0-cm intrasellar pituitary adenoma.

In addition to stopping levothyroxine, which of the following is the best next step in this patient's management?

A. Refer to genetics to evaluate for thyroid receptor β pathogenic variants
B. Refer to neurosurgery for tumor removal
C. Repeat thyroid function tests in 4 weeks
D. Start methimazole, 5 mg twice daily
E. Start octreotide, 50 mcg 3 times daily

94 A 26-year-old woman (G0, P0) with a 10-year history of type 1 diabetes, hypoglycemia unawareness, peripheral neuropathy, and nonproliferative diabetic retinopathy presents for preconception planning. She has used a hybrid closed-loop insulin pump for the past 2 years. She would like to become pregnant within the next 6 months. Her menses are generally regular, but she does occasionally skip months. Her most recent eye examination was 18 months ago. She has no history of cardiovascular disease or gastroparesis. She is currently using barrier contraception. She exercises 5 times per week for 30 minutes.

On physical examination, her blood pressure is 112/72 mm Hg and pulse rate is 68 beats/min. Her height is 66 in (167.6 cm), and weight is 152 lb (68.9 kg) (BMI = 24.5 kg/m^2). Her feet have mild decreased sensation to monofilament testing to the base of her toes. There is normal vibration sense, ankle reflexes, and distal pulses.

Laboratory test results:

Hemoglobin A_{1c} = 6.8% (4.0%-5.6%) (50.8 mmol/mol [20-38 mmol/mol])
Creatinine = 0.72 mg/dL (0.6-1.1 mg/dL) (SI: 63.6 µmol/L [53.0-97.2 µmol/L])
Urinary albumin-to-creatinine ratio = 35 mg/g creat (<30 mg/g creat)
Total cholesterol = 153 mg/dL (<200 mg/dL [optimal]) (SI: 3.96 mmol/L [<5.18 mmol/L])
LDL cholesterol = 80 mg/dL (<100 mg/dL [optimal]) (SI: 2.07 mmol/L [<2.59 mmol/L])
TSH = 0.5 mIU/L (0.5-5.0 mIU/L)
Celiac screen, negative

In addition to improved glycemic control, which of the following referrals is the best next step?

A. Diabetes education to learn about multiple daily insulin injections in anticipation of stopping hybrid closed-loop insulin pump therapy
B. Nutrition for weight management to improve fertility
C. Nephrology in anticipation of worsening albuminuria in pregnancy
D. Ophthalmology in anticipation of worsening retinopathy in pregnancy
E. Reproductive endocrinology in anticipation of infertility

95 A 56-year-old man presents for evaluation of mild hypercalcemia noted on recent laboratory testing ordered by his primary care physician. He has no polyuria, polydipsia, constipation, fatigue, myalgias, or difficulty concentrating. He has never passed a kidney stone. He states that he has had mildly elevated calcium levels for many years. He consumes 2 to 3 servings of dietary calcium daily. Medical history is notable for hypertension, a 20-year history of bipolar disease (stable on lithium), and asthma. There is no family history of kidney stones, hypercalcemia, or bipolar disorder. Medications include an albuterol inhaler as needed, amlodipine, and lithium carbonate.

On physical examination, his blood pressure is 128/72 mm Hg and pulse rate is 82 beats/min. His height is 68 in (172.7 cm), and weight is 170 lb (77.1 kg) (BMI = 25.8 kg/m^2). Examination findings are unremarkable, and there are no palpable neck masses.

Laboratory test results:

Total calcium = 11.4 mg/dL (8.2-10.2 mg/dL) (SI: 2.9 mmol/L [2.1-2.6 mmol/L])
Ionized calcium = 5.63 mg/dL (4.60-5.08 mg/dL) (SI: 1.41 mmol/L [1.15-1.27 mmol/L])
25-Hydroxyvitamin D = 32 ng/mL (30-80 ng/mL [optimal]) (SI: 79.9 nmol/L [74.9-199.7 nmol/L])
PTH = 112 pg/mL (10-65 pg/mL) (SI: 11.9 pmol/L [1.1-6.9 pmol/L])
Phosphate = 2.5 mg/dL (2.3-4.7 mg/dL) (SI: 0.8 mmol/L [0.7-1.5 mmol/L])
Albumin = 4.0 g/dL (3.5-5.0 g/dL) (SI: 40 g/L [35-50 g/L])
Creatinine = 0.8 mg/dL (0.7-1.3 mg/dL) (SI: 70.7 µmol/L [61.9-114.9 µmol/L])
Alkaline phosphatase = 125 U/L (50-120 U/L) (SI: 2.09 µkat/L [0.84-2.00 µkat/L])
TSH = 2.05 mIU/L (0.5-5.0 mIU/L)
Urinary calcium excretion = 86 mg/24 h (100-300 mg/24 h) (SI: 2.2 mmol/d [2.5-7.5 mmol/d])

Neck exploration reveals 2 parathyroid adenomas that are excised with biochemical cure. Twelve months after surgery, he presents with recurrent hypercalcemia (calcium = 10.9 mg/dL [2.7 mmol/L]). Findings on parathyroid technetium ^{99m}Tc technetium sestamibi imaging are normal.

Which of the following is the best next step in this patient's care?

A. Order genetic testing for *CASR* pathogenic variants
B. Resect 2 remaining parathyroid glands with forearm autotransplant of one-half gland
C. Start cinacalcet
D. Stop dietary calcium
E. Stop lithium and remeasure calcium in 6 months

96 A 56-year-old woman with type 2 diabetes and hypertension is referred by her primary care provider for evaluation and management of thyroid cancer and postoperative hypothyroidism. She has a history of American Joint Committee on Cancer stage I papillary thyroid cancer, first diagnosed 8 years ago during a workup for primary hyperparathyroidism. She previously underwent left inferior parathyroidectomy and total thyroidectomy due to large bilateral thyroid nodules, with surgical pathology demonstrating a 2-g parathyroid adenoma and a 0.4-cm papillary thyroid cancer confined to the right thyroid lobe with negative surgical margins, multiple benign thyroid nodules, and 1 benign lymph node. She did not receive postoperative radioactive iodine therapy.

The patient has not seen an endocrinologist for many years but continued taking levothyroxine as prescribed. She recently established care with a new primary care physician who ordered routine laboratory testing:

Hemoglobin A_{1c} = 6.9% (4.0%-5.6%) (52 mmol/mol [20-38 mmol/mol])
Basic metabolic panel, normal
TSH = 0.5 mIU/L (0.5-5.0 mIU/L)
Thyroglobulin (immunometric assay) = 250 ng/mL (SI: 250 µg/L)
Thyroglobulin antibodies = <4.0 IU/mL (≤4.0 IU/mL) (SI: <4.0 kIU/L [≤4.0 kIU/L])
Repeated serum thyroglobulin (liquid chromatography/tandem mass spectrometry) = 228 ng/mL (SI: 228 µg/L)

The patient reports some difficulty swallowing and discomfort in her upper neck radiating to her left ear when she attempts to swallow. She has had no voice changes, cough, or shortness of breath. Current medications include metformin, empagliflozin, levothyroxine, and hydrochlorothiazide.

On physical examination, her blood pressure is 125/79 mm Hg and pulse rate is 89 beats/min. Her height is 64 in (163 cm), and weight is 181 lb (82.1 kg) (BMI = 31.1 kg/m^2). There is an anterior neck scar. There are no nodules in the thyroid bed or cervical lymphadenopathy, but a 3-cm, firm mass is palpated in the upper neck near the midline.

Neck ultrasonography and neck CT with contrast are performed (*see images*).

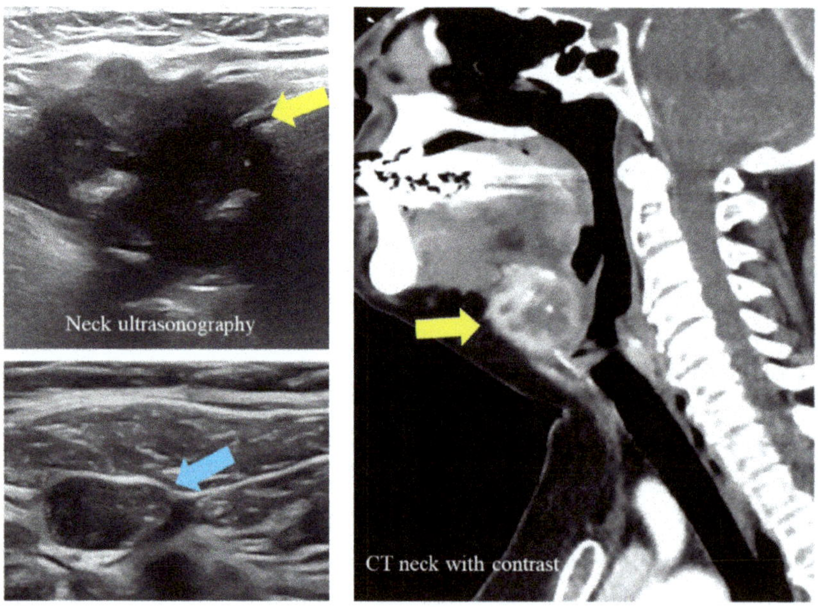

Left panel, Neck ultrasonography showing no mass or nodule in the thyroid bed, a midline zone 1 lesion (*yellow arrow*), and a left zone 3 lymph node (*blue arrow*). *Right panel*, Neck CT showing a midline zone 1 lesion (*yellow arrow*).

Which of the following is the most likely etiology of the patient's elevated serum thyroglobulin?

A. Heterophile antibody interference in the thyroglobulin assay
B. Primary thyroid cancer arising from a thyroglossal duct cyst
C. Recurrent papillary thyroid cancer in cervical lymph nodes
D. Thyroglossal duct cyst
E. Thyroid remnant tissue

97 A 62-year-old man with type 2 diabetes and dyslipidemia is seen for follow-up after having undergone 3-vessel coronary artery bypass grafting 3 months ago. He has a strong family history of premature coronary artery disease and was experiencing increased exertional dyspnea, which led to cardiac evaluation and diagnosis of triple-vessel disease. Type 2 diabetes was diagnosed 10 years ago, and his glycemic control improved after starting dulaglutide 3 years ago. He also takes metformin and insulin glargine, 25 units at bedtime, without any significant hypoglycemic episodes. He has been taking rosuvastatin, 40 mg daily, and an over-the-counter fish oil capsule. He quit smoking cigarettes 5 years ago and has restricted alcohol intake to 1 or 2 glasses of wine once weekly.

On physical examination, his blood pressure is 124/70 mm Hg and pulse rate is 60 beats/min. His height is 66.1 in (168 cm), and weight is 216 lb (98 kg) (BMI = 34.8 kg/m^2). He has truncal obesity and diminished sensation to 10-g monofilament over the great toes. There are no skin breaks or calluses.

Laboratory test results:

Fasting plasma glucose = 108 mg/dL (70-99 mg/dL) (SI: 6.0 mmol/L [3.9-5.5 mmol/L])
Hemoglobin A_{1c} = 6.8% (4.0%-5.6%) (51 mmol/mol [20-38 mmol/mol])
ALT = 47 U/L (10-40 U/L) (SI: 0.78 µkat/L [0.17-0.67 µkat/L])
AST = 29 U/L (20-48 U/L) (SI: 0.48 µkat/L [0.33-0.80 µkat/L])
Serum creatinine = 1.5 mg/dL (0.7-1.3 mg/dL) (SI: 132.6 µmol/L [61.9-114.9 µmol/L])
Total cholesterol = 134 mg/dL (<200 mg/dL [optimal]) (SI: 3.47 mmol/L [<5.18 mmol/L])
Triglycerides = 346 mg/dL (<150 mg/dL [optimal]) (SI: 3.91 mmol/L [<1.70 mmol/L])
HDL cholesterol = 42 mg/dL (>60 mg/dL [optimal]) (SI: 1.09 mmol/L [>1.55 mmol/L])
LDL cholesterol = <30 mg/dL (<100 mg/dL [optimal]) (SI: <0.78 mmol/L [<2.59 mmol/L])

Which of the following is the best option to further reduce this patient's risk of atherosclerotic cardiovascular disease?

A. Add fenofibrate, 145 mg daily
B. Decrease rosuvastatin dosage to 20 mg daily and add fenofibrate, 48 mg daily
C. Increase over-the-counter fish oil to 4 capsules a day
D. Replace over-the-counter fish oil with icosapent ethyl, 4 g daily
E. Start prandial insulin

98 A 69-year-old man is referred for evaluation of a right adrenal nodule that was discovered on abdominal MRI when he presented with abdominal pain. MRI shows a homogeneous, 2.3-cm right adrenal lesion with low T2 signal intensity, consistent with intracellular fat. His medical history includes Parkinson disease, glaucoma, and vitamin D deficiency. His personal and family history is negative for malignancies. Current medications are carbidopa, levodopa, entacapone, rasagiline, latanoprost, and cholecalciferol. He has no history of cigarette smoking and consumes no alcohol.

On physical examination, his blood pressure is 140/60 mm Hg and pulse rate is 68 beats/min. His height is 69 in (175 cm), and weight is 145 lb (65.8 kg) (BMI = 21.4 kg/m^2). He has no plethora, bruises, or striae. Muscle strength is normal.

Laboratory test results:

Serum aldosterone = 8.7 ng/dL (4-21 ng/dL) (SI: 240 pmol/L [111.0-582.5 pmol/L])
Plasma renin activity = 1.1 ng/mL per h (0.6-4.3 ng/mL per h)

Urinary normetanephrine = 191 μg/24 h (<733 μg/24 h) (SI: 1040 nmol/d [<4000 nmol/d])
Urinary metanephrine = 321 μg/24 h (<394 μg/24 h) (SI: 1630 nmol/d [<2000 nmol/d])
Urinary 3-methoxytyramine = 67,270 nmol/d (<2300 nmol/d)

The serum cortisol value following a 1-mg dexamethasone-suppression test (8 AM) is 1.8 μg/dL (SI: 49 nmol/L).

Which of the following is the best next step in this patient's management?

A. ^{68}Ga-DOTATATE PET
B. Measurement of plasma metanephrines and 3-methoxytyramine
C. Neck MRI
D. No further workup or follow-up
E. Right adrenalectomy

99 A 70-year-old woman presents to the emergency department with dizziness. She has a history of coronary artery disease, atrial fibrillation, hyperlipidemia, and hypothyroidism. She underwent resection of a "calcified pituitary mass" followed by radiation more than 40 years ago.

Laboratory test results from 2 years ago (blood drawn at 8 AM):
Cortisol = 2.4 μg/dL (5-25 μg/dL) (SI: 66.2 nmol/L [137.9-689.7 nmol/L])
Free T_4 = 0.9 ng/dL (0.8-1.8 ng/dL) (SI: 11.6 pmol/L [10.30-23.17 pmol/L])
TSH = 0.09 mIU/L (0.5-5.0 mIU/L)

She stopped coming to clinic after her husband's death 2 years ago.

Current medications include aspirin, 325 mg daily; metoprolol XL, 50 mg daily; atorvastatin, 10 mg daily; hydrocortisone, 10 mg daily; levothyroxine, 100 mcg daily; and a multivitamin daily. In the past few months, she reports taking all of her medications at once (with dinner, her only meal of the day) to avoid forgetting to take her pills. She does not think she is missing doses. Pharmacy records indicate that levothyroxine and hydrocortisone dosages have not changed and that the patient regularly picks up her prescriptions.

She has had no weight changes, tremor, change in bowel movements, or heat or cold intolerance. She has experienced poor appetite, nausea, and, more recently, vomiting and dizziness.

On physical examination, she appears anxious and has slight bilateral leg edema. Her blood pressure is 92/60 mm Hg, and pulse rate is 55 beats/min with an irregular rhythm. Her height is 63 in (160 cm), and weight is 158 lb (71.7 kg) (BMI = 28 kg/m^2; body surface area = 2.1 m^2). Laboratory evaluation is performed a few hours before she takes her medications. Complete blood cell count and basic chemistry panel are normal.

Which the following patterns of laboratory test results is most likely in this patient?

Answer	Free T_4	TSH	Cortisol
A.	0.65 ng/dL (SI: 8.37 pmol/L)	0.1 mIU/L	22.0 μg/dL (SI: 606.9 nmol/L)
B.	0.92 ng/dL (SI: 11.84 pmol/L)	0.1 mIU/L	7.5 μg/dL (SI: 206.9 nmol/L)
C.	1.90 ng/dL (SI: 24.45 pmol/L)	0.1 mIU/L	0.5 μg/dL (SI: 13.8 nmol/L)
D.	0.65 ng/dL (SI: 8.37 pmol/L)	0.1 mIU/L	7.5 μg/dL (SI: 206.9 nmol/L)
E.	1.00 ng/dL (SI: 12.87 pmol/L)	7.7 mIU/L	7.5 μg/dL (SI: 206.9 nmol/L)
Reference ranges	0.8-1.8 ng/dL (SI: 10.30-23.17 pmol/L)	0.5-5.0 mIU/L	2-14 μg/dL (PM) (SI: 55.2-386.2 nmol/L)

100 A 35-year-old pregnant woman (G2, P1) with a 4-year history of type 2 diabetes presents for a follow-up visit. She has a history of moderately increased albuminuria but no other known microvascular or macrovascular diabetes-related complications. In her pregnancy 6 years ago, she had mild preeclampsia that was well-managed and not associated with any peripartum complications. At an appointment 6 months ago, she shared that she and her partner were thinking about expanding their family. Empagliflozin, semaglutide, and ramipril were discontinued, and she was prescribed insulin degludec and insulin lispro. Her metformin dosage was kept the same.

Her blood pressure has been well controlled at 112/70 mm Hg. She was able to successfully decrease her hemoglobin A_{1c} level to 6.8% (51 mmol/mol) 3 months ago. Her last eye examination was 6 months ago, and no abnormalities were noted.

At today's visit, she is about 8 weeks pregnant. She checks her blood glucose before meals 3 times daily using capillary blood glucose monitoring.

Regarding use of aspirin, which of the following is the best recommendation for this pregnant patient?

A. Aspirin is not indicated in this patient
B. Start 81 mg daily at week 14 and continue until delivery
C. Start 81 mg daily now and continue until delivery
D. Start 81 mg daily now and stop at the start of third trimester
E. Start 325 mg daily at week 14 and continue until delivery

101 A 72-year-old man with hypertension, type 2 diabetes (hemoglobin A_{1c} = 9.0% [75 mmol/mol]), and chronic obstructive pulmonary disease is found to have an incidental thyroid nodule on screening carotid ultrasonography. Subsequent dedicated thyroid ultrasonography shows that there is a solitary well-demarcated, solid, hypoechoic thyroid nodule in the midpole of the right lobe that measures 1.1 cm (sagittal) × 0.9 cm (anterior/posterior) × 0.7 cm (transverse). The nodule has no microcalcifications or intranodular hypervascularity, and there are no abnormal-appearing lymph nodes on ultrasonography. In accordance with the American Thyroid Association guidelines for the management of thyroid nodules, FNA biopsy is recommended, which reveals a Bethesda VI cytologic result (papillary thyroid cancer).

He does not wish to proceed with thyroid surgery and states that he would be interested in having the thyroid nodule followed closely instead. He is offered the option of active surveillance of this biopsy-proven papillary thyroid cancer, to which he agrees. Through shared decision-making, a plan is formulated for him to return every 6 months for neck ultrasonography.

Which of the following scenarios would be the best indication for proceeding to thyroid surgery?

A. Development of a second thyroid nodule
B. Growth of the thyroid nodule by ≥3 mm in any 1 of the 3 dimensions measured
C. Increase in serum thyroglobulin by >2.0 ng/mL (>2.0 μg/L) while TSH is suppressed to <0.1 mIU/L
D. New diagnosis of papillary thyroid cancer with a positive *BRAF* V600E pathogenic variant in an immediate family member
E. Patient's inability to attend follow-up appointments more than once per year

102 A 33-year-old woman with a history of mild asthma presents with acute mid-back pain after lifting her 4-month-old son from his car seat. She heard a "pop" that was followed by sudden and severe mid-back pain. She tried rest and ibuprofen with some relief. She had an uneventful pregnancy, and her son was delivered vaginally at 39 weeks' gestation. She is solely breastfeeding, and her menses have not yet resumed. She has been pumping between feeds to store milk in anticipation of returning to work next week.

Menarche occurred at age 13 years, and she has regular cycles. She did not have any issues with conception. She adheres to a vegetarian diet and tries to consume 2 servings of dairy products per day. She fractured her left humerus at age 8 years after a fall. She had asthma exacerbations in childhood; however, her asthma is currently well controlled. She does not smoke cigarettes or use recreational drugs. Before pregnancy and breastfeeding, she drank wine 2 to 3 times a week. Her mother has osteopenia but no fractures. The patient intends to breastfeed for 1 year and desires future pregnancies. She currently takes elemental calcium citrate, 315 mg daily, and vitamin D, 1000 international units daily (25 mcg daily).

On physical examination, her blood pressure is 107/68 mm Hg, and pulse rate is 76 beats/min. Her height is 61 in (154.9 cm), and weight is 107 lb (48.5 kg) (BMI = 20.2 kg/m^2). Her thyroid gland is palpable, nontender, and not enlarged. No nodules are appreciated. Reflexes are 2+ at the biceps and patella. Spinal and paraspinal tenderness is noted in the mid-back. The rest of her examination findings are normal.

Spine radiographs show an acute compression fracture with 30% height loss at the T6 vertebral body.

Laboratory test results:

- Hemoglobin = 13.8 g/dL (12.1-15.1 g/dL) (SI: 138 g/L [121-151 g/L])
- Calcium = 9.5 mg/dL (8.2-10.2 mg/dL) (SI: 2.4 mmol/L [2.1-2.6 mmol/L])
- Phosphate = 2.5 mg/dL (2.3-4.7 mg/dL) (SI: 0.8 mmol/L [0.7-1.5 mmol/L])
- Creatinine = 0.8 mg/dL (0.6-1.1 mg/dL) (SI: 70.7 µmol/L [53.0-97.2 µmol/L])
- Albumin = 4.0 g/dL (3.5-5.0 g/dL) (SI: 40 g/L [35-50 g/L])
- Intact PTH = 20 pg/mL (10-65 pg/mL) (SI: 2.1 pmol/L [1.1-6.9 pmol/L])
- 25-Hydroxyvitamin D = 32 ng/mL (30-80 ng/mL [optimal]) (SI: 79.9 nmol/L [74.9-199.7 nmol/L])
- Alkaline phosphatase = 115 U/L (50-120 U/L) (SI: 1.92 µkat/L [0.84-2.00 µkat/L])
- TSH = 1.21 mIU/L (0.5-5.0 mIU/L)
- Urinary calcium = 125 mg/24 h (100-300 mg/24 h) (SI: 3.1 mmol/d [2.5-7.5 mmol/d])
- Prolactin = 28 ng/mL (10-200 ng/mL [lactating females]; 4-30 ng/mL [nonlactating females]) (SI: 1.22 nmol/L [0.43-8.70 nmol/L] [lactating females]; 0.17-1.30 nmol/L [nonlactating females])

DXA documents the following values:

- Lumbar spine Z score (L1-L4) = −3.0
- Femoral neck Z score = −1.8
- Total hip Z score = −1.3

Which of the following is the best management advice for this patient?

A. Increase calcium intake to 1200-1500 mg daily
B. Recommend use of a thoracolumbosacral orthosis
C. Start alendronate, 70 mg weekly
D. Start cabergoline, 0.5 mg twice weekly
E. Stop breastfeeding

103 A 59-year-old man with type 1 diabetes presents to the diabetes clinic for evaluation and management. His medical history is notable for hypertension, diabetic retinopathy, congestive heart failure (New York Heart Association class II), end-stage kidney disease, diabetic neuropathy, and dyslipidemia. His treatment consists of insulin glargine, 25 units daily; insulin aspart, 8 units with each meal; and correctional insulin, 1 unit for every 30 mg/dL (1.7 mmol/L) glucose is above 130 mg/dL (>7.2 mmol/L).

In the past, the patient worked with a nutritionist for instruction on counting carbohydrates. However, he finds carbohydrate counting to be extremely difficult. After discussing this further with the nutritionist, she informs you that the patient struggles to understand the difference between large-, medium-, and low-carbohydrate meals. He is frustrated that he frequently experiences hypoglycemia, sometimes 3 times per week. The episodes can occur after meals, overnight, or after increased activity, including exercise or even gardening. The patient has hypoglycemia unawareness, which has led to 2 hospitalizations in the last 12 months. Although he has a cell phone, he does not use a smartphone because he finds them too complicated.

Laboratory test results:

- Hemoglobin = 11.0 g/dL (13.8-17.2 g/dL) (SI: 110 g/L [138-172 g/L])
- Hemoglobin A_{1c} = 9.4% (4.0%-5.6%) (79 mmol/mol [20-38 mmol/mol])
- Creatinine = 6.1 mg/dL (0.7-1.3 mg/dL) (SI: 539.4 µmol/L [61.9-114.9 µmol/L])

Which of the following is the most appropriate management plan for this patient?

A. Decrease basal and bolus insulin dosages by 30%
B. Decrease basal insulin dosage by 30%
C. Recommend insulin pump initiation and use a hybrid automated insulin delivery system
D. Refer the patient for islet-cell transplant
E. Refer the patient for pancreas and kidney transplant

104 A 79-year-old woman presents to discuss weight management before undergoing knee replacement surgery. She reports numerous weight-loss attempts over the years but has not been able to maintain her weight. She has been asked to lose 25 lb (11.3 kg) before the knee replacement and has been advised to consider bariatric surgery. Her weight has been the same for about a year. She has been monitoring portion sizes and does not consume caloric beverages.

She can only mobilize with a cane and cannot stand for longer than 10 minutes because of knee pain. She performs physical therapy exercises 3 times per week. She has a large hiatal hernia for which she takes pantoprazole daily. DXA performed this year shows normal bone density at the hips (spine was nondiagnostic).

On physical examination, her blood pressure is 158/72 mm Hg and pulse rate is 82 beats/min. Her height is 66.1 in (167.9 cm), and weight is 270 lb (122.5 kg) (BMI = 43.4 kg/m^2). She has symmetric weight distribution between her upper and lower body. She uses her arms to stand from a seated position. She is unable to get on and off the examination table without assistance.

Laboratory test results:

25-Hydroxyvitamin D = 57 ng/mL (30-80 ng/mL [optimal]) (SI: 142.3 nmol/L [74.9-199.7 nmol/L])
TSH = 5.9 mIU/L (0.5-5.0 mIU/L)
Serum creatinine = 0.6 mg/dL (0.6-1.1 mg/dL) (SI: 53.0 μmol/L [53.0-97.2 μmol/L])

Which of the following should be recommended to help this patient achieve her weight-loss goal?

A. Increase dietary protein intake
B. Refer for sleeve gastrectomy
C. Start a meal replacement program
D. Start levothyroxine
E. Start phentermine

105 A 38-year-old man is admitted to the emergency department with sudden and severe abdominal pain, vomiting, and headache. He recently received his first dose of the ChAdOx1 nCoV-19 vaccine. He has no notable medical history and takes no regular medications. He smokes 15 cigarettes per day and consumes no alcohol. He has no family history of clotting disorders.

On physical examination, his blood pressure is 133/90 mm Hg and pulse rate is 110 beats/min. His height is 69 in (175 cm), and weight is 181 lb (82 kg) (BMI = 26.8 kg/m^2). He has diffuse abdominal tenderness.

Abdominal CT shows retroperitoneal fat stranding and high-density fluid around the adrenal glands (*see image, arrows*).

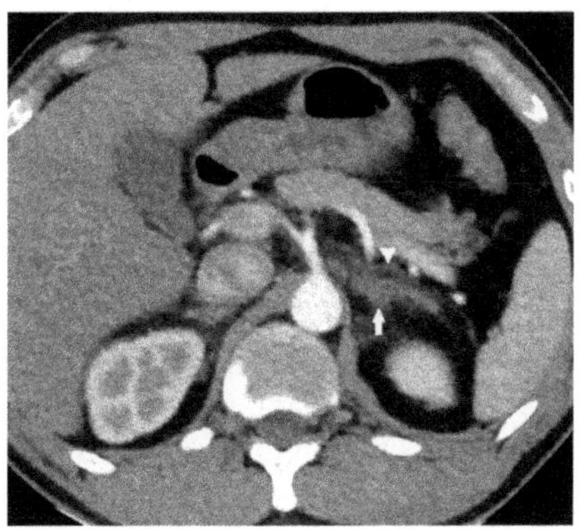

Reprinted from Taylor P et al. *Clin Endocrinol*, 2022; 97(1):26-27. © The Authors. Published by John Wiley & Sons Ltd.

Laboratory test results:

Electrolytes, normal
Serum creatinine, normal
Amylase = 51 U/L (26-102 U/L) (SI: 0.85 μkat/L [0.43-1.70 μkat/L])
Lactate = 45.0 mg/dL (5.4-20.7 mg/dL) (SI: 5.0 mmol/L [0.6-2.3 mmol/L])
Hemoglobin = 12.9 g/dL (13.8-17.2 g/dL) (SI: 129 g/L [138-172 g/L])
Platelet count = 139 × 10^3/μL (150-450 × 10^3/μL) (SI: 139 × 10^9/L [150-450 × 10^9/L])
Prothrombin time = 12.1 s (9.0-13.0 s)
Activated partial thromboplastin time = 28.7 s (24.0-36.0 s)

Clauss fibrinogen = 3.5 g/L (2.0-4.0 g/L)
D-dimer = >20,000 μg/L (<500 μg/L)
C-reactive protein = 117 mg/L (0.8-3.1 mg/L) (SI: 1114.3 nmol/L [7.62-29.52 nmol/L])

Which of the following is the best next step in this patient's management?

A. Administer intravenous hydrocortisone
B. Administer intravenous immunoglobulins
C. Perform a cosyntropin-stimulation test
D. Perform heparin-induced thrombocytopenia antibody screen
E. Recommend laparotomy

106 A 52-year-old man returns for continued management of type 2 diabetes, which was diagnosed 10 years ago. He was initially treated with metformin monotherapy, but oral semaglutide was prescribed 3 years ago to maintain his hemoglobin A_{1c} below 7.0% (<53 mmol/mol). He has painful peripheral neuropathy that is well controlled with pregabalin. He also has hypertension and hyperlipidemia. He reports taking a medication (he cannot recall the name) for high cholesterol for many years, but he stopped because of muscle aches. He currently takes metformin, oral semaglutide, valsartan, and chlorthalidone.

On physical examination, his blood pressure is 124/80 mm Hg and pulse rate is 72 beats/min. His height is 68 in (172.7 cm), and weight is 225 lb (102.3 kg) (BMI = 34 kg/m^2). He has 1+ pitting edema and 2+ dorsalis pedis pulse in the bilateral lower extremities. His lungs are clear to auscultation, and his heart has a regular rate and rhythm. There is decreased sensation to 10-g monofilament testing on the distal plantar aspect of his feet bilaterally.

Hemoglobin A_{1c} = 6.8% (4.0%-5.6%) (51 mmol/mol [20-38 mmol/mol])
Total cholesterol = 188 mg/dL (<200 mg/dL [optimal]) (SI: 4.87 mmol/L [<5.18 mmol/L])
HDL cholesterol = 51 mg/dL (>60 mg/dL [optimal]) (SI: 1.32 mmol/L [>1.55 mmol/L])
LDL cholesterol = 116 mg/dL (<100 mg/dL [optimal]) (SI: 3.00 mmol/L [<2.59 mmol/L])
Triglycerides = 157 mg/dL (<150 mg/dL [optimal]) (SI: 1.77 mmol/L [<1.70 mmol/L])
Serum creatinine = 1.4 mg/dL (0.7-1.3 mg/dL) (SI: 123.8 μmol/L [61.9-114.9 μmol/L])
Estimated glomerular filtration rate = 54 mL/min per 1.73 m^2 (>60 mL/min per 1.73 m^2)
Potassium = 5.2 mEq/L (3.5-5.0 mEq/L) (SI: 5.2 mmol/L [3.5-5.0 mmol/L])

Which of the following is the most appropriate recommendation in this patient's management?

A. Bempedoic acid
B. Fenofibrate
C. Icosapent ethyl
D. Nutrition counseling
E. Rosuvastatin

107 A 19-year-old woman presents for evaluation of primary amenorrhea. She has no anosmia, galactorrhea, headaches, hirsutism, acne, or vasomotor symptoms. She does not exercise excessively, has good nutritional status, and has no history of an eating disorder.

On physical examination, her blood pressure is 112/62 mm Hg and pulse rate is 69 beats/min. Her height is 62 in (157.5 cm), and weight is 115 lb (52.2 kg) (BMI = 21.0 kg/m^2). Breast development is Tanner stage 5, and she has small areolae and nipple dysplasia. She has no axillary hair. There is no lymphadenopathy, and her thyroid gland is normal. On pelvic examination, she has no pubic hair, and on speculum examination she has a short and shallow vaginal cavity with no uterus or ovaries. These findings are confirmed by transvaginal ultrasonography.

Baseline laboratory test results:

Total testosterone = 873 ng/dL (SI: 30.3 nmol/L) (male range: 300-900 ng/dL [10.4-31.2 nmol/L]; female range: 8-60 ng/dL [0.3-2.1 nmol/L])

Estradiol = 27 pg/mL (SI: 99.1 pmol/L) (male range: 10-40 pg/mL [36.7-146.8 pmol/L];
premenopausal female range: 30-400 pg/mL [110.1-1468.4 pmol/L] [varies widely throughout the menstrual cycle];
postmenopausal female range: <20 pg/mL [<73.4 pmol/L])
LH = 43.0 mIU/mL (SI: 43.0 IU/L)
FSH = 13.0 mIU/mL (SI: 13.0 IU/L)

Her karyotype is 46,XY. Genetic testing identifies a pathogenic variant in the ligand-binding domain of the androgen receptor gene.

In addition to gonadal and pituitary hormone assessment, which of the following is the best next step in this patient's management?

A. Assess bone density with DXA
B. Measure α-fetoprotein
C. Measure hCG
D. Perform abdominal ultrasonography
E. Start estrogen therapy

108 A 60-year-old man presents for ongoing care of type 2 diabetes diagnosed 15 years ago. He has no known macrovascular diabetes complications. Two years ago, he was noted to have moderately increased albuminuria (urinary albumin-to-creatinine ratio = 42 mg/g creat [<30 mg/g creat]). This has since been well controlled after initiating candesartan, 8 mg once daily. His most recent urinary albumin-to-creatinine ratio is 25 mg/g creat, and blood pressure is 112/75 mm Hg.

He is concerned because his younger brother recently had a myocardial infarction at age 45 years. He mentions that coronary artery disease is common in his family; both parents and his paternal grandfather had myocardial infarctions in their 50s. He has a sedentary job but tries to stay active and walks at least 10,000 steps per day at a moderately brisk pace without dyspnea or chest discomfort. His medical history includes type 2 diabetes, dyslipidemia, hypertension, and peptic ulcer disease, He has a remote history of cigarette smoking but he quit more than 15 years ago (12 pack-year history).

Current medications are empagliflozin, 10 mg daily; oral semaglutide, 14 mg daily; metformin, 850 mg 3 times daily; candesartan, 8 mg daily; and rosuvastatin, 20 mg daily.

Laboratory test results:
Hemoglobin A_{1c} = 6.5% (4.0%-5.6%) (48 mmol/mol [20-38 mmol/mol])
LDL cholesterol = 65 mg/dL (<100 mg/dL [optimal]) (SI: 1.68 mmol/L [<2.59 mmol/L])
HDL cholesterol = 42 mg/dL (>60 mg/dL [optimal]) (SI: 1.09 mmol/L [>1.55 mmol/L])
Triglycerides = 225 mg/dL (<150 mg/dL [optimal]) (SI: 2.54 mmol/L [<1.70 mmol/L])
Estimated glomerular filtration rate = 55 mL/min per 1.73 m^2 (>60 mL/min per 1.73 m^2)

Which of the following is the best next step in this patient's care?

A. Continue the current regimen
B. Start aspirin, 81 mg daily
C. Start icosapent ethyl, 2 g twice daily
D. Start micronized fenofibrate, 90 mg daily
E. Switch from oral semaglutide to injectable semaglutide, 1 mg weekly

109 A 54-year-old woman with pulmonary sarcoidosis, degenerative joint disease, American Joint Committee on Cancer (AJCC) stage I follicular thyroid carcinoma, and postoperative hypothyroidism presents for follow-up. Follicular thyroid carcinoma was first diagnosed 3 years ago following left hemithyroidectomy for an indeterminate thyroid nodule (Bethesda IV cytopathology, follicular neoplasm). Molecular testing was not performed. Surgical pathology demonstrated a 4.5-cm follicular thyroid carcinoma with minimal capsular invasion and at least 4 foci of angioinvasion. There was no extrathyroidal extension and surgical margins were negative. Completion thyroidectomy demonstrated benign thyroid tissue. The patient subsequently received 125 mCi ^{131}I therapy 2 months after surgery. Posttherapy whole-body scan showed radioiodine uptake in the left thyroid bed and no iodine-avid metastases. Neck ultrasonography performed 6 months after radioactive iodine therapy and repeated 1 year ago demonstrated no nodules in the thyroid bed or suspicious cervical lymph nodes.

The patient is currently feeling well. Levothyroxine is her only prescribed medication. Four months ago, she began taking a hair and nail supplement containing 300 mcg of biotin (vitamin B_7).

On physical examination, her blood pressure is 130/74 mm Hg and pulse rate is 90 beats/min. Her height is 68.5 in (174 cm), and weight is 249 lb (112.9 kg) (BMI = 37.3 kg/m^2). There is an anterior neck scar. There are no palpable neck masses or cervical lymphadenopathy. The rest of the examination findings are normal.

Laboratory test results:

Measurement	3 Years ago (after rhTSH injection for ^{131}I therapy)	2 Years ago	6 Months ago	Today	Reference range
TSH	67.0 mIU/L	0.8 mIU/L	0.5 mIU/L	0.18 mIU/L	0.5-5.0 mIU/L
Free T_4	1.4 ng/dL (SI: 18.02 pmol/L)	...	...	1.7 ng/dL (SI: 21.88 pmol/L)	0.8-1.8 ng/dL (SI: 10.30-23.17 pmol/L)
Thyroglobulin	7.8 ng/mL (SI: 7.8 μg/L)	0.4 ng/mL (SI: 0.4 μg/L)	1.2 ng/mL (SI: 1.2 μg/L)	6.9 ng/mL (SI: 6.9 μg/L)	
Thyroglobulin antibodies	<4.0 IU/mL (SI: <4.0 kIU/L)	<4.0 IU/mL (SI: <4.0 kIU/L)	<4.0 IU/mL (SI: <4.0 kIU/L)	<4.0 IU/mL (SI: <4.0 kIU/L)	≤4.0 IU/mL (SI: ≤4.0 kIU/L)

Which of the following is the most appropriate next step in this patient's management?

A. Perform FDG PET/CT scan
B. Perform neck ultrasonography
C. Perform rhTSH-stimulated radioactive iodine (^{123}I) whole-body scan
D. Reassure the patient and measure serum thyroglobulin in 6 months
E. Remeasure serum thyroglobulin after instructing the patient to stop the biotin-containing supplement

110 A 42-year-old woman is evaluated in the emergency department after a near-fainting episode associated with nausea. In the past 3 weeks, she has had low-grade fever, cough, and fatigue. Several years ago, she was evaluated for daily throbbing headaches, vision obscurations, and tinnitus. After evaluation by a neurologist and ophthalmologist, she underwent brain MRI, which indicated findings of "empty sella" and a lumbar puncture. She was subsequently prescribed acetazolamide, 500 mg twice daily, and was referred to a nutritionist for weight loss. Her headaches improved, and she lost 21 lb (9.5 kg).

Review of systems in the emergency department is remarkable for oligomenorrhea that is attributed to polycystic ovary syndrome. Over the past year, menses occurred every 6 to 8 weeks, whereas they previously occurred every 4 to 6 weeks.

On physical examination, she is pale. Her decubitus blood pressure is 105/60 mm Hg and pulse rate is 100 beats/min. After standing, her blood pressure is 107/68 mm Hg, and pulse rate is 105 beats/min. Her height is 60 in (152.4 cm), and weight is 200 lb (90.7 kg) (BMI = 39.0 kg/m^2). She has generalized fat distribution. The thyroid gland is palpable, symmetric, and without nodules. Coarse rhonchi are noted on lung auscultation. Visual fields are grossly normal on confrontation, and neurologic findings are nonfocal. She has no lower-extremity edema.

Laboratory test results:

Sodium = 128 mEq/L (136-142 mEq/L) (SI: 128 mmol/L [136-142 mmol/L])
Other chemistry panel results, normal
TSH = 2.5 mIU/L (0.5-5.0 mIU/L)

Review of records from the patient's annual visit 1 year ago documents a serum sodium value of 135 mEq/L (135 mmol/L).

Which of the following is the best next step in this patient's management?

A. Measure serum cortisol now
B. Measure serum cortisol, free T_4, IGF-1, prolactin, and FSH the next morning
C. Perform cosyntropin-stimulation testing the next morning
D. Perform MRI of the brain and pituitary gland
E. Stop acetazolamide

111 A 62-year-old man is undergoing kidney pretransplant evaluation. He has hypertensive nephropathy (has received hemodialysis for the past 3 years) and hepatitis C. Current medications are olmesartan, amlodipine, metoprolol, atorvastatin, furosemide, vitamin D, and sevelamer. He has a history of a thyroid nodule with benign results on FNA biopsy 2 years ago, and he reports that the nodule has not changed in size. He has no history of fractures. He has a family history of type 2 diabetes and hypertension in both parents. He drinks 4 alcoholic beverages per week and does not smoke cigarettes.

On physical examination, his blood pressure is 148/78 mm Hg and pulse rate is 86 beats/min. His height is 73 in (185.4 cm), and weight is 245 lb (111.1 kg) (BMI = 32.3 kg/m^2). Findings on cardiac and pulmonary examinations are normal.

Laboratory test results:

Hemoglobin = 11.0 g/dL (13.8-17.2 g/dL) (SI: 111 g/L [138-172 g/L])
Platelet count = 365 × 10^3/µL (150-450 × 10^3/µL) (SI: 365 × 10^9/L [150-450 × 10^9/L])
Creatinine = 4.7 mg/dL (0.7-1.3 mg/dL) (SI: 415.5 µmol/L [61.9-114.9 µmol/L])
Total cholesterol = 112 mg/dL (<200 mg/dL [optimal]) (SI: 2.90 mmol/L [<5.18 mmol/L])
LDL cholesterol = 65 mg/dL (<100 mg/dL [optimal]) (SI: 1.68 mmol/L [<2.59 mmol/L])
Triglycerides = 132 mg/dL (<150 mg/dL [optimal]) (SI: 1.49 mmol/L [<1.70 mmol/L])
TSH = 3.6 mIU/L (0.5-5.0 mIU/L)
PTH = 268 pg/mL (10-65 pg/mL) (SI: 28.4 pmol/L [1.1-6.9 pmol/L])
Calcium = 8.6 mg/dL (8.2-10.2 mg/dL) (SI: 2.2 mmol/L [2.1-2.6 mmol/L])
Phosphate = 4.5 mg/dL (2.3-4.7 mg/dL) (SI: 1.5 mmol/L [0.7-1.5 mmol/L])
Albumin = 3.8 g/dL (3.5-5.0 g/dL) (SI: 38 g/L [35-50 g/L])
AST = 25 U/L (20-48 U/L) (SI: 0.42 µkat/L [0.33-0.80 µkat/L])
ALT = 21 U/L (10-40 U/L) (SI: 0.35 µkat/L [0.17-0.67 µkat/L])
Hemoglobin A_{1c} = 5.9% (4.0%-5.6%) (41 mmol/mol [20-38 mmol/mol])

DXA documents the following:

Location	T-score
Lumbar spine (L1-L4)	1.2
Left total hip	−1.0
Left femoral neck	−0.8

This patient is at highest risk to develop which of the following after transplant?

A. Hepatic cirrhosis
B. Hypertriglyceridemia
C. Osteoporosis
D. Posttransplant diabetes
E. Thyroid cancer

112 A 57-year-old woman is found to have multiple vertebral compression fractures (T10, T12-L5) during evaluation for back pain without a clear precipitating incident. She has always been physically active in her job, which includes lifting. She has noticed that her torso appears shorter than it used to be. No secondary causes of osteoporosis are apparent, including genetic evaluation for bone fragility disorders and workup for myeloma. She takes no glucocorticoids.

On physical examination, her blood pressure is 134/79 mm Hg and pulse rate is 68 beats/min. Dentition is normal. She has kyphosis and has only 1 cm distance between her lower ribs and pelvic bones.

Based on severe osteoporosis with multiple low-trauma compression fractures and very low T-scores, she is treated with teriparatide for 2 years, after which bone mineral density increases. Because she has no contraindications to oral bisphosphonates, she is treated with alendronate for 5 years, followed by a drug holiday. During these 7 years of osteoporosis treatment, she also had a non–ST-elevation myocardial infarction. Kidney function is normal.

Bone density measurements over time:

	Timing			
Site	**Baseline**	**After 2 years of teriparatide**	**After 5 years of alendronate**	**After 2 years of drug holiday**
Spine BMD, g/cm^2	0.679	0.847	0.903	0.930
Spine T-score	−3.3	−1.8	−1.3	−1.1
Left femoral neck BMD, g/cm^2	0.596	0.693	0.716	0.675
Left femoral neck T-score	−2.3	−1.4	−1.2	−1.6
Left total hip BMD, g/cm^2	0.787	0.838	0.860	0.858
Left total hip T-score	−1.3	−0.9	−0.7	−0.7

Two years after stopping alendronate, she again develops severe back pain after slipping and falling, and MRI indicates an acute T11 compression fracture, as well as worsened compression deformities of L2 through L4 to grade 3 and stable grade 1 chronic compression deformities of T12 and L5 (compared with imaging from initial presentation). Updated evaluation does not identify any new secondary causes for osteoporosis or fractures.

Which of the following is the best next step in this patient's management?

A. Continue drug holiday
B. Restart alendronate
C. Restart teriparatide
D. Start romosozumab
E. Start zoledronate

113 A 70-year-old man is evaluated for the management of papillary thyroid cancer that was initially treated 2 years ago with total thyroidectomy and postoperative radioactive iodine ablation. The post–^{131}I therapy whole-body scan showed uptake in the thyroid bed and scattered foci bilaterally at the lung bases. In the 3 years since his initial treatment, he has maintained minimally positive serum thyroglobulin concentrations ranging from 2.8 to 3.6 ng/mL (2.8-3.6 μg/L). He has been taking a stable levothyroxine dosage (125 mcg daily) since surgery, and his TSH concentration is less than 0.1 mIU/L. He currently follows a regular diet.

A diagnostic ^{123}I whole-body uptake and scan is performed to further evaluate his thyroid cancer, and it shows no appreciable uptake.

Which of the following aspects could account for false-negative radioiodine uptake in this patient?

A. Dietary iodine intake >50 mcg daily in the week before administration of the radioiodine tracer
B. Discontinuing levothyroxine 6 weeks before administration of the radioiodine tracer
C. Failure to ensure that the serum TSH is >60.0 mIU/L on the day that the radioiodine tracer is administered
D. Receipt of iodinated contrast media 4 months before administration of the radioiodine tracer
E. Receipt of rhTSH (2 consecutive daily doses) concluding the day before administration of the radioiodine tracer

114

A 63-year-old man is referred for an incidentally detected adrenal mass during CT chest for lung cancer screening. He has smoked Hookah for many years. A subsequent adrenal protocol CT demonstrates a well-defined right adrenal mass measuring 8.4 × 9.6 × 8.7 cm, with areas of heterogeneity and foci of calcification in the wall (*see noncontrast images, arrows*). Most of the nodule does not enhance following contrast administration, except small areas that demonstrate retained contrast in the delayed phase. The pancreas, spleen, left adrenal gland, and both kidneys are unremarkable.

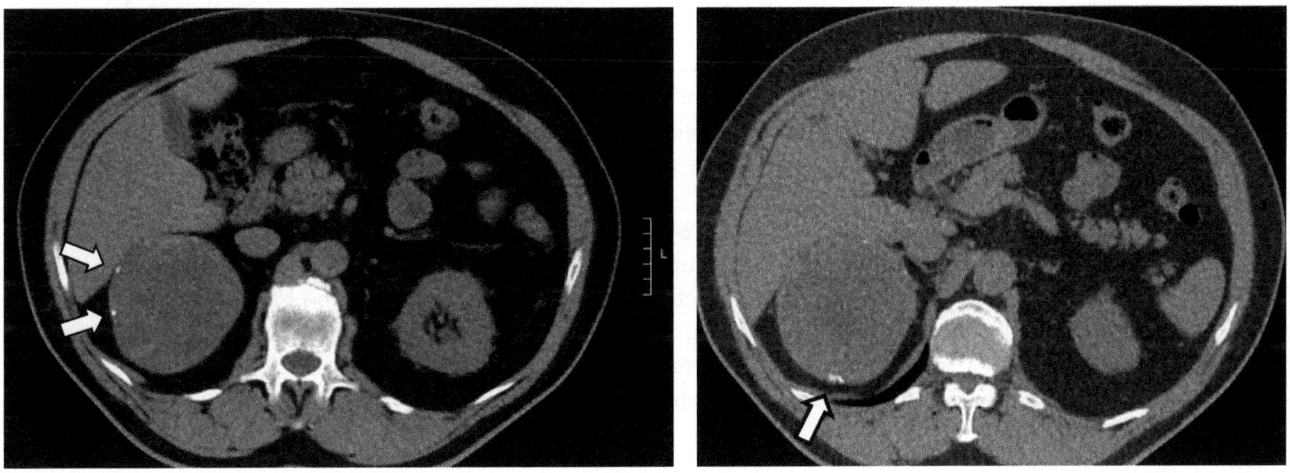

His medical history includes type 2 diabetes treated with metformin, 500 mg twice daily, and semaglutide, 0.5 mg weekly. He lost several pounds after initiation of semaglutide. He has had hypertension for more than 10 years, which is well controlled with amlodipine, 10 mg daily. He has no easy bruising, muscle weakness, or palpitations. He rarely has headaches. He has episodic anxiety.

On physical examination, his blood pressure is 133/87 mm Hg and pulse rate is 68 beats/min. His height is 71 in (180.3 cm), and weight is 228 lb (103.4 kg) (BMI = 31.8 kg/m^2). He has no plethora, bruises, striae, or dorsocervical adiposity. Muscle strength is within normal limits, and he has no edema.

Laboratory test results:

Comprehensive metabolic panel, normal
Complete blood cell count, normal
Hemoglobin A_{1c} = 6.0% (4.0%-5.6%) (SI: 42 mmol/mol [20-38 mmol/mol])
Serum cortisol (8 AM) = 13 µg/dL (5-25 µg/dL) (SI: 358.6 nmol/L [137.9-689.7 nmol/L])
ACTH (8 AM) = 55 pg/mL (10-60 pg/mL) (SI: 12.1 pmol/L [2.2-13.2 pmol/L])
Serum aldosterone = 5.5 ng/dL (4-21 ng/dL) (SI: 152.6 pmol/L [111.0-582.5 pmol/L])
Plasma renin activity = 0.5 ng/mL per h (0.6-4.3 ng/mL per h)
Serum DHEA-S = 40 µg/dL (25-131 µg/dL) (SI: 1.08 µmol/L [0.68-3.55 µmol/L])
Serum 11-deoxycortisol = 70 ng/dL (10-79 ng/dL)
Urinary metanephrines and catecholamines, normal

FDG PET-CT shows a large right adrenal mass measuring 8.5 × 8 × 7.5 cm. There are areas of calcification peripherally with central photopenia. Uptake along the periphery is at physiologic levels. No hypermetabolic abdominal lymph nodes are identified.

The patient undergoes laparoscopic right adrenalectomy. The pathologic findings are consistent with a mature teratoma.

Which of the following is the best management approach for this patient?

A. Initiate mitotane
B. Perform a cosyntropin-stimulation test
C. Perform follow-up abdominal CT in 3 months
D. Recommend no hormonal or imaging follow-up
E. Refer to radiation oncology

115 A 38-year-old man is referred for evaluation of gynecomastia. He reports breast tenderness that started 6 months ago. At a visit 3 months ago, his primary care physician told him that breast tenderness "might be due to a testosterone problem and he should see a specialist." The breast tenderness has since resolved, but the patient is concerned about developing breasts. He reports no change in soaps, laundry detergents, colognes, or diet. His weight has not changed. He reports no vision changes, diplopia, or eye irritation. He does not use lavender-containing products. Neither his profession nor hobbies involve chemicals. His wife uses a transdermal hormonal contraceptive patch. He reports that she has not changed lotions or other skin products. He takes no medications.

On physical examination, his blood pressure is 132/84 mm Hg, pulse rate is 68 beats/min, and respiratory rate is 12 breaths/min. He height is 69 in (175.3 cm), and weight is 210 lb (95.3 kg) (BMI = 31.0 kg/m^2). He has no proptosis, lid lag, chemosis, or edema of the medial rectus. There is no visible goiter, and he has a normal-sized thyroid gland with no palpable nodules. He has 3-cm nontender gynecomastia. Jugular venous pressure is normal, and heart rate is regular with no gallop. He has no edema. His penis is normal, and testes are 25 cc bilaterally. There is no acropachy or waxy induration on the lower extremities.

Laboratory test results:

Hematocrit = 43% (41%-51%) (SI: 0.43 [0.41-0.51])
Serum total testosterone = 250 ng/dL (264-916 ng/dL) (SI: 8.7 nmol/L [9.2-31.8 nmol/L]) (assay validated and harmonized by the CDC)
Repeat serum testosterone (sample drawn at 7:30 AM) = 260 ng/dL (SI: 9.0 nmol/L)
(Calculated free testosterone, normal)
SHBG = 1.7 µg/mL (1.1-6.7 µg/mL) (SI: 15 nmol/L [10-60 nmol/L])
FSH = 4.0 mIU/mL (1.0-13.0 mIU/mL) (SI: 4.0 IU/L [1.0-13.0 IU/L])
LH = 2.0 mIU/mL (1.0-9.0 mIU/mL) (SI: 2.0 IU/L [1.0-9.0 IU/L])
Free T_4 = 1.2 ng/dL (0.8-1.8 ng/dL) (SI: 15.4 pmol/L [10.30-23.17 pmol/L])
TSH = 0.5 mIU/L (0.5-5.0 mIU/L)

Which of the following is the best next step in this patient's evaluation?

A. Breast ultrasonography
B. No further evaluation
C. Serum β-hCG measurement
D. Serum estradiol measurement
E. Serum prolactin measurement

116 A 65-year-old man with a 20-year history of type 2 diabetes (generally suboptimally controlled with hemoglobin A_{1c} values >9.0% [>75 mmol/mol] for many years) presents for ongoing assessment. He mentions he has been having significant right-sided shoulder pain for the last 2 months. The pain is worse at night and makes it difficult to sleep. He has had no falls and no trauma to the right shoulder. His fasting blood glucose values have been running higher than usual. He has thus had to increase the basal insulin dosage by 30% in the past few days, but his glucose values have still remained elevated. He describes the pain as diffuse (not limited to anterior shoulder) and without radiation. During the day, it prevents him from being able to reach overhead or move his right arm across his chest. He does not engage in vigorous physical activity and has a sedentary job. A physiotherapist told him that the pain is related to diabetes.

On physical examination, there is restriction of both active and passive range of motion on testing for external rotation and abduction of the shoulder on the right. There is no weakness noted on resistance testing. He is unable to place his hand on his buttock. There is no point tenderness over the acromioclavicular joint.

Which of the following etiologies most likely accounts for his presentation?

A. Acromion fracture
B. Adhesive capsulitis
C. Bicipital tendonitis
D. Cervical disk disease
E. Rotator cuff tear

117 A 46-year-old transgender woman has identified as being female since childhood, but she did not have the language or understanding to realize what she was feeling until the last few years during the COVID pandemic when she explored online social media platforms and spent more time reflecting on her own life's goals. She is particularly uncomfortable with her deep voice and hair growth. She asks about gender-affirming hormone therapy to achieve a more feminine appearance, particularly breast development. She is eager to have full feminization as quickly as possible. She expresses a deep concern about thinning hair on the top of her head because balding is prevalent in her family. She worries about how balding would affect her confidence in presenting as a woman.

On physical examination, blood pressure is 115/75 mm Hg and pulse rate is 76 beats/min. Her height is 68 in (172.7 cm), and weight is 160 lb (72.6 kg) (BMI = 24.3 kg/m^2). She is well masculinized with normal body hair distribution, receding temporal hair line, and thinning hair on the vertex. There are normal, bilaterally descended testes (testicular volume, 15 cc bilaterally).

Baseline laboratory test results:

Total testosterone = 650 ng/dL (SI: 22.6 nmol/L) (male range: 300-900 ng/dL [10.4-31.2 nmol/L]; female range: 8-60 ng/dL [0.3-2.1 nmol/L])

Estradiol = 13 pg/mL (SI: 47.7 pmol/L) (male range: 10-40 pg/mL [36.7-146.8 pmol/L]; premenopausal female range: 30-400 pg/mL [110.1-1468.4 pmol/L] [varies widely throughout the menstrual cycle]; postmenopausal female range: <20 pg/mL [<73.4 pmol/L])

She is started on estradiol valerate, 4 mg subcutaneously every 7 days. After 3 months, the following laboratory test results are documented (sample drawn midcycle, while fasting at 8 AM):

Total testosterone = 26 ng/dL (SI: 0.9 nmol/L) (male range: 300-900 ng/dL [10.4-31.2 nmol/L]; female range: 8-60 ng/dL [0.3-2.1 nmol/L])

Estradiol = 196 pg/mL (SI: 719.5 pmol/L) (male range: 10-40 pg/mL [36.7-146.8 pmol/L]; premenopausal female range: 30-400 pg/mL [110.1-1468.4 pmol/L] [varies widely throughout the menstrual cycle]; postmenopausal female range: <20 pg/mL [<73.4 pmol/L])

At her follow-up visit, she reports feeling affirmed and has noticed decreased libido, decreased spontaneous erections, and recent onset of breast tenderness. However, she continues to worry about balding and requests treatment to prevent scalp hair loss.

Which of the following is the best treatment for this patient now?

A. Biotin (vitamin B_7)
B. Finasteride
C. Leuprolide
D. Oral minoxidil
E. Topical minoxidil

118 A 72-year-old man with a history of hypertension, hyperlipidemia, and myocardial infarction (status post stent placement 5 years ago) presents for an annual follow-up appointment. Lately, he reports feeling unsteady on his feet but has not had any falls. In addition, he has also noted a gradual decline in erectile function and loss of muscle mass. He is active and walks 2.5 miles 5 days a week. He has no other symptoms and no fracture history. Current medications are lisinopril, atorvastatin, clopidogrel, and metoprolol.

On physical examination, his blood pressure is 163/78 mm Hg and pulse rate is 71 beats/min. His height is 67 in (170 cm), and weight is 210 lb (95.3 kg) (BMI = 32.9 kg/m^2). Spinal curvature is normal, and there is no spinal tenderness.

Laboratory test results:

Total testosterone (8 AM) = 248 ng/dL (300-900 ng/dL) (SI: 8.3 nmol/L [10.4-32.1 nmol/L])
Repeat total testosterone (8 AM) = 252 ng/dL (SI: 8.7 nmol/L)
LH = 1.0 mIU/mL (1.0-9.0 mIU/mL) (SI: 1.0 IU/L [1.0-9.0 IU/L])
Hematocrit = 44% (41%-51%) (SI: 0.47 [0.41-0.51])
Chemistry panel (including TSH, 25-hydroxyvitamin D, calcium, and creatinine), normal
PSA = 2.0 ng/mL (<7.0 ng/mL) (SI: 2.0 μg/L [<7.0 μg/L])

DXA documents the following values:

Lumbar spine T-score = −2.3
Femoral neck T-score = −2.0
Total hip T-score = −2.4

Recent chest and abdominal CT shows no vertebral compression fractures. His FRAX score is calculated to be 8.7% for major fractures and 2.8% for hip fracture.

He is not keen on taking oral bisphosphonates or anabolic agents and inquiries about the recent study that evaluated the long-term benefits of testosterone.

Which of the following is the best next step in the management of this patient's osteoporosis?

A. Calcium and vitamin D plus referral to physical therapy
B. Denosumab
C. Teriparatide
D. Testosterone therapy
E. Zoledronic acid

119 A 41-year-old woman with a history of secondary progressive multiple sclerosis, depression, and anxiety disorder presents for evaluation of abnormal thyroid function. Symptoms prompting recent laboratory testing include fatigue, 30-lb (13.6-kg) weight gain over the past year, heat and cold intolerance, constipation, and dysphoric mood. There is subjective weakness of the lower extremities and gait instability. Review of systems is otherwise normal.

Current medications include ocrelizumab; vitamin D_3, 5000 international units daily (125 mcg daily); and biotin, 10 mg 3 times daily. There is no family history of thyroid disease.

On physical examination, her blood pressure is 124/76 mm Hg and pulse rate is 55 beats/min. Her height is 66 in (168 cm), and weight is 159 lb (72.1 kg) (BMI = 25.7 kg/m^2). The thyroid gland is 20 g without thyroid nodules. There are no tremors of the upper extremities, and deep tendon reflexes are 2+ throughout.

Laboratory test results from 2 weeks ago:

TSH = 0.02 mIU/L (0.5-5.0 mIU/L)

Free T_4 = 5.2 ng/dL (0.8-1.8 ng/dL) (SI: 66.93 pmol/L [10.30-23.17 pmol/L])

Which of the following is the most appropriate next step in this patient's management?

A. Evaluate for heterophile antibodies
B. Measure TRAb
C. Measure TSH and free T_4 again after holding biotin for 3 days
D. Prescribe methimazole
E. Perform thyroid uptake and scan

120 A 52-year-old postmenopausal woman with type 2 diabetes and obesity presents for further evaluation. She reports a 3-year history of progressively worsening glycemic control and weight gain despite implementation of diet and exercise. Initially, she was treated with metformin monotherapy. Her current regimen consists of metformin, glipizide, insulin lispro at fixed doses before meals, and insulin degludec. Since starting insulin, she has gained 20 lb (44.1 kg). She uses the Plate Method for meal composition and portions. She exercises 4 to 6 times per week by walking for 15 minutes and lifting weights for 15 minutes. She has significant glycemic variability with postprandial hypoglycemia 3 to 4 times per week. She drinks 1 to 2 glasses of wine with dinner and is trying to cut down. She quit smoking cigarettes 2 years ago. Her most recent hemoglobin A_{1c} value is 9.2% (77 mmol/mol). She plans to stop insulin, lose weight, and improve her glycemic control.

On physical examination, her blood pressure is 123/84 mm Hg and pulse rate is 84 beats/min. Her height is 62 in (157.5 cm), and weight is 208 lb (94.3 kg) (BMI = 38.0 kg/m^2)

If insulin is discontinued, which of the following classes of medications is the best option to achieve her goals of weight loss and improved glycemic control?

A. Dual receptor agonist (GLP-1 and glucose-dependent insulinotropic polypeptide)
B. GLP-1 receptor agonist
C. SGLT-2 inhibitor
D. Thiazolidinedione
E. Triple receptor agonist (glucagon, GLP-1, and glucose-dependent insulinotropic polypeptide)

ANSWERS

ENDOCRINE SELF-ASSESSMENT PROGRAM 2025

Part II

1 ANSWER: D) Refer to surgery for right adrenalectomy

The incidental finding of a large, heterogenous adrenal mass mandates consideration of adrenal cortical carcinoma (ACC), pheochromocytoma, and other more rare adrenal tumors. Hormonal testing and imaging studies are most often sufficient for preoperative diagnosis. Unequivocal elevation of plasma and urinary catecholamines and/or normetanephrines, correlating with tumor size, is diagnostic of a pheochromocytoma. ACCs often overproduce steroid hormones, particularly cortisol and/or androgens and precursors (such as 11-deoxycortisol and 17-hydroxypregnenolone). When the hormonal workup is unremarkable, as in this vignette, dedicated imaging can often establish the malignant potential of an adrenal mass. Adrenal protocol CT (Answer A) is the imaging of choice for homogenous adrenal masses. Pheochromocytomas and ACCs are richly vascularized, rendering noncontrast CT enhancement greater than 20 Hounsfield units and poor contrast washout. However, these imaging characteristics cannot be accurately assessed in heterogenous masses, so adrenal protocol CT is not the best approach in this case.

Indeterminate adrenal masses should be evaluated with ^{18}F-FDG PET-CT, as was done in this patient. Increased FDG uptake is commonly observed in malignant adrenal masses. Rarely, benign adrenal adenomas can also display increased FDG uptake. An optimal standardized uptake value (SUV) cutoff that accurately distinguishes between malignant and benign adrenal masses has not been established. Nevertheless, uptake at the background level is suggestive of a benign mass. For nonfunctional adrenal tumors with unequivocal benign imaging characteristics, tumor size alone is not a criterion for surgery. The patient in this vignette, however, underwent surgical resection of the tumor (Answer D) because it demonstrated indeterminate imaging features (internal heterogeneity and calcifications), and its large size was associated compressive symptoms. Pathological examination revealed that it was a ganglioneuroma.

Adrenal ganglioneuromas are large, encapsulated tumors, with an average size of 7 cm. Ganglioneuromas are slow-growing, typically benign neuroblastic tumors, and they are related to neuroblastomas and ganglioneuroblastomas. Like pheochromocytomas and paragangliomas, ganglioneuromas are rare, and they are derived from embryonic neural crest cells. They can be located anywhere along the sympathetic chain, such as in the mediastinum, retroperitoneum, and adrenal glands. Ganglioneuromas contain ganglion cells, neurites, and Schwann cells, which help distinguish them histologically from pheochromocytomas/paragangliomas (composed primarily of chief cells and sustentacular cells) and from neuroblastomas (composed of undifferentiated neuroblasts).

Ganglioneuromas tend to occur more frequently in females, and 60% develop in persons younger than 20 years. Most tumors are sporadic, although rare familial forms exist. Ganglioneuromas are typically asymptomatic and are identified incidentally on imaging, as in this vignette. When ganglioneuromas are large, they can cause local compressive symptoms such as dyspnea, constipation, or radicular pain. A minority of ganglioneuromas produce catecholamines, typically in modest amounts, and these cases are usually not associated with symptoms of catecholamine excess. This is in contrast with pheochromocytomas, in which the catecholamine production typically correlates closely with tumor size.

On cross-sectional imaging, ganglioneuromas are large masses with regular margins and a distinct capsule. They are often heterogenous with internal calcifications. Their ^{18}F-FDG uptake is variable, and many ganglioneuromas, but not all, demonstrate increased uptake. MIBG uptake (Answer C) cannot differentiate pheochromocytomas, paragangliomas, and neuroblastomas (commonly MIBG-positive) from ganglioneuromas (50%-60% display MIBG avidity).

Large ganglioneuromas are surgically removed whenever feasible. The capsule can be adherent to important structures, and total excision might not always be possible. Preoperative preparation with α-adrenergic blockers, β-adrenergic blockers, and intravascular volume expansion (Answer E) is necessary for pheochromocytomas and secretory paragangliomas, but not in this patient who showed minimal elevation of normetanephrine.

Observation and periodic surveillance with imaging (Answer B) could be considered in an asymptomatic patient who is older and in whom surgery poses complication risks.

Educational Objective

Consider ganglioneuroma in the differential diagnosis of large adrenal masses and guide its management.

Reference(s)

Fliedner SMJ, Winkelmann PER, Wesley R, Vonthein R, Lehnert H. Ganglioneuromas across age groups: systematic review of individual patient data. *Clin Endocrinol* (Oxf). 2021;94(1):12-23. PMID: 32702779

Decarolis B, Simon T, Krug B, et al. Treatment and outcome of ganglioneuroma and ganglioneuroblastoma intermixed. *BMC Cancer.* 2016;16:542. PMID: 27465021

Fassnacht M, Tsagarakis S, Terzolo M, et al, European Society of Endocrinology clinical practice guidelines on the management of adrenal incidentalomas, in collaboration with the European Network for the Study of Adrenal Tumors. Eur J Endocrinol. 2023;189(1):G1-G42. PMID: 37318239

2 **ANSWER: D) Add ezetimibe, 10 mg daily**

Effective lipid-lowering therapy is an important step in reducing the high risk of atherosclerotic cardiovascular disease (ASCVD) in patients with type 2 diabetes. Based on the beneficial effect seen in multiple primary and secondary prevention trials, there is general agreement that most patients with type 2 diabetes must be on statin therapy of appropriate intensity. High-intensity statin therapy has been recommended for patients with established ASCVD or for those at high risk. Over the last decade, guidelines have shifted their focus from target LDL-cholesterol concentrations to appropriate statin intensity, but more recently the recommendations have refocused on LDL-cholesterol thresholds at which further intensification of lipid-lowering therapy should be considered. Accordingly, recent guidelines from the American Diabetes Association and the American College of Cardiology/American Heart Association guidelines recognize a threshold LDL-cholesterol concentration of 70 mg/dL (1.81 mmol/L) for primary prevention and 55 mg/dL (1.42 mmol/L) for secondary prevention to intensify lipid-lowering therapy.

The patient in this vignette does not have established ASCVD but does have longstanding type 2 diabetes and other risk factors with LDL-cholesterol values above 70 mg/dL (>1.81 mmol/L). Making no changes to his lipid-lowering therapy (Answer E) is therefore inappropriate. He is already on high-intensity statin therapy, so one must now choose the best add-on therapy to lower LDL cholesterol and ASCVD risk. Of the 4 choices, ezetimibe (Answer D) would be the most cost-effective option that has been shown to reduce ASCVD risk when added to statin therapy. Subgroup analysis of the IMPROVE-IT trial showed that risk reduction with addition of ezetimibe to statin therapy in patients with type 2 diabetes was nearly double compared with risk reduction in those without diabetes (13% vs 7%). Ezetimibe is relatively inexpensive and well tolerated and should be the first option as add-on therapy to statins.

Both bempedoic acid (Answer A) and evolocumab (Answer C) reduce LDL cholesterol and ASCVD risk when added to statin therapy in patients with diabetes. But given the higher cost, they should be considered after a trial of ezetimibe if the LDL-cholesterol concentration remains above 70 mg/dL (>1.81 mmol/L). Both bempedoic acid and evolocumab are currently approved only for secondary prevention or in patients with heterozygous familial hypercholesterolemia who have elevated LDL cholesterol despite maximally tolerated statin therapy. Further, use of bempedoic acid in this patient could potentially increase uric acid and the risk for recurrent gout.

Colesevelam (Answer B) is a bile acid–binding resin that is also an effective LDL-cholesterol–lowering agent, but there are limited data on ASCVD outcomes when it is used as add-on therapy. It could also potentially worsen hypertriglyceridemia.

Educational Objective

Guide therapy based on the new LDL-cholesterol treatment target threshold for intensification of lipid-lowering therapy in patients at risk for atherosclerotic cardiovascular disease.

Reference(s)

American Diabetes Association Professional Practice Committee. 10. Cardiovascular disease and risk management: standards of care in diabetes-2024. *Diabetes Care.* 2024;47(Suppl 1):S179-S218. PMID: 38078592

Virani SS, Newby LK, Arnold SV, et al. 2023 AHA/ACC/ACCP/ASPC/NLA/PCNA guideline for the management of patients with chronic coronary disease: a report of the American Heart Association/American College of Cardiology Joint Committee on Clinical Practice Guidelines. *Circulation.* 2023;148(9):e9-e119. PMID: 37471501

Cannon CP, Blazing MA, Giugliano RP, et al; IMPROVE-IT Investigators. Ezetimibe added to statin therapy after acute coronary syndromes. *N Engl J Med.* 2015;372(25):2387-2397. PMID: 26039521

3 **ANSWER: E) Refer for diabetes education**

The patient has type 1 diabetes. Her detectable antibodies and the 2-hour postprandial blood glucose value greater than 200 mg/dL (>11.1 mmol/L), confirmed via a second oral glucose tolerance test, are consistent with type 1 diabetes. Given this new diagnosis, she should be referred for diabetes education (Answer E).

Teplizumab (Answer A) is an anti-CD3 antibody that has been shown to delay progression to clinical type 1 diabetes in participants at high risk. In the trial performed by the Type 1 Diabetes TrialNet Study Group, participants at high risk for development of clinical diabetes were nondiabetic relatives of patients with type 1 diabetes, 8 years or older at the time of random assignment, that had 2 or more diabetes-related autoantibodies detected in 2 samples obtained within 6 months before randomization. In addition, participants had to have evidence of dysglycemia during an oral glucose tolerance test. Dysglycemia was defined as a fasting glucose concentration of 110 to 125 mg/dL (6.1-6.9 mmol/L); a 2-hour postprandial plasma glucose concentration of at least 140 mg/dL (≥7.8 mmol/L) and less than 200 mg/dL (<11.1 mmol/L); or an intervening postprandial glucose concentration at 30, 60, or 90 minutes greater than 200 mg/dL (>11.1 mmol/L) on 2 occasions, within 52 days before enrollment. The protocol was amended in 2014 to allow enrollment of participants younger than 18 years who had a single abnormal oral glucose-tolerance test result because the rates of type 1 diabetes progression were similar with or without a confirmatory oral glucose tolerance test in this age group.

This patient already had 2 positive antibody tests, so assessing additional diabetes-related autoantibodies (Answers C and D) is not necessary to be eligible for teplizumab therapy. However, she was found to have a 2-hour postprandial glucose value greater than 200 mg/dL (>11.1 mmol/L) on 2 occasions. A 2-hour postprandial glucose value greater than 200 mg/dL during a standard oral glucose tolerance test is consistent with a diagnosis of diabetes, not "dysglycemia" (see *table* for a summary of type 1 diabetes staging). Accordingly, this patient has type 1 diabetes and is not eligible for teplizumab therapy (Answer A). Her reported low-carbohydrate diet and active lifestyle may have kept the diagnosis from being recognized (ie, her blood glucose or hemoglobin A_{1c} were not very high at this time, and she was also asymptomatic). However, over time, as more β-cell function is lost, her blood glucose control would have slowly deteriorated until the formal diagnosis was recognized.

Repeating the hemoglobin A_{1c} measurement (Answer B) would not change the results of the 2-hour oral glucose tolerance tests, which were consistent with the diagnosis of diabetes.

A recent study investigated the role of teplizumab therapy in patients with recent-onset type 1 diabetes (Recent-Onset Type 1 Diabetes Trial Evaluating Efficacy and Safety of Teplizumab [PROTECT]). In this phase 3, randomized, placebo-controlled trial, β-cell preservation was evaluated in children and adolescents who were assigned to receive teplizumab or placebo for two 12-day courses. The primary end point was the change from baseline in β-cell function, as measured by stimulated C-peptide levels at week 78. The key secondary end points were the insulin doses that were required to meet glycemic goals, hemoglobin A_{1c} levels, time in the target glucose range, and clinically important hypoglycemic events. Patients treated with teplizumab (n = 217) had significantly higher stimulated C-peptide levels than those of patients receiving placebo (n = 111) at week 78 (least-squares mean difference, 0.13; 95% CI, 0.09-0.17; $P < .001$). Of patients treated with teplizumab, 94.9% (95% CI, 89.5-97.6) maintained a clinically meaningful peak C-peptide concentration of 0.2 pmol/mL or greater, as compared with 79.2% (95% CI, 67.7-87.4) of those receiving placebo. Unfortunately, while the trial met its primary end point of C-peptide preservation with teplizumab at week 78 vs placebo, the groups did not differ significantly regarding the key secondary end points. As of 2024, teplizumab is not approved for the treatment of patients with recent-onset type 1 diabetes (ie, stage 3 of type 1 diabetes) (*see table*).

Table. Staging of Type 1 Diabetes

	Stage 1	Stage 2	Stage 3
Characteristics	• Autoimmunity • Normoglycemia • Presymptomatic	• Autoimmunity • Dysglycemia • Presymptomatic	• Autoimmunity • Overt hyperglycemia • Symptomatic
Diagnostic criteria	• Multiple islet autoantibodies • No IGT or IFG	• Islet autoantibodies (usually multiple) • Dysglycemia: IFG and/or IGT • FPG 100–125 mg/dL (5.6–6.9 mmol/L) • 2-h PG 140–199 mg/dL (7.8–11.0 mmol/L) • A1C 5.7–6.4% (39–47 mmol/mol) or ≥10% increase in A1C	• Autoantibodies may become absent • Diabetes by standard criteria

FPG, fasting plasma glucose; IFG, impaired fasting glucose; IGT, impaired glucose tolerance; 2-h PG, 2-h plasma glucose.

Educational Objective

Determine when treatment with teplizumab therapy is appropriate for at-risk relatives of patients with type 1 diabetes.

Reference(s)

Herold KC, Bundy BN, Long SA, et al; Type 1 Diabetes TrialNet Study Group. An anti-CD3 antibody, teplizumab, in relatives at risk for type 1 diabetes. *N Engl J Med.* 2019;381(7):603-613. PMID: 31180194

Ramos EL, Dayan CM, Chatenoud L, et al. Teplizumab and β-cell function in newly diagnosed type 1 diabetes. *N Engl J Med.* 2023;389(23):2151-2161. PMID: 37861217

ElSayed NA, Aleppo G, Aroda VR, et al American Diabetes Association. 2. Classification and diagnosis of diabetes: standards of care in diabetes-2023. *Diabetes Care.* 2023;46(Suppl 1):S19-S40. PMID: 36507649

ElSayed NA, Aleppo G, Aroda VR, et al. Addendum. 2. Classification and diagnosis of diabetes: standards of care in diabetes-2023. *Diabetes Care.* 2023;46(Suppl 1):S19-S40. *Diabetes Care.* 2023;dc23ad08. PMID: 37356047

4 **ANSWER: A) Bisphosphonate-induced orbital inflammation**

Bisphosphonates are potent antiresorptive agents that inhibit osteoclastic bone resorption and are widely used for the treatment of osteoporosis. The nitrogen-containing aminobisphosphonates trigger an immune cascade via activation of gamma delta T cells and the release of cytokines, including interleukin-1, interleukin-6, and acute-phase proteins. While acute-phase reactions can occur in up to one-third of individuals receiving the first dose of bisphosphonate therapy, inflammatory adverse effects involving the eyes and periocular structures (eg, anterior uveitis, episcleritis, and scleritis) are not commonly reported. In the HORIZON pivotal fracture trial, eye pain was noted in 0.2% of patients taking bisphosphonates (vs 0% of control participants), and "eye inflammation" was noted in 0.4% (vs 0.1% of control participants). While the exact mechanism is not well understood, one theory is that orbital fibroblasts, unlike other fibroblasts, are of neural ectodermal origin and have a greater inflammatory response to stimulation than fibroblasts in the skin and lung, for example. Bisphosphonate-associated orbital inflammation usually occurs within 6 to 72 hours of treatment. In addition to pain and restricted eye movement, other symptoms include lid edema, conjunctival hyperemia, and chemosis.

Most reported cases of bisphosphonate-induced orbital inflammation have been with zoledronic acid, probably because of its higher potency, but it has been reported with alendronate as well. There is no consensus on rechallenge with a bisphosphonate after an adverse ocular event and this should be assessed on an individual basis. Intraocular or topical nonsteroidal anti-inflammatory agents, and sometimes systemic glucocorticoids, can be recommended in addition to ophthalmology consultation.

In this vignette, there is no evidence of metastatic disease, serum calcium is not elevated based on the normal metabolic panel, and imaging does not show masses. In addition, the strong correlation of symptom onset after drug exposure makes metastatic breast cancer (Answer C) an unlikely diagnosis. Orbital imaging can be very useful in supporting a diagnosis of bisphosphonate-associated orbital inflammation and can confirm the presence of orbital inflammatory changes and exclude other diagnoses, such as orbital metastases.

The absence of clinical infectious parameters makes orbital cellulitis (Answer D) an unlikely diagnosis. There is also a strong correlation between drug administration and symptom onset. In orbital cellulitis, CT shows thickening of the fat (best appreciated in the preseptal space), fat infiltration, and contrast enhancement.

This patient presents with flulike symptoms and headache, which raises concern for giant-cell arteritis (Answer B). Other symptoms associated with this condition include fatigue, weight loss, and jaw pain caused by mastication. Ocular involvement seen in giant-cell arteritis varies from transient vision loss (amaurosis fugax) to permanent vision loss caused by anterior or posterior ischemic neuropathy or central retinal artery occlusion. Fundoscopic examination can help identify the cause of vision loss. A characteristic laboratory abnormality in many patients with giant-cell arteritis is a high erythrocyte sedimentation rate, which can reach 100 mm/h. This patient's near-normal erythrocyte sedimentation rate and symptom onset after administration of a new drug make giant-cell arteritis a less likely diagnosis.

Thyroid eye disease (Answer E) is an autoimmune condition of the orbit, most often associated with Graves hyperthyroidism. This patient received radioactive iodine therapy to treat Graves disease, and radioactive iodine has been shown to increase the risk of Graves orbitopathy. However, the normal TSH and free T_4 concentrations and the occurrence of symptoms after administration of bisphosphonates makes thyroid eye disease unlikely.

Educational Objective

Explain the association between bisphosphonate use and orbital inflammation, a condition that is typically self-limiting.

Reference(s)

Reid IR, Gamble GD, Mesenbrink P, Lakatos P, Black DM. Characterization of and risks for the acute-phase response after zoledronic acid. *J Clin Endocrinol Metab.* 2010;95(9):4380-4387. PMID: 20554708

Chang HY, Chi JT, Dudoit S, et al. Diversity, topographic differentiation, and positional memory in human fibro- blasts. *Proc Natl Acad Sci U S A.* 2002;99(20):12877-12882. PMID: 12297622

van Steensel L, Paridaens D, Dingjan GM, et al. Platelet- derived growth factor-BB: a stimulus for cytokine production by orbital fibroblasts in Graves' ophthalmopathy. *Invest Ophthalmol Vis Sci.* 2010;51(2):1002-1007. PMID: 19797221

5 ANSWER: B) Sellar MRI

This patient has symptoms suggestive of hypogonadism (decreased libido and decreased early-morning erections), and his testicles are large (≥30 cc). He has low to low-normal serum total and calculated free testosterone concentrations measured with a high-quality assay. He also has markedly elevated serum FSH, low-normal serum LH, and slightly elevated serum prolactin. This constellation of symptoms, large testes, and laboratory results suggests hypogonadism due to an FSH-secreting macroadenoma that is causing mild hyperprolactinemia by compressing the pituitary stalk. Sellar MRI (Answer B) would confirm the diagnosis.

The most common pituitary macroadenomas are nonfunctional gonadotropinomas—tumors that do not typically secrete FSH and LH (and what is secreted has low bioactivity). However, gonadotropinomas rarely produce large amounts of FSH and/or LH with variable bioactivity. Functional gonadotropinomas in men often secrete FSH and/or LH that is bioinactive. These men present similarly to those with classic nonfunctional pituitary tumors, typically macroadenomas with mild hyperprolactinemia due to stalk effect and variable degrees of hypopituitarism, with hypogonadotropic hypogonadism being most common. In men, FSH-producing gonadotropinomas are more common than LH-producing gonadotropinomas. Men with gonadotropinomas that secrete bioactive FSH or LH usually present with macroadenomas that may cause hyperprolactinemia and variable pituitary deficiency. Men with FSH-producing gonadotropinomas may present with large testes (as this patient did) due to FSH trophic effects on the seminiferous tubules of the testes. Serum FSH concentrations are typically much higher than those of serum LH in patients with FSH-producing gonadotropinomas. LH-producing gonadotropinomas are rare in men and may present with elevated serum testosterone.

Seminal fluid analysis (Answer C) in men with functional gonadotropinomas usually demonstrates low sperm concentrations, but the results could be normal.

Measurement of serum estradiol (Answer D) and inhibin B (Answer E) is seldom useful in men and would not be helpful in this patient's evaluation. He needs sellar imaging independent of the results of these hormone measurements. Commonly used estradiol fluoroimmunoassays are not accurate at the low serum estradiol concentrations typical in most men. With an accurate assay, his serum estradiol concentration might be low-normal or normal and would not help discern the etiology of his hypogonadism, large testes, and isolated elevation of serum FSH. Inhibin B assays have not been standardized and provide inconsistent results. With an accurate assay, his serum inhibin B concentration might be low, normal, or high depending on the bioactivity of the FSH in his serum. A high serum inhibin B concentration would be consistent with an FSH-producing adenoma, but a low or normal serum inhibin B concentration would not exclude an FSH-producing adenoma.

Although serum FSH concentrations are higher than serum LH concentrations in men with Klinefelter syndrome, both gonadotropins are elevated in men with this syndrome (unlike in this man who has a low-normal serum LH value). In addition, men with Klinefelter syndrome do not present with large testes. Thus, karyotype analysis (Answer A) is not necessary.

Educational Objective

Evaluate for an FSH-secreting gonadotropinoma in a man with large testes, elevated FSH, low-normal serum testosterone and LH, and slightly elevated prolactin concentrations.

Reference(s)

Chamoun R, Layfield L, Couldwell WT. Gonadotroph adenoma with secondary hypersecretion of testosterone. *World Neurosurg.* 2013;80(6):900.e7-e11. PMID: 23201183

Ntali G, Capatina C. Updating the landscape for functioning gonadotroph tumors. *Medicina (Kaunas).* 2022;58(8):1071. PMID: 36013538

Ntali G, Capatina C, Grossman A, Karavitaki N. Clinical review: Functioning gonadotroph adenomas. *J Clin Endocrinol Metab.* 2014;99(12):4423-4433. PMID: 25166722

Young Jr WF, Scheithauer BW, Kovacs KT, Horvath E, Davis DH, Randall RV. Gonadotroph adenoma of the pituitary gland: a clinicopathologic analysis of 100 cases. *Mayo Clin Proc.* 1996;71(7):649-656. PMID: 8656706

6 **ANSWER: C) Measure IGF-1**

Most patients with acromegaly have acral enlargement and facial changes at diagnosis; however, these symptoms develop insidiously, and diagnosis is usually delayed after onset of clinical manifestations. In some patients, the typical physical changes related to GH excess are not easily recognized. In this patient diagnosed in the context of mass effects, coarsened facial features were apparent only after reviewing old photos. MRI identified a large pituitary adenoma and frontal bossing, with the latter best visualized on sagittal images. Several comorbidities of acromegaly were present (hypertension, prediabetes, osteoarthritis), although these are also common in this age group in the general population. Because delayed diagnosis of GH excess is associated with an increased number of comorbidities, screening for acromegaly with IGF-1 measurement is recommended for all patients with pituitary tumors (per Endocrine Society and Pituitary Society recommendations). Because GH secretion is pulsatile, the GH concentration alone is not sufficient to confirm or rule out GH hypersecretion. Measurement of GH at baseline and every 30 minutes during a 120-minute glucose tolerance test is used to confirm the diagnosis of acromegaly.

Measuring IGF-1 (Answer C) to complete the hormonal evaluation for a pituitary tumor is the best next step in this case. His IGF-1 concentration was 483 ng/mL (63.3 nmol/L), corresponding to a Z-score of 4.2. Acromegaly was confirmed with an oral glucose tolerance test, which revealed the following GH values:

Baseline = 2.1 ng/mL (2.1 μg/L)
Postglucose administration
- 30 minutes = 1.54 ng/mL (1.54 μg/L)
- 60 minutes = 1.35 ng/mL (1.35 μg/L)
- 90 minutes = 1.38 ng/mL (1.38 μg/L)
- 120 minutes = 1.24 ng/mL (1.24 μg/L)

A normal response would have been a GH nadir below 0.4 ng/mL (<0.4 μg/L).

The patient underwent transsphenoidal surgery, and histopathology revealed strong, diffuse nuclear PIT-1 immunoreactivity in the adenoma cells. The pituitary adenoma had mixed populations of cells: one demonstrating predominantly strong immunoreactivity for prolactin and another smaller population of cells with staining for GH, supporting the diagnosis of a mixed somatotroph-lactotroph adenoma.

Cabergoline (Answer E) is not the best next step in management because the diagnostic evaluation has not been completed. Current recommendations from Pituitary Society suggest following acromegaly treatment guidelines in patients with prolactinomas who have autonomous GH secretion. Histological differentiation of mammosomatotroph (combined GH and prolactin secretion from the same cells) and somatotroph-lactotroph adenomas (both cell types present) has prognostic implications. In this case, prolactin significantly improved and IGF-1 normalized postoperatively. Low-dosage cabergoline was started postoperatively, with subsequent prolactin normalization.

When acromegaly is suspected, GH must be measured at baseline and every 30 minutes after glucose administration during oral glucose tolerance testing. Measuring GH only 120 minutes after oral administration of 75 g glucose (Answer B) is not sufficient due to lower sensitivity than multiple measurements after glucose administration.

Endocrine evaluation of the tumor is not yet complete, and surgery (Answer D) would not be indicated if the patient had only prolactin excess. In patients with prolactinoma, dopamine agonists are recommended as the primary therapy. Rarely, patients with acute apoplexy in the setting of large macroadenomas develop severe neurological and ophthalmological consequences and urgent surgery is necessary as the primary therapy.

The patient has no clinical features of hypercortisolism, and his ACTH elevation is minimal. Although this is not a screening test for Cushing syndrome, his morning cortisol value is not elevated. Measurement of 24-hour urinary cortisol excretion (Answer A) is not necessary. Corticotroph and lactotroph cells require different transcription factors for differentiation. Rarely pituitary tumors arise from early pluripotent stem cells, but pituitary surgery would still have been required. In this case, repeated cortisol and ACTH measurements were normal.

Educational Objective

Describe the presentation of mixed GH- and prolactin-secreting pituitary tumors.

Reference(s)

Giustina A, Barkan A, Beckers A, et al. A consensus on the diagnosis and treatment of acromegaly comorbidities: an update. *J Clin Endocrinol Metab.* 2020;105(4):dgz096. PMID: 31606735

Katzelson L, Atkinson JLD, Cook DM, Ezzat SZ, Hamrahian AH, Miller KK; AACE Acromegaly Task Force. American Association of Clinical Endocrinologists medical guidelines for clinical practice for the diagnosis and treatment of acromegaly--2011 update: executive summary. *Endocr Pract.* 2011;17(4):636-646. PMID: 21846619

Giustina A, Barkhoudarian G, Beckers A, et al. Multidisciplinary management of acromegaly: a consensus. *Rev Endocr Metab Disord.* 2020;21(4):667-678. PMID: 32914330

Petersenn S, Fleseriu M, Casanueva FF, et al. Diagnosis and management of prolactin-secreting pituitary adenomas: a Pituitary Society international consensus statement. *Nat Rev Endocrinol.* 2023;19(12):722-740. PMID: 37670148

7 ANSWER: D) *HNF1A*

Monogenic diabetes accounts for less than 5% of all causes of diabetes. Previously referred to as maturity-onset diabetes of the young (MODY), monogenic diabetes is typically diagnosed before age 25 years in the absence of type 1 diabetes autoantibodies and in the presence of a strong family history of diabetes (typically autosomal dominant inheritance). Phenotypes associated with monogenic diabetes are characterized by decreased insulin secretion and impaired glucose sensing with normal insulin sensitivity. Determining the presence of monogenic diabetes significantly influences treatment and prognosis. To date, more than 40 different genetic subtypes of monogenic diabetes have been identified. The 3 most common types of monogenic diabetes in adults are *HNF1A*-related MODY (MODY type 3), *GCK*-related MODY (MODY type 2), and *HNF4A*-related MODY (MODY type 1).

The 2024 American Diabetes Association Standards of Care recommend screening for monogenic diabetes in the following instances:

- Diabetes diagnosed in the first 6 months of life
- Diabetes in children or young adults who have negative type 1 antibodies and absence of obesity in the setting of a strong family history of diabetes
- Mild, stable hyperglycemia (100-150 mg/dL [5.6-8.5 mmol/L]) in the absence of obesity

A recent systematic review, however, identified additional criteria to screen for monogenic diabetes:

- Diabetes diagnosed before age 12 months
- Women with gestational diabetes and fasting glucose >99 mg/dL (>5.5 mmol/L) (screen for *GCK* variants)
- Persistent, mild hyperglycemia at any age in the absence of obesity (screen for *GCK* variants)
- All patients younger than 30 years without obesity with negative type 1 antibodies or detectable C-peptide

These latter recommendations were based on data from 100 publications mainly describing European populations, which may limit their application in more diverse populations. The criteria using absence of obesity in nonpregnant adults increases the yield for positive results. In other words, if screening were performed in those with obesity, the percentage of patients positive for monogenic diabetes would be lower but not zero, so determining when to screen patients with obesity requires higher clinical suspicion. Many monogenic diabetes risk calculators have been published in the literature (see Table 2 by Murphy et al). A simple model using clinical variables was developed by the University of Exeter (https://www.diabetesgenes.org/exeter-diabetes-app/ModyCalculator). If the predicted probability is greater than 5%, testing for monogenic diabetes is recommended. Using this calculator, the patient in this vignette has a 75.5% probability of having monogenic diabetes and testing is recommended.

HNF1A (Answer D) encodes the transcription factor HNF1A, which is expressed in hepatocytes and pancreatic islet cells. HNF1A promotes insulin transcription and is also important for β-cell development and function. As such, persons with *HNF1A* pathogenic variants often have impaired insulin secretion, which is exacerbated in pregnancy. They respond well to low-dosage sulfonylureas but can progress to requiring insulin over time. *HNF1A* variants are also associated with lowering of the renal glucose threshold, leading to glycosuria at normal glucose concentrations. Individuals with *HNF1A* pathogenic variants are at increased risk of microvascular and macrovascular disease, in particular cardiovascular mortality.

GCK encodes glucokinase, which phosphorylates glucose to glucose-6-phosphate and regulates glucose sensing. Pathogenic variants in *GCK* increase the glucose threshold for insulin secretion, leading to mild, stable hyperglycemia. Affected individuals are not at increased risk of diabetes-related complications. In most cases, treatment consists of diet alone, except in the setting of pregnancy. If the fetus does not carry the *GCK* variant, maternal hyperglycemia results in fetal hyperinsulinemia and leads to macrosomia. In this case, treatment of maternal hyperglycemia is warranted. However, if the fetus also carries the *GCK* variant, treatment of maternal hyperglycemia is not indicated as the fetus is not affected by maternal hyperglycemia because the fetus's glucose threshold is similarly increased as the mother's.

HNF4A pathogenic variants are associated with impaired insulin secretion presenting in childhood or adolescence. Treatment consists of sulfonylureas, although insulin may be required over time. There is increased risk of both microvascular and macrovascular disease.

Pathogenic variants in *CFTR* (Answer A) cause cystic fibrosis (CF), a multisystemic disease, and cystic fibrosis–related diabetes (CFRD). Screening for CF is now part of newborn screening in the United States (since 2010). CFRD is rarely the sole presentation of CF. CFRD usually presents insidiously in a person with known CF as failure to thrive or a decrease in pulmonary function, or it can occur in the setting of recurrent CF-related pancreatitis or pancreatic exocrine insufficiency. Given the absence of other CF symptoms in this vignette, testing for *CFTR* pathogenic variants is unlikely to reveal the cause of this patient's hyperglycemia.

Pathogenic variants in *FOXP3* (Answer B) are associated with IPEX syndrome (immune dysregulation, polyendocrinopathy, enteropathy, X-linked). IPEX is a rare cause of neonatal diabetes presenting with failure to thrive, dermatitis, and severe diarrhea.

HLA-DQA1 pathogenic variants (Answer C) are associated with autoimmune disease and type 1 diabetes. Testing for *HLA-DQA1* is not currently recommended to diagnose type 1 diabetes, but it has been included in polygenic risk scores for type 1 diabetes.

Pathogenic variants in *TCF7L2* (Answer E) are associated with an increased risk of type 2 diabetes.

The patient in this vignette is a young, healthy woman without obesity but with prediabetes, as well as and diabetes during pregnancy. She has a very strong family history of diabetes, including her younger sister who was recently diagnosed. She most likely has monogenic diabetes, and of the presented answer options, *HNF1A* is the only gene associated with monogenic diabetes. Type 1 diabetes should be excluded; however, autoantibody testing was not listed as an option. Testing for pathogenic variants in *HLA-DQA1* or *TCF7L2* would not establish the cause of her diabetes. While identifying pathogenic variants in either *CFTR* or *FOXP3* would establish a cause, they are unlikely in this patient given her lack of other symptoms and age of disease onset.

Educational Objective

Determine when screening for monogenic diabetes is appropriate.

Reference(s)

American Diabetes Association Professional Practice Committee. 2. Diagnosis and classification of diabetes: standards of care in diabetes-2024. *Diabetes Care.* 2024;47(Suppl 1):S20-S42. PMID: 38078589

Murphy R, Colclough K, Pollin TI, et al. The use of precision diagnostics for monogenic diabetes: a systematic review and expert opinion. *Commun Med.* 2023;3(1):136. PMID: 37794142

8 ANSWER: D) 100 mcg estradiol patch and medroxyprogesterone acetate, 10 mg cycled days 1 to 12 of each month

This patient has primary ovarian insufficiency. Primary ovarian insufficiency is characterized by hypergonadotropic hypogonadism in a woman younger than 40 years who has had amenorrhea for 4 months or more, with 2 serum FSH values (measured at least 1 month apart) in the menopausal range. This condition was once referred to as premature ovarian failure or premature menopause; however, about 50% of such women actually have varying and unpredictable ovarian function, with 5% to 10% conceiving and delivering a child after they are diagnosed. The underlying mechanism of primary ovarian insufficiency is either follicle dysfunction or follicle depletion. Affected women have lower serum estradiol concentrations than other women of similar age, and therefore they need to be treated with physiologic estrogen and progestogen replacement until they reach menopause. Physiologic estrogen and progestogen replacement mitigates the long-term effects of estrogen deprivation on various organ systems (eg, cardiovascular system and bone health).

A dosage of 100 mcg of estradiol daily, administered by transdermal patch, achieves average serum estradiol concentrations in the range of 100 to 200 pg/mL (367-734 pmol/L), which is the average estradiol range in a premenopausal woman. This is considered physiologic replacement of estrogen. Cyclic medroxyprogesterone acetate, 10 mg daily for 12 days of the luteal phase of the month (Answer D), is supported by evidence as the preferred progestin to use for endometrial protection because it fully induces secretory endometrium and, with withdrawal bleeding, provides protection against endometrial cancer while taking exogenous estrogen.

This patient would be fine with an unintended pregnancy, and therefore oral combined hormonal contraception is not indicated. Furthermore, compared with transdermal estrogen, oral estrogen use (Answers A and B) is associated with higher rates of venous thromboembolism, which makes it less favorable as a treatment option unless contraception in the form of oral combined hormonal contraception is needed. If pregnancy is not desired, oral combined hormonal contraception or an intrauterine device, for example, is suggested to prevent the unlikely occurrence of an unintentional pregnancy. Transdermal administration bypasses the liver and avoids the first-pass effect with the associated increase in clotting factors and increased risk of thromboembolic events.

A 50-mcg estradiol patch and micronized progesterone, 100 mg nightly (Answer C), would be a good choice if this woman were older and postmenopausal. Given that she is 37 years old with a diagnosis of primary ovarian insufficiency, she needs physiologic replacement of estradiol, and this regimen would be insufficient. Finally, a 150-mcg estradiol patch and micronized progesterone, 200 mg cycled days 1 to 12 of each month (Answer E), represents an estradiol dosage that is too high, and this answer option also proposes use of cyclical micronized progesterone for uterine protection, which is not supported by evidence.

Educational Objective

Recommend treatment of primary ovarian insufficiency.

Reference(s)

Nelson LM. Clinical practice. Primary ovarian insufficiency. *N Engl J Med.* 2009;360(6):606-614. PMID: 19196677

Kapoor E. Primary ovarian insufficiency: a toolkit for the busy clinician. *Menopause.* 2024;31(1):65-67. PMID: 38086010

Giri R, Vincent AJ. Prevalence and risk factors of premature ovarian insufficiency/early menopause. *Semin Reprod Med.* 2020;38(4-5):237-246. PMID: 33434933

9 ANSWER: A) Increase the calcium citrate dosage to 400 mg 3 times daily

Low bone density on DXA is not always due to osteoporosis, and acquired osteomalacia in an adult may also present with low bone density. This patient has low bone density and stress fractures with biochemical evidence of osteomalacia, including elevated alkaline phosphatase and secondary hyperparathyroidism. Osteomalacia is a common consequence of malabsorptive bariatric procedures. Roux-en-Y gastric bypass leads to the enteral intake bypassing the duodenum and part of the jejunum, which are the main locations of active calcium transport stimulated by 1,25-dihydroxyvitamin D. Thus, even with normal serum 25-hydroxyvitamin D (or vitamin D stores), and in this case high 1,25-dihydroxyvitamin D, calcium absorption is still impaired, as evidenced by low 24-hour urinary calcium excretion. PTH is elevated in response to low intestinal calcium absorption driving the elevation in 1,25-dihydroxyvitamin D and also leading to mild hypophosphatemia due to downregulation of renal phosphate transporters. Alkaline phosphatase is elevated in osteomalacia in response to impaired osteoid mineralization, to limit pyrophosphate accumulation and bring the phosphate-to-pyrophosphate ratio back into balance to promote mineralization. Over time, the incidence of secondary hyperparathyroidism increases if vitamin D and calcium are not adequately supplemented.

In a study from Olmstead County, Minnesota, over a 15-year follow-up period after bariatric surgery (mean, 8.9 years), 58% of persons had fractures, of which 9.8% were spontaneous stress fractures. Relative risk for fracture is increased 2.3-fold after bariatric surgery, with vitamin D deficiency and calcium deficiency as important factors, although calcium deficiency did not reach statistical significance in this study. Bone density loss is greatest in the first year after surgery; early after surgery, elevations in bone resorption markers are apparent. The prevalence of low bone density also increases with subsequent years after surgery. However, this patient's underlying osteomalacia must be managed before addressing her risk for fracture with medications.

In this patient, increasing calcium intake (Answer A), both in terms of dosage and frequency throughout the day, would lead to improved calcium absorption primarily by increasing absorption along the later portions of the intestine dependent on passive transport mechanisms. The absorption of calcium carbonate is more influenced by gastric acid than that of calcium citrate, and calcium citrate may be better absorbed in states of impaired gastric acid secretion. Some studies suggest decreased gastric acid secretion after gastric bypass. Thus, calcium citrate is often recommended for these patients, with better absorption supported by some, but not all, studies of calcium supplementation after gastric bypass surgery. Roux-en-Y gastric bypass is also associated with hyperoxaluria, hypocitraturia, and risk of calcium oxalate nephrolithiasis. Thus, supplementation with calcium citrate may further help address this risk.

Her 25-hydroxyvitamin D concentration is nearly in the normal range on her current intake, which was recently increased. Thus, she should not increase the cholecalciferol dosage to 50,000 international units daily (1250 mcg daily) (Answer B).

The hypophosphatemia is secondary to the high PTH level, and while administering phosphate salts (Answer E) may increase the phosphate concentration, it would not address calcium deficiency and may make it worse by binding calcium and decreasing its absorption. Treating the calcium deficit will eventually improve the serum phosphate concentration by decreasing PTH.

Importantly, osteoporosis medications (Answers C and D) do not treat osteomalacia, and bisphosphonates are mineralization inhibitors that may worsen osteomalacia. Clinically significant improvement in bone mineral density can occur over time after resolution of the osteomalacia.

Educational Objective

Identify osteomalacia as a complication of gastric bypass surgery.

Reference(s)

Nakamura KM, Haglind EG, Clowes JA, et al. Fracture risk following bariatric surgery: a population-based study. *Osteoporos Int.* 2014;25(1):151-158. PMID: 23912559

Sperb LF, Leotti VB, Silveiro SP, de Azevedo MJ, Viana LV. Long-term changes in bone density and bone metabolism after gastric bypass surgery: a retrospective cohort study. *Obes Surg.* 2023;33(3):911-919. PMID: 36609743

10 ANSWER: B) Counsel on the potential drawbacks of using a non–FDA-approved device (DIY AID) and institute a back-up plan in case of device failure

DIY or open-source automated insulin delivery systems combine a pump and a real-time continuous glucose monitor with a controller and an algorithm designed to automate insulin delivery. A high degree of effort is required to start and maintain this type of therapy, and the user must always remain aware of daily activities that might affect blood glucose. These systems are not approved by the US FDA, although efforts are underway to obtain regulatory clearance. The patient should be counseled that the safety and effectiveness of these systems depend on the user being able to appropriately program and adjust the devices. Since the devices have not been reviewed by regulators, there is the potential for increased risk with their use. The current American Diabetes Association (2023) standards of care guidelines state that, "Individuals with diabetes may be using systems not approved by the U.S. Food and Drug Administration, such as do-it-yourself closed-loop systems and others; health care professionals cannot prescribe these systems but should assist in diabetes management to ensure the safety of people with diabetes." Hence, although these systems cannot be prescribed by providers, it is important to keep patients safe if they are using these methods for AID. Part of this entails ensuring patients have a back-up plan in case of pump failure (Answer B). Discharging the patient from the clinical practice without further guidance (Answer C) is incorrect.

Online DIY forums can assist with how to set up and manage these systems, but should not be relied on for individual management. Contributors on the forums may not be health care providers, and thus posted information cannot be interpreted as medical advice for insulin titration (Answer A). In most DIY systems, insulin doses are adjusted based on the pump settings for basal rates, carbohydrate ratios, correction doses, and insulin activity. Therefore, these settings can be evaluated and changed based on the patient's insulin requirements.

There is no indication to switch back to multiple daily injections (Answer D) for a patient who prefers using continuous subcutaneous insulin infusion therapy.

Current literature suggests significant improvement in time-in-range when patients with type 1 diabetes are randomly assigned to the use of DIY pumps compared with sensor-augmented insulin pumps (Answer E). In this patient, the previous use of sensor-augmented pump therapy led to suboptimal glycemic control.

Another alternative, although not listed as an option in this vignette, would be to discuss the possibility of switching to an FDA-approved hybrid closed-loop pump. Evidence suggests that these systems may reduce hemoglobin A_{1c} levels, improve time-in-range, lower the risk of exercise-related hypoglycemia, and have psychosocial benefits. The use of these systems depends on patient preference and appropriate selection of individuals (and/or caregivers) who are capable of safely and effectively using the devices.

Educational Objective

Assist in the diabetes management of patients who use do-it-yourself automated insulin delivery pump systems.

Reference(s)

ElSayed NA, Aleppo G, Aroda VR, et al; American Diabetes Association. 7. Diabetes technology: *standards of medical care in diabetes-2023. Diabetes Care.* 2023;46(Suppl 1):S111-S117. PMID: 36507635

Diabetes Canada Clinical Practice Guidelines Expert Working Group, Halperin IJ, Chambers A, et al. Do-it-yourself automated insulin delivery: a position statement. *Can J Diabetes.* 2023;47(5):381-388. PMID: 37532365

Burnside MJ, Lewis DM, Crocket HR, et al. Open-source automated insulin delivery in type 1 diabetes. *N Engl J Med.* 2022;387(10):869-881. PMID: 36069869

Heinemann L, Lange K. "Do it Yourself" (DIY)-automated insulin delivery (AID) systems: current status from a German point of view. *J Diabetes Sci Technol.* 2020;14(6):1028-1034. PMID: 31875681

11 ANSWER: E) The absorption test was appropriately performed; she has decreased ability to absorb oral levothyroxine and should be referred to a gastroenterologist for further evaluation

Levothyroxine is one of the most widely prescribed medications and is the most common form of thyroid hormone replacement used to correct hypothyroidism. This medication has a half-life of 6 to 7 days and is absorbed in the small intestine within hours after oral ingestion and enters the circulation via various solute carrier transporters. Levothyroxine dosing is based on an individual's weight, etiology of hypothyroidism, degree of serum TSH elevation, TSH goal, impending pregnancy plans, and age. A full replacement dosage in an individual who is completely reliant on thyroid hormone (ie, after total thyroidectomy or in longstanding Hashimoto thyroiditis) is estimated to be 1.6 mcg/kg per day. Recent data demonstrate that switching between different generic formulations of levothyroxine, or whether the individual takes generic or branded levothyroxine, is not associated with significant differences in TSH concentration. TSH values should not be used to assess the appropriateness of levothyroxine dosing in patients with central hypothyroidism; free T_4 levels should be targeted to the upper half of the reference range in such scenarios.

In an individual with primary hypothyroidism, the response to levothyroxine is assessed by measuring serum TSH 4 to 6 weeks after the drug's initiation or a dosage change. However, up to 30% of patients who take an appropriate dosage of levothyroxine do not maintain a stable serum TSH concentration. Confirming with pharmacy records that the patient has been dispensed levothyroxine at appropriate intervals is recommended to assess for potential medication nonadherence. The patient should be counseled to ingest the levothyroxine 60 minutes before breakfast or at bedtime (3 or more hours after the evening meal) and separated at least 4 hours from any use of potentially interfering medications and supplements (eg, calcium, iron) to optimize consistent absorption. If the cause of fluctuating TSH values still cannot be ascertained, an oral levothyroxine absorption test can help delineate the reason for difficult-to-manage hypothyroidism.

An oral levothyroxine absorption test is performed by administering a supraphysiologic single dose (usually 600-1000 mcg) of levothyroxine by mouth after an overnight fast. This high dose is required to achieve a significantly increased serum T_4 concentration above baseline (thus, Answer A is incorrect). Serum free T_4 or total T_4 is measured regularly (every 30-60 minutes up to 6 hours) to assess for peak level, which is usually seen 2 to 5 hours after the levothyroxine loading dose. Serum TSH concentrations do not change over the short duration of the test, and T_3 concentrations have not been well-studied to correlate with absorption (thus, Answers B and C are incorrect).

Poor adherence to levothyroxine is suspected when greater than 60% absorption is observed during the test (ie, correlating to a peak total T_4 value >6 µg/dL [>77 nmol/L] or a minimal rise of free T_4 >0.40 ng/dL [>5.14 pmol/L] above baseline). In contrast, malabsorption of levothyroxine should be suspected when less than 60% absorption is observed during the test (ie, correlating to a peak total T_4 value <6 µg/dL [<77 nmol/L] or the failure of free T_4 to increase >0.40 ng/dL [>5.14 pmol/L] above baseline). This patient's oral levothyroxine absorption test produced a free T_4 rise of only 0.3 ng/dL (3.86 pmol/L), consistent with malabsorption, and she should be referred to gastroenterology for further evaluation (Answer E). In the meantime, her levothyroxine dosage should be increased above her weight-based dosage to overcome the problem of impaired absorption. The inappropriate rise in the free T_4 level after the oral challenge makes Answer D incorrect.

Educational Objective

Interpret the results of an oral levothyroxine absorption test in a patient with primary hypothyroidism who has difficulty maintaining a stable serum TSH concentration.

Reference(s)

Brito JP, Deng Y, Ross JS, et al. Association between generic-to-generic levothyroxine switching and thyrotropin levels among US adults. *JAMA Intern Med.* 2022;182(4):418-425. PMID: 35226058

Brito JP, Ross JS, Sangaralingham L, et al. Comparative effectiveness of generic vs brand-name levothyroxine in achieving normal thyrotropin levels. *JAMA Netw Open.* 2020;3(9):e2017645. PMID: 32997127

Caron P, Declèves X. The use of levothyroxine absorption tests in clinical practice. *J Clin Endocrinol Metab.* 2023;108(8):1875-1888. PMID: 36916146

Jonklaas J, Bianco AC, Bauer AJ, et al; American Thyroid Association Task Force on Thyroid Hormone Replacement. Guidelines for the treatment of hypothyroidism: prepared by the American Thyroid Association task force on thyroid hormone replacement. *Thyroid.* 2014;24(12):1670-1751. PMID: 25266247

Gonzales KM, Stan MN, Morris JC 3rd, Bernet V, Castro MR. The levothyroxine absorption test: a four-year experience (2015-2018) at the Mayo Clinic. *Thyroid.* 2019;29(12):1734-1742. PMID: 31680654

12 ANSWER: C) Continue current regimen

Obesity is a chronic disease that is best managed with a multimodal strategy, including dietary intervention, behavioral modification, and, when necessary, pharmacotherapy or surgical intervention. Several FDA-approved antiobesity medications are available that, with the exception of orlistat, decrease caloric intake by suppressing appetite (*see table*). Current prescribing guidelines state that patients with a BMI of 27 kg/m^2 or greater with a weight-related comorbidity and patients with a BMI of 30 kg/m^2 or greater without a weight-related comorbidity are eligible for an antiobesity medication. Choice of medication should be based on patient factors such as comorbid conditions, drug-drug interactions, acceptability of the adverse effect profile, cost, and accessibility.

Table. FDA-Approved Weight-Loss Medications

Drug and dosage	Year approved	Mean placebo-subtracted weight loss, % in randomized controlled trials
Orlistat, 120 mg 3 times daily	1999	4.0
Phentermine/topiramate, 15/92 mg once daily	2012	8.7
Naltrexone/bupropion, 8/90 mg two tablets twice daily	2014	4.8
Liraglutide, 3 mg once weekly subcutaneously	2014	5.4
Semaglutide, 2.4 mg once weekly subcutaneously	2021	12.4
Setmelanotide, 3 mg once daily subcutaneously	2020, 2022 (Bardet-Biedl syndrome)	NA
Tirzepatide, 15 mg once weekly subcutaneously	2023	17.8

An appropriate response to an antiobesity medication is defined as at least 5% weight loss 3 months after starting the medication. Qualitatively, patients on antiobesity medications often report improved satiety, satiation, and fewer cravings, all of which contribute to a sustainable decrease in caloric intake. The patient in this vignette lost approximately 8% of her body weight 3 months after starting the medication, which categorizes her as a responder. However, her weight loss has slowed significantly, and she now appears to be at a weight plateau.

Weight plateaus often occur approximately 6 months after starting a weight-loss intervention because of biological metabolic adaptation and/or behavioral drift. The patient continues to describe appetite suppression and control of eating behaviors. Therefore, the medication is still exerting a therapeutic effect and should be continued (Answer C). Stopping the medication (Answer D) would most likely lead to weight regain and would not be advisable in the absence of adverse effects or objective signs of intolerance (eg, tachycardia and hypertension).

Augmentation of therapy with metformin (Answer A) or orlistat (Answer B) is incorrect because neither drug has been studied in combination with phentermine/topiramate ER or been shown to offer additional weight loss. While metformin could be considered in light of this patient's hemoglobin A_{1c} value in the prediabetes range, it has not been shown to be beneficial for patients with prediabetes and a BMI less than 35 kg/m^2.

Naltrexone/bupropion (Answer E) is a less-potent appetite suppressant than phentermine/topiramate; therefore, switching to this drug would be inappropriate.

A reasonable alternative that is not presented here would be to increase the dosage of phentermine/topiramate to afford greater appetite suppression and therefore greater weight loss. The patient should be counseled that adverse effects are dose-dependent and are more likely to occur at a higher dosage. However, in this patient, maintaining the same drug dosage is also very reasonable because she describes adequate appetite control.

Setting realistic weight goals before initiating an antiobesity medication is an essential component of counseling patients. This discussion helps patients to be accepting of naturally occurring phenomena such as weight plateaus, which can then be managed expectantly without interrupting therapy.

Educational Objective

Manage antiobesity medications during a weight plateau.

Reference(s)

Chakhtoura M, Haber R, Ghezzawi M, Rhayem C, Tcheroyan R, Mantzoros CS. Pharmacotherapy of obesity: an update on the available medications and drugs under investigation. *EClinicalMedicine.* 2023;58:101882. PMID: 36992862

13 ANSWER: B) Follow-up with plasma metanephrine measurement in 3 to 6 months

Multiple endocrine neoplasia type 2 (MEN 2) refers to a set of autosomal dominant disorders caused by gain-of-function variants in the *RET* proto-oncogene and consequent predisposition to multicentric tumor formation in organs where the gene is expressed. MEN 2 is further subclassified into 2A and 2B. MEN 2A (as this patient has) enhances the risk of medullary thyroid cancer (MTC), pheochromocytoma, and primary parathyroid hyperplasia. MEN 2B predisposes to MTC, pheochromocytoma, mucosal neuromas, and intestinal ganglioneuromas and is associated with a marfanoid habitus.

Virtually all individuals with MEN 2 develop MTC, often early in life. Persons with pathogenic variants in codon 634 (as in this patient) often develop MTC in the first 2 to 5 years of life, and risk-reducing thyroidectomy is recommended before age 5 years.

Pheochromocytoma occurs in approximately 50% of persons with MEN 2, and the penetrance varies according to the specific *RET* pathogenic variant. Of persons with *RET* codon 634 variants, 25% develop pheochromocytoma by age 30 years, 52% by age 50 years, and 88% by age 77 years. Yearly screening for pheochromocytoma is recommended in all individuals with MEN 2, and measurement of plasma metanephrines is often sufficient for this purpose.

Pheochromocytomas associated with MEN 2 preferentially produce metanephrines, and affected individuals often develop hypertension and suggestive symptoms, such as episodic headaches, sweating, palpitations, and anxiety. Unlike sporadic pheochromocytomas that are typically unilateral, pheochromocytomas associated with MEN 2 are often bilateral. However, MEN 2A–associated pheochromocytomas are almost always benign, and prophylactic bilateral adrenalectomy is not recommended. Because the progression of pheochromocytoma is commonly very slow, and a contralateral pheochromocytoma might not develop for many years, unilateral adrenalectomy (Answer E) is usually pursued.

In patients with bilateral pheochromocytomas on presentation, cortical-sparing surgery (Answer C) can be attempted when feasible to preserve some residual adrenal cortical function and delay iatrogenic primary adrenal insufficiency. However, because patients with MEN 2 often have diffuse medullary disease, cortical-sparing adrenal surgery is associated with an increased risk of pheochromocytoma recurrence, which should be discussed with the patient. The timing of surgery is personalized. In this patient, given that the pheochromocytoma is very small and her metanephrine elevation is minimal, close monitoring (Answer B) is reasonable, with a plan for surgical intervention when the disease becomes clinically relevant. While the patient reports symptoms that could be attributed to a pheochromocytoma, they were present when monitoring for pheochromocytoma was negative and they have remained disproportionate relative to the small tumor size and hormonal abnormality. As targeted therapy for migraine headaches and anxiety have not yet been initiated, these issues can be addressed by her primary care physician.

Preoperative α-adrenergic blockade, followed by β-adrenergic blockade, should be initiated in patients with pheochromocytomas and secretory paragangliomas to target low-normal blood pressure. Calcium channel blockers are acceptable alternatives when the former therapy is not tolerated or not available. Oral hydration and salt intake are needed to overcome the catecholamine-induced volume contraction and orthostasis associated with α-adrenergic blockade.

Decreasing the levothyroxine dosage (Answer A) while the TSH concentration is low-normal is not mandatory in a young patient, but it should be considered in older patients treated with levothyroxine because of the risk of tachyarrhythmia and bone loss. An abnormally low TSH value would indicate overreplacement with levothyroxine, which could contribute to sweating, anxiety, and weight loss—symptomatology also observed in individuals with pheochromocytomas.

Performing a ^{68}Ga-DOTATATE scan (Answer D) is not necessary in this patient with MEN 2A who has a high pretest probability of pheochromocytoma and in whom cross-sectional imaging and plasma metanephrine results are both indicative of a small tumor. ^{68}Ga-DOTATATE is sensitive for the detection of primary tumors in patients with clinical and biochemical evidence of pheochromocytoma/paraganglioma who have negative abdominal cross-sectional imaging or when metastases are suspected.

Educational Objective

Guide the surveillance for and management of pheochromocytoma in patients with multiple endocrine neoplasia type 2A.

Reference(s)

Thosani S, Ayala-Ramirez M, Palmer L, et al. The characterization of pheochromocytoma and its impact on overall survival in multiple endocrine neoplasia type 2. *J Clin Endocrinol Metab.* 2013;98(11):E1813-E1819. PMID: 24030942

Neumann HP, Berger DP, Sigmund G, et al. Pheochromocytomas, multiple endocrine neoplasia type 2, and von Hippel-Lindau disease. *N Engl J Med.* 1993;329(21):1531-1538. PMID: 8105382

14 ANSWER: A) Hepatitis B, pneumococcal, inactive influenza, zoster

Part of a comprehensive medical visit for patients with diabetes is providing routine vaccinations. As was seen during the COVID-19 pandemic, individuals with diabetes are at higher risk for severe infections requiring hospitalization and ventilation, as well as increased risk of death from infection. Vaccination reduces mortality and severity of infectious diseases. The CDC has a list of recommended vaccinations for persons with diabetes based on age.

This patient requires many vaccines based on CDC recommendations. The COVID-19 vaccine booster should be given 8 weeks after the last dose of their COVID-19 vaccine. Either Moderna or Pfizer boosters are recommended. Tdap should be given every 10 years. Hepatitis B vaccination is recommended for all adults younger than 60 years. Persons with diabetes are at higher risk of hepatitis B infections and should be fully vaccinated. Hepatitis B is a 2- or 3-dose series. In adults who are unsure whether they received hepatitis B vaccination previously or lack evidence of immunity, the CDC recommends administering 2 to 3 doses of hepatitis B vaccination to achieve immunity. There is no guidance on routine hepatitis A vaccination in persons with diabetes or in cruise ship travelers.

Persons with diabetes are at high risk for bacterial pneumonia infections and have a 50% mortality rate for nosocomial pneumonia. Pneumococcal vaccination is recommended even if a person with diabetes was previously infected with pneumonia. Currently, there are several pneumonia vaccines available. Pneumococcal conjugate vaccines are PCV13, PCV15, and PCV20. The pneumococcal polysaccharide vaccine is PCV23. Children with diabetes are recommended to receive 4 doses of PCV13 or PCV15 before age 15 months followed by 1 dose of PCV23 between ages 6 and 18 years. Adults younger than 65 years should receive PCV15 or PCV20. Adults 65 years or older should receive 1 dose of PCV20. For adults who received PCV13 or PCV15, PCV23 should then be administered 1 year after. If an adult only receives PCV23, they should receive PCV20 one year after. In this case, adult vaccine status for pneumonia is unknown, so the patient should receive PCV20. If he receives either PCV13 (Answer B) or PCV23 (Answer D), he should follow-up in 1 year with PCV23 or PCV20, respectively. The table summarizes the vaccine schedule.

<table>
<tr><th></th><th>PCV13</th><th>PCV15</th><th>PCV20*</th><th>PCV23</th></tr>
<tr><th>Vaccine Type</th><th>Conjugate</th><th>Conjugate</th><th>Conjugate</th><th>Polysaccharide</th></tr>
<tr><td colspan="5">Administration guidelines by age</td></tr>
<tr><td>0-15 months</td><td colspan="2">4 doses of either</td><td></td><td></td></tr>
<tr><td>6-17 years</td><td></td><td></td><td></td><td>1 dose</td></tr>
<tr><td>18-64 years</td><td></td><td colspan="2">1 dose of either</td><td>1 dose 1 year after receiving PCV13 or PCV15</td></tr>
<tr><td>≥65 years</td><td></td><td></td><td>1 dose</td><td></td></tr>
</table>

*1 dose of PCV20 is recommended 1 year after PCV23 in adults aged 18 to 64 years if they never received any other pneumococcal vaccine before PCV23.

Influenza vaccination is recommended for everyone 6 months or older. In persons with diabetes, influenza vaccination is associated with decreased numbers of hospitalizations and all-cause mortality. Influenza vaccination is recommended annually. The active influenza vaccine is administered via nasal spray and is recommended for patients aged 2 to 49 years. In persons with diabetes, however, the CDC and the American Diabetes Association recommend the inactive influenza vaccine (thus, Answer C is incorrect). In persons 65 years and older, the high-dose quadrivalent vaccine is recommended. In this case, therefore, the patient should receive the inactive influenza vaccine.

Zoster vaccination with 2 doses is recommended for all adults 50 years and older regardless of diabetes status or previous zoster vaccination. This patient should therefore receive zoster vaccination. Varicella vaccination is recommended for everyone who has never had chickenpox and never been vaccinated. He had chickenpox as a child, so varicella vaccination (Answer E) is not necessary.

Not included in the options is the respiratory syncytial virus (RSV) vaccine which is recommended for all adults 60 years and older.

Given these recommendations, this patient should receive hepatitis B, PCV20, inactive influenza, and zoster vaccination (Answer A).

Educational Objective

Recommend appropriate immunizations for patients with diabetes.

Reference(s)

Centers for Disease Control. Recommended adult immunization schedule for ages 19 years or older. 2024. https://www.cdc.gov/vaccines/schedules/downloads/adult/adult-combined-schedule.pdf

American Diabetes Association Professional Practice Committee. 4. Comprehensive medical evaluation and assessment of comorbidities: standards of care in diabetes-2024. *Diabetes Care.* 2024;47(Suppl 1):S52-S76. PMID: 38078591

15 ANSWER: A) Avoid high-fat diet and weight gain

This patient has hypocholesterolemia with hepatic steatosis and a family history of liver cirrhosis. Even though lipid levels in other family members are not available, this clinical picture is strongly suggestive of familial hypobetalipoproteinemia.

Hypocholesterolemia occurs in a variety of acquired and inherited disorders. Low lipid concentrations are most often secondary to other systemic illnesses and catabolic states such as malnutrition, malabsorption, chronic infections, and malignancies. Hyperthyroidism and advanced liver disease are other common causes of hypocholesterolemia. When these acquired causes are ruled out, primary genetic disorders causing hypocholesterolemia must be considered. These include heterozygous and homozygous hypobetalipoproteinemia due to apolipoprotein B pathogenic variants, and abetalipoproteinemia due to biallelic *MTTP* (microsomal triglyceride transfer protein) pathogenic variants. Other rare conditions include chylomicron retention disease due to *SAR1B* pathogenic variants and activating *PCSK9* variants causing hypocholesterolemia (*see table*).

Heterozygous familial hypobetalipoproteinemia is a common autosomal codominant disorder with an estimated prevalence of 1 in 3000. Most affected patients are asymptomatic but are at risk for fatty liver disease. Due to defective apolipoprotein B_{100} synthesis, VLDL secretion is impaired, thus causing hepatic accumulation of triglycerides. Fat malabsorption can similarly occur due to defective apolipoprotein B_{48} and chylomicron synthesis in the gut, but this is rare and is more often seen in patients with homozygous variants in the gene encoding apolipoprotein B or in those with abetalipoproteinemia.

Hypobetalipoproteinemia should be suspected in patients with fatty liver disease who do not have obesity and have low cholesterol and apolipoprotein B concentrations. Obtaining lipid profiles in a patient's parents and other first-degree relatives is very helpful for making the diagnosis. No specific therapy is required, but patients should be counseled to avoid both weight gain and a high-fat diet (Answer A) because these factors can increase the risk for fatty liver disease. Supplementation with fat-soluble vitamins should be considered in patients with evidence of fat malabsorption and low vitamin levels.

This patient does not have steatorrhea or evidence of vitamin deficiencies, and therefore starting vitamin supplements (Answer E) is less important now. However, high-dose, fat-soluble vitamins are essential in treating patients with homozygous hypobetalipoproteinemia, abetalipoproteinemia, and chylomicron retention disease who have undetectable apolipoprotein B levels and are at risk for neurological and retinal manifestations developing early in life. This patient is not at increased risk for cancer or cardiovascular disease, and therefore statin initiation (Answer D) or early screening for colon cancer (Answer B) is not required.

Table. Clinical Features and Causes of Primary Hypocholesterolemia

Disorder	Gene	Inheritance	Pathophysiology	Clinical features
Hypobetalipoproteinemia	APOB (apolipoprotein B)	Autosomal dominant	Defective hepatic (and to lesser extent intestinal) export of triglyceride-rich particles	Low cholesterol and apolipoprotein B, heterozygotes often asymptomatic, fatty liver disease. Homozygotes have features similar to those of persons with abetalipoproteinemia
Abetalipoproteinemia	MTTP (microsomal triglyceride transfer protein)	Autosomal recessive	Defective synthesis of VLDL and chylomicrons	Markedly low/absent apolipoprotein B, fatty liver disease, steatorrhea, neurological and retinal manifestations early in life
Chylomicron retention disease	SAR1B (secretion associated Ras related GTPase 1B)	Autosomal recessive	Defective synthesis of chylomicrons due to effects on intracellular lipid transport, normal hepatic VLDL synthesis	Steatorrhea, early neurological and retinal manifestations, no liver abnormality
PCSK9 activating pathogenic variants	PCSK9 (proprotein convertase subtilisin/kexin type 9)	Autosomal dominant	Increased clearance of LDL particles	No apparent clinical manifestations

GLP-1 receptor agonist therapy (Answer C) reduces hepatic fat accumulation in patients with obesity who have nonalcoholic fatty liver disease, but it has not been studied in patients with hypobetalipoproteinemia. GLP-1 receptor agonists can potentially also cause fat malabsorption and should not be the initial therapy, especially in the absence of diabetes and obesity.

Educational Objective

Identify clinical features of hypobetalipoproteinemia and recommend appropriate management to prevent complications.

Reference(s)

Welty FK. Hypobetalipoproteinemia and abetalipoproteinemia. *Curr Opin Lipidol.* 2014;25(3):161-168. PMID: 24751931

Burnett JR, Hooper AJ, Hegele RA. *APOB*-related familial hypobetalipoproteinemia. 2021 In: Adam MP, Feldman J, Mirzaa GM, et al, eds. *GeneReviews.* Published online May 13, Accessed November 28, 2023. https://www.ncbi.nlm.nih.gov/books/NBK570370/

Buryska S, Ahn JC, Allen AM, Simha V, Simonetto DA. Familial hypobetalipoproteinemia: an underrecognized cause of lean NASH. *Hepatology.* 2021;74(5):2897-2898. PMID: 34091928

16 ANSWER: B) Neck ultrasonography

This patient's clinical and laboratory findings are consistent with overt hyperthyroidism. Thyroid uptake and scan have been completed, demonstrating heterogeneous uptake throughout the enlarged thyroid gland, the presence of a pyramidal lobe (*blue arrow*), and elevated 6- and 24-hour radioiodine uptake values. These findings are compatible with a diagnosis of Graves disease. Measurement of serum TRAb would have been an alternative method to confirm the diagnosis. The thyroid scan also shows a large photopenic defect in the left thyroid lobe (*orange arrow*).

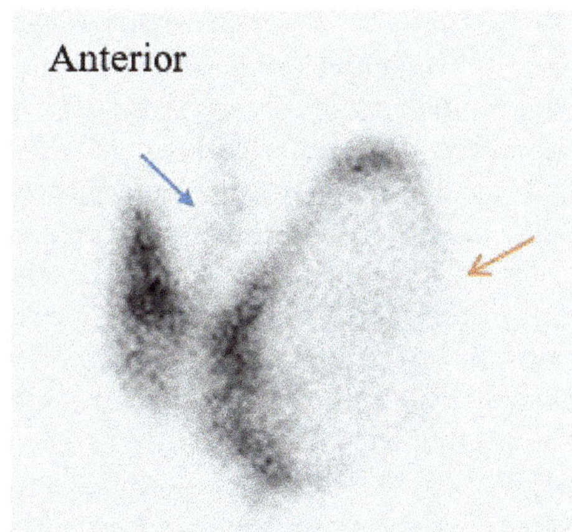

To further evaluate the photopenic area seen on scintigraphy, neck ultrasonography (Answer B) is indicated to confirm the presence of a left thyroid nodule, evaluate its grayscale features, and assess for the presence of abnormal cervical lymph nodes. In this case, neck ultrasonography showed a solid nodule measuring up to 4.7 cm in the left lobe corresponding to the photopenic area on the nuclear medicine study.

The grayscale sonographic features that should be evaluated for all thyroid nodules are composition (cystic, solid, or mixed), echogenicity, shape (measured in the transverse plane), margins, and the presence or absence of echogenic foci. These 5 features are the basis of every sonographic risk stratification system used worldwide, including those from the American Thyroid Association and the American College of Radiology. By combining these grayscale features into discrete risk of malignancy categories, sonographic risk stratification systems can be used to determine a nodule's likelihood of malignancy and to inform patient management, including FNA.

Performing neck CT (Answer A) is not the best next step in this patient's management. She does report a possible compressive symptom (globus sensation), and neck CT may help to delineate the anatomic relationship between the nodule and the esophagus. However, barium esophagography would have the potential to provide more direct evidence that the patient's symptom is the result of extrinsic compression of the esophagus. Moreover, CT does not provide information about the grayscale features of the nodule, which are needed to evaluate its risk of malignancy. CT is not as sensitive as neck ultrasonography for the detection of cervical lymph node metastases from thyroid cancer.

Referral for radioactive iodine therapy (Answer C) or thyroidectomy (Answer E) is not recommended now. This patient's dysphagia is likely related to the thyroid nodule. Because the nodule is nonfunctioning, radioactive iodine therapy would be unlikely to lead to a reduction in its size and resolve her symptoms. Moreover, a malignant left thyroid nodule would need to be ruled out for radioactive iodine therapy to be considered a viable therapeutic option. Although this patient may ultimately choose surgery for management of both Graves disease and the left thyroid nodule, it would be premature to refer for thyroidectomy before performing neck ultrasonography. Ultrasonography of the cervical lymph nodes is needed to evaluate for metastasis in all patients with thyroid nodules. Moreover, perioperative treatment with a thionamide medication (methimazole) would be recommended given the patient's advanced age and overt hyperthyroidism.

Finally, measurement of serum calcitonin (Answer D) is not routinely recommended in the preoperative evaluation of thyroid nodules unless medullary thyroid carcinoma is suspected.

Educational Objective

Identify nonfunctioning thyroid nodule(s) on thyroid scintigraphy in patients with Graves disease and recommend appropriate management.

Reference(s)

Haugen BR, Alexander EK, Bible KC, et al. 2015 American Thyroid Association management guidelines for adult patients with thyroid nodules and differentiated thyroid cancer: the American Thyroid Association Guidelines Task Force on Thyroid Nodules and Differentiated Thyroid Cancer. *Thyroid.* 2016;26(1):1-133. PMID: 26462967

Ross DS, Burch HB, Cooper DS, et al. 2016 American Thyroid Association guidelines for diagnosis and management of hyperthyroidism and other causes of thyrotoxicosis. *Thyroid.* 2016;26(10):1343-1421. PMID: 27521067

Tessler FN, Middleton WD, Grant EG, et al. ACR Thyroid Imaging, Reporting and Data System (TI-RADS): white paper of the ACR TI-RADS Committee. *J Am Coll Radiol.* 2017;14(5):587-595. PMID: 28372962

17 **ANSWER: D) Osteopetrosis**

High bone mineral density (BMD) on routine DXA is not uncommon. While artifactual increases or degenerative disease are most often the cause of high BMD on DXA, increases in BMD can also arise secondary to an underlying disorder with skeletal effects. When evaluating high BMD, it is important to first consider whether it is (1) localized or generalized and (2) acquired or inherited. The most common localized acquired causes include Paget disease, hypophosphatasia, tumors, chronic infectious osteomyelitis, and osseous tuberous sclerosis.

This young man has increased BMD on radiographic evaluation for back pain. High BMD is seen at the spine and hip with sparing of the radius. Osteopetrosis (Answer D) is a rare genetic condition due to reduced osteoclastic bone resorption. It is also known as marble bone disease because of abnormally dense and brittle sclerotic bone. Historically, osteopetrosis is further classified into autosomal dominant type 1 and type 2. Autosomal dominant osteopetrosis type 1 is due to *LRP5* pathogenic variants; however, it is not considered a primary osteoclast disease with increased bone fragility and therefore is not considered to be an osteopetrosis. Autosomal dominant osteopetrosis type 2 (also called Albers-Schnoberg disease) is caused by pathogenic variants in the *CLCN7* gene. The clinical phenotype encompasses facial nerve palsy, visual loss, carpal tunnel syndrome, osteoarthritis of hip, increased fracture risk and delayed fracture healing, osteomyelitis, dental abscesses, and deep decay, and, in extreme cases, bone marrow failure. The clinical presentation is varied due to variable penetrance of 60% to 80%. Classic radiographic features include vertebral end-plate thickening (rugger jersey spine), bone-within-bone (particularly in the pelvis), and transverse sclerotic bands within the distal femur. The radiographs shown in this vignette illustrate a classic example of rugger jersey spine with vertebral end-plate thickening. Genetic testing in this patient documented a *CLCN7* pathogenic variant. The differential diagnosis of rugger jersey spine on radiographs includes osteopetrosis, fluorosis, and renal osteodystrophy.

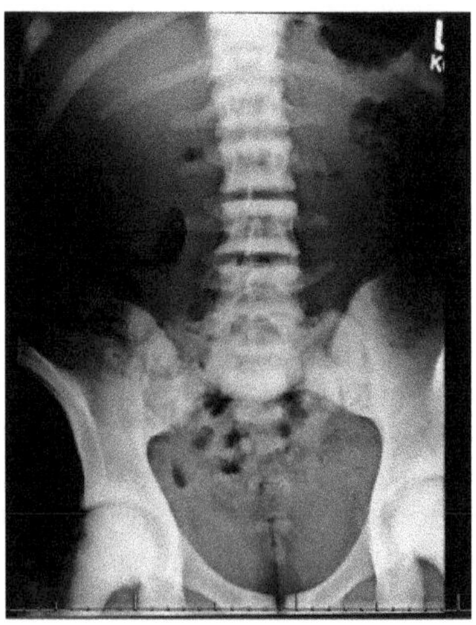

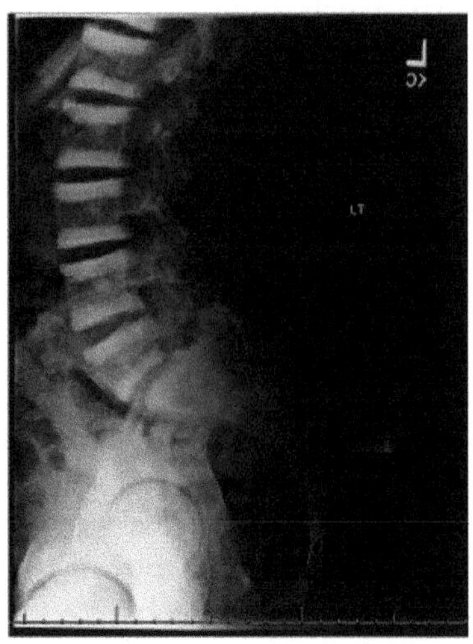

While fluorosis (Answer B) is a possibility, this patient has no history of excess ingestion of tea or well water with high fluoride content and there is no evidence of difluoroethane abuse/huffing.

Normal liver function and absence of risk factors for hepatitis C exclude hepatitis C–associated osteosclerosis (Answer C).

Ochronosis (Answer A), a bluish-black discoloration of tissue, is caused by alkaptonuria, a rare metabolic disorder due to a pathogenic variant in the homogentisate oxidase gene (*HGD*). Deficiency of homogentisate oxidase leads to accumulation of homogentisic acid in the blood and urine and deposition of homogentisic acid in soft tissues, tendons, cartilage, large joints, and intervertebral disks. Ochronosis typically affects the lumbar region and causes narrowing and calcification of the intervertebral spaces. The entire disk is usually sclerotic, and the classic rugger jersey spine appearance is not observed. Skin-lightening creams usually contain hydroquinone and can result in exogenous ochronosis with long-term use. Ochronosis is deemed endogenous when related to alkaptonuria and exogenous when related to the use of skin-lightening agents.

Sclerosteosis (Answer E) is a progressive disease causing generalized enhanced bone formation with increased bone strength and resistance to fracture. Loss-of-function pathogenic variants in the *SOST* gene reduce sclerostin concentrations, thereby activating the osteoblastic Wnt pathway. Sclerosteosis causes gigantism, mandible enlargement, torus palatinus and mandibularis, and cranial and facial nerve compression due to calvarial overgrowth. Cutaneous syndactyly of fingers and toes is an important defining feature. This patient's physical examination findings are not consistent with these features.

Educational Objective

Differentiate among acquired and inherited causes of high bone mass.

Reference(s)

Wu CC, Econs MJ, DiMeglio LA, et al. Diagnosis and management of osteopetrosis: consensus guidelines from the osteopetrosis working group. *J Clin Endocrinol Metab.* 2017;102(9):3111-3123. PMID: 28655174

Whyte MP. Osteopetrosis: discovery and early history of "marble bone disease." *Bone.* 2023;171:116737. PMID: 36933855

Paccou J, Michou L, Kolta S, Debiais F, Cortet B, Guggenbuhl P. High bone mass in adults. *Joint Bone Spine.* 2018;85(6):693-699. PMID: 29407041

18 ANSWER: D) Discuss proper continuous glucose monitor placement techniques with the patient

Continuous glucose monitoring (CGM) devices are important systems for diabetes management, especially among patients who are treated with multiple daily insulin injection regimens. Several CGM systems are commercially available, and some have received FDA clearance for treatment decisions without needing point-of-care fingerstick glucose testing. Unlike point-of-care fingerstick testing (a relatively painful method of glucose measurement as it requires manual penetration of fingertips with lancets to obtain blood), CGM represents a much less painful method of glucose evaluation.

CGM has many other benefits, including assessment of glucose values as frequently as every 1 to 5 minutes. Such frequent monitoring provides a significant amount of information that can be used either retrospectively or in real-time for immediate treatment decisions. High- or low-glucose alarms allow patients with diabetes to take immediate treatment action when necessary. In the scenario of hyperglycemia, when the glucose concentration reaches a value above the higher glucose alarm threshold, patients are alerted so they can administer additional short-acting insulin to correct hyperglycemia. Similarly, CGM devices have low-glucose alarms, which notify patients of impending hypoglycemia. However, to be effective in preventing hypoglycemia, these low-glucose value alarm thresholds should be higher than hypoglycemic values (>70 mg/dL [SI: >3.9 mmol/L]), as patients need time to prevent hypoglycemia before it occurs. Most patients with diabetes who use CGM devices set low-glucose alarm thresholds at 80 mg/dL (4.4 mmol/L).

The combined availability of low-glucose alarms and frequent glucose measurements means that CGM can be used during sensitive periods such as during exercise, where rapid glucose changes can occur, or overnight while patients are sleeping. Although CGM devices are superior to point-of-care blood glucose measurement overnight when patients sleep, they are not without limitations. Importantly, the sensor can be compressed by body positioning (most commonly during sleep), which leads to rapid and sharp decreases in glucose values, creating falsely "low" glucose measurements. When the glucose values fall below the hypoglycemia threshold, the low-glucose alarm is activated. These falsely low glucose values are usually easily corrected when patients simply change position. Patients could check point-of-care glucose fingerstick values to evaluate whether the low glucose values are accurate. As the compression effect is released following repositioning, the glucose concentration rises and usually reaches the same or higher values as before, which can be identified as a "V shape" in the CGM graph (as observed in this patient's CGM tracing). Therefore, discussing proper CGM placement techniques with the patient (Answer D) is the best management option in this vignette. Based on their sleeping pattern, patients can be instructed to place the CGM on areas of the body that are less likely to be affected by compression, assuming that the device has received regulatory clearance to be used on those body areas.

Changing the low-glucose threshold (Answer A) would not help because the compression effect is the main cause of the "low" glucose values. This would also be a relatively unsafe practice, as it would remove one of the main benefits of CGM devices: to prevent hypoglycemia, especially overnight.

One major benefit of CGM is that it has increased our understanding of glucose dynamics in relation to insulin or other conditions (nutrition, exercise, etc). In this case, the rapid decrease of glucose values manifested early in the morning cannot be attributed to basal insulin glargine. Although hypoglycemia can occur with insulin glargine, this type of insulin does not usually create the observed pattern of rapid glucose decrease. If glargine were the cause of this patient's hypoglycemia, her CGM pattern should be different: CGM glucose values would start to decrease 1 to 2 hours (11 PM-12 AM) after insulin administration, eventually leading to low glucose values. However, as this is evident from the CGM report, the patient's drop in glucose values is sudden and occurs many hours after insulin administration. Thus, decreasing the insulin glargine dosage (Answer B) is incorrect.

Although this glycemic pattern could be secondary to short-acting insulin, the patient's last short-acting insulin dose was administrated at 8 PM. As the duration of action of short-acting insulin is 4 to 5 hours, this cannot be the cause of hypoglycemia, which occurs several hours later according to the CGM tracing. Thus, decreasing the short-acting insulin dose with dinner (Answer C) is incorrect.

Although initiating insulin pump therapy (Answer E) is an appropriate option for patients with type 1 diabetes, this action would not correct the false-positive hypoglycemia that occurs secondary to compression of the CGM device.

Educational Objective

Identify compression of continuous glucose monitors as a cause of false hypoglycemia alarms.

Reference(s)

Mensch BD, Wisniewski NA, Neil BM, Burnett DR. Susceptibility of interstitial continuous glucose monitor performance to sleeping position. *J Diabetes Sci Technol.* 2013;7(4):863-870. PMID: 23911167

19 ANSWER: B) Lymphoplasmacytic infiltrate with positive IgG-4 staining

This patient's presentation with anterior hypopituitarism and a mass involving the hypothalamus contiguous to a thickened infundibulum suggests an inflammatory condition. A predominantly nodular mass of the infundibulum would suggest a tumoral etiology, such as Rathke cleft cyst, craniopharyngioma, and rarely metastasis. Inflammatory conditions of the hypothalamus, infundibulum, and pituitary can be primary (usually autoimmune, affecting women more often than men) or secondary to local or systemic processes such as tumors (craniopharyngioma, Rathke cleft cyst), sarcoidosis, histiocytic disorders, Wegener granulomatosis, and infections (tuberculosis, HIV, fungal). This 70-year-old man without a history of neoplastic or infectious disease is unlikely to have lymphocytic hypophysitis. The diagnosis of IgG4-related hypophysitis (Answer B) is suggested by his recent episode of pancreatitis. IgG4-related disease is an immune-mediated fibroinflammatory condition that can affect multiple organs. The clinical spectrum includes autoimmune pancreatitis, sclerosing cholangitis, lymphadenopathy, sialadenitis, and retroperitoneal fibrosis. Up to 8% of patients with IgG4-related disease develop hypophysitis. However, the possibility of IgG4-related hypophysitis should be considered even in the absence of IgG4 systemic disease as shown in a retrospective study that identified IgG4-related hypophysitis in 41% of specimens initially diagnosed as primary autoimmune hypophysitis. Clinically, affected individuals present with multiple anterior hormone deficiencies. They can also develop arginine vasopressin deficiency, which may become apparent after glucocorticoid replacement. These patients can also exhibit mass effect symptoms such as headaches or vision changes from optic nerve compression.

Diagnosis of IgG4-related hypophysitis relies on documenting a serum IgG4 concentration greater than 140 mg/dL and/or anatomic pathological diagnosis of lymphoplasmacytic infiltrate with greater than 10 IgG4-positive cells per high-power field. In this case, the biopsy of the suprasellar mass identified a dense lymphoplasmacytic infiltrate and fibrosis, with IgG4 immunohistochemistry positive in most of the plasma cells.

Recommended treatment for IgG4-related hypophysitis is prednisone, 30 to 40 mg daily, with progressive taper to physiologic dosages. In this patient, after 4 weeks of treatment with 40 mg daily, the nodular component of the suprasellar mass decreased in size and the infundibulum became thinner. This is typical of IgG4-related hypophysitis, but relapse can occur when prednisone is tapered, and rituximab and other steroid-sparing agents can be used.

Necrotizing granuloma (Answer C) should be considered in geographical regions where tuberculosis is endemic and usually occurs in patients with history of extracranial tuberculosis. This diagnosis however is unlikely in this case.

Sarcoidosis affecting the hypothalamus, infundibulum, or pituitary is unlikely in this patient who does not have lung or sinonasal disease. Other manifestations of sarcoidosis such as hypercalcemia are also lacking. The patient's serum ACE concentration is also normal; however, this is not always elevated in individuals with neurosarcoidosis. The gold standard for diagnosis consists of histological demonstration of noncaseating granulomas (Answer D).

Langerhans cell histiocytosis (Answer A) is unlikely in this patient who does not have disease elsewhere (eg, bone, lung, skin) or constitutional symptoms. This is a rare granulomatous disease characterized by abnormal expansion of the dendritic cells with onset usually in childhood and less frequently in young adulthood. *BRAF* V600E pathogenic variants are present in 50% of affected individuals and can be detected in histiocytosis lesions or plasma cell-free DNA.

The patient has infundibular thickening contiguous to the nodular mass, thus papillary craniopharyngioma (Answer E) is an unlikely diagnosis. Craniopharyngioma presents as a nodular mass in the suprasellar region, which can be solid or solid with cystic components. Infundibular thickening is not present.

Educational Objective

Construct the differential diagnosis of inflammatory conditions involving the hypothalamus and pituitary.

Reference(s)

Donegan D, Honegger J. Hypophysitis. *Endocr Pract.* 2022;28(9):901-910. PMID: 35779833

Amirbaigloo A, Esfahanian F, Mouodi M, Rakhshani N, Zeinalizadeh M. IgG4-related hypophysitis. *Endocrine.* 2021;73(2):270-291. PMID: 33837927

Langlois F, Varlamov EV, Fleseriu M. Hypophysitis, the growing spectrum of a rare pituitary disease. *J Clin Endocrinol Metab.* 2022;107(1):10-28. PMID: 34528683

Bando H, Iguchi G, Fukuoka H, et al. The prevalence of IgG4-related hypophysitis in 170 consecutive patients with hypopituitarism and/or central diabetes insipidus and review of the literature. *Eur J Endocrinol.* 2013;170(2):161-172. PMID: 24165017

20 ANSWER: D) Decrease the dosage of testosterone cypionate to 75 mg weekly

Erythrocytosis is a common adverse effect of testosterone therapy in older men. The effects of testosterone on hematocrit are dosage-dependent and might relate more to peak serum testosterone concentrations. Testosterone increases erythropoiesis by direct stimulation of the bone marrow, stimulation of erythropoietin secretion, and suppression of hepcidin that is secreted by the liver. Suppression of hepcidin results in increased iron absorption in the small intestine, release of iron from reticuloendothelial system macrophages, and increased delivery of iron for erythropoiesis to the bone marrow.

Due to changes in metabolism, older men typically require lower testosterone dosages than younger men. Although intramuscular testosterone cypionate 100 mg weekly or 200 mg every 14 days is a typical dosage for young men, many men older than 50 years require only 75 mg weekly or 150 mg every 14 days. This man with erythrocytosis and high-normal serum total testosterone (measured at mid-interval between injections) while taking intramuscular testosterone cypionate 100 mg weekly should have his testosterone dosage reduced (Answer D).

Transdermal testosterone therapy (Answer C) causes erythrocytosis less commonly than intramuscular testosterone. For example, erythrocytosis occurred in 0.2% and 1.5% of testosterone-treated men in the Testosterone Trials and TRAVERSE trial, respectively, whereas 22% of testosterone-treated men in the T4DM study developed erythrocytosis. Transdermal testosterone was used in the Testosterone Trials, and intramuscular testosterone undecanoate was used in the T4DM trial. However, reduction of the intramuscular dosage generally suffices to normalize serum hematocrit, and intramuscular testosterone cypionate is much less expensive than transdermal testosterone. A reduced dosage of intramuscular testosterone cypionate should be tried for this patient who does not want to risk secondary transfer of testosterone to his grandchildren and wife, and who is otherwise doing well on that formulation.

Sleep apnea should be considered in men who develop erythrocytosis on a low or a modest testosterone dosage. It is the most common cause of erythrocytosis in men who develop polycythemia while on a typical dosage of testosterone. This patient has no symptoms of sleep apnea, and there is evidence that his testosterone dosage is excessive. His dosage should be reduced before outpatient polysomnography (Answer A) is performed.

Genetic testing for *JAK2* pathogenic variants (Answer B) causing polycythemia vera should be considered only if erythrocytosis persists after his testosterone dosage has been reduced and sleep apnea has been excluded.

Phlebotomy and blood donation (Answer E) is not usually necessary (and might not be permissible in some regions of the world) for the treatment of testosterone-induced erythrocytosis.

Educational Objective

Recommend a reduction in testosterone dosage to treat testosterone-induced erythrocytosis.

Reference(s)

Coviello AD, Kaplan B, Lakshman KM, Chen T, Singh AB, Bhasin S. Effects of graded doses of testosterone on erythropoiesis in healthy young and older men. *J Clin Endocrinol Metab.* 2008;93(3):914-919. PMID: 18160461

Bachman E, Travison TG, Basaria S, et al. Testosterone induces erythrocytosis via increased erythropoietin and suppressed hepcidin: evidence for a new erythropoietin/hemoglobin set point. *J Gerontol A Biol Sci Med Sci.* 2014;69(6):725-735. PMID: 24158761

Roy CN, Snyder PJ, Stephens-Shields AJ, et al. Association of testosterone levels with anemia in older men: a controlled clinical trial. *JAMA Intern Med.* 2017;177(4):480-490. PMID: 28241237

Pencina KM, Travison TG, Artz AS, et al. Efficacy of testosterone replacement therapy in correcting anemia in men with hypogonadism: a randomized clinical trial. *JAMA Network Open.* 2023;6(10):e2340030. PMID: 37889486

Wittert G, Bracken K, Robledo KP, et al. Testosterone treatment to prevent or revert type 2 diabetes in men enrolled in a lifestyle programme (T4DM): a randomised, double-blind, placebo-controlled, 2-year, phase 3b trial. *Lancet Diabetes Endocrinol.* 2021;9(1):32-45. PMID: 33338415

21 ANSWER: A) 8-AM cortisol and ACTH measurement

This patient has symptoms and signs of hypercortisolism, with thin skin, bruising, and facial plethora. Her medication list includes itraconazole, which is a potent inhibitor of the cytochrome p450 3A4 (CYP3A4) enzyme. Such drugs inhibit the metabolism of some synthetic glucocorticoids (eg, fluticasone, budesonide, triamcinolone, and mometasone), leading to increased glucocorticoid exposure. In the long-term, this leads to the apparent clinical paradox of Cushing syndrome and secondary adrenal insufficiency as a consequence of adrenocortical suppression. Measurement of 8-AM cortisol and ACTH (Answer A) would confirm this diagnosis. Measurement of synthetic glucocorticoid drug concentrations may also help, but these assays are not widely available.

Several drugs are recognized as potent inhibitors of CYP3A4 with the potential to cause Cushing syndrome and adrenal insufficiency. Most commonly, this interaction has been described with protease inhibitors, such as ritonavir, used in combination with inhaled steroid administration, although interactions with intra-articular triamcinolone have also been described. Ritonavir has the greatest inhibitory effect on CYP3A4 and is frequently used in low dosages to "boost" the concentrations of other protease inhibitors, thus allowing less frequent dosing with greater potency. Cobicistat is another antiretroviral agent with potent CYP3A4 inhibitory effects. In contrast to protease inhibitors, the potential interaction of synthetic glucocorticoids with antifungal agents is less well appreciated. Co-administration of itraconazole, which is widely used in the treatment of allergic bronchopulmonary aspergillosis, substantially increases systemic concentrations of inhaled fluticasone. Other potent inhibitors of CYP3A4 include the azole antifungals (voriconazole, ketoconazole, and posaconazole) and the commonly administered antibiotic clarithromycin.

As with other causes of adrenal insufficiency, management should include patient education about stress dosing during intercurrent illness, including access to oral and injectable supplies of hydrocortisone, provision of a steroid emergency card, and education on when to seek emergency care. Physiological replacement dosages of hydrocortisone may also be needed as exogenous glucocorticoid concentrations decrease. Educating the patient and their health care providers is also important in safeguarding against these drug interactions. Additional treatment options include alternative agents to address the underlying disease that do not affect CYP3A4 (eg, non–steroid-interacting antiretrovirals such as integrase inhibitors) and/or reviewing the need for, and dosage of, synthetic glucocorticoid required to control the underlying disease (eg, asthma). Endocrinologists and other specialists must work closely together to avoid these drug interactions.

Screening tests for Cushing syndrome, such as 24-hour urinary cortisol measurement (Answer B), late-night salivary cortisol measurement (Answer D), and the overnight dexamethasone-suppression test (Answer E), are useful when endogenous Cushing syndrome is suspected, but they have no utility in the workup of a patient with features of Cushing syndrome secondary to excess synthetic glucocorticoid exposure, in whom endogenous cortisol production will be suppressed. Dexamethasone is metabolized via CYP3A4, and this has practical implications when interpreting the results of dexamethasone-suppression testing. False-positive results can occur from concomitant treatment with CYP3A4 inducers (eg, carbamazepine, phenytoin, rifampicin, St. John's wort). Simultaneous measurement of dexamethasone along with 9-AM cortisol measurement helps interpretability in such circumstances.

Adrenal antibody assessment (Answer C) would not be indicated in this case given the clear etiology of excess synthetic glucocorticoid exposure as a cause of secondary adrenal insufficiency. Adrenal antibody measurement is recommended in patients with a confirmed diagnosis of primary adrenal insufficiency when this has been confirmed by an elevated ACTH concentration.

Educational Objective

Distinguish adrenal insufficiency from corticosteroid use and coprescription of a cytochrome P450 3A4 inhibitor.

Reference(s)

Naef R, Schmid C, Hofer M, Minder S, Speich R, Boehler A. Itraconazole comedication increases systemic levels of inhaled fluticasone in lung transplant recipients. *Respiration.* 2007;74(4):418-422. PMID: 17003527

Bornstein SR, Allolio B, Arlt W, et al. Diagnosis and treatment of primary adrenal insufficiency: an Endocrine Society clinical practice guideline. *J Clin Endocrinol Metab.* 2016;101(2):364-389. PMID: 26760044

Fleseriu M, Auchus R, Bancos I, et al. Consensus on diagnosis and management of Cushing's disease: a guideline update. *Lancet Diabetes Endocrinol.* 2021;9(12):847-875. PMID: 34687601

22 ANSWER: B) Effect of denosumab cessation

The first and most important thing to consider in this situation is recurrence or worsening of breast cancer (Answer E). However, this patient had a thorough evaluation that did not demonstrate new malignancy or worsening of existing bone lesions. Denosumab and zoledronate are both antiresorptive agents used in the setting of breast cancer to treat or prevent bone metastases and bone complications (eg, from metastases or from effects of anticancer therapies such as aromatase inhibitors). The appropriate long-term duration of treatment with antiresorptive agents in the setting of cancer is uncertain, with studies generally only extending to 5 years of treatment. Bisphosphonates such as zoledronate have a long skeletal half-life, and a rapid rebound elevation of bone turnover is not typical. However, denosumab is a monoclonal antibody to RANKL that does not incorporate into bone. Denosumab loses its effect on resorption as the antibody is metabolized, leading to a rapid rebound of increased bone turnover (Answer B). The elevated bone resorption markers in this patient support a high turnover state with increased resorption as the cause of hypercalcemia.

When dosing for osteoporosis, denosumab (60 mg) is administered every 6 months. When denosumab is continued at this dosage for 2 years and then therapy is stopped, bone resorption increases above baseline for up to 2 years after cessation. When dosing for cancer, it is administered at a much higher dosage of 120 mg every month, resulting in greater suppression of bone resorption, and potentially a higher rebound increase in bone turnover, which may continue to be evident for a longer period. After stopping denosumab, the increased bone turnover is accompanied by decreases in bone density and increased risk of vertebral compression fracture. In clinical studies in children, this effect has also been associated with hypercalcemia during the rebound phase. In adults, hypercalcemia appears to be less common but has been reported.

In response to hypercalcemia, administration of denosumab or intravenous bisphosphonates is typically required to address elevated bone resorption (in addition to usual treatment with intravenous saline).

Hydrochlorothiazide (Answer D) can decrease kidney calcium excretion, but it is unlikely to be the cause of acute hypercalcemia after long-term use without calcium problems.

Granulomatous disease (Answer C) causes hypercalcemia through activation of 1,25-dihydroxyvitamin D. However, this patient's 1,25-dihydroxyvitamin D concentration is low, making granulomatous disease less likely to be the cause.

This patient has an acute increase in creatinine. Acute kidney insufficiency (Answer A) may itself cause hypercalcemia or be caused by hypercalcemia. However, acute kidney insufficiency—if it was the sole cause of hypercalcemia—would be expected to demonstrate low urinary calcium excretion, rather than elevated as seen in this patient. Despite acute kidney insufficiency, she has elevated urinary calcium excretion, along with high (not low) excretion of NTX as a bone resorption marker.

Educational Objective

Identify increased bone turnover and hypercalcemia as complications of denosumab withdrawal.

Reference(s)

Horiuchi K, Kobayashi E, Mizuno T, Susa M, Chiba K. Hypercalcemia following discontinuation of denosumab therapy: a systematic review. *Bone Rep.* 2021;15:101148. PMID: 34825020

Koldkjær Sølling AS, Harsløf T, Kaal A, Rejnmark L, Langdahl B. Hypercalcemia after discontinuation of long-term denosumab treatment. *Osteoporos Int.* 2016;27(7):2383-2386. PMID: 27098536

Bone HG, Bolognese MA, Yuen CK, et al. Effects of denosumab treatment and discontinuation on bone mineral density and bone turnover markers in postmenopausal women with low bone mass. *J Clin Endocrinol Metab.* 2011;96(4):972-980. PMID: 21289258

23 ANSWER: A) Abdominal MRI

This patient has adult-onset consumptive hypothyroidism resulting from a hepatic tumor overexpressing type 3 deiodinase (D3). Consumptive hypothyroidism is a rare paraneoplastic syndrome that was first described in 2000. Approximately 50 cases have been reported in the literature, mostly in children with infantile hepatic hemangiomas that are rich in type 3 deiodinase (D3). High expression of D3 in these individuals converts T_3 and T_4 to the inactive thyroid hormone metabolites T_2 and reverse T_3, respectively, leading to the presentation of usually profound primary hypothyroidism and its associated clinical manifestations. Hemangiomatosis arising from cutaneous or extrahepatic visceral lesions has also been reported in some patients and may serve as a diagnostic clue. The presentation of consumptive hypothyroidism in adults is much rarer, but it has been reported in some patients with gastrointestinal stromal tumors and hepatic hemangiomas that express D3.

Treatment with extremely high thyroid hormone dosages may be required to meet the demands of the increased thyroid hormone metabolic state (ie, the initial description of this condition required 9-fold the typical thyroid hormone dosage for an infant). In children, most tumors are benign and self-involute by age 3 to 9 years, but approximately 10% are aggressive and may be life-threatening. Hemangiomas themselves are usually responsive to shrinkage with steroids and propranolol, which can decrease the disease burden, although surgery, liver transplant, or ligation of the hepatic artery may be required in refractory cases. Liver transplant was required to address the disease burden in this patient, and hypothyroidism and associated clinical stigmata subsequently fully resolved.

Abdominal MRI (Answer A) is the best test to assess for hepatic hemangiomas, which have a typical radiologic appearance as T2-hyperintense masses containing central hypointense regions. In this case, abdominal MRI demonstrated hepatomegaly and the presence of multiple hepatic multifocal hemangiomas, consistent with this diagnosis.

Genetic testing (Answer B) would be indicated for suspicion of a thyroid hormone–resistant state, which would not explain the physical examination findings or be consistent with the thyroid function test abnormalities in this patient.

Measuring serum thyroglobulin (Answer C) has no role in the evaluation of a patient with an intact thyroid gland and should primarily be used to monitor differentiated thyroid cancer after total thyroidectomy. Serum thyroglobulin measurement could be helpful if there were suspicion of exogenous thyroid hormone ingestion, but this is not suggested by the pattern of the thyroid function values in this patient. Similarly, his clinical presentation and laboratory pattern are not consistent with central hypothyroidism, so pituitary-directed MRI (Answer D) would not be helpful. Radioiodine uptake (Answer E) could be performed if there were biochemical evidence of hyperthyroidism, but this is also not the case.

Educational Objective

Diagnose and manage consumptive hypothyroidism resulting from high D3 expression in a case of hepatic hemangiomatosis.

Reference(s)

Luongo C, Trivisano L, Alfano F, Salvatore D. Type 3 deiodinase and consumptive hypothyroidism: a common mechanism for a rare disease. *Front Endocrinol (Lausanne).* 2013;4:115. PMID: 24027558

Huang SA, Tu HM, Harney JW, et al. Severe hypothyroidism caused by type 3 iodothyronine deiodinase in infantile hemangiomas. *N Engl J Med.* 2000 Jul 20;343(3):185-9. PMID: 10900278

Siano MA, Ametrano O, Barbato F, et al. Consumptive hypothyroidism due to hepatic hemangiomas: a case series and review of the literature. *JPGN Rep.* 2022;3(4):e270. PMID: 37168485

Huang SA, Fish SA, Dorfman DM, et al. A 21-year-old woman with consumptive hypothyroidism due to a vascular tumor expressing type 3 iodothyronine deiodinase. *J Clin Endocrinol Metab.* 2002;87(10):4457-4461. PMID: 12364418

Maynard MA, Marino-Enriquez A, Fletcher JA, et al. Thyroid hormone inactivation in gastrointestinal stromal tumors. *N Engl J Med.* 2014;370(14):1327-1334. PMID: 24693892

24 ANSWER: C) Presence of lipohypertrophy

Lipohypertrophy (Answer C) is a common complication of subcutaneous insulin injections and is characterized by abnormal accumulation of fat underneath the skin at injection sites. It presents as thickened, rubbery tissue swelling that can be either soft or firm and may not always be visible. The most common independent risk factors associated with lipohypertrophy are reuse of pen needles, incorrect injection site rotation, and longer duration of insulin therapy. The likelihood of lipohypertrophy in patients who demonstrate appropriate site rotation is 5%, compared with 98% in those who demonstrate incorrect site rotation. A recent study showed that most insulin users (61%) reported injecting into only 1 site. About 48% of patients with lipohypertrophy develop it 1 to 5 years after initiating insulin therapy. Patients with lipohypertrophy are 7 times more likely than patients without lipohypertrophy to have glycemic variability. Patients should be asked about the sites they use to inject, and those sites should be examined. Lipohypertrophy is often mistaken for a lipoma or may be missed completely.

Although inaccurate carbohydrate counting (Answer A) can lead to glycemic variability, it would not be applicable in this patient who is adept at carbohydrate counting.

Lifting skin before administering insulin (Answer B) is not needed for 4-mm needles, but pinching would be unlikely to contribute to glycemic variability. The use of 4-mm needles (Answer E) is unlikely to contribute to increased glycemic variability.

Intramuscular injection (Answer D) can lead to glycemic variability because of unpredictable insulin absorption, and this may lead to episodes of hypoglycemia. Risk of accidental intramuscular injection is directly proportionate to needle length. However, use of 4-mm needles has the lowest likelihood of unintentional intramuscular injection across a wide range of BMIs (19-65 kg/m^2).

Optimal needle insertion technique is shown in the figure.

Figure. Optimal insulin injection technique

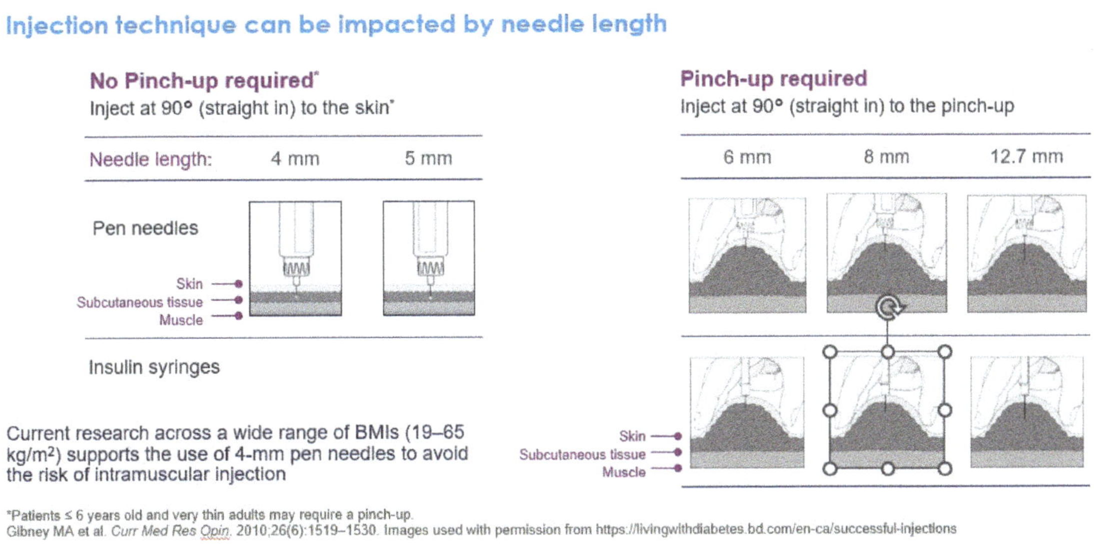

Reprinted from Gibney MA et al. Curr. Med. Res. Opin, 2010; 26(6): 1519-1530. Images used with permission from Embecta, https://livingwithdiabetes.bd.com/en-ca/successful-injections

Educational Objective

Review correct insulin injection techniques and outcomes associated with faulty injection techniques in patients with diabetes.

Reference(s)

Frid AH, Hirsch LJ, Menchior AR, Morel DR, Strauss KW. Worldwide injection technique questionnaire study: population parameters and injection practices. *Mayo Clin Proc.* 2016;91(9):1212-1223. PMID: 27594185

Blanco M, Hernández MT, Strauss KW, Amaya M. Prevalence and risk factors of lipohypertrophy in insulin-injecting patients with diabetes. *Diabetes Metab.* 2013;39(5):445-453. PMID: 23886784

Gibney MA, Arce CH, Byron KJ, Hirsch LJ. Skin and subcutaneous adipose layer thickness in adults with diabetes at sites used for insulin injections: implications for needle length recommendations. *Curr Med Res Opin.* 2010;26(6):1519-1530. PMID: 20429833

Bari B, Corbeil MA, Farooqui H, et al. Insulin injection practices in a population of Canadians with diabetes: an observational study. *Diabetes Ther.* 2020;11(11):2595-2609. PMID: 32893337

25 ANSWER: A) Copper

The frequency of bariatric or metabolic surgery continues to increase worldwide, consistent with the increasing global prevalence of obesity. Bariatric surgery is a safe and effective long-term treatment for obesity when combined with a comprehensive lifestyle intervention program and adequate monitoring of nutritional parameters.

Roux-en-Y gastric bypass is an operation that creates a small gastric pouch (30 mL) and bypass of most of the stomach and first part of the duodenum (*see figure*). As a result, food and drink that is ingested traverses the gastric pouch and directly enters the small intestine before it is digested and absorbed. Anatomical changes resulting from the Roux-en-Y gastric bypass predispose individuals to nutritional deficiencies (eg, iron and vitamin D), which can occur despite adherence to a consistent multivitamin and mineral supplementation program.

Figure. Roux-en-Y gastric bypass.

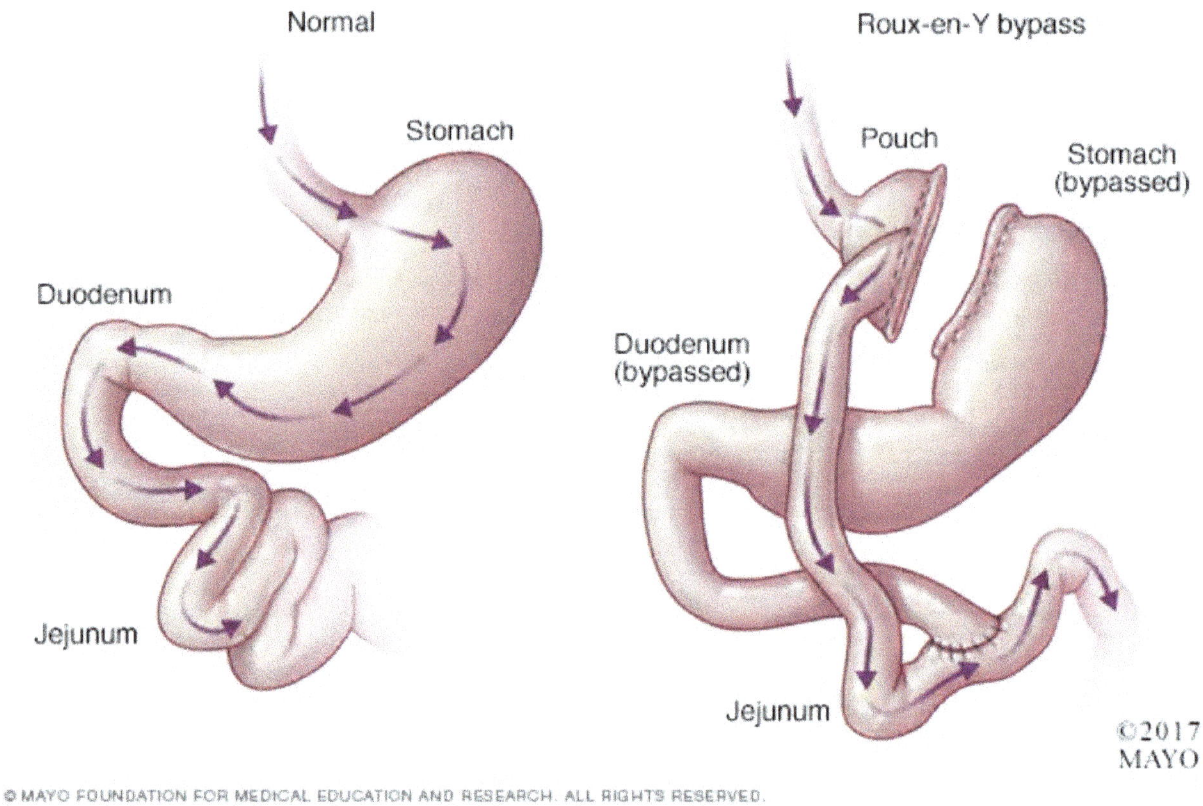

Copper deficiency (Answer A) is a rare but potentially serious sequela of Roux-en-Y gastric bypass. Copper is primarily absorbed in the stomach and proximal duodenum via the copper transporter CTR1. Zinc interferes with intestinal copper absorption by increasing intestinal synthesis of metallothionein, which binds copper and promotes its excretion into the intestinal lumen. Hence, risk factors for copper deficiency include upper gastrointestinal surgery, defects in CTR1, and zinc toxicity. Zinc toxicity (Answer E) is not this patient's diagnosis because she does not present with the typically vague gastrointestinal complaints associated with this condition such as abdominal pain, nausea, and diarrhea. The most common hematological manifestations of copper deficiency are anemia (microcytic, macrocytic, and normocytic) and leukopenia. Thrombocytopenia is rare. Neurological manifestations of copper deficiency include myelopathy and sensory-motor polyneuropathy. Following parenteral copper supplementation, this patient experienced improvement in anemia, leukopenia, and neurological symptoms.

Although iron deficiency (Answer B) is a common complication of Roux-en-Y gastric bypass, it would not be expected to cause febrile neutropenia. Moreover, measuring ferritin 1 month after receiving an iron infusion is too soon to detect persistent iron deficiency; it should be measured about 6 to 8 weeks after the infusion to minimize interference with iron assays.

Vitamin B_{12} deficiency (Answer C) can manifest with pancytopenia and neurological symptoms. However, because the patient takes a parenteral form of B_{12} that does not require gut absorption, she is unlikely to be B_{12} deficient.

Excess or deficient vitamin E (Answer D) is rare after Roux-en-Y gastric bypass, and it does not need to be addressed for most patients because of uncertain clinical significance.

Educational Objective

Diagnose copper deficiency as a rare but serious consequence of bariatric surgery.

Reference(s)

Griffith DP, Liff DA, Ziegler TR, Esper GJ, Winton EF. Acquired copper deficiency: a potentially serious and preventable complication following gastric bypass surgery. *Obesity (Silver Spring)*. 2009;17(4):827-831. PMID: 19148115

Kumar N, McEvoy KM, Ahlskog JE. Myelopathy due to copper deficiency following gastrointestinal surgery. *Arch Neurol*. 2003;60(12):1782-1885. PMID: 14676057

26 ANSWER: B) Measure estradiol

This patient has primary ovarian insufficiency, which is characterized by hypergonadotropic hypogonadism in a woman younger than 40 years who has had amenorrhea for 4 or more months, with 2 serum FSH values (measured at least 1 month apart) in the menopausal range. Previous estimates indicated that 1% of women have primary ovarian insufficiency; however, newer estimates are as high as 3%. This condition was once referred to as premature ovarian failure or premature menopause, but about 50% of such women actually have varying and unpredictable ovarian function, with 5% to 10% conceiving and delivering a child after they are diagnosed. The underlying mechanism of primary ovarian insufficiency is either follicle dysfunction or follicle depletion. These women have lower serum estradiol concentrations than other women of similar age, and therefore they need to be treated with physiologic estrogen and progestin replacement until they reach menopause.

This patient's physiologic dosage of estradiol (100 mcg daily, transdermally) is expected to achieve goal average serum estradiol concentrations in the range of 100 to 200 pg/mL (367-734 pmol/L), which is the average estradiol range in a premenopausal woman. Such estradiol levels maintain bone mineral density and mitigate the increased risk of cardiovascular disease and cognitive impairment with aging observed in women not treated with hormone therapy. However, this patient continues to have breakthrough vasomotor symptoms, low libido, and vaginal dryness. These persistent symptoms, which are secondary to estrogen insufficiency, raise the question of adequate estradiol absorption through her transdermal patch. At this point, the best next step would be to measure estradiol (Answer B). She may need to add an additional estradiol patch to achieve goal serum estradiol levels. Of note, in clinical practice, when instructing patients who are on an estradiol patch to check estradiol levels, many clinicians advise measuring estradiol on a day other than the day they change their patch (or the day before a patch change) to avoid potentially high or low levels related to the patch delivery system (although there are no guidelines stating this).

Adding vaginal estradiol (Answer A) could be a good option if vaginal dryness, most likely related to vaginal atrophy, continues after vasomotor symptoms are treated.

Measuring testosterone (Answer C) is not indicated in this woman because her presenting symptoms are consistent with estrogen insufficiency, and ensuring adequate estrogen replacement should be addressed first.

If she continues to have low libido after all her hormones are optimized and vaginal atrophy is treated, consultation with a sex therapist (Answer D) could be considered as adjuvant care for sexual health, but this is not the best next step.

Vaginal laser treatment (Answer E) to address genitourinary syndrome of menopause (for vaginal atrophy symptoms) is supported by evidence. Laser treatment leads to thermal necrosis, which induces collagen remodeling and synthesis, neovascularization, and elastin formation and may restore premenopausal and estrogen-rich epithelial function. The first step in this case is to address why she continues to have systemic symptoms of primary ovarian insufficiency (breakthrough vasomotor symptoms and low libido). Once resolved, if the symptoms of vaginal atrophy continue then laser treatment can be considered.

Educational Objective
Assess adequate physiologic hormone replacement in a woman with primary ovarian insufficiency.

Reference(s)

Nelson LM. Clinical practice. Primary ovarian insufficiency. *N Engl J Med.* 2009;360(6):606-614. PMID: 19196677

Kapoor E. Primary ovarian insufficiency: a toolkit for the busy clinician. *Menopause.* 2024;31(1):65-67. PMID: 38086010

Giri R, Vincent AJ. Prevalence and risk factors of premature ovarian insufficiency/early menopause. *Semin Reprod Med.* 2020;38(4-5):237-246. PMID: 33434933

Popat VB, Calis KA, Kalantaridou SN, et al. Bone mineral density in young women with primary ovarian insufficiency: results of a three-year randomized controlled trial of physiological transdermal estradiol and testosterone replacement. *J Clin Endocrinol Metab.* 2014;99(9):3418-3426. PMID: 24905063

27

ANSWER: D) Sequential low-dose/high-dose dexamethasone-suppression test

Rare forms of ACTH-independent hypercortisolism with apparently normal adrenal imaging include primary pigmented nodular adrenocortical disease (PPNAD) and nonpigmented forms associated with pathogenic variants in the phosphodiesterase 11A isoform 4 gene (*PDE11A*) or the phosphodiesterase 8B gene (*PDE8B*). On thin-section high-resolution CT, PPNAD might appear as a string of beads, representing multiple subcentimeter nodules spread across an atrophic cortex. Histologically, the adrenal cortex harbors nodules smaller than 4 mm and is composed of cells with lipofuscin content, a coarsely granular, brown pigment.

This patient's clinical presentation (weight gain with central distribution, hyperglycemia, high-normal blood pressure for a young woman) mandates consideration of Cushing syndrome and polycystic ovary syndrome (PCOS). Her hormonal workup is consistent with mild ACTH-independent hypercortisolism (low baseline ACTH and DHEA-S; incompletely suppressed cortisol following 1 mg of dexamethasone, despite full suppression of ACTH). Most cases of mild autonomous cortisol excess are diagnosed during hormonal evaluation of incidentally found adrenal adenomas. Cortisol-producing adenomas are typically sizeable, measuring at least 1 to 1.5 cm. In this patient, however, the adrenal glands appear morphologically normal on cross-sectional imaging.

PPNAD is diagnosed in fewer than 1% of patients with Cushing syndrome. Unlike ACTH-independent hypercortisolism associated with adrenal incidentalomas, which typically presents after the fifth decade of life, PPNAD presents in young patients, including adolescents and children. After puberty, PPNAD affects more females than males. Two-thirds of PPNAD cases are familial and are caused by Carney complex or isolated familial PPNAD. The clinical presentation of hypercortisolism is often subtle and develops gradually over years. Cyclical hypercortisolism has also been reported.

Carney complex is an autosomal dominant multiple neoplasia syndrome characterized by spotty pigmentation, endocrine tumors (PPNAD, acromegaly, thyroid carcinomas), and nonendocrine tumors (cardiac myxomas, breast myxomatosis, blue nevi). The hyperpigmented lentigines are distributed on lips, conjunctiva, inner or outer canthi, and vaginal and penile mucosa.

Most patients with PPNAD have a paradoxical increase in urinary free cortisol during sequential low-dose/high-dose dexamethasone-suppression testing (the Liddle test [Answer D]). The test consists of 24-hour urinary free cortisol measurement at baseline and during dexamethasone suppression at doses of 0.5 mg every 6 hours for 2 days, followed by 2 mg every 6 hours for 2 days. An increase in urinary free cortisol excretion occurs in 69% to 75% of patients with PPNAD. This response is due to glucocorticoid receptor overexpression in the nodules.

Pituitary-directed MRI (Answer C) and ^{68}Ga-DOTATATE scanning (Answer A) are imaging modalities relevant for ACTH-dependent Cushing syndrome, such as Cushing disease or ectopic Cushing syndrome.

In patients with clinically suspected Cushing syndrome due to exogenous glucocorticoids, ACTH and serum cortisol are typically fully suppressed. Most synthetic glucocorticoids are not detected by serum or urinary cortisol assays. As an exception, hydrocortisone is converted to cortisol via the enzyme 11β-hydroxysteroid dehydrogenase type 1 and is captured in the usual cortisol assay. In contrast, cortisol immunoassays have low cross-reactivity with prednisone and prednisolone and no cross-reactivity with dexamethasone. If exogenous glucocorticoid use is suspected, a synthetic glucocorticoid screen (Answer E) should be ordered.

Oral contraceptive pills with third-generation progestin (Answer B) are first-line treatment for women with PCOS if conception is not being pursued. The clinical presentation of PCOS overlaps greatly with that of mild hypercortisolism, including excess adiposity, metabolic syndrome features, and/or signs of androgen excess. However, PCOS is a diagnosis of exclusion, and in this patient, the hormonal evaluation is indicative of ACTH-independent hypercortisolism.

Educational Objective

Explain the limitations of adrenal imaging in patients with ACTH-independent Cushing syndrome caused by primary pigmented nodular adrenocortical disease.

Reference(s)

Louiset E, Stratakis CA, Perraudin V, et al. The paradoxical increase in cortisol secretion induced by dexamethasone in primary pigmented nodular adrenocortical disease involves a glucocorticoid receptor-mediated effect of dexamethasone on protein kinase A catalytic subunits. *J Clin Endocrinol Metab.* 2009;94(7):2406-2413. PMID: 19383776

Almeida MQ, Stratakis CA. Carney complex and other conditions associated with micronodular adrenal hyperplasias. *Best Pract Res Clin Endocrinol Metab.* 2010;24(6):907-914. PMID: 21115159

28 ANSWER: E) Stop octreotide LAR

This patient has radiological and biochemical evidence of residual tumor causing GH excess, with a marginally elevated IGF-1 concentration while being treated with octreotide LAR. However, in normal pregnancy, placental GH production increases during the second trimester, leading to a rise in IGF-1 but suppressed pituitary-derived GH through negative feedback. IGF-1 levels may rise above the age-specific nonpregnancy reference range; hence, the Endocrine Society recommends against measuring GH and IGF-1 during pregnancy. Pregnancy might be expected to worsen disease control because of this placental mechanism and continued pituitary-derived GH secretion from a lack of negative feedback, but studies have indicated biochemical and clinical stability during most pregnancies, especially when adenomas have been treated effectively before conception. Experience with somatostatin analogue use in pregnancy is very limited. However, octreotide can cross the placenta and may affect fetal development in tissues where somatostatin receptors are expressed, including the brain. Intrauterine growth restriction has also been reported, although the risk of congenital malformation appears to be low. Because of the limited safety data, the Endocrine Society recommends discontinuation of long-acting somatostatin analogues ideally 2 months in advance of planned conception, with short-acting octreotide being used if necessary for disease control until pregnancy is achieved (thus, Answer E is correct and Answer A is incorrect). Short-term cessation of therapy is unlikely to affect long-term patient outcomes given the protracted nature of the disease.

Subfertility is common in patients with acromegaly, affecting 37% patients at diagnosis in a referral-center study. Contributory factors include menstrual disturbances (which are more prevalent in patients with higher IGF-1 concentrations), uterine leiomyoma, ovarian cysts, and polycystic ovarian morphology. Menstrual disturbance may be improved with successful disease control. Despite these challenges, 73% of patients were able to conceive at least once in this study, with 87% of deliveries occurring at term. In a systematic review of 273 pregnancy outcomes, tumor growth occurred in 9%. Thus, patients must be monitored clinically for headache and visual symptoms. There were no fetal or maternal deaths, but new-onset gestational diabetes or worsening of preexisting diabetes occurred in 9%, most likely driven by GH-induced insulin resistance. Patients can also experience worsening of preexisting hypertension. The frequencies of preterm birth (9%), miscarriage (4%), small-for-gestational-age babies (5%), and congenital malformations (1%) appear to be similar to frequencies in the background population.

The significance of the prolactin value is uncertain because prolactin is expected to rise naturally during pregnancy. Thus, measurement during pregnancy is not usually helpful. Even if this mixed somatotroph-lactotroph tumor was continuing to secrete prolactin in addition to GH, there is no indication for treatment with either bromocriptine (Answer C) or cabergoline (Answer D) at this time because dopamine agonists are only required for patients with clinically relevant adenoma growth, particularly if this extends suprasellarly with the potential to threaten the optic chiasm. For this reason, visual field evaluation is recommended every 3 months in patients with macroprolactinomas. Transsphenoidal surgery (Answer B) is not indicated because the tumor extends into the cavernous sinus only and the visual fields are normal on confrontation testing.

Educational Objective

Guide the management of acromegaly during pregnancy.

Reference(s)

Katznelson L, Laws ER Jr, Melmed S, et al; Endocrine Society. Acromegaly: an Endocrine Society clinical practice guideline. *J Clin Endocrinol Metab.* 2014;99(11):3933-3951. PMID: 25356808

Pirchio R, et al. Fertility in acromegaly: a single-center experience of female patients during active disease and after disease remission. *J Clin Endocrinol Metab.* 2023;108(8):e583-e593. PMID: 36790068

Bandeira DB, Olivatti TOF, Bolfi F, Boguszewski CL, Nunes-Nogueira VDS. Acromegaly and pregnancy: a systematic review and meta-analysis. *Pituitary.* 2022;25(3):352-362. PMID: 35098440

Petersenn S, Fleseriu M, Casanueva FF, et al. Diagnosis and management of prolactin-secreting pituitary adenomas: a Pituitary Society international consensus statement. *Nat Rev Endocrinol.* 2023;19(12):722-740. PMID: 37670148

29 **ANSWER: B) Perform DXA**

Overt hyperthyroidism is a well-established cause of bone loss and secondary osteoporosis. However, data on bone loss and fracture risk in subclinical hyperthyroidism are unclear. Studies in postmenopausal women with subclinical hyperthyroidism have shown loss in bone mineral density at the femoral neck and radius. Fewer studies with conflicting data are available for premenopausal women and men. Overall, study findings support evaluating bone health in postmenopausal women and in men older than 70 years. However, the role of DXA in premenopausal women and in men younger than 70 years is unclear. Since the detrimental effect of hyperthyroidism on bone health in these populations is uncertain, DXA assessment can be considered. This patient has a TSH concentration less than 0.1 mIU/L and a family history of hip fracture in his mother. Thus, performing DXA (Answer B) is the best next step.

Given his risk factors for bone loss and fracture, annual TSH monitoring without addressing his bone mineral density (Answer C) is inappropriate. Also, the recent guidelines from the Korean Endocrine Society recommend optimizing calcium and vitamin D, even in individuals with normal bone mineral density.

Pharmacotherapy should not be started empirically in this individual. While FRAX can predict the probability of a fracture based on risks alone, this patient's FRAX score is 8.7% for major fractures and 0.4% for hip fractures, which is below the treatment threshold. Thus, starting alendronate (Answer D) is incorrect.

Testosterone replacement (Answer E) should be considered if a man's testosterone concentration is less than 200 ng/dL (<6.9 nmol/L), based on evidence that there is higher bone turnover and bone loss with values lower than this threshold. Also, the diagnosis of hypogonadism should not be based on a single laboratory value, and testosterone measurement should be repeated fasting between 8 and 9 AM.

Subclinical hyperthyroidism increases his risk of atrial fibrillation; however, in the absence of symptoms and or physical signs, a loop recorder (Answer A) is not indicated.

Educational Objective

Explain the effect of subclinical hyperthyroidism on bone mineral density and recommend the frequency of DXA in men younger than 70 years.

Reference(s)

Blum MR, Bauer DC, Collet T-H, et al. Subclinical thyroid dysfunction and fracture risk: a meta-analysis. *JAMA.* 2015;313(20):2055-2065. PMID: 26010634

Leader A, Ayzenfeld RH, Lishner M, et al. Thyrotropin levels within the lower normal range are associated with an increased risk of hip fractures in euthyroid women, but not men, over the age of 65 years. *J Clin Endocrinol Metab.* 2014;99(8):2665-2673. PMID: 24885627

Hong AR, Kang H-C. Evaluation and management of bone health in patients with thyroid diseases: a position statement of the Korean Thyroid Association. *Endocrinol Metab (Seoul).* 2023;38(2):175-189. PMID: 37150514

30 ANSWER: A) Fecal T_4 measurement

This patient has many signs and symptoms of thyrotoxicosis, including eyelid retraction, which is due to sympathetic overactivity and can occur in any form of thyrotoxicosis. Periorbital edema, chemosis, and proptosis are findings that are specific for thyroid eye disease (TED) associated with Graves disease. The 24-hour thyroid uptake is very low (0.93%), and the corresponding scan shows essentially no radiotracer accumulation. The differential diagnosis of these findings includes inflammation or destruction of thyroid tissue with release of stored hormone into circulation (thyroiditis) or an extrathyroidal source of thyroid hormone. The latter includes struma ovarii and factitious thyrotoxicosis. In this case, factitious thyrotoxicosis should be suspected given undetectable serum thyroglobulin in the absence of potentially interfering thyroglobulin antibodies. In contrast, both thyroiditis and struma ovarii would be associated with elevated, or at least detectable, serum thyroglobulin. Fecal T_4 measurement (Answer A) can confirm a diagnosis of factitious thyrotoxicosis when in doubt. Surreptitious administration of thyroid hormone is associated with high fecal T_4 concentrations. One study demonstrated that fecal T_4 concentrations were 12 to 24 times higher in patients with factitious thyrotoxicosis than in control participants. In contrast, fecal T_4 was only 2 times higher in patients with hyperthyroidism due to Graves disease than in control participants. Decreased Doppler flow of the thyroid parenchyma on ultrasonography can also support a diagnosis of factitious thyrotoxicosis.

Factitious thyrotoxicosis results from the exogenous administration of thyroid hormone and may be intentional (factitious disorder; Munchausen syndrome) or unintentional. The unintentional administration of thyroid hormone can occur in patients taking dietary and herbal supplements, including those intended to promote weight loss, or through the consumption of tainted food products. Hamburger thyrotoxicosis occurs when beef neck meat containing bovine thyroid gland is consumed.

Pelvic radioimaging using ^{123}I (Answer C) would be recommended if struma ovarii were suspected. However, the low serum thyroglobulin value rules out this diagnosis.

In Graves disease, serum thyroglobulin would be elevated, and the 24-hour uptake would be elevated or inappropriately normal, with a thyroid scan showing diffuse uptake. Thus, TRAb measurement (Answer E) is not indicated.

Thyrotoxicosis due to thyroiditis has also been ruled out, so repeating thyroid function tests in 4 to 6 weeks (Answer D) may not be needed.

Methimazole (Answer B) is indicated only for the management of hyperthyroidism due to overproduction of thyroid hormone by the thyroid gland, as with Graves disease or autonomously functioning thyroid nodules, and it has no role in the management of factitious thyrotoxicosis or thyroiditis.

Educational Objective

Diagnose factitious thyrotoxicosis caused by exogenous administration of thyroid hormone.

Reference(s)

Bouillon R, Verresen L, Staels F, Bex M, De Vos P, De Roo M. The measurement of fecal thyroxine in the diagnosis of thyrotoxicosis factitia. *Thyroid.* 1993;3(2):101-103. PMID: 8369649

Kang GY, Parks JR, Fileta B, et al. Thyroxine and triiodothyronine content in commercially available thyroid health supplements. *Thyroid.* 2013;23(10):1233-1237. PMID: 23758055

Bernet VJ. Thyroid hormone misuse and abuse. *Endocrine.* 2019;66(1):79-86. PMID: 31617167

31 ANSWER: C) Polysomnography

Fatigue is a nonspecific symptom that is frequently used to describe multiple states, including inability to perform physical activity (either initiating or maintaining), inability to concentrate, or excessive sleepiness. Sleep-disordered breathing is prevalent in persons with type 2 diabetes (up to 58%). Obstructive sleep apnea (OSA) is the most common type of sleep-disordered breathing in this group of patients, with a prevalence of 24% to 86% in population-based cohorts. The American Diabetes Association recommends assessing for OSA as part of the initial comprehensive medical evaluation for persons with diabetes. OSA and sleep disorders increase the risk of developing type 2 diabetes and gestational diabetes. In addition, sleep disruption is associated with higher hemoglobin A_{1c} concentrations, increased cardiovascular complications (eg, atrial fibrillation, hypertension, heart failure), and increased cardiovascular mortality.

In a patient with fatigue and symptoms of daytime sleepiness (as evidenced by her need to walk to avoid falling asleep), evaluation for OSA (Answer C) is recommended. Other symptoms and signs suggestive of OSA are snoring, witnessed apneic episodes or periods of silence between snoring, dry mouth on waking, morning headaches, nocturia, and restlessness while sleeping. Persons with OSA are more likely to also have obesity, hypertension, metabolic dysfunction-associated steatotic liver disease, heart failure, and pulmonary hypertension. Other pertinent physical examination clues are large neck size (>16 in [>40.6 cm] in women and >17 in [>43.2 cm] in men), increased waist circumference (>38 in [>95.5 cm] in women and >40 in [>100.5 cm] in men), and a crowded oropharyngeal airway (Mallampati class ≥3).

Increased risk of moderate to severe OSA is indicated by:

- Presence of excessive daytime sleepiness

AND at least 2 of the following:

- Habitual loud snoring
- Witnessed apnea, gasping, or choking
- Diagnosed hypertension

Polysomnography is the gold-standard diagnostic test for OSA. Questionnaires such as STOP-Bang (Answer E) or the Epworth Sleepiness Scale are not diagnostic of OSA and have poor accuracy compared with polysomnography. The sensitivity and specificity of the STOP-Bang questionnaire is 84% and 56%, respectively, for mild OSA. Similarly, nocturnal oximetry (Answer B) has decreased accuracy to diagnose mild OSA with varying sensitivities and specificities reported due to variation in the pulse oximetry device used, the person interpreting the data, and whether it was worn correctly overnight. In addition, nocturnal oximetry is unable to differentiate OSA from other sleep disorders. The American Academy of Sleep Medicine recommends polysomnography (either in laboratory or at home with a technically adequate device) to diagnose OSA in persons with a high pretest probability for OSA, which this patient has. Unfortunately, to date, no studies show a significant reduction in primary prevention of cardiovascular events with continuous positive airway pressure treatment. Currently published studies, however, have limited generalizability to all persons with continuous positive airway pressure due to specific inclusion/exclusion criteria and low treatment adherence in their cohorts.

A complete blood cell count with ferritin measurement (Answer A) should be performed if anemia is suspected. This laboratory investigation is recommended as part of the initial evaluation of fatigue if there are no identifiable causes.

The patient has no signs or symptoms to suggest adrenal insufficiency, so cortisol measurement (Answer D) is not indicated.

Educational Objective

Assess sleep health in patients with diabetes.

Reference(s)

Kapur VK, Auckley DH, Chowdhuri S, et al. Clinical practice guideline for diagnostic testing for adult obstructive sleep apnea: an American Academy of Sleep Medicine clinical practice guideline. *J Clin Sleep Med.* 2017;13(3):479-504. PMID: 28162150

American Diabetes Association Professional Practice Committee. 5. Facilitating positive health behaviors and well-being to improve health outcomes: standards of care in diabetes-2024. *Diabetes Care.* 2024;47(Suppl 1):S77-S110. PMID: 38078584

32 **ANSWER: D) Intravenous zoledronate during this hospitalization**

This patient is at risk for being lost to follow-up, as well as further fractures. He has had evidence of osteoporosis (vertebral fractures) for at least 2 years, but he has not yet been treated. The current acute fracture event is an opportunity to treat this patient, and a fracture liaison service can help ensure initiation of osteoporosis treatment. Several osteoporosis medications demonstrate fracture risk reduction in this setting. While one could argue that with a history of hip fracture and vertebral compression fracture anabolic therapy (Answer A) would be beneficial, treatment cannot generally start during a hospitalization (these agents are often not on inpatient formularies), and PTH receptor agonists require ongoing daily administration after hospitalization for benefit. Similarly, use of oral alendronate (Answer B) also requires ongoing, long-term administration.

In the Horizon Trial, zoledronate was administered up to 90 days after hip fracture and demonstrated reduction of subsequent fracture risk. However, since bisphosphonates do not impair fracture healing, there is no need to delay this therapy until his fractures have healed. If a patient is at risk for being lost to follow-up, treatment for osteoporosis during the acute hospitalization represents an important opportunity. Treatment with zoledronate (Answer D) is safe during hospitalization for acute fracture, with low risk of hypocalcemia in properly selected patients (normal vitamin D concentration, creatinine clearance greater than 35 mL/min per 1.73 m^2, and no acute kidney injury). While fever does occur, treatment with scheduled acetaminophen minimizes this effect, and zoledronate treatment does not increase length of stay or 30-day readmission rate.

Transdermal testosterone (Answer E) has not been demonstrated to consistently improve bone mineral density, especially when the testosterone concentration is already greater than 200 ng/dL (>6.9 nmol/L). Importantly, for this patient, fracture risk reduction is the treatment goal, and recent randomized controlled trials failed to demonstrate fracture risk reduction in men with hypogonadism treated with transdermal testosterone.

Given that this patient's 25-hydroxyvitamin D value is already 26 ng/mL (64.9 nmol/L), there is no added benefit of increasing the cholecalciferol dosage to 50,000 units weekly (1250 mcg weekly) (Answer C). However, ensuring adequate daily supplementation on an ongoing basis is recommended (recommended daily allowance of 800 units daily). Long-term treatment with high-dosage vitamin D in the absence of deficiency may increase fall and fracture risk.

Educational Objective

Initiate osteoporosis treatment in the acute setting after fragility fracture.

Reference(s)

Lyles KW, Colon-Emeric CS, Magaziner JS, et al; HORIZON Recurrent Fracture Trial. Zoledronic acid and clinical fractures and mortality after hip fracture. *N Engl J Med.* 2007;357(18):1799-1809. PMID: 17878149

Fan W, Leder BZ, Mannstadt M, Ly TV, Franco-Garcia E, Bolster MB. Safety of inpatient zoledronic acid in the immediate postfracture setting. *J Clin Endocrinol Metab.* 2023;108(11):e1282-e1288. PMID: 37227016

Fan W, Machado M, Leder BZ, et al. Inpatient zoledronic acid and integrated orthopedic and fracture liaison services improve osteoporosis treatment rates. *J Clin Endocrinol Metab.* 2022;108(1):191-197. PMID: 36056816

Snyder PJ, Bauer DC, Ellenberg SS, et al. Testosterone treatment and fractures in men with hypogonadism. *N Engl J Med.* 2024;390(3):203-211. PMID: 38231621

33

ANSWER: C) Recombinant human leptin replacement therapy

This patient has had severe hypertriglyceridemia from a young age in association with marked insulin resistance and diminished subcutaneous fat. These features are consistent with congenital generalized lipodystrophy (CGL), a rare autosomal recessive monogenic cause of severe hypertriglyceridemia. Patients with CGL have marked paucity of subcutaneous fat all over the body from birth, muscular prominence, acromegaloid appearance, and acanthosis nigricans. In addition, they develop metabolic complications related to insulin resistance such as diabetes, dyslipidemia, steatohepatitis, and polycystic ovary syndrome. Much progress has been made in recent years in understanding the genetic basis of these disorders, and at least 4 genetic loci have been identified. The most common forms of CGL are due to pathogenic variants in the *AGPAT2* gene (CGL type 1) and the *BSCL2* gene (CGL type 2). Besides generalized fat loss, lipodystrophy can also manifest as partial lipodystrophy, which is generally characterized by loss of peripheral subcutaneous fat in early childhood with preserved or excess truncal fat. Similar to CGL, familial partial lipodystrophy has been noted to display considerable phenotypic and genotypic heterogeneity. In addition, both generalized and partial lipodystrophy can occur as an acquired disorder, usually in association with other autoimmune disorders or secondary to medications such as HIV antiretroviral therapy and immune checkpoint inhibitors. Despite varying clinical features, metabolic complications among the different lipodystrophy syndromes are similar, although the severity may depend on extent of fat loss. The cause of these metabolic complications is both the absence of adipose tissue storage depots leading to fat accumulation in nonadipose tissues such as liver and muscle and the lack of adipokines such as leptin and adiponectin.

Leptin replacement therapy (Answer C) dramatically improves metabolic complications of generalized lipodystrophy, including hyperglycemia, hyperlipidemia, and steatohepatitis. Often these patients are on very high dosages of traditional glucose- and lipid-lowering medications with suboptimal response, and introduction of leptin therapy usually facilitates reduction and even discontinuation of these medications. The benefits of leptin therapy in patients with partial lipodystrophy is less clear, and currently it is approved by the US FDA only for patients with generalized lipodystrophy.

In Europe and Japan, it is approved for use in patients with partial lipodystrophy and low leptin levels. Leptin therapy is generally well tolerated, but there are rare reports of hematological malignancies in children with acquired generalized lipodystrophy treated with leptin. Some patients develop neutralizing antibodies, leading to loss of efficacy. Leptin is available as a lyophilized powder that is reconstituted before injection and is administered subcutaneously daily.

This patient had compound heterozygous pathogenic variants in the *AGPAT2* gene, a critical enzyme involved in triglyceride synthesis and a known locus for CGL. Her leptin concentration was undetectable, and recombinant human leptin therapy was initiated. She responded remarkably well, with decreased hyperphagia and improved metabolic parameters. Over the next 3 to 6 months, her hemoglobin A_{1c} level declined to 6.0% (42 mmol/mol), serum triglycerides were consistently below 150 mg/dL (<1.70 mmol/L), liver enzymes normalized, and eruptive xanthomas resolved (*see image*).

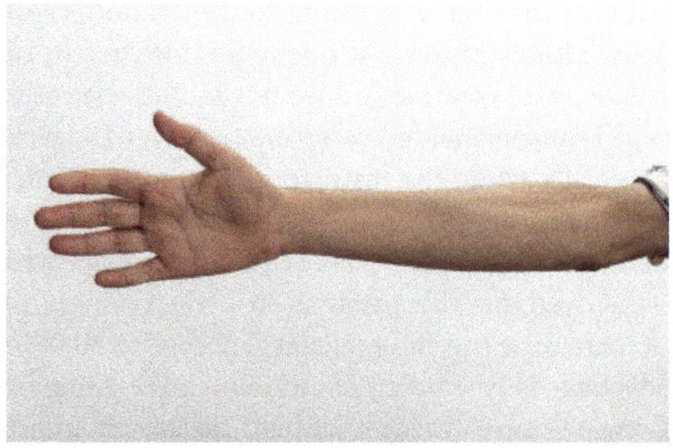

Such a response would have been unlikely with either intensification of insulin therapy (Answer A) or the addition of semaglutide (Answer D).

Patients with CGL do not have deficiency of lipoprotein lipase and therefore replacing it (Answer B) is incorrect. Patients with lipoprotein lipase deficiency also present with severe hypertriglyceridemia and recurrent pancreatitis starting in early childhood, but they do not have insulin resistance or phenotypic features such as fat loss and muscular prominence as seen in this patient.

Volanesorsen (Answer E) is a novel antisense oligonucleotide to apolipoprotein CIII, which is an inhibitor of lipoprotein lipase, and it reduces triglycerides by both lipoprotein lipase–mediated and non–lipoprotein lipase–mediated pathways. Volanesorsen is effective in patients with severe hypertriglyceridemia, including those with lipoprotein lipase deficiency, but it has not been studied in patients with CGL. Further, due to the high incidence of thrombocytopenia, volanesorsen is currently not approved by the US FDA for clinical use.

Educational Objective

Identify congenital lipodystrophy as a monogenic cause of severe hypertriglyceridemia and explain the role of leptin replacement therapy.

Reference(s)

Brown RJ, Araujo-Vilar D, Cheung PT, et al. The diagnosis and management of lipodystrophy syndromes: a multi-society practice guideline. *J Clin Endocrinol Metab.* 2016;101(12):4500-4511. PMID: 27710244

Oral EA, Simha V, Ruiz E, et al. Leptin-replacement therapy for lipodystrophy. *N Engl J Med.* 2002;346(8):570-578. PMID: 11856796

Musso C, Major ML, Andres E, Simha V. Metreleptin treatment in three patients with generalized lipodystrophy. *Clin Med Insights Case Rep.* 2017;9:123-127. PMID: 28096701

34 ANSWER: B) Measure bone mineral density

Patients with prolactinoma causing hypogonadism have an increased risk of osteoporosis. Primary treatment of prolactinoma consists of dopamine agonists, and cabergoline is the preferred choice because it is more effective than bromocriptine. Treatment with cabergoline decreases the prolactin concentration and reduces tumor size in most patients, with a favorable effect on hypothalamic-pituitary-gonadal axis function. However, in patients with large prolactin-secreting adenomas, central hypogonadism may persist with resulting bone density abnormalities. In a study in 40 men and 60 women with prolactinoma, prevalence of impaired bone mineral density at a median of 79 months after onset of medical treatment or pituitary surgery remained high in men (37% in men vs 7% in women). In this study, most patients had adequate control of hyperprolactinemia and hypogonadism.

This patient has had persistent hypogonadism despite taking cabergoline. Also, he takes a total daily hydrocortisone dose of 30 mg, which may contribute to his increased risk of osteoporosis. Current prolactinoma guidelines recommend that baseline bone mineral density be assessed (Answer B) in all patients with prolactinoma and longstanding hypogonadism. Data on recovery of hypogonadism after cabergoline treatment are sparse, and osteoporosis drugs should be considered in patients with osteoporosis.

Increasing the cabergoline dosage to 1.5 mg twice weekly (Answer A) is incorrect because of the patient's favorable response to the current dosage of cabergoline with almost normalization of the prolactin concentration and reduction in tumor size. A study of 124 patients with microprolactinoma and 61 patients with macroprolactinoma found that tumor volume reduction after 3 to 4 months of medical therapy was a good predictor of future tumor shrinkage. Also, increasing the cabergoline dosage above a weekly total dose of 2 mg may be associated with long-term complications, including possible valvulopathy.

This patient has already undergone baseline echocardiography, which did not show significant valve disease, and he does not have a murmur on heart examination. Thus, performing echocardiography now (Answer E) is incorrect. Current recommendations indicate that echocardiography should be repeated every 2 to 3 years in patients receiving more than 2.0 mg of cabergoline per week.

Total testosterone concentrations are significantly low in a fasting early-morning blood draw. Free testosterone (Answer C) is unlikely to be normal in the absence of severe obesity or known protein-binding disorders and would not have any influence on the next management steps. After 6 months of dopamine agonist therapy, testosterone replacement should be considered for men with ongoing hypogonadism who are not interested in fertility.

Measuring total testosterone in 12 months (Answer D) is incorrect because there is a risk that the patient might develop significant bone loss in the absence of testosterone replacement.

Prolactin concentration and tumor size should be monitored after testosterone replacement. Cases of enlarging tumors after testosterone replacement have been reported and are attributed to aromatization to estradiol. Therefore, a short-acting testosterone formulation, such as daily topical administration, is preferred to injections. Off-label use of aromatase inhibitors has a favorable effect on tumor size (in retrospective studies), but this therapy requires further study and would have detrimental effects on bone density.

Educational Objective

Recommend bone density screening in a patient with medically treated macroprolactinoma and persistent hypogonadism.

Reference(s)

Petersenn S, Fleseriu M, Casanueva FF, et al. Diagnosis and management of prolactin-secreting pituitary adenomas: a Pituitary Society international consensus statement. *Nat Rev Endocrinol.* 2023;19(12):722-740. PMID: 37670148

Andereggen L, Frey J, Andres RH, et al. Persistent bone impairment despite long-term control of hyperprolactinemia and hypogonadism in men and women with prolactinomas. *Sci Rep.* 2021;4;11(1):5122. PMID: 33664388

Biagetti B, Sarria-Estrada S, Ng-Wong YK, et al. Shrinkage by the third month predicts long-term response of macroprolactinoma after cabergoline. *Eur J Endocrinol.* 2021;185(4):587-595. PMID: 34374649

Melmed S, Casanueva FF, Hoffman AR, et al; Endocrine Society. Diagnosis and treatment of hyperprolactinemia: an Endocrine Society clinical practice guideline. *J Clin Endocrinol Metab.* 2011;96(2):273-288. PMID: 21296991

35

ANSWER: C) Struma ovarii

Causes of elevated serum thyroid hormone levels (thyrotoxicosis) include thyroidal and extrathyroidal etiologies. The term hyperthyroidism refers to the condition of increased thyroid hormone arising specifically from the thyroid gland. When encountering a patient with persistently abnormal laboratory values consistent with thyrotoxicosis (ie, decreased TSH and increased T_3 and/or T_4 values), obtaining thyroid antibody levels, checking for thyroid eye disease (which is present in 40% of patients with Graves disease), and eliciting personal or family history of autoimmune conditions may help increase the suspicion of Graves disease, particularly if thyroid ultrasonography shows a diffusely hypervascular gland with heterogeneous echotexture throughout. If the etiology of hyperthyroidism cannot be determined by these data, a radioiodine uptake and scan should be obtained.

When considering thyroidal causes of thyrotoxicosis, the most common cause of hyperthyroidism worldwide is Graves disease (Answer A), which would produce diffusely increased thyroid radioiodine uptake, a result that is not seen in this patient. Other common etiologies of hyperthyroidism include autonomously functioning thyroid nodule(s) and the thyrotoxic phase of thyroiditis. Autonomously functioning thyroid nodule(s) (also termed hot nodule[s] or toxic nodule[s]) result in a picture of biochemical hyperthyroidism and focally increased uptake on radioiodine uptake and scan at the site of the autonomous nodule, which should be confirmed structurally by thyroid ultrasonography. The overall uptake of the thyroid gland may be low, normal, or high in such patients.

In all forms of thyroiditis, including subacute thyroiditis (also known as de Quervain thyroiditis) (Answer D), painless thyroiditis, postpartum thyroiditis, acute suppurative thyroiditis, drug-induced thyroiditis, and radiation-induced thyroiditis, the thyrotoxic phase can appear to be hyperthyroidism, including the presentation of hyperthyroid clinical symptoms and laboratory test results showing hyperthyroidism, due to the leakage of preformed thyroid hormone from the thyroid gland into circulation. The diagnosis of a thyrotoxic phase resulting from thyroiditis is determined by low or absent thyroidal uptake with a diagnostic radiotracer such as ^{99m}Tc or ^{123}I. The chronicity of this patient's symptoms and persistently suppressed TSH are not suggestive of subacute thyroiditis. Very rare causes of hyperthyroidism include constitutively activating (gain-of-function) pathogenic variants in the TSH receptor gene (*TSHR*), resulting in a form of nonautoimmune familiar or sporadic hyperthyroidism that may also show diffusely elevated radioiodine uptake values. This diagnosis is confirmed by genetic testing.

Extrathyroidal causes of thyrotoxicosis usually produce low or absent radiotracer uptake in the thyroid gland, as in this patient. These etiologies are less common than thyroidal causes and include thyrotoxicosis factitia, defined as the intentional or unintentional ingestion of exogenous thyroid hormone (Answer E), functional thyroid cancer metastasis, TSH-producing pituitary adenoma, gestational trophoblastic tumor, and struma ovarii (Answer C). Additional imaging may be needed to localize some of these conditions, such as a whole-body radioiodine uptake and scan to detect functional cancer metastasis and struma ovarii and pituitary MRI or CT to detect a TSH-producing pituitary adenoma. In addition, a serum thyroglobulin concentration (reflecting endogenous thyroid hormone production) can be measured, which would be elevated in all cases of extrathyroidal thyrotoxicosis except for exogenous thyroid hormone ingestion. In this case, serum thyroglobulin would be low or absent, reflecting the suppression of endogenous thyroid hormone production when there is exogeneous thyroid hormone intake.

hCG-producing tumors (Answer B) can arise both in pregnancy and in nonpregnancy and range from premalignant to malignant conditions. Examples include benign trophoblastic lesions, hydatidiform moles, malignant invasive moles, choriocarcinomas, placental site trophoblastic tumors, and epithelioid trophoblastic tumors. Thyrotoxicosis may be a feature of some hCG-producing tumors, due to the weak stimulatory activity of hCG. However, this patient's condition is inconsistent with an hCG-producing tumor, as thyroid radioiodine uptake is not used for diagnosis (and even if obtained, it would not be low or absent); hCG-producing tumors are localized by cross-sectional imaging and share the hallmark feature of markedly elevated serum hCG concentrations.

Struma ovarii accounts for only 0.3% to 1.0% of all ovarian tumors and is a type of teratoma composed of 50% or more thyroid tissue that is located in the ovary. A whole-body radioiodine uptake and scan would reveal low or absent thyroidal uptake and instead demonstrate a focus of increased uptake in the pelvis corresponding to the tumor. Affected women are usually asymptomatic, but they may report pain in the lower abdomen, have a palpable abdominal/pelvic mass, or present with vaginal bleeding. More than 95% of struma ovarii tumors are benign, and the overall prognosis of this rare condition is excellent following surgery of the ovarian mass. The thyrotoxicosis is expected to resolve with surgical resection, although in rare cases, postoperative adjuvant radioactive iodine therapy may also be needed for metastatic or recurrent disease.

Educational Objective

Diagnose struma ovarii based on clinical, biochemical, and imaging characteristics.

Reference(s)

Leung AM. Thyrotoxicosis of extrathyroidal origin. In: *Werner & Ingbar's The Thyroid: A Fundamental and Clinical Text*, 11th Edition, eds: Braverman LE, Cooper D, Kopp P. Philadelphia: Lippincott, Williams, and Wilkins, 2020.

Ang LP, Avram AM, Lieberman RW, Esfandiari NH. Struma ovarii with hyperthyroidism. *Clin Nucl Med.* 2017;42(6):475-477. PMID: 28394842

Addley S, Mihai R, Alazzam M, Dhar S, Soleymani Majd H. Malignant struma ovarii: surgical, histopathological and survival outcomes for thyroid-type carcinoma of struma ovarii with recommendations for standardising multi-modal management. A retrospective case series sharing the experience of a single institution over 10 years. *Arch Gynecol Obstet.* 2021;303(4):863-870. PMID: 33512594

Yoo S-C, Chang K-H, Lyu M-O, Chang S-J, Ryu H-S, Kim H-S. Clinical characteristics of struma ovarii. *J Gynecol Oncol.* 2008;19(2):135-138. PMID: 19471561

Oikonomou C, Spathari N, Doumoulaki S, Koutela A, Stagkoglou C, Keramidaris D. Recurrent struma ovarii presented with high levels of thyroglobulin. *Case Rep Surg.* 2021;2021:8868095. PMID: 33824772

Giovanella L, Avram AM, Ovčariček PP, Clerc J. Thyroid functional and molecular imaging. *Presse Med.* 2022;51(2):104116. PMID: 35124101

36 ANSWER: E) Transdermal testosterone gel plus annual infusion of zoledronic acid

This man has symptomatic hypogonadism with 2 documented low early-morning serum testosterone concentrations. He also has osteoporosis based on his recent history of a fragility fracture of the lower vertebral spine. The etiology of his hypogonadotropic hypogonadism is likely neck irradiation; this is a common adverse effect of radiotherapy for head or neck cancer. He would benefit from testosterone replacement therapy to improve his sexual function, resolve breast tenderness and breast growth, and increase bone density and bone strength. However, testosterone therapy has not been shown to prevent fragility fractures, and he should be prescribed a medication that has been shown to decrease major osteoporotic fractures (eg, a bisphosphonate). Of the choices offered, testosterone therapy plus zoledronic acid (Answer E) is the best management. There is insufficient evidence to prescribe testosterone therapy alone (Answers B and D) to a hypogonadal man who has a high risk of major osteoporotic fracture.

Although testosterone therapy increases areal and volumetric bone density and bone strength in men with low serum testosterone concentrations, the recently published TRAVERSE study demonstrated no difference in major osteoporotic fractures in men treated with testosterone vs those treated with placebo for a mean duration of 22 months. Curiously, the overall fracture rate was higher in testosterone-treated men; the increase in fractures was mostly due to an increase of ankle and rib fractures that are typically associated with trauma. The TRAVERSE study was designed to test the cardiovascular safety of testosterone therapy. It was not designed to test whether testosterone therapy would reduce fractures. The limitations of the TRAVERSE study include not recruiting men at high risk of fracture (eg, this patient with a history of low-trauma fracture), lack of systematic collection of data on the possible effects of testosterone on behavior that might increase trauma-induced fractures, and the lack of bone densitometry at baseline or at the end of treatment.

Although this man has at least 2 risk factors for cardiovascular disease, the TRAVERSE study demonstrated that testosterone therapy for up to 4 years does not increase the risk of major cardiovascular events in men with (or at high risk of) cardiovascular disease, compared with placebo. Likewise, this study showed no increase in lower urinary tract symptoms in men such as this patient who had baseline benign prostatic hyperplasia without severe lower urinary tract symptoms. Findings from the Testosterone Trials and the TRAVERSE study suggest that testosterone therapy for this patient is likely to improve libido and sexual function, modestly improve mood and energy, and normalize hematocrit. Testosterone therapy would most likely be beneficial and safe for this hypogonadal man with osteoporosis, and treatment to prevent fragility fractures without testosterone therapy (Answers A and C) would be incorrect. Of note, romosozumab has not been approved for use in men (as of 2024).

Educational Objective

Recommend the best treatment for men with a history of a fragility fracture and newly diagnosed hypogonadism.

Reference(s)

Thirumalai A, Anawalt BD. Epidemiology of male hypogonadism. *Endocrinol Metab Clin North Am.* 2022;51(1):1-27. PMID: 35216709

Cunningham GR, Ellenberg SS, Bhasin S, et al. Prostate-specific antigen levels during testosterone treatment of hypogonadal older men: data from a controlled trial. *J Clin Endocrinol Metab.* 2019;104(12):6238-6246. PMID: 31504596

Grossmann M, Jayasena CN, Anawalt BD. Approach to the patient: the evaluation and management of men ≥50 years with low serum testosterone concentration. *J Clin Endocrinol Metab.* 2023;108(9):e871-e884. PMID: 36995891

Grossmann M, Anawalt BD. Breaking news - testosterone treatment and fractures in older men. *N Engl J Med.* 2024;390(3):267-268. PMID: 38231628

Ng Tang Fui M, Hoermann R, et al. Effect of testosterone treatment on bone microarchitecture and bone mineral density in men: a 2-year RCT. *J Clin Endocrinol Metab.* 2021;106(8):e3143-e3158. PMID: 33693907

Snyder PJ, Kopperdahl DL, Stephens-Shields AJ, et al. Effect of testosterone treatment on volumetric bone density and strength in older men with low testosterone: a controlled clinical trial. *JAMA Intern Med.* 2017;177(4):471-479. PMID: 28241231

Thirumalai A, Anawalt BD. Epidemiology of male hypogonadism. *Endocrinol Metab Clin North Am.* 2022;51(1):1-27. PMID: 35216709

37 ANSWER: B) Perform a 2-hour oral glucose tolerance test

This patient is at risk for developing type 1 diabetes. His history and positive antibody tests warrant further evaluation with a 2-hour oral glucose tolerance test (Answer B) to identify dysglycemia, which is required, in addition to 2 positive antibody markers of type 1 diabetes, to be eligible for teplizumab therapy. This patient is presently not eligible for teplizumab therapy (Answer E); dysglycemia must be confirmed first. This patient has not been thoroughly evaluated for the presence of dysglycemia. Although measuring hemoglobin A_{1c} (Answer D) and fasting glucose is helpful in the evaluation, a 2-hour oral glucose tolerance test is often the best test to identify early abnormalities of glucose metabolism. The first glucose abnormality detected is often a rise in the postprandial glucose concentration because of reduced first-phase insulin secretion. Fasting glucose levels rise as β-cell activity deteriorates over time. Diabetes eventually develops with a further reduction in insulin secretion. A 2-hour plasma glucose value less than 140 mg/dL (<7.8 mmol/L) is considered normal, 140 to 199 mg/dL (7.8-11.0 mmol/L) indicates impaired glucose tolerance, and 200 mg/dL or greater (≥11.1 mmol/L) indicates diabetes. There are advantages and limitations for each of the recommended screening tests to identify diabetes or individuals with increased risk of diabetes (*see table*).

He is clearly making sufficient insulin given his hemoglobin A_{1c} and fasting glucose values, so serum C-peptide (Answer A) would certainly be detectable. C-peptide measurement would not help to stratify his risk of developing type 1 diabetes. In addition, he has already tested positive for 2 diabetes-related autoimmune antibodies, so assessing additional diabetes-related autoantibodies (Answer C) is not necessary to be eligible for teplizumab therapy.

Table. Advantages and Limitations of Current Screening Tests for Prediabetes and Diabetes

Current screening tests for prediabetes/diabetes – advantages and limitations.

Screening test	Advantages	Limitations
Fasting Plasma Glucose (FPG)	Can be performed as a single blood draw.	Requires overnight fast. Less sensitive than the OGTT.
Oral Glucose Tolerance Test (OGTT)	Includes assessment of both fasting plasma glucose and the 2-hour glucose after the oral glucose load. Allows assessment of the glucose response after an oral glucose challenge. Identifies more individuals with dysglycemia than the FPG or HbA1c.	Requires overnight fast. Associated nausea in a subset of individuals after ingestion of 75 g glucose load. Two-hour test duration. Sensitive to day-to-day differences due to diet and/or physical activity. Can vary according to time of day of testing. Reproducibility is not as good as the FPG or HbA1c.
HbA1c	Reflects integrated glucose levels over preceding 3 months. Convenient. Does not require fasting. Can be performed as a single blood draw. High reproducibility (precision). Less day-to-day perturbations during stress and illness. Standardized worldwide. Quality assurance in place.	Less sensitive than the FPG and OGTT. Interpretation and accuracy can be affected by presence of hemoglobin variants (i.e., sickle cell trait), chronic renal failure, iron deficiency anemia, differences in red blood cell lifespan, and differences with age and race. May be high or low relative to underlying average glucose levels (accuracy – HbA1c "mismatches" as a reflection of average glucose levels).
Random Plasma Glucose (RPG)	Convenient. Does not require fasting. Can be performed as a single blood draw. Often included in "metabolic profile" panels Very specific when elevated.	Levels which (a) should be followed by confirmatory diagnostic tests, or (b) indicate a low likelihood of dysglycemia, have not been established.

Reprinted from Bergman M et al. *Diabetes Res Clin Pract*, 2020:165:108233. Published by Elsevier B.V.

Educational Objective

Explain the advantages and limitations of each of the recommended screening tests for prediabetes and diabetes.

Reference(s)

Bergman M, Abdul-Ghani M, DeFronzo RA, et al. Review of methods for detecting glycemic disorders [published correction appears in *Diabetes Res Clin Pract.* 2021;180:108632]. *Diabetes Res Clin Pract.* 2020:165:108233. PMID: 32497744

Herold KC, Bundy BN, Long SA, et al. An anti-CD3 antibody, teplizumab, in relatives at risk for type 1 diabetes. *N Engl J Med.* 2019;381:603-613. PMID: 31180194

American Diabetes Association Professional Practice Committee. 2. Diagnosis and classification of diabetes: standards of care in diabetes-2024. *Diabetes Care.* 2024;47(Suppl 1):S20-S42. PMID: 38078589

38 **ANSWER: D) Provide reassurance that treatment with thyroid hormone is not required now**

This patient's laboratory test results demonstrate a serum TSH value that is above the standard reference range on 2 separate occasions with normal free T_4 results. These findings are consistent with a diagnosis of subclinical hypothyroidism. A modest rise in serum TSH may not be abnormal in older adults and may instead reflect a change in the TSH setpoint with aging.

The key to answering this question is to assess the potential risks and benefits of initiating treatment of mild subclinical hypothyroidism in older adults. The TRUST Study was a randomized controlled trial that included 737 older adults (>65 years) with persistent subclinical hypothyroidism who were treated with levothyroxine or placebo. The mean age was 74.4 years and the mean baseline TSH concentration was 6.4 mIU/L. After 1 year of treatment, there was no significant difference in the coprimary outcomes of hypothyroid symptoms or fatigue scores. Similarly, a large systematic review and meta-analysis of 2192 adult patients with subclinical hypothyroidism found that there was no improvement in overall quality of life or thyroid-related symptoms associated with treatment with thyroid hormone. Finally, a predefined TRUST substudy found that after 1 year of treatment with levothyroxine or placebo, depressive symptoms did not significantly differ between groups.

When the serum TSH concentration is above 7.0 mIU/L, some studies have demonstrated an association between subclinical hypothyroidism and an increased risk of coronary artery disease and cardiac mortality, while serum TSH values above 10.0 mIU/L have been associated with heart failure. However, not all studies have demonstrated a clear link between subclinical hypothyroidism and increased cardiovascular risk, especially in older adults. Furthermore, no large, randomized trials have demonstrated that treatment with levothyroxine significantly improves cardiovascular outcomes in patients with subclinical hypothyroidism.

The lack of clear benefit of treating mild subclinical hypothyroidism in older adults must be weighed against the potential for harm. Overtreatment with levothyroxine occurs in more than one-third of patients older than 65 years and is most common in older women. Initiation of treatment with levothyroxine or other thyroid hormone preparations is not recommended for patients 65 years or older when the serum TSH concentration remains below 7.0 to 10.0 mIU/L. Offering the patient reassurance that treatment with thyroid hormone is not required now (Answer D) is the most appropriate management. Recommending further treatment by prescribing a branded levothyroxine gel cap (Answer C) or the branded levothyroxine tablet prescribed by her primary care provider (Answer E) is incorrect, given the lack of clear benefit and potential for harm.

Although serum thyroglobulin antibodies (Answer A) can be detected in a subset of patients with autoimmune hypothyroidism, their measurement is unlikely to provide added value to this patient's care, especially given the fact that TPO antibodies were undetectable.

Finally, measuring serum total T_3 (Answer B) is incorrect due to the higher sensitivity of serum T_4 in diagnosing overt hypothyroidism.

Educational Objective

Determine when to initiate levothyroxine in patients with subclinical hypothyroidism.

Reference(s)

Stott DJ, Rodondi N, Kearney PM, et al; TRUST Study Group. Thyroid hormone therapy for older adults with subclinical hypothyroidism. *N Engl J Med.* 2017;376(26):2534-2544. PMID: 28402245

Feller M, Snel M, Moutzouri E, et al. Association of thyroid hormone therapy with quality of life and thyroid-related symptoms in patients with subclinical hypothyroidism: a systematic review and meta-analysis. *JAMA.* 2018;320(13):1349-1359. PMID: 30285179

Wildisen L, Feller M, Del Giovane C, et al. Effect of levothyroxine therapy on the development of depressive symptoms in older adults with subclinical hypothyroidism: an ancillary study of a randomized clinical trial. *JAMA Netw Open.* 2021;4(2):e2036645. PMID: 33566107

Gencer B, Moutzouri E, Blum MR, et al. The impact of levothyroxine on cardiac function in older adults with mild subclinical hypothyroidism: a randomized clinical trial. *Am J Med.* 2020;133(7):848-856.e5. PMID: 32171774

39 ANSWER: B) Cosyntropin-stimulation test

The differential diagnosis for this clinical presentation includes polycystic ovary syndrome (PCOS), nonclassic congenital adrenal hyperplasia, and Cushing syndrome. The phenotypes of PCOS and nonclassic congenital adrenal hyperplasia overlap closely. Because PCOS is an exclusionary diagnosis, testing for nonclassic congenital adrenal hyperplasia should routinely be pursued in women presenting with infertility and signs of androgen excess. Most cases of congenital adrenal hyperplasia are caused by pathogenic variants in the *CYP21A2* gene, leading to defective 21-hydroxylase activity (P450c21, CYP21A2). Screening for CYP21A2 deficiency is done by measuring serum 17-hydroxyprogesterone. Rare forms of congenital adrenal hyperplasia that might display features similar to those of 21-hydroxylase deficiency include 11β-hydroxylase (CYP11B1) deficiency and 3β-hydroxysteroid dehydrogenase type 2 (HSD3B2) deficiency.

HSD3B2 catalyzes the steroidogenic step that precedes the one executed by CYP21A2, and this reaction occurs both in the adrenal glands and in the gonads (*figure*). Severe forms of HSD3B2 deficiency impair both mineralocorticoid and glucocorticoid synthesis and result in steroid flux diversion to DHEA overproduction. DHEA is metabolized to active androgens in peripheral tissues that express HSD3B1 and downstream enzymes. The resultant androgen quantities produced peripherally are excessive for females, leading to mild virilization in newborn girls, but are insufficient for males, who are undervirilized. Nonclassic HSD3B2 deficiency is very rare, and it is difficult to distinguish clinically from CYP21A2 deficiency or CYP11B1 deficiency. All 3 forms present with hirsutism, menstrual irregularities, and/or acne, often in adolescents or young women. These patients have normal cortisol and aldosterone production because of mild ACTH elevation. The combination of HSD3B2 deficiency and compensatory ACTH stimulation leads to disproportionate elevation of steroids above the enzymatic defect, including 17-hydroxypregnenolone and DHEA. These elevations are further amplified with cosyntropin-stimulation testing (Answer B). A cosyntropin-stimulated 17-hydroxypregnenolone concentration greater than 5000 ng/dL (>150 nmol/L) is suggestive of HSD3B2 deficiency. Intermediate elevations of 17-hydroxypregnenolone and DHEA were previously thought to represent nonclassic forms of HSD3B2 deficiency, but such patients were later proved to harbor normal *HSD3B2* gene sequencing and a PCOS phenotype. Genetic testing for *HSD3B2* pathogenic variants (Answer C) can be considered when hormonal data are ambiguous.

Figure. Pathways of adrenal steroid hormone synthesis

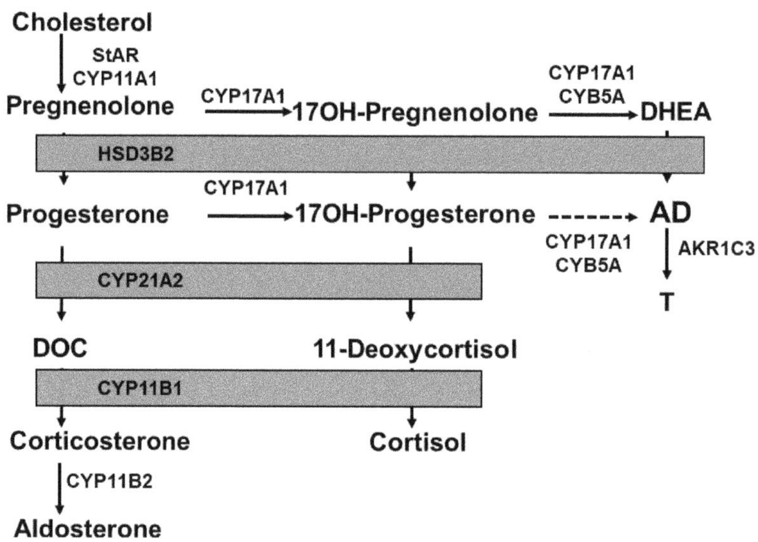

Abbreviations: StAR, steroidogenic acute regulatory protein; CYP11A1, cytochrome P450 cholesterol side-chain cleavage; HSD3B2, 3β-hydroxysteroid dehydrogenase type 2; CYP17A1, 17α-hydroxylase/17,20-lyase; CYB5A, cytochrome b_5; CYP21A2, 21-hydroxylase; CYP11B1, 11β-hydroxylase; CYP11B2, aldosterone synthase; AKR1C3, 17β-hydroxysteroid dehydrogenase type 5; DOC, 11-deoxycorticosterone; AD, androstenedione; T, testosterone.

Adrenal CT (Answer A) would be instrumental when evaluating for adrenal cortical carcinoma. These tumors can produce large amounts of DHEA-S, usually concentrations greater than 700 µg/dL (>19.0 µmol/L), and sometimes cortisol. Cortisol-producing adrenal tumors lead to ACTH suppression. If androgens are not coproduced from an adrenal tumor, DHEA-S is also low for sex and age, reflecting its ACTH dependency. This patient has a moderate DHEA-S elevation, and her clinical features and the insidious timeline are not suggestive of a malignancy.

Cushing disease, a term used for ACTH-producing pituitary adenomas, can present with weight gain, irregular menses, acne, and hirsutism, and it should be considered in a woman with this clinical presentation. However, this patient has a disproportionately high ACTH value compared with her serum cortisol value, which is suggestive of nonclassic congenital adrenal hyperplasia. Pituitary MRI (Answer E) is indicated in patients with hormonal workup consistent with ACTH-dependent Cushing syndrome (at least 2 of 3 abnormal results: 24-hour urinary cortisol, midnight salivary cortisol, and/or dexamethasone-suppression test).

Ovarian ultrasonography (Answer D) is performed when ovarian pathology is suspected. In women, androstenedione and testosterone are produced roughly equally by the adrenal glands and the ovaries. In this patient, both hormones are normal. Characteristic ovarian morphology of PCOS includes an antral follicle count greater than 20 and/or increased ovarian volume. Multiple subcentimeter follicles located at the periphery of the ovary, and increased central stroma are also common. Presence of ovarian cysts is a supportive finding but is not a mandatory diagnostic criterion for PCOS. Thus, ovarian ultrasonography is not the most helpful next step for this patient.

Educational Objective

Diagnose nonclassic congenital adrenal hyperplasia due to 3β-hydroxylase deficiency.

Reference(s)

Al Alawi AM, Nordenström A, Falhammar H. Clinical perspectives in congenital adrenal hyperplasia due to 3β-hydroxysteroid dehydrogenase type 2 deficiency. *Endocrine.* 2019;63(3):407-421. PMID: 30719691

Sharma A, Welt CK. Practical approach to hyperandrogenism in women. *Med Clin North Am.* 2021;105(6):1099-1116. PMID: 34688417

Turcu AF, Auchus RJ. The next 150 years of congenital adrenal hyperplasia. *J Steroid Biochem Mol Biol.* 2015;153:63-71. PMID: 26047556

40 ANSWER: B) Continue hybrid closed-loop insulin pump therapy on auto mode and change correction factor settings to decrease insulin delivered through correction boluses

Type 1 diabetes is a demanding disease from the patient perspective. Unfortunately, data from the T1D Exchange Registry have shown that only a few patients with type 1 diabetes achieve the recommended glycemic targets. Among the different treatment options, insulin pumps are important tools for the management of type 1 diabetes. Insulin pumps can administer insulin continuously, alleviating the need for multiple daily insulin injections. Insulin pumps have evolved over the years and can now communicate with continuous glucose monitoring devices. Insulin pumps use sophisticated algorithms to auto-adjust insulin based on continuous glucose monitoring values.

Several hybrid closed-loop insulin pump systems (automated insulin delivery) are available and have different characteristics. In this case, the patient is using a system that automatically adjusts basal rates and administers autocorrection bolus insulin. This algorithm uses autocorrection boluses when glucose values are predicted to be greater than 180 mg/dL (>10.0 mmol/L). The autocorrection bolus dose is calculated using the insulin sensitivity (correction) factor and a target as low as 110 mg/dL (6.1 mmol/L) and delivers 60% of that value. One issue that might arise with this automated insulin delivery system is that patients can experience hypoglycemia when they receive an autocorrection bolus and the insulin sensitivity factor (correction factor) is aggressive (higher than normal). Indeed, in this clinical vignette, the patient experienced several episodes of hypoglycemia. Although there is limited information on the first episode that occurred at 1:30 AM, there is sufficient information about the other 2 episodes that occurred at 6:30 PM and 11:30 PM. Before the second episode, the patient had a snack or meal at 4 PM, perhaps with the goal of preventing impending hypoglycemia. As her glucose values started to rise, the automated insulin delivery system provided an autocorrection bolus when it predicted that the glucose concentration would be greater than 180 mg/dL (>10.0 mmol/L). As a result, the patient became hypoglycemic at 6:30 PM, which led to self-treatment. However, she probably overtreated and a second autocorrection bolus was given, which led to another episode of hypoglycemia around 11:30 PM.

This patient should continue hybrid closed-loop insulin pump therapy on auto mode, but she should change correction factor settings to decrease insulin delivered through correction boluses (Answer B) with the goal of avoiding these autocorrection-induced hypoglycemic episodes. Her hypoglycemic events would not be prevented by decreasing the basal rate or changing the insulin-to-carbohydrate ratio to decrease insulin given with meals (Answers A, C, and D). Discontinuing the insulin pump and placing the patient on a multiple daily injection regimen (Answer E) would not correct the problem of aggressive correction factor settings, and the patient would lose many of the benefits conferred by insulin pump therapy.

Educational Objective

Recommend the appropriate correction factor settings for patients using closed-loop insulin pump systems that administer autocorrection bolus insulin.

Reference(s)

Foster NC et al. State of type 1 diabetes management and outcomes from the T1D exchange in 2016-2018. *Diabetes Technol Ther.* 2019;21(2):66-72. PMID: 30657336

Phillip M, Nimri R, Bergenstal RM, et al. Consensus recommendations for the use of automated insulin delivery technologies in clinical practice. *Endocr Rev.* 2023;44(2):254-280. PMID: 36066457

41 **ANSWER: C) Recommend that she contact the clinic immediately upon confirmed pregnancy**
Adequate maternal thyroid hormone is essential for normal fetal growth and neurodevelopment, and untreated or undertreated maternal hypothyroidism is associated with a host of adverse obstetric and neonatal outcomes. During pregnancy, thyroid hormone requirements are increased compared with requirements for nonpregnant individuals because of several physiologic changes that occur during gestation. These changes include nearly 2-fold higher serum thyroid-binding globulin levels secondary to rising maternal estradiol levels (especially in early pregnancy); increased degradation of thyroid hormones (T_4 and T_3) by the type 3 inner ring deiodinase that is abundantly expressed in the placenta, chorion, and amnion; higher volume of T_4 distribution due to increased plasma volume during pregnancy; and transplacental transfer of T_4 from the mother to the fetus. As such, women with hypothyroidism, including those due to Hashimoto thyroiditis, require a levothyroxine dosage increase of up to 30% once pregnant. When seen during preconception counseling, this patient should be advised to call the clinic immediately upon confirmed pregnancy (Answer C), so guidance can be provided to increase her levothyroxine dosage and a plan can be arranged for thyroid function monitoring throughout gestation. The Endocrine Society recommends adjusting the levothyroxine dosage to trimester-specific TSH values if available, or alternatively to maintain the TSH concentration less than 2.5 mIU/L during the first trimester and less than 3.0 mIU/L during the second and third trimesters. If this patient with Hashimoto thyroiditis (ie, TPO antibodies are positive) did not have preexisting hypothyroidism, the American Thyroid Association advises starting levothyroxine if the TSH concentration is greater than 4.0 mIU/L and considered if the TSH concentration is between 2.5 and 4.0 mIU/L during preconception and pregnancy.

In women with hypothyroidism who are taking levothyroxine to target a stable TSH concentration, there is no need to supplement with additional iodine (Answer D), although the other components of a prenatal multivitamin (ie, folate) are advised for routine prenatal care. The role of iodine supplementation in preconception, pregnancy, and lactation is only to provide adequate thyroid hormone reserve to a woman without preexisting hypothyroidism, given her anticipated increased thyroid hormone requirements upon a confirmed pregnancy.

The possible benefits of selenium supplementation (Answer E) and adoption of a gluten-free diet (Answer A) to improve fertility and/or decrease serum thyroid autoantibody titers in pregnancy have been inconsistent and are not standardly recommended.

There is no role for any liothyronine-containing forms of thyroid hormone replacement (Answer B) during pregnancy, which includes both synthetic liothyronine and all desiccated thyroid extract preparations. The fetus is reliant on thyroid hormone for development, particularly neurodevelopment, during early pregnancy (beginning as early as week 3 of gestation). As the fetal thyroid gland does not start to form until weeks 11 to 12 of pregnancy (and remains mostly immature until at least mid to late pregnancy), the fetus depends on the transplacental availability of adequate maternal thyroid hormone concentrations throughout the entire gestational period. In fact, at birth, nearly half of the thyroid hormone concentrations in cord blood is still that from the mother. However, as the fetal blood-brain barrier is relatively impermeable to T_3, any maternal use of liothyronine for the treatment of hypothyroidism would not be able to supply the fetal brain and is thus absolutely contraindicated.

Educational Objective

Formulate a plan to manage hypothyroidism associated with Hashimoto thyroiditis during pregnancy.

Reference(s)

De Groot L, Abalovich M, Alexander EK, et al. Management of thyroid dysfunction during pregnancy and postpartum: an Endocrine Society clinical practice guideline [published correction appears in *J Clin Endocrinol Metab.* 2021;106(6):e2461]. *J Clin Endocrinol Metab.* 2012;97(8):2543-2565. PMID: 22869843

Alexander EK, Pearce EN, Brent GA, et al. 2017 guidelines of the American Thyroid Association for the diagnosis and management of thyroid disease during pregnancy and the postpartum [published correction appears in *Thyroid.* 2017;27(9):315-389]. *Thyroid.* 2017;27(3):315-389. PMID: 28056690

Abalovich M, Amino N, Barbour LA, et al. Management of thyroid dysfunction during pregnancy and postpartum: an Endocrine Society clinical practice guideline [published correction appears in *J Clin Endocrinol Metab.* 2021;106(7):e2844]. *J Clin Endocrinol Metab.* 2007;92(8 Suppl):S1-S47. PMID: 17948378

Mandel SJ, Larsen PR, Seely EW, Brent GA. Increased need for thyroxine during pregnancy in women with primary hypothyroidism. *N Engl J Med.* 1990;323(2):91-96. PMID: 2359428

Minnetti M, Sada V, Feola T, et al. Selenium supplementation in pregnant women with autoimmune thyroiditis: a practical approach. *Nutrients.* 2022;14(11):2234. PMID: 35684035

Stazi AV, Mantovani A. A risk factor for female fertility and pregnancy: celiac disease. *Gynecol Endocrinol.* 2000;14(6):454-463. PMID: 11228068

42 **ANSWER: C) Intravenous bisphosphonate**

This patient has polyostotic fibrous dysplasia of bone, with a technetium bone scan showing uptake in the right cranial bones, vertebral column, right ribs, right acetabulum, and right femur. Fibrous dysplasia is a mosaic disorder that can occur in monostotic or polyostotic forms and is frequently accompanied by other features of McCune-Albright syndrome (MAS) as demonstrated in this patient who has a history of a functional ovarian cyst as a young child and café-au-lait macules. Other common features of MAS include hyperfunctioning thyroid nodules and pituitary adenomas. Fibrous dysplasia and MAS are caused by postzygotic pathogenic variants in the *GNAS* gene, leading to constitutive activity of the associated G-protein–coupled receptors in target tissues. The bone in fibrous dysplasia lesions is replaced by disorganized fibro-osseous tissue resulting in woven bone and high turnover with active osteoclastogenesis. Bone formation markers, such as total and bone-specific alkaline phosphatase, and resorption markers, such as N-telopeptide, are typically elevated when fibrous dysplasia lesions are metabolically active.

As fibrous dysplasia is a high–bone turnover state, inhibition with antiresorptive agents has been used. Treatment with intravenous bisphosphonates (Answer C) can improve bone pain, especially when lesions involve long bones, but they do not eliminate pain, and most studies indicate that there is not an improvement in lesion size. Daily oral bisphosphonates (Answer D) did not improve bone pain in a randomized controlled trial in participants with fibrous dysplasia. Treatment of fibrous dysplasia with denosumab has been shown to be associated with decreased bone pain in case reports, a retrospective cohort, and a small prospective clinical trial. Interestingly while bisphosphonates do not appear to decrease lesion volume, limited data suggest a decrease in lesion volume with denosumab. However, discontinuation of monthly doses of denosumab in a prospective trial was associated with severe rebound elevation of bone turnover markers, and 3 of 8 participants developed rebound hypercalcemia, which was severe in 1 participant. Because of these risks, denosumab requires caution, frequent follow-up, and laboratory monitoring (at least every 2 weeks), and it should be considered an experimental treatment to be conducted under research protocols.

Resection of fibrous dysplasia lesions is typically associated with lesion regrowth, which may be rapid or delayed, and complete resection is usually not possible. Thus, surgery (Answer E) is reserved for situations when vital structures are compressed. Although this patient has hearing loss, her vision is not impaired; resection of the skull lesion was not pursued due to concerns that potential expansion of the skull lesion after surgery could compromise intracranial structures.

Radiation therapy (Answer B) is contraindicated because of risks of fibrous dysplasia transforming into osteosarcoma.

Fibrous dysplasia lesions produce FGF-23. Some individuals with fibrous dysplasia develop FGF-23–mediated renal phosphate losses and resultant hypophosphatemia. The likelihood of hypophosphatemia is proportionate to the total body fibrous dysplasia lesion burden. Burosumab (Answer A), a monoclonal antibody to FGF-23, is FDA approved for treating hypophosphatemia due to X-linked hypophosphatemia and tumor-induced osteomalacia. There are case reports of using burosumab to manage hypophosphatemia of fibrous dysplasia, although it is not approved for that indication. However, this patient does not have hypophosphatemia.

Educational Objective

Medically manage metabolic bone disease of fibrous dysplasia.

Reference(s)

Hartley I, Zhadina M, Collins MT, Boyce AM. Fibrous dysplasia of bone and McCune-Albright syndrome: a bench to bedside review. *Calcif Tissue Int.* 2019;104(5):517-529. PMID: 31037426

de Castro LF, Michel Z, Pan K, et al. Safety and efficacy of denosumab for fibrous dysplasia of bone. *N Engl J Med.* 2023;388(8):766-768. PMID: 36812441

43 **ANSWER: C) Bempedoic acid**

This patient has established atherosclerotic cardiovascular disease (ASCVD), requiring percutaneous coronary intervention, and he is at high risk for future events. Effective lipid-lowering therapy would be very beneficial in reducing this risk. Unfortunately, he is intolerant of high-intensity statin therapy. While taking maximally tolerated statin therapy and ezetimibe, his LDL-cholesterol concentration is still elevated at 112 mg/dL (2.90 mmol/L), well above the recommended threshold of 55 mg/dL (1.42 mmol/L) for further intensification of therapy. It would be prudent to add another lipid-lowering agent to his current regimen, and of the choices presented, bempedoic acid (Answer C) is most likely to be effective and well tolerated.

Bempedoic acid is a relatively new cholesterol-lowering medication that has a similar mechanism of action to that of statins. It reduces endogenous cholesterol synthesis, leading to increased LDL-receptor expression and clearance of LDL particles. Unlike statins, which inhibit HMG CoA reductase, bempedoic acid inhibits ATP citrate lyase, which catalyzes a more proximal step in cholesterol biosynthesis. Further, it is a prodrug, and it must be activated by a long-chain acyl CoA synthetase, which is present only in the liver and not in the skeletal muscle. Therefore, its effects are largely limited to the liver, and it is less likely to cause muscle adverse effects seen with statin therapy that may result from statin-induced reduction of cholesterol and other biological intermediates in the skeletal muscle. Indeed, bempedoic acid has been extensively studied in statin-intolerant patients. In the CLEAR Serenity and CLEAR Tranquility studies, it reduced LDL cholesterol by about 25% and high-sensitivity C-reactive protein by about 30%. The recently completed CLEAR Outcomes study also provided evidence for ASCVD risk reduction. In this randomized, double-blind, placebo-controlled study of nearly 14,000 statin-intolerant patients at high risk, bempedoic acid reduced the relative risk of a composite 4-point major adverse cardiovascular event by 13% and fatal and nonfatal myocardial infarction by 23%. Bempedoic acid was approved by the US FDA in 2023 for use in patients with either established ASCVD or those with heterozygous familial hypercholesterolemia who have elevated LDL cholesterol on maximally tolerated statin therapy. Although bempedoic acid is generally well tolerated, it may increase the risk of hyperuricemia and gout, which should be monitored in patients with previously elevated uric acid levels. It has not been associated with hyperglycemia and increased risk of diabetes, and there is a tendency for lower incidence of new-onset diabetes compared with incidence in placebo groups.

In this patient with prediabetes, ASCVD, elevated LDL cholesterol and high-sensitivity C-reactive protein, and intolerance to high-dosage statin therapy, bempedoic acid is indicated for further risk reduction. Switching rosuvastatin to pravastatin (Answer A) is unlikely to lead to more robust lipid-lowering. Similarly, adding a stanol ester supplement (Answer E) is unlikely to lower LDL cholesterol more than 10%, and it has not been shown to reduce ASCVD risk. A PCSK9 inhibitor would certainly be a reasonable option, but this was not offered as an answer choice.

He has prediabetes, but given the improvement in his body weight and hemoglobin A_{1c}, restarting metformin (Answer B) for diabetes prevention is not required. Metformin has not been shown to reduce risk of either microvascular or macrovascular complications in patients with prediabetes.

Similarly, GLP-1 receptor agonists such as semaglutide (Answer D) are indicated in patients with type 2 diabetes and high ASCVD risk, with little data supporting their use in prediabetes for ASCVD risk reduction.

Educational Objective

Explain the role of bempedoic acid in the treatment of hypercholesterolemia in statin-intolerant patients.

Reference(s)

Nissen SE, Lincoff AM, Brennan D, et al; CLEAR Outcomes Investigators. Bempedoic acid and cardiovascular outcomes in statin-intolerant patients. *N Engl J Med.* 2023;388(15):1353-1364. PMID: 36876740

Nissen SE, Menon V, Nicholls SJ, et al. Bempedoic acid for primary prevention of cardiovascular events in statin-intolerant patients. *JAMA.* 2023;330(2):131-140. PMID: 37354546

Ballantyne CM, Bays H, Catapano AL, Goldberg A, Ray KK, Saseen JJ. Role of bempedoic acid in clinical practice. *Cardiovasc Drugs Ther.* 2021;35(4):853-864. PMID: 33818688

44 ANSWER: B) Measure ACTH

This patient has evidence of overt hypercortisolism (Cushing syndrome) based on his history, examination findings, and radiology report. Suggestive features include abdominal distention (which could be due to central weight gain), spontaneous Achilles tendon rupture (recognized as a complication of Cushing syndrome), deep vein thrombosis, and hypertension. The examination findings of bruising and proximal myopathy are consistent with hypercortisolism. The loss of vertebral height may be due to osteoporotic fractures, and it is notable that he has an elevated hemoglobin A_{1c} value in the absence of any family history of diabetes. The clinical likelihood of Cushing syndrome is therefore high, and it is supported by the results of the overnight dexamethasone-suppression test, which shows clear failure of suppression in the face of adequate dexamethasone concentrations, as well as the elevated late-night salivary cortisol concentrations. Guidelines from the Pituitary Society recommend proceeding to ACTH measurement (Answer B) in patients with a high index of suspicion of Cushing syndrome and abnormal findings on 2 to 3 screening tests, especially if these are confirmed on repeat testing.

Additional measurement of 24-hour urinary free cortisol (Answer A) is not necessary in this clinical setting and would only add to diagnostic delay.

There is no need to delay management by reassessing the aldosterone-to-renin ratio after discontinuing spironolactone (Answer C). Endocrine Society guidelines recommend discontinuation of mineralocorticoid receptor antagonists for at least 4 weeks before assessing the aldosterone-to-renin ratio; spironolactone would be expected to raise plasma renin activity.

The suppressed aldosterone in this instance provides an additional clue for the presence of hypercortisolism, since cortisol excess saturates the 11β-hydroxysteroid dehydrogenase 2 enzyme leading to spillover to the mineralocorticoid receptor.

The European guidelines on management of adrenal incidentalomas emphasize the importance of undertaking a careful clinical evaluation for symptoms and signs of adrenal hormone excess. Furthermore, in the context of an adrenal incidentaloma, cortisol values higher than 5 µg/dL (>138 nmol/L) are often suggestive of overt Cushing syndrome. Adrenalectomy (Answer D) is recommended in patients with an adrenal incidentaloma, autonomous cortisol secretion, and potential cortisol-related comorbidities, but ACTH independence has to be demonstrated by a suppressed or low basal morning plasma ACTH value before this can be contemplated. In this instance, there are features that suggest ACTH dependence, including the biochemical finding of a nonsuppressed DHEA-S concentration and the radiologic appearance of the contralateral left adrenal gland that is easily visible and not atrophic on CT. This patient's ACTH concentration was elevated, with workup confirming Cushing disease and subsequent transsphenoidal surgery leading to biochemical remission. The coexistence of ACTH-dependent Cushing disease with an adrenal adenoma has been recognized. Reduction in adenoma size after successful transsphenoidal surgery can occur, suggesting that adrenal tumor growth is at least partially ACTH dependent in some cases. Testing for ongoing autonomous adrenal cortisol secretion following surgical cure is nevertheless prudent.

Medical therapy for Cushing disease includes drugs that target adrenal steroidogenesis (such as metyrapone [Answer E]), the glucocorticoid receptor, and somatostatin and dopamine receptors in the pituitary gland. However, drug therapy is usually reserved for patients who decline or are not candidates for surgery, who have persistent or recurrent hypercortisolism after initial surgery, or who need control of hypercortisolism while radiotherapy takes effect.

Educational Objective

Recognize that an incidental adrenal mass may be the initial presentation of Cushing disease.

Reference(s)

Fleseriu M, et al. Consensus on diagnosis and management of Cushing's disease: a guideline update. *Lancet Diabetes Endocrinol.* 2021;9(12):847-875. PMID: 34687601

Funder JW, Carey RM, Mantero F, et al. The management of primary aldosteronism: case detection, diagnosis, and treatment: an Endocrine Society clinical practice guideline. *J Clin Endocrinol Metab.* 2016;101(5):1889-1916. PMID: 26934393

Fassnacht M, Tsagarakis S, Terzolo M, et al. European Society of Endocrinology clinical practice guidelines on the management of adrenal incidentalomas, in collaboration with the European Network for the Study of Adrenal Tumors. *Eur J Endocrinol.* 2023;189(1):G1-G42. PMID: 37318239

Shakir MKM, Ebrahim IC, Spiro A, Mai VQ, Hoang TD. Coexistence of Cushing disease with a solitary adrenocorticotropic hormone-dependent adrenal adenoma. *AACE Clinical Case Rep.* 2022;8:41-44. PMID: 35097202

45 ANSWER: A) Perform clinical assessment, including neck exam, in 12 months

Although this patient's thyroid nodule was incidentally detected on PET-CT imaging, it was not noted to be FDG-avid. Conversely, thyroid nodules with focal FDG uptake are associated with a malignancy rate of approximately 30%. Although a higher SUVmax (maximum standardized uptake value) has been associated with a higher likelihood of malignancy, it is not easy to distinguish malignant from benign nodules based on SUVmax alone. The observation that a thyroid nodule is FDG-avid should be incorporated into decision-making about FNA.

The presence of high-risk sonographic findings at baseline or the subsequent development of suspicious features during follow-up represent the strongest indications for a repeat FNA of thyroid nodules after an initial benign biopsy result. Fortunately, the development of new suspicious features is a rare event, with only 3% to 4% of cytologically benign thyroid nodules developing suspicious sonographic features during surveillance. Growth is another consideration in the follow-up of benign thyroid nodules. A prospective multicenter study from Italy of 1567 benign thyroid nodules demonstrated that only 11.1% of nodules grew during follow-up and that growth was typically slow. Furthermore, nodule growth is a relatively poor predictor of subsequent malignant diagnosis following negative FNA. Nodule growth should not be considered a strong indication for repeating FNA of a nodule with benign cytology. However, in the case described here, another FNA has already been performed, again demonstrating benign cytology.

Thyroid nodules that have undergone ultrasound-guided FNA with benign results on 2 occasions have an extremely low probability of malignancy (<1%), and ultrasonography follow-up on a routine basis to evaluate for missed malignancy is not warranted. Therefore, performing neck ultrasonography in 12 or 24 months (Answers C and D) is incorrect. Patients with larger thyroid nodules should continue to be assessed clinically for the development of signs or symptoms that could be related to their thyroid nodule (dysphagia, dysphonia, dyspnea, cough, pain, and symptoms of hyperthyroidism). Performing a clinical assessment, including neck examination, in 12 months (Answer A) would be appropriate management for this patient. Repeated neck ultrasonography would be indicated if the clinical assessment suggested that significant growth had occurred that was symptomatic or likely to result in compressive symptoms. Fortunately, most cytologically benign nodules that are classified as asymptomatic remain that way over time.

There is no indication for neck CT with contrast (Answer B). Neck CT would be potentially indicated if there were clinical suspicion for substernal extension of the thyroid/nodule or to further evaluate the relationship between the thyroid nodule and nearby anatomical structures (eg, trachea or esophagus) because of clinical compressive symptoms. The patient described in this vignette is asymptomatic.

Similarly, there is no current indication to refer the patient for a right thyroid lobectomy (Answer E).

Educational Objective

Recommend appropriate follow-up of a thyroid nodule with 2 negative FNA biopsies.

Reference(s)

Durante C, Costante G, Lucisano G, et al. The natural history of benign thyroid nodules. *JAMA.* 2015;313(9):926-35. PMID: 25734734

Haugen BR, Alexander EK, Bible KC, et al. 2015 American Thyroid Association management guidelines for adult patients with thyroid nodules and differentiated thyroid cancer: the American Thyroid Association Guidelines Task Force on Thyroid Nodules and Differentiated Thyroid Cancer. *Thyroid.* 2016;26(1):1-133. PMID: 26462967

Singh Ospina N, Maraka S, Espinosa DeYcaza A, et al. Diagnostic accuracy of thyroid nodule growth to predict malignancy in thyroid nodules with benign cytology: systematic review and meta-analysis. *Clin Endocrinol.* 2016;85(1):122-31. PMID: 26562828

46 ANSWER: C) Evaluate for causes of medication nonadherence

Achieving a glycemic target hemoglobin A_{1c} value below 7.0% (<53 mmol/mol) is not always easy for patients with type 2 diabetes. In the natural course of type 2 diabetes, most patients will experience higher glucose concentrations over time, which prompts modification of existing treatment regimens. Various treatment approaches can be used, and these methods differ based on, among other things, the degree of hyperglycemia, the risk of hypoglycemia, and/or underlying comorbidities.

Several novel medications have received FDA approval for the treatment of type 2 diabetes. Among them are the new GLP-1 receptor agonists, the dual-acting GLP-1 and gastric inhibitory peptide receptor agonists, and the SGLT-2 inhibitors, which have not only antihyperglycemic properties but also important cardiovascular and kidney protective benefits. In addition, GLP-1 receptor agonists are important weight-loss medications that can be used to address obesity, a condition that can be a risk factor for the development of not only diabetes but also many other diseases (cardiovascular, pulmonary, orthopedic) and malignancies.

Managing diabetes is a significant economic burden for both patients and society. The total cost of diabetes in the United States continues to rise. In 2022, the total cost was $412.9 billion, including $306.6 billion in direct medical costs and $106.3 billion in indirect costs attributable to diabetes. Seventeen percent of these total direct medical costs are secondary to glucose-lowering medications and expenses of diabetes supplies. Many patients with diabetes face significant socioeconomic challenges. For some patients, nutrition options are less than optimal, and fast-food becomes an affordable meal. Although some newer diabetes medications are very efficient and have pleiotropic benefits, they are also relatively expensive, and some patients cannot afford them.

Physicians should consider the whole picture when a patient states that they are adherent to their medication regimen, especially in a case such as this one, in which the patient has not had any improvement in hemoglobin A_{1c}. His blood pressure did not improve, even after the initiation of an SGLT-2 inhibitor, a medication class that has blood pressure–lowering properties. This suggests that he is probably not taking the medication as prescribed. Physicians should also be aware that patients might not be willing to disclose sensitive information about their personal lives, such as possible financial burdens. This is especially a concern when a patient sees a new provider with whom they have not yet established a patient-physician relationship, as in this clinical vignette. This patient is currently unemployed and might be uninsured or have limited funds to support his medications or personal needs. Such information might not be easily disclosed. Establishing a rapport and a relationship with patients is fundamental in managing chronic diseases, such as diabetes, but this requires time. Two very potent medications were prescribed to this patient (ie, a GLP-1 receptor agonist and an SGLT-2 inhibitor), with no improvement in his hemoglobin A_{1c}. Despite the patient stating that he adheres to his treatment regimen, further evaluation is needed to rule out medication nonadherence (Answer C). One option would be to call the patient's pharmacy to determine whether prescriptions were filled.

A thiazolidinedione (Answer B) or a sulfonylurea (Answer A) should not be added to the existing regimen without further evaluating why his hyperglycemia is not improving. Thiazolidinediones are also not recommended for this patient because he has history of congestive heart failure. If he had disclosed from the beginning that the cost of medications was a significant concern, adding a sulfonylurea or a thiazolidinedione or increasing the basal insulin dosage (Answer D) could have been considered as an initial therapeutic step. These options are more affordable than SGLT-2 inhibitors or GLP-1 receptor agonists. Adding prandial insulin (Answer E) and placing the patient on a multiple daily injection regimen might make his management significantly more complex, as he might require 3 or 4 insulin injections per day, and this could decrease adherence. Furthermore, a multiple daily injection regimen might not be needed, as his hemoglobin A_{1c} value is only mildly elevated, and noninsulin medications can be added to achieve glycemic targets.

Educational Objective

Identify causes of medication nonadherence in patients with diabetes, including socioeconomic challenges.

Reference(s)

Parker ED, Lin J, Mahoney T, et al. Economic costs of diabetes in the U.S. in 2022. *Diabetes Care.* 2024;47(1):26-43. PMID: 37909353

American Diabetes Association Professional Practice Committee. 9. Pharmacologic approaches to glycemic treatment: standards of care in diabetes-2024. *Diabetes Care.* 2024;47(Suppl 1):S158-S178. PMID: 38078590

47 **ANSWER: E) Repeat ACTH measurement using a separate immunoassay method**

The clinical picture and biochemical findings in this vignette are supportive of mild autonomous cortisol secretion (MACS), with no overt clinical signs of Cushing syndrome (pale striae are well recognized in Ehlers-Danlos syndrome), normal midnight salivary cortisol, and normal urinary free cortisol. The suppressed DHEA-S concentration and comparatively large adenoma are consistent with MACS arising from the right adrenal adenoma independently of ACTH. The image shows a normal, nonhyperplastic contralateral adrenal gland, which would be inconsistent with ACTH dependence. The elevated ACTH concentration is thus at odds with the other findings. ACTH immunoassay interference offers the best explanation for this inconsistency and repeating the ACTH measurement with an alternative immunoassay platform (Answer E) is the most appropriate next step in this patient's management. All immunoassays are potentially subject to assay interference, including ACTH immunoassays. Interference by heterophile antibodies, human antianimal antibodies, rheumatoid factor, and biotin is well known and may affect decision-making if unrecognized. The M-protein in monoclonal gammopathy can occur in high concentrations and may also cause significant assay interference. A repeat measurement on a different analytical platform in this instance resulted in an appropriately suppressed ACTH concentration of 6 pg/mL (1.3 pmol/L). Serial dilution could also be considered, either with the recommended diluent or a patient sample with undetectable ACTH, to establish whether the sample dilutes linearly. In patients with assay interference, a nonlinear dose response would be suggestive of analytical interference.

While Cushing disease may be associated with adrenal adenomas, these are rarely large. Furthermore, the DHEA-S concentration is normal or elevated in the setting of Cushing disease, and there is often radiologic evidence of a hypertrophied contralateral adrenal gland. Given these inconsistencies, the radiologic investigations used in localizing a central or peripheral source of ACTH excess, namely CT of the thorax, abdomen, and pelvis (Answer A), bilateral inferior petrosal sinus sampling (Answer B), and MRI pituitary (Answer C), would not be indicated in the absence of a confirmatory high ACTH concentration. Moreover, clinical signs of hypercortisolism are absent and findings on endocrine workup are normal, with the exception of the suppressed DHEA-S and nonsuppressed cortisol following 1-mg overnight dexamethasone. These observations argue strongly against ectopic ACTH production and Cushing disease, even if it were cyclical.

Right adrenalectomy (Answer D) might be considered in this patient in view of her young age, but only after ACTH independence is confirmed. Patients with MACS have an increased risk of morbidity and mortality, with the greatest risk being apparent in young women. The recently updated European guidelines on the management of adrenal incidentalomas recommend considering adrenalectomy for all patients with unilateral tumors and clinically significant hormone excess. In MACS, adrenalectomy may be considered dependent on the presence of potential cortisol-related comorbidities, condition of the patient, patient preference, and age.

Educational Objective

Guide the management of mild autonomous cortisol excretion and recognize immunoassay interference as a cause of falsely elevated ACTH.

Reference(s)

Donegan DM, Algeciras-Schimnich A, Hamidi O, Young WF, Nippoldt T, Bancos I. Corticotropin hormone assay interference: a case series. *Clin Biochem.* 2019;63:143-147. PMID: 30423322

Deutschbein T, Reimondo G, Di Dalmazi G, et al. Age-dependent and sex-dependent disparity in mortality in patients with adrenal incidentalomas and autonomous cortisol secretion: an international, retrospective, cohort study. *Lancet Diabetes Endocrinol.* 2022;10(7):499-508. PMID: 35533704

Fassnacht M, Tsagarakis S, Terzolo M, et al. European Society of Endocrinology clinical practice guidelines on the management of adrenal incidentalomas, in collaboration with the European Network for the Study of Adrenal Tumors. *Eur J Endocrinol.* 2023;1891:G1-G42. PMID: 37318239

Greene LW, Geer EB, Page-Wilson GB, Findling JW, Raff H. Assay-specific spurious ACTH results lead to misdiagnosis, unnecessary testing, and surgical misadventure-a case series. *J Endocr Soc.* 2019;3(4):763-772. PMID: 30963134

48 ANSWER: D) Start transdermal estrogen with cyclic oral progesterone

In the setting of anorexia nervosa, bone health is impaired primarily by the caloric deficit and the resulting hormonal changes, including limitations in estradiol production (through hypothalamic hypogonadism) and a state of relative GH resistance leading to decreased IGF-1 (*see figure*). Additionally, the physiologic stress of starvation activates the hypothalamic-pituitary-adrenal axis and hypercortisolism, which further contributes to impairments in bone mass. Hyponatremia is common in anorexia nervosa and is associated with lower bone mineral density. Hyponatremia is also associated with osteoporosis and fractures in persons without anorexia nervosa (such as postmenopausal women and elderly men).

Because anorexia nervosa typically begins during puberty, which is the period of the greatest accumulation of an individual's peak bone mass, there are decrements in bone mineral accrual, as well as potential loss of bone mass over time in adulthood, as the disorder persists. The most important factor in improving bone health is adequate weight gain with the resulting resumption of spontaneous menstrual cycles as evidence of sufficient weight attainment.

Oral combined hormonal contraceptives (Answer B) are generally not effective in improving bone density in the setting of anorexia nervosa, with most studies suggesting no benefit. This is likely in part because the first-pass effect of estrogens on the liver leads to lower IGF-1 production, and this impedes bone mass gains. Transdermal estrogen (Answer D) avoids this first-pass effect, and in one study it led to improvements in bone density compared with placebo. For uterine protection, transdermal estrogens must be combined with a form of progestin, which, if given cyclically, will induce menstrual cycles. However, artificially inducing menstrual cycles has the disadvantage of removing the utility of resumption of spontaneous menses as a marker of adequate weight gain.

Calcium (Answer A) and vitamin D supplementation is important if adequate amounts are not present in the diet. Persons with anorexia nervosa demonstrate decreased intestinal calcium absorption and increased urinary excretion, likely related to hypercortisolemia. However, individuals with anorexia nervosa are often already supplementing with calcium and vitamin D, and this patient's calcium and vitamin D intake is adequate. Merely increasing these dosages will likely result in hypercalciuria. Benefit from further increases in calcium and vitamin D supplementation has not been demonstrated and would not overcome the issues caused by caloric deficit and hypothalamic hypogonadism.

Figure. Neuroendocrine adaptations to starvation contributing to loss of bone mass in anorexia nervosa

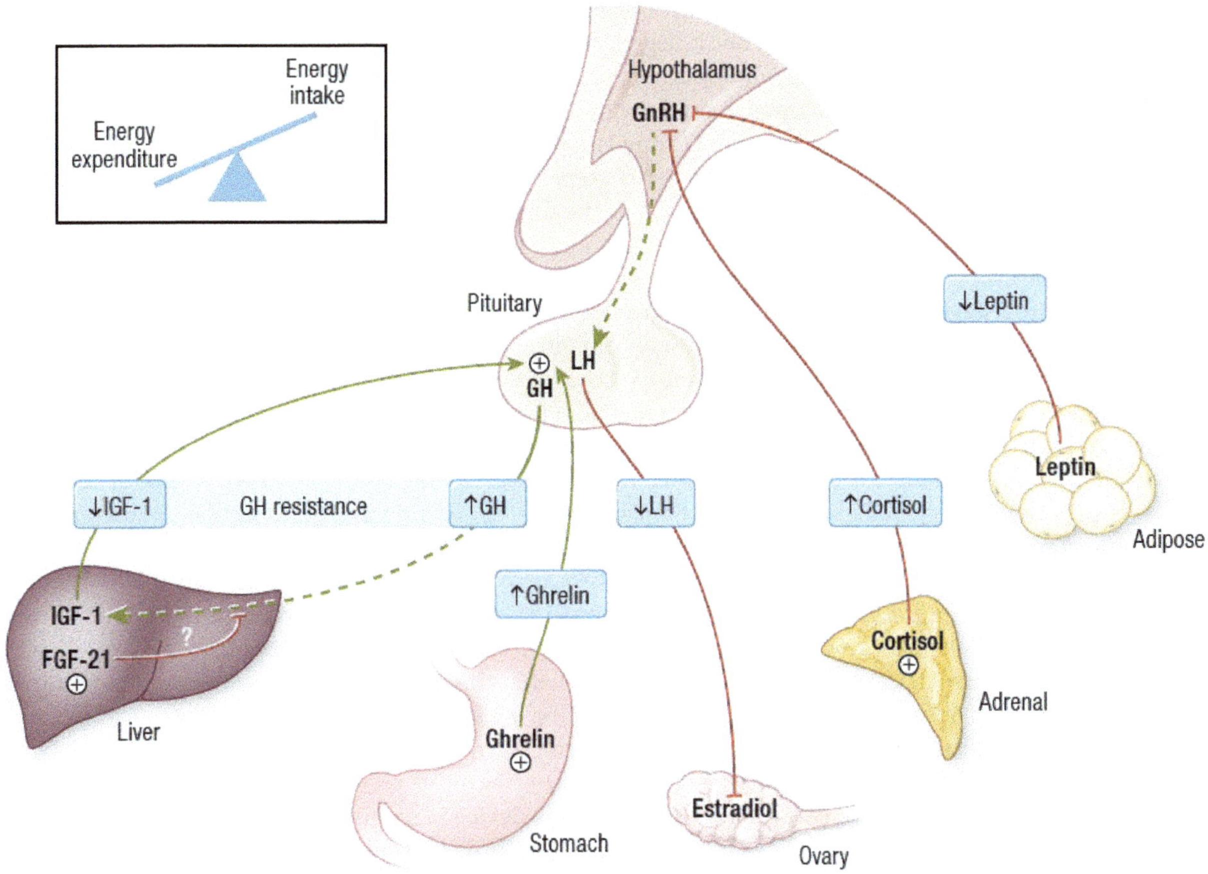

Reprinted from Fazeli PK & Klibanski A. *Endocr Rev*, 2018;39(6):895-910. © 2018 Illustration Presentation Endocrine Society.

Androgens (DHEA [Answer C] or testosterone [Answer E]) have also been evaluated for their effects on bone density in this situation and do not increase bone density over placebo, although one study indicated stabilization of bone density (rather than improvement) when administering DHEA and a combined oral contraceptive together, compared with outcomes with placebo.

Studies evaluating the use of bisphosphonates or teriparatide for bone treatment in anorexia nervosa have shown some increases in bone density when treating adults (by 2%-4% for bisphosphonates and 6%-10% for teriparatide). Use of bone-active drugs is primarily reserved for patients in whom other approaches have failed and who are deemed to be at high risk (eg, frequent fragility fractures). However, due to the suppressive effect of bisphosphonates on bone turnover, these agents are not generally effective in preventing stress fractures and might actually increase that risk. In general, the bone density benefit of any approach in anorexia nervosa is more limited if not accompanied by improvements in body weight.

Educational Objective

Evaluate and manage bone health in the setting of anorexia nervosa and hypothalamic amenorrhea.

Reference(s)

Indirli R, Lanzi V, Mantovani G, Arosio M, Ferrante E. Bone health in functional hypothalamic amenorrhea: what the endocrinologist needs to know. *Front Endocrinol (Lausanne)*. 2022;13:946695. PMID: 36303862

Fazeli PK, Klibanski A. Effects of anorexia nervosa on bone metabolism. *Endocr Rev.* 2018;39(6):895-910. PMID: 30165608

Robinson L, Aldridge V, Clark EM, Misra M, Micali N. Pharmacological treatment options for low bone mineral density and secondary osteoporosis in anorexia nervosa: a systematic review of the literature. *J Psychosom Res.* 2017;98:87-97. PMID: 28554377

49 ANSWER: B) Prescribe a low-dosage oral combined hormonal contraceptive pill

This woman is in perimenopause, the time during which variations in menstrual cycle length occur. The total duration of perimenopause varies, but it begins with early perimenopause when the menstrual cycle is variable in length, with a 7- or more day difference in length of consecutive cycles. Early perimenopause then progresses to late perimenopause during which time the interval of amenorrhea is 60 or more days. Perimenopause ends 12 months after the final menstrual period. During late perimenopause, women can also begin to experience menopause-related symptoms and declining reproductive capacity because of ovarian aging. However, hormonally normal and ovulatory menstrual cycles can occur up until the final menstrual period. Therefore, contraception is needed if pregnancy is not desired and the person is engaging in sexual behavior that can lead to conception.

For a woman nearing the average age of menopause (51 years) who has irregular menses and classic symptoms, there is no benefit to measuring estradiol and FSH (Answer A) and this is not recommended. Estradiol and FSH values would not change the best next step in her management.

She is currently perimenopausal with symptoms of estrogen deficiency, and she would benefit from menopausal hormone therapy with combination estradiol and progesterone if she does not have contraindications. Her mother had a provoked deep vein thrombosis, and the patient has recently been told she has an *MTHFR* variant (Ala1298Cys), which raises additional concern for increased clotting risk. The *MTHFR* gene encodes the methylenetetrahydrofolate reductase enzyme that has a role in converting the amino acid homocysteine to methionine. Therefore, variants in the *MTHFR* gene can lead to elevated homocysteine levels. These polymorphisms are common in the general population. There is concern for theoretical abnormal blood clotting; however, data from large meta-analyses do not support the association of these variants with increased clotting risk. In the case of an incidentally detected *MTHFR* variant, repeating thrombophilia screening (Answer E) is not indicated. While functional medicine providers could recommend a methylated B-complex vitamin (Answer C), there is no evidence to support this intervention and studies show a lack of effect.

Given that she is sexually active with ovum meeting sperm, she needs perimenopausal contraception. Of the listed answer choices, a low-dosage oral combined hormonal contraceptive pill (Answer B) is the best option. Another contraceptive option would be a progestin intrauterine device in combination with estrogen replacement therapy. However, this is off-label use in the United States. Transdermal estradiol and micronized progesterone (Answer D) would not protect this patient against conception given that menopausal hormone therapy is not currently a reliable method because of the low hormone dosing.

Educational Objective

Explain the implications of *MTHFR* variants when prescribing menopausal hormone therapy.

Reference(s)

The 2022 hormone therapy position statement of The North American Menopause Society Advisory Panel. The 2022 hormone therapy position statement of The North American Menopause Society. *Menopause.* 2022;29(7):767-794. PMID: 35797481

Voedisch AJ, Ariel D. Perimenopausal contraception. *Curr Opin Obstet Gynecol.* 2020;32(6):399-407. PMID: 33002952

Dhawan A, Eng C. Is the *MTHFR* gene mutation associated with thrombosis? *Cleve Clin J Med.* 2023;90(11):661-663. PMID: 37914198

50 ANSWER: D) Switch from dulaglutide to tirzepatide at the highest tolerated dosage

Individuals with diabetes who have a BMI of 25 kg/m^2 or greater may benefit from modest or marked weight loss. Weight management is an important component of glucose lowering in patients with type 2 diabetes. Selection of the glucose-lowering regimen should consider approaches that support weight management goals in persons with overweight or obesity. Although this patient is at goal for glycemic control, he has obesity, which could compound other comorbidities (dyslipidemia, hypertension, sleep apnea and osteoarthritis). Current diabetes management guidelines suggest that in addition to glycemic control, the other core components of managing diabetes include weight management, cardiovascular risk factor management, and cardiorenal protection.

Figure. Four core components of care of patients with type 2 diabetes

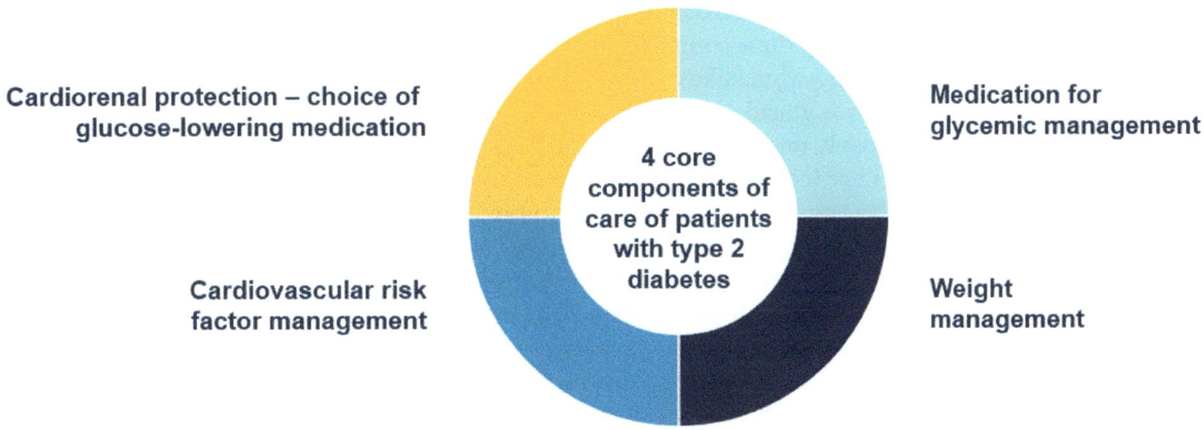

Adapted from ElSayed NA et al. *Diabetes Care*, 2023; 46(Suppl 1):S19-S40. © The American Diabetes Association.

This is a case of primary prevention; the patient's regimen addresses cardiovascular risk factor management, and he is taking glucose-lowering medications that address cardiorenal protection. Even though he does not need further glucose reduction, comprehensive management of his diabetes as per current standard of care includes helping him optimize weight management. Tirzepatide (Answer D) is the first dual GLP-1 receptor agonist and glucose-dependent insulinotropic polypeptide (GIP) receptor agonist approved to treat type 2 diabetes. At week 40, tirzepatide, 15 mg, was associated with mean weight loss of up to 27.3 lb (12.4 kg) from baseline. Thus, switching from dulaglutide to tirzepatide would help with further weight loss and alleviate knee pain.

Although this patient is already at his glycemic goal with a hemoglobin A_{1c} value less than 7.0% (<53 mmol/mol), continuing current management (Answer A) would not address his underlying obesity, which has not improved despite optimizing lifestyle modifications.

Increasing the metformin dosage (Answer C) would not be helpful because he is already at his glycemic goal, and increasing the dosage would be unlikely to help with weight loss. Similarly, switching from dulaglutide, 1.5 mg weekly, to dulaglutide, 3.0 mg weekly, (Answer B) is unlikely to induce any substantial additional weight loss.

Initiation of naltrexone/bupropion (Answer E) would not be appropriate in this patient with uncontrolled hypertension. Additionally, the weight loss seen with naltrexone/bupropion would be much less compared to that achieved with tirzepatide.

Educational Objective

Prioritize weight management in patients with diabetes and obesity even if glycemic targets are already met.

Reference(s)

American Diabetes Association Professional Practice Committee. 9. Pharmacologic approaches to glycemic treatment: standards of care in diabetes-2024. *Diabetes Care.* 2024;47(Suppl 1):S158-S178. PMID: 36507650

Lingvay I, Sumithran P, Cohen RV, le Roux CW. Obesity management as a primary treatment goal for type 2 diabetes: time to reframe the conversation. *Lancet.* 2022;399(10322):394-405. PMID: 34600604

Muzurović E, Dragnić S, Medenica S, et al. Weight-centric pharmacological management of type 2 diabetes mellitus - an essential component of cardiovascular disease prevention. *J Diabetes Complications.* 2020;34(8):107619. PMID: 32499116

Inagaki N, Takeuchi M, Oura T, Imaoka T, Seino Y. Efficacy and safety of tirzepatide monotherapy compared with dulaglutide in Japanese patients with type 2 diabetes (SURPASS J-mono): a double-blind, multicentre, randomised, phase 3 trial. *Lancet Diabetes Endocrinol.* 2022;10(9):623-633. PMID: 35914543

Wysham C, Blevins T, Arakaki R, et al. Efficacy and safety of dulaglutide added onto pioglitazone and metformin versus exenatide in type 2 diabetes in a randomized controlled trial (AWARD-1). *Diabetes Care.* 2014;37(8):2159-2167 PMID: 24879836

Boulé NG, Prud'homme D. Canadian Adult Obesity Clinical Practice Guidelines: Physical Activity in Obesity Management. Available at: https://obesitycanada.ca/guidelines/physicalactivity. Accessed January 2024

51 ANSWER: C) hCG subcutaneously 3 times weekly

Abuse of high dosages of androgenic steroids causes marked suppression of spermatogenesis and fertility. Chronic abuse (>1 year) of high dosages of these compounds may suppress the hypothalamus-pituitary-testicular (HPT) axis for years. For this couple who would like to conceive now, the most reliable and rapid approach for restoring spermatogenesis and fertility is gonadotropin replacement therapy. LH replacement therapy in the form of hCG (Answer C) is likely to effectively raise sperm concentrations enough for fertility within 6 to 12 months in this man who has testes that are 12 cc (combined volume is 24 cc) and some degree of spermatogenesis. Normal adult testes are 15 cc or greater with a combined volume of 30 cc or greater.

For men who have gone through puberty and have combined testicular volumes of 16 cc or more (ie, each testis is at least 8 cc), hCG monotherapy is often adequate to restore spermatogenesis to levels that allow conception. FSH replacement therapy (typically with recombinant human FSH) may be added if there is an inadequate response to hCG alone. With gonadotropin replacement therapy, conception often occurs at sperm concentrations below normal (3-15 million) if the female partner is ovulating. This man's wife is most likely ovulating because she has regular periods. Regular periods with intervals of 28 to 35 days are associated with ovulation in more than 95% of women.

"Watchful waiting" with episodic reassessment of serum testosterone and gonadotropins (Answer D) after cessation of androgenic steroid abuse is a reasonable option, particularly for men who have abused androgens for 1 year or less. Recovery of spermatogenesis typically occurs within 12 to 18 months for these men. However, men who have abused androgens for longer than 1 year may have symptoms of hypogonadism for many months (or even years) and may have suppressed spermatogenesis for years (or longer). A recent study suggested that fertility reverted to normal within 1 to 3 years after cessation of androgen abuse that lasted 3 to 4 years. This patient's partner is 34 years old, and a delay of 1 to 3 years might significantly reduce their chances of conception; watchful waiting is not a good option.

There are case reports of aromatase inhibitor therapy (Answer A) being effective in treating secondary hypogonadism and azoospermia or severe oligospermia in men with class 3 obesity, but there are no conclusive data for men with abnormal spermatogenesis due to a suppressed HPT axis from chronic androgenic steroid abuse. Furthermore, aromatase inhibitor therapy causes adverse effects on bone health and body composition (increased body fat) in men.

Similarly, selective estrogen receptor modulators (eg, clomiphene [Answer B]) might restore HPT function and fertility for these men, but the evidence is weak for the effectiveness and timeliness of response for these drugs compared with gonadotropin replacement therapy. All data for clomiphene are based on case series. In addition, clomiphene has been associated with central retinal vein thrombosis.

Spermatogenesis should be optimized before any surgical extraction procedure (Answer E) is considered. Surgical extraction is likely to be unnecessary for this man.

Educational Objective

Recommend subcutaneous hCG as the initial treatment for men with secondary hypogonadism, azoospermia (or severe oligospermia), and infertility due to chronic androgenic steroid abuse if there is a goal to conceive within the next 2 years.

Reference(s)

Anawalt BD. Diagnosis and management of anabolic androgenic steroid use. *J Clin Endocrinol Metab.* 2019;104(7):2490-2500. PMID: 30753550

Anawalt BD. Male fertility after androgenic steroid use: how little we know. *J Clin Endocrinol Metab.* 2021;106(7):e2813-e2815. PMID: 33861859

Anawalt BD. Approach to male infertility and induction of spermatogenesis. *J Clin Endocrinol Metab.* 2013;98(9):3532-3542. PMID: 24014811

Thirumalai A, Anawalt BD. Androgenic steroids use and abuse: past, present, and future. *Urol Clin North Am.* 2022;49(4):645-663. PMID: 36309421

52 ANSWER: D) Spinal epidural lipomatosis

The examination findings of paraparesis with hypertonicity and sensory loss below the midthoracic region point to spinal cord compression as the cause of this patient's presentation. Spine MRI shows generalized excess fat in the extradural space, with signal characteristics of fat on all sequences (T1: high signal; T2: high signal; T2-STIR: hypointense). This indicates spinal epidural lipomatosis (Answer D) as the underlying cause (*Figure 2*). *Figure 3* shows MRI of a normal spine for comparison.

Figure 2. MRI showing epidural lipomatosis

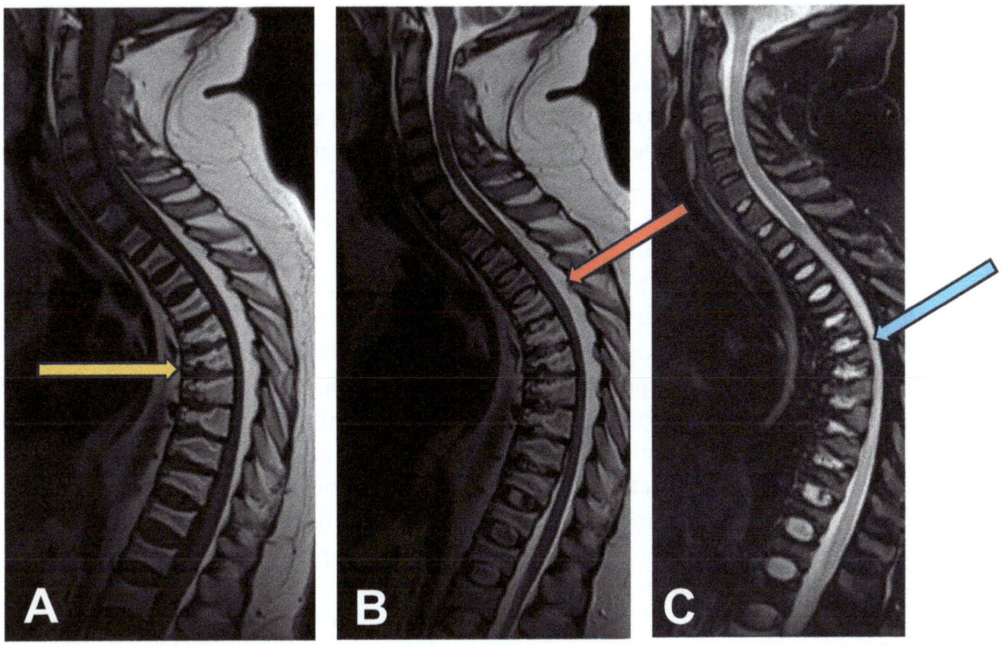

Panel A, sagittal T1-weighted; *Panel B*, sagittal T2-weighted; and *Panel C*, T2-STIR sequence (short Tau inversion recovery). Yellow arrow indicates multiple compression fractures; red arrow indicates spinal epidural lipomatosis; blue arrow indicates compressed cord. Adapted with permission of Bioscientifica Ltd. from Ahmad S et al. *Endocrinol Diabetes Case Rep*, 2020; 2020:20-0111.

Figure 3. MRI showing a normal spine

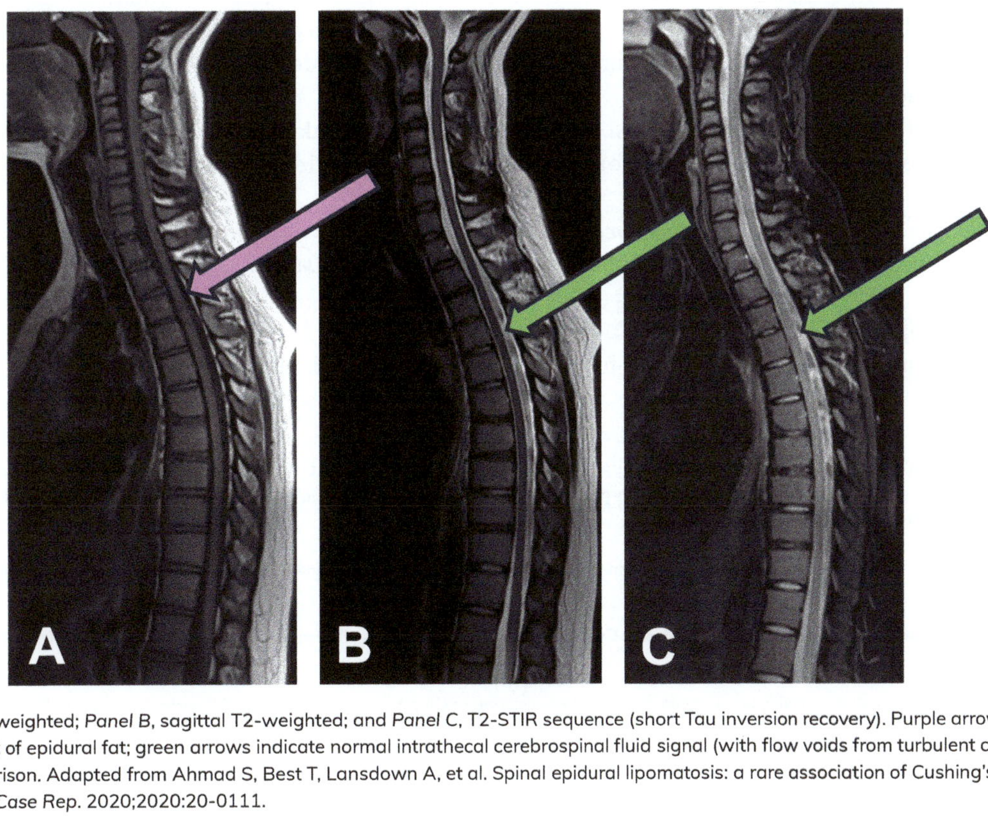

Panel A, sagittal T1-weighted; *Panel B*, sagittal T2-weighted; and *Panel C*, T2-STIR sequence (short Tau inversion recovery). Purple arrow indicates normal small amount of epidural fat; green arrows indicate normal intrathecal cerebrospinal fluid signal (with flow voids from turbulent cerebrospinal fluid flow) for comparison. Adapted from Ahmad S, Best T, Lansdown A, et al. Spinal epidural lipomatosis: a rare association of Cushing's disease. *Endocrinol Diabetes Case Rep*. 2020;2020:20-0111.

Spinal epidural lipomatosis is defined as excessive accumulation of fat within the epidural space leading to compression of the thecal sac, and it should be considered in the differential diagnosis of myopathy in Cushing syndrome. Spinal epidural lipomatosis may be asymptomatic, but in severe cases it can cause pain, weakness, radicular symptoms, and/or paresthesia. The lower lumbar and/or lower thoracic levels are the most commonly affected regions. Risk factors include glucocorticoid excess, most commonly from long-term steroid administration, and obesity. Endogenous glucocorticoid excess from Cushing syndrome is also recognized as a risk factor for spinal epidural lipomatosis. No specific intervention, beyond weight loss and a review of exogenous steroid therapy, is required in most cases, although surgical decompression may occasionally be necessary. In patients with Cushing syndrome, control of hypercortisolism by surgery or medical therapy is expected to gradually improve symptoms. Spinal epidural lipomatosis is likely to have been present for many months preoperatively in this patient, with his back pain and lower-limb weakness attributed to osteoporotic fractures and proximal myopathy, respectively. It is also conceivable that some degree of spinal cord ischemia arising in an already compromised spinal canal (from spinal epidural lipomatosis and wedge fractures) may have accounted for the acute postoperative presentation.

Spontaneous spinal cord infarction (Answer C) is thought to result from atheromatous processes as in cerebral stroke, with overlapping risk factors of hypertension, diabetes, hyperlipidemia, and cigarette smoking. Cushing syndrome is associated with an increased risk of thromboembolic events, and prophylactic anticoagulation is typically recommended perioperatively. Spinal cord infarction usually presents with sudden and severe back pain, followed by rapid bilateral flaccid limb weakness and sensory loss, particularly for temperature and pain. While the rapidity of symptom onset should prompt consideration of cord infarction as a potential diagnosis, the clinical presentation is not typical. Furthermore, the radiological hallmark of spinal cord infarction is the presence of abnormal T2 signal within the cord, which is absent in this case.

Osteoporotic fractures of the spine (Answer A) are common in individuals with Cushing syndrome. Indeed, this patient's spine MRI shows loss of height and compression fractures from T1 to L1. However, fractures are not expected to account for the neurological findings, except in rare instances where there is retropulsion. Management of osteoporosis secondary to glucocorticoid excess is nevertheless important, and it usually requires calcium and vitamin D supplementation. In the presence of fragility fractures, as in this patient, treatment with anabolic or antiresorptive agents is also indicated, alongside specific therapy to restore eucortisolism.

Proximal myopathy (Answer B) is characteristic of Cushing syndrome, but this would not explain the patient's wider neurological signs. The pathophysiology includes protein breakdown, muscle atrophy secondary to inactivity, and accumulation of intramuscular fat. Proximal myopathy may persist despite biochemical remission, but it usually improves over time.

Vitamin D deficiency (Answer E) can also contribute to muscle weakness and is common in individuals with Cushing syndrome, but it does not account for the full extent of this patient's signs.

**Note: Images in the stem are reprinted from Ahmad S, Best T, Lansdown A, et al. Spinal epidural lipomatosis: a rare association of Cushing's disease.* Endocrinol Diabetes Case Rep. *2020;2020:20-0111.*

Educational Objective

Differentiate among the causes of myopathy in the setting of Cushing disease.

Reference(s)

van Haalen FM, Kaya M, Pelsma ICM, et al; Endo-ERN Cushing and Thrombosis study group. Current clinical practice for thromboprophylaxis management in patients with Cushing's syndrome across reference centers of the European Reference Network on Rare Endocrine Conditions (Endo-ERN). *Orphanet J Rare Dis.* 2022;17(1):178. PMID: 35505430

Fleseriu M, Auchus R Bancos I, et al. Consensus on diagnosis and management of Cushing's disease: a guideline update. *Lancet Diabetes Endocrinol.* 2021;9(12):847-875. PMID: 34687601

Frara S, Allora A, di Filippo L, et al. Osteopathy in mild adrenal Cushing's syndrome and Cushing disease. *Best Pract Res Clin Endocrinol Metab.* 2021;35(2):101515. PMID: 33795196

Ahmad S, Best T, Lansdown A, et al. Spinal epidural lipomatosis: a rare association of Cushing's disease. *Endocrinol Diabetes Case Rep.* 2020;2020:20-0111. PMID: 33434165

53 ANSWER: B) Dapagliflozin

This patient has type 2 diabetes, chronic kidney disease, and severely increased albuminuria. Dapagliflozin (Answer B), an SGLT-2 inhibitor, is approved to reduce the risk of sustained decline in estimated glomerular filtration rate (eGFR), end-stage kidney disease, cardiovascular death, and hospitalization for heart failure in adults with chronic kidney disease at risk of progression. In the Dapagliflozin in Patients with Chronic Kidney Disease study, 4304 participants with an eGFR of 25 to 75 mL/min per 1.73 m^2 of body-surface area and a urinary albumin-to-creatinine ratio (with albumin measured in milligrams and creatinine measured in grams) of 200 to 5000 mg/g were randomly assigned to receive dapagliflozin (10 mg once daily) or placebo. The primary outcome was a composite of a sustained decline in the eGFR of at least 50%, end-stage kidney disease, or death of kidney-related or cardiovascular causes. Over a median of 2.4 years, a primary outcome event occurred in 197 of 2152 participants (9.2%) in the dapagliflozin group and in 312 of 2152 participants (14.5%) in the placebo group (hazard ratio, 0.61 [95% CI, 0.51-0.72]; $P < .001$; number needed to treat to prevent 1 primary outcome event, 19 [95% CI, 15-27]). The hazard ratio for the composite of sustained decline in the eGFR of at least 50%, end-stage kidney disease, or death of kidney-related causes was 0.56 (95% CI, 0.45-0.68; $P < .001$), and the hazard ratio for the composite of death of cardiovascular causes or heart failure was 0.71 (95% CI, 0.55-0.92; $P = .009$). The effects of dapagliflozin were similar in participants with or without type 2 diabetes. Of note, dapagliflozin, when not being used to improve glycemic control, can be initiated for patients with an eGFR as low as 25 mL/min per 1.73 m^2, and it can be continued up to the start of dialysis. Dapagliflozin is not recommended for use to improve glycemic control in adults with type 2 diabetes with an eGFR rate less than 45 mL/min per 1.73 m^2. Dapagliflozin, like most SGLT-2 inhibitors, is less effective in improving glycemic control at lower eGFR rates based on its mechanism of action. Lastly, hyperkalemia is not a contraindication to initiating an SGLT-2 inhibitor. In patients with baseline electrolyte abnormalities, particularly if also receiving other medications that can affect electrolyte levels (diuretics, RAAS inhibitors, etc), closer monitoring of electrolytes and volume status would be advised upon initiation of an SGLT-2 inhibitor. The fact that the patient's hemoglobin A_{1c} is already at goal (<7.0% [<53 mmol/mol]) has no bearing on the decision to initiate an SGLT-2 inhibitor for one of its secondary indications, which in this case is to reduce the risk of sustained decline in eGFR rate, end-stage kidney disease, cardiovascular death, and heart failure in adults with chronic kidney disease at risk of progression. It is important to assess for other reasons to intensify therapy in patients with type 2 diabetes, including the initiation of additional type 2 diabetes therapies, beyond the patient's hemoglobin A_{1c} goal attainment status.

Similar data would also support the use of other SGLT-2 inhibitors that have been demonstrated to reduce the risk of chronic kidney disease progression. While empagliflozin is approved to do so in patients with and without type 2 diabetes, like dapagliflozin, canagliflozin is only approved to do so in patients with type 2 diabetes and chronic kidney disease.

Initiating acarbose (Answer A), insulin glargine (Answer B), or pioglitazone (Answer C) is incorrect. The patient's hemoglobin A_{1c} is at target (<7.0% [<53 mmol/mol]), and there is currently no need for intensification of therapy to improve glycemic control. Rather, the addition of therapy at this time should be focused on addressing his risk of chronic kidney disease progression.

Finerenone (Answer C), a nonsteroidal mineralocorticoid receptor antagonist, has also demonstrated kidney benefits in patients with type 2 diabetes. The FIDELIO-DKD trial was conducted to evaluate the effect of finerenone on chronic kidney disease outcomes in persons with type 2 diabetes. This trial included 5734 patients with chronic kidney disease and type 2 diabetes. Patients were required to have a urinary albumin-to-creatinine ratio (UACR) of 30 to 300 mg/g and an eGFR of 25 to 75 mL/min per 1.73 m^2 and diabetic retinopathy, or UACR of 300 to 5000 mg/g and an eGFR of 25 to 75 mL/min per 1.73 m^2. All patients were treated with RAS blockade. The primary composite outcome, assessed in a time-to-event analysis, was kidney failure, a sustained decrease of at least 40% in the eGFR from baseline, or death of kidney-related causes. During a median follow-up of 2.6 years, finerenone reduced the risk of the primary composite outcome (hazard ratio 0.82 [95% CI, 0.73-0.93]; $P < .001$). Although compared with placebo, finerenone was associated with a higher overall risk of hyperkalemia, discontinuation of the trial regimen due to hyperkalemia was infrequent in patients who received finerenone vs placebo (2.3% vs 0.9%). Of note, the risk of hyperkalemia with finerenone is lower than that observed with use of steroidal mineralocorticoid antagonists (spironolactone, eplerenone). The risk of hyperkalemia associated with the steroidal mineralocorticoid antagonists often precludes their use, particularly in patients already receiving the maximum tolerated dosage of ACE/ARB therapy. While the risk of hyperkalemia is lower with finerenone than with traditional steroidal mineralocorticoid antagonists, it should not be initiated if the serum potassium concentration is greater than 5.0 mEq/L (>5.0 mmol/L). Given this patient's potassium concentration of 5.2 mEq/L (5.2 mmol/L), starting finerenone is incorrect. If serum potassium concentrations are greater than 4.8 to 5.0 mEq/L, initiation of finerenone may be considered with additional serum potassium monitoring during the first 4 weeks of treatment, based on clinical judgment and serum potassium values. Use of potassium binders may also be considered.

Educational Objective
Identify patients with type 2 diabetes and chronic kidney disease who may benefit from SGLT-2 inhibitor therapy.

Reference(s)

Heerspink HJL, Stefánsson BV, Correa-Rotter R, et al. Dapagliflozin in patients with chronic kidney disease. *N Engl J Med.* 2020;383:1436-1446. PMID: 32970396

The EMPA-KIDNEY Collaborative Group; Herrington WG, Staplin N, et al. Empagliflozin in patients with chronic kidney disease. *N Engl J Med.* 2023;388(2):117-127. PMID: 36331190

Perkovic V, Jardine MJ, Neal B, et al; CREDENCE Trial Investigators. Canagliflozin and renal outcomes in type 2 diabetes and nephropathy. *N Engl J Med.* 2019;380(24):2295-2306. PMID: 30990260

Neal B, Perkovic V, Mahaffey KW, et al; CANVAS Program Collaborative Group. Canagliflozin and cardiovascular and renal events in type 2 diabetes. *N Engl J Med.* 2017;377(7):644-657. PMID: 28605608

Bakris GL, Agarwal R, Anker SD, et al; FIDELIO-DKD Investigators. Effect of finerenone on chronic kidney disease outcomes in type 2 diabetes. *N Engl J Med.* 2020;383(23):2219-2229. PMID: 33264825

54 ANSWER: B) Perform pituitary-directed imaging

Hypogonadism is more common in men with obesity than in men without obesity. There is a bidirectional relationship between hypogonadism and obesity, in that obesity decreases fertility while hypogonadism contributes to increased fat mass accumulation. Obesity is a recognized cause of male reproductive dysfunction. In at least one bariatric surgery cohort, half of male patients were found to have hypogonadism. Weight loss improves hypogonadism and fertility in both men and women.

Lower serum SHBG concentrations in men with obesity largely account for lower total testosterone concentrations. However, there is also an inverse relationship between total and free testosterone concentrations, and BMI; as weight increases, both total and free testosterone concentrations decrease.

The patient in this vignette has class 3 obesity and hypogonadism. Hypogonadism has been confirmed on repeat morning testosterone measurement. However, the significantly low testosterone concentration is unlikely to be explained by obesity alone. Although there is no accepted weight-adjusted range for total testosterone, concentrations below 180 ng/dL (<6.2 nmol/L) have rarely been reported in men with obesity. Therefore, an alternative etiology for the patient's hypogonadism should be explored.

In light of the low gonadotropins, the best next step to evaluate this patient's markedly low total testosterone is pituitary-directed imaging (Answer B) to assess for a pituitary mass.

Restarting clomiphene (Answer E) would not be recommended before the patient has been evaluated for a pituitary mass.

If there were clinical suspicion for Cushing syndrome, a 1-mg overnight dexamethasone-suppression test (Answer A) could be performed. For suppression testing, a 1-mg dose of dexamethasone shows good clinical performance in patients with obesity. However, this would not be the best next test for this patient and would not explain his significant hypogonadotropic hypogonadism.

Treatment of obesity with a GLP-1 receptor agonist (Answer C) could certainly be considered, but the etiology of the patient's low total testosterone concentration should be pursued first.

Buprenorphine is a partial μ-opioid agonist used as pharmacotherapy for opioid dependence, and although it acts centrally, it suppresses the hypothalamic-pituitary-adrenal axis more than other opioids. Communicating with the pain clinic to help the patient wean off buprenorphine (Answer D) may be necessary in the long term, but it should be done after a pituitary etiology has been excluded.

Educational Objective
Identify alternative etiologies for hypogonadism in a man with obesity.

Reference(s)

Hofstra J, Loves S, van Wageningen B, Ruinemans-Koerts J, Jansen I, de Boer H. High prevalence of hypogonadotropic hypogonadism in men referred for obesity treatment. *Neth J Med.* 2008;66(3):103-109. PMID: 18349465

Calderon B, Gomez-Martin JM, Vega-Pinero B, et al. Prevalence of male secondary hypogonadism in moderate to severe obesity and its relationship with insulin resistance and excess body weight. *Andrology.* 2016;4(1):62-67. PMID: 26663756

55 ANSWER: A) Refer to hepatology

Evaluation for metabolic dysfunction-associated steatotic liver disease (MASLD) is recommended in patients with type 2 diabetes, cardiometabolic comorbidities, and/or elevated liver enzymes. More than 70% of persons with type 2 diabetes have MASLD. MASLD is a spectrum that includes steatotic liver disease, steatohepatitis, fibrosis, and cirrhosis. The presence of clinically significant fibrosis (defined as F2 based on histological findings) increases the risk of progression to cirrhosis, liver failure, and hepatocellular carcinoma. MASLD is one of the leading causes of liver transplant and is also associated with increased cardiovascular morbidity and mortality. The goal of early screening for MASLD is to prevent and/or delay the progression of fibrosis to decrease the risk of end-stage liver disease.

The Fibrosis-4 score (FIB-4) is endorsed by several guidelines as one of the first noninvasive steps in the evaluation of MASLD (*see equation*).

$$\text{FIB-4} = \text{age (years)} \times [\text{AST (U/L)} / \sqrt{(\text{PLT } (10^9/\text{L}) \times \text{ALT (U/L)}}]$$

A FIB-4 score less than 1.3 is deemed low risk for hepatic fibrosis, 1.3 to 2.67 is indeterminate risk, and greater than 2.67 is high risk. The area under the curve for FIB-4 in diagnosing hepatic fibrosis is 0.78 to 0.8; further diagnostic testing is needed in patients who are not at low risk. In this vignette, the patient's FIB-4 score falls in the indeterminate-risk category. The next step per the American Diabetes Association Standards of Care is to perform further testing either by vibration-controlled transient elastography or by measuring fibrosis markers. In this case, vibration-controlled transient elastography showed elevated controlled attenuation parameter and liver stiffness measurement scores. The controlled attenuation parameter assesses the degree of steatosis. A threshold of 288 dB/m or higher is associated with clinically significant steatosis; however, the presence of steatosis alone does not predict fibrosis. Liver stiffness measurement by elastography is the most commonly used method to assess liver stiffness associated with fibrosis. A threshold of less than 8.0 kPa rules out clinically significant fibrosis, while greater than 12 kPa has a high likelihood of fibrosis. In this case, elastography indicated an intermediate risk of fibrosis. Given that advanced fibrosis has not been effectively ruled out, the best next step is referral to hepatology (Answer A) for further risk stratification either by MRI or liver biopsy.

The Mediterranean diet has been extensively studied in cardiometabolic disease, including in persons with type 2 diabetes and MASLD. Implementation of the Mediterranean diet (Answer B) would benefit this patient's overall health, but assessing for liver fibrosis is more important for risk stratification.

Dapagliflozin (Answer C), an SGLT-2 inhibitor, would help lower hemoglobin A_{1c}. There have been small studies using SGLT-2 inhibitors in patients with type 2 diabetes and MASLD with improvement in liver parameters on imaging.

Pioglitazone (Answer D) improves steatosis and steatohepatitis in persons with type 2 diabetes and MASLD, and it also improves glycemic control. Whether this patient has clinically significant fibrosis is unknown, so initiation of additional medications at this time is premature.

MASLD is associated with increased cardiovascular morbidity and mortality. Stopping rosuvastatin (Answer E) would further increase this risk and cause harm. The American Association for the Study of Liver Diseases recommends continuing statins even in compensated cirrhosis because of the cardiovascular benefit. Statins can be continued in the setting of decompensated cirrhosis in patients with high cardiovascular risk, but careful monitoring is necessary because adverse events are higher in this population. Persistent elevations of liver enzymes occur in 0.5% to 3% of statin users. Statins should not be discontinued in the setting of mild liver enzyme elevation, but discontinuation could be considered in patients with a confirmed ALT concentration more than 3 times the upper normal limit.

Educational Objective

Determine when to refer patients with metabolic dysfunction-associated steatotic liver disease to gastroenterology.

Reference(s)

American Diabetes Association Professional Practice Committee. 4. Comprehensive medical evaluation and assessment of comorbidities: standards of care in diabetes-2024. *Diabetes Care.* 2024;47(Suppl 1):S52-S76. PMID: 38078591

Rinella ME, Neuschwander-Tetri BA, Siddiqui MS, et al. AASLD practice guidance on the clinical assessment and management of nonalcoholic fatty liver disease. *Hepatology.* 2023;77(5):1797-1835. PMID: 36727674

56 **ANSWER: B) Initiate oral hydrocortisone twice daily**

The United States faces a concerning opioid epidemic, with a rising number of opioid prescriptions over the past 2 decades. Opioid use is associated with well-recognized adverse effects, including constipation, sedation, and nausea. Endocrine complications, such as opioid-induced adrenal insufficiency (OIAI), are less known, in part due to the overlapping symptomatology of adrenal insufficiency, chronic pain, and opioids effects. Such symptoms include fatigue, nausea, vomiting, weight loss, dizziness, and muscle aches/weakness. Adrenal insufficiency is associated with increased morbidity and risk of death, and its timely recognition and treatment are critical.

Prolonged opioid use and a high dosage are associated with opioid-induced secondary hypogonadism, but risk factors for developing OIAI remain inadequately studied. Opioids suppress the hypothalamic-pituitary-adrenal axis. As in other forms of secondary/tertiary adrenal insufficiency, mineralocorticoid deficiency and associated electrolyte abnormalities are not present.

Available diagnostic tests for adrenal insufficiency have limitations and require interpretation within the appropriate clinical context. In a patient without major alterations of circadian rhythm who presents with signs and/or symptoms suggestive of adrenal insufficiency, a low morning serum cortisol value and a low or inappropriately normal ACTH value are suggestive of secondary or tertiary adrenal insufficiency. In patients with recent-onset central adrenal insufficiency, however, the adrenal cortex is not yet atrophied, and supraphysiologic synthetic corticotropin doses might mount normal levels of cortisol. Thus, a standard high-dose cosyntropin-stimulation test (250 mcg) cannot exclude recent central adrenal insufficiency. The low-dose (1 mcg) corticotropin-stimulation test does not provide better diagnostic accuracy. The low-dose test is also logistically limited because there are no commercially available 1-mcg preparations (requiring dilution from 250 mcg/L), and it is not well validated. In such patients, an insulin tolerance test would uncover central adrenal insufficiency. Insulin-induced hypoglycemia is a significant stressor, prompting the release of ACTH and, subsequently, cortisol. When the serum glucose concentration falls below 50 mg/dL (<2.8 mmol/L), serum cortisol should rise to 18 µg/dL or higher (≥498 nmol/L). Although the insulin tolerance test is considered the gold standard for diagnosing central adrenal insufficiency, it is logistically complex, requiring an endocrine testing unit and close supervision. Moreover, the resultant hypoglycemia can pose risks, particularly if comorbidities are present.

For both standard high-dose cosyntropin-stimulation testing (250 mcg) and insulin tolerance testing, improvement in cortisol assays (specific monoclonal antibody immunoassays or mass spectrometry assays) have led to the more recently proposed cortisol cutoff of 14 µg/dL or greater (≥387 nmol/L) as an indicator of normal response.

In light of these limitations and variations in cortisol cutoffs, low concentrations of morning serum cortisol, ACTH, and DHEA-S may be considered sufficient for the diagnosis of OIAI when clinical suspicion is high. A morning cortisol value less than 5 µg/dL (<138 nmol/L) predicts a poor response to the 250-mcg cosyntropin-stimulation test. In such situations, a therapeutic trial of physiologic glucocorticoid hormone replacement can be considered and continued if sustained symptom improvement is observed.

Hydrocortisone is used most frequently for glucocorticoid replacement and is typically administered in 2 to 3 divided doses daily, with the highest dose taken upon waking (Answer B). Alternative glucocorticoid preparations with longer half-lives, such as prednisone or prednisolone, can used once daily, but they are associated with greater risk of glucocorticoid adverse effects.

When secondary adrenal insufficiency is suspected to be caused by exogenous glucocorticoids or opioids, these medications should be tapered when possible. This approach should always be discussed with the treating physician, and glucocorticoid or opioid therapy should never be stopped abruptly. Thus, stopping oxycodone and measuring 8-AM cortisol and ACTH in 2 months (Answer E) is not appropriate. Suppression of the hypothalamic-pituitary-adrenal axis is also possible from inhaled or transcutaneous glucocorticoids. Discussing use of a nonglucocorticoid inhaler with her treating physician (Answer A) is reasonable, but alternative therapies are not effective in all patients. In addition, the concern about OIAI would persist in this patient.

To complete the hormonal evaluation of this newly diagnosed pituitary microadenoma, prolactin and IGF-1 should be measured (Answer C), but this should not delay treatment initiation for secondary adrenal insufficiency.

Sleep apnea is associated with chronic day-time fatigue and circadian rhythm disruption. However, more often, nocturnal hypoxia causes sympathetic and hypothalamic-pituitary-adrenal axis activation, and pseudo-Cushing syndrome. A sleep study (Answer D) should be pursued in patients with an unclear etiology of fatigue, but it should not delay the evaluation for adrenal insufficiency in this patient.

Educational Objective

Evaluate and manage opioid-induced adrenal insufficiency.

Reference(s)

Li T, Cunningham JL, Gilliam WP, et al. Prevalence of opioid-induced adrenal insufficiency in patients taking chronic opioids. *J Clin Endocrinol Metab.* 2020 Oct 1;105(10):e3766-75. PMID: 32866966

Gadelha MR, Karavitaki N, Fudin J, Bettinger JJ, Raff H, Ben-Shlomo A. Opioids and pituitary function: expert opinion. *Pituitary.* 2022;25(1):52-63. PMID: 35066756

Gruber LM, Bancos I. Secondary adrenal insufficiency: recent updates and new directions for diagnosis and management. *Endocr Pract.* 2022;28(1):110-117. PMID: 34610473

Javorsky BR, Raff H, Carroll TB, et al. New cutoffs for the biochemical diagnosis of adrenal insufficiency after ACTH stimulation using specific cortisol assays. *J Endocr Soc.* 2021;5(4):bvab022. PMID: 33768189

57 ANSWER: D) Recurrence of the same condition with future pregnancies

In women who have given birth, postpartum thyroiditis occurs in 5% to 8% in the first year after delivery. This condition is thought to arise from underlying thyroid autoimmunity that manifests after the physiologic period of immune tolerance, conferred by pregnancy, has ended. Postpartum thyroiditis is a painless type of thyroiditis. Although mild thyromegaly is possible, there is no local neck pain or tenderness (Answer B).

The phases of postpartum thyroiditis are similar to those of the other thyroiditis subtypes: in the classic scenario, there is an initial thyrotoxic phase for 2 to 3 months followed by a hypothyroid phase in the next 2 to 3 months before biochemical euthyroidism is restored. Symptoms of postpartum thyroiditis can be those typical of either hyperthyroidism or hypothyroidism, depending on the phase of the condition in which the patient comes to attention. The thyrotoxic phase of thyroiditis results from the leakage of preformed thyroid hormone from the thyroid gland, and therefore there is no role for the use of antithyroid drugs during this phase. The hypothyroidism phase is characterized by the depletion of stored thyroid hormone from the thyroid gland and may require the short-term use of thyroid hormone replacement before endogenous normal thyroid function is reestablished.

Postpartum thyroiditis is more common in women with serum thyroid-antibody positivity and in those with a personal or family history of autoimmune conditions. Following an episode, there is an up to 70% risk of recurrent postpartum thyroiditis with future pregnancies (Answer D). Affected women should be counseled to monitor for hyperthyroid or hypothyroid symptoms after any future pregnancies.

The thyrotoxic phase of postpartum thyroiditis may resemble Graves disease, and both are possible in this patient. Distinguishing features are summarized in the *table*, and the diagnosis can usually be confirmed without radioiodine uptake, which would require temporary cessation of breastfeeding. In this vignette, the time frame of symptom onset, absence of pathognomonic signs of Graves disease, absent TRAb and thyroid-stimulating immunoglobulin titers, and relatively low ratio of serum T_3 to T_4 concentrations support the diagnosis of postpartum thyroiditis.

Finally, postpartum thyroiditis is not associated with risk of thyroid eye disease (Answer E), future infertility (Answer A), or the time course of lactation should a woman breastfeed her newborn (Answer C).

Table. Distinguishing Features of Postpartum Thyroiditis and Graves Disease

Features	Thyrotoxic phase of postpartum thyroiditis	Graves disease
Etiology	Autoimmune	Autoimmune
Time course	Onset usually 2-4 months after delivery	May appear any time after delivery
Hyperthyroid symptoms	Mild to moderate	Moderate to severe
Thyroid eye disease	Absent	Possible
Thyroid bruit	Absent	Possible
Goiter	Absent	Possible
Serum TPO antibodies	Positive	Positive
Serum TRAb	Negative	Usually positive
Serum thyroid-stimulating immunoglobulin	Negative	Usually positive
Serum T_3-to-T_4 ratio	Low (ie, <20)	High (ie, >20)
Radioiodine uptake[a]	Low	High

[a] Thyroid scintigraphy is contraindicated during active lactation. If the test must be performed, the woman should be advised to temporarily discontinue breastfeeding for 2 to 3 days following the radiotracer administration. During this period, the nursing infant may be given stored breastmilk that was previously pumped.

Educational Objective

Diagnose postpartum thyroiditis and identify its associated risks.

Reference(s)

Nguyen CT, Mestman JH. Postpartum thyroiditis. *Clin Obstet Gynecol.* 2019;62(2):359-364. PMID: 30844908

Lee SY, Pearce EN. Assessment and treatment of thyroid disorders in pregnancy and the postpartum period. *Nat Rev Endocrinol.* 2022;18(3):158-171. PMID: 34983968

Amino N, Arata N. Thyroid dysfunction following pregnancy and implications for breastfeeding. *Best Pract Res Clin Endocrinol Metab.* 2020;34(4):101438. PMID: 32651061

De Groot L, Abalovich M, Alexander EK, et al. Management of thyroid dysfunction during pregnancy and postpartum: an Endocrine Society clinical practice guideline [published correction appears in *J Clin Endocrinol Metab.* 2021;106(6):e2461]. *J Clin Endocrinol Metab.* 2012;97(8):2543-2465. PMID: 22869843

Alexander EK, Pearce EN, Brent GA, et al. 2017 guidelines of the American Thyroid Association for the diagnosis and management of thyroid disease during pregnancy and the postpartum [published correction appears in *Thyroid.* 2017;27(9):1212]. *Thyroid.* 2017;27(3):315-389. PMID: 28056690

58 ANSWER: E) Start micronized progesterone to induce ductal side-branching and acinar development

In general, gender-affirming therapy is associated with decreased gender dysphoria in both transgender men and women. The goal of feminizing gender-affirming hormone therapy is to reduce endogenous testosterone levels and increase estrogen levels. Standard feminizing gender-affirming hormone therapy regimens for transgender women typically consist of an estrogen combined with an antiandrogen. Estrogen therapy induces feminine characteristics and reduces testosterone due to suppression of the hypothalamic-pituitary-gonadal axis, and antiandrogens such as spironolactone are used to decrease testosterone synthesis and block at the receptor level, targeting a testosterone concentration less than 50 ng/dL (<1.7 nmol/L). The woman in this vignette is interested in additional breast development. The data on progestogens in feminizing gender-affirming therapy are limited; however, many transgender women have sought the addition of a progestogen (progesterone or other synthetic progesterone called progestins) in response to anecdotal reports of improved breast development, libido, insomnia, mood, and more rapid feminization. Some adverse effects documented with progestogen use in the transgender population are fatigue, drowsiness, nausea, vomiting, breast pain and tenderness, mood swings, weight gain, and hyperlipidemia. While the World Professional Association for Transgender Health (WPATH) Standards of Care 8 indicate there are no quality data to support use of progestins and that there can be potential harm with extended exposure, they state, "If, after a discussion of the risks and benefits of progesterone treatment, there is a collaborative decision to begin a trial of progesterone therapy, the prescriber should evaluate the patient within a year to review the patient's response to this treatment."

Currently, based on clinical experience in girls with Turner syndrome (beginning cyclic progesterone treatment after 2 years of estrogen), in the setting of feminizing gender-affirming hormone therapy, progesterone may be considered after at least 12 months of estradiol treatment for a patient who expresses desire for additional breast development. Progesterone is responsible for side-branching in breast development and using it too soon may lead to tubular breast development. Discussion with patient should address the potential risks and benefits of progesterone and the many unknowns. Currently, high-quality data are lacking regarding progestogens in transfeminine populations in terms of any outcome. Four existing studies provide some data around potential risks and benefits:

- Meyer et al (1986) evaluated 15 transgender women taking progestogens and found no difference in breast hemicircumference between progestogen vs nonprogestogen users.
- Jain et al (2019) evaluated 39 transgender women taking progestogens, and 26 of 39 (67%) on medroxyprogesterone acetate self-reported improvement of breast development.
- Bahr et al (2024) evaluated 29 transgender women taking oral micronized progesterone and reported higher rates of satisfaction with breast development at 6 and 9 months and improvements in mental health at 6 months compared with outcomes in transgender women not taking micronized progesterone.
- Patel et al (2022) evaluated adverse effects related to progestogen use in a large retrospective analysis and demonstrated that recent progestogen prescriptions (medroxyprogesterone acetate, micronized progesterone) were associated with venous thromboembolism, lower HDL cholesterol, and variable effect on blood pressure.

The Women's Health Initiative study that evaluated 16,000 postmenopausal cisgender women on conjugated equine estrogens and medroxyprogesterone acetate vs placebo found that those on treatment had increased risks of coronary heart disease (by 24%), stroke (by 31%), venous thromboembolism (by 107%), and invasive breast cancer (by 29%). However, it must be emphasized that the Women's Health Initiative was a study of postmenopausal cisgender women with a median age of 63 years, with use of both synthetic estrogen and progestogen.

In the case of this 28-year-old transgender woman, shared decision-making with individualization, education, and good listening is the best approach. She has no inherent risk factors for progestogen use (no family history of breast cancer, stroke, or cardiovascular disease), and she is experiencing dysphoria related to the small size of her breasts. As such, starting micronized progesterone to induce ductal side-branching and acinar development (Answer E) is the best next step, although there is no clear consensus/standard of care.

Increasing the estradiol dosage to induce further ductal and lobular formation (Answer A) is incorrect because the patient has already achieved goal estradiol levels and a higher dosage would increase the risk of adverse events such as venous thromboembolism.

The onset of expected breast growth is 3 to 6 months after starting feminizing gender-affirming hormone therapy, with a maximum period of 2 to 3 years for total growth. Referring for affirming breast augmentation surgery (Answer B) could be considered, but many surgeons prefer to wait a longer duration for the effects of feminizing gender-affirming hormone therapy.

Referring her for gender expression care (Answer C) is incorrect because it is not directly addressing her desire for additional breast development. Gender expression care augments traditional medical care with professional guidance and practical support centered around specific components of social transition, including individualization of self-expression, gender-affirming movement and exercise, gender-affirming presentation, and vocal training in real-world environments.

Referring her for mental health services to address persisting dysphoria (Answer D) is not the best next step, although Endocrine Society guidelines indicate that it is good practice to have the patient working with a mental health practitioner, so they can receive support around the psychosocial impact of potential changes on one's life.

Educational Objective

Describe the limited data available about progestin use for breast development in transgender women.

Reference(s)

Hembree WC, Cohen-Kettenis PT, Gooren L, et al. Endocrine treatment of gender-dysphoric/gender-incongruent persons: an Endocrine Society clinical practice guideline. *J Clin Endocrinol Metab.* 2017;102(11):3869-3903. PMID: 28945902

Center of Excellence for Transgender Health, Department of Family and Community Medicine, University of California, San Francisco. *Guidelines for the Primary and Gender-Affirming Care of Transgender and Gender Nonbinary People.* 2nd ed. Deutsch MB, ed. June 2016. Available at: www.transhealth.ucsf.edu/guidelines.

Patel KT, Adeel S, Miragaya JR, Tangpricha V. Progestogen use in gender-affirming hormone therapy: a systematic review. *Endocr Pract.* 2022;28(12):1244-1252. PMID: 36007714

Jain J, Kwan D, Forcier M. Medroxyprogesterone acetate in gender-affirming therapy for transwomen: results from a retrospective study. *J Clin Endocrinol Metab.* 2019;104(11):5148-5156. PMID: 31127826

Bahr C, Ewald J, Dragovich R, Gothard MD. Effects of progesterone on gender affirmation outcomes as part of feminizing hormone therapy. *J Am Pharm Assoc.* 2024;64(1):268-272. PMID: 37549733

59 ANSWER: E) Zoledronate

This patient has osteoporosis with very low bone density and a history of multiple fractures, including a vertebral compression fracture. Her osteoporosis must be treated to reduce her fracture risk. However, she also has pain in the right hip and thigh region and elevated alkaline phosphatase, suggesting underlying pathology that may include insufficiency fracture, osteoarthritis, osteomalacia, or bone lesions. There are radiographic findings indicating bony abnormalities in the right femur, including irregular cortical sclerosis and a lateral linear lucency. On technetium bone scan, the entire femur demonstrates tracer uptake, with an area of linear focus indicating an insufficiency fracture. These findings are consistent with Paget disease, with a fracture through the pagetic bone.

Paget disease affects 1% to 2% of the population, and it is often diagnosed incidentally on radiographs or during follow-up of a high alkaline phosphatase value on laboratory testing. However, Paget disease can result in mixed osteolytic and sclerotic bony lesions that are expansile and deforming, which can cause pain, insufficiency fracture, as well as nerve compression and hearing loss depending on lesion location. Lesions have high uptake on technetium scans. Pathologic fracture can also occur in these lesions. While Paget disease is often detected by elevated alkaline phosphatase (a bone formation marker), the hallmark is abnormally active osteoclasts, followed by active bone formation with deposition of disorganized woven bone. These metabolically active lesions can cause overlying warmth of the skin and sometimes redness. Fibrous dysplasia of bone similarly presents in one or more bones with high turnover and mixed lytic and sclerotic expansile bony lesions and elevated bone alkaline phosphatase.

Osteomalacia from various causes may lead to insufficiency fractures in the femur and elsewhere, and these may present as low bone mineral density on DXA. Hypophosphatasia is a genetic disorder of deficient alkaline phosphatase that presents in adults with fractures and bone pain. Hypophosphatasia, by itself, can predispose to atypical femur fractures, even in the absence of treatment with antiresorptive agents. The defining biochemical feature of hypophosphatasia is low serum total and bone-specific alkaline phosphatase. This patient's alkaline phosphatase concentration is elevated, however, which suggests osteomalacia or other bone lesions, and these must be assessed before attempting therapy with osteoporosis medications. Osteomalacia most often results from vitamin D deficiency or hypophosphatemia. Her concentrations of calcium, phosphate, and 25-hydroxyvitamin D are normal, arguing against osteomalacia. Bone mineral density is usually not low in the genetic osteomalacia disorder X-linked hypophosphatemia.

Burosumab (Answer B) is a monoclonal antibody binding FGF-23, which is used to treat hypophosphatemia in patients with X-linked hypophosphatemia. Based on her biochemical and radiographic findings, she is not hypophosphatemic and does not have X-linked hypophosphatemia.

Asfotase alfa (Answer A) is an enzyme replacement therapy for hypophosphatasia. Her alkaline phosphatase concentration is high, not low, so asfotase alfa is not indicated.

Physical examination should evaluate for signs of vertebral compression fractures. In most healthy individuals, the occiput touches the wall when they stand with their back to the wall. If the distance is more than 2 cm, it is considered abnormal and consistent with kyphosis. The distance between the lowest ribs and the pelvis may also be a sign of vertebral compression fracture, especially lumbar fractures. This is measured as the distance between the inferior rib margin and the superior pelvis at the midaxillary line. As published by Siminoski et al, distances of 2 fingerbreadths or less (<3.6 cm) are suggestive of vertebral compression fractures (although with poor specificity [47%]), while distances of 1 fingerbreadth or less (<1.8 cm) have a specificity of 88%. This is a practical and simple measurement to make during the clinical visit. Given the variability in adult fingerbreadth, measurement in cm may be more reliable.

While this patient's severe osteoporosis may justify anabolic agents, teriparatide (Answer D) (or abaloparatide) would be contraindicated in the setting of concurrent Paget disease or fibrous dysplasia, as there is some risk of sarcomatous transformation in these lesions. The effect or safety of anabolic treatment with romosozumab in patients with Paget disease is unknown.

While raloxifene (Answer C) might be useful to treat her osteoporosis, it does not have randomized controlled trial evidence that it reduces risk of nonvertebral fractures, and it does not treat Paget disease of bone.

The treatment of choice for the high-turnover lesions of Paget disease is intravenous or oral bisphosphonates (Answer E) to target bone resorption, and which would also target her osteoporosis. A lateral transverse femur shaft insufficiency fracture may suggest an atypical femur fracture, raising concerns about use of antiresorptive agents. However, by definition, fractures occurring in pathologic bone lesions are not classified as atypical femur fractures. The cause of elevated alkaline phosphatase should be identified before starting osteoporosis therapy, as it may guide treatment.

Educational Objective

Explain the influence of concurrent Paget disease of bone on osteoporosis treatment decisions.

Reference(s)

Siminoski K, Warshawski RS, Jen H, Lee KC. Accuracy of physical examination using the rib-pelvis distance for detection of lumbar vertebral fractures. *Am J Med.* 2003;115(3):233-236. PMID: 12935830

Green AD, Colon-Emeric CS, Bastian L, Drake MT, Lyles KW. Does this woman have osteoporosis? *JAMA.* 2004;292(23):2890-2900. PMID: 15598921

Cook SJ, Wall C. Paget's disease of bone: a clinical update. *Aust J Gen Pract.* 2021;50(1-2):23-29. PMID: 33543158

60 ANSWER: A) Dapagliflozin

This patient's clinical condition deteriorated after undergoing an elective surgical procedure. Her white blood cell count, hemoglobin, hematocrit, body temperature, and normal blood pressure are all reassuring findings and are not consistent hypovolemic shock (ie, an infectious process, bleeding complication, or adrenal insufficiency). The patient's blood glucose is currently not out of control, with a postoperative random value of 164 mg/dL (9.10 mmol/L). However, she is noted to have elevated ketones, acidemia, normal lactate level, and an elevated anion gap, which should be concerning for diabetic ketoacidosis (DKA), particularly euglycemic DKA given her history of SGLT-2 inhibitor therapy. Although the risk of euglycemic DKA is higher in persons with type 1 diabetes, it can be seen in those with type 2 diabetes as well.

The diagnosis of euglycemic DKA requires the triad of less marked hyperglycemia (blood glucose <250 mg/dL [<13.9 mmol/L]), an elevated anion gap metabolic acidosis, and ketosis. Patients receiving SGLT-2 inhibitor therapy are at risk of developing euglycemic DKA, where the glucose levels are not as high as typically observed in patients with "classic" DKA because of the ongoing hyperglycosuria associated with SGLT-2 inhibitor therapy. Accordingly, the diagnosis of DKA can often be overlooked, particularly in the postoperative setting, when glucose values are not frankly elevated and when patients have a history of type 2, rather than type 1, diabetes.

SGLT-2 inhibitor therapy is known to raise serum levels of glucagon and ketones. The underlying pathophysiology of euglycemic DKA includes an absolute or, more commonly, a relative insulin deficiency in the setting of severe insulin resistance. This deficit can be further exacerbated by the metabolic stress associated with surgery or infection, particularly in the setting of background SGLT-2 inhibitor therapy, which can cause further increases in glucagon production and release of free fatty acids, leading to ketogenesis and acidemia.

Patients receiving SGLT-2 inhibitor therapy who undergo a period of restricted oral intake and/or a surgical procedure are at risk of developing euglycemic DKA. Accordingly, in 2020, the United States approved safety labeling changes to SGLT-2 inhibitors, recommending temporary discontinuation of SGLT-2 inhibitors before a scheduled surgery (or any scheduled fast) due to the greater risk of patients developing ketoacidosis. Canagliflozin, dapagliflozin, and empagliflozin should be discontinued 3 days before scheduled surgery, and ertugliflozin should be stopped at least 4 days before.

Blood glucose levels should be carefully monitored after discontinuation of SGLT-2 inhibitor therapy and appropriately managed before surgery. SGLT-2 inhibitor therapy can be resumed once the patient's oral intake is back to baseline and any other risk factors for ketoacidosis have resolved.

While euglycemic DKA has been reported after the initiation of GLP-1 receptor agonist therapy in patients receiving background SGLT-2 inhibitor therapy, this patient has been reported to be receiving stable doses of her current type 2 diabetes medications for the past few years. Thus, it is unlikely to be the etiology in this case. The proposed mechanism for euglycemic DKA after initiation of a GLP-1 receptor agonist in patients receiving background SGLT-2 inhibitor therapy is thought to be related to the decrease in oral intake, and in some cases severe nausea/vomiting and/or diarrhea, that can sometimes be observed shortly after initiation of a GLP-1 receptor agonist. In this case, dulaglutide (Answer B) is not the best choice.

Metformin (Answer E) is rarely associated with lactic acidosis, which is a recognized etiology of metabolic acidosis with an elevated anion gap. However, this patient's lactic acid level was normal.

Etomidate (Answer C) and lisinopril/hydrochlorothiazide (Answer D) are not associated with the development of an elevated anion gap metabolic acidosis. A well-documented adverse effect of etomidate is suppression of adrenal synthesis of cortisol. In a dose-dependent fashion, it inhibits adrenal mitochondrial 11β-hydroxylase, the enzyme responsible for the final conversion of 11-deoxycortisol to cortisol. Thus, adrenal suppression or adrenal insufficiency resulting in hemodynamic and metabolic changes can occur in some patients. However, the clinical presentation and laboratory findings presented in this case are not consistent with adrenal insufficiency.

Educational Objective

Identify SGLT-2 inhibitor use as a risk factor for euglycemic diabetic ketoacidosis in the postoperative setting and the need to hold SGLT-2 inhibitor therapy before a scheduled surgery due to the greater risk developing ketoacidosis.

Reference(s)

Peters AL, Buschur EO, Buse JB, Cohan P, Diner JC, Hirsch IB. Euglycemic diabetic ketoacidosis: a potential complication of treatment with sodium-glucose cotransporter 2 inhibition. *Diabetes Care.* 2015;38(9):1687-1693. PMID: 26078479

Patel K, Nair A. A literature review of the therapeutic perspectives of sodium-glucose cotransporter-2 (SGLT2) inhibitor-induced euglycemic diabetic ketoacidosis. *Cureus.* 2022;14(9):e29652. PMID: 36320965

Franki L. FDA Announcement: FDA says to stop SGLT2 inhibitors for diabetes prior to surgery. https://www.medscape.com/viewarticle/927047?form=fpf

Alduraibi RK, Alrebdi YM, Altowayan YF. Euglycemic diabetic ketoacidosis after the initiation of dulaglutide in patient with type 2 diabetes. *Medicine (Baltimore).* 2023;102(23):e34027. PMID: 37335652

Thompson Bastin ML, Baker SN, Weant KA. Effects of etomidate on adrenal suppression: a review of intubated septic patients. *Hosp Pharm.* 2014;49(2):177-183. PMID: 24623871

61 ANSWER: C) Change evolocumab to inclisiran

Combination lipid-lowering therapy is often required in patients with heterozygous familial hypercholesterolemia (FH) who generally have baseline LDL-cholesterol concentrations in the range of 200 to 400 mg/dL (5.18-10.36 mmol/L). In the past, the goal of therapy was to achieve at least 50% reduction from the baseline concentration, and this often required LDL apheresis in addition to high-dosage statin therapy. The advent of PCSK9 monoclonal antibodies, which reduce cholesterol concentrations by 50% to 60% when added to statin therapy, has facilitated discontinuation of LDL apheresis in many patients. PCSK9 inhibitors have also made it possible to aim for more aggressive lipid-lowering goals. Considering treatment intensification in patients with FH even after 50% reduction has been achieved would be reasonable if the LDL-cholesterol concentration were still greater than 100 mg/dL (>2.59 mmol/L) for primary prevention or greater than 70 mg/dL (>1.81 mmol/L) for secondary prevention.

This patient not only has known heterozygous FH but also has established atherosclerotic cardiovascular disease (ASCVD). He has received combination lipid-lowering therapy with a high-intensity statin, ezetimibe, and the PCSK9 monoclonal antibody evolocumab. His untreated baseline LDL-cholesterol concentration was 320 mg/dL (8.29 mmol/L), and he had previously been treated with LDL apheresis. With the addition of evolocumab, his LDL-cholesterol concentrations had been consistently below 100 mg/dL (<2.59 mmol/L) (even without LDL apheresis) and he had had no recent ASCVD events. However, he has recently been noted to have intermittent elevations in LDL cholesterol despite no apparent change in medications or lifestyle habits, prompting reevaluation of therapeutic options to maintain good lipid control. Many postmarketing studies have documented that adherence to PCSK9 monoclonal antibody therapy is a main determinant of therapeutic response. This patient's intermittent elevations in LDL cholesterol may have resulted from missed doses of evolocumab. High medication burden and possible dementia can contribute to missed medication doses, especially in patients taking intermittent doses such as biweekly evolocumab. Given his previous excellent response to the introduction of PCSK9 inhibitor therapy, every effort must be made to ensure that it is optimally administered. In addition to monoclonal antibodies, small interfering RNA–based therapy such as inclisiran (Answer C) is now available, and this can be administered in a health care facility every 6 months. Inclisiran would be the most convenient and effective therapy for this patient to help maintain low LDL-cholesterol concentrations and prevent ASCVD events.

Inclisiran is a double-stranded small interfering RNA that is taken up by hepatocytes, and it activates an RNA-induced silencing complex targeting PCSK9 mRNA. Inclisiran leads to prolonged lowering of PCSK9 levels in hepatocytes and a subsequent increase in LDL-receptor activity. A single 300 mg dose maintains a nearly 50% LDL-cholesterol reduction for up to 6 months. In the ORION-9 trial that enrolled 482 patients with heterozygous FH, inclisiran therapy achieved LDL-cholesterol concentrations below 100 mg/dL (<2.59 mmol/L) in 65% of patients and below 70 mg/dL (<1.81 mmol/L) in 40% of patients. While outcomes data are not yet available, inclisiran was approved by the US FDA in December 2021 for patients with heterozygous FH or those with established ASCVD who have elevated LDL-cholesterol concentrations on maximally tolerated statin therapy. It is available as a single-dose prefilled syringe (284 mg/1.5 mL) and is administered subcutaneously by a health care provider. Initial injection is followed by a second injection in 3 months and subsequently every 6 months. While there are no head-to-head trials comparing inclisiran's efficacy with that of PCSK9 monoclonal antibodies, the magnitude of LDL-cholesterol lowering is similar and inclisiran also promotes long-term adherence.

Changing from one high-intensity statin to another (atorvastatin 80 mg to rosuvastatin 40 mg [Answer A]) or from one PCSK9 monoclonal antibody to another (evolocumab to alirocumab [Answer B]) can be considered if patients are intolerant of a medication, but such regimen changes are unlikely to significantly affect lipid levels or address adherence issues.

Increasing the evolocumab dosage to 420 mg every 4 weeks (Answer D) is a reasonable option if the patient finds this more convenient, but it is unlikely to be as effective as directly administered therapy.

Reintroduction of LDL apheresis (Answer E) is also a less desirable option, as it is very time- and labor-intensive, requiring vascular access and visits to an apheresis center every 2 weeks.

Educational Objective

Explain the role of the novel small interfering RNA–based therapy inclisiran in the treatment of patients with severe hypercholesterolemia.

Reference(s)

Raal FJ, Kallend D, Ray KK, et al; ORION-9 Investigators. Inclisiran for the treatment of heterozygous familial hypercholesterolemia. *N Engl J Med.* 2020;382(16):1520-1530. PMID: 32197277

Levin AA. Treating disease at the RNA level with oligonucleotides. *N Engl J Med.* 2019 3;380(1):57-70. PMID: 30601736

Ray KK, Troquay RPT, Visseren FLJ, et al. Long-term efficacy and safety of inclisiran in patients with high cardiovascular risk and elevated LDL cholesterol (ORION-3): results from the 4-year open-label extension of the ORION-1 trial. *Lancet Diabetes Endocrinol.* 2023;11(2):109-119. PMID: 36620965

62 ANSWER: A) Discuss how to recognize symptoms of hyperglycemia and ketoacidosis

The US FDA approved short-acting subcutaneous pasireotide (in 2012) and long-acting intramuscular pasireotide injections (in 2018) for patients with Cushing disease in whom surgery is not an option or is not curative. Pasireotide is a somatostatin receptor multiligand that binds with high affinity to somatostatin receptor type 5, which is usually abundantly expressed in ACTH-secreting tumors. A phase 3 randomized, double-blind clinical trial with pasireotide LAR demonstrated clinical improvement in most patients (weight, waist circumference, quality of life, and blood pressure), normalized urinary free cortisol in 40% of patients after 7 months, and reduced tumor size in 50% of patients. Urinary free cortisol was more likely to normalize in patients with baseline urinary free cortisol values 1.5 to 2 times the upper normal limit than in those with baseline values 2 to 5 times the upper normal limit. The patient in this vignette has relatively mild biochemical hypercortisolism and residual tumor visible on postoperative MRI and is thus a good candidate for pasireotide treatment.

In patients with Cushing disease, a frequent adverse event of pasireotide is perturbation of glucose metabolism. In the phase 3 study, almost 50% of patients had hyperglycemia and 24% had diabetes mellitus. Fifty percent of patients experiencing hyperglycemia required treatment with a new antidiabetes medication. In the extension study, the biochemical and clinical improvements were sustained, and mean fasting glucose and hemoglobin A_{1c} values were stable, with diabetes medications initiated or escalated in some patients. In real-world studies, hyperglycemia is noted in the first few weeks of therapy, and it stabilizes over time. The mechanism of pasireotide-induced hyperglycemia is the expression of somatostatin receptor type 5 in the pancreatic β cells and enteroendocine cells, with decreased insulin and incretin secretion. Studies in healthy volunteers receiving pasireotide indicate that DPP-4 inhibitors and GLP-1 receptor agonists are effective in decreasing the degree of hyperglycemia after oral glucose tolerance testing.

When pasireotide is prescribed, patients must be counseled about monitoring blood glucose concentrations at home, as well as symptoms of hyperglycemia and ketoacidosis (Answer A). Hemoglobin A_{1c} should be measured before initiation of pasireotide LAR, not after 3 injections (Answer B). Patients taking pasireotide require weekly blood glucose monitoring in the first 3 months after initiation and in the first 4 to 6 weeks after dosage escalation. If diabetes develops, standard medical therapy is recommended, usually starting with metformin and intensifying the regimen to achieve glycemic control. Incretin-based therapies (DPP-4 inhibitors or GLP-1 receptor agonists) should be considered. In a randomized, open-label, phase 4 study, 25 patients with Cushing disease treated with pasireotide were randomly assigned to sitagliptin (replaced by liraglutide if needed) or insulin. There was a trend of better hemoglobin A_{1c} control and fewer hypoglycemic events for patients in the incretin-based arm compared with patients in the insulin arm. Because DPP-4 inhibitors augment endogenous incretin action, they are less effective than GLP-1 receptor agonists in treating pasireotide-induced hyperglycemia.

Treatment of hyperglycemia (Answers C and D) should not be started based on a single abnormal glucose value of 115 mg/dL (6.4 mmol/L).

Stopping hydrochlorothiazide (Answer E) would not address pasireotide's effect on glucose metabolism. Adjusting the patient's blood pressure medication could be considered depending on the potassium concentration and blood pressure measurements.

Educational Objective

Explain the pathophysiology of pasireotide-induced hyperglycemia in a patient with persistent postoperative hypercortisolism and recommend appropriate management.

Reference(s)

Fleseriu M, Auchus R, Bancos I, et al. Consensus on diagnosis and management of Cushing's disease: a guideline update. *Lancet Diabetes Endocrinol.* 2021;9(12):847-875. PMID: 34687601

Lacroix A, Gu F, Gallardo W, et al; Pasireotide G2304 Study Group. Efficacy and safety at once-monthly pasireotide in Cushing's disease: a 12 month clinical trial. *Lancet Diabetes Endocrinol.* 2018;6(1):17-26. PMID: 29032078

Fleseriu M, Petersenn S, Biller BMK, et al. Long-term efficacy and safety of once-monthly pasireotide in Cushing's disease: a phase III extension study. *Clin Endocrinol (Oxf).* 2019;91(6):776-785. PMID: 31465533

Samson SL, Gu F, Feldt-Rasmussen U, et al. Managing pasireotide-associated hyperglycemia: a randomized, open-label, phase IV study. *Pituitary.* 2021;24(6):887-903. PMID: 34275099

Breitschaft A, Hu K, Hermosillo Resendiz K, Darstein C, Golor G. Management of hyperglycemia associated with pasireotide (SOM230): healthy volunteer study. *Diabetes Res Clin Pract.* 2014;103(3):458-465. PMID: 24461109

63 **ANSWER: A) Alprostadil (intracavernous injection)**

This patient has erectile dysfunction that is not responding adequately to oral phosphodiesterase (PDE) inhibitor therapy, the typical first-line treatment choice. A man must try an oral PDE inhibitor at least 4 to 6 times before the clinician concludes that PDE inhibitors are ineffective. Many clinicians prescribe a different PDE inhibitor before choosing an alternative treatment. The key elements of choosing the next therapy after failure of a PDE inhibitor are the likelihood of effectiveness, convenience, acceptability, cost, and adverse effects.

This patient should be offered the most effective medical therapy: intracavernosal treatment with alprostadil alone (Answer A) or in combination of phentolamine, papaverine, or vasoactive intestinal peptide. It has an overall satisfactory response rate of 80% to 90% compared with 60% to 70% with oral PDE inhibitors. Alprostadil has a different mechanism of action than that of PDE inhibitors, and it is often effective when PDE inhibitors are not. Alprostadil increases production of cyclic AMP, whereas PDE inhibitors inhibit the metabolism of cyclic GMP; both cyclic AMP and GMP are important for penile vascular engorgement and tumescence.

Intraurethral alprostadil (Answer B) is effective in only 40% to 50% of patients and causes severe penile pain in up to 30% of applications. It could be offered as a trial to this patient if intracavernosal injections are effective. Lower response rates to all medical therapies are seen in patients with erectile dysfunction due to neuropathy and vasculopathy (eg, diabetes mellitus), spinal cord injury, radical prostatectomy (with injury to nerves to the penis), multiple sclerosis, or severe vascular disease.

A third trial of a PDE inhibitor (Answer C) is sometimes effective, but this is likely to lead to frustration for this man who tried 2 oral medications already.

Penile pump therapy (with a constrictive band placed over the base of the penis after tumescence has been achieved) (Answer D) can be effective and is covered as an inexpensive option by many medical systems and third-party payers. However, most men do not like using these pumps because they interrupt spontaneity, often result in suboptimal tumescence, and cause erections that are cold, bluish purple, and/or feel numb.

Penile prostheses (Answer E) are generally reserved for patients who do not respond to medical therapies. Penile protheses have greater than 90% satisfaction rates and last for 10 to 15 years. The worst adverse effect of penile prosthesis implantation is infection that can occur in approximately 2% of men when the prosthesis is implanted by a surgeon who frequently performs the procedure.

Men with erectile dysfunction should be encouraged to exercise regularly to improve endothelial function, increase synthesis of nitric oxide (the vasodilator that causes penile tumescence), improve self-esteem, and decrease anxiety. For men with obesity, weight loss (due to lifestyle changes) is associated with improved erectile function.

A growing number of alternative therapies are being offered to treat erectile dysfunction, including herbal remedies, L-arginine, and extracorporeal shock-wave therapy. The evidence for herbal remedies has been inconclusive because there are no large placebo-controlled studies. L-arginine therapy is likely safe, but its effectiveness has not been confirmed. A 2019 meta-analysis found only 7 studies of extracorporeal shock-wave therapy (with 522 patients) that included sham therapy as a control; there was a modest benefit on erectile function (about half that typically seen with PDE inhibitors). Other therapies under investigation include injection of stem cells and platelet-rich plasma.

Educational Objective

Recommend intracavernosal therapy (eg, alprostadil or a combination of alprostadil/papaverine/phentolamine) as the most effective medical therapy for erectile dysfunction.

Reference(s)

Allen MS. Physical activity as an adjunct treatment for erectile dysfunction. *Nat Rev Urol.* 2019;16(9):553-562. PMID: 31239541

Dong L, Chang D, Zhang X, et al. Effect of low-intensity extracorporeal shock wave on the treatment of erectile dysfunction: a systematic review and meta-analysis. *Am J Mens Health.* 2019;13(2):1557988319846749. PMID: 31027441

Maiorino MI, Bellastella G, Esposito K. Lifestyle modifications and erectile dysfunction: what can be expected? *Asian J Androl.* 2015;17(1):5-10. PMID: 25248655

64 ANSWER: E) Discontinue U500 insulin, metformin, and semaglutide and start basal-bolus U100 insulin at significantly decreased doses compared with the outpatient regimen

Insulin is an important medication used for the management of both type 1 and type 2 diabetes. Most patients with diabetes use U100 insulin regimens that contain 100 units of insulin per mL. For most patients, these regimens—if used correctly with or without other diabetes medications—effectively manage hyperglycemia. However, in severely insulin-resistant patients, this approach might not be sufficient for the management of hyperglycemia. Generally, if patients require more than 200 units of U100 insulin per day, U500 insulin regimens should be considered instead. These contain 500 units per mL of regular insulin. Such concentrated insulin regimens deliver more potent amounts of insulin in smaller volumes, which leads to better insulin absorption. This, in turn, leads to improved effectiveness. The use of U500 insulin is becoming less common given the availability of other insulin-sparing medications, especially GLP-1 receptor agonists. These medications can induce significant weight loss and therefore address obesity in addition to improving glycemic control. This patient, despite being treated with both a GLP-1 receptor agonist and metformin, still requires U500 insulin, and her diabetes remains suboptimally controlled as evidenced by her hemoglobin A_{1c} values.

U500 insulin can cause significant hypoglycemia because it is extremely potent. This is especially relevant for patients who have decreased appetite or an altered diet (eg, consume meals that contain lower amounts of carbohydrates than their usual meals). Hospitalized patients, with or without diabetes, can have decreased appetite due to the underlying illness that necessitated their hospitalization. As a result, the use of U500 insulin regimens in hospitalized patients carries an increased risk of hypoglycemia. In addition, diet can be significantly different in the hospital, as many institutions use carbohydrate-controlled diets for patients with diabetes; these diets typically contain 60 to 80 g of carbohydrates per meal. More importantly, patients with diabetes are frequently placed on nothing-by-mouth status before procedures/surgeries or on diets that have very few or no carbohydrates (clear-liquid diet). On all these occasions, either because appetite is decreased or because the diet is significantly different from the diet followed in the outpatient setting, patients with diabetes can develop significant hypoglycemia if they continue their U500 insulin regimen and use the same outpatient dosage during hospitalization. Although higher dosages might eventually be needed during hospitalization, the total insulin dose should be significantly reduced as a first step. Indeed, studies have shown that most patients taking U500 require conversion to U100 along with a 50% to 75% reduction in the total daily insulin dose. Therefore, continuing U500 at the current dosage (Answer A), discontinuing U500 and placing the patient on the equivalent dose of U100 (Answer D), or continuing U500 at slightly decreased dosage (Answer B) is incorrect, as these actions would place the patient at high risk for hypoglycemia.

Discontinuing U500 insulin and having the patient continue either metformin (Answer C) or semaglutide is not the best option because these medications by themselves will not be sufficient to achieve appropriate glycemic control. There is limited information about the safety and efficacy of metformin or GLP-1 receptor agonists in the hospital setting.

The best treatment option is to discontinue U500 insulin, metformin, and semaglutide and initiate a basal-bolus U100 insulin regimen with a significantly decreased total daily dose (Answer E). After this initial decrease, insulin could be further adjusted as clinically needed to achieve normoglycemia.

Educational Objective

Describe limitations of using U500 insulin in the hospital setting.

Reference(s)

Davidson MB et al. U-500 regular insulin: clinical experience and pharmacokinetics in obese, severely insulin-resistant type 2 diabetic patients. *Diabetes Care.* 2010;33(2):281-283. PMID: 19910500

Paulus AO, Colburn JA, True MW, et al. Evaluation of total daily dose and glycemic control for patients taking U-500 regular insulin admitted to the hospital. *Endocr Pract.* 2016;22(10):1187-1191. PMID: 27359287

Kedia R, Desouza C, Smith LM, Shivaswamy V. A retrospective review of insulin requirements in patients using U-500 insulin hospitalized to a Veterans Affairs Hospital. *J Diabetes Complications.* 2017;31(5):874-879. PMID: 28274680

American Diabetes Association Professional Practice Committee. 16. Diabetes care in the hospital: standards of care in diabetes-2024. *Diabetes Care.* 2024;47(Suppl 1):S295-S306. PMID: 38078585

Korytkowski MT. Management of hyperglycemia in hospitalized adult patients in non-critical care settings: an Endocrine Society clinical practice guideline. *J Clin Endocrinol Metab.* 2022;107(8):2101-2128. PMID: 35690958

65 ANSWER: B) Ectopic Cushing syndrome

This patient presents with signs, symptoms, and hormonal values indicative of profound Cushing syndrome. Her severe hypokalemia is mediated via activation of the mineralocorticoid receptors by cortisol. Cortisol has mineralocorticoid receptor affinity equivalent to that of aldosterone, and its physiologic concentrations exceed aldosterone by 3 orders of magnitude. In mineralocorticoid receptor–expressing tissues, however, cortisol is inactivated to cortisone by 11β-dehydrogenase isozyme 2 (HSD11B2). In Cushing syndrome, the HSD11B2 enzyme capacity is saturated, allowing cortisol to overactivate mineralocorticoid receptors with consequent hypertension and hypokalemia. Profound hypokalemia is commonly observed in severe forms of Cushing syndrome, as seen with ectopic ACTH-producing tumors (Answer B) or cortisol-producing adrenal carcinomas.

Ectopic ACTH or, rarely, corticotropin-releasing hormone secretion occurs from neuroendocrine tumors located in the lungs, pancreas, thymus, or appendix. Medullary thyroid carcinomas and pheochromocytomas also rarely produce ACTH. The incidence of ectopic Cushing syndrome increases after age 50 years, as does the incidence of lung carcinoma, with small differences between sexes. Although the clinical severity is heterogeneous and is influenced by the malignant potential of the underlying tumor and disease burden, ectopic Cushing syndrome is often fulminant, predisposing to thromboembolism, opportunistic infections, and sepsis, and it has the highest mortality of all forms of Cushing syndrome. Prophylaxis with low–molecular weight heparin (for venous thromboembolism) and trimethoprim-sulfamethoxazole (160/800 mg orally administered daily or 3 times weekly for pneumocystis pneumonia) should be considered in patients with severe Cushing syndrome who are hospitalized. Due to the low incidence of clinically profound Cushing syndrome, initiation of prophylactic therapy is not well standardized and is considered on an individual basis.

Cushing disease (Answer A) is caused by ACTH-producing pituitary adenomas. Cushing disease is most common in young women (25-45 years) and is typically more indolent than ectopic Cushing syndrome. Cushing disease should always be considered in patients who have evidence of ACTH-dependent hypercortisolism. A pituitary tumor larger than 6 mm and/or an unequivocal central-to-peripheral plasma ACTH gradient during inferior petrosal sinus sampling indicate a pituitary source of ACTH. Cortisol suppression following administration of 8 mg of dexamethasone is suggestive of Cushing disease rather than ectopic ACTH secretion, but the sensitivity and specificity of this test are suboptimal. Localization tests should be conducted in this patient, but presence of a small pituitary adenoma should not deter one from considering ectopic Cushing syndrome given the severe clinical phenotype.

In glucocorticoid-remediable aldosteronism (GRA), mineralocorticoid receptor activation by aldosterone promotes hypertension and hypokalemia. Aldosterone synthase (CYP11B2) is expressed in the zona glomerulosa, and it executes the final steps in aldosterone synthesis. In the adrenal zona fasciculata, steroid 11β-hydroxylase (CYP11B1) converts 11-deoxycortisol to cortisol. CYP11B1 activity is governed by ACTH, while CYP11B2 function is primarily regulated by the renin-angiotensin system. GRA is an autosomal dominant disorder characterized by the presence of a chimeric gene containing the promoter region of *CYP11B1* and the coding sequences of *CYP11B2*. This results in ACTH-dependent aldosterone synthesis, following a typical circadian pattern. Patients with GRA are susceptible to hypokalemia when exposed to thiazide diuretics, and they are at risk of hemorrhagic strokes. A biochemical hallmark of GRA is elevation of "hybrid steroids," 18-hydroxycortisol and 18-oxocortisol, resulting from the coexpression of 17α-hydroxylase (CYP17A1) and CYP11B2 in the zona fasciculata. GRA (Answer C) is not suspected in this patient, who has a low morning concentration of aldosterone and hormonal and clinical evidence of Cushing syndrome.

Primary pigmented nodular adrenal disease (Answer D) is a rare cause of ACTH-independent hypercortisolism. In this condition, the adrenal glands are commonly described as normal on routine CT. High-resolution CT might reveal multiple subcentimeter nodules, resembling beads on a string. Bilateral adrenal hypertrophy can be seen in ACTH-dependent Cushing syndrome, as sustained ACTH excess has a trophic effect on the adrenal glands.

Stress-induced hypercortisolism (Answer E) with robust elevation of both ACTH and cortisol can be observed during acute illness, hospitalization, and surgery. Hypothalamic-adrenal axis activation with more modest ACTH and cortisol values can also be caused by sleep apnea, major depressive disorders, alcoholism, severe malnutrition, or obesity. Features specific to Cushing syndrome are not usually present in these circumstances.

Educational Objective

Consider ectopic Cushing syndrome as a cause of profound hypokalemia.

Reference(s)

Reincke M, Fleseriu M. Cushing syndrome: a review. *JAMA*. 2023;330(2):170-181. PMID: 37432427

Ferriere A, Tabarin A. Biochemical testing to differentiate Cushing's disease from ectopic ACTH syndrome. *Pituitary*. 2022;25(5):705-708. PMID: 35759056

66 **ANSWER: D) Radiofrequency ablation is recommended; refer for a procedural consultation**

There is increasing interest in the use of ultrasound-guided ablation procedures, including radiofrequency ablation (RFA), laser thermal ablation, and ethanol ablation, as an alternative to surgery for the treatment of benign thyroid nodules. Both RFA and laser thermal ablation are thermal ablation techniques, while ethanol ablation is a chemical ablation technique. RFA uses a high-frequency alternating electric current to generate heat and ablate the target tissue. In laser thermal ablation, a laser is used to deliver thermal energy. The ethanol ablation procedure involves the localized injection of ethanol leading to cellular dehydration and localized vascular damage.

Clinical indications for ultrasound-guided ablation of benign thyroid nodules include the presence of compressive symptoms and/or cosmetic concerns. Larger nodules (>20 mL volume) may require multiple procedures. Autonomously functioning thyroid nodules may also be considered for thermal ablation procedures;

however, the efficacy for resolving hyperthyroidism (up to 70% success rate) is lower than for surgery or radioactive iodine therapy. Successful achievement of euthyroidism is higher for smaller autonomously functioning nodules, such that a cutoff of 10 to 12 mL has been suggested as the thyroid nodule volume upper limit for this procedure. It is recommended that a subjective voice assessment be performed before any ultrasound-guided ablation procedure. A laryngeal examination is recommended for all patients with dysphonia or a history of neck surgery. Patients without evidence of voice impairment do not require preprocedural flexible laryngoscopy (Answer E).

A 2020 meta-analysis reported a pooled nodule volume reduction of 64.5%, 76.9%, and 92.2% at 6, 12, and 36 months, respectively, following RFA. Ultrasound follow-up is recommended for all nodules treated with RFA because up to 24% of nodules can demonstrate regrowth after 3 to 5 years. Thermal ablation procedures have been associated with improvement in thyroid nodule–related symptoms, cosmetic findings, and health-related quality of life. In comparison to surgery, thermal ablation is associated with a shorter postprocedural recovery time and a much lower risk of subsequent hypothyroidism. RFA does not result in an anterior neck scar. However, like surgery, thermal ablation procedures are not without risk. Complications may include recurrent laryngeal nerve injury (1%-2%), thyroid nodule rupture, transient thyrotoxicosis from destructive thyroiditis, hematoma, superficial burns, and tracheal or esophageal injury. Thyroidectomy remains the preferred treatment modality for large, symptomatic multinodular goiters. However, if a bilateral ablation procedure is chosen (eg, patient not fit for surgery), a staged procedure is recommended with laryngeal examination performed before ablation of the contralateral side to avoid the rare complication of bilateral recurrent laryngeal nerve injury. Relative contraindications to RFA of benign thyroid nodules include pregnancy; bleeding diathesis; sonographic findings concerning for malignancy; indeterminate cytopathology results; significant substernal extension; vocal cord paralysis on the contralateral side; and the presence of a pacemaker, automated implantable defibrillator, or cochlear implant.

This case describes an ideal candidate for an ultrasound-guided thermal ablation procedure. She is euthyroid with a large, solid thyroid nodule and associated compressive symptoms. She has undergone ultrasound-guided FNA biopsy twice with benign cytopathology and subjective assessment of her voice is normal. Consequently, RFA can be recommended, and it is appropriate to refer her for a procedural consultation (Answer D) given her interest in exploring this option. Although right thyroid lobectomy (Answer A) would also be a reasonable management approach, the patient has expressed an interest in learning about RFA and has no contraindications to the procedure. Conversely, radioactive iodine therapy (Answer B) would not be a preferred management strategy for a nonfunctioning thyroid nodule. Finally, ethanol ablation (Answer C) is most effective for treating predominantly cystic lesions, including cystic thyroid nodules and lymph nodes, and is not recommended as the sole intervention for the treatment of an entirely solid or predominantly solid thyroid nodule, such as this one.

Educational Objective

List the indications and contraindications for thyroid nodule radiofrequency ablation.

Reference(s)

Cho SJ, Baek JH, Chung SR, Choi YJ, Lee JH. Long-term results of thermal ablation of benign thyroid nodules: a systematic review and meta-analysis. *Endocrinol Metab.* 2020;35(2):339-350. PMID: 32615718

Orloff LA, Noel JE, Stack BC Jr, et al. Radiofrequency ablation and related ultrasound-guided ablation technologies for treatment of benign and malignant thyroid disease: an international multidisciplinary consensus statement of the American Head and Neck Society Endocrine Surgery Section with the Asia Pacific Society of Thyroid Surgery, Associazione Medici Endocrinologi, British Association of Endocrine and Thyroid Surgeons, European Thyroid Association, Italian Society of Endocrine Surgery Units, Korean Society of Thyroid Radiology, Latin American Thyroid Society, and Thyroid Nodules Therapies Association. *Head Neck.* 2022;44(3):633-660. PMID: 34939714

Sinclair CF, Baek JH, Hands KE, et al. General principles for the safe performance, training, and adoption of ablation techniques for benign thyroid nodules: an American Thyroid Association statement. *Thyroid.* 2023;33(10):1150-1170. PMID: 37642289

67 ANSWER: B) Direct the patient to measure point-of-care fingerstick glucose values several times between 2 and 6 PM

Continuous glucose monitoring (CGM) devices are important systems for diabetes management, especially among patients on insulin pump therapy or multiple daily injection regimens, such as the individual in this vignette. One of the benefits of CGM devices is that they can check glucose values every 1 to 15 minutes. By using specific software applications, patients with diabetes can monitor glucose values in "real-time" or periodically receive reports on their glucose control, even daily. These can be discussed with their health care providers to address any issues identified. For example, the patient in this vignette self-identified a period of hyperglycemia after 1:30 to 2 PM that persists until 6 or 6:30 PM.

When a period of persistent hyperglycemia is identified, targeted changes in the insulin regimen can be made. For example, if a patient is experiencing fasting hyperglycemia upon waking, the basal insulin dosage should be up-titrated. If a patient is experiencing hyperglycemia after a meal, then up-titration of prandial premeal insulin should be considered.

When determining the best treatment strategy, one must consider whether the glucose values obtained from the CGM device are accurate. Although newer CGM systems have improved accuracy, some medications can cause inaccurate readings, including mannitol, tetracycline, alcohol, ascorbic acid (dosages >500 mg daily), acetaminophen, and hydroxyurea, among others. The interference of medications with different types of CGM systems is not the same, as some medications can affect one CGM type/brand and have no effect on another.

The patient in this vignette is taking hydroxyurea, which could be affecting her CGM sensor and leading to falsely elevated glucose values. The effect of hydroxyurea on glucose levels is not immediate and usually occurs several hours later. Thus, she might not be experiencing true hyperglycemia. The best next step is to recommend that she check several point-of-care fingerstick glucose values between 2 and 6 PM (Answer B).

Continuing current management (Answer A) is not appropriate because the significant hyperglycemia documented on her CGM tracing requires further evaluation.

Increasing the prelunch insulin dose (Answer C) would be the next step only if the point-of-care glucose values obtained correlate with the CGM glucose values, proving that she is experiencing true hyperglycemia.

The degree of this patient's hyperglycemia (ie, rising from 140 to 300 mg/dL [8.3 to 16.7 mmol/L]) is clinically significant and cannot be explained by the timing of insulin glargine administration or by a decreased dose of basal insulin glargine. Sometimes patients with diabetes experience severe fasting hyperglycemia overnight and upon waking (dawn phenomenon), which is not the case here. Therefore, switching insulin glargine administration from 12 PM to bedtime (Answer E) is incorrect.

Recommending an automated insulin delivery system (insulin pump therapy) (Answer D) could potentially be dangerous if the glucose values are falsely elevated, as these systems administer insulin based on CGM glucose readings.

Educational Objective

Identify medications that can affect accuracy of continuous glucose monitoring devices.

Reference(s)

Battelino T, Alexander CM, Amiel SA, et al. Continuous glucose monitoring and metrics for clinical trials: an international consensus statement. *Lancet Diabetes Endocrinol.* 2023;11(1):42-57. PMID: 36493795

Szmuilowicz ED, Aleppo G. Interferent effect of hydroxyurea on continuous glucose monitoring. *Diabetes Care.* 2021;44(5):e89-e90. PMID: 33653823

68 **ANSWER: E) Start phentermine/topiramate ER**

Hypothalamic obesity can accompany panhypopituitarism following surgery and radiation treatment to the sella and suprasellar region, as in this patient. The hypothalamus centrally coordinates complex peripheral and external appetite signals, which lead to the initiation, continuation, and cessation of ingestive behaviors. When these appetite signals are disrupted because of surgery or radiation treatment to the hypothalamus and surrounding areas, appetite regulation is adversely affected.

Hypothalamic obesity has been described as a sequela of surgical and/or radiation treatment of craniopharyngioma and brain tumors. Affected patients describe hyperphagia, loss of satiety after eating and between meals, cravings, and difficulty knowing when to stop eating. Weight gain is typically rapid, particularly in the early postoperative stages (which may also be associated with supraphysiologic dosages of glucocorticoids) before slowing down and eventually plateauing.

The patient in this vignette has hypothalamic obesity secondary to surgery and radiation for a mixed germ-cell tumor of the suprasellar area. While caloric restriction is the ultimate goal, dietary counseling alone (Answer B) is unlikely to produce weight loss in this patient who has iatrogenic appetite dysregulation. Similarly, increasing physical activity (Answer A) would most likely not decrease body weight, although it could mitigate some of the changes in body composition that occur with rapid weight gain.

It is important to ensure that the patient is on the right dosages of anterior pituitary hormone replacement. He is showing some signs of excess glucocorticoid exposure and is currently on a slightly supraphysiologic dosage of glucocorticoid. Reducing his glucocorticoid dosage to a more physiologic dosage (Answer C) is appropriate but is unlikely to affect his weight-management efforts. GH replacement (Answer D) may be necessary to optimize body composition and can be considered if tumor growth is not a concern, but GH deficiency will likely need to be confirmed with repeat testing with or without provocation. GH replacement is also unlikely to significantly affect his weight trajectory.

Initiation of an appetite suppressant such as phentermine/topiramate ER (Answer E) is the best step for this particular patient. Hypothalamic obesity is notoriously difficult to treat, and no single obesity treatment option is superior to another. Treatment of hypothalamic obesity often requires a multimodal approach using behavioral strategies to modulate food intake, increased physical activity, different classes of pharmaceutical agents, and potentially bariatric surgery. Despite the comprehensive management approach, reported weight outcomes are variable, and patient expectations should be managed accordingly.

Educational Objective

Manage hypothalamic obesity in the context of panhypopituitarism.

Reference(s)

Dimitri P. Treatment of acquired hypothalamic obesity: now and the future. *Front Endocrinol (Lausanne).* 2022;13:846880. PMID: 35464063

69 ANSWER: A) Continue current management

In this patient with longstanding type 2 diabetes, macrovascular disease, and good glycemic control (hemoglobin A_{1c} 7.0%-8.0% [53-64 mmol/mol]), one must consider the indication and safety of medications that can be used for cardiac protection.

Dapagliflozin is an SGLT-2 inhibitor that is approved to reduce the risk of cardiovascular death and hospitalization for heart failure in adults with heart failure (New York Heart Association classes II-IV) with reduced ejection fraction, regardless of the presence or absence of type 2 diabetes. The DAPA-HF trial (Dapagliflozin in Patients with Heart Failure and Reduced Ejection Fraction) found that among patients with heart failure with reduced ejection fraction (left ventricular ejection fraction ≤40%), the risk of worsening heart failure or death of cardiovascular causes was lower among those who received dapagliflozin than among those who received placebo. In addition, in the DAPA-CKD trial (Dapagliflozin in Patients with Chronic Kidney Disease), the risk of the composite outcome of a sustained decline in estimated glomerular filtration rate (eGFR) of at least 50%, end-stage kidney disease, or death of kidney or cardiovascular causes among patients with chronic kidney disease was lower in those who received dapagliflozin than in those who received placebo. Dapagliflozin can be initiated for patients with an eGFR as low as 25 mL/min per 1.73 m^2 and can be continued up to the start of dialysis to reduce the risk of estimated GFR decline, end-stage kidney disease, or cardiovascular disease. Accordingly, stopping dapagliflozin (Answer D) is incorrect. Only the 10 mg dosage of dapagliflozin was used in the DAPA-HF and DAPA-CKD trials, and there is no indication to reduce the dosage to 5 mg daily (Answer B) based on eGFR. Of note, dapagliflozin and empagliflozin have also been shown to reduce the combined risk of cardiovascular death or hospitalization for heart failure in patients with heart failure and a preserved ejection fraction, regardless of the presence or absence of diabetes.

At the patient's current eGFR, the glucose-lowering effect of SGLT-2 inhibitor therapy is reduced. Accordingly, use of an SGLT-2 inhibitor for the primary effect of inducing glucosuria to improve blood glucose control is not indicated, and the use of these medications for the primary intent of lowering hemoglobin A_{1c}, depending on the specific SGLT-2 inhibitor, is limited to patients with an eGFR greater than 45 mL/min per 1.73 m^2. In this patient, the primary goal with dapagliflozin is continued risk reduction for both cardiovascular/heart failure and chronic kidney disease.

Finerenone (Answer C) was studied in the FIDELIO-DKD and FIGARO-DKD randomized placebo-controlled trials. The FIDELIO-DKD trial was conducted to evaluate the effect of finerenone on chronic kidney disease outcomes in type 2 diabetes. The trial included 5734 patients with chronic kidney disease and type 2 diabetes. Patients were required to have:

Urinary albumin-to-creatinine ratio ≥30 to <300 mg/g *and* eGFR ≥25 to <75 mL/min per 1.73 m^2 *and* diabetic retinopathy

or

Urinary albumin-to-creatinine ratio ≥300 to ≤5000 mg/g *and* eGFR ≥25 to <75 mL/min per 1.73 m^2

All patients were treated with RAAS blockade as tolerated. The primary composite outcome, assessed in a time-to-event analysis, was kidney failure, a sustained decrease of at least 40% in the eGFR from baseline, or death of kidney-related causes. During a median follow-up of 2.6 years, finerenone reduced the risk of the primary composite outcome (hazard ratio, 0.82; 95% CI, 0.73-0.93; $P < .001$). The FIGARO-DKD trial was conducted to evaluate the effect of finerenone on cardiovascular outcomes. It included 7437 patients with chronic kidney disease and type 2 diabetes. Patients were required to have:

Urinary albumin-to-creatinine ratio ≥30 to < 300 mg/g *and* eGFR ≥25 to <75 mL/min per 1.73 m^2

or

Urinary albumin-to-creatinine ratio ≥300 to ≤5000 mg/g *and* eGFR ≥60 mL/min per 1.73 m^2

All patients were treated with RAAS blockade as tolerated. The primary outcome, assessed in a time-to-event analysis, was a composite of death of cardiovascular causes, nonfatal myocardial infarction, nonfatal stroke, or hospitalization for heart failure. During a median follow-up of 3.4 years, finerenone reduced the risk of the primary composite outcome (hazard ratio, 0.87 [95% CI, 0.76-0.98]; $P = .03$). Given that this patient's potassium concentration is already slightly elevated, finerenone would not be the best choice. It is not recommended to initiate finerenone if the serum potassium concentration is greater than 5.0 mEq/L. If the serum potassium concentration is 4.8 mEq/L or less, it should be measured 4 weeks after initiating treatment. If the serum potassium concentration is greater than 4.8 to 5.0 mEq/L, initiation of finerenone can be considered with additional serum potassium monitoring within the first 4 weeks based on clinical judgement. The patient's current eGFR is less than 25 mL/min per 1.73 m^2, and the initiation of finerenone is not recommended in this case. The starting finerenone dosage is 10 mg once daily in patients with an eGFR 25 mL/min per 1.73 m^2 or greater (but less than 60 mL/min per 1.73 m^2), and 20 mg once daily in patients with an eGFR greater than 60 mL/min per 1.73 m^2.

Lastly, while this patient has hyperkalemia, it is mild and in a range frequently encountered in patients with stage 4 chronic kidney disease, particularly in those receiving RAAS-blocking therapy. The patient's current potassium concentration is slightly elevated, but stable, and it is therefore not a reason to stop losartan therapy (Answer E). Moreover, the patient's blood pressure is well-controlled.

The best course of action for this patient is to continue current management (Answer A).

Educational Objective

Recommend the most appropriate management of SGLT-2 inhibitors and related therapies in patients with heart failure with reduced ejection fraction and chronic kidney disease with or without type 2 diabetes.

Reference(s)

McMurray JJV, Solomon SD, Inzucchi SE, et al; DAPA-HF Trial Committees and Investigators. Dapagliflozin in patients with heart failure and reduced ejection fraction. *N Engl J Med.* 2019;381(21):1995-2008. PMID: 31535829

Heerspink HJL, Stefánsson BV, Correa-Rotter R, et al; DAPA-CKD Trial Committees and Investigators. Dapagliflozin in patients with chronic kidney disease. *N Engl J Med.* 2020;383(15):1436-1446. PMID: 32970396

Bakris GL, Agarwal R, Anker SD, et al; FIDELIO-DKD Investigators. Effect of finerenone on chronic kidney disease outcomes in type 2 diabetes. *N Engl J Med.* 2020;383(23):2219-2229. PMID: 33264825

Pitt B, Filippatos G, Agarwal R, et al; FIGARO-DKD Investigators. Cardiovascular events with finerenone in kidney disease and type 2 diabetes. *N Engl J Med.* 2021;385(24):2252-2263. PMID: 34449181

70 ANSWER: C) Hyperparathyroidism–jaw tumor syndrome

Hyperparathyroidism–jaw tumor syndrome (HPT-JT) (Answer C) is a rare autosomal dominant disorder characterized by the development of parathyroid tumors, ossifying fibromas of the mandible and maxilla, cystic and neoplastic kidney abnormalities, and hyperplastic and neoplastic uterine involvement. Primary HPT is the most common feature, followed by uterine tumors (can occur in 50% of women with this syndrome) and jaw tumors that occur in about one-third of affected individuals. Parathyroid carcinoma is also common. HPT-JT is caused by germline loss-of-function pathogenic variants in the *CDC73* gene, which encodes the parafibromin nuclear protein. There is incomplete disease penetrance, and 10% to 35% of *CDC73* variant carriers do not manifest overt symptoms. Correctly diagnosing HPT-JT is crucial for optimal clinical and surgical management, and this is accomplished with genetic testing. Testing is generally indicated in the presence of familial primary HPT; sporadic primary HPT with age of onset less than 40 years; multiglandular involvement; cystic, atypical, or malignant parathyroid involvement; or coexistence of ossifying jaw fibromas, kidney tumors, or uterine tumors. The patient in this vignette has hypercalcemia and a uterine leiomyosarcoma. Relevant family history includes hypercalcemia in her daughter and kidney cysts in her maternal aunt. Therefore, testing for a loss-of-function pathogenic variant in *CDC73* would be helpful.

Loss-of-function pathogenic variants in the *MEN1* gene are associated with multiple endocrine neoplasia type 1 (MEN 1) (Answer D). MEN 1 is an autosomal dominant disorder characterized by parathyroid, pituitary, and pancreatic tumors. Other tumors such as thymic and bronchial carcinoids and adrenal cortical tumors have also been observed. Similar to HPT-JT, primary HPT is the most common manifestation. Hypercalcemia is usually mild and parathyroid carcinoma is rare. MEN 1 is not associated with uterine tumors or kidney manifestations.

MEN 2A (Answer E) is characterized by HPT, pheochromocytoma, and medullary thyroid cancer. Unlike in MEN 1 and HPT-JT, HPT occurs in 30% of patients with MEN 2A, and it typically occurs in the third or fourth decade of life. MEN 2A is caused by gain-of-function pathogenic variants in the *RET* proto-oncogene. There is no history suggestive of pheochromocytoma or medullary thyroid cancer in this vignette, making MEN 2A an unlikely diagnosis.

Because there is a history of hypercalcemia in both the patient and her daughter, familial hypocalciuric hypercalcemia (FHH) (Answer A) is a consideration. There are several variants of FHH with subtle differences in phenotype. Most affected individuals have mild to moderate hypercalcemia. The most common form is inherited in an autosomal dominant manner due to pathogenic variants in the *CASR* gene, which encodes the calcium-sensing receptor. Considerable overlap exists between primary HPT and FHH; the calcium-to-creatinine clearance ratio is the main distinguishing factor. The family history of other genitourinary tumors raises concern for HPT-JT; therefore, testing for loss-of-function pathogenic variants in the *CASR* gene would not be helpful.

Gain-of-function pathogenic variants in the *GCM2* gene are seen in familial isolated HPT (Answer B). The hallmark is the absence of clinical manifestations of other associated hereditary syndromes. The diagnosis of familial isolated HPT is often made after evaluation for other syndromes, given the overlap with certain variants with incomplete penetrance. The patient in this vignette has other clinical manifestations, so *GCM2* genetic testing would not be helpful.

Educational Objective

Confirm the diagnosis of hyperparathyroidism–jaw tumor syndrome with genetic testing.

Reference(s)

Newey PJ. Hereditary primary hyperparathyroidism. *Endocrinol Metab Clin North Am.* 2021;50(4):663-681. PMID: 34774240

Torresan F, Iacobone M. Clinical features, treatment, and surveillance of hyperparathyroidism-jaw tumor syndrome: an up-to-date and review of the literature. *Int J Endocrinol.* 2019;2019:1761030. PMID: 31929790

Minisola S, Arnold A, Belaya Z, et al. Epidemiology, pathophysiology, and genetics of primary hyperparathyroidism. *J Bone Miner Res.* 2022;37(11):2315-2329. PMID: 36245271

71 ANSWER: C) Perform vibration-controlled transient elastography

This woman has polycystic ovary syndrome (PCOS). PCOS is the most common endocrine disorder among reproductive-aged women. The prevalence of PCOS is between 10% and 13% in the general population. The 2003 Rotterdam diagnostic criteria state that 2 of the 3 following findings must be present: (1) oligoovulation and/or anovulation; (2) clinical and/or biochemical signs of hyperandrogenism; and (3) polycystic ovaries (≥12 follicles measuring 2-9 mm in diameter and/or ovarian volume >10 mL in at least 1 ovary on ultrasonography).

The presentation of PCOS is heterogeneous and requires multisystem care, including considerations in reproduction, metabolism, psychological health, and cosmetics. In addition to menstrual irregularities, hyperandrogenism, and polycystic ovaries, metabolic issues are very common in affected women. Women with PCOS have a higher risk of insulin resistance, hypertension, dyslipidemia, obesity, glucose intolerance, type 2 diabetes, fatty liver, sleep apnea, and cardiovascular disease. To address the increased metabolic risk associated with PCOS, screening with oral glucose tolerance testing or hemoglobin A_{1c} measurement, lipid profile, and transaminase measurements is recommended. In this case, the patient has a normal hemoglobin A_{1c} value, elevated triglycerides on lipid profile, obesity, and elevated transaminase levels.

Given her transaminitis and the known increased risk of metabolic dysfunction-associated steatotic liver disease (MASLD) in women with PCOS, the best next step is vibration-controlled transient elastography (Answer C). Vibration-controlled transient elastography is the most commonly used noninvasive method to assess for hepatic fibrosis by estimating liver stiffness. A threshold less than 8.0 kPa rules out clinically significant hepatic fibrosis, while a value greater than 12 kPa has a high likelihood of advanced hepatic fibrosis. MASLD is a spectrum that includes steatotic liver disease, steatohepatitis, fibrosis, and cirrhosis. It is characterized by excessive fat accumulation in the liver and can progress to metabolic dysfunction-associated steatohepatitis (MASH), which is characterized by hepatocyte injury and apoptosis. With time and further inflammation, MASH can progress to cirrhosis. The prevalence of MASLD in the PCOS population ranges from 15% to 55%, and it can affect women with and without obesity. The presence of clinically significant fibrosis increases the risk of progression to cirrhosis, liver failure, and hepatocellular carcinoma. MASLD is one of the leading causes of liver transplant and is also associated with increased cardiovascular morbidity and mortality. The goal of early screening for MASLD is to prevent and/or delay the progression of fibrosis to decrease the risk of end-stage liver disease. Once MASLD is diagnosed, the recommendation is for patients to lose weight, screen for cardiovascular risk factors, and treat identified risk factors. First-line therapy for MASLD in patients with obesity is weight loss (5%-10% of initial body weight).

Liver ultrasonography (Answer A) is insufficient to assess for hepatic fibrosis and is therefore not the best next step.

Regardless of age and BMI, women with PCOS have an increased risk of impaired fasting glucose, impaired glucose tolerance, and type 2 diabetes. Therefore, glycemic status should be assessed in every woman with PCOS at time of diagnosis, and then reassessed every 1 to 3 years, based on additional individual risk factors for diabetes. While a 2-hour oral glucose tolerance test (Answer B) is often the best test to identify early abnormalities of glucose metabolism, this patient has already had hemoglobin A_{1c} measured, and therefore oral glucose tolerance testing is not the next best step in her overall health management.

This patient has a normal hemoglobin A_{1c} value and is lean (normal BMI). According to the 2023 International Evidence-based Guideline for the Assessment and Management of Polycystic Ovary Syndrome, "Metformin alone should be considered in adults with PCOS and a BMI >25 kg/m^2 for anthropometric and metabolic outcomes including insulin resistance, glucose, and lipid profiles." Metformin (Answer E) could be considered to support the goal of pregnancy in this lean woman with PCOS and no dysglycemia, as metformin can increase the number of ovulatory menstrual cycles in the setting of PCOS. However, she hopes to conceive in 2 to 3 years, and this is not the next best step in her overall management.

Inositol (Answer D) is a sugar made in the body and found in foods. Some dieticians recommend myo-inositol, a dietary supplement, for the treatment of obesity and/or impaired glucose tolerance, and some women may express interest in inositol due to individual values and preferences. However, inositol cannot currently be recommended for women with PCOS due to a lack of high-quality evidence. Using shared decision-making, physicians should inform women with PCOS about limited harm, theoretical potential for improvement in metabolic measures, yet limited clinical benefits (regarding ovulation, hirsutism, or weight). In addition, patients should be advised that regulatory status and quality control of inositol in any form (like other nutrient supplements) vary and are different from those of FDA-approved pharmacological agents.

Educational Objective

Diagnose metabolic dysfunction-associated steatotic liver disease in a lean woman with polycystic ovary syndrome.

Reference(s)

Kelly CE, Brown AJ, Diehl AM, Setji TL. Review of nonalcoholic fatty liver disease in women with polycystic ovary syndrome. *World J Gastroenterol.* 2014;20(39):14172-14184. PMID: 25339805

Teede HJ, Tay CT, Laven JJE, et al. Recommendations from the 2023 international evidence-based guideline for the assessment and management of polycystic ovary syndrome. *J Clin Endocrinol Metab.* 2023;108(10):2447-2469. PMID: 37580314

Legro RS, Arslanian SA, Ehrmann DA, et al; Endocrine Society. Diagnosis and treatment of polycystic ovary syndrome: an Endocrine Society clinical practice guideline. *J Clin Endocrinol Metab.* 2013;98(12):4565-4592. PMID: 24151290

Morley LC, Tang T, Yasmin E, Norman RJ, Balen AH. Insulin-sensitising drugs (metformin, rosiglitazone, pioglitazone, D-chiro-inositol) for women with polycystic ovary syndrome, oligo amenorrhoea and subfertility. *Cochrane Database Syst Rev.* 2017;2017(11):CD003053. PMID: 29183107

Afdhal NH. Fibroscan (transient elastography) for the measurement of liver fibrosis. *Gastroenterol Hepatol (N Y).* 2012;8(9):605-607. PMID: 23483859

72 ANSWER: B) Instructing the patient to keep a detailed food record

The first series describing patients experiencing hyperinsulinemic hypoglycemia following gastric bypass was published in 2005. Current estimates show the prevalence of hypoglycemia after gastric bypass to be approximately 10%, with 1% of patients developing severe hypoglycemia defined as a blood glucose concentration of 40 mg/dL or less (≤2.2 mmol/L).

The postprandial nature of post–gastric bypass hypoglycemia is unique. Hypoglycemia that occurs in the fasting state should not be assumed to be secondary to gastric bypass. Hypoglycemia is diagnosed when the patient has fulfilled the Whipple triad (low venous glucose associated with typical symptoms of hypoglycemia that improve with carbohydrate ingestion). Patients often describe hypoglycemic symptoms occurring after ingestion of simple carbohydrates, such as the patient in this vignette, while mixed meals or protein-rich foods do not trigger symptoms. This patient has fulfilled the Whipple triad and has inappropriate endogenous hyperinsulinemia.

Several changes that occur after gastric bypass contribute to the pathophysiology of this condition. Rapid transit of nutrients through the gastric pouch into the Roux limb (small bowel) triggers a rapid rise in glucose, followed by higher GLP-1 secretion and insulin release. Additionally, there is evidence that postoperative impairments in α-cell function (glucagon) and delayed clearance of insulin contribute to hypoglycemia. Currently, there is no predictive tool to identify patients who are most at risk of developing post–gastric bypass hypoglycemia.

Meeting with an experienced bariatric dietitian is the first step in managing hypoglycemia, and keeping a food record (Answer B) can help identify dietary triggers and eating patterns to target. Typical eating patterns that trigger hypoglycemic symptoms include eating too quickly, eating and drinking at the same time or not allowing at least 20 minutes between eating solids and drinking liquids, and consuming carbonated beverages. Alcohol and excessive caffeine consumption increase the risk of postprandial hypoglycemia by impairing hepatic glucose release and should therefore be avoided.

Because the patient's symptoms are always triggered by eating and do not occur during fasting, there is no need to pursue evaluation for insulinoma. Therefore, a 72-hour fast (Answer E), CT of the abdomen with pancreas protocol (Answer A), and selective arterial calcium stimulation (Answer D) are incorrect.

After the ingestion of liquid carbohydrates, such as during an oral glucose tolerance test (Answer C), individuals who have undergone gastric bypass are at high risk of experiencing dumping syndrome due to the rapid transit of nutrients into the small bowel. In one series of patients who had undergone gastric bypass, approximately 10% developed hypoglycemia (by plasma glucose) after oral glucose tolerance testing, although there was no correlation between the development of hypoglycemia and subjective symptoms. Thus, oral glucose tolerance testing should not be used to diagnose hypoglycemia after Roux-en-Y gastric bypass.

Educational Objective

Diagnose hypoglycemia as a rare but serious complication of gastric bypass surgery.

Reference(s)

Pigeyre M, Vaurs C, Raverdy V, Hanaire H, Ritz P, Pattou F. Increased risk of OGTT-induced hypoglycemia after gastric bypass in severely obese patients with normal glucose tolerance. *Surg Obes Relat Dis.* 2015;11(3):573-577. PMID: 25892342

Salehi M, Vella A, McLaughlin T, Patti ME. Hypoglycemia after gastric bypass surgery: current concepts and controversies. *J Clin Endocrinol Metab.* 2018;103(8):2815-2826. PMID: 30101281

73 **ANSWER: D) Stop metformin and valsartan, start daily exercise and labetalol**

This patient is high risk for preeclampsia given her age, comorbidities, and previous pregnancy outcomes. Preconception counseling and planning are recommended for all women with diabetes who can conceive. Preconception counseling and planning include assessing comorbidities, nutrition, and glycemic targets; updating immunizations; discussing congenital risks; ensuring contraception until glycemic targets are achieved; ensuring adequate folic acid supplementation; and reviewing medications and their use in pregnancy. A hemoglobin A_{1c} target less than 6.5% (<48 mmol/mol) before conception is ideal.

This patient's hemoglobin A_{1c} value is above target. She currently takes metformin and insulin glargine with an acceptable hemoglobin A_{1c} level for someone who does not desire pregnancy. She is not likely achieving the pregnancy goals of fasting glucose between 70 and 95 mg/dL (SI: 3.9-5.3 mmol/L), 1-hour postprandial glucose between 110 and 140 mg/dL (SI: 6.1-7.8 mmol/L), or 2-hour postprandial glucose between 100 and 120 mg/dL (SI: 5.6-6.7 mmol/L). Use of metformin in pregnancy is controversial, and it is not currently recommended as first-line treatment for diabetes in pregnancy. In the MitY study (Metformin in Women with Type 2 Diabetes in Pregnancy), newborns of mothers who were assigned to metformin had lower rates of large-for-gestational-age babies and macrosomia, but higher rates of small-for-gestational-age babies. This was not seen in the MOMPOD study (Metformin Plus Insulin for Preexisting Diabetes or Gestational Diabetes in Early Pregnancy). However, in the Early Metformin in Gestational Diabetes study, a similar finding was noted of increased rate of small-for-gestational-age infants in the metformin group. Current guidance by the American Diabetes Association 2024 recommends insulin as the preferred treatment of type 2 diabetes in pregnancy. The American College of Obstetricians and Gynecologists recommends limiting use of agents other than insulin and individualization of therapy if noninsulin agents are needed. Given that this patient has hypertension and a history of preeclampsia (and thus increased risk for placental insufficiency) and previously had 2 babies who were small-for-gestational-age, continuing metformin (Answers A and B) is not recommended.

Use of insulin glargine in pregnancy has not been associated with any adverse maternal or neonatal outcomes. In addition, transplacental transfer studies show that insulin glargine does not cross the placenta. Stopping insulin glargine (Answers A and B) is not recommended. She will most likely require premeal insulin to achieve glycemic targets because insulin sensitivity will decrease as her pregnancy progresses.

Risk of preeclampsia is increased in women with diabetes. Low-dosage aspirin (100-150 mg daily) starting at 12 to 16 weeks' gestation is recommended in all women with diabetes to lower the risk of preeclampsia. However, she is currently 8 weeks pregnant, so low-dosage aspirin (Answers A, C, and E) would not be recommended now.

ACE inhibitors, angiotensin receptor blockers, spironolactone, and statins are commonly used medications in patients with diabetes. In women who plan to conceive and in all sexually active women of childbearing age who are not using reliable contraception, medications in these classes should be stopped. Valsartan is an angiotensin receptor blocker that is associated with congenital anomalies, including fetal kidney dysplasia, pulmonary hypoplasia, oligohydramnios, and intrauterine growth restriction. This patient's blood pressure while on valsartan is above pregnancy goals. Target blood pressure in pregnancy is less than 140/90 mm Hg, ideally 110-135/<85 mm Hg. If valsartan is stopped, it should be replaced with an antihypertensive agent that is safe in pregnancy because her blood pressure will further increase when off valsartan (thus, Answers A, B, and C are wrong). Options are methyldopa, nifedipine, labetalol, diltiazem, clonidine, and prazosin. Valsartan should be stopped and switched to labetalol (Answer D), which is safe in pregnancy. Hydralazine (Answer E) is reserved for acute management of hypertension or severe preeclampsia. Atenolol (Answers A and C) is not recommended during pregnancy because it crosses the placenta and can cause adverse fetal and neonatal events, including bradycardia, hypotension, and low birth weight.

Exercise is important in the management of weight, glycemic targets, and high blood pressure. Exercise reduces the risk of preeclampsia and cesarean delivery. Exercise has been shown to delay initiation of insulin in women with gestational diabetes. However, more studies are needed in women with preexisting type 2 diabetes. The American College of Obstetrics and Gynecology recommends 20 to 30 minutes of moderate-intensity exercise on most or all days of the week during pregnancy. Exercise would likely aid in reducing glucose concentrations during the first trimester while insulin sensitivity is higher; however, she will most likely need intensification of insulin therapy with premeal insulin in her second and third trimesters when insulin resistance significantly increases. This patient should stop metformin, replace valsartan with labetalol, and start daily exercise.

Educational Objective

Recommend blood pressure treatment goals in pregnant women with diabetes.

Reference(s)

American Diabetes Association Professional Practice Committee. 15. Management of diabetes in pregnancy: standards of care in diabetes-2024. *Diabetes Care.* 2024;47(Suppl 1):S282-S294. PMID: 38078583

Pollex EK, Feig DS, Lubetsky A, Yip PM, Koren G. Insulin glargine safety in pregnancy: a transplacental transfer study. *Diabetes Care.* 2010;33(1):29-33. PMID: 19808914

Wang J, Ji X, Liu T, Zhao N. Maternal and neonatal outcomes with the use of long acting, compared to intermediate acting basal insulin (NPH) for managing diabetes during pregnancy: a systematic review and meta-analysis. *Diabetol Metab Syndr.* 2022;14(1):154. PMID: 36271431

ACOG Committee Opinion No. 650: physical activity and exercise during pregnancy and the postpartum period. *Obstet Gynecol.* 126(6):p e135-e142. PMID: 26595585

74 **ANSWER: D) Switch sevelamer to lanthanum carbonate**

In the setting of hypoparathyroidism, persistent hypocalcemia despite dosage escalation requires evaluation of potential causes. One of the most obvious and common causes of persistent hypocalcemia is missed doses of calcium and activated vitamin D analogues. Patients often state they are taking all of their medicines and may even believe they are taking them reliably when they are actually missing doses. When required to take medications several times a day, missing doses is certainly easy to do, even with the best intentions. When there is a language barrier, extra effort is required to ensure the importance of the medication regimen is understood, as well as the medical problems that arise when not followed. However, this patient is able to describe dosing properly and has reliable help from his family. Thus, reminding him to not miss doses (Answer B), although important, is unlikely to solve his issue, as hypocalcemia is persisting while high doses are being administered in the hospital.

Calcium carbonate (Answer C) is likely to be less well absorbed than calcium citrate, and such a formulation switch would probably not help. Switching to intravenous calcium (Answer E) may transiently improve serum calcium, but it would not address the home regimen.

Doubling the calcitriol dosage (Answer A) may be useful, but he is on a very high dosage already, raising questions about absorption. It is uncommon to chronically require calcitriol dosages of more than 2 or 3 mcg per day to maintain normocalcemia, and in this person taking 6 mcg of calcitriol daily, increasing the dosage further increases the pill burden, but is uncertain whether it would normalize his calcium. Evaluating for causes of gastrointestinal malabsorption, such as celiac disease, is reasonable in a situation such as this one. However, the patient is taking sevelamer and calcitriol at the same time. Sevelamer binds to calcitriol and inhibits its absorption, thus limiting the calcitriol effect. This is the likely reason that he needs a higher calcitriol dosage. Lanthanum, however, does not interfere with calcitriol absorption. After switching phosphate binders from sevelamer to lanthanum (Answer D), he was able to decrease the calcitriol dosage (to 1 mcg twice daily) required to maintain a calcium concentration in the target range and also decrease his calcium citrate dosage (to 600 mg 3 times daily).

Educational Objective

Explain how sevelamer can interfere with calcitriol absorption.

Reference(s)

Pierce D, Hossack S, Poole L, Robinson A, et al. The effect of sevelamer carbonate and lanthanum carbonate on the pharmacokinetics of oral calcitriol. *Nephrol Dial Transplant.* 2011;26(5):1615-1621. PMID: 20921291

75 **ANSWER: D) Primary hypothyroidism**

Hyperprolactinemia is variably associated with primary hypothyroidism (Answer D), and up to 58% of individuals with untreated or undertreated hypothyroidism have elevated serum prolactin concentrations. The mechanism is multifactorial but is primarily due to the physiologic stimulation of prolactin release from lactotrophs, which is caused by increased thyrotropin-releasing hormone levels in individuals with untreated or undertreated primary hypothyroidism.

This patient was referred to endocrinology for mild hyperprolactinemia and a pituitary T1 postcontrast MRI that showed a 2.1 × 1.3 × 1.0-cm, homogeneously enhancing soft-tissue mass in the region of the sella associated with suprasellar extension, mass effect, and upward lifting of the optic chiasm (*see images*). Further workup with thyroid function testing then revealed untreated primary hypothyroidism, which is the underlying etiology of her symptoms and presentation. She later confirmed that she was skipping most levothyroxine doses over the past several months due to nausea associated with the medication.

Figure. Soft-tissue sellar mass with suprasellar extension

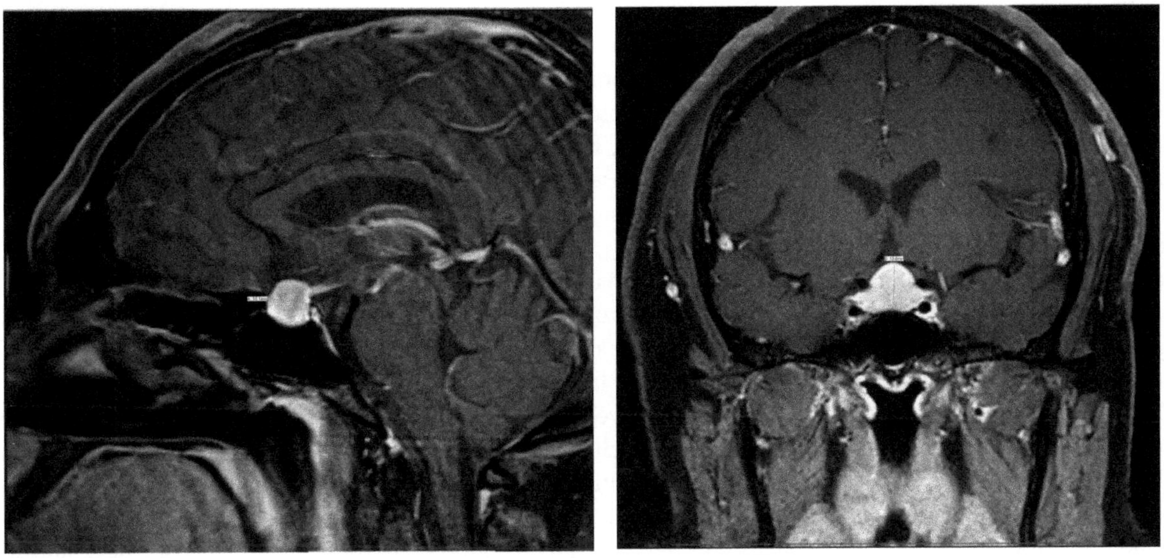

Image courtesy of Anu Sharma.

Hyperprolactinemia can occur in individuals with Cushing disease (Answer A) because of concomitant production and secretion of prolactin and cortisol from corticotroph adenomas (ie, cosecretion leading to stalk effect). However, diagnosis of this condition should not rely on random serum cortisol and ACTH measurements and would require measurement of 24-hour urinary free cortisol or late-night salivary cortisol or a dexamethasone-suppression test. The mildly elevated serum cortisol in this patient is consistent with oral contraceptive pill use and not Cushing disease.

This patient's chronicity of symptoms, hyperprolactinemia, and abnormal thyroid function test results are inconsistent with pregnancy (Answer B), and the pregnancy test is also negative.

Functional pituitary adenomas, including those that produce TSH, prolactin, and/or GH (Answers C and E), arise from precursor or stem cells, as evidenced by overexpression of pituitary-specific transcriptional factor (Pit-1) in these tumors. These answer choices are inconsistent with this patient's findings. Individuals with TSH-secreting pituitary adenomas would present with clinical signs and symptoms of hyperthyroidism, elevated serum T_3 and T_4 levels, and inappropriately normal or elevated serum TSH values. The degree of hyperprolactinemia in patients with a prolactinoma would be proportionate to the size of the pituitary adenoma (thus inconsistent with this patient's presentation). GH-secreting pituitary adenomas are diagnosed with elevated serum IGF-1 values and would demonstrate signs and symptoms of GH excess, all of which are absent in this patient.

Educational Objective

Identify untreated primary hypothyroidism as an etiology of hyperprolactinemia.

Reference(s)

Sharma LK, Sharma N, Gadpayle AK, Dutta D. Prevalence and predictors of hyperprolactinemia in subclinical hypothyroidism. *Eur J Intern Med.* 2016;35:106-110. PMID: 27473607

Honbo KS, van Herle AJ, Kellett KA. Serum prolactin levels in untreated primary hypothyroidism. *Am J Med.* 1978;64(5):782-787. PMID: 645742

Meier C, Christ-Crain M, Guglielmetti M, Huber P, Staub JJ, Müller B. Prolactin dysregulation in women with subclinical hypothyroidism: effect of levothyroxine replacement therapy. *Thyroid.* 2003;13(10):979-985. PMID: 14611708

Nieman LK, Biller BM, Findling JW, et al. The diagnosis of Cushing's syndrome: an Endocrine Society clinical practice guideline. *J Clin Endocrinol Metab.* 2008;93(5):1526-1540. PMID: 18334580

76 ANSWER: D) Start testosterone replacement

This patient has primary (hypergonadotropic) hypogonadism, as indicated by the low testosterone and high gonadotropin levels. In primary hypogonadism, FSH is usually more elevated than LH, which is the case here. When he was asked about any relevant past testicular pathology, he mentioned a remote history of seminoma treated with radiation. Other causes of primary hypogonadism (eg, Klinefelter syndrome, cryptorchidism, varicocele) should be considered in the setting of primary hypogonadism, and a testicular examination is necessary. This patient also reported that he was an avid biker throughout his life, which may have increased his risk of osteoporosis. Interpretation of laboratory test results is complicated because of his history of pituitary macroadenoma and transsphenoidal surgery with pathology indicative of gonadotroph adenoma. Postoperatively, FSH and LH remained elevated from persistent primary hypogonadism. FSH and LH concentrations do not represent markers of pituitary tumor activity. In rare cases of functioning gonadotroph pituitary adenomas, laboratory data reveal hypergonadotropic hypergonadism (ie, high testosterone concentration along with inappropriately high FSH or LH, or both).

Both clinically nonfunctioning and functioning gonadotroph adenomas are often positive for FSH, LH, or α-subunit and for steroidogenic factor 1 (SF-1) by immunohistochemistry. In a man who does not desire fertility, management of primary hypogonadism consists of testosterone replacement, and therapy should be initiated for this patient (Answer D). Treatment is also likely to also have a favorable effect on his anemia.

While pituitary radiotherapy (Answer A) is effective for postoperative pituitary tumor control in most cases, its use must be balanced with the risk of complications, including hypopituitarism. In a multicenter single-center study of gamma radiation, 21% of patients developed hypopituitarism at a median follow-up of 51 months. Radiotherapy is usually reserved for sizeable residual tumor that cannot be surgically removed, when pathology markers associated with risk of tumor growth are present (eg, increased proliferation index [>3%] and mitoses), and when there is evidence of tumor progression on serial scans. In this case, because the 3-month postoperative scan was not done, it is unclear whether the residual tumor is a stable postoperative finding or represents tumor progression.

Reoperation (Answer B) is indicated for pituitary adenomas with sizeable residual tumor, especially if mass effect manifestations are present (eg, headaches, optic neuropathy), or if there is evidence of tumor growth during serial MRI.

This patient has only mild hyperprolactinemia, which is unlikely to be significantly contributing to his hypogonadism. Repeating prolactin measurement after opioid treatment has been completed would help distinguish between stalk effect and drug-induced hyperprolactinemia. However, treatment with a dopamine agonist (Answer C) is unlikely to improve his testosterone, which is low on a primary and not secondary basis.

Opioid treatment has only been used recently in this patient, and his biochemical picture is still compatible with hypergonadotropic hypogonadism. Thus, waiting to initiate therapy until opioids are stopped and hormones are reevaluated (Answer E) is incorrect.

Educational Objective

Differentiate among the laboratory and histopathologic characteristics of pituitary adenomas and construct the differential diagnosis of elevated gonadotroph hormones.

Reference(s)

Ntali G, Capatina C, Grossman A, Karavitaki N. Clinical review: functioning gonadotroph adenomas. *J Clin Endocrinol Metab.* 2014;99(12):4423-4433. PMID: 25166722

Fong KY, Jun Rui Lim M, Fu S, et al. Postsurgical outcomes of nonfunctioning pituitary adenomas: a patient-level meta-analysis. *Pituitary.* 2023;26(4):461-473. PMID: 37389776

Chanson P, Wolf P. Clinically non-functioning pituitary adenomas. *Presse Med.* 2021;50(4)104086. PMID: 34718111

Cordeiro D, Xu Z, Mehta GU, et al. Hypopituitarism after Gamma Knife radiosurgery for pituitary adenomas: a multicenter, international study. *J Neurosurg.* 2018;131(4):1188-1196. PMID: 31369225

77 ANSWER: A) Acute kidney failure

Internists, hospitalists, and critical care and emergency department physicians frequently encounter patients with acid-base disorders. Endocrinologists are often asked to evaluate patients with diabetes or endocrine diseases who have acid-base disorders. Patients with diabetes, either type 1 or type 2, are at risk of developing diabetes-related ketoacidosis (DKA). Predisposing factors for DKA include acute illness (most commonly secondary to infections) or medications. DKA can also be the initial manifestation of new-onset type 1 diabetes. SGLT-2 inhibitors can cause euglycemic ketoacidosis, with glucose values usually less than 250 mg/dL (<13.9 mmol/L). SGLT-2 inhibitor–induced euglycemic DKA is thought to be caused by SGLT-2 inhibitor–induced glucosuria, which leads to decreased plasma glucose concentration/carbohydrate deficit, decreased insulin production, and increased glucagon release. This results in increased lipolysis and ketogenesis.

Because DKA is associated with increased morbidity and mortality, endocrinologists should be able to diagnose and treat this condition early in its course. Treatment sometimes requires management in the intensive care unit. In this vignette, euglycemic DKA (Answer C) is a reasonable consideration, especially given this patient's recent initiation of an SGLT-2 inhibitor. One should also consider DKA due to suboptimally controlled diabetes (Answer B) and metformin-induced acidosis (secondary to lactic acid production) (Answer D) because of the history of metformin use. DKA and euglycemic DKA can both cause metabolic acidosis with an increased anion gap due to the increased production of ketones. Similarly, metformin-induced lactic acidosis, a rare condition, can cause metabolic acidosis with an increased anion gap. In this case, however, the patient is experiencing metabolic acidosis with a normal, not increased, anion gap.

Acute kidney failure (Answer A) can induce metabolic acidosis, with either an elevated or normal anion gap. When impairment of kidney function is mild or moderate (glomerular filtration rate >15 or 20 mL/min per 1.73 m^2), there is decreased hydrogen excretion, which leads to normal anion gap acidosis. When kidney function is severe, an accumulation of anions elevates the anion gap. Acute kidney failure is the most likely cause of this patient's acid-base disorder.

In contrast to the other answer options, type 4 renal tubular acidosis (Answer E) causes normal anion gap metabolic acidosis. Type 4 renal tubular acidosis can occur among patients with diabetes and kidney impairment (diabetic nephropathy). It is believed that this is secondary to reduced aldosterone production, which is secondary to low renin production. The low renin production is caused by deceased conversion of prorenin to renin. The acidosis is primarily due to decreased urinary ammonium excretion. In addition to metabolic acidosis, mild hyperkalemia can be observed. This patient's potassium concentration is normal. In addition, type 4 renal tubular acidosis does not develop acutely, but rather it is a chronic condition that develops gradually.

Educational Objective

Differentiate among metabolic acidosis conditions in patients with diabetes.

Reference(s)

Taylor SI, Blau JE, Rother KI. SGLT2 inhibitors may predispose to ketoacidosis. *J Clin Endocrinol Metab.* 2015;100(8):2849-2852. PMID: 26086329

Berend K. Review of the diagnostic evaluation of normal anion gap metabolic acidosis. *Kidney Dis.* 2017;3(4):149-159. PMID: 29344509

78 ANSWER: D) Referral to specialist for testicular sperm extraction

For patients with normal serum gonadotropins and testosterone, infertility, and azoospermia, surgical sperm extraction (Answer D) is the only proven effective therapy. Although this man does not have evidence of obstructive azoospermia on ultrasonography, imaging is not 100% accurate. Furthermore, nests of sperm can be found and surgically extracted (40%-60%) in men who have nonobstructive azoospermia on seminal fluid analysis. Even in men with Klinefelter syndrome, 35% to 50% have some sperm for surgical extraction. Intraoperative microscopy (microscopic testicular extraction [mTESE]) can enhance detection rates.

Although this man has mild hyperprolactinemia, dopamine agonist therapy (Answer A) is not effective for men who are infertile and azoospermic (or oligospermic) but who have normal serum gonadotropin and testosterone concentrations.

Although a 2015 meta-analysis concluded that clomiphene (Answer B) might increase sperm concentrations in men with oligospermia, this meta-analysis included only 2 randomized controlled trials and one cohort study with a total of 115 clomiphene-treated men and 85 control participants. There are no randomized, placebo-controlled studies demonstrating that clomiphene is effective for infertile men with azoospermia, but one study provided preliminary data that clomiphene might increase the yield for surgical extraction of sperm for men with nonobstructive azoospermia.

There is moderately strong evidence that increased seminal plasma oxidant concentrations (eg, reactive oxygen species) contribute to decreased male fertility. Eating a diet rich in vegetables, nuts, and seeds provides antioxidants, including arginine, carnitines, carotenoids, folate, selenium, zinc, magnesium, myo-inositol polyunsaturated fatty acids, resveratrol, and vitamin B complex. A 2022 Cochrane review concluded the following about antioxidant supplements (Answer C): "The side-effect profile of antioxidant supplements appears to be low and mild... Subfertile couples should be advised that overall, the current evidence is inconclusive based on serious risk of bias due to poor reporting of methods of randomisation, failure to report on the clinical outcomes of live birth rate and clinical pregnancy, often unclear or even high attrition, and also imprecision due to often low event rates and small overall sample sizes. Further large well-designed randomised placebo-controlled trials studying infertile men and reporting on pregnancy and live births are still required to clarify the exact role of antioxidants." Currently, it seems best to recommend a well-balanced diet that includes vegetables, nuts, and seeds. It is probably harmless to take antioxidant supplements, but there is inadequate evidence to support recommending them.

There is a theoretical benefit of FSH therapy (Answer E) for men with idiopathic infertility, oligospermia (and/or decreased numbers of normally shaped and/or motile sperm), and certain polymorphisms in the FSH β unit and FSH receptor genes, but the evidence is too weak to recommend this therapy. Further study with randomized, placebo-controlled studies must done to determine which, if any, patients might benefit and the appropriate dosage for FSH therapy.

Educational Objective

Recommend surgical sperm extraction as the only proven effective method to obtain sperm for conception in men who have nonobstructive azoospermia and do not have hypogonadotropic hypogonadism.

Reference(s)

Bridges N, Trofimenko V, Fields S, Carrell D, Aston K, Hotaling J. Male factor infertility and clomiphene citrate: a meta-analysis-the effect of clomiphene citrate on oligospermia. *Urol Pract.* 2015;2(4):199-205. PMID: 37559276

de Ligny W, Smits RM, Mackenzie-Proctor R, et al. Antioxidants for male subfertility. *Cochrane Database Syst Rev.* 2022;5(5):CD007411. PMID: 35506389

Gundewar T, Kuchakulla M, Ramasamy R. A paradoxical decline in semen parameters in men treated with clomiphene citrate: a systematic review. *Andrologia.* 2021;53(1):e13848. PMID: 33108678

Hussein A, Ozgok Y, Ross L, Rao P, Niederberger C. Optimization of spermatogenesis-regulating hormones in patients with non-obstructive azoospermia and its impact on sperm retrieval: a multicentre study. *BJU Int.* 2013;111(3 Pt B):E110-E114. PMID: 22958644

Simoni M, Brigante G, Rochira V, Santi D, Casarini L. Prospects for FSH treatment of male infertility. *J Clin Endocrinol Metab.* 2020;105(7):dgaa243. PMID: 32374828

79 ANSWER: E) Semaglutide

Use of GLP-1 receptor agonists in patients with chronic kidney disease is generally safe and effective, but there have been postmarketing reports of acute kidney injury and worsening of chronic kidney disease, which sometimes require hemodialysis. Similar adverse events were also observed during the clinical trial programs that evaluated the safety and efficacy of GLP-1 receptor agonist therapies. Some of these events were reported in patients without known underlying kidney disease. Most reported events occurred in patients who had experienced nausea, vomiting, diarrhea, or dehydration, which, if severe, can lead to a prerenal state of azotemia (ie, acute kidney injury). Because these reactions may worsen kidney function, caution should be exercised when initiating or escalating dosages of GLP-1 receptor agonists in patients with kidney impairment. Kidney function should be monitored in patients with kidney impairment who report severe adverse gastrointestinal reactions, as well as in patients with significant reductions in appetite and oral intake, as reported by this patient and supported by her 18-lb (8.2-kg) weight loss over 3 months. Weight loss related to GLP-1 receptor agonist therapy is mediated through caloric restriction. Significant reduction in oral intake and the associated weight loss are often not just related to reduced caloric intake but also to reduced fluid intake. The patient's serum urea nitrogen-to-creatinine ratio greater than 20 (0.081 in IU) is consistent with prerenal azotemia (ie, acute kidney injury). Accordingly, intravenous fluid should be initiated. While the patient's background therapy with empagliflozin (Answer A) and enalapril (Answer B) may have contributed to her development of acute kidney injury, she tolerated this regimen without any issues until semaglutide was initiated. Thus, the therapy most likely responsible for her current clinical condition is semaglutide (Answer E). Resuming GLP-1 receptor agonist therapy in the future may be considered: either semaglutide at a lower dosage or a trial of an alternative GLP-1 receptor agonist; however, this would need to be performed cautiously and with close follow-up.

No dosage adjustments are recommended in patients with kidney impairment, including end-stage kidney disease, with the most used long-acting GLP-1 receptor agonists (liraglutide, dulaglutide, semaglutide) or the gastric inhibitor polypeptide/GLP-1 receptor dual agonist tirzepatide. The only 2 commercially available GLP-1 receptor agonists that have recommendations regarding limitation of use in patients with chronic kidney disease are lixisenatide and exenatide-based formulations. These formulations are not recommended for patients with an estimated glomerular filtration rate less than 15 mL/min per 1.72 m^2 (lixisenatide) or less than 30 mL/min per 1.72 m^2 (exenatide-based formulations).

While euglycemic diabetic ketoacidosis has been reported after the initiation of GLP-1 receptor agonists in patients already receiving SGLT-2 inhibitors, this patient's laboratory findings are not consistent with euglycemic diabetic ketoacidosis; rather, they are consistent with acute kidney injury (prerenal etiology). The proposed mechanism for euglycemic diabetic ketoacidosis after initiation of GLP-1 receptor agonist therapy in patients receiving background SGLT-2 inhibitor therapy is thought to be related to decreased oral intake, and in some cases severe nausea/vomiting and/or diarrhea, that can sometimes be observed shortly after the initiation of the GLP-1 receptor agonist. This patient had been doing very well on the SGLT-2 inhibitor therapy until semaglutide was initiated and titrated to 1 mg weekly. Thus, given the case presented, laboratory results, and timing of symptom onset, empagliflozin (Answer A) is not the correct answer.

Although metformin (Answer D) and enalapril (Answer B) should be stopped in this patient with acute kidney injury (at least temporarily since her estimated glomerular filtration rate is low), neither of these medications is the root cause.

Her symptoms are not attributable to hypoglycemia, and her acute kidney injury is not related to insulin glargine therapy (Answer C).

Educational Objective

Explain the role of GLP-1 receptor agonist therapy in patients with kidney impairment and the risk of worsening kidney function when initiating or escalating dosages of GLP-1 receptor agonists.

Reference(s)

Dong S, Sun C. Can glucagon-like peptide-1 receptor agonists cause acute kidney injury? An analytical study based on post-marketing approval pharmacovigilance data. *Front Endocrinol (Lausanne)*. 2022;13:1032199. PMID: 36583004

Leehey DJ, Rahman MA, Borys E, Picken MM, Clise CE. Acute kidney injury associated with semaglutide. *Kidney Med.* 2021;3(2):282-285. PMID: 33851124

Patoulias D, Boulmpou A, Papadopoulos CE, Doumas M. Glucagon-like peptide-1 receptor agonists and the risk of acute kidney injury: alarming, or not? *Kidney Med.* 2021;3(4):674-675. PMID: 34401735

80 ANSWER: B) Cholesterol efflux capacity

Patients with longstanding type 1 diabetes are at increased risk for atherosclerotic cardiovascular disease (ASCVD) despite the absence of significant dyslipidemia. Statin therapy is therefore recommended for most patients with an LDL-cholesterol cholesterol greater than 70 mg/dL (>1.81 mmol/L) if they are older than 40 years or have had diabetes for more than 20 years. Screening for subclinical atherosclerosis with cardiac CT is generally not recommended for individuals at high risk, including those with diabetes, but if obtained, the presence of a non-zero coronary artery calcium score further supports statin initiation, as recommended for this patient.

While patients with type 2 diabetes have characteristic atherogenic dyslipidemia consisting of elevated triglycerides and low HDL-cholesterol levels, patients with type 1 diabetes generally do not have elevated triglycerides unless they have marked hyperglycemia from inadequate insulin therapy. In addition, their HDL-cholesterol levels are usually normal or even elevated. Nonetheless there are qualitative defects in HDL function that likely contribute to increased risk for atherogenesis. While the exact cause of this is not known, it is speculated that posttranslational modification of HDL proteins due to abnormal glucose and insulin levels may have a role. One of the most important roles of HDL is reverse cholesterol transport, which involves the delivery of cholesterol from peripheral macrophages to the liver for excretion. The first step in this process is ATP-binding cassette transporter (ABCA1)–mediated transfer of free cholesterol from macrophages to apolipoprotein A1 leading to formation of nascent HDL particles. Subsequently, other proteins such as lecithin cholesterol acyl transferase (LCAT) help in the maturation of nascent HDL to mature spherical HDL particles that deliver cholesterol to the liver either directly by docking on the scavenger receptors or indirectly through a cholesteryl ester transfer protein (CETP)–mediated process. The efficiency of apolipoprotein A1 to accept cholesterol from macrophages, which initiates this whole process, can be assessed by measuring the cholesterol efflux capacity (Answer B).

Cholesterol efflux capacity is assessed by measuring the efflux of labeled cholesterol from cultured macrophages to apolipoprotein B–depleted plasma of the patient, which contains the HDL particles. Cholesterol efflux capacity has been shown to inversely correlate with prevalence of ASCVD independent of HDL-cholesterol levels. The longitudinal Dallas Heart Study also demonstrated that there was a 67% lower risk of ASCVD events in patients in the highest quartile of cholesterol efflux capacity compared with those in the lowest quartile after adjustment for other traditional risk factors, including HDL-cholesterol concentration. The failure of medications such as

CETP inhibitors to significantly reduce ASCVD risk despite markedly increasing the HDL-cholesterol levels also indicates the importance of assessing HDL function over the HDL-cholesterol concentration. There are limited studies in patients with type 1 diabetes that have demonstrated impaired cholesterol efflux and other HDL functions in patients with either well-controlled or poorly controlled type 1 diabetes. Improved glycemic control has also been shown to improve cholesterol efflux capacity concomitant with reduced posttranslational modification of apolipoprotein A1. Unfortunately, this test is still not widely available for clinical use and is mostly restricted to research settings.

While the LDL-cholesterol concentration is the most commonly used lipid measure to assess ASCVD risk, it may not accurately reflect this risk in certain situations, and advanced lipoprotein profile testing may be required. In patients with elevated triglycerides, such as those with type 2 diabetes and metabolic syndrome, LDL cholesterol is often not elevated but there is a preponderance of small, dense LDL particles that confers high risk. Nuclear magnetic resonance spectroscopy (Answer C) helps to quantify the number and size of LDL particles. Besides small-dense LDL particles, triglyceride-rich remnant lipoprotein particles, which are generated after initial hydrolysis of chylomicrons and VLDL particles, are known to be atherogenic and are increased in patients with moderate hypertriglyceridemia. It is possible to measure the cholesterol content of these remnant lipoprotein particles both directly and indirectly (Answer E), and this is a good risk marker. Another way of assessing the total burden of atherogenic lipoproteins is to measure apolipoprotein B (Answer A), as all lipoproteins, including LDL and remnants, carry one molecule of apolipoprotein B. However, all these measures are superior to LDL-cholesterol measurement only in patients with hypertriglyceridemia. In those with normal triglyceride levels, they do not provide any additional information beyond that learned from the LDL-cholesterol concentration.

Lipoprotein (a) (Answer D) is an independent risk factor that should be measured in patients who have a personal or family history of premature ASCVD or familial hypercholesterolemia. It should also be measured in individuals who do not demonstrate the expected LDL-cholesterol lowering with statin therapy and in those who continue to demonstrate progressive atherosclerosis despite well-controlled cholesterol levels. Since this patient does not have any of these features, lipoprotein (a) measurement is unnecessary.

Educational Objective

Explain the utility of an HDL functional assay over HDL-cholesterol measurement in assessing the risk of atherosclerotic cardiovascular disease.

Reference(s)

Rohatgi A, Khera A, Berry JD, et al. HDL cholesterol efflux capacity and incident cardiovascular events. *N Engl J Med.* 2014;371(25):2383-2393. PMID: 25404125

Manjunatha S, Distelmaier K, Dasari S, Carter RE, Kudva YC, Nair KS. Functional and proteomic alterations of plasma high density lipoproteins in type 1 diabetes mellitus. *Metabolism.* 2016;65(9):1421-1431. PMID: 27506748

Denimal D, Monier S, Simoneau I, Duvillard L, Vergès B, Bouillet B. HDL functionality in type 1 diabetes: enhancement of cholesterol efflux capacity in relationship with decreased HDL carbamylation after improvement of glycemic control. *Cardiovasc Diabetol.* 2022;21(1):154. PMID: 35962339

81 ANSWER: A) Add hydrochlorothiazide

The goal for long-term management of hypoparathyroidism is the amelioration of symptoms and maintaining a low to near-normal calcium concentration. Calcium supplements and vitamin D are the treatment mainstays. Active vitamin D has also been used to minimize the amount of calcium needed. Supplementation of calcium via diet alone is not adequate. Calcium carbonate has 40% elemental calcium, while calcium citrate has 21%. Calcium carbonate is better absorbed with acid in the stomach, so when an individual is on proton-pump inhibitor therapy, calcium citrate is the preferred choice. While the patient in this vignette is frustrated by the amount of calcium she must take, she does not seem to have problems with calcium carbonate absorption. In addition, if she were to switch from calcium carbonate to calcium citrate (Answer E), she would need to take more calcium citrate tablets, adding to the pill burden. Although higher amounts of calcium supplements can be used, the occurrence of hypercalciuria limits further increases in the calcium carbonate dosage (Answer C).

Active vitamin D formation in the kidney is impaired in persons with hypoparathyroidism due to hyperphosphatemia and the lack of PTH action. With its short half-life, active vitamin D must be administered at least twice daily. The patient's calcium concentration is already at the lower end of the reference range and reducing the calcitriol dosage further (Answer B) could result in additional symptoms.

Hypomagnesemia causes hypocalcemia because it reduces the sensitivity of PTH; however, this patient's magnesium concentration is normal. Thus, starting magnesium (Answer D) is not the best next step.

Hydrochlorothiazide (Answer A), although not frequently used in this scenario, could correct this patient's hypercalciuria and in turn improve the serum calcium concentration. Thiazides promote increased kidney calcium absorption and can therefore help with hypercalciuria. Higher dosages of thiazide diuretics (hydrochlorothiazide, 25 mg twice daily or 50 mg once daily; chlorthalidone, 25 or 50 mg daily; indapamide, 1.25-2.5 mg once daily) are usually needed to achieve this effect. Hypokalemia is a concern with long-term use, and potassium supplementation maybe required.

PTH receptor agonists have potential utility. PTH (1-84) replacement was effective in maintaining calcium within the reference range, reducing the required dosages of calcium and active vitamin D, and reducing urinary calcium excretion. However, this agent is no longer available for use in the United States. Palopegteriparatide consists of PTH (1-34) transiently linked to a branched methoxypolyethylene glycol carrier through a proprietary linker. This formulation allows for once-daily administration to achieve stable PTH levels. In the PaTHway trial, 79% of individuals on the study drug achieved normal calcium concentrations, independence from conventional therapy (defined as no active vitamin D supplementation and <600 mg daily of calcium), and no need for increased dosage of the study drug over 4 weeks. Additionally, palopegteriparatide also showed improved SF-36 scores. Palopegteriparatide is now approved for use in Europe and the United States.

Educational Objective

Describe the conventional treatment of hypoparathyroidism and the role of PTH replacement for hypocalcemia.

Reference(s)

Bilezikian JP. Hypoparathyroidism. *J Clin Endocrinol Metab.* 2020;105(6):1722-1736. PMID: 32322899

Khan AA, Rubin MR, Schwarz P, et al. Efficacy and safety of parathyroid hormone replacement with TransCon PTH in hypoparathyroidism: 26-week results from the phase 3 PaTHway Trial. *J Bone Miner Res.* 2023;38(1):14-25. PMID: 36271471

82 ANSWER: E) Repeat neck ultrasonography and measure serum thyroglobulin and thyroglobulin antibodies in 12 months

This patient has been diagnosed with American Joint Committee on Cancer (8th edition) stage I papillary thyroid carcinoma (PTC) and was initially stratified to have an intermediate risk (approximately 20%) for structural recurrence due to the presence of clinical lymph node metastases. Pathologic findings were consistent with the diffuse sclerosing variant of PTC, which is often found in younger patients and is characterized by diffuse involvement of the thyroid gland and a higher frequency of lymph node and distant (especially lung) metastases. Although the diffuse sclerosing variant of PTC is associated with lower disease-free survival than classic PTC, overall mortality remains low.

Following initial surgery, the patient was treated with adjuvant radioactive iodine therapy, and the posttherapy whole-body scan demonstrated scant residual normal and/or malignant thyroid tissue in the left thyroid bed and no evidence of metastatic disease. The patient has subsequently been followed for 2 years with neck ultrasonography and measurement of serum thyroglobulin and thyroglobulin antibodies. Serum thyroglobulin cannot be measured using an immunoassay in many patients with positive thyroglobulin antibodies due to negative interference. However, a declining thyroglobulin antibody titer over time is typically seen in patients who are disease-free after initial therapy and is considered a good prognostic marker. After total thyroidectomy and radioactive iodine therapy, the median time for resolution of thyroglobulin antibodies is approximately 3 years, but they may persist indefinitely in some patients who are free of disease. In this patient's case, the observation that the thyroglobulin antibody titer is declining when measured using the same assay coupled with normal findings on neck ultrasonography should lead to the conclusion that additional diagnostic testing is not needed now (Answer C), and the patient should be advised to repeat neck ultrasonography and measure serum thyroglobulin and thyroglobulin antibodies in 12 months (Answer E).

The patient has been tolerating a generic levothyroxine tablet well without clinical or biochemical evidence of malabsorption. A retrospective study of nearly 17,600 hypothyroid patients demonstrated that treatment with generic or branded levothyroxine was associated with similar rates of normal and stable TSH levels. There is no evidence-based clinical indication to change this patient's prescription to a branded levothyroxine formulation (Answer A), which could incur additional cost to the patient and health system. Furthermore, for this patient with an indeterminate response to therapy (positive but declining thyroglobulin antibody titer) 2 years after initial treatment for thyroid cancer, a continued TSH goal of 0.1 to 0.5 mIU/L is appropriate, and TSH suppression below this range (Answer B) is both unnecessary and is more likely than lesser degrees of TSH suppression to cause adverse effects.

Interference in immunoassays by heterophile antibodies may result in false-negative or false-positive results. One of the methods for detecting heterophile antibody interference is to reanalyze the sample after pretreatment with a commercially available heterophile-blocking antibody reagent (Answer D). A significant deviation from the initial results following pretreatment with the "blocker" provides evidence of heterophile antibody interference. Other options to demonstrate the presence of interfering heterophile antibodies include: (1) evaluating linearity after serial dilutions; (2) analyzing samples using a different immunoassay platform; or (3) analyzing samples using a radioimmunoassay or liquid chromatography/tandem mass spectrometry–based thyroglobulin assay. Although it is important to consider the possibility of heterophile antibody interference in the interpretation of laboratory test results, in this case there is a high likelihood that the undetectable serum thyroglobulin is resulting from thyroglobulin antibody interference, which is often seen in patients with thyroid cancer and concomitant autoimmune thyroid disease (eg, Hashimoto thyroiditis). The fact that the serum thyroglobulin antibody titer is downtrending in this patient is a good prognostic marker and thus pursuing alternative methods of measuring serum thyroglobulin (liquid chromatography/tandem mass spectrometry or radioimmunoassay) or testing for heterophile antibody interference is superfluous.

Educational Objective

Assess response to therapy in a patient with papillary thyroid cancer and positive thyroglobulin antibodies.

Reference(s)

Brito JP, Ross JS, Sangaralingham L, et al. Comparative effectiveness of generic vs brand-name levothyroxine in achieving normal thyrotropin levels. *JAMA Netw Open.* 2020;3(9):e2017645. PMID: 32997127

Giovanella L, Suriano S, Ceriani L, Verburg FA. Undetectable thyroglobulin in patients with differentiated thyroid carcinoma and residual radioiodine uptake on a postablation whole-body scan. *Clin Nucl Med.* 2011;36(2):109-112. PMID: 21220971

Haugen BR, Alexander EK, Bible KC, et al. 2015 American Thyroid Association management guidelines for adult patients with thyroid nodules and differentiated thyroid cancer: the American Thyroid Association Guidelines Task Force on Thyroid Nodules and Differentiated Thyroid Cancer. *Thyroid.* 2016;26(1):1-133. PMID: 26462967

83 ANSWER: A) Measure serum IGF-1

Incidentally discovered lesions in the pituitary gland are increasingly reported with the rise in modern imaging methods undertaken for investigation of headache, trauma, or other indications. On the surface, this case appears to be most consistent with a nonfunctioning pituitary adenoma, since there are no clinical signs of endocrine hyperfunction and the patient's hormone profile is normal. However, serum IGF-1 (Answer A) has not been reported and should be measured next, in line with Endocrine Society guidance and other recommendations. The Endocrine Society recommends screening for GH and prolactin excess in all incidentally discovered adenomas, regardless of the presence or absence of symptoms. The symptoms of acromegaly may be subtle, and a careful symptom inquiry (sweating, symptoms of sleep apnea, headaches, and peripheral numbness) should be undertaken even when the patient is apparently asymptomatic. Similarly, the physical signs of acromegaly are insidious and often go unrecognized. Nevertheless, clinically silent somatotroph adenomas have been reported, and they potentially become symptomatic if left untreated. Since these tumors are biologically active, definitive therapy is preferred over surveillance, not only to prevent symptom development but also to reduce the risk of comorbidities associated with long-term exposure to GH excess. Silent pituitary adenomas with positive immunostaining for GH but without biochemical evidence of GH hypersecretion have also been described. These adenomas usually present with mass effects or as incidental findings, and they are managed by following recommendations for nonfunctioning pituitary adenomas.

In contrast to its recommendation for IGF-1 measurement, the Endocrine Society recommends testing for hypercortisolism (Answer B) only when this is suspected clinically, as there is a high false-positive rate and low true-positive rate in unselected populations. However, silent corticotroph adenomas are not uncommon findings on histologic analysis following surgery for apparent nonfunctioning tumors. These demand more intense surveillance because of their increased growth potential.

In this vignette, there is no radiologic evidence of threat to the optic apparatus, so visual field testing (Answer C) is not immediately required. For similar reasons, immediate transsphenoidal surgery (Answer D) would not be indicated in the absence of chiasmal compression or hormone excess.

Another pituitary MRI (Answer E) would be needed to survey for tumor growth in the absence of surgical plans. However, the endocrine workup has not been completed here, and a nonfunctioning adenoma cannot be presumed until the IGF-1 concentration is documented to be in the reference range. The frequency of radiological surveillance for nonfunctioning pituitary adenomas should be tailored according to the size of the lesion and rate of growth evident on any previous imaging. The Endocrine Society recommends MRI of the pituitary 6 months after the initial scan for patients with incidentally discovered macroadenomas and 1 year after the initial scan for patients with microlesions. Subsequent imaging is recommended annually for macroadenomas. Recent data confirm a low probability that nonfunctioning pituitary microadenomas will grow; thus, after the initial 1-year follow-up MRI, subsequent imaging can be delayed to 3 years.

Educational Objective

Select appropriate endocrine tests for a patient with a pituitary incidentaloma and explain the importance of IGF-1 measurement even when there are no clinical features of GH excess.

Reference(s)

Freda PU, Beckers AM, Katznelson L, et al; Endocrine Society. Pituitary incidentaloma: an endocrine society clinical practice guideline. *J Clin Endocrinol Metab.* 2011;96(4):894-904. PMID: 21474686

Giraldi E, Allen JW, Ioachimescu A. Pituitary incidentalomas: best practices and looking ahead. *Endocr Pract.* 2023;29(1):60-68. PMID: 36270609

Hamblin R, Fountas A, Lithgow K, et al. Natural history of non-functioning pituitary microadenomas: results from the UK non-functioning pituitary adenoma consortium. *Eur J Endocrinol.* 2023;189(1):87-95. 2023;189(1):87-95. PMID: 37345849

84 ANSWER: B) DPP-4 inhibitor

Before 1993, there was limited evidence to support the importance or benefit of achieving better glycemic control in patients with diabetes. The DCCT/EDIC trial (Diabetes Control and Complications Trial/Epidemiology of Diabetes Interventions and Complications) for patients with type 1 diabetes in 1993 and the UKPDS (United Kingdom Prospective Diabetes Study) for patients with type 2 diabetes in 1998 showed that intensive glycemic control can lead to reduced diabetes-related complications, especially microvascular complications. However, one of the findings was that the intensively treated group had a higher risk of hypoglycemia compared with risk in the control group. Several studies have shown an association between hypoglycemia and adverse clinical outcomes such as an increased risk of cardiovascular disease and mortality, altered mental status, and falls. Very low glucose values, if left untreated, can cause loss of consciousness and even seizures among patients with diabetes.

Achieving tighter glycemic control, especially when targeting hemoglobin A_{1c} values lower than 7.0% (<53 mmol/mol), can increase risk of hypoglycemia. Professional organizations therefore recommend higher glycemic targets, with hemoglobin A_{1c} values greater than 7.0% (>53 mmol/mol), especially among older patients with diabetes-related complications, comorbid conditions, or limited life expectancy. These higher glycemic targets, while appropriate in individuals who are particularly vulnerable to hypoglycemia, should not be applied universally to all patients with diabetes. Most patients with diabetes should have a target hemoglobin A_{1c} value less than 7.0% (<53 mmol/mol). The patient in this clinical scenario is a relatively healthy individual with no major underlying diseases, comorbidities, or diabetes-related complications. Although she is 67 years old, average life expectancy for women in the United States is 77.28 years (in 2020 before the COVID-19 pandemic). If this patient has increased hemoglobin A_{1c} for a prolonged period, she might be at higher risk of developing diabetes-related complications over the next 10 years.

This patient can be managed with medications that do not cause hypoglycemia and therefore do not place her at a higher risk of hypoglycemia-induced adverse events. In addition to metformin and empagliflozin, she can start a DPP-4 inhibitor (Answer B), which could lower her hemoglobin A_{1c} level below 7.0% (<53 mmol/mol). DPP-4 inhibitors inhibit the enzyme DPP-4, which inactivates many peptides, among them GLP-1 and glucose-dependent insulinotropic polypeptide (GIP). Overall, the effect on glycemic control is only modest, usually resulting in a hemoglobin A_{1c} reduction of 0.5%. This goal, however, is appropriate for this patient with mild hyperglycemia.

Basal or prandial insulin (Answers A and C) or a sulfonylurea (Answer D) are not appropriate because they can cause hypoglycemia.

Adding no medication (Answer E) is not appropriate in this relatively healthy patient with diabetes who has no significant underlying diseases and good life expectancy. Her diabetes can be managed with medications that do not carry an increased risk of hypoglycemia.

Educational Objective

Safely achieve tighter glucose control in older patients by prescribing medications with a low risk for hypoglycemia.

Reference(s)

Diabetes Control and Complications Trial Research Group, Nathan DM, Genuth S, et al. The effect of intensive treatment of diabetes on the development and progression of long-term complications in insulin-dependent diabetes mellitus. *N Engl J Med.* 1993;329(14):977-986. PMID: 8366922

Intensive blood-glucose control with sulphonylureas or insulin compared with conventional treatment and risk of complications in patients with type 2 diabetes (UKPDS 33). UK Prospective Diabetes Study (UKPDS) Group. *Lancet.* 1998;352(9131):837-853. PMID: 9742976

Cryer PE. Symptoms of hypoglycemia, thresholds for their occurrence, and hypoglycemia unawareness. *Endocrinol Metab Clin North Am.* 1999;28(3):495-500, v-vi. PMID: 10500927

Lee AK, Warren B, Lee CJ, et al. The association of severe hypoglycemia with incident cardiovascular events and mortality in adults with type 2 diabetes. *Diabetes Care.* 2018;41(1):104-111. PMID: 29127240

American Diabetes Association Professional Practice Committee. 6. Glycemic goals and hypoglycemia: standards of care in diabetes-2024. *Diabetes Care.* 2024;47(Suppl 1):S111-S125. PMID: 38078586

85 ANSWER: B) Plasma fluoride

This young woman has gradually worsening pain in multiple joints. Wrist pain from carpal tunnel syndrome can occur in acromegaly. She has no history of recurrent/frequent headaches, excessive sweating, or oligomenorrhea, and physical examination did not document frontal bossing, prognathism, macroglossia, or wide spacing of the teeth. She has no history of diabetes or hypertension. Serum IGF-1 measurement (Answer A) is the screen used when GH excess is suspected. In the absence of a constellation of symptoms, signs, and conditions that maybe associated with acromegaly, this is an unlikely diagnosis.

Elevated plasma fluoride concentrations (Answer B) are observed in individuals with fluorosis. Skeletal fluorosis is associated with generalized osteosclerosis with diffuse periosteal new bone formation (which is painful) in the hands, forearms, and lower extremities. High bone mineral density at the spine and hips is noted. A high fluoride concentration can occur with increased tea ingestion/well water and is endemic in some countries. Long-term use of certain medications such as voriconazole can also cause skeletal fluorosis. In the United States, long-term use of inhalants has been associated with skeletal fluorosis. Inhalants are easily obtained and induce euphoria; they are often used or substituted for other recreational drugs because of their availability. Compressed air used for computer cleaning contains hydrofluorocarbons such as 1,1-difluoroethane. In this patient, previous alcohol use disorder places her at risk for ongoing substance abuse. She began inhaling computer cleaners when she stopped drinking alcohol 3 years ago. The high alkaline phosphate, high osteocalcin, and radiographic findings are suggestive of a diagnosis of fluorosis. Discontinuation is the primary form of treatment; however, radiographic abnormalities may take a long time to resolve because of the prolonged half-life of fluoride in bone.

Mastocytosis is a disorder of increased mast-cell proliferation with both cutaneous and systemic manifestations. The classic clinical features include flushing, pruritus, abdominal pain, diarrhea, hypotension, musculoskeletal pain, and the pathognomonic Darier sign. Radiographs show diffuse osteolysis and osteosclerosis (in severe disease) mainly affecting the axial skeleton and the ends of long bones. DXA imaging can show osteopenia or osteoporosis. The diagnosis is suspected when an adult presents with one or more of the following: unexplained ulcer disease or malabsorption, radiographic or 99Tc technetium bone scan abnormalities, organomegaly, lymphadenopathy, unexplained flushing, or hypotension. Serum tryptase concentrations (Answer E) are often elevated greater than 20 ng/mL (>20 µg/L). While systemic mastocytosis can occur in the absence of cutaneous manifestations, it is an unlikely diagnosis in this patient.

Hepatitis C–associated osteosclerosis can cause generalized osteosclerosis with elevated T-scores. Other than substance abuse, she does not have any other risk factors for hepatitis C, and her clinical presentation along with normal liver function tests makes it unlikely. Therefore, assessing for reactive hepatitis C antibodies (Answer C) is not necessary.

While phalangeal involvement and enlargement along with joint pain is seen in rheumatoid arthritis (Answer D), the patient's bone scan findings are not typical of this condition. Ankylosing spondylitis—not rheumatoid arthritis—can cause artifactual elevations in spine bone mineral density due to syndesmophyte formation at the vertebral margins.

Educational Objective

Construct the differential diagnosis for osteosclerosis, including skeletal fluorosis due to inhalant abuse of a difluoroethane.

Reference(s)

Tucci JR, Whitford GM, McAlister WH, et al. Skeletal fluorosis due to inhalation abuse of a difluoroethane-containing computer cleaner. *J Bone Miner Res.* 2017;32(1):188-195. PMID: 27449958

Cook FJ, Seagrove-Guffey M, Mumm S, et al. Non-endemic skeletal fluorosis: causes and associated secondary hyperparathyroidism (case report and literature review). *Bone.* 2021;145:115839. PMID: 33418099

Fawell J, Bailey K, Chilton J, Dahi E, Fewtrell L, Magara Y. *Fluoride in Drinking-Water.* World Health Organization; 2006.

86 **ANSWER: C) Human menopausal gonadotropins (hMG) followed by hCG**

This patient has panhypopituitarism. She is an excellent candidate for ovulation induction given her normal antimullerian hormone concentration and normal pelvic anatomy (uterus and patent fallopian tubes). She is likely very fertile and will probably respond well to either intrauterine insemination or in vitro fertilization following ovulation induction. In such a patient, collaboration with colleagues in reproductive endocrinology and infertility is needed to administer and closely monitor ovulation induction, especially given her increased risk of ovarian hyperstimulation. Moreover, there are many complexities in her case because of panhypopituitarism and likely GH deficiency. Currently, there are no recommendations to replace GH to achieve conception, although GH replacement therapy may be needed for fertility in selected women with hypopituitarism.

For ovulation induction in this patient, both human menopausal gonadotropins (hMG) and hCG are needed (Answer C). hMG is a highly purified urinary preparation containing both FSH and LH, which are required for normal follicular development. LH is required to stimulate the theca cells to produce androstenedione, which is then aromatized by the granulosa cells into estradiol under the influence of FSH. This allows for follicular development and recruitment of a dominant follicle. hCG is used to treat female infertility by inducing ovulation. hCG shares an α subunit with LH and therefore the hCG pulse or surge replaces the endogenous LH surge to stimulate ovulation. Using only injections of hMG (Answer B) would help to stimulate follicle development, but without hCG, ovulation would not occur.

Clomiphene citrate (Answer A) is a selective estrogen receptor modulator that blocks the estrogen receptor at the level of the hypothalamus and leads to a compensatory increase in GnRH secretion and subsequent rise in LH and FSH. Clomiphene citrate would not be effective in a woman with panhypopituitarism.

Likewise, letrozole (Answer D) would not be effective in this patient. Letrozole is an aromatase inhibitor that works by inhibiting the cytochrome P450 isoenzymes 2A6 and 2C19 of the aromatase enzyme complex, and therefore down-regulating estradiol production. This decrease in estradiol results in a lack of negative feedback to the hypothalamus and pituitary gland, which therefore leads to stronger GnRH pulses. Elevated GnRH levels stimulate the pituitary gland to produce more FSH, which induces follicle development in the ovaries. However, in a person with either GnRH or LH/FSH deficiency, letrozole is ineffective.

Finally, given that the patient has normal ovarian reserve and wants to be the egg source and carry the pregnancy, the best next step is to try ovulation induction. If ovulation induction attempts are unsuccessful, then discussion about oocyte donation and in vitro fertilization (Answer E) would be appropriate.

Educational Objective

Manage fertility goals in a woman with panhypopituitarism.

Reference(s)

Vila G, Felseriu M. Fertility and pregnancy in women with hypopituitarism: a systematic literature review. *J Clin Endocrinol Metab.* 2020;105(3):e55-e67. PMID: 31652320

Von Hofe J, Bates GW. Ovulation induction. *Obstet Gynecol Clin North Am.* 2015;42(1):27-37. PMID: 25681838

87 ANSWER: E) Refer to behavioral psychology

Weight regain is a phenomenon that occurs in approximately 25% to 30% of patients who have undergone bariatric surgery, and it is associated with relapse of weight-related comorbidities. The prevalence of weight regain following bariatric surgery varies depending on the type of bariatric procedure performed, but it is higher after sleeve gastrectomy than after gastric bypass. Factors that predispose to the development of post–bariatric surgery weight regain include changes in dietary habits, decline in physical activity, weight-promoting medications, and alterations to mood, as described in this patient. The ideal approach to post–bariatric surgery weight regain should be multifaceted to comprehensively address dietary, behavioral, and anatomic variables.

The patient started pharmacotherapy for the management of weight regain and has experienced weight loss. However, the current concern is that she has lost a relatively large amount of weight quickly and continues to lose weight. Based on her weight trajectory, her estimated oral intake is less than 1000 kilocalories per day, which is equivalent to a very-low-calorie diet. In the long-term, very-low-calorie diets lead to a negative metabolic adaptation and lowering of the basal metabolic rate. Such diets are therefore not a preferred long-term strategy for weight management.

The patient's significant appetite suppression is a result of semaglutide. However, discontinuing semaglutide (Answer A) would lead to a rebound increase in hunger, which can be quite distressing for patients. A reasonable alternative not presented here would be to decrease the dosage of semaglutide.

An assessment of her nutrition status is important to determine next steps. However, a 24-hour urine collection for protein excretion (Answer B) would not provide data on the adequacy of protein stores, muscle mass, or function. Measuring gonadotropins and estradiol (Answer C) would not provide additional information about the etiology or sequelae of her rapid weight loss.

In this patient who has no clinical features suggestive of a distended sleeve (excessive portion sizes, not feeling full), the yield of a double contrast upper gastrointestinal fluoroscopic study (Answer D) is likely to be low.

The patient has a personal goal of losing more weight, despite the risk of malnutrition. Understanding her health goals and differentiating these from her personal goals is best addressed with a psychologist (Answer E) to manage expectations and provide optimal care.

Educational Objective

Manage rapid weight loss associated with antiobesity medications.

Reference(s)

Lucas E, Simmons O, Tchang B, Aronne L. Pharmacologic management of weight regain following bariatric surgery. *Front Endocrinol (Lausanne)*. 2022;13:1043595. PMID: 36699042

Ames GE, Koball AM, Clark MM. Behavioral interventions to attenuate driven overeating and weight regain after bariatric surgery. *Front Endocrinol (Lausanne)*. 2022;13:934680. PMID: 35923629

88 ANSWER: D) Start metyrapone

Adrenocortical carcinomas (ACCs) are rare tumors, with an incidence of 1 to 2 cases per million per year. ACC has a bimodal age distribution, with peaks occurring before age 5 years and in the fourth and fifth decades of life. In general, the disease progresses more rapidly and aggressively in adults than in children. ACC is more common in women than in men by a factor of 1.5 to 2.5.

Approximately 60% of ACC cases are associated with overt hormonal excess. In adults, 45% of ACCs overproduce cortisol only, while 25% overproduce both cortisol and androgens. Less than 10% of ACCs present with virilization alone, and even fewer patients overproduce aldosterone or estrogens.

Clinical symptoms linked to glucocorticoid excess, such as central weight gain, weakness, and insomnia, usually develop rapidly over 3 to 6 months. Patients with concurrent hypersecretion of adrenal androgens may not experience the typical catabolic effects of glucocorticoid excess such as muscle and skin atrophy. In patients with profound hypercortisolism, medical therapy should be initiated promptly to decrease the risk of glucocorticoid-related complications, such as hypercoagulability, opportunistic infections, hyperglycemia, hypertension, and hypokalemia.

Systematic data comparing steroidogenic enzyme inhibitors in ACC-associated Cushing syndrome are lacking, but expert centers primarily use metyrapone (Answer D). Metyrapone inhibits 11β-hydroxylase (CYP11B1), the enzyme that catalyzes the last step in cortisol synthesis (*figure*). Metyrapone is initiated at 250 mg 4 times daily, and the dosage can be titrated up to 6 g daily. Metyrapone therapy often mitigates cortisol overproduction within 3 to 7 days. When metyrapone alone is insufficient to control hypercortisolemia, ketoconazole may be introduced, initially at a dosage of 200 mg 3 times daily, and further titration up to 400 mg 3 times daily when necessary. Higher dosages are rarely effective. Addition of mitotane (Answer E) or mifepristone, a glucocorticoid receptor antagonist, may be necessary to achieve adequate control of cortisol excess.

Figure. Medical therapies for Cushing syndrome

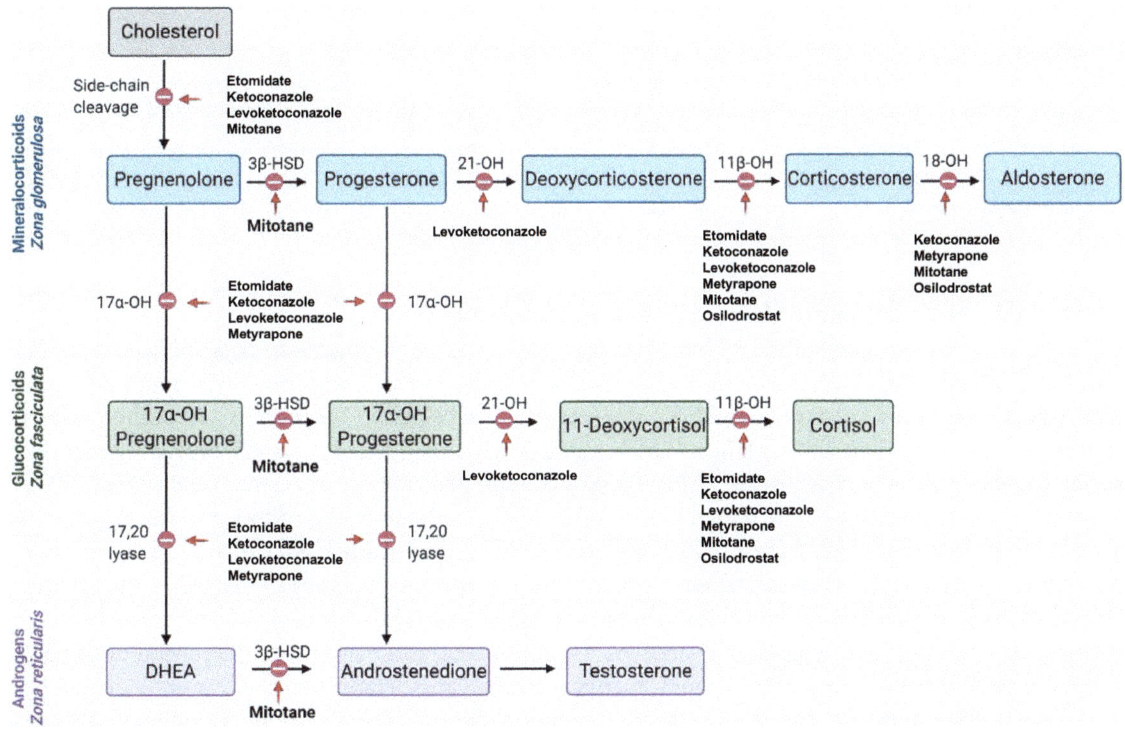

Adapted from Varlamov EV et al. *Best Pract Res Clin Endocrinol Metab*, 2021;35(1):101490.

Similar to metyrapone, osilodrostat inhibits CYP11B1. Osilodrostat was approved by the US FDA in 2020 for adults with Cushing disease who either are ineligible for pituitary surgery or have residual disease following transsphenoidal surgery. In a placebo-controlled randomized study of osilodrostat in 137 patients with Cushing disease, 86% of patients taking osilodrostat maintained a complete response. Data on adequate dosage and effectiveness of osilodrostat for cortisol-producing ACC are lacking. The choice of therapy can be influenced by drug availability in various countries.

In patients who are critically ill and unable to take oral medications, intravenous etomidate (Answer A) (which also blocks CYP11B1) can be initiated at a low, nonhypnotic dosage of 0.3 mg/kg per hour. Etomidate requires supervision in an intensive care unit and thus is only used when oral therapy is not feasible.

ACC can metastasize to the liver, lungs, lymph nodes, and bone. Chest and liver CT imaging, as well as a bone scan, are typically part of ACC staging. Although FDG PET-CT (Answer B) is more sensitive than CT or radiographic bone scans for detecting distant metastases across various clinical scenarios, it has a lower resolution than cross-sectional imaging. In this patient, medical treatment for hypercortisolism should be prioritized over ACC staging.

Surgical resection is the only potentially curative approach for localized ACC, and it is the preferred primary treatment. Right adrenalectomy (Answer C) is indicated in this patient, but only after stabilizing the acute medical problems, including hypercortisolemia.

Adjuvant therapy with mitotane (Answer E), chemotherapy, and/or radiotherapy is considered for unresectable or recurrent ACC, following multidisciplinary considerations of risks vs benefits for each patient. Mitotane has adrenolytic effects via mitochondrial destruction and adrenocortical cell necrosis. In addition, mitotane inhibits steroidogenesis by blocking cholesterol side-chain cleavage (CYP11A1) and 11β-hydroxylase (CYP11B1). Close clinical and hormonal monitoring of hypercortisolism is needed during therapy with mitotane. In patients with nonfunctional ACC, hydrocortisone is initiated along with mitotane in anticipation that adrenal insufficiency will develop. In patients with cortisol-producing ACC, hydrocortisone therapy is delayed until hypercortisolism is controlled. Mitotane amplifies hepatic cytochrome P450 enzymatic metabolism, and its use mandates higher than physiologic glucocorticoid replacement dosages.

Educational Objective

Recommend medical therapy options for cortisol-producing adrenal cortical carcinoma.

Reference(s)

Fassnacht M, Dekkers O, Else T, et al. European Society of Endocrinology clinical practice guidelines on the management of adrenocortical carcinoma in adults, in collaboration with the European Network for the Study of Adrenal Tumors. *Eur J Endocrinol.* 2018;179(4):G1-G46. PMID: 30299884

Berruti A, Fassnacht M, Libe R, et al. First randomized trial on adjuvant mitotane in adrenocortical carcinoma patients: the Adjuvo study. *J Clin Oncol.* 2022;40:(6).

Varlamov EV, Han AJ, Felseriu M. Updates in adrenal steroidogenesis inhibitors for Cushing's syndrome – a practical guide. *Best Pract Res Clin Endocrinol Metab.* 2021;35(1):101490. PMID: 3370708

89 **ANSWER: D) Resistance to thyroid hormone (β subtype)**

This patient has a constellation of symptoms that are typical of resistance to thyroid hormone (RTH) (Answer D). First described in 1967, the condition is usually inherited in an autosomal dominant pattern with an estimated prevalence of up to 1 in 40,000, and both sexes are affected equally. More than 240 pathogenic variants have been described in more than 1000 families (>3000 cases worldwide). Most cases are caused by pathogenic variants in the thyroid hormone receptor β (*THRB)* gene, although additional pathogenic variants in *THRA* have also been identified over the past decade, which has broadened the understanding of this condition. Affected individuals have impaired sensitivity to thyroid hormone and characteristically have elevated serum T_3 and T_4 levels and normal or elevated TSH levels. As peripheral thyroid hormone levels are increased in the setting of an abnormal thyroid hormone receptor, clinical manifestations range from those of thyroid hormone deficiency and thyroid hormone excess, although many individuals are asymptomatic and may not be diagnosed until adulthood. The constellation of features in this patient (goiter, tachycardia, hearing deficit, hyperactive behavior, depression, mild developmental delay, tremors, and brisk reflexes) is a hallmark of RTHβ. There is evidence of increased cardiovascular morbidity risk and decreased survival in affected individuals. No specific treatment is currently available to fully reverse the RTHβ defect, and many patients may not require any treatment at all, as the resistance to thyroid hormone may be appropriately matched by compensatory increased T_3 and T_4 levels.

Usual management consists of an individualized approach to the patient's symptoms resulting either from tissue thyroid hormone excess or deficiency. Symptoms of thyroid hormone excess such as tachycardia may be addressed by β-adrenergic blockade, while goiter and cognitive issues can be treated with intermittent high-dosage liothyronine or 3,5,3'-triiodothyroacetic acid (Triac), a thyroid hormone analogue with thyromimetic effects in the pituitary and liver that has the ability to suppress TSH levels. Symptoms of thyroid hormone deficiency such as delayed growth and neurodevelopment can be treated with supraphysiologic doses of levothyroxine. There is no role for thyroidectomy or radioactive iodine in patients with RTHβ.

RTH may mimic a TSH-secreting pituitary adenoma (Answer E), particularly in the absence of relevant family history. Patients with a TSHoma may be distinguished by an often elevated serum pituitary α-subunit concentration and the inability to achieve TSH suppression after receiving supraphysiologic doses of T_3.

In contrast, familial dysalbuminemic hyperthyroxinemia (Answer A) is an autosomal dominant condition in which there is increased affinity of T_4 to an abnormal albumin molecule (due to gain-of-function variants in the albumin gene [*ALB*]), resulting in its increased affinity to T_4 but not T_3. Affected individuals have elevated serum free and total T_4 concentrations that are not associated with clinical symptoms. Patients with familial dysalbuminemic hyperthyroxinemia may be mistaken for being hyperthyroid, but they are asymptomatic and do not require treatment.

This patient's thyroid panel is not consistent with Graves disease, which would usually show suppressed serum TSH levels, elevated peripheral thyroid hormone levels, and positive thyroid antibody titers.

The presence of heterophile antibodies to various components of the thyroid laboratory panel such as T_3, T_4, and TSH (Answers B and C) can falsely increase thyroid function measurements. Heterophilic antibodies are endogenous serum antibodies that can develop from mice and mice product exposures, autoimmune diseases, dialysis, maternal transfer during pregnancy, or receipt of immunizations and blood transfusions. Their interference with immunoassays usually results in falsely elevated values of the analyte through their bridging of the capture and labeled antibodies of the assay platform independently of an antigen. However, the spurious laboratory results would be due to assay interference and would not produce any clinical symptoms as was seen in this patient.

Educational Objective

Identify the clinical and biochemical features of thyroid hormone resistance.

Reference(s)

Pappa T, Refetoff S. Resistance to thyroid hormone beta: a focused review. *Front Endocrinol (Lausanne).* 2021;12:656551. PMID: 33868182

Brent GA. Mechanisms of thyroid hormone action. *J Clin Invest.* 2012;122(9):3035-3043. PMID: 22945636

Tagami T. An overview of thyroid function tests in subjects with resistance to thyroid hormone and related disorders. *Endocr J.* 2021;68(5):509-517. PMID: 33827995

Okosieme OE, Usman D, Taylor PN, et al. Cardiovascular morbidity and mortality in patients in Wales, UK with resistance to thyroid hormone β (RTHβ): a linked-record cohort study. *Lancet Diabetes Endocrinol.* 2023;11(9):657-666. PMID: 37475119

Favresse J, Burlacu MC, Maiter D, Gruson D. Interferences with thyroid function immunoassays: clinical implications and detection algorithm. *Endocr Rev.* 2018;39(5):830-850. PMID: 29982406

90 ANSWER: B) Add metformin

Insulin is the preferred medication for treating hyperglycemia in patients with gestational diabetes. However, there are some people with gestational diabetes requiring medical therapy who may not be able to use insulin safely or effectively during pregnancy due to cost, language barriers, comprehension, personal choice, or cultural influences. Oral agents may be an alternative for these individuals after discussing the known risks and the need for more long-term safety data in offspring. Use of metformin (Answer B) (with or without insulin) has been shown to have similar perinatal complications as those associated with use of insulin by itself. The risks of macrosomia (odds ratio, 0.59 [95% CI, 0.46-0.77]; $P < .001$) and large-for-gestational age (odds ratio, 0.78 [95% CI, 0.62-0.99]; $P = .04$) are lower following maternal treatment with metformin than the risks associated with insulin. Despite lower average birth weight, metformin-exposed children appear to experience accelerated postnatal growth, resulting in heavier infants and higher BMI by mid-childhood compared with BMI of children whose mothers were treated with insulin. In mid-childhood, BMI is significantly higher (mean difference, 0.78 kg/m^2 [95% CI, 0.23-1.33]; $I_2 = 7\%$; $P = .005$) following metformin exposure than following insulin exposure, although the difference in absolute weight between the groups is not significantly different ($P = .09$). Tolerating metformin is sometimes an issue because of gastrointestinal adverse effects, and there is uncertainty whether metformin is associated with an increased risk of premature birth. Metformin monotherapy cannot usually attain the strict glycemic targets recommended during pregnancy. However, given this patient's situation (limitations taking subcutaneous therapy because of fear of needles), some improvement gained by metformin would still benefit maternal and fetal outcomes. When discussing pharmacotherapy, it is important to address the risks of uncontrolled gestational diabetes vs the known risks of metformin use.

Glipizide (Answer A) is not approved for use in the management of gestational diabetes. Sulfonylureas are known to cross the placenta and have been associated with increased neonatal hypoglycemia. However, glyburide, another sulfonylurea, can be used in pregnancy. Glyburide can increase the risk of hypoglycemia and has been associated with multiple adverse neonatal outcomes (neonatal intensive care unit admissions, large-for-gestational-age infants, hypoglycemia, respiratory distress, and birth injury) vs outcomes with insulin therapy.

Pharmacotherapy for the management of gestational diabetes is indicated if blood glucose values are persistently elevated despite optimization of lifestyle/medical nutrition therapy. The American Diabetes Association and the American College of Obstetricians and Gynecologists recommend the following targets for glucose values in patients with gestational diabetes (while acknowledging that these thresholds have been extrapolated from recommendations proposed for pregnant patients with preexisting diabetes):

- Fasting and preprandial blood glucose: <95 mg/dL (<5.3 mmol/L)
- One-hour postprandial blood glucose: <140 mg/dL (<7.8 mmol/L)
- Two-hour postprandial glucose: <120 mg/dL (<6.7 mmol/L)

Semaglutide (Answer C) is not approved for use in pregnancy because it is unknown whether it has teratogenic effects. This is true for all medications in the GLP-1 receptor agonist class. Patients on semaglutide (either taken for the management of type 2 diabetes and/or for the management of obesity) should discontinue the medication 8 weeks before attempting to conceive.

Although insulin pump therapy (Answer D) and subcutaneous insulin therapy (Answer E) are reasonable options, these may not be helpful for this patient who has a fear of needles.

Educational Objective

Counsel patients about the role of noninsulin pharmacotherapy in the management of gestational diabetes.

Reference(s)

American Diabetes Association Professional Practice Committee. 15. Management of diabetes in pregnancy: standards of care in diabetes-2024. *Diabetes Care.* 2024;47(Suppl 1):S282-S294). PMID: 38078583

Tarry-Adkins JL, Aiken CE, Ozanne SE. Neonatal, infant, and childhood growth following metformin versus insulin treatment for gestational diabetes: a systematic review and meta-analysis. *PLoS Med.* 2019;16(8):e1002848. PMID: 31386659

Gui J, Liu Q, Feng L. Metformin vs insulin in the management of gestational diabetes: a meta-analysis. *PLoS One.* 2013;8(5):e64585. PMID: 23724063

Rowan JA, Hague WM, Gao W, Battin MR, Moore MP; MiG Trial Investigators. Metformin versus insulin for the treatment of gestational diabetes. *N Engl J Med.* 2008;358(19):2003-2015. PMID: 18463376

91 ANSWER: E) Treat with denosumab

This patient has hypercalcemia of malignancy, which is associated with high morbidity and mortality. This condition may respond to effective treatment of the underlying malignancy, but affected patients generally require intervention with medication to lower calcium, at least in the interim.

This man has hypoxia, congestive heart failure, and volume issues that limit his ability to tolerate an increased intravenous fluid rate (Answer A). Adding a loop diuretic in this situation would be appropriate to aid with fluid management, but it would not primarily address serum calcium. Loop diuretics do not effectively manage hypercalcemia.

Calcitonin (Answer C) has a modest and only very brief effect on the serum calcium concentration, lasting only 24 to 48 hours despite continued dosing. This patient needs a treatment with more sustained response.

He has suppressed PTH but elevated PTHrP, and hypercalcemia is due to malignancy. Cinacalcet (Answer D) is useful to add to manage hypercalcemia in the setting of parathyroid carcinoma or primary hyperparathyroidism, but it would not be expected to have a significant effect on hypercalcemia mediated by PTHrP or other causes.

In 2023, the Endocrine Society published a new clinical practice guideline for treatment of hypercalcemia of malignancy in which they make a strong recommendation to treat with either intravenous bisphosphonates (Answer B) or denosumab (Answer E). Both agents potently inhibit osteoclastic bone resorption. While hypercalcemia often responds well to intravenous fluids followed by intravenous bisphosphonates, some patients with cancer have persistent hypercalcemia despite bisphosphonates, or require frequent doses. Zoledronate affects serum calcium in 48 to 72 hours, but a week may be needed to determine whether the target calcium concentration can be reached. After that, additional doses may be given if the calcium concentration is above target. This patient has persistent significant hypercalcemia (albumin-corrected calcium of 13.1 mg/dL [3.3 mmol/L]) after 2 doses of zoledronate administered 10 days apart, along with worse kidney function compared with baseline function. There is also potential risk for worsening kidney function with doses of zoledronate, which may be transient or sustained, favoring denosumab (Answer E) in the setting of chronic kidney disease for hypercalcemia. The guidelines provide a conditional recommendation to treat patients with refractory hypercalcemia with denosumab. For hypercalcemia, the denosumab dose is 120 mg. Denosumab lowers serum calcium 3 to 10 days after dosing but may take 3 weeks for complete response. Additional considerations with denosumab are the higher cost compared with zoledronate and the potential for rebound increased bone resorption requiring additional dosing.

Educational Objective

Treat hypercalcemia of malignancy refractory to intravenous bisphosphonates.

Reference(s)

El-Hajj Fuleihan G, Clines GA, Hu MI, et al. Treatment of hypercalcemia of malignancy in adults: an Endocrine Society clinical practice guideline. *J Clin Endocrinol Metab.* 2023;108(3):507-528. PMID: 36545746

92 **ANSWER: A) Condoms for vaginal intercourse with a sex partner with a penis**
Contraception for transgender patients should be approached with the same goals as those for cisgender patients: (1) contraceptive effectiveness; (2) patient preference; (3) minimization of adverse effects; (4) potential noncontraceptive benefits; and (5) future family planning preferences. For this polyamorous patient who has receptive vaginal intercourse with multiple sex partners, the contraceptive plan should include barrier methods that prevent sexually transmitted infections. When used correctly, condoms (Answer A) offer effective contraception and protection against sexually transmitted infections. Although condoms are effective when used correctly (98% effectiveness during 1 year of use), there are misconceptions about their use that result in 13% failure rates during the first year of use as sole contraception. It is important to counsel the patient that the condom must be placed before there is any vaginal-penile contact and that they cannot be re-used.

Although oral micronized progesterone (Answer C) likely has contraceptive effects, it has not been systematically studied in large contraceptive trials. Furthermore, it offers no protection against sexually transmitted infections.

Subcutaneous progestin implants (Answer D), progestin injections (eg, depot medroxyprogesterone injections), and progestin-eluting intrauterine devices offer effective contraception, may decrease irregular menstrual bleeding, and have additive androgenic effects to testosterone therapy for transmen However, these progestin-containing contraceptive options do not offer protection against sexual transmitted infections and possibly may have a permissive role for HIV infection. Furthermore, progestin implants and intrauterine devices provide contraception for at least 3 to 5 years and would be a poor choice for a patient who might wish to conceive in 2 to 3 years; these long-acting reversible contraceptives (LARCs) require removal at least 3 months before the patient wishes to start conceiving.

Testosterone therapy typically causes amenorrhea in transmen within 12 months after initiation. However, testosterone therapy does not reliably induce anovulation, and there have been case reports of unplanned pregnancies in transmen taking testosterone therapy (thus, Answer E is incorrect). Suppression of ovulation likely correlates with duration of testosterone use. An effective contraceptive must be used while he is sexually active and does not plan to have children. It is estimated that 40% or more of transmen do not use effective contraception while taking testosterone, and many of them believe that testosterone-induced amenorrhea is evidence of contraceptive effectiveness. There is no evidence to indicate that long-term testosterone therapy affects long-term fertility. It is likely that—similar to ciswomen—fertility is affected by aging for transmen, but it is not affected by hormone-based suppression of the hypothalamic-pituitary-ovarian axis. This patient is taking a testosterone dosage that is producing testosterone concentrations in the mid-normal range (at the mid-interval of injection), and there is no evidence to indicate that there would be greater contraceptive or virilizing benefit with a supraphysiological dosage of testosterone (Answer B). Testosterone therapy may be monitored and adjusted based on testosterone concentration measurements, similarly to cismen (eg, midinterval for injections and 4 to 6 hours after application for transdermal formulations).

This patient should also be informed of emergency contraceptive options (ulipristal or levonorgestrel).

Educational Objective

Recommend an effective contraceptive for transmen with testosterone-induced amenorrhea based on their sexual activity and plans for potential conception.

Reference(s)

Hembree WC, Cohen-Kettenis PT, Gooren L, et al. Endocrine treatment of gender-dysphoric/gender-incongruent persons: an Endocrine Society clinical practice guideline. *J Clin Endocrinol Metab.* 2017;102(11):3869-3903. PMID: 28945902

Mancini I, Alvisi S, Gava G, Seracchioli R, Meriggiola MC. Contraception across transgender. *Int J Impot Res.* 2020;33(7):710-719. PMID: 33558672

Rowlands S, Amy J-J. Preserving the reproductive potential of transgender and intersex people. *Eur J Contracept Reprod Health Care.* 2018;23(1):58-63. PMID: 29323576

Sundaram A, Vaughan B, Kost K, et al. Contraceptive failure in the United States: estimates from the 2006-2010 National Survey of Family Growth. *Perspect Sex Reprod Health.* 2017;49(1):7-16. PMID: 28245088

Taub RL, Ellis SA, Neal-Perry G, Magaret AS, Prager SW, Micks EA. The effect of testosterone on ovulatory function in transmasculine individuals. *Am J Obstet Gynecol.* 2020;223(2):229.e1-229.e8. PMID: 32044312

93 **ANSWER: E) Start octreotide, 50 mcg 3 times daily**

The differential diagnosis of hyperthyroidism with an inappropriately high TSH concentration includes TSH-secreting pituitary adenoma (thyrotropinoma); thyroid hormone resistance (caused most frequently by an autosomal dominant pathogenic variant in the thyroid hormone β receptor gene); and spurious elevation of thyroid hormone levels in the context of assay methodological interference (familial dysalbuminemic hyperthyroxinemia, T_4 antibodies, or heterophilic antibodies). History and laboratory evaluation are important in evaluating patients with a biochemical picture of central hyperthyroidism. In patients with a thyrotropinoma, TSH concentrations range between normal and extremely high due to variable biological activity and do not respond to thyrotropin-releasing hormone stimulation. Thyrotropin-releasing hormone cannot be used for diagnosis because it is not currently available in the United States. α-subunit is elevated in 50% to 85% cases; also in this setting, the α-subunit-to-TSH ratio is usually greater than 1 (α-subunit [μg/L] / TSH [mIU/L] × 10). The result must be interpreted with caution in women because α-subunit levels fluctuate during the menstrual cycle and increase at menopause. Also, the α-subunit range increases with age. Pituitary imaging most frequently shows a macroadenoma.

This patient's TSH, free T_4, and free T_3 concentrations normalized after a 3-week course of subcutaneous octreotide. He was interested in achieving definitive remission and subsequently underwent surgery. Histopathological examination confirmed the diagnosis. Most tumor cells stained positive for TSH, and a small cell population stained positive for GH. The tumor also diffusely and strongly stained positive for pituitary-specific transcription factor 1 (Pit-1). Co-staining for other Pit-1–derived pituitary hormones is not unusual in patients with thyrotropinomas, although it is less frequently clinically or biochemically apparent. In a meta-analysis, surgical removal of the thyrotropinoma was shown to resolve hyperthyroidism in 66% of patients. Postoperatively, biochemical remission is apparent in the first few days when both TSH and thyroid hormone levels decrease. The somatostatin receptor ligands octreotide (Answer E) or lanreotide can be used as primary or adjuvant treatment postoperatively, with control of central hyperthyroidism in 75% of patients. Biochemical response to short- or long-acting somatostatin receptor ligand therapy is usually evident within a month, which makes a somatostatin receptor ligand useful as a preoperative diagnostic trial. More than half of patients treated with a somatostatin receptor ligand as primary therapy also experience reduction in tumor size. The favorable effects of somatostatin receptor ligands in treating thyrotropinomas are explained by the high tumoral expression of type 2 somatostatin receptors. Some tumors express dopamine receptors, and cases of thyrotropinomas with biochemical control achieved by bromocriptine and cabergoline have been reported. Postoperative pituitary radiotherapy is effective in patients with postoperative residual tumor, but this treatment carries a risk of hypopituitarism.

Genetics referral to evaluate for thyroid receptor β pathogenic variants (Answer A) is not the best next step in this patient's management, as he has relatively recent onset of hyperthyroid clinical manifestations and no family history of a thyroid disorder. De novo pathogenic variants in the thyroid hormone receptor genes and rare cases of thyrotropinomas developing in patients with such variants have been reported. Therefore, genetic testing would be indicated if an octreotide therapeutic trial is ineffective.

Repeating thyroid function tests (Answer C) is incorrect because the patient has symptomatic hyperthyroidism requiring treatment.

Starting methimazole (Answer D) is incorrect because methimazole is not a primary therapy for thyrotropinomas. In fact, long-term use of antithyroid therapies can promote the tumor growth. Antithyroid medications can be used postoperatively or after radiation if thyroid hormones remain elevated, with careful monitoring of pituitary imaging.

Neurosurgery for tumor removal (Answer B) is incorrect because the diagnosis of thyrotropinoma has not been clearly demonstrated and the imaging abnormality could represent a nonfunctioning incidentaloma. In addition, the patient's thyroid status requires optimization before surgery.

Educational Objective

Recommend therapeutic options for patients with TSH-secreting adenomas.

Reference(s)

Socin HV, Chanson P, Delemer B, et al. The changing spectrum of TSH-secreting pituitary adenomas: diagnosis and management in 43 patients. *Eur J Endocrinol.* 2003;148(4):433-442. PMID: 12656664

Cossu G, Daniel RT, Pierzchala K, et al. Thyrotropin-secreting pituitary adenomas: a systematic review and meta-analysis of postoperative outcomes and management. *Pituitary.* 2019;22(1):79-88. PMID: 30390277

Illous F, Chanson P, Sonnet E, et al. Somatostatin receptor ligands induce TSH deficiency in thyrotropin-secreting pituitary adenoma. *Eur J Endocrinol.* 2021;184(1):1-8. PMID: 33112257

Fukuhara N, Horiguchi K, Nishioka H, et al. Short-term preoperative octreotide treatment for TSH-secreting pituitary adenoma. *Endocr J.* 2015;62(1):21-27. PMID: 25273395

Azzalin A, Appin CL, Schniederjan MJ, et al. Comprehensive evaluation of thyrotropinomas: single-center 20-year experience. *Pituitary.* 2016;19(2):183-193. PMID: 26689573

94 ANSWER: D) Ophthalmology in anticipation of worsening retinopathy in pregnancy

A goal hemoglobin A_{1c} value less than 6.5% (<48 mmol/mol) is recommended before conception because risk of diabetic embryopathy is directly proportional to hemoglobin A_{1c} elevation, especially during the first 10 weeks of pregnancy. Reliable contraception should be used until the goal hemoglobin A_{1c} is achieved.

Lifestyle and behavioral management are important in the care of pregnant women with diabetes and will aid in achieving this patient's preconception glycemic goal and pregnancy targets after she conceives. While referral to a nutritionist is important in her overall care, the goal of weight management (Answer B) will not help her currently because she has a normal BMI. In addition, she has not attempted to conceive, so she does not meet criteria for infertility. Infertility is diagnosed if pregnancy is not achieved after 12 months of regular, unprotected intercourse. While diabetes increases infertility rates, this patient has regular menses most months, suggesting normal ovulation. Eighty-five percent of women conceive in the first 12 months of attempting pregnancy. Referral to reproductive endocrinology now (Answer E) is premature.

The presence of microvascular complications (nephropathy and retinopathy) is associated with adverse pregnancy outcomes. Diabetic retinopathy is known to worsen during pregnancy, and closer monitoring is required. In addition, rapid correction of hyperglycemia can worsen retinopathy, again requiring very close monitoring. In a recent meta-analysis, the pooled progression rate per 100 pregnancies for new diabetic retinopathy was 15.0 (95% CI, 9.9-20.8), worsened nonproliferative diabetic retinopathy was 31.0 (95% CI, 23.2-39.2), progression from nonproliferative diabetic retinopathy to proliferative diabetic retinopathy was 6.3 (95% CI, 3.3-10.0), and worsened proliferative diabetic retinopathy was 37.0 (95% CI, 21.2-54.0). Given her known history of nonproliferative diabetic retinopathy and the fact that her last eye exam was 18 months ago, the best next step is referral to ophthalmology (Answer D). Regarding preexisting diabetic nephropathy (as defined by ICD codes), there are insufficient data on significantly worsening kidney disease. In this case, she has moderately increased albuminuria, which is associated with increased risk of preeclampsia but no report of worsening albuminuria independent of preeclampsia or hypertension. Referral to nephrology (Answer C) is not needed for her mild albuminuria.

Currently, there is only one hybrid closed-loop system that has been studied in pregnant women with type 1 diabetes. Women using the hybrid closed-loop system had more time-in-range (68.2% vs 55.6%), lower time-above-range, and minimal hypoglycemia. At this time, this system is only available in the United Kingdom (Cambridge model predictive control algorithm using the Dana Diabecare RS insulin pump and the Dexcom G6 continuous glucose monitor). There are no currently approved hybrid closed-loop systems for use in pregnancy in the United States pregnancy due to lack of customizable glucose targets. Women who use a hybrid closed-loop system can discontinue the closed-loop system and use the insulin pump device to deliver insulin. Switching to multiple daily injections (Answer A) is not necessary. Although many of the current hybrid closed-loop systems are not designed for pregnancy, some women choose to continue using their hybrid closed-loop systems during pregnancy (off-label use) because they note better glycemic control when using the system than when they do not have closed-loop functionality. The choice of whether to use a hybrid closed-loop system during pregnancy should be personalized. Use of continuous glucose monitoring during pregnancy has been studied, and it is associated with improved glycemic control with no difference if they were on insulin pump or multiple daily injections. In addition, women who use continuous glucose monitoring during pregnancy have less neonatal hypoglycemia and lower risk for large-for-gestational-age babies than women who monitor their capillary glucose via fingerstick testing (CONCEPTT trial). This patient should continue her hybrid closed-loop system until conception and then switch to just the insulin pump with continuous glucose monitoring or to a hybrid closed-loop system that allows for pregnancy glucose target modifications.

Educational Objective

Recommend screening for worsening retinopathy in pregnancy.

Reference(s)

Relph S, Patel T, Delaney L, Sobhy S, Thangaratinam S. Adverse pregnancy outcomes in women with diabetes-related microvascular disease and risks of disease progression in pregnancy: a systematic review and meta-analysis. *PLoS Med.* 2021;18(11):e1003856. PMID: 34807920

Widyaputri F, Rogers SL, Kandasamy R, Shub A, Symons RCA, Lim LL. Global estimates of diabetic retinopathy prevalence and progression in pregnant women with preexisting diabetes: a systematic review and meta-analysis. *JAMA Ophthalmol.* 2022;140(5):486-494. PMID: 35357410

Lee TTM, Collett C, Bergford S, et al; AiDAPT Collaborative Group. Automated insulin delivery in women with pregnancy complicated by type 1 diabetes. *N Engl J Med.* 2023;389(17):1566-1578. PMID: 37796241

Feig DS, Donovan LE, Corcoy R, et al; CONCEPTT Collaborative Group. Continuous glucose monitoring in pregnant women with type 1 diabetes (CONCEPTT): a multicentre international randomised controlled trial. *Lancet.* 2017;390(10110):2347-2359. PMID: 28923465

95 **ANSWER: B) Resect 2 remaining parathyroid glands with forearm autotransplant of one-half gland**

Hypercalcemia is a well-known complication of lithium therapy with an incidence between 8% and 15%. The risk of lithium-induced hypercalcemia increases with treatment duration. Lithium has a direct effect on calcium-regulated PTH release. Lithium stimulates the growth of the parathyroid cells through the Wnt pathway and can cause both parathyroid hyperplasia and multiple adenomas. Since multigland involvement is common with lithium, recurrence risk following surgery is deemed to be higher than with parathyroidectomy for primary hyperparathyroidism. While the hypercalcemia caused by lithium can be reversible, with longer exposure to the medication the chances of parathyroid function normalizing are low. This patient has been taking lithium for 20 years and therefore reversibility is unlikely. In addition, he has evidence of parathyroid adenomas, which decreases the likelihood of response to lithium discontinuation. Stopping lithium and remeasuring calcium in 6 months (Answer E) is incorrect.

Since this patient is young and lithium therapy is likely to be continued, another neck surgery with removal of the remaining 2 parathyroid glands and autotransplant of one-half of 1 gland into the forearm (Answer B) is the most appropriate management strategy. The goal of a repeated surgery in this man is surgical cure. During his initial surgery, 2 parathyroid glands were removed. Multiple adenomas are not uncommon in lithium-induced hypercalcemia. Due to long duration of lithium exposure, the remaining glands could have undergone hyperplasia, possibly explaining why the repeated sestamibi study was negative.

His 24-hour urinary calcium concentration (<100 mg/24 h [<2.5 mmol/d]) raises concern for familial hypocalciuric hypercalcemia (FHH). Lithium enhances kidney absorption of calcium from the thick limb of the loop of Henle. Due to this additional effect on renal tubule calcium handling, urinary calcium excretion can biochemically mimic that which is observed with FHH. Given the patient's long-term treatment with lithium, lack of relevant family history, recently noted hypercalcemia, and 2 parathyroid adenomas removed during initial surgery, the likelihood of FHH is low and genetic testing for *CASR* pathogenic variants (Answer A) is not needed.

Cinacalcet (Answer C) is a calcimimetic agent that acts on the calcium-sensing receptors and reduces PTH secretion by increasing the sensitivity of the calcium-sensing receptors on parathyroid cells. Cinacalcet lowers calcium and PTH concentrations. Commonly reported adverse effects include nausea, vomiting, and muscle weakness. Evidence of long-term efficacy of cinacalcet in reducing calcium and PTH compared with effectiveness of parathyroidectomy has not been demonstrated. Another consideration is cinacalcet's lack of effect on bone mineral density, unlike what is seen after surgical parathyroidectomy. Lastly, long-term treatment with cinacalcet may not be cost-effective. While there is a role for cinacalcet after surgery for hyperparathyroidism, including lithium-induced hyperparathyroidism, the choice in this case is surgery with a goal for cure.

Stopping dietary calcium (Answer D) would not be beneficial. About 30% to 40% of dietary calcium is absorbed in the intestinal tract. Calcium acts on the calcium-sensing receptor and inhibits parathyroid proliferation. Low calcium intake can stimulate PTH production and secretion. Data from the Nurses' Health Study show that increased calcium intake (>500 mg daily) is independently associated with reduced risk of primary hyperparathyroidism. Similarly, data in men and women have shown that dietary calcium intake reduces the risk of kidney stones. While a similar study is not available in individuals who have lithium-induced hypercalcemia, the data from the Nurses' Health Study can be extrapolated here for guidance.

Educational Objective

Evaluate and manage lithium-induced hyperparathyroidism.

Reference(s)

Paik JM, Curhan GC, Taylor EN. Calcium intake and risk of primary hyperparathyroidism in women: prospective cohort study. *BMJ.* 2012;345:e6390. PMID: 23080543

Broome JT, Solorzano CC. Lithium use and primary hyperparathyroidism. *Endocr Pract.* 2011;17(Suppl 1):31-35. PMID: 21247849

96 **ANSWER: B) Primary thyroid cancer arising from a thyroglossal duct cyst**

This patient has a remote history of low-risk microscopic papillary thyroid cancer (PTC) that was detected incidentally during a preoperative workup for primary hyperparathyroidism. Based on pathology alone, she would have been considered to be at low risk for structural recurrence according to dynamic risk stratification using the American Thyroid Association's 2015 system. However, postoperative serum thyroglobulin was never measured in this patient, even though this would have been standard practice following total thyroidectomy for thyroid cancer. Had postoperative serum thyroglobulin been measured, it may have yielded an inappropriately elevated level, prompting further workup at that time. Following her initial surgery, the patient did not have close follow-up for a few years, but she eventually completed laboratory testing with a new primary care physician that revealed a markedly elevated serum thyroglobulin value. The elevated serum thyroglobulin by immunoassay was subsequently confirmed by liquid chromatography/tandem mass spectroscopy, which rules out heterophile antibody interference (Answer A) as a potential cause.

Neck ultrasonography (*see images*) revealed a mixed cystic and solid mass with lobular borders and macroscopic calcifications, sonographic features that, taken together, are suspicious for malignancy. Ultrasonography also demonstrated a left zone 3 (lateral neck) lymph node (*left panel*) with loss of fatty hilum, rounded shape, and a few punctate echogenic foci (*yellow arrows*), indicative of cervical lymph node metastasis. For comparison, a normal cervical lymph node (*right panel*) with intact fatty hilum (*blue arrow*) is also shown.

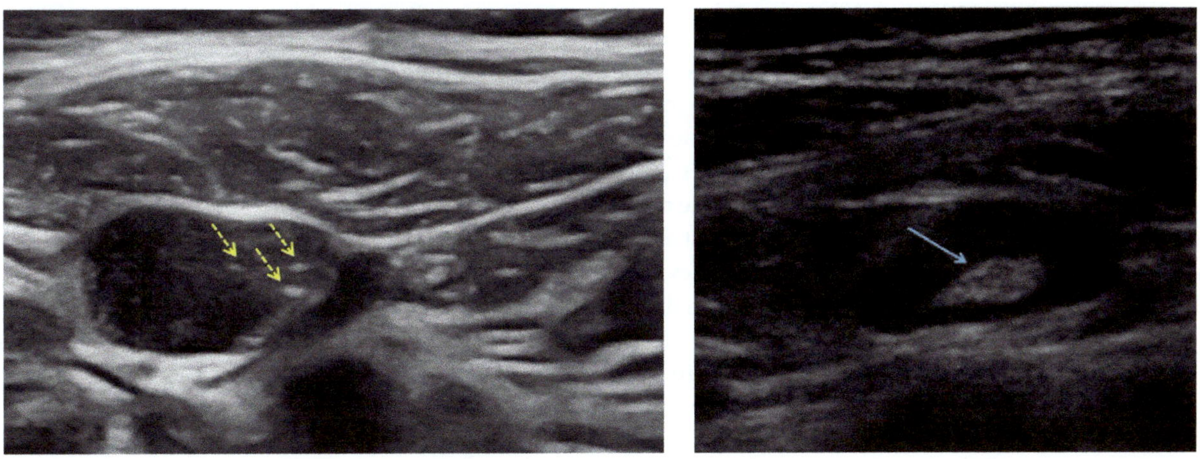

Based on ultrasonography images alone, discerning the exact location of the suspicious cystic and solid neck mass is difficult, and thus recurrent PTC in cervical lymph nodes (Answer C) is possible. However, neck CT with contrast indicates that an alternative diagnosis is more likely. The CT images show that the mass is in the upper midline neck between the hyoid bone and cricoid cartilage, a location that is highly characteristic of a thyroglossal duct cyst. Furthermore, CT shows likely invasion of the mass into the hyoid bone and laryngeal soft tissues anterior to the epiglottis. These imaging findings plus the markedly elevated serum thyroglobulin concentration and suspicious left lateral neck lymph node seen on ultrasonography are consistent with a diagnosis of primary thyroid cancer arising from a thyroglossal duct cyst (Answer B), rather than a benign thyroglossal duct cyst (Answer D). The imaging findings are not consistent with thyroid remnant tissue (Answer E), as the expected location of remnant tissue would be in the thyroid bed (lower neck) and should resemble benign thyroid parenchyma on ultrasonography.

Malignancy within thyroglossal duct cysts is rare, occurring in only 1% to 2% of cases. PTC is the most frequent cancer type diagnosed in this setting. Findings on imaging that suggest the presence of thyroid cancer within a thyroglossal duct cyst include a mural solid component within the cyst, irregular margins of the solid component, calcifications on ultrasonography or CT, invasion of nearby anatomical structures, increased FDG uptake on PET/CT, and locoregional cervical lymphadenopathy.

Through FNA of the midline upper neck mass and left zone 3 lymph node, this patient was confirmed to have a second primary PTC within a thyroglossal duct cyst and left lateral neck lymph node metastasis. She subsequently underwent a Sistrunk procedure, an operation that removes the entire thyroglossal duct cyst and the central portion of the hyoid bone, and a left lateral neck dissection. Intraoperatively, the thyroid cancer was found to have invaded the hyoid bone and right lamina of the thyroid cartilage, the base of tongue musculature, and the right anterior and posterior bellies of the digastric muscle. The patient's surgical pathology revealed a 3-cm PTC infiltrating skeletal muscle and bone. Metastatic PTC was documented in 5 of 44 resected lymph nodes, the largest measuring 2 cm without extranodal extension. A benign pyramidal lobe was also resected. Unfortunately, this patient's serum thyroglobulin reached a nadir of only 103 ng/mL (103 µg/L) following surgery, and she was subsequently diagnosed with distant metastatic disease on whole-body imaging.

Educational Objective

Diagnose papillary thyroid cancer arising from thyroglossal duct remnant tissue.

Reference(s)

Haugen BR, Alexander EK, Bible KC, et al. 2015 American Thyroid Association management guidelines for adult patients with thyroid nodules and differentiated thyroid cancer: the American Thyroid Association Guidelines Task Force on Thyroid Nodules and Differentiated Thyroid Cancer. *Thyroid.* 2016;26(1):1-133. PMID: 26462967

Hou T, Liu Z, Gan Q, Debnam JM, Kirshnamurthy S. Clinical and cytopathological features of suspected thyroglossal duct cysts and neoplasms arising from them: a large series from a referral cancer center. *Cancer Cytopathol.* 2022;130(1):72-79. PMID: 34529338

Iftikhar H, Ikram M, Rizwan Nathani K, Yar Muhammad A. Papillary thyroid carcinoma within thyroglossal duct cyst: case series and literature review. *Int Arch Otorhinolaryngol.* 2018;22(3):253-255. PMID: 29983764

97 **ANSWER: D) Replace over-the-counter fish oil with icosapent ethyl, 4 g daily**

Type 2 diabetes is associated with atherogenic dyslipidemia characterized by seemingly normal total and LDL-cholesterol levels, elevated triglycerides, and low HDL-cholesterol levels. In addition, there is preponderance of small, dense LDL particles that likely contributes to increased atherogenicity. Since insulin resistance is the primary driver of these lipid abnormalities, lifestyle changes, including caloric restriction and regular physical exercise to maintain ideal body weight, are the first steps in management of diabetic dyslipidemia. Statins have been shown in multiple trials and subgroup analyses to help with both primary and secondary prevention, and they are the first-line pharmacotherapy in patients with diabetes. The role of additional nonstatin therapies in patients with diabetes and hypertriglyceridemia has been a subject of much investigation and debate. The only intervention that has demonstrated risk reduction when added to statin therapy in this patient population is icosapent ethyl (eicosapentaenoic acid [EPA]) at a dosage of 4 g daily (Answer D).

The REDUCE-IT trial enrolled more than 8000 patients who either had established ASCVD or were at high risk, were on statin therapy, and had a median triglyceride concentration of 216 mg/dL (2.44 mmol/L). Treatment with 4 g of EPA daily was associated with a 25% relative risk reduction of the primary composite end point. Interestingly, the risk reduction did not seem to be related to triglyceride reduction, and EPA is postulated to have additional antithrombotic or anti-inflammatory effects. A similar study (STRENGTH) that used a combination of EPA and docosahexaenoic acid (DHA), however, failed to show any benefit. Whether these discrepant results were due to the unique effects of EPA, or due to a potentially atherogenic effect of the comparator mineral oil used in REDUCE-IT, is not clear. Nonetheless, available trial data only support adding EPA, 4 g daily, and not over-the-counter fish oil (Answer C), which contains both DHA and EPA.

The role of additional fibrate therapy (Answers A and B) is also not clear. While some of the older placebo-controlled trials did show the benefit of fibrates for ASCVD risk reduction in patients with hypertriglyceridemia, the combination of statin and fibrate did not lower risk in the ACCORD study compared with risk lowering with statins only. However, only about one-fifth of patients in the ACCORD trial had hypertriglyceridemia, and these patients did show benefit with combination therapy. Other subgroup analyses also suggest a benefit of adding fibrates to statins in patients with high triglycerides and low HDL cholesterol. A recent large trial (PROMINENT) showed no benefit of adding pemafibrate to statins in patients with type 2 diabetes and moderate hypertriglyceridemia. There is currently no evidence supporting the addition of fibrates to statin therapy for ASCVD risk reduction. There is also no reason to decrease the statin dosage because of low LDL cholesterol as long as patients are tolerating the current dosage. In patients with hypertriglyceridemia, calculated LDL cholesterol does not accurately reflect the overall atherogenic lipid burden, and it is better to assess this by measuring non–HDL cholesterol or apolipoprotein B.

Good glycemic control is essential for managing hypertriglyceridemia and diabetic dyslipidemia, and this may sometimes require the institution of a basal-bolus insulin regimen. However, this patient has fairly good glycemic control with a hemoglobin A_{1c} value of 6.8% (51 mmol/L) and no hypoglycemia on current therapy with basal insulin, GLP-1 receptor agonist, and metformin. Addition of prandial insulin (Answer E) to further improve diabetes control is unlikely to have a significant effect on dyslipidemia or reduce overall ASCVD risk.

Educational Objective

Explain the role of fibrates and omega-3 fatty acids in the management of hypertriglyceridemia.

Reference(s)

Bhatt DL, Steg PG, Brinton EA, et al; REDUCE-IT Investigators. Rationale and design of REDUCE-IT: reduction of cardiovascular events with icosapent ethyl-intervention trial. *Clin Cardiol.* 2017;40(3):138-148. PMID: 28294373

Das Pradhan A, Glynn RJ, Fruchart JC, et al; PROMINENT Investigators. Triglyceride lowering with pemafibrate to reduce cardiovascular risk. *N Engl J Med.* 2022;387(21):1923-1934. PMID: 36342113

Doi T, Langsted A, Nordestgaard BG. A possible explanation for the contrasting results of REDUCE-IT vs STRENGTH: cohort study mimicking trial designs. *Eur Heart J.* 2021;42(47):4807-4817. PMID: 34455435

98 **ANSWER: D) No further workup or follow-up**

This patient presents with an incidental finding of an adrenal adenoma. Clinical examination does not indicate any signs of hormone excess, and MRI findings are consistent with benign disease. Results of the endocrine workup are unremarkable, with the exception of the markedly elevated urinary 3-methoxytyramine (3-MT) concentration. The key feature in this patient's presentation is his history of Parkinson disease and its treatment with levodopa, which accounts for the elevated 3-MT concentration since 3-MT is a metabolite of dopamine. For these reasons, the patient does not require additional evaluation or follow-up (Answer D).

Measurement of 3-MT (Answer B) marginally improves the diagnostic sensitivity of plasma metanephrines for pheochromocytoma and paraganglioma by identifying rare tumors that solely secrete dopamine (particularly tumors located in the head and neck). Plasma 3-MT measurement may also be helpful in predicting metastatic pheochromocytoma and paraganglioma, although its sensitivity (50%) is limited and little better than germline *SDHB* variant status used for the same purpose. Recent data indicate that the predictive value for malignancy may be significantly enhanced by the use of machine-learning models that incorporate additional patient, tumor, and biochemical characteristics. Although elevated 3-MT can occur in isolation, the degree of elevation in this vignette is marked and explained entirely by the patient's treatment with levodopa.

MRI (Answer C) and ^{68}Ga-DOTATATE PET (Answer A) are useful imaging modalities for head and neck paragangliomas but are not indicated when there is such a clear explanation for the biochemical findings. MRI is preferred over CT in suspected neck/skull base paragangliomas because it offers better spatial resolution. Functional imaging provides better specificity than anatomical imaging for pheochromocytomas and paragangliomas and is especially useful in staging metastatic or multifocal disease. ^{68}Ga-DOTATATE PET has high diagnostic accuracy in head and neck paragangliomas and in *SDHB*-associated metastatic disease.

The Endocrine Society guidelines recommend initial biochemical testing for pheochromocytoma and paraganglioma by measuring plasma free or urinary fractionated metanephrines. However, metanephrine measurement is unnecessary in a patient with an incidentally discovered adrenal lesion when the imaging findings are clearly consistent with a homogeneous, lipid-rich adenoma. Plasma free metanephrines combined with plasma 3-MT may provide a slightly greater sensitivity than urinary fractionated metanephrines, but this difference is marginal (99% vs 95%), and plasma measurement offers limited additional value when urinary values are available. Plasma free metanephrines have a very high negative predictive value such that pheochromocytoma and paraganglioma can largely be excluded when the result is normal, with the exception of rare, biochemically silent tumors harboring an *SDHB* pathogenic variant. Compared with urinary metanephrines, measurement of plasma metanephrines is associated with fewer false-positive results (specificity 89% and 96%, respectively). Causes include analytical or pharmacological interference from drugs such as tricyclic antidepressants or sympathomimetics.

Borderline elevations may also reflect inappropriate sampling and needing repeat measurement in the supine position. Dietary considerations (especially catecholamine-rich foods such as nuts and fruits) may be important when measuring 3-MT. The clinical context is also an important consideration: critical illness (eg, needing intensive care) can cause marked metanephrine elevation, while milder elevations may be observed in the setting of obstructive sleep apnea because of increased nocturnal sympathetic drive. The clonidine-suppression test, or a wait-and-retest approach, is sometimes necessary when there is low probability of a tumor and an underlying cause is not immediately apparent.

The recently updated European Society of Endocrinology guidelines for management of adrenal incidentalomas have helped clarify the indications for adrenalectomy (Answer E), which should account for the risk of malignancy, extent of any hormonal excess, patient preference, and general health. Adrenalectomy is not indicated here given the benign features on MRI and normal findings on endocrine workup. Repeated hormonal evaluation is not required in patients with a normal hormonal evaluation at initial assessment unless new signs of hormone excess emerge or there is worsening of endocrine-associated comorbidities.

Educational Objective

Identify levodopa therapy as a cause of elevated 3-methoxytyramine and describe the causes of false-positive results in patients with pheochromocytoma or paraganglioma.

Reference(s)

Pamporaki C, Berends AMA, Filippatos A, et al. Prediction of metastatic pheochromocytoma and paraganglioma: a machine learning modelling study using data from a cross-sectional cohort. *Lancet Digit Health.* 2023;5(9):e551-e559. PMID: 37474439

Ryder SJ, Love AJ, Duncan EL, Pattison DA. PET detectives: molecular imaging for pheochromocytomas and paragangliomas in the genomics era. *Clin Endocrinol (Oxf).* 2021;95(1):13-28. PMID: 33296100

Lenders JWM, Duh Q-Y, Eisenhofer G, et al; Endocrine Society. Pheochromocytoma and paraganglioma: an Endocrine Society clinical practice guideline. *J Clin Endocrinol Metab.* 2014;99(6):1915-1942. PMID: 24893135

Fassnacht M, Tsagarakis S, Terzolo M, et al. European Society of Endocrinology clinical practice guidelines on the management of adrenal incidentalomas, in collaboration with the European Network for the Study of Adrenal Tumors. *Eur J Endocrinol.* 2023;189(1):G1-G42. PMID: 37318239

99 ANSWER: D) Free T_4, 0.65 ng/dL (SI: 8.37 pmol/L); TSH, 0.1 mIU/L; cortisol, 7.5 μg/dL (SI: 206.9 nmol/L)

This patient has longstanding hypopituitarism as a result of surgery and radiation to treat a World Health Organization grade I craniopharyngioma. Clinical evaluation of thyroid status can be challenging in elderly patients, especially in the presence of β-adrenergic blockers. The key element is the patient's reports of taking levothyroxine with meals and the multivitamin, which diminishes its intestinal absorption. Therefore, the free T_4 concentration is expected to be lower than normal. The TSH concentration is expected to be lower than normal in patients taking long-term thyroid hormone replacement for central hypothyroidism, and its measurement is not helpful for dosage adjustments. Before starting levothyroxine, patients with pituitary hypothyroidism have normal or slightly elevated TSH levels.

She also has manifestations of cortisol deficiency (nausea, vomiting, poor appetite, low blood pressure) and takes a small dose of hydrocortisone once daily with dinner. Because she took hydrocortisone earlier in the evening of hospital admission (when the blood sample was drawn for laboratory tests), the cortisol concentration is not expected to be low. A higher-than-normal cortisol value would not be expected because the patient only took 10 mg of hydrocortisone, and her previous morning cortisol value was less than 3.0 μg/dL (<82.8 nmol/L).

The pattern of laboratory test results most likely to be observed in this patient is low TSH and free T_4 and normal cortisol (Answer D).

Endocrine management in this case entails a stress dose of intravenous hydrocortisone because of hypotension and vomiting. The patient should be counseled about correct administration of levothyroxine on an empty stomach and apart from the multivitamin. Since the endogenous production of cortisol appears to be very low and the patient is symptomatic, the long-term replacement dose of hydrocortisone should be increased to a total of 20 mg daily. This is calculated based on the patient's body surface area of 2.1 m^2 and can be achieved by taking 15 mg in the morning and 5 mg 6 to 8 hours later.

Educational Objective

Identify the caveats of laboratory tests when evaluating a patient with hypopituitarism treated with cortisol and thyroid replacement.

Reference(s)

Fleseriu M, Hashim IA, Karavitaki N, et al. Hormonal replacement in hypopituitarism in adults: an Endocrine Society clinical practice guideline. *J Clin Endocrinol Metab.* 2016;101(11):3888-3921. PMID: 27736313

Liu H, Wei Li, Zhang W, Sun S, Chen C. Levothyroxine: conventional and novel drug delivery formulations. *Endocr Rev.* 2023;44(3):393-416. PMID: 36412275

Scavone C, Sportiello L, Cimmaruta D, et al. Medication adherence and the use of new pharmaceutical formulations: the case of levothyroxine. *Minerva Endocrinol.* 2016;41(2): 279-289. PMID: 27015567

100 **ANSWER: B) Start 81 mg daily at week 14 and continue until delivery**

Preeclampsia is a disease defined by hypertension (office-based blood pressure ≥140/90 mm Hg on 2 separate occasions during the second half of pregnancy [>20 weeks]) accompanied by proteinuria. Proteinuria is defined as urinary protein excretion greater than 300 mg/24 h, a single voided urine protein-to-creatinine ratio of 0.3 or greater, or a urine dipstick reading of 2+ (used only if other quantitative methods are not available).

Preeclampsia is one of the most serious health problems that affects pregnant persons. It is a complication in approximately 4% of pregnancies in the United States, and it contributes to both maternal and infant morbidity and mortality. Persons at the highest risk for preeclampsia are those who have a history of preeclampsia in a previous pregnancy, type 1 or type 2 diabetes, and chronic hypertension. Additional conditions that place a person at high risk for preeclampsia include multifetal gestation, conception using assisted reproductive technology, autoimmune disease, and kidney disease. Other risk factors are nulliparity, high prepregnancy BMI, family history of preeclampsia, and advanced maternal age (35 years or older). In addition, Black individuals have higher rates of preeclampsia and are at increased risk for serious complications due to various societal and health inequities.

Low-dosage aspirin prophylaxis (81 mg daily) is recommended for women at high risk of preeclampsia, and this should be initiated between 12 and 28 weeks' gestation (optimally before 16 weeks) and continued until delivery (Answer B).

Starting aspirin at week 8 (Answers C and D) may be too early based on the current guideline recommendations. Although there was a single study in low-income and medium-income countries showing that initiation of aspirin as early as 6 weeks' gestation could potentially reduce the incidence of preterm delivery, this was only in nulliparous women and it failed to demonstrate a difference in overall risk of preeclampsia compared with outcomes in women who started aspirin between 6 weeks 0 days and 13 weeks 6 days vs outcomes in those who do not start aspirin. As such, earlier initiation of aspirin for reducing the risk of preeclampsia is not currently a guideline recommendation. Also, as the purpose of using aspirin is to prevent preeclampsia, it should be taken until delivery. Discontinuation at the start of third trimester would be incorrect.

High-dosage aspirin (325 mg daily) (Answer E) has not been studied in this population for prevention of preeclampsia.

This patient has at least a moderate risk of preeclampsia (risk factors: age ≥35 years, type 2 diabetes, and moderate albuminuria), so not using aspirin (Answer A) is incorrect.

Educational Objective

Recommend appropriate use of aspirin for the prevention of preeclampsia in the management of preexisting diabetes mellitus in pregnancy.

Reference(s)

Low-dose aspirin use during pregnancy. ACOG Committee Opinion No. 743: low-dose aspirin use during pregnancy. *Obstet Gynecol.* 2018;132(1):e44-e52. PMID: 29939940

LeFevre ML, US Preventive Services Task Force. Low-dose aspirin use for the prevention of morbidity and mortality from preeclampsia: U.S. Preventive Services Task Force recommendation statement. U.S. Preventive Services Task Force. *Ann Intern Med.* 2014;161:819-826. PMID: 25200125

US Preventive Services Task Force, Davidson KW, Barry MJ, et al. Aspirin use to prevent preeclampsia and related morbidity and mortality: US Preventive Services Task Force Recommendation Statement. *JAMA.* 2021;326(12):1186-1191. PMID: 34581729

Hoffman MK, Goudar SS, Kodkany BS, et al; ASPIRIN Study Group. Low-dose aspirin for the prevention of preterm delivery in nulliparous women with a singleton pregnancy (ASPIRIN): a randomised, double-blind, placebo-controlled trial. *Lancet.* 2020;395(10220):285-293. PMID: 31982074

101 **ANSWER: B) Growth of the thyroid nodule by ≥3 mm in any 1 of the 3 dimensions measured**

Active surveillance is being increasingly adopted as an alternative management option for patients with low-risk papillary thyroid cancer (PTC), in whom long-term nonsurgical observation of this usually indolent malignancy has excellent outcomes. In the early- to mid-1990s, active surveillance was initiated for low-risk PTC at 2 centers in Japan, around the same time that active surveillance for some prostate cancers was also adopted in several centers globally. Patients enrolled in the active surveillance programs for PTC in Japan, and other programs developed since, are usually followed up with neck ultrasonography every 6 to 12 months. Thus, the patient's inability to attend follow-up appointments only once per year (Answer E) is not an indication to proceed to thyroid surgery. Ideal candidates are patients with low-risk PTC and usually cancers that (1) are smaller than 1 cm; (2) have no high-risk histologic features; (3) are not adjacent to the trachea or recurrent laryngeal nerve, with ≥2 mm of surrounding normal thyroid parenchyma; (4) have no suggestion of extrathyroidal extension on ultrasonography; and (5) have no evidence of nodal or distant metastases. Patients must also be willing to return for regular sonographic monitoring.

Indications for terminating active surveillance and proceeding with thyroid surgery include the appearance of a new suspicious lymph node(s); greater than 3-mm increase from baseline of the biopsy-proven PTC (Answer B); or the preference of the patient, clinician, or both to transition the care plan to thyroid surgery. There is no evidence that development of new thyroid nodules (Answer A) or relevant family history (Answer D) would affect long-term outcomes. Serum thyroglobulin concentrations (Answer C) are not used to monitor thyroid cancer in individuals with an intact thyroid gland.

In the nearly 30 years of follow-up in Japan, similar risks of recurrence, distant metastases, disease-specific and all-cause mortality, and other outcomes have been observed when comparing active surveillance to immediate thyroid surgery for the treatment of low-risk PTC. Several systematic reviews and meta-analyses have confirmed these data of other programs initiated worldwide (Korea, United States, Canada, Brazil, Columbia, Argentina, and Italy), including some that have expanded inclusion to PTC 2.0 cm or smaller, although these have had substantially shorter durations of follow-up.

Educational Objective

Assess important factors in the consideration of active surveillance for the management of low-risk papillary thyroid cancer.

Reference(s)

Ito Y, Miyauchi A. Active surveillance as first-line management of papillary microcarcinoma. *Annu Rev Med.* 2019;70:369-379. PMID: 30230964

Haugen BR, Alexander EK, Bible KC, et al. 2015 American Thyroid Association management guidelines for adult patients with thyroid nodules and differentiated thyroid cancer: the American Thyroid Association guidelines task force on thyroid nodules and differentiated thyroid cancer. *Thyroid.* 2016;26(1):1-133. PMID: 26462967

Chou R, Dana T, Haymart M, et al. Active surveillance versus thyroid surgery for differentiated thyroid cancer: a systematic review. *Thyroid.* 2022;32(4):351-367. PMID: 35081743

Brito JP, Ito Y, Miyauchi A, Tuttle RM. A clinical framework to facilitate risk stratification when considering an active surveillance alternative to immediate biopsy and surgery in papillary microcarcinoma. *Thyroid.* 2016;26(1):144-149. PMID: 26414743

Kim MJ, Moon JH, Lee EK, et al. Active surveillance for low-risk thyroid cancers: a review of current practice guidelines. *Endocrinol Metab (Seoul).* 2024;39(1):47-60. PMID: 38356210

102 **ANSWER: A) Increase calcium intake to 1200-1500 mg daily**

Fragility fractures are uncommon during pregnancy or breastfeeding. During breastfeeding there is a temporary increase in fracture risk due to the obligatory loss in the skeletal mineral content and strength. On average, a breastfeeding woman loses 210 mg of calcium in the milk daily. Intestinal calcium absorption is increased during pregnancy, but it returns to nonpregnant absorption following delivery. Therefore, the additional calcium needed for production of breast milk is provided by the maternal skeleton. The mechanism of maternal skeleton loss is explained by the brain-breast-bone circuit. Suckling and prolactin inhibit hypothalamic GnRH, thereby suppressing FSH and LH and reducing the ovarian sex steroids estradiol and progesterone. PTHrP production and release from the breast tissue are stimulated by increased prolactin levels. This, in turn, results in bone resorption and provides calcium in the breast milk. Osteoclast-mediated bone resorption especially affects trabecular bone (5%-10% loss in bone mineral density over 3 to 6 months of lactation), while cortical and whole-body sites are less affected. The typical rate of bone mineral density loss is 1% to 3% per month. Calcium-rich diets reduce the amount of bone loss during lactation in animal models. Following weaning, the maternal skeleton undergoes significant remodeling and remineralization and returns to the prepregnancy state within 6 to 12 months, thus supporting that this is a temporary decrease in bone mineral density and bone strength. Parity and lactation do not increase the long-term risk of low bone mineral density, fractures, or osteoporosis.

Treatment is focused on optimizing calcium intake (1200-1500 mg daily) (Answer A) and vitamin D intake to achieve 25-hydroxyvitamin D concentrations greater than 20 ng/mL (>50 nmol/L); correcting any nutritional deficiency; and implementing weight-bearing exercises.

While weaning from breastfeeding (Answer E) might be beneficial from the bone perspective, respecting the mother's preferences about breastfeeding is important. It is imperative to emphasize that the skeleton eventually restores the lost bone mineral density upon weaning and the risk of fractures in subsequent pregnancies and lactation is not increased.

Although hyperprolactinemia can cause bone loss, this patient's prolactin concentration is increased because of breastfeeding, and cabergoline treatment (Answer D) is not indicated.

Some advocate for a temporary supportive corset for pain reduction, but thoracolumbosacral orthosis (Answer B) can cause paraspinal muscle weakness and increased risk of fractures and therefore should be used with caution.

While pharmacologic therapies such as alendronate (Answer C) have been studied in pregnancy and lactation-associated osteoporosis, their use is individualized, especially given the lack of data on long-term safety concerns. In addition, this patient desires future fertility, and there is concern for bisphosphonate crossing the placenta and interfering with the fetal skeleton in future pregnancies. Since bone mineral density improves upon weaning, treatment may not be indicated.

Educational Objective

Explain the reversible impact of breastfeeding on the skeleton and limitations of available therapies in women of childbearing age.

Reference(s)

Kovacs CS, Ralston SH. Presentation and management of osteoporosis presenting in association with pregnancy or lactation. *Osteoporos Int.* 2015;26(9):2223-2241. PMID: 25939309

Kirby BJ, Ma Y, Martin HM, Buckle Favaro KL, Karaplis AC, Kovacs CS. Upregulation of calcitriol during pregnancy and skeletal recovery after lactation do not require parathyroid hormone. *J Bone Miner Res.* 2013;28(9):1987-2000. PMID: 23505097

Lampropoulou-Adamidou K, Trovas G, Triantafyllopoulos IK, et al. Teriparatide treatment in patients with pregnancy- and lactation-associated osteoporosis. *Calcif Tissue Int.* 2021;109(5):554-562. PMID: 34132853

Anagnostis P, Lampropoulou-Adamidou K, Bosdou JK, et al, Comparative effectiveness of therapeutic interventions in pregnancy and lactation-associated osteoporosis: a systematic review and meta-analysis. *J Clin Endocrinol Metab.* 2024;109(3):879-901. PMID: 37708365

103 ANSWER: E) Refer the patient for pancreas and kidney transplant

Before the discovery of insulin almost one century ago, type 1 diabetes was considered a terminal illness, as the life expectancy following this diagnosis was only a few months. With technological advances in the management and treatment of patients with type 1 diabetes, life expectancy has increased significantly. Despite this progress, managing type 1 diabetes can be challenging, which places a significant clinical and emotional burden on patients. One of the notable causes of hyperglycemia is nutrition and its effect on glucose concentration. Prandial (short-acting) insulin is used to prevent postprandial hyperglycemia, but this strategy is only successful if a patient can either correctly count carbohydrates or estimate the amount of carbohydrates in each meal and categorize them into low-, medium-, and heavy-carbohydrate meals. Another option for these patients is to be consistent and consume similar types of meals, although this approach is more difficult to achieve. Overall, if there is a mismatch between the amount and type of food consumed and the amount of insulin used, the patient will develop either hyperglycemia (if less insulin is used) or hypoglycemia (if more insulin is used).

In this clinical vignette, the patient has significant hyperglycemia as evidenced by the increased hemoglobin A_{1c} level. He also experiences significant hypoglycemia 3 times per week, at different times of the day. More importantly, this patient has hypoglycemia unawareness, and the episodes are so severe that they have led to 2 hospitalizations in the last 12 months. Although reducing the basal insulin dosage (Answer B) or both the basal and bolus insulin dosages (Answer A) might reduce the number of hypoglycemic episodes, this approach will worsen hyperglycemia. Therefore, this does not represent the best management option for the patient.

Another strategy is to use a subcutaneous insulin infusion system (insulin pump) (Answer C). The evolution of insulin pumps has led to the development of automated insulin delivery systems. These systems adjust insulin continuously by using algorithms that consider different factors, such as glucose values from continuous glucose monitors, insulin on board (residual insulin), and whether the patient is exercising or sleeping. Although in the past carbohydrate counting was a skill that a patient with diabetes was required to have to use insulin pumps, the most recent systems can be used even if a patient is less proficient in carbohydrate-counting techniques. However, the currently available automated insulin delivery systems require some technological expertise. Considering this patient's challenges with carbohydrate counting and his lack of knowledge in using technology (eg, finds using a smartphone complicated), a hybrid automated insulin delivery system is not the best management option.

Pancreas and islet-cell transplant should be used only in select patients with type 1 diabetes. These options are usually reserved for patients with severe metabolic complications (eg, hypoglycemia [especially when it is accompanied by unawareness], diabetic ketoacidosis, incapacitating problems with exogenous insulin delivery, and failure of insulin-based options to prevent acute complications). Although there are no direct trials comparing pancreas transplant with islet-cell transplant, the former is usually associated with the rate of insulin independence and also with higher morbidity because of the significant surgery. In considering which long-term approach is best for this patient who also has end-stage kidney disease, most providers would recommend proceeding with combined pancreas and kidney transplant (Answer E), as this will improve glycemia, reverse metabolic complications, and address underlying kidney disease. Referring the patient for islet-cell transplant alone (Answer D) would not address the underlying significant kidney disease, which is associated with significant mortality, especially among patients with diabetes.

Educational Objective

Identify scenarios for which pancreas-kidney transplantation is recommended.

Reference(s)

Holt RIG, DeVries JH, Hess-Fischl A, et al. The management of type 1 diabetes in adults. A consensus report by the American Diabetes Association (ADA) and the European Association for the Study of Diabetes (EASD). *Diabetes Care.* 2021;44(11):2589-2625. PMID: 34593612

Phillip M, Nimri R, Bergenstal RM, et al. Consensus recommendations for the use of automated insulin delivery technologies in clinical practice. *Endocr Rev.* 2023;44(2):254-280. PMID: 36066457

104 ANSWER: C) Start a meal replacement program

The patient in this vignette has a modest weight-loss goal of approximately 10% of her body weight before becoming a candidate for knee replacement surgery. While several strategies could be used to help her achieve this weight goal, the pros and cons of each should be carefully considered before making a recommendation.

In elderly patients with obesity, the risk of worsening sarcopenia should be weighed against the benefit of the weight-loss intervention. Sarcopenia is defined as the age-related loss of muscle mass and strength or physical function. During a weight-loss intervention, approximately 25% of the weight that is lost consists of fat-free mass (includes skeletal muscle and bone), although proportions are influenced by activity, dietary composition, and age. The patient in this vignette has baseline decreased functionality and cannot stand from a seated position without assistance. She is therefore at higher-than-average risk of worsening sarcopenia and frailty following marked or rapid weight loss.

Bariatric surgery can be safe and effective in patients older than 70 years. However, in this patient who has a large hiatal hernia, sleeve gastrectomy (Answer B) is relatively contraindicated due to the risk of worsening gastroesophageal reflux. Moreover, she does not require a bariatric procedure to lose 10% of her body weight and the risks of surgery in this case outweigh the benefits.

Ensuring that the patient is consuming an adequate amount of protein (Answer A) is important, but this would not lead to weight loss without a concomitant decrease in overall caloric intake. To that end, starting a meal replacement program (Answer C) is a convenient, short-term strategy to achieve calorie restriction without significantly altering dietary patterns or meal timing. It is also an appropriate strategy to achieve the intended amount of weight loss over a reasonable period.

Levothyroxine supplementation (Answer D) should not be initiated with this mild degree of TSH elevation solely for the purpose of weight loss.

Phentermine (Answer E) is not a good choice for appetite suppressant in a patient who has hypertension, and it should generally be used with caution in elderly patients.

Educational Objective

Choose the appropriate weight-management strategy in a postmenopausal woman based on weight-loss goals and risk of sarcopenia.

Reference(s)

Villareal DT, Chode S, Parimi N, et al. Weight loss, exercise, or both and physical function in obese older adults. *N Engl J Med.* 2011;364(13):1218-1229. PMID: 21449785

Dowgiallo-Gornowicz N, Lech P, Major P; Collaborative Study G. Bariatric and metabolic surgery in patients older than 65 years - a multicenter study. *Obes Surg.* 2023;33(10):3106-3111. PMID: 37566339

105 ANSWER: A) Administer intravenous hydrocortisone

The CT appearance of retroperitoneal fat stranding and high-density fluid are typical of the early stages of adrenal hemorrhage before a hematoma has organized. Adrenal congestion is also a vital clue to early radiological diagnosis. While electrolytes are normal in this patient, it is clear that he is critically unwell, with tachycardia and elevated lactate. The best next step in management is thus intravenous hydrocortisone (Answer A). Bilateral adrenal hemorrhage is usually accompanied by adrenal insufficiency, and hydrocortisone may be lifesaving. Hyponatremia and other electrolyte disturbances are not universal in adrenal crisis, especially when the presentation is acute. Treatment should be given promptly and should never be delayed while awaiting the results of diagnostic tests such as cosyntropin stimulation (Answer C). The Endocrine Society guidelines recommend prompt administration of 100 mg intravenous hydrocortisone in adults, followed by fluid resuscitation and 200 mg hydrocortisone per 24 hours (given either as continuous intravenous infusion or injection every 6 hours).

The adrenal glands are particularly vulnerable to hemorrhagic infarction because of their precarious blood supply. The adrenals have a rich arterial supply from the superior, middle, and inferior adrenal arteries. In contrast, venous drainage is limited to a single adrenal vein that drains into the renal vein on the left side and into the inferior vena cava on the right side. This creates a "vascular dam" that may predispose to venous congestion and hemorrhage under conditions of stress, driven by an increase in adrenal vascularity from elevated ACTH and catecholamine-induced vasoconstriction of the draining venules.

Adrenal hemorrhage has a range of etiologies, including blunt abdominal trauma, sepsis, anticoagulant use, and procoagulant states (eg, antiphospholipid syndrome, essential thrombocytosis, and other coagulopathies). Most cases are unilateral and self-limiting, although an underlying tumor, especially pheochromocytoma, should also be considered. The presentation is often nonspecific, ranging from abdominal pain to varying degrees of adrenal insufficiency. A high index of suspicion is therefore needed to make a prompt diagnosis.

More recently, adrenal hemorrhage has been reported as a complication of COVID-19, most likely due to the increased thrombotic risk associated with this condition. Several cases of adrenal hemorrhage have also been reported secondary to vaccine-induced immune thrombocytopenia and thrombosis. The pathophysiology of this rare complication of adenoviral-based COVID-19 vaccines (ChAdOx1 and Ad26.COV2.S) closely resembles that of heparin-induced thrombocytopenia, in which antibodies against platelet factor 4 lead to platelet activation, thrombocytopenia (this patient's platelet count fell rapidly to $14 \times 10^3/\mu L$), and an increased risk of thromboembolism. A heparin-induced thrombocytopenia antibody screen (Answer D) helps confirm the diagnosis of vaccine-induced immune thrombocytopenia and thrombosis, but it should not take priority over intravenous hydrocortisone. Vaccine-induced immune thrombocytopenia and thrombosis classically has a symptom onset of 5 to 30 days after vaccination and is characterized by thrombocytopenia, thrombosis, elevated D-dimer, and positive antiplatelet factor 4 antibodies. Mortality is high, and treatment includes intravenous immunoglobulin (Answer B). Although systemic glucocorticoid therapy should be administered first given the risk for adrenal crisis, plasma exchange and direct thrombin inhibitors such as argatroban are also used in this condition.

Management of adrenal hemorrhage is centered around timely administration of hydrocortisone when adrenal insufficiency is suspected, accompanied by appropriate fluid resuscitation and supportive therapy, including blood transfusion if necessary. The hemorrhage itself is usually managed conservatively. Endovascular embolization is minimally invasive and may be needed in cases of ongoing hemorrhage with hemodynamic instability. Exploratory laparotomy (Answer E) is usually only required in the absence of facilities for interventional radiology or where the bleeding vessel cannot be identified. Recovery of adrenal function after bilateral adrenal hemorrhage is uncommon since the adrenals typically become atrophic. Most patients therefore require lifelong glucocorticoid replacement, although mineralocorticoid loss may be relatively spared, perhaps due to the distance of the zona glomerulosa from the vascular dam.

**Note: Image reprinted from Taylor P, Allen L, Shrikrishnapalasuriyar N, Stechman M, Rees A. Vaccine-induced thrombosis and thrombocytopenia with bilateral adrenal haemorrhage.* Clin Endocrinol (Oxf). *2022;97(1):26-27.*

Educational Objective

Diagnose and manage adrenal hemorrhage as a presentation of vaccine-induced thrombosis and thrombocytopenia.

Reference(s)

Bornstein SR, Allolio B, Arlt W, et al. Diagnosis and treatment of primary adrenal insufficiency: an Endocrine Society clinical practice guideline. *J Clin Endocrinol Metab.* 2016;101(2):364-389. PMID: 26760044

Elhassan YS, Ronchi C, Wijewickrama P, Baldeweg SE. Approach to the patient with adrenal hemorrhage. *J Clin Endocrinol Metab.* 2023;108(4):995-1006. PMID: 36404284

Elhassan YS, Iqbal F, Arlt W, et al. COVID-19-related adrenal hemorrhage: multicentre UK experience and systematic review of the literature. *Clin Endocrinol (Oxf).* 2023;98(6):766-778. PMID: 36710422

106 ANSWER: E) Rosuvastatin

This patient has type 2 diabetes, which is a cardiovascular risk equivalent that warrants the initiation of statin therapy, regardless of his LDL-cholesterol concentration. The patient reports previous use of medication for cholesterol management, presumably statin therapy given the muscle aches he noted. While his history suggests possible statin intolerance, he is not sure which medication he took. This history should not preclude a statin trial now (Answer E), particularly one such as rosuvastatin at a low dosage, which tends to be associated with less intolerability than many statins.

While a referral to nutrition counseling (Answer D) may benefit patients in terms of their glycemic control, lipid control, blood pressure management, and weight-loss efforts, it is not a substitute for a trial of statin therapy.

Bempedoic acid (Answer A) is an ATP citrate lyase inhibitor that reduces LDL-cholesterol levels and is associated with a low incidence of muscle-related adverse events. Among statin-intolerant patients, bempedoic acid is associated with a lower risk of major adverse cardiovascular events (death of cardiovascular causes, nonfatal myocardial infarction, nonfatal stroke, or coronary revascularization) (hazard ratio, 0.87 [95% CI, 0.79-0.96]; $P = .004$). However, determining whether this patient has a real intolerance to statin therapy is warranted before considering bempedoic acid therapy.

Icosapent ethyl (Answer C) has been demonstrated to reduce the risk of cardiovascular events (death of cardiovascular disease, nonfatal myocardial infarction, nonfatal stroke, coronary revascularization, or unstable angina) in patients with fasting triglyceride concentrations of 135 to 499 mg/dL (1.53-5.64 mmol/L) and LDL-cholesterol concentrations ranging from 41 to 100 mg/dL (1.06-2.59 mmol/L) despite statin therapy and who had established cardiovascular disease (70%) or diabetes plus other high-risk factors (30%). The primary results demonstrated a 25% reduction in the primary end point of death of cardiovascular disease, nonfatal myocardial infarction, nonfatal stroke, coronary revascularization, or unstable angina with icosapent ethyl vs placebo ($P < .001$). Adding icosapent ethyl is incorrect because the patient's lack of statin therapy should be addressed first. A moderate-intensity statin is recommended in patients with type 2 diabetes, even if their LDL-cholesterol value is not above target. If the patient had established cardiovascular disease, then a high-intensity statin would be recommended, as tolerated. Patients included in the REDUCE-IT trial had a triglyceride concentration greater than 135 mg/dL (>1.53 mmol/L). It is unclear whether icosapent ethyl is associated with reduction of cardiovascular disease risk in patients with triglyceride concentrations less than 135 mg/dL (<1.53 mmol/L).

There is no role for fenofibrate therapy (Answer B) in this patient. Fenofibrate should only be considered for severe hypertriglyceridemia (triglycerides ≥500 mg/dL [≥5.65 mmol/L]) to reduce the risk of acute pancreatitis. There is no conclusive evidence that fibrate therapy reduces cardiovascular risk. Even among patients with chronic coronary disease receiving statin therapy, adding niacin, fenofibrate, or dietary supplements containing omega-3 fatty acids is not beneficial in reducing cardiovascular risk.

Educational Objective

Explain the role of statin therapy for patients with type 2 diabetes who have an unclear history of statin intolerance.

Reference(s)

Grundy SM, Stone NJ, Baily AL, et al. 2108 AHA/ ACC/AACVPR/AAPA/ABC/ACPM/ADA/AGS/APhA/ASPC/NLA/PCNA Guideline on the management of blood cholesterol: a report of the American College of Cardiology/American Heart Association Task Force on Clinical Practice Guidelines. *Circulation.* 2019;139(25):e1082-e1143. PMID: 30586774

Bhatt DL, Steg PG, Miller M, et al; REDUCE-IT Investigators. Cardiovascular risk reduction with icosapent ethyl for hypertriglyceridemia. *N Engl J Med.* 2019;380(1):11-22. PMID: 30415628

Nissen SE, Lincoff AM, Brennan D, et al; CLEAR Outcomes Investigators. Bempedoic acid and cardiovascular outcomes in statin-intolerant patients. *N Engl J Med.* 2023;388(15):1353-1364. PMID: 36876740

Virani SS, Virani LK, Arnold SV, et al. 2023 AHA/ACC/ACCP/ASPC/NLA/PCNA Guideline for the management of patients with chronic coronary disease: a report of the American Heart Association/American College of Cardiology Joint Committee on Clinical Practice Guidelines. *Circulation.* 2023;148(9):e9-e119. PMID: 37471501

107 ANSWER: D) Perform abdominal ultrasonography

Primary amenorrhea is defined as the absence of menarche in girls aged 13 years or older without signs of pubertal development (ie, delayed puberty) or absence of menarche in girls aged 15 years or older with developed secondary sexual characteristics and normal growth. The evaluation of primary amenorrhea should be initiated 3 years after thelarche or within 5 years of thelarche if breast development started before age 10 years.

This patient's clinical presentation is consistent with complete androgen insensitivity syndrome, and both her karyotype and genetic testing result confirm this diagnosis. Complete androgen insensitivity syndrome is a difference of sexual development. Affected individuals have a 46,XY karyotype and testicular gonads, but they are completely insensitive to androgens due to a pathogenic variant in the androgen receptor gene (Xq11-q12 [the ligand-binding domain]) and therefore have a female phenotype (no male external genitalia). These individuals often have testes in the inguinal canal presenting as hernias, or masses in the labioscrotal folds, but the testes are most commonly found intra-abdominally. The testes make testosterone and antimullerian hormone. Antimullerian hormone leads to regression of the mullerian structures: the uterus, fallopian tubes, and upper two-thirds of the vagina. Therefore, female internal genitalia are absent, and the vagina is short and shallow with a blind end. Breasts are generally well-developed because testosterone is aromatized to estrogen, but nipple development can be incomplete given that this process is dependent on androgens. Furthermore, in patients with complete androgen insensitivity syndrome, there is typically absence of axillary and pubic hair due to androgen resistance. The typical hormone profile in a postpubertal patient with complete androgen insensitivity syndrome and intact gonads is a testosterone concentration in the adult male range or higher, an increased LH concentration (despite elevated testosterone), and a normal or slightly elevated FSH concentration.

Persons with complete androgen insensitivity syndrome have an increased risk of gonadal tumors that is associated with dysfunctional spermatogenesis and a history of undescended testes. Given that the testes are most commonly found intra-abdominally, abdominal ultrasonography (Answer D) to detect undescended gonads is the best next step in this patient's management. Historically, patients with complete androgen insensitivity syndrome were managed with removal of gonadal tissue at various ages to avoid the risk of gonadal malignancy. Germ-cell tumors and gonadoblastomas may develop in the cryptorchid testes (1.5% to 2% of undescended testes); however, this is rare and early prophylactic gonadal tissue removal is no longer recommended. Rather, the patient should be followed up closely and engaged in shared decision-making to discuss the potential risks, benefits, and alternatives of gonadectomy vs active surveillance of gonadal tissue. To date, there are no clinical practice guidelines on the optimal screening approaches and timing of recommended gonadectomy for patients with undescended gonads.

Patients who do undergo gonadectomy require hormone replacement with estradiol, but women with complete androgen insensitivity have reported generally feeling "unwell" after gonadectomy despite receiving estrogen. This may be due to their bodies experiencing very high levels of estradiol from the aromatization of testosterone to estradiol; therefore, the replacement dosage of estradiol may not suffice. Some patients report feeling better after the addition of testosterone. For patients who choose gonadal retention, baseline transabdominal ultrasonography and MRI are recommended to locate the position of the gonads, visualize their features, and assess for common neoplastic findings. If the gonads appear normal without features suggestive of neoplastic disease, then annual monitoring with transabdominal ultrasonography is recommended.

hCG (Answer C) and α-fetoprotein (Answer B) are not good screening instruments for these tumors because of variable secretion from seminomas.

DXA (Answer A) is not indicated because bone mineral density is normal when gonads are intact and testosterone levels are in the male range (300-900 ng/dL [10.4-31.2 nmol/L]).

Starting estrogen therapy (Answer E) is not indicated while the patient has intact gonads. If she chooses to undergo gonadectomy, then estrogen replacement therapy would be indicated.

Educational Objective

Recommend the best monitoring approach for malignancy in a patient with complete androgen insensitivity syndrome.

Reference(s)

Guo M, Huan J-C, Li C-F, Liu Y-Y. Complete androgen insensitivity syndrome: a case report and literature review. *J Int Med Res.* 2023;51(2):1-7. PMID: 36851849

Hughes IA, et al. Androgen resistance. *Best Pract Res Clin Endocrinol Metab.* 2006;20(4):557-598. PMID: 17161333

Weidler EM, Linnaus ME, Baratz AB, et al. A management protocol for gonad preservation in patients with androgen insensitivity syndrome. *J Pediatr Adolesc Gynecol.* 2019;(32):605-611. PMID: 31233832

Seppa S, Kuiri-Hanninen T, Holopainen E, Voutilainen R. Management of endocrine disease: diagnosis and management of primary amenorrhea and female delayed puberty. *Eur J Endocrinol.* 2021;184(6):R225-R242. PMID: 33687345

Hughes IA, Davies JD, Bunch TI, Pasterski V, Mastroyannopoulou K, MacDougall J. *Lancet.* 2012;380(9851):1419-1428. PMID: 22698698

Melo KFS, Mendonca BB, Billerbeck AEC, et al. Clinical, hormonal, behavioral, and genetic characteristics of androgen insensitivity syndrome in a Brazilian cohort: five novel mutations in the androgen receptor gene. *J Clin Endocrinol Metab.* 2003;88(7):3241-3250. PMID: 12843171

108 ANSWER: C) Start icosapent ethyl, 2 g twice daily

The REDUCE-IT trial showed significant reduction (hazard ratio, 0.75 [95% CI, 0.68-0.83]; P < .001) in the occurrence of major adverse cardiovascular events (MACE) in secondary prevention, as well as high-risk primary prevention, in patients with type 2 diabetes when icosapent ethyl was added to standard of care in those with fasting triglyceride concentrations of 150 mg/dL or greater (≥1.70 mmol/L) but less than 500 mg/dL (<5.65 mmol/L). The high-risk primary prevention group in this trial was defined as patients with diabetes and 1 or more of the following:

- Age ≥55 years (men) or ≥65 years (women)
- Hypertension
- Active cigarette smoker or quit within the past 3 months
- HDL cholesterol ≤40 mg/dL (≤1.04 mmol/L) (men) or ≤50 mg/dL (≤1.30 mmol/L) (women)
- High-sensitivity C-reactive protein >3 mg/L (>28.6 nmol/L)
- Creatinine clearance >30 and <60 mL/min per 1.73 m^2
- Retinopathy
- Moderately or severely increased albuminuria (previously called microalbuminuria and macroalbuminuria, respectively)
- Ankle brachial index <0.9 without symptoms of claudication

Given the magnitude of benefit seen with use of icosapent ethyl in this group, not making any changes to this patient's regimen (Answer A) would be incorrect.

Although adding aspirin (Answer B) would be helpful in reducing cardiovascular risk for individuals with type 2 diabetes who are at high risk, it also caused major bleeding events when evaluated in the ASCEND trial.

Current guidelines encourage having a comprehensive discussion with the patient on the benefits vs the comparable increased risk of bleeding when aspirin is being considered as a primary prevention strategy. Given this patient's history of peptic ulcer disease, aspirin might increase the risk of gastrointestinal bleeding, which may outweigh the cardiovascular benefit.

Addition of fibrate therapy to statins (Answer D) has not been shown to improve atherosclerotic cardiovascular disease outcomes and is generally not recommended.

Among GLP-1 receptor agonist therapies, cardiovascular outcome trials showed reduction of 3-point MACE with dulaglutide, liraglutide, and injectable semaglutide in secondary prevention, and a reduction of 3-point MACE with dulaglutide in high-risk primary prevention. There are no data to suggest that switching from oral semaglutide to injectable semaglutide (Answer E) would have any additional cardiovascular benefits in primary prevention in patients at high risk.

Educational Objective

Manage cardiovascular risk reduction in primary prevention in patients with diabetes and hypertriglyceridemia.

Reference(s)

Bhatt DL, Steg PG, Miller M, et al; REDUCE-IT Investigators. Cardiovascular risk reduction with icosapent Ethyl for Hypertriglyceridemia. *N Engl J Med.* 2019;380(1):11-22. PMID: 30415628

ASCEND Study Collaborative Group; Bowman L, Mafham M, et al. Effects of aspirin for primary prevention in persons with diabetes mellitus. *N Engl J Med.* 2018;379(16):1529-1539. PMID: 30146931

American Diabetes Association. 10. Cardiovascular disease and risk management: standards of medical care in diabetes-2019. *Diabetes Care.* 2019;42(Suppl 1):S103-S123. PMID: 30559236

109 **ANSWER: C) Perform rhTSH-stimulated radioactive iodine (^{123}I) whole-body scan**

Follicular thyroid carcinoma (FTC) is the second most common thyroid cancer type after papillary thyroid carcinoma. FNA typically demonstrates indeterminate cytology (most often follicular neoplasm or atypia of undetermined significance). A diagnosis of FTC is usually rendered based on histopathologic findings following surgical resection, which includes complete evaluation of the tumor capsule and assessment for angioinvasion. FTC can be classified into 3 subgroups to reflect prognosis according to the 2022 World Health Organization Classification of Thyroid Tumors. These subgroups are (1) minimally invasive (capsular invasion only); (2) encapsulated angioinvasive; and (3) widely invasive (obliterated or only focally intact tumor capsule and gross invasion throughout the thyroid gland). According to the 2015 American Thyroid Association guidelines, tumors with extensive vascular invasion (>4 foci) are also considered to have a high risk of structural recurrence (30%-55%).

The patient described in this case was found to have a 4.5-cm FTC of the encapsulated angioinvasive subtype with at least 4 foci of angioinvasion (American Thyroid Association high risk of recurrence). She was subsequently treated with completion thyroidectomy and radioactive iodine therapy. Her serum thyroglobulin reached a nadir of 0.4 ng/mL (0.4 μg/L) 1 year after radioactive iodine therapy but increased over the next 2 years to reach the current concentration of 6.9 ng/mL (6.9 μg/L). Given the current serum thyroglobulin value and the observed thyroglobulin trend over time (rising), additional imaging is indicated to evaluate for structural disease. Most studies evaluating the significance of thyroglobulin doubling times have found that a doubling time of less than 1 year, as in this patient, is associated with disease recurrence and reduced survival. Simply reassuring the patient and measuring serum thyroglobulin again in 6 months (Answer D) would be inappropriate.

Of the remaining answer choices, 3 imaging modalities are listed: FDG PET/CT (Answer A), neck ultrasonography (Answer B), and rhTSH-stimulated radioactive iodine (^{123}I) whole-body scan (Answer C). The patient's last imaging assessment was neck ultrasonography that was performed 1 year ago and was negative. Cervical lymph node metastases are uncommon in FTC, affecting 5% to 10% of patients. Given this observation, normal findings on ultrasonography only 1 year ago, and the sheer rise in the patient's serum thyroglobulin concentration, repeating neck ultrasonography (Answer B) now would be of low utility. Rather, more extensive body imaging should be performed. Although disagreement remains regarding how best to define radioactive iodine–refractory thyroid cancer, proposed definitions include: (1) no evidence of radioactive iodine uptake on whole-body scan; (2) no uptake in some lesions on whole-body scan; (3) clear growth within 12 to 16 months of radioactive iodine (^{131}I) therapy; and/or (4) a lifetime dose of 600 mCi ^{131}I. Based on the data provided, there is insufficient evidence to label this patient as having radioactive iodine–refractory disease, and therefore performing an rhTSH-stimulated radioactive iodine (^{123}I) whole-body scan (Answer C), which could provide both diagnostic and therapeutic information, is the best next step in this patient's management. If the diagnostic ^{123}I whole-body scan is negative, then FDG PET/CT (Answer A) could be considered.

Biotin (vitamin B_7) is known to cause spurious laboratory test results with immunoassay platforms that incorporate streptavidin-biotin binding. Elevated serum biotin can lead to falsely high values in competitive immunoassays and falsely low values in noncompetitive immunometric (sandwich) assays. As a large molecule, thyroglobulin is measured using a sandwich assay. Therefore, biotin interference should cause a falsely low, not high, serum thyroglobulin result. In addition, spurious laboratory results from biotin interference are typically seen in patients taking high-dosage biotin supplements (5-10 mg daily), not 300 mcg daily. Remeasuring serum thyroglobulin after instructing the patient to stop the biotin-containing supplement (Answer E) is therefore unnecessary.

Educational Objective

Evaluate a patient with follicular thyroid cancer and rising serum thyroglobulin levels.

Reference(s)

Baloch ZW, Asa SL, Barletta JA, et al. Overview of the 2022 WHO Classification of Thyroid Neoplasms. *Endocr Pathol.* 2022;33(1):27-63. PMID: 35288841

Giovanella L, Garo ML, Albano D, Gorges R, Ceriani L. The role of thyroglobulin doubling time in differentiated thyroid cancer: a meta-analysis. *Endocr Connect.* 2022;11(4):e210648. PMID: 35245206

Ylli D, Soldin SJ, Stolze B, et al. Biotin interference in assays for thyroid hormones, thyrotropin and thyroglobulin. *Thyroid.* 2021;31(8):1160-1170. PMID: 34042535

110 ANSWER: A) Measure serum cortisol now

This patient's history is suggestive of idiopathic intracranial hypertension (IIH), which predominantly affects premenopausal women. Clinical manifestations include daily headaches, vision changes, and bilateral papilledema. Pulse synchronous pulsatile tinnitus and diplopia also occur in some patients. Lumbar puncture with an opening pressure greater than 250 mm H_2O confirms the diagnosis of IIH in the absence of other causes of increased intracranial hypertension on imaging. Weight-loss interventions and acetazolamide are usually successful in managing IIH, as described in this case. Surgery is only required in patients with resistant cases. Brain MRI with and without contrast shows empty sella in up to 80% of affected patients.

Empty sella was first described in 1951 and is defined as the herniation of subarachnoid space within the sella turcica in the context of a congenital or acquired defect of the sellar diaphragm, with cerebrospinal fluid occupying most of the sella and flattening of the pituitary gland. The diagnosis of primary empty sella entails absence of pituitary pathology and is generally a radiologic finding without clinical implications. Secondary empty sella can occur in patients with congenital pituitary hypoplasia, pituitary infarction, masses, or infections and after surgery or radiation. An evaluation of pituitary function is necessary in patients with empty sella syndrome (ie, clinical manifestations suggestive of mass effects or hormone abnormalities), which can unveil isolated or multiple hormone deficits and/or mild hyperprolactinemia. A recent study of 60 patients with primary empty sella indicates that 50% have some degree of hypopituitarism.

This patient with a history suggestive of empty sella syndrome became symptomatic in the context of acute illness, raising suspicion for inadequate cortisol response. Serum cortisol (Answer A) measured at 4 PM was 2.4 µg/dL (66.2 nmol/L). After cosyntropin stimulation (250 mcg) in the emergency department, cortisol values remained low at 30 minutes (5.1 µg/dL [140.7 nmol/L]) and at 60 minutes (8.5 µg/dL [234.5 nmol/L]). Free T_4 was also low (0.65 ng/dL [8.4 pmol/L]). She was diagnosed with ACTH and TSH deficiencies, which were contributory to the moderate hyponatremia. Management in the emergency department included 50 mg intravenous hydrocortisone and azithromycin. She was discharged on hydrocortisone, 25 mg twice daily, and levothyroxine, 100 mcg daily, with plans to follow up with endocrinology in 1 week.

The patient requires timely evaluation and treatment, so waiting until the following morning (Answers B and C) is inappropriate. Additionally, while baseline serum cortisol concentrations are higher in the early morning, ACTH-stimulated concentrations are usually informative regardless of the time of the day.

She has no acute neurological manifestations, so performing MRI of the brain and pituitary gland (Answer D) is incorrect.

Stopping acetazolamide (Answer E) is incorrect because it is not responsible for the acute manifestations suggestive of untreated hypocortisolism in the context of acute illness.

Educational Objective

Recommend the appropriate hormone evaluation in patients with empty sella syndrome.

Reference(s)

Lu M, Ye J, Gao F. Analysis of clinical features of primary empty sella. *Ann Endocrinol (Paris).* 2023;84(2):249-253. PMID: 36642370

Fleseriu M, Hashim IA, Karavitaki N, et al. Hormonal replacement in hypopituitarism in adults: an Endocrine Society clinical practice guideline. *J Clin Endocrinol Metab.* 2016;101(11):3888-3921. PMID: 27736313

Chiloiro S, Giampietro A, Bianchi A, De Marinis L. Empty sella syndrome: multiple endocrine disorders. *Handb Clin Neurol.* 2021:161:29-40. PMID: 34238465

De Marinis L, Bonadonna S, Bianchi A, Maira G, Giustina A. Primary empty sella. *J Clin Endocrinol Metab.* 2005;90(9):5471-5477. PMID: 15972577

Chiloiro S, Giampietro A, Bianchi A, et al. Diagnosis of endocrine disease: primary empty sella: a comprehensive review. *Eur J Endocrinol.* 2017;177(6):R275-R285. PMID: 28780516

111 ANSWER: D) Posttransplant diabetes

The number of organ transplants increases every year. In the United States, more than 40,000 organs were transplanted in 2022 compared with approximately 25,000 in 2002. Care of patients after transplant requires a multidisciplinary approach given higher risks of multiple comorbidities, including endocrinopathies. Metabolic complications increase risk of organ rejection, cardiovascular morbidity and mortality, cancer-related mortality, and overall survival. Due to effective screening and management of complications, 1- and 5-year survival rates have improved. In kidney transplant recipients, 1-year survival was 97% or greater and 5-year survival was 86% or greater, with higher rates in living donor recipients. However, more than 70% of kidney transplant mortality is due to cardiovascular disease.

Posttransplant diabetes (PTD) (Answer D) is defined as the presence of diabetes after transplant that was not diagnosed before transplant. There are many cases of "missed" type 2 diabetes during the pretransplant period because many centers do not routinely screen for diabetes during the pretransplant evaluation. PTD is diagnosed based on the American Diabetes Association criteria for diabetes if fulfilled after 6 weeks posttransplant:

- Symptoms of diabetes + random plasma glucose ≥200 mg/dL (SI: ≥11.1 mmol/L) *or*
- Fasting plasma glucose ≥126 mg/dL (SI: ≥7.0 mmol/L) *or*
- 2-hour postprandial glucose ≥200 mg/dL (SI: ≥11.1 mg/dL) during a 75-g oral glucose tolerance test *or*
- Hemoglobin A_{1c} ≥6.5% (≥48 mmol/mol)

Kidney transplant recipients should be screened (regardless of risk factors) with one of the following:

- Fasting plasma glucose weekly for 4 weeks post-discharge, then 3 months, 6 months, and annually
- Hemoglobin A_{1c} at 3 months, 6 months, 12 months, and annually
- 75-g oral glucose tolerance test at 3 months, 6 months, 12 months, and annually

While oral glucose tolerance testing is the gold standard to diagnose PTD, it is cumbersome and time consuming. Most centers use fasting plasma glucose and hemoglobin A_{1c} measurement. Of note, hemoglobin A_{1c} accuracy is affected by anemia and changes in red blood cell turnover due to immunosuppressants. PTD is found in 10% to 20% of solid-organ transplant recipients and is associated with decreased graft and overall survival rates. Pretransplant risk factors for PTD are age 40 years and older, African American and Asian races, Hispanic ethnicity, history of gestational diabetes, history of hepatitis C or cytomegalovirus infection, family history of type 2 diabetes in a first-degree relative, obesity, and pretransplant impaired glucose tolerance. The patient in this case has many of these risk factors and is at high risk for PTD (Answer D). Other risk factors for PTD include perioperative hyperglycemia, HLA mismatching, allografts from male and deceased donors, and the immunosuppressive agents used after transplant (high-dosage glucocorticoids, calcineurin inhibitors [tacrolimus more than cyclosporine], and mTOR inhibitors [sirolimus and everolimus]).

Posttransplant dyslipidemia is also very common. Forty-four to seventy-eight percent of kidney transplant recipients have dyslipidemia. Given the high cardiovascular mortality in this population, the goal is to lower the LDL-cholesterol concentration to less than 100 mg/dL (SI: <2.59 mmol/L). If triglyceride concentrations are still elevated, then the triglyceride and non–HDL-cholesterol concentrations should be decreased to less than 130 mg/dL (SI: <3.36 mmol/L). Pretransplant risk factors for dyslipidemia driven by LDL cholesterol posttransplant are age 50 years or older, excessive dietary intake, obesity, kidney insufficiency, genetic predisposition, and antihypertensive medications (diuretics and β-adrenergic blockers). In this vignette, the patient's pretransplant lipid profile while on a statin is excellent with a triglyceride concentration in the reference range, making his risk for posttransplant hypertriglyceridemia (Answer B) less likely. Other risk factors for posttransplant dyslipidemia are immunosuppressive agents (glucocorticoids, calcineurin inhibitors [cyclosporine more than tacrolimus], and mTOR inhibitors), nephrotic syndrome, hypothyroidism, type 2 diabetes, chronic liver disease, and excessive alcohol intake.

Posttransplant osteoporosis (Answer C) is prevalent, and bone loss is highest in the first 3 to 6 months after transplant. Fracture rate is highest in patients who had low bone mass before transplant. Glucocorticoids contribute significantly to bone loss after transplant, and calcineurin inhibitors (cyclosporine and tacrolimus) are also associated with bone loss. Pretransplant risk factors for osteoporosis are those associated with osteoporosis in general, with the underlying etiology for transplant being a contributor (eg, renal osteodystrophy, hepatic osteodystrophy, heart failure, diabetes, end-stage pulmonary disease). For pretransplant kidney candidates, hyperparathyroidism should be treated. All patients should undergo bone density measurement and screening for vertebral fractures before transplant. In patients who do not meet criteria for treatment (normal bone mineral density, no history of fragility fractures, etc), no pretransplant treatment for osteoporosis is required. The patient in this case has normal electrolytes, PTH concentration within goal range for a hemodialysis recipient (goal range less than 2 to 9 times the upper limit of the PTH assay), no history of fractures, and normal bone density. His overall risk for posttransplant osteoporosis is low.

Malignancy after transplant is common and regular screening is required. The 3 most frequently observed malignancies are Kaposi sarcoma, skin (nonmelanoma, nonepithelial), and nonHodgkin lymphoma. Immunosuppressive agents have been linked to increased posttransplant malignancy risk, more so with antibody therapy, calcineurin inhibitors, and azathioprine. Coexisting viral infection also poses an increased risk, particularly Epstein-Barr virus, human herpesvirus-8, human papillomavirus, and Merkel cell polyomavirus. Rarely, there can be donor transmission of tumor cells. Risk of thyroid cancer is highest with kidney transplant. In this case, the patient had a negative biopsy of his nodule. His overall risk of thyroid cancer (Answer E), while higher than that of the general population, is lower than the risk of developing PTD.

The prevalence of hepatitis C ranges between 10% and 65% in patients undergoing dialysis and between 6% and 46% in kidney transplant recipients. Pretransplant hepatitis B and hepatitis C infections increase the risk of posttransplant liver disease, with higher rates in those with hepatitis B. Pretransplant kidney recipients and/or patients undergoing dialysis are screened for hepatitis C routinely with a hepatitis C virus antibody test. Hepatitis C infection is typically treated before transplant, although it can be treated after transplant as well. In this case, the patient has no evidence of cirrhosis before transplant and the decision to treat before or after transplant will be dependent on timing of transplant. Early studies show similar outcomes in recipients who achieved virological clearance compared with outcomes in those with no infection. While his risk for hepatic cirrhosis is higher than the risk of an individual without hepatitis C, with effective therapy for hepatitis C (sustained virological response >95%) and lack of current fibrosis/liver involvement, his risk of hepatic cirrhosis (Answer A) after transplant is lower than his risk of PTD.

Educational Objective

Determine individual risk for posttransplant diabetes.

Reference(s)

Bhat M, Usmani SE, Azhie A, Woo M. Metabolic consequences of solid organ transplantation. *Endocr Rev.* 2021;42(2):171-197. PMID: 33247713

Kidney Disease: Improving Global Outcomes (KDIGO) Transplant Work Group. KDIGO clinical practice guideline for the care of kidney transplant recipients. *Am J Transplant.* 2009;9(Suppl 3):S1-S155. PMID: 19845597

Kitahara CM, Yanik EL, Ladenson PW, et al. Risk of thyroid cancer among solid organ transplant recipients. *Am J Transplant.* 2017;17(11):2911-2921. PMID: 28397388

Yuan Q, Hong S, Perez-Ortiz Aet al. Effect of recipient hepatitis C status on outcomes of deceased donor kidney transplantation. *J Am Coll Surg.* 2020;230(6):853-886. PMID: 32035979

112 **ANSWER: C) Restart teriparatide**

This patient initially had significant improvement in bone density with teriparatide followed by an antiresorptive agent. Treatment sequences that start with anabolic agents followed by antiresorptive agents are associated with the largest increases in bone density. Based on bone density response and lack of fracture during treatment, she was placed on a drug holiday. However, she currently has evidence that she is at high risk of fragility fracture, as emphasized by her multiple new or worsening (and symptomatic) vertebral compression fractures without a clear traumatic mechanism, even after 7 years of osteoporosis therapy.

Using the Genant semiquantitative criteria, the severity of vertebral compression deformities is graded by degree of reduction in vertebral body height in the anterior, middle, or posterior portions of the vertebrae as: grade 1 mild (20-25% height reduction), grade 2 moderate (25-40% height reduction), and grade 3 severe (40% or more height reduction). Worsening of vertebral body deformities over time are consistent with incident further fracture of previously fractured vertebrae. New fragility fractures are an indication to restart osteoporosis therapy rather than continue a drug holiday (Answer A). An incident fracture in patients during a drug holiday or otherwise not undergoing treatment should trigger re-evaluation of the need for pharmacotherapy, as well as a search for new risk factors. Although this patient's bone density values do indicate declines at the femoral neck of 5.7%, the International Society for Clinical Densitometry does not recommend monitoring femoral neck alone in individual patients due to greater potential for effects of orientation and rotation on this measured region of interest that could artificially cause the appearance of change. In this patient's case, the current bone density would also underestimate her fracture risk.

Although the initial FDA label for teriparatide restricted use to 2 years due to theoretical risk for osteosarcoma in animal models (resulting in a boxed warning), 15 years of postmarketing studies did not demonstrate an increased risk in humans. Thus, the FDA label was modified in 2021 to allow consideration for treating longer than 2 years with teriparatide, although only in those remaining at high fracture risk. The FDA 2023 label for the other PTH receptor agonist, abaloparatide, still indicates that treatment with abaloparatide beyond 2 years is not recommended, despite also removing the boxed warning. Anabolic therapies are expensive, and judicious use would apply extension of teriparatide treatment to patients clearly demonstrating high fracture risk either through severely low or worsening bone density despite therapy, or as in the case of this patient, having additional fragility fractures despite bone density improvements. While romosozumab (Answer D) is also a reasonable choice for potent anabolic therapy in a person who has fragility fractures despite previous treatment, she also has evidence of coronary artery disease, which raises some concern for risk for cardiovascular events while taking romosozumab. In the randomized clinical trial of romosozumab vs alendronate, there was a slight increase in cardiovascular events in the romosozumab arm compared with events in the alendronate arm, although this was not observed in a placebo-controlled trial.

Restarting alendronate (Answer B) or starting an intravenous bisphosphonate (zoledronate [Answer E]) are both reasonable, although less-potent, options for fracture risk reduction in this patient who has high fracture risk and recent vertebral fracture. The VERO trial in postmenopausal women with moderate to severe vertebral compression fractures is relevant to this patient. In that study, incidental vertebral fractures occurred in 5.4% of women treated with teriparatide vs 12% of women treated with risedronate (hazard ratio, 0.44 [95% CI, 0.29-0.68]) and clinical fractures in 4.8% vs 9.8%, respectively (hazard ratio, 0.48 [95% CI, 0.32-0.74]). Thus, restarting teriparatide (Answer C) is the best next step for anabolic therapy in this woman who has ongoing high risk for vertebral fractures.

Educational Objective

List indications for treating high-risk osteoporosis with more than 2 years of teriparatide.

Reference(s)

Shoback D, Rosen CJ, Black DM, Cheung AM, Murad MH, Eastell R. Pharmacological management of osteoporosis in postmenopausal women: an Endocrine Society guideline update. *J Clin Endocrinol Metab.* 2020;105(3):587-594. PMID: 32068863

Eli Lilly and Company. Forteo Prescribing Information. Eli Lilly and Company 2021. http://pi.lilly.com/us/forteo-pi.pdf. Accessed September 23, 2021

Shuhart C, Cheung A, Gill R, Gani L, Goel H, Szalat A. Executive summary of the 2023 Adult Position Development Conference of the International Society for Clinical Densitometry: DXA reporting, follow-up BMD testing and trabecular bone score application and reporting. *J Clin Densitom.* 2024;27(1):101435. PMID: 38007332

Kendler DL, Marin F, Zerbini CAF, et al. Effects of teriparatide and risedronate on new fractures in post-menopausal women with severe osteoporosis (VERO): a multicentre, double-blind, double-dummy, randomised controlled trial. *Lancet.* 2018;391(10117):230-240. PMID: 29129436

113 ANSWER: A) Dietary iodine intake >50 mcg daily in the week before administration of the radioiodine tracer

Patients with differentiated thyroid cancer who continue to have positive serum thyroglobulin concentrations despite total thyroidectomy and ^{131}I ablation may require a diagnostic ^{123}I whole-body uptake and scan. Preparation for a radioiodine uptake and scan (using ^{123}I or ^{131}I) requires a low-iodine diet (<50 mcg daily) for several days, ideally up to a week. Dietary iodine intake greater than 50 mcg daily in the week before administration of the radioiodine tracer (Answer A) could cause a false-negative radioiodine uptake result.

Preparation for the test also includes no receipt of iodinated contrast media in the 1 to 2 months before the radioiodine tracer is administered. Receiving iodinated contrast media 4 months before the test (Answer D) would not cause a false-negative uptake result.

Results of the radioiodine uptake and scan are most reliable when there is concomitant stimulation of any thyroid tissue present, which is confirmed with a serum TSH concentration greater than 30.0 mIU/L. A TSH value greater than 60.0 mIU/L (Answer C) is not required. Stimulation can be accomplished by 2 consecutive injections of rhTSH in the 2 days before the test. Receipt of rhTSH (2 consecutive daily doses) concluding the day before administration of the radioiodine tracer (Answer E) would not cause a false-negative radioiodine uptake result. Thyroid hormone withdrawal may be preferred if there is a reasonable chance that a treatment dose of ^{131}I may be needed shortly after the diagnostic uptake and scan. Thyroid hormone withdrawal is accomplished by discontinuing levothyroxine for 4 to 6 weeks and then initiating a temporary course of liothyronine, 25 mcg twice daily, which is also discontinued in the 2 weeks before the test. Thus, discontinuing levothyroxine 6 weeks before administration of the radioiodine tracer (Answer B) would not cause a false-negative uptake result.

Educational Objective

Identify the reasons for a false-negative diagnostic radioiodine uptake in a patient receiving levothyroxine for differentiated thyroid cancer after thyroidectomy.

Reference(s)

Haugen BR, Alexander EK, Bible KC, et al. 2015 American Thyroid Association management guidelines for adult patients with thyroid nodules and differentiated thyroid cancer: the American Thyroid Association guidelines task force on thyroid nodules and differentiated thyroid cancer. *Thyroid.* 2016;26(1):1-133. PMID: 26462967

Van Nostrand D. Radioiodine refractory differentiated thyroid cancer: time to update the classifications. *Thyroid.* 2018;28(9):1083-1093. PMID: 30105931

Liu H, Yang D, Li L, Tu Y, Chen C, Sun S. Appraisal of radioiodine refractory thyroid cancer: advances and challenges. *Am J Cancer Res.* 2020;10(7):1923-1936. PMID: 32774993

114 ANSWER: D) Recommend no hormonal or imaging follow-up

Teratomas are germ-cell tumors originating from pluripotent cells with multidirectional potential of differentiation. In adults, germ-cell tumors occur most commonly in the gonads. Extragonadal germ-cell tumors can be found in the retroperitoneum, mediastinum, head, and neck areas. In infants and young children, germ-cell tumors are often sacrococcygeal and intracranial. Adrenal teratomas are very rare. Men with Klinefelter syndrome (karyotype 47,XXY) have an increased risk of germ-cell tumors, particularly mediastinal.

Most teratomas contain more than one embryonic germ-cell layer, although elements derived from ectoderm are most common. Teratomas tend to have a heterogenous composition, and they might contain elements of fat, fluid, bone, teeth, and/or hair. Some of these elements are easily recognizable on CT. Calcifications are very common.

Teratomas are often found incidentally but can cause pain or other symptoms due to compression of adjacent structures. Adrenal teratomas present as large, nonfunctional tumors, with an average diameter of 8 to 10 cm. Most teratomas are benign but some have potential for malignant degeneration and can be very aggressive. Mature teratomas are surgically resected, and these tumors are benign and do not carry recurrence risk. Consequently, no additional therapy or follow-up is needed (Answer D). Follow-up CT (Answer C) would be appropriate for tumors with potential for recurrence.

The optimal treatment for immature teratomas is uncertain, although radical resection should be pursued whenever possible. Cisplatin-based chemotherapy has been used for mediastinal and retroperitoneal immature teratomas, but such data are not available for adrenal tumors. Malignant germ-cell tumors (seminomas or nonseminomatous) presenting in adulthood affect mostly men in the third decade of life. Malignant germ-cell tumors are exquisitely sensitive to both cisplatin-based chemotherapy and radiotherapy, regardless of location. A multidisciplinary approach, including oncology and radiation oncology (Answer E), is desirable for such patients.

Mitotane (Answer A) is an adrenocorticolytic drug used in patients with ACC when surgery is incomplete or not feasible or when there is risk of recurrence. Mitotane does not have a role in the treatment of teratomas.

Patients with cortisol-producing tumors and ACTH suppression have contralateral adrenal insufficiency, and they require glucocorticoid replacement until the hypothalamic-pituitary-adrenal axis recovers. In such patients, glucocorticoids are administered perioperatively, and they are continued following surgery, with tapering in the outpatient setting. Adrenal insufficiency is not otherwise a concern following unilateral adrenalectomy, and there is no role for performing cosyntropin-stimulation testing (Answer B).

Educational Objective

Consider teratomas in the differential diagnosis of adrenal incidentalomas with calcifications and guide their management.

Reference(s)

Gatcombe HG, Assikis V, Kooby D, Johnstone PA, Primary retroperitoneal teratomas: a review of the literature. *J Surg Oncol.* 2004;86(2):107-113. PMID: 15112254

Zhou L, Pan X, He T, et al. Primary adrenal teratoma: a case series and review of the literature. *Mol Clin Oncol.* 2018;9(4):437-442. PMID: 30214733

115 ANSWER: B) No further evaluation

This man has chronic, nontender gynecomastia and normal serum testosterone and gonadotropins. Further evaluation is not necessary or useful (thus, Answer B is correct).

Gynecomastia results from excess estrogen or estrogen effect, deficiency of androgen or androgen effect, or a combination. Gynecomastia smaller than 2.5 cm is common; nontender gynecomastia that is smaller than 2.5 cm may be considered a normal variant. Most causes of gynecomastia have some component of androgen deficiency because causes of excess estrogen lead to gynecomastia directly and indirectly via suppression of pituitary gonadotropins, which suppresses Leydig-cell secretion of testosterone. For most men with gynecomastia, the cause is not definitively determined. For men with gynecomastia that is 4 cm or smaller and who have no symptoms of an endocrinopathy such as hyperthyroidism, the evaluation can be limited to a careful history to review medications and possible environmental exposures to compounds with estrogenic or antiandrogenic effects; symptoms and signs of hyperthyroidism, hypogonadism, and hyperprolactinemia; and examination focused on findings of hyperthyroidism and hypogonadism. Generally, measurement of serum testosterone and gonadotropins is the sole additional useful evaluation.

Distinguishing chronic gynecomastia (present for 6 months or longer) from new-onset gynecomastia is crucial. New-onset gynecomastia is characterized by pain and tenderness to palpation, and it is much more likely to be related to an endocrinopathy that perturbs the balance between circulating androgens and estrogens. New-onset gynecomastia is best treated by resolving the underlying cause, but some boys and men with tender, new breast growth benefit from 3 to 6 months of oral tamoxifen. Chronic gynecomastia requires no endocrine evaluation and cannot be treated medically.

The most important elements of the physical examination are to distinguish breast tissue (characteristically a disk of rubbery tissue, with its center just below the nipple) from fat tissue. This disk of breast tissue has a firm edge, and the entire disk can be "flipped" or "rocked" by the examiner's fingers. Evaluation for breast cancer by imaging with either ultrasonography (Answer A) or mammography should be restricted to hard, fixed, and irregular breast masses that are often eccentric to the nipple.

Serum testosterone and gonadotropin values generally identify or exclude hormonal causes of gynecomastia. For example, neoplastic production (or an exogenous source) of hCG (Answer C) suppresses serum LH and FSH but results in low-normal to elevated serum testosterone. (When a neoplastic source of hCG is suspected, an assay that measures all forms of hCG and its β subunit should be used.)

Neoplastic production (or an exogenous source) of estrogen suppresses serum gonadotropins and testosterone. Measurement of serum estradiol (Answer D) is almost never useful in the assessment of gynecomastia because serum estradiol concentrations are not accurately measured at the low serum concentrations present in men. In addition, slightly elevated serum estradiol concentrations are a common, nondiagnostic, normal variant in men.

Hyperprolactinemia (Answer E) causes gynecomastia by suppressing serum gonadotropins and testosterone and causing secondary hypogonadism. This man is eugonadal. He has a low total testosterone concentration but a normal calculated free testosterone concentration. His total testosterone is low because of low SHBG (likely due to BMI >30 kg/m^2).

Educational Objective

Explain that nontender gynecomastia 4.0 cm or smaller is unlikely to be caused by a clinically significant endocrinopathy.

Reference(s)

Abalovich M, Levalle O, Hermes R, et al. Hypothalamic-pituitary-testicular axis and seminal parameters in hyperthyroid males. *Thyroid.* 1999;9(9):857-863. PMID: 10524563

Gronowski AM. Clinical assays for human chorionic gonadotropin: what should we measure and how? *Clin Chem.* 2009;55(11):1900-1904. PMID: 19797713

Klang E, Kanana N, Grossman A, et al. Quantitative CT assessment of gynecomastia in the general population and in dialysis, cirrhotic, and obese patients. *Acad Radiol*, 2018;25(5):626-635. PMID: 29326049

Thirumalai A, Anawalt BD. Androgenic steroids use and abuse: past, present, and future. *Urol Clin North Am.* 2022;49(4):645-663. PMID: 36309421

116 ANSWER: B) Adhesive capsulitis

Adhesive capsulitis (Answer B) or frozen shoulder is a common musculoskeletal complication of diabetes, occurring in up to 29% of persons with type 1 or type 2 diabetes. It is directly proportional to hemoglobin A_{1c} levels and duration of diabetes. Adhesive capsulitis is characterized by diffuse shoulder pain that is worse at night and associated with stiffness and decreased range of motion (both active and passive) in 2 or more planes, compared with the contralateral side. It is generally spontaneous in onset without an underlying cause, but it can also occur subsequent to prolonged immobilization after shoulder surgery or injury. Treatment of adhesive capsulitis in patients with diabetes is the same as in those without diabetes—conservative treatment with pain relief, glucocorticoid injection, and physiotherapy as first-line management and surgery reserved for those who do not respond.

Acromion fracture (Answer A) occurs due to trauma or overuse injury, and associated pain persists throughout the day and night. Physical examination findings are notable for direct point tenderness over the acromioclavicular joint and pain elicited on passive cross-body adduction of the arm to cause compression of the acromioclavicular joint. There is no associated increased risk of acromion fracture in people with diabetes.

Bicipital tendonitis (Answer C) pain is focused in the anterior shoulder with radiation distally over the biceps. The pain is aggravated by lifting, pulling, or performing repetitive overhead activities. There is no known association with diabetes.

Cervical disk disease (Answer D) is associated with neck pain/stiffness that worsens with neck movement. The pain may radiate to the shoulder, arm, or hand and is associated with tingling, numbness, or burning sensation.

Rotator cuff tear (Answer E) is typically associated with repetitive overhead motion, generally due to sports such as baseball, tennis, swimming, gymnastics, etc. With rotator cuff pathology, passive range of motion is usually greater than active range of motion.

Educational Objective

Identify common musculoskeletal complications of diabetes leading to shoulder pain.

Reference(s)

Balci N, Balci MK, Tüzüner S. Shoulder adhesive capsulitis and shoulder range of motion in type II diabetes mellitus: association with diabetic complications. *J Diabetes Complications.* 1999;13(3):135-140. PMID: 10509873

Juel NG, Brox JI, Brunborg C, Holte KB, Berg TJ. Very high prevalence of frozen shoulder in patients with type 1 diabetes of ≥45 years' duration: the Dialong Shoulder Study. *Arch Phys Med Rehabil.* 2017;98(8):1551-1559. PMID: 28219686

Chan JH, Ho BS, Alvi HM, Saltzman MD, Marra G. The relationship between the incidence of adhesive capsulitis and hemoglobin A_{1c}. *J Shoulder Elbow Surg.* 2017;26(10):1834-1837. PMID: 28495575

117 **ANSWER: B) Finasteride**

This transgender woman is seeking feminizing gender-affirming hormone therapy. In general, gender-affirming therapy is associated with decreased gender dysphoria in both transgender men and women. The goal of feminizing gender-affirming hormone therapy is to reduce the endogenous testosterone levels and increase estrogen. It is important to counsel transgender women about the varying effects of estrogen on the body, as well as the potential risks. The approach to feminizing gender-affirming therapy includes estrogen and may include an antiandrogen, and sometimes a 5α-reductase inhibitor and/or progesterone. Estrogen therapy induces feminine characteristics; reduces testosterone due to hypothalamic-pituitary-gonadal feedback; and can be administered in oral, transdermal, or injectable forms. While on estrogen therapy, the goal serum estradiol target is in the average premenopausal female range of 100 to 200 pg/mL (367-734 pmol/L). An antiandrogen can be necessary when estrogen is not sufficient to lower the testosterone into the goal range of less than 50 ng/dL (<1.7 nmol/L).

This patient has achieved both goal estradiol and total testosterone levels with an injectable estradiol ester (estradiol valerate, 4 mg subcutaneously every 7 days). There is no indication to add an antiandrogen such as spironolactone, which blocks the action and downstream effects of testosterone on target tissues by blocking androgens during their interaction with the androgen receptor. Mechanistically, spironolactone also blocks testosterone synthesis within the testicle and adrenal gland via inhibition of 17α-hydroxylase and 17,20-lyase. Likewise, there is also no benefit of adding leuprolide, a GnRH agonist (Answer C) at this time given that estradiol valerate alone is adequately blocking the hypothalamic-pituitary-gonadal axis and decreasing endogenous production by the natal testes.

Androgenetic alopecia is a hair loss disorder mediated by dihydrotestosterone, the potent form of testosterone. Dihydrotestosterone is an androgen that is converted from testosterone by 5α-reductase, and it is responsible for external male genitalia development, androgenetic alopecia, prostate growth, and increased sebaceous gland activity (acne). Dihydrotestosterone induces miniaturization of hair follicles, causing transformation of terminal hair into vellus hair (miniaturized hairs), which leads to hair thinning and loss. Without treatment, patients experience progressive hair loss.

5α-Reductase inhibitors such as finasteride (Answer B), block conversion of testosterone to dihydrotestosterone. Therefore, a 5α-reductase inhibitor lowers dihydrotestosterone and can be used to prevent scalp hair loss. In studies evaluating the effect of finasteride, 1 mg, compared with placebo in cisgender men, finasteride led to a significant change in hair-density per cm^2 from baseline to follow-up. This improvement was more than that gained from topical minoxidil and low-level laser light therapy in cisgender men. Finally, finasteride may help further suppress this patient's libido and erections, which she has expressed as positive changes thus far.

Minoxidil is a piperidino-pyrimidine derivative that affects follicular cells by enhancing hair growth and reducing hair loss. It has been used for hair-loss treatment for several decades. In a meta-analysis, topical minoxidil (Answer E) at all concentrations provided superior results in reducing hair loss in cisgender men as compared with results achieved in the placebo group. However, this effect is smaller than that of finasteride. Specifically, hair growth was improved by 8 to 12 hairs/cm^2 with topical minoxidil 2% and by 114 hairs/cm^2 with topical minoxidil 5% (compared with 18 hairs/cm^2 with finasteride). In addition, the most common adverse effect of minoxidil is irritant contact dermatitis with the typical symptoms of itching and scaling; some patients may not tolerate this.

While topical minoxidil is an effective treatment option for hair loss, adherence is variable because it must be applied twice daily, it can cause scalp irritation, and it has a potential undesirable effect on hair texture. The topical lotions can make the hair greasy, and the topical foams can make the hair sticky—both can be difficult to apply to the scalp. Oral minoxidil (Answer D) was introduced in the 1970s as a treatment of severe refractory hypertension and was incidentally found to lead to hair growth. At low dosages (<5 mg), studies have shown tolerability, with few and mild adverse effects in cisgender men. However, due to reports of postural hypotension/dizziness, lower limb edema, mild blood pressure changes, and electrocardiographic changes, it is not the recommended first-line treatment.

Finally, biotin (vitamin B_7))Answer A) is a water-soluble vitamin that serves as a cofactor for several carboxylases involved in fatty acid synthesis, gluconeogenesis, and amino acid metabolism. Biotin is commonly included at high dosages in hair and nail vitamins; however, there is no evidence to suggest that biotin directly affects hair follicle cycling or development. Biotin has also never been studied independently as a supplement for hair loss, and there is no published evidence supporting that oral biotin supplements promote hair growth in the absence of biotin deficiency. In addition, biotin is known to cause spurious laboratory test results with immunoassay platforms that incorporate streptavidin-biotin binding. Elevated serum biotin can lead to falsely high values in competitive immunoassays and falsely low values in noncompetitive immunometric (sandwich) assays.

Educational Objective

Address concerns of balding in a transgender woman while affirming her gender identity.

Reference(s)

Hembree WC, Cohen-Kettenis PT, Gooren L, et al. Endocrine treatment of gender-dysphoric/gender-incongruent persons: an Endocrine Society clinical practice guideline. *J Clin Endocrinol Metab.* 2017;102(11):3869-3903. PMID: 28945902

Center of Excellence for Transgender Health, Department of Family and Community Medicine, University of California, San Francisco. *Guidelines for the Primary and Gender-Affirming Care of Transgender and Gender Nonbinary People.* 2nd ed. Deutsch MB, ed. June 2016. Available at: www.transhealth.ucsf.edu/guidelines.

Suchonwanit P. Thammarucha S. Leerunyakul K. Minoxidil and its use in hair disorders: a review. *Drug Des Devel Ther.* 2019;13:2777–2786. PMID: 31496654

Motofei DL, Rowland DL, Baconi M, et al. Androgenetic alopecia; drug safety and therapeutic strategies, *Expert Opin Drug Saf.* 2018;17(4):407-412. PMID: 29363345

Adil A, Godwin M. The effectiveness of treatments for androgenetic alopecia: a systematic review and meta-analysis. *J Am Acad Dermatol.* 2017;77(1):136-141.e5. PMID: 28396101

Randolph M, Tosti A. Oral minoxidil treatment for hair loss: a review of efficacy and safety. *J Am Acad Dermatol.* 2021;84(3):737-746. PMID: 32622136

Adelman MJ, Bedford LM, Potts FA. Clinical efficacy of popular oral hair growth supplement ingredients. *Int J Dermatol.* 2021;60(10):1199-1210. PMID: 33296077

118 **ANSWER: A) Calcium and vitamin D plus referral to physical therapy**

Falling is a risk factor for fractures, and one of the limitations of the FRAX calculator is that it does not account for the number of falls. FRAX was developed in 2008, and it is widely used despite this limitation. FRAX has been validated in 26 studies across 9 countries. The Garvan fracture risk calculator accounts for the number of falls in the last 12 months, although it has been validated in only 6 studies in 3 countries. For individuals with increased fall risk as in this clinical vignette (patient reports feeling unsteady in his gait), physical therapy to focus on strength and balance training must be incorporated. This patient's T-scores are within the osteopenic range, and the FRAX prediction model shows numbers that are typically below the treatment threshold. Therefore, the best course of action is to optimize calcium and vitamin D and refer him to physical therapy for strength and balance training (Answer A).

He does not wish to take oral bisphosphonates, and based on his FRAX score and DXA findings, treatment is not indicated. Therefore, denosumab (Answer B), teriparatide (Answer C), and zoledronic acid (Answer E) are incorrect.

Serum testosterone levels decrease annually with age. Testosterone replacement therapy (Answer D) can aid in maintaining and improving bone mineral density in hypogonadal men. However, the data about efficacy of testosterone replacement in bone health are conflicting. In a meta-analysis of 55 studies, there was a significant association between hypogonadism and low bone mineral density. Testosterone therapy was not associated with reduced risk of falls in the Testosterone Trials. In the recent TRAVERSE study, a multicenter, randomized, double-blind, placebo-controlled noninferiority trial of 5204 men aged 45 to 80 years with at least 1 symptom of hypogonadism, 2 fasting serum testosterone values less than 300 ng/dL (<10.4 nmol/L), and preexisting cardiovascular disease or elevated cardiovascular risk were randomly assigned, 1:1, to transdermal 1.62% testosterone gel or matching placebo gel daily. While the primary and secondary end points showed cardiovascular noninferiority, there were 154 confirmed fractures in the testosterone group compared with 97 in the placebo group. In this study of middle-aged and older men with hypogonadism, testosterone treatment did not lower the incidence of overall or major osteoporotic clinical fractures. In fact, the group randomly assigned to testosterone had numerically higher overall fracture incidence than in the placebo group. This finding was surprising, especially given the previously published Testosterone Trials showing improved bone mineral density with testosterone. The divergence between the 2 groups was seen early in the trial and persisted throughout the study. However, it should be noted that the study population was men aged 45 to 80 years who had 1 symptom of hypogonadism that did not include osteoporosis (was not a criterion for study entry), and assessment of bone mineral density and fracture risk was not done at baseline. In addition, information about falls, physical activity, and risk-taking was not assessed.

Educational Objective

Summarize the data from TRAVERSE study that showed increased fracture risk in men on long-term testosterone therapy.

Reference(s)

Burnett-Bowie S-AM, McKay EA, Lee H, Leder BZ. Effects of aromatase inhibition on bone mineral density and bone turnover in older men with low testosterone levels. *J Clin Endocrinol Metab.* 2009;94(12):4785-4792. PMID: 19820017

Lincoff AM, Bhasin S, Flevaris P, et al. Cardiovascular safety of testosterone-replacement therapy. *N Engl J Med.* 2023;389(2):107-117. PMID: 37326322

Snyder PJ, Bauer DC, Ellenberg SS, et al. Testosterone treatment and fractures in men with hypogonadism. *N Engl J Med.* 2024;390(3):203-211. PMID: 38231621

119 **ANSWER: C) Measure TSH and free T_4 again after holding biotin for 3 days**

Biotin (vitamin B_7) is a water-soluble vitamin that serves as a cofactor for several carboxylases involved in fatty acid synthesis, gluconeogenesis, and amino acid metabolism. Dietary sources of biotin include chicken, salmon, eggs, dairy products, nuts, whole-grain cereals and breads, and some fruits and vegetables. A second source of biotin is through de novo synthesis by bacteria in the large intestine. The Institute of Medicine recommends a daily biotin intake of 30 mcg for all adults, which can be easily achieved through dietary sources. True biotin deficiency is rare and is mostly due to genetic causes (biotinidase deficiency, carboxylase deficiencies, or biotin transport defects). However, consumption of 2 or more raw eggs per day over several months can cause biotin malabsorption through binding to avidin, which is present in egg whites, leading to biotin deficiency. This does not occur with cooked eggs.

Hypothesized to affect remyelination, high-dosage biotin (up to 100 mg 3 times daily) has been investigated in the treatment of progressive multiple sclerosis. Although a small randomized controlled study demonstrated improved multiple sclerosis–related disability for patients treated with high-dosage biotin, a larger phase 3 trial (SPI2) demonstrated no benefit. Despite these negative results, some patients with multiple sclerosis continue to take biotin supplements. Biotin is also widely marketed as an over-the-counter dietary supplement to promote hair, nail, and skin health. Although biotin supplementation is generally well-tolerated without many clinical adverse effects, it is known to cause spurious laboratory test results with immunoassay platforms that incorporate streptavidin-biotin binding.

Biotin is rapidly absorbed after ingestion, with peak serum levels occurring in 1 to 2 hours and a half-life of 1.8 hours. Steady state is achieved after 3 days of regular ingestion. Elevated serum biotin concentrations can lead to falsely high results in competitive immunoassays, including total and free T_4 and total and free T_3, and falsely low results in noncompetitive immunometric (sandwich) assays, including serum TSH. Consequently, thyroid function test results in the context of biotin interference may mimic hyperthyroidism biochemically (eg, low serum TSH, elevated free T_4). Clinically significant effects on laboratory test results are not seen with low biotin dosages (<300 mcg daily), while dosages of 10 mg daily may affect some but not all biotin-based assays. In patients taking moderate to high-dosage biotin supplementation (>5 to 10 mg daily), it is necessary to first rule out the possibility of biotin interference before pursuing additional diagnostic testing or treatment for "hyperthyroidism," which may prove to be unnecessary, costly, and potentially harmful. One method to rule out biotin interference is to remeasure serum TSH and free T_4 in the same assay after holding biotin for 3 days (Answer C). This would likely be feasible even in patients with partial biotinidase deficiency, but it may not be possible for patients with the most severe forms. Alternative approaches to confirm biotin interference may also be used, including: (1) performing serial dilution; (2) repeating testing with an alternative immunoassay that uses antianimal antibodies instead of biotinylated antibodies, or through a liquid chromatography-tandem mass spectrometry assay; (3) using depletion protocols in which materials to bind and remove biotin are added to the sample before retesting; or (4) preconjugating the biotinylated capture antibody to streptavidin.

Laboratory interference should be suspected in this case given the absence of signs or symptoms suggestive of hyperthyroidism. In fact, thyroid function testing was ordered because hypothyroidism was suspected clinically (weight gain, depression, fatigue, and constipation). In a patient such as this one, ruling out laboratory interference is important before moving forward with further diagnostic testing for hyperthyroidism (measuring TRAb [Answer B] or performing thyroid uptake and scan [Answer E]).

Prescribing methimazole (Answer D) would also be inappropriate unless a definitive diagnosis of hyperthyroidism was made.

Although an evaluation for heterophile antibodies (Answer A) should be pursued when a patient's clinical and biochemical findings are discordant, this is less likely than biotin interference given the reported history of high-dosage biotin supplementation.

Educational Objective

Suspect biotin interference as a cause of spurious thyroid function test results.

Reference(s)

Angelopoulos N, Paparodis RD, Androulakis I, et al. The clinical significance of low dose biotin supplements (<300μg/day) in the treatment of patients with hypothyroidism: crucial or overestimated? *Thyroid Res.* 2023;16(1):18. PMID: 37455308

Bowen R, Benavides R, Colón-Franco JM, et al. Best practices in mitigating the risk of biotin interference with laboratory testing. *Clin Biochem.* 2019;74:1-11. PMID: 31473202

Cree BAC, Cutter G, Wolinsky JS, et al; SPI2 investigative teams. Safety and efficacy of MD1003 (high-dose biotin) in patients with progressive multiple sclerosis (SPI2): a randomised, double-blind, placebo-controlled, phase 3 trial. *Lancet Neurol.* 2020;19(12):988-997. PMID: 33222767

120 ANSWER: E) Triple receptor agonist (glucagon, GLP-1, and glucose-dependent insulinotropic polypeptide)

Therapies to treat type 2 diabetes and obesity are constantly evolving. The first SGLT-2 inhibitor was approved in 2013. In persons with type 2 diabetes, this medication class (Answer C) has significant cardiovascular and kidney benefits, in addition to helping with weight loss and hemoglobin A1c lowering. Inhibition of the SGLT-2 cotransporter does not result in hypoglycemia. In a pooled analysis of SGLT-2 inhibitors, these agents were associated with a mean hemoglobin A1c reduction of 0.5% and mean weight loss of 5.5 lb (2.5 kg) after 1 year of therapy.

Thiazolidinediones (Answer D) bind to peroxisome proliferator–activated receptors (PPARs), mainly PPAR-γ, to improve insulin sensitivity in adipose tissue, muscle, and the liver (PPAR-α). Due to the variable expression of PPARs in tissue and each thiazolidinedione differing with receptor activity, the effects of each thiazolidinedione differ. Pioglitazone is the only thiazolidinedione available in the United States and Europe. Pioglitazone has both PPAR-γ and PPAR-α effects and can decrease hemoglobin A_{1c} up to 1.4%; however, it is associated with fluid retention, weight gain, and increased fracture risk. The weight gain associated with thiazolidinediones is dose- and time-dependent, with an average of 12-lb (5.3-kg) weight gain by 36 months with pioglitazone.

The first GLP-1 receptor agonist (Answer B) to be approved was exenatide in 2005. Since then, an additional 10 incretin-based medications have been studied in patients with type 2 diabetes. With each new medication, the reported weight loss and glycemic control have been better than that described with the prior drug. Activation of insulin secretion by stimulating the GLP-1 receptor is glucose dependent, thus hypoglycemia is rare. GLP-1 receptor agonists are available in oral and injectable formulations. The longer-acting and newer formulations are associated with greater hemoglobin A1c reductions and weight loss. Oral semaglutide (14 mg) is associated with a 1.1% reduction in hemoglobin A_{1c} and 5.1-lb (2.3-kg) weight loss, whereas the 50-mg dose (not yet commercially available) is associated with a 2% reduction in hemoglobin A_{1c} and 17.6-lb (8-kg) weight loss. Orforglipron (not commercially available) is an oral nonpeptide GLP-1 receptor agonist studied in persons with type 2 diabetes. In a phase 2 study, orforglipron was associated with a 2.1% reduction in hemoglobin A_{1c} and a 22.3-lb (10.1-kg) weight loss at week 26. Injectable once-weekly semaglutide is associated with slightly greater hemoglobin A_{1c} and weight reductions compared with outcomes of dulaglutide and liraglutide. In persons with type 2 diabetes, injectable semaglutide at a dosage of 2.0 mg resulted in a 2.2% reduction in hemoglobin A_{1c} and a 15.2-lb (6.9-kg) weight loss at week 40.

GLP-1 is produced by the L cells in the small and large intestine, whereas glucose-dependent insulinotropic polypeptide is produced by K cells. Tirzepatide is the first dual GLP-1 and glucose-dependent insulinotropic polypeptide (GIP) receptor agonist (Answer A) approved to treat type 2 diabetes. At week 40, tirzepatide, 15 mg, was associated with a 2.07% reduction in hemoglobin A_{1c} and a 20.9 lb (9.5 kg) weight loss from baseline. Retatrutide (not commercially available) is a triple-receptor agonist targeting GLP-1, GIP, and glucagon receptors (Answer E). At week 24, retatrutide, 12 mg, resulted in a 2.02% hemoglobin A_{1c} reduction and a 37.9 lb (17.18 kg) weight loss. While hemoglobin A_{1c} reductions are similar with the highest dosages of these various incretin therapies, weight loss is greatest with the triple-receptor agonist retatrutide.

The rationale behind targeting these various receptors is due to their effects on metabolism as noted in the table below. There is a synergistic net effect to promote both improved glycemic control and weight loss when multiple receptors are stimulated.

Table. Effects of GLP-1, GIP, and Glucagon on Metabolism

Hormone	Food intake	Gastric emptying	Adipose lipolysis	Energy expenditure
GLP-1	↓	↓	↑	↑
GIP	No effect	↑	↑	No effect
Glucagon	↓	↓	↑	↑

A concern can be increased adipose tissue lipolysis. While adipose tissue lipolysis results in increased free fatty acid release, insulin inhibits lipolysis and promotes triglyceride storage. Clinical data show mild reductions in concentrations of LDL and total cholesterol with no change in triglycerides when patients with type 2 diabetes are treated with semaglutide, tirzepatide, or retatrutide. GIP stimulates glucagon release and potentiates the action of GLP-1, which in turn increases insulin secretion in a glucose-dependent manner. In addition, postprandial lipid concentrations are lower with GLP-1 receptor agonism. The net effect of all these actions therefore results in clinically insignificant changes to the lipid profile.

In this case, a triple receptor agonist would have the highest likelihood of achieving the goals of eliminating the need for insulin, improving glycemic control, and improving weight loss.

Educational Objective

Explain the physiology of glucagon receptor agonism in glycemic control.

Reference(s)

Liu X-Y, Zhang N, Chen R, Zhao J-G, Yu P. Efficacy and safety of sodium-glucose cotransporter 2 inhibitors in type 2 diabetes: a meta-analysis of randomized controlled trials for 1 to 2 years. *J Diabetes Complications.* 2015;29(8):1295-303. PMID: 26365905

Frias JP, Hsia S, Eyde S, et al. Efficacy and safety of oral orforglipron in patients with type 2 diabetes: a multicentre, randomised, dose-response, phase 2 study. *Lancet.* 2023;402(10400):472-483. PMID: 37369232

Frías JP, Auerbach P, Bajaj HS, et al. Efficacy and safety of once-weekly semaglutide 2·0 mg versus 1·0 mg in patients with type 2 diabetes (SUSTAIN FORTE): a double-blind, randomised, phase 3B trial. *Lancet Diabetes Endocrinol.* 2021;9(9):563-574. PMID: 34293304

Rosenstock J, Wysham C, Frías JP, et al. Efficacy and safety of a novel dual GIP and GLP-1 receptor agonist tirzepatide in patients with type 2 diabetes (SURPASS-1): a double-blind, randomised, phase 3 trial. *Lancet.* 2021;398(10295):143-155. PMID: 34186022

Rosenstock J, Frias J, Jastreboff AM, et al. Retatrutide, a GIP, GLP-1 and glucagon receptor agonist, for people with type 2 diabetes: a randomised, double-blind, placebo and active-controlled, parallel-group, phase 2 trial conducted in the USA. *Lancet.* 2023;402(10401):529-544. PMID: 37385280

Question-Mapping Index

ENDOCRINE SELF-ASSESSMENT PROGRAM 2025

Part III

This question-mapping index groups question topics according to the 8 umbrella sections of ESAP (Adrenal, Bone-Calcium, Diabetes, Lipids-Obesity, Pituitary, Reproduction [Female], Reproduction [Male], and Thyroid). Relevant ***question numbers*** *follow each topic.*

ADRENAL

CALCIUM-BONE

DIABETES

LIPIDS & OBESITY

PITUITARY

REPRODUCTION, FEMALE

REPRODUCTION, MALE

THYROID